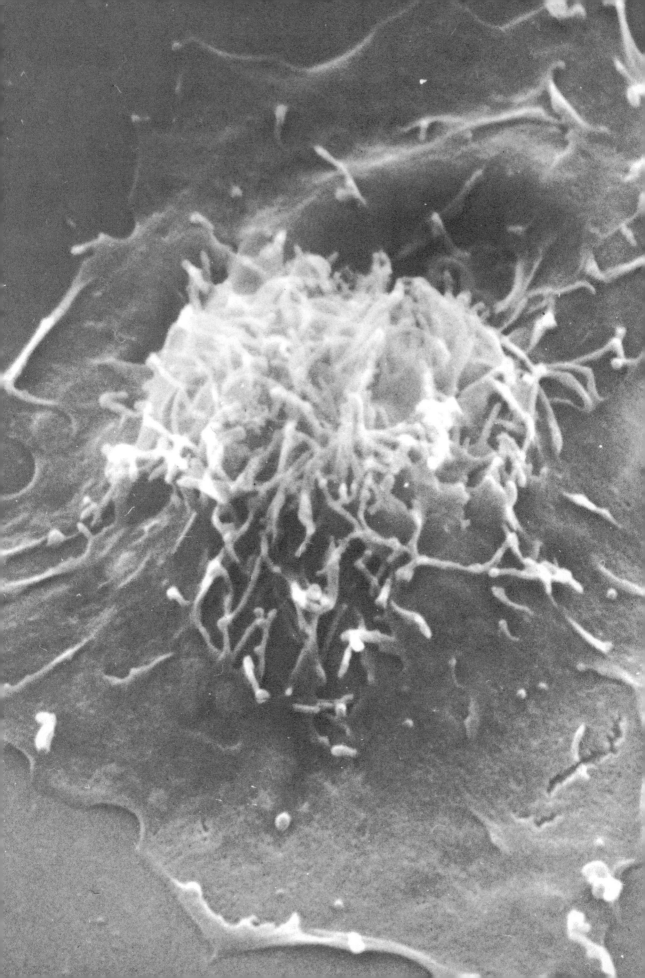

Cell and Molecular Biology

Seventh Edition

E.D.P. De Robertis, M.D.
University of Buenos Aires
Buenos Aires, Argentina

E.M.F. De Robertis, Jr., M.D.,Ph.D.
MRC Laboratory of Molecular Biology
Cambridge, England

SAUNDERS COLLEGE
Philadelphia

Saunders College
West Washington Square
Philadelphia, PA.   19105

Listed here is the latest translated edition of this book together
with the language of the translation and publisher.

Portuguese (*1st Edition*) — El Ateneo, Buenos Aires, Argentina

French (*1st Edition*) — University of Laval, Quebec, Canada

Hungarian (*2nd Edition*) — Akadeniai Kiodo, Budapest, Hungary

Italian (*2nd Edition*) — Editore Nicola Zannichelli,
                  Bologna, Italy

Japanese (*2nd Edition*) — Asakura, Tokyo, Japan

Polish (*2nd Edition*) — Panstwowe Wydawnictwo, Navkowe
                  Warsaw, Poland

There is also a Russian translation, *2nd Edition.*

cover photo: Histone-depleted metaphase chromosome from a human Hela cell. A scaffold or core,
having the shape characteristic of a metaphase chromosome, is surrounded by a halo of DNA.
The halo consists of many loops of DNA, each anchored in the scaffold at its base; most of the DNA
exists in loops at least 10-30 μm long. (From J. R. Paulson and U. K. Laemmli, *Cell* **12:**817–828, 1977,
Copyright © M.I.T.)

Cell and Molecular Biology   ISBN 0-03-056749-1

This edition is a tribute to the memory of Prof. Francisco A. Saez, our former co-author, who passed away in 1975. He was a great scientist and a master who contributed with unfailing dedication to the development of cytogenetics throughout the Latin-American countries. It is most unfortunate that he is no longer with us to continue in the wonderful experience of writing about discoveries in cell and molecular biology.

PREFACE

Biological life has an immense variety of forms that arose by the process of evolution, but all living organisms share a master plan of structural and functional organization. This book is about the building blocks — *cells* and *molecules* — that constitute the unity of the living world.

Progress in our field of science has been so rapid that during the lifetime of this book, first published in 1946, we have witnessed the most revolutionary discoveries, such as those involving the ultrastructure and macromolecular organization of cell components and the molecular basis of the genetic code and gene expression. This tremendous progress necessitated making extensive revisions every five years and changing the original title, *General Cytology*, to *Cell Biology* in 1965 and to *Cell and Molecular Biology* in the present edition. This text has been entirely rewritten and should be considered a new book rather than a new edition.

The present title reflects the fact that molecular biology, which deals with the fundamental roles of nucleic acids and proteins, cannot be separated from cell biology, which deals with the cell as the structural and functional unit of life. In this edition we have attempted to integrate the most recent advances in molecular biology with our knowledge of the structure and function of cells, while taking into account the work of classical cytologists — often forgotten these days — which laid the foundations of our understanding of the living cell.

E. De Robertis
E. M. De Robertis, Jr.
Sept. 1979

ACKNOWLEDGMENTS

We have been stimulated in our task by the good reception that previous editions have received in their Spanish, Portuguese, Italian, French, Hungarian, Polish, Russian, and Japanese translations. We want especially to thank Professors A. Bajer, E. Carlson, G. A. O'Donovan, S. Gerbi, L. J. Kleinsmith, H. Massover, G. E. Palade, B. Strauss, and other reviewers who wish to remain anonymous. All of them, through their suggestions and criticisms of various editions, have greatly contributed to the improvement of this book.

Among the illustrations, we have included the original micrographs and designs of many experiments that led to major discoveries in cell biology. It is fortunate that the organization of the cell can now be visualized even at the macromolecular level. We hope that these experiments will help the students to recognize the way in which scientific progress is made. We especially wish to express our gratitude to our colleagues who have contributed to the illustrations of this edition:

K. Akert, J. M. Anderson, P. A. Armond, C. J. Arntzen, G. F. Bahr, A. D. Bangham, W. Barnes, B. Barrell, E. K. F. Bautz, H. Beevers, G. Bennett, M. L. Birnstiel, G. Blobel, B. R. Brinkley, R. J. Britten, H. G. Callan, S. J. Chan, S. Cohen, D. P. Costello, A. C. Cuello, S. De Petris, J. T. Finch, W. W. Franke, A. Globus, R. D. Goldman, E. González, R. G. Goss, P. Greengard, U. Groeschel-Stewart, J. B. Gurdon, M. D. Hatch, J. Herkovits, F. Hucho, H. Huxley, J. D. Jamieson, M. Jamrich, J. Kezer, D. E. Kohne, G. Kreibich, U. K. Laemmli, S. Latt, R. A. Laskey, C. P. Leblond, P. Leder, R. C. Lewandowski, J. R. McIntosh, V. A. Mc Kusick, O. L. Miller, Jr., C. Milstein, H. H. Mollenhauer, D. J. Morré, J. A. Nathanson, E. Neher, G. L. Nicolson, T. W. O'Brien, Y. Okada, A. L. Olins, M. Osborn, M. P. Osborn, G. E. Palade, B. Pearse, J. Pecci Saavedra, A. Pellegrino de Iraldi, D. J. Pitts, K. R. Porter, D. M. Prescott, E. Racker, M. C. Raff, S. W. Rasmussen, N. R. Ringertz, D. Röhme, D. D. Sabatini, O. Sanchez, R. E. Savage, U. Scheer, W. Schiebler, J. B. Schvartzman, L. A. Staehelin, T. L. Steck, D. F. Steiner, E. F. Stevens, D. L. Taylor, J. O. Thomas, H. Thoenen, P. N. Uwin, L. Van Deenen, K. Weber, A. Worcel, J. J. Yunis.

Mr. Robert Lakemacher and all the staff of W. B. Saunders Company have contributed immensely to improving the presentation of this textbook. We are particularly happy with the work of the Art Department.

CONTENTS

Part 1

INTRODUCTION TO CELL BIOLOGY—METHODS OF STUDY

Chapter 1

INTRODUCTION AND HISTORY OF CELL BIOLOGY 3

1–1 Levels of Organization in Biology 3
    Levels of Organization and Instrumental Resolving Power 5
    Summary: Levels of Organization 7
1–2 History of Cell Biology 7
    The Cell Theory 8
    Cytogenetics—Cytology and Genetics 8
    Cell Physiology 9
    Cytochemistry—Chemical and Physicochemical Cellular
        Analysis 9
    Ultrastructure and Molecular Biology 10
    Summary: Modern Cell Biology 11
1–3 Literary Sources in Cell Biology 11

Chapter 2

PROKARYOTIC AND EUKARYOTIC CELLS—GENERAL STRUCTURE 13

2–1 General Structure of Prokaryotic Cells 13
    Escherichia coli (*E. coli*)—The Most Studied Prokaryote 13
    Smallest Living Mass—DNA, RNAs, and a Plasma Membrane 16
2–2 General Structure of Eukaryotic Cells 16
    Cell Size—Generally Microscopic 16
    Cell Shape—Specific for Each Cell Type 16
    Living Cell—Only Some Structural Components Visible 17
    Fixed Cell—Complex Structural Organization 18
    Summary: Prokaryotic and Eukaryotic Cells 20
2–3 The Nucleus, Chromosomes, and the Cell Cycle—
    General Concepts 21
2–4 Mitosis and Meiosis—Essentials 22
    Mitosis—Maintenance of Chromosomal Continuity
        and Diploid Number 24
    Meiosis—Reduction of Chromosomal Number to Haploid Set 26
    Summary: Essentials About Nucleus and Chromosomes 26

Chapter 3

INSTRUMENTAL ANALYSIS OF BIOLOGICAL STRUCTURES 28

3–1 Various Types of Light Microscopy 28
    Phase Microscopy—Detects Small Differences in
        Refractive Index 29
    Interference and Nomarski Miscroscopy—Detect Continuous
        Changes in Refractive Index of Cell Structures 31

Darkfield Microscopy—Based on Light Scattering at
Cell Boundaries 31
Polarization Microscopy—Detects Anisotrophy with
Polarized Light 32
3–2 Electron Microscopy (EM) 34
Thin Specimens—Essential for EM Study 35
Freeze Fracturing—Membranes Split on Cleavage Planes 35
Preparation of Thin Sections—Epoxy Resins and
Ultramicrotomes 35
Shadow Casting or Negative Staining—Increased Contrast 36
Tracers—Opaque Macromolecules Used 37
High Voltage EM—Allows Study of Thicker Specimens 38
Scanning EM—Surface View of Cell Structures 38
3–3 X-ray Diffraction 39
Summary: Microscopy 42

Chapter 4

CYTOLOGIC AND CYTOCHEMICAL METHODS **45**

4–1 Cell Culture and Microsurgery 45
4–2 Fixation 46
Osmium Tetroxide and Electron Microscopy 47
Freeze-Drying and Freeze-Substitution 48
Microtomes and Embedding 48
4–3 Chemical Basis of Staining 49
Metachromasia—A Change in Original Dye Color 50
Summary: Observation of Living and Fixed Cells 51
4–4 Cytochemistry 51
Cell Fractionation—Separation of Subcellular Fractions 53
Differential or Gradient Centrifugation—Separation of
Cell Particles and Macromolecules 53
Micro- and Ultramicromethods—Detection of
Minute Quantities 54
4–5 Cytochemical and Histochemical Staining Methods 55
Schiff's Reagent—Detection of Aldehydes 55
Lipids—Detection by Lipid-soluble Stains 57
Enzymes—Detection by Incubation with Substrates 57
4–6 Cytochemical Methods Using Physical Techniques 59
Cytophotometric Methods 59
Fluorescence Microscopy—Autofluorescence and
Fluorochrome Dyes 60
Immunocytochemistry—Detection of Antigens with
Labeled Antibodies 61
Radioautography—Interaction of Radioisotopes with
Photographic Emulsions 62
Summary: Cytochemistry 63

Part 2

MOLECULAR COMPONENTS AND METABOLISM OF THE CELL

Chapter 5

BIOCHEMISTRY OF THE CELL **69**

5–1 Chemical Components of the Cell 69
Water is the Most Abundant Cell Component 70
Salts and Ions are Essential 70
Macromolecules—Polymers of Repeating Monomers 70
5–2 Proteins 71
Proteins are Chains of Amino Acids Linked by Peptide Bonds 71

Primary Structure—Amino Acid Sequence.................................. 72
Secondary Structure—α-Helix or Pleated Sheet
Conformation.. 72
Tertiary Structure—Three-dimensional Folding 73
Quaternary Structure—Protein Subunits................................... 74
Weak Interactions are Essential for Protein Structure.................... 75
Electric Charges of Proteins and the Isoelectric Point 76
Separation of Cell Proteins—Isoelectric Focusing and
SDS Electrophoresis 76
5-3 Carbohydrates.. 78
Complex Polysaccharides 78
Glycoproteins—Two-step Carbohydrate Addition....................... 78
5-4 Lipids 80
Triglycerides—Three Fatty Acids Bound to Glycerol..................... 80
Compound Lipids—Phospholipids and Biological Membranes 80
Summary: Molecular Components of the Cell............................ 82
5-5 Nucleic Acids.. 83
Nucleic Acids—A Pentose, Phosphate, and Four Bases................. 84
Regularities in DNA Base Composition—A = T and G = C........... 86
DNA is a Double Helix.. 86
Denaturation and Annealing of DNA....................................... 88
Circular DNA—Supercoiled Conformation................................ 89
RNA Structure—Ribose and Uracil instead of Deoxyribose
and Thymine... 90
The Simplest Infectious Agent—A Circular RNA Chain................. 91
Summary: Nucleic Acids.. 93

Chapter 6
ENZYMES, BIOENERGETICS, AND CELL RESPIRATION **95**

6-1 Enzymes... 95
Enzymes are Proteins 95
Enzymes are Highly Specific 96
Some Enzymes Require Cofactors... 96
Substrates Bind to the Active Site 97
Enzyme Kinetics—Km and Vmax Define Enzyme Behavior........... 97
Enzyme Inhibitors can be Very Specific 99
Isoenzymes—Multiple Molecular Forms 100
The Cell is not Simply a Bag Full of Enzymes 100
Allosteric Enzymes have Multiple Interacting Subunits 100
6-2 Metabolic Regulation.. 101
Enzymes are Regulated at the Catalytic and the Genetic Levels 101
Enzyme Interconversions Also Regulate Metabolism 102
Cyclic AMP—The Second Messenger in Hormone Action 102
Summary: Enzymes in The Cell 105
6-3 Bioenergetics.. 106
Entropy is Related to the Degree of Molecular Disorder................. 106
Photosynthesis is Essential in the Biological Energy Cycle 107
Cells Utilize Chemical Energy.. 107
ATP has High Energy Bonds 108
6-4 Cell Respiration 109
Anaerobic Glycolysis Yields only 2 ATPs per Glucose
Molecule.. 109
Aerobic Respiration Produces 36 ATPs per Glucose Molecule 110
The Krebs Cycle—The Common Final Pathway in the
Degradation of Fuel Molecules.. 110
The Respiratory Chain—Stepwise Release of Energy by
Electron Pairs 111
Oxidative Phosphorylation—The Energy Released by
Electron Pairs Produces ATP... 112
Summary: Bioenergetics and Cell Respiration 112

Part 3

SUPRAMOLECULAR STRUCTURE AND THE CELL SURFACE

Chapter 7

SUPRAMOLECULAR ORGANIZATION AND THE ORIGIN OF CELLS............ 117

7-1 The Shape of Protein Molecules 118
7-2 The Assembly of Macromolecules.. 118
    Assembly of Viruses—Nucleic Acids and Proteins 119
    Collagen Fibers—Supramolecular Assemblies of Tropocollagen....... 120
    Fibrinogen and Thrombin in Blood Clotting 120
    Glycogen Particles—Three Levels of Organization 122
7-3 Elementary Membranous Structures.. 123
    Lipids Tend to Form Monolayer Films 123
    Artificial Lipid Bilayers—Important Model Systems 123
    Bulk Phospholipid-Water Systems Form Hexagonal and
        Lamellar Structures 124
    Liposomes and Phospholipid Vesicles—Possible
        Applications in Biology and Medicine 126
    Summary: Supramolecular Organization.................................... 126
7-4 The Origin of Cells ......:.. 127
    Chemical Evolution Produced Carbon-Containing Molecules.......... 127
    Mechanisms of Assembly Were at Work to Form Primitive
        Proteinoids 128
    Prokaryotic Cells Preceded the Eukaryotic Cells.......................... 129
    Summary: The Origin of Cells 129

Chapter 8

THE CELL MEMBRANE AND PERMEABILITY; INTERCELLULAR
INTERACTIONS 131

8-1 Molecular Organization of the Cell Membrane 132
    The Cell Membrane—Composed of Proteins, Lipids, and
        Carbohydrates... 132
    Lipids are Asymmetrically Distributed within the Bilayer.............. 133
    Carbohydrates—In the Form of Glycolipids and Glycoproteins....... 134
    Membrane Proteins—Peripheral or Integral 135
    Polypeptides of the Red Cell Membrane.................................... 135
    Every Protein of the Cell Membrane is Distributed
        Asymmetrically.. 135
    Major Polypeptides of the Red Cell Membrane are Well
        Characterized 137
    Asymmetrical Distribution of Enzymes 138
    Summary: Molecular Organization of the Cell Membrane 139
8-2 Molecular Models of the Cell Membrane.. 139
    The Unit Membrane Model—Re-evaluation of the EM Image 140
    The Fluid Mosaic Model is Now Generally Accepted..................... 140
    Membrane Fluidity—Studied with Physical and Biological
        Techniques 142
    Membrane Fluidity and Coupling of Receptors to
        Adenylate Cyclase.. 143
    The Myelin Sheath and the Photoreceptors—Special
        Multi-layered Membranes 144
    Summary: Membrane Molecular Models.................................... 147
8-3 Cell Permeability 148
    Different Ionic Concentrations across the Membrane Create
        Electrical Potentials 148
    Passive Permeability—Dependent on the Concentration
        Gradient and the Partition Coefficient.................................... 149

Passive Ionic Diffusion—Dependent on Concentration and
Electrical Gradients 150
Active Transport of Ions Uses Energy 151
A "Sodium Pump" is Postulated in the Active Efflux of Na$^+$.............. 151
Ionic Transport through Charged Pores in the Membrane............. 153
Anion Transport in Erythrocytes Involves the Special
Band-3 Polypeptide.. 153
The Vectorial Function of Na$^+$ K$^+$ ATPase—The Carrier
Hypothesis 154
Various Substances are Transported by a Carrier Mechanism......... 155
Selectivity of Transport—Dependent on Permease Systems 156
Penetration by Large Molecules—Various Mechanisms.................. 156
Summary: Cell Permeability 157
8–4 Differentiations At The Cell Surface and Intercellular
Communications.. 158
Microvilli—A Greatly Increased Cell Membrane Surface Area 158
Desmosomes and Intermediary and Tight Junctions—
Intercellular Attachments 158
Gap Junctions (Nexus) and Intercellular Communications............. 161
Electrical Coupling between Cells Depends on Gap Junctions 161
Gap Junctions—Channels Permeable to Ions and Small Molecules... 163
Coupling between Cells Enables Metabolic Cooperation............... 165
Altered Coupling in Cancer Cells 165
Summary: Differentiations of the Cell Membrane and
Intercellular Communications 166
8–5 Coats of the Cell Membrane and Cell Recognition 166
Numerous Functions are Attributed to the Cell Coat.................... 168
Cell-cell Recognition—Specific Cell Adhesion and
Contact Inhibition... 168
Cancer Cells—Many Changes in the Cell Surface Properties........... 169
Transformation of Cells—Produced by Certain Viruses.................. 173
Summary: Cell Coats and Cell Recognition 173

Part 4

THE CYTOPLASM AND CYTOPLASMIC ORGANELLES

Chapter 9

THE CYTOSKELETON AND CELL MOTILITY: MICROTUBULES
AND MICROFILAMENTS **179**

9–1 The Cytosol and the Cytoskeleton 179
The Cytoskeletal Fabric—Microtubules and Microfilaments.............. 181
Summary: The Cytoskeleton 183
9–2 Microtubules.. 183
Tubulin—The Main Protein of Microtubules............................... 183
Microtubules—Assembled from Tubulin Dimers 185
Detection of Microtubules in Culture Cells by Anti-tubulin
Antibodies... 185
Functions of Cytoplasmic Microtubules...................................... 186
Summary: Properties of Microtubules 187
9–3 Microtubular Organelles: Cilia, Flagella, and Centrioles...................... 187
Ciliary and Flagellar Motions—Present in Cells and in Tissues....... 188
The Ciliary Apparatus—The Cilium, Basal Bodies, and Ciliary
Rootlets.. 188
The Axoneme Contains Microtubular Doublets 188
Basal Bodies (Kinetosomes) and Centrioles Contain
Microtubular Triplets... 190
Ciliary Movement—A Sliding of Microtubular Doublets That
Involves Dynein 191

Kartagener's Syndrome—A Mutation Involves a Lack of Dynein..... 192
Photoreceptors are Derived from Cilia 193
Cilia and Flagella Originate from Basal Bodies 193
Summary: Structure, Motion, and Origins of Cilia and Flagella....... 196
9–4 Microfilaments 197
Microtrabecular Lattice in Cytosol—Revealed by
High-Voltage EM.. 197
Cytochalasin B Impairs Several Cellular Activities Involving
Microfilaments 197
Actin, Myosin, and Other Contractile Proteins Present in
Non-muscle Cells... 198
Contractile and Regulatory Proteins—Detected by Specific
Antibodies; Stress Fibers 199
Two Recognizable Types of Microfilaments 199
Microfilaments—Involved in All Motility in Non-muscle Cells 200
Cytoplasmic Streaming (Cyclosis)—Observed in Large Plant Cells... 200
Ameboid Motion—Characteristic of Amebae and Many Free Cells... 201
Summary: Microfilaments, Cyclosis, and Ameboid Motion 203

Chapter 10

THE ENDOPLASMIC RETICULUM AND CELL SECRETION I **206**

10–1 General Morphology of the Endomembrane System........................... 208
The Rough ER—Ribosomes and Protein Synthesis........................ 209
Ribosomal Binding to the ER—60S Subunit and Ribophorins
Involved 210
The Smooth ER Lacks Ribosomes... 210
Summary: Endoplasmic Reticulum 213
10–2 Microsomes—Biochemical Studies... 213
Microsomal Membranes—A Complex Lipid and Protein
Composition 214
Two Microsomal Electron Transport Systems—Flavoproteins
and Cytochromes b and P-450 Involved 216
Microsomal Enzymes—Glycosidation and Hydroxylation of
Amino Acids.. 217
Microsomal Enzymes—Asymmetry Across the Membrane 217
Summary: Microsomes... 217
10–3 Functions of the Endoplasmic Reticulum..................................... 218
Membrane Biogenesis Involves a Multi-Step Mechanism 218
ER-Membrane Fluidity and Flow Through the Cytoplasm.............. 219
Ions and Small Molecules—Transport Across ER Membranes 219
Special Functions of the Smooth ER—Detoxification, Lipid
Synthesis, and Glycogenolysis 219
Summary: Functions of Endoplasmic Reticulum 220
10–4 The Endoplasmic Reticulum and Synthesis of Exportable Proteins 221
Special Initial Codons for Signal Peptides—In mRNA in
ER-bound Polysomes... 221
The Hydrophobic Signal Peptide is Removed by a Signal
Peptidase... 221
Membrane Proteins are Made and Assembled in Different
Compartments... 222
Summary: Synthesis of Exportable Proteins—The Signal
Hypothesis 223

Chapter 11

GOLGI COMPLEX AND CELL SECRETION II.. **226**

11–1 Morphology of the Golgi Complex (Dictyosomes) 226
Dictyosomes—A Forming Face, and a Maturing Face Near
the GERL... 227

Polarization of Dictyosomes and Membrane Differentiation 228
Summary: Morphology of the Golgi Complex.............................. 232
11-2 Cytochemistry of the Golgi Complex 232
Chemical Composition of the Golgi Complex – Intermediate
Between Those of the ER and the Plasma Membrane 233
Glycosyl Transferases are Concentrated in the Golgi 233
Summary: Cytochemistry of the Golgi Complex 234
11-3 Functions of the Golgi Complex.. 235
Synthesis of Glycosphingolipids and Glycoproteins – A
Major Role of the Golgi.. 235
Cancer Cells – Changes in Glycosidation of Lipids and Proteins 236
Secretion – The Main Function of the Golgi Complex 236
The Secretory Cycle – Continuous or Discontinuous...................... 236
The Secretory Process in the Pancreas – Six Consecutive Steps........ 239
The GERL Region and Lysosome Formation 244
Insulin Biosynthesis – A Good Example of the Molecular
Processing of Secretion 245
Summary: Functions of the Golgi Complex................................. 245

Chapter 12

MITOCHONDRIA AND OXIDATIVE PHOSPHORYLATION.............................. **249**

12-1 Morphology of Mitochondria.. 250
Summary: Mitochondrial Morphology 252
12-2 Mitochondrial Structure 252
The Mitochondrial Matrix Contains Ribosomes and a
Circular DNA... 253
F_1 Particles – On the M Side of the Inner Mitochondrial
Membrane 253
Mitochondrial Structural Variations in Different Cell Types 256
Mitochondrial Sensitivity to Cell Injury and Resultant
Degeneration... 259
Summary: Structure of Mitochondria 259
12-3 Isolation of Mitochondrial Membranes... 259
Outer and Inner Membranes – Structural and Chemical
Differences... 261
Mitochondrial Enzymes are Highly Compartmentalized 261
The Inner Membrane – Regional Structural and
Enzymatic Differences.. 263
Summary: Isolation of Mitochondrial Membranes 263
12-4 Molecular Organization and Function of Mitochondria...................... 263
Electrons Flow Along a Cytochrome Chain Having a
Gradient of Redox Potential... 264
The Respiratory Chain – Four Molecular Complexes...................... 265
Electron Transport – Coupled to the Phosphorylation at
Three Points.. 267
Mitochondrial ATPase – A Structurally Complex Proton Pump........ 268
Topological Organization of the Respiratory Chain and
the Phosphorylation System 268
The Chemiosmotic Hypothesis – An Electrochemical Link
Between Respiration and Phosphorylation.............................. 269
The Chemical-Conformational Hypothesis Involves
Short-Range Interactions... 270
Summary: Molecular Organization and Function 270
12-5 Permeability of Mitochondria 271
ADP, ATP, and Pi – Transport by Specific Carriers 272
Mitochondrial Conformation – Changes with Stages of
Oxidative Phosphorylation... 272
Mitochondrial Swelling and Contraction – Agent-Induced.............. 272
Mitochondrial Accumulation of Ca^{2+} and Phosphate.................... 274
Summary: Mitochondrial Permeability 275

12–6 Biogenesis of Mitochondria 275
    Mitochondria are Semiautonomous Organelles............................ 276
    Mitochondrial DNA—Circular and Localized in the Matrix 276
    Mitochondrial Ribosomes—Smaller than Cytoplasmic Ribosomes ... 277
    Mitochondria Synthesize Mainly Hydrophobic Proteins 278
    The Symbiont Hypothesis—Mitochondria and Chloroplasts are
        Intracellular Prokaryotic Parasites... 278
    Summary: Mitochondrial Biogenesis... 279

Chapter 13

LYSOSOMES, THE CELL DIGESTIVE SYSTEM, AND PEROXISOMES............ **282**

13–1 Major Characteristics of Lysosomes 282
    Various Cytochemical Procedures for Identifying Lysosomes.......... 285
    Lysosomes are Very Polymorphic... 285
    Primary Lysosomes and Three Types of Secondary Lysosomes 287
    Lysosomal Enzymes—Synthesized in the ER, Packaged in
        the Golgi 288
    Summary: Lysosomal Morphology and Cytochemistry................... 288
13–2 Endocytosis... 288
    Phagocytosis—The Process of Cellular Ingestion of
        Solid Material.. 289
    Pinocytosis—The Ingestion of Fluids 289
    Micropinocytosis and the Ingestion and Transport of Fluids 290
    Micropinocytosis—Frequently Associated with the Formation
        of Coated Vesicles.. 292
    Endocytosis—An Active Mechanism Involving the Contrac-
        tion of Microfilaments 293
    Summary: Endocytosis and Lysosomes 293
13–3 Functions of Lysosomes—Intracellular Digestion 294
    Autophagy by Lysosomes—The Renovation and Turnover
        of Cell Components... 294
    Lysosomal Removal of Cells and Extracellular Material
        and Developmental Processes 294
    Release of Lysosomal Enzymes to the Medium for
        Extracellular Action 294
    Lysosomal Enzyme Involvement in Thyroid Hormone
        Release and in Crinophagy... 295
    Leukocyte Granules are of a Lysosomal Nature 295
    Lysosomes are Important in Germ Cells and Fertilization 295
    Lysosomal Involvement in Human Diseases and Syndromes.......... 295
    Storage Diseases—Caused by Mutations That Affect
        Lysosomal Enzymes 296
    Lysosomes in Plant Cells and Their Role in Seed Germination 296
    Summary: Functions of Lysosomes 296
13–4 Peroxisomes.. 297
    Peroxisomes Contain Nucleoids: Microperoxisomes
        Lack Them.. 297
    Peroxisomal Enzymes—Made in the Rough ER 297
    Peroxisomes Contain Enzymes Related to the
        Metabolism of H_2O_2... 298
    Plant Peroxisomes are Involved in Photorespiration 298
    Summary: Peroxisomes... 298

Chapter 14

THE PLANT CELL AND THE CHLOROPLAST **300**

14–1 The Cell Wall of Plant Cells 301
    The Cell Wall—A Network of Cellulose Microfibrils and
        a Matrix... 301

Development of Primary and Secondary Walls and Plant
Cell Differentiation 302
Cell Wall Components—Synthesis Associated with the Golgi
or the Plasma Membrane 302
Plasmodesmata Establish Communication Between
Adjacent Cells 302
Summary: The Plant Cell Wall 305
14–2 Plant Cell Cytoplasm 305
The Endoplasmic Reticulum of Plants—Formation of Protein
Bodies, Glyoxysomes, and Vacuoles....................................... 305
Development—Cell Division, Cell Expansion, and a Drying Phase... 305
Seed Germination Involves Synthesis of Hydrolytic Enzymes......... 307
Glyoxysomes—Organelles Related to Triglyceride Metabolism........ 307
Dictyosomes are Involved in Several Secretion Processes 307
Mitochondria can be Distinguished from Proplastids 307
Summary: Plant Cell Cytoplasm 309
14–3 The Chloroplast and Other Plastids 309
Life on Earth is Dependent on the Function of
Chloroplasts in Photosynthesis 310
Chloroplast Morphology Varies in Different Cells........................ 310
Chloroplasts are Motile Organelles and Undergo Division 310
The Envelope, Stroma, and Thylakoids are the Main
Components of Chloroplasts.. 311
Freeze-fracturing Best Reveals the Substructure of the
Thylakoid Membrane 312
Chloroplast Development—The Formation of Granal and
Stroma Thylakoids 314
14–4 Molecular Organization of Thylakoids... 314
Several Chlorophyll-protein Complexes in the
Thylakoid Membrane 317
The Photophosphorylation Coupling Factor and the Photo-
systems—Vectorially Disposed Across the Thylakoid Membrane... 318
Summary: Structure and Molecular Organization of Chloroplasts..... 318
14–5 Photosynthesis 319
The Primary Reaction of Photosynthesis—The Photochemical
Reaction 320
Photosynthetic Carbon Reduction—The Major Group of
Chemical Reactions in Photosynthesis 322
The C_4 Pathway is Found in Some Angiosperms.......................... 324
Summary: Photosynthesis and the Chloroplast 324
14–6 A Structural-Functional Model of the Chloroplast membrane............... 325
Ion Fluxes and Conformational Changes—Caused by Light
and Darkness 326
Summary: A Structural—Functional Model and
Conformational Changes... 327
14–7 Chloroplasts as Semiautonomous Organelles.................................... 327
Summary 328

Part 5

THE NUCLEUS AND CHROMOSOMES

Chapter 15

**THE INTERPHASE NUCLEUS, CHROMATIN, AND THE
CHROMOSOMES** **333**

15–1 The Nuclear Envelope.. 334
Nuclear Pores are not Wide-Open Channels; Pore Complexes 334
Annulated Lamellae—Cytoplasmic Stores of Pore Complexes 335
Nuclear Pores—Selective Diffusion Barriers Between
Nucleus and Cytoplasm 335

Nuclear Proteins Accumulate Selectively in the Nucleus 337
Summary: The Nuclear Envelope.. 339
15-2 Chromatin 340
Nuclei Contain a Constant Amount of DNA 340
Thin-section Electron Microscopy did not Reveal Details
of Chromatin Structure 340
Chromatin is a Complex of DNA and Histones 342
Electron Microscopy of Chromatin Spreads Revealed
a Beaded Structure 343
The Nucleosome—An Octamer of Four Histones (H2A, H2B,
H3, and H4) Complexed with DNA 343
The 20 to 30 nm Fiber—A Further Folding of the
Nucleosome Chain 347
Summary: Chromatin 347
15-3 The Chromosomes 348
Chromosomal Shape is Determined by the Position of
the Centromere... 348
Chromosomal Nomenclature—Chromatid, Centromere,
Kinetochore, Telomere, Satellite, Secondary Constriction,
Nucleolar Organizer 348
Karotype—All the Characteristics of a Particular
Chromosomal Set... 350
Each Chromatid has a Single DNA Molecule............................. 350
Metaphase Chromosomes are Symmetrical 351
Interphase Chromosomal Arrangement—Random or
Non-random?... 351
Chromosomes Undergo Condensation-Decondensation Cycles 352
A Chromosomal Scaffold of Non-histone Proteins 354
15-4 Heterochromatin.. 354
Heterochromatin—Chromosomal Regions That do not
Decondense During Interphase 354
Heterochromatin can be Facultative or Constitutive 354
Heterochromatin is Genetically Inactive 356
Constitutive Heterochromatin Contains Repetitive DNA
Sequences.. 357
Summary: The Chromosomes and Heterochromatin.................... 358
15-5 The Nucleolus 359
Electron Microscopy of the Nucleolus—A Fibrillar and
a Granular Zone 359
The Nucleolus is Disassembled During Mitosis.......................... 361
Summary: The Nucleolus 361

Chapter 16

THE CELL CYCLE AND DNA REPLICATION.. **364**

16-1 The Cell Cycle.. 364
Interphase—The G_1, S, and G_2 Phases 365
G_1—The Most Variable Period of the Cell Cycle 365
Visualization of Chromosomes During G_1, S, and G_2 by
Premature Condensation... 366
Condensed Chromosomes do not Synthesize RNA 368
Several Molecular Events Occur at Defined Stages of
the Cell Cycle 369
Summary: The Cell Cycle 370
16-2 DNA Replication 370
DNA Replication is Semiconservative 371
Replication of the *E. coli* Chromosome is Bidirectional 371
DNA Synthesis is Discontinuous 373
Reverse Transcriptase can Copy RNA into DNA......................... 374
Eukaryotic Chromosomes Have Multiple Origins of Replication...... 375
Eukaryotic DNA Synthesis is Bidirectional 376
The Number of Replication Units is Developmentally Regulated..... 376

Reinitiation within the Same Cycle is Prevented in
  Eukaryotic Cells.. 377
New Nucleosomes are Assembled Simultaneously with
  DNA Replication 377
RNA Synthesis Continues During DNA Replication 378
DNA Repair Enzymes Remove Thymine Dimers Induced
  by UV Light... 378
Xeroderma Pigmentosa Patients Have Defective DNA Repair......... 379
Summary: DNA Replication 380

Chapter 17

MITOSIS AND CELL DIVISION... **383**

17–1 A General Description of Mitosis 383
      Prophase—Chromatid Coiling, Nucleolar Disintegration,
        and Spindle Formation 384
      Metaphase—Chromosomal Orientation at the Equatorial Plate 385
      Anaphase—Movement of the Daughter Chromosomes Toward
        the Poles.. 385
      Telophase—Reconstruction of the Daughter Nuclei...................... 385
      Summary: Events in Mitosis.. 387
17–2 Molecular Organization and Functional Role of the Mitotic Apparatus... 387
      Centromere Association with the Kinetochore, Where Spindle
        Microtubules are Implanted...................................... 391
      Spindle Microtubules—Kinetochore, Polar, and Free 392
      Assembly of Spindle Microtubules—Control by Poles and
        Kinetochores 392
      Metaphase Chromosomal Motion—Caused by Kinetochore and
        Polar Microtubular Interaction 393
      Anaphase Chromosomal Movement—The Dynamic
        Equilibrium and the Sliding Mechanism Hypothesis 394
      Cytokinesis in Animal Cells—Contractile Ring of Actin
        and Myosin... 395
      Cytokinesis in Plant Cells—The Phragmoplast and Cell Plate 395
      Summary: The Mitotic Apparatus.. 397

Chapter 18

MEIOSIS AND SEXUAL REPRODUCTION **400**

18–1 A Comparison of Mitosis and Meiosis 401
18–2 A General Description of Meiosis...................................... 402
      Leptotene Chromosomes Appear to be Single and Have
        Chromomeres... 402
      Zygonema—Pairing of Homologous Chromosomes and
        Synaptonemal Complex Formation 405
      Pachynema—Crossing Over and Recombination Between
        Homologous Chromatids 408
      The Synaptonemal Complex—Homologue Alignment and
        Recombination 409
      The Recombination Nodule is Probably Related to Crossing Over... 409
      Diplonema—Separation of the Paired Chromosomes Except
        at the Chiasmata.. 409
      Diakinesis—Reduction in the Number of Chiasmata.................... 410
      Meiotic Division I—Separation of the Homologous Centromeres..... 410
      Meiotic Division II—Separation of the Sister Centromeres............ 410
      Chromosome Distribution in Mitosis and Meiosis—
        Dependent on Kinetochore Orientation................................ 410
      Summary: Events in Meiosis 412
18–3 Genetic Consequences of Meiosis and Types of Meiosis...................... 413
      Meiosis is Intermediary or Sporic in Plants.............................. 414

Meiosis may Last for 50 Years in the Human Female 415
Meiosis Starts after Puberty in the Human Male.......................... 416
Fertilization – A Species-specific Interaction Between the
    Gametes 416
Summary: More About Meiosis.. 416
18–4 Biochemistry of Meiosis 417
Summary: 418

Chapter 19

CYTOGENETICS – CHROMOSOMES AND HEREDITY **420**

19–1 Mendelian Laws of Inheritance 420
The Law of Segregation – Genes are Distributed without Mixing 421
Independent Assortment – Genes in Different Chromosomes
    are Distributed Independently During Meiosis......................... 422
Genes are Linked When They are on the Same Chromosome 422
Linkage May be Broken by Recombination During Meiotic
    Prophase 424
Neurospora – Ideal for Studying Recombination and Gene
    Expression 424
Summary: Fundamental Genetics 424
19–2 Chromosomal Changes and Cytogenetics 426
Euploidy – A Change in the Number of Chromosome Sets;
    Aneuploidy – A Loss or Gain of Chromosomes 426
Aneuploidy – The Result of Non-disjunction at Meiosis or Mitosis... 427
Chromosomal Aberrations – Structural Alterations; Gene
    Mutations – Molecular Changes.. 428
Radiation and Chemical Mutagens Act Mainly on the
    DNA Molecule... 429
Germ Cell Mutations are Transmittable, Somatic are not............... 430
Chromosomal Aberrations – Deletion, Duplication, Trans-
    location, Inversion, and Sister Chromatid Exchange 430
Sister Chromatid Exchanges are Increased in Diseases
    with Altered DNA Repair 432
19–3 Chromosomes Play A Fundamental Role in Evolution 433
Summary: Chromosomal Aberrations, Action of Mutagens,
    and Cytogenetics and Evolution.. 435

Chapter 20

HUMAN CYTOGENETICS **438**

20–1 The Normal Human Karotype.. 439
Karotype Preparation – Chromosomal Ordering by Size
    and Centromere Position 439
Banding Techniques Reveal Structural Details of Chromosomes...... 440
Banding Provides New Features for Identification of Human
    Chromosomes 440
Summary: The Human Karotype 443
20–2 Sex Chromosomes and Sex Determination 444
X and Y Sex Chromatin in Interphase Nuclei 444
X Chromatin – Facultative Heterochromatin Bodies Equaling
    nX − 1 in Number 445
Y Chromatin – The Heterochromatic Region of the Y
    Chromosome... 446
Sex is Determined by Sex Chromosomes, but Hormones
    Influence Differentiation 446
Summary: Sex Chromosomes and Sex Determination 448
20–3 Human Chromosomal Abnormalities 448
Aneuploidy – Caused by Non-disjunction of Chromosomes 449
Reciprocal Translocations – Identified by Banding....................... 450
Sex Chromosome Abnormalities in Human Syndromes 450

Mongolism—The Best Known Autosomal Abnormality 452
Structural Aberrations May Cause Syndromes Besides
    Aneuploidy... 454
Banding Techniques Have Detected More Than 30
    New Syndromes 454
Certain Human Tumors Show Specific Chromosomal
    Aberrations.. 454
Summary: Chromosomal Abnormalities 454
20–4 Human Chromosomes and The Genetic Map 455
Daltonism and Hemophilia—The Best Known Sex-Linked
    Diseases 455
Chromosome Mapping—Linkage Analysis, Somatic Cell
    Hybridization, and In Situ Hypridization............................. 457
Localization of Genes in Chromosomes and Regions of
    Chromosomes 458
Summary: The Human Genetic Map.. 458

Part 6

GENE EXPRESSION

Chapter 21

THE GENETIC CODE AND GENETIC ENGINEERING 463

21–1 The Genetic Code 464
Genes Code for Proteins—Inborn Errors of Metabolism.................. 464
Mutations Produce Amino Acid Changes 465
Genes are Made of DNA—Genetic Transformation 467
Three Nucleotides Code for One Amino Acid 468
Artificial mRNAs Were Used to Decipher the Genetic Code........... 468
Synonyms in the Code—61 Codons for 20 Amino Acids............... 470
AUG—The Initiation Codon; UAG, UAA, and UGA—
    Termination Codons.. 470
The Genetic Code is Universal.. 471
Summary: The Genetic Code 471
21–2 Mutations and the Genetic Code 472
Suppressor Mutations—Reversal of the Effect of Point
    Mutations 472
Chemical Mutagens Can be Very Specific................................... 473
Summary: Mutations and the Genetic Code................................ 473
21–3 DNA Sequencing and the Genetic Code...................................... 474
The 5375-Nucleotide Sequence of ϕ X 174 Revealed
    Overlapping Genes 475
Nucleic Acid Sequencing Confirmed the Genetic Code 476
Summary: DNA Sequencing and the Confirmation of the
    Genetic Code 478
21–4 Genetic Engineering—Restriction Endonucleases............................ 478
Restriction Enzymes Recognize Specific DNA Sequences.............. 479
Eukaryotic Genes can be Introduced into Plasmids and
    Cloned in E. coli 480
Summary: Genetic Engineering.. 482

Chapter 22

TRANSCRIPTION AND PROCESSING OF RNA 484

22–1 Messenger RNA in Prokaryotes.. 484
Prokaryotic RNAs are Transcribed by a Single RNA Polymerase..... 486
Transcription has Three Stages—Initiation at Promoters,
    Elongation, and Termination 486
In Prokaryotes Transcription is Coupled to Protein Synthesis........ 489
Summary: RNA Transcription in Prokaryotes............................... 489

22–2 Transcription in Eukaryotes.. 490
    Three Different RNA Polymerases Transcribe the Various
      Eukaryotic RNAs 490
    Transcription May be Visualized by Electron Microscopy 491
    Summary: Transcription in Eukaryotes.................................... 492
22–3 Eukaryotic Messenger RNA 492
    The 5′ End of Eukaryotic mRNA is Blocked by 7-Methyl-G........... 493
    Eukaryotic Genes Frequently Contain Insertions of Non-
      Coding DNA.. 494
    The 3′ End of Eukaryotic mRNA – A Poly A Segment 495
    Heterogeneous Nuclear RNAs may be mRNA Precursors
      Containing Intervening Sequences 497
    Eukaryotic mRNAs are Associated with Proteins........................ 497
    Summary: Eukaryotic Messenger RNA 498

Chapter 23

RIBOSOMES AND NUCLEOLAR FUNCTION **501**

23–1 Ribosomes.. 501
    Ribosomes may be Free or Membrane-bound 501
    Ribosome Composition – A Large and a Small Subunit 502
    Ribosomal RNAs – 18S, 28S, 5.8S, and 5S in Eukaryotes................ 503
    Summary: Structure of Ribosomes and rRNAs 505
23–2 Ribosomal Proteins 506
    Dissociation and Reconstitution of Ribosomal Proteins 506
    Prokaryotic and Eukaryotic Ribosomes – Little Homology
      but the Same Basic Function... 507
    Summary: Organization of Ribosomes 507
23–3 Biogenesis of Ribosomes and Nucleolar Function 508
    The Nucleolar Organizer Contains Ribosomal DNA..................... 508
    Ribosomal Genes are Tandemly Repeated and Separated
      by Spacer DNA 509
    5S Genes – Random Repeats with Spacers, but outside the
      Nucleolar Organizer 511
    Ribosomal DNA is Highly Amplified in Oöcytes........................ 511
    Ribosomal RNAs Undergo a Complex Processing in the
      Nucleolus 513
    Ribosomal Biogenesis can be Followed Under the Electron
      Microscope 514
    Summary: Biogenesis of Ribosomes.. 515

Chapter 24

PROTEIN SYNTHESIS.. **519**

24–1 Transfer RNA.. 519
    tRNA Transcription – Long Precursors with Complex Post-
      transcriptional Modifications 521
    The Fidelity of Protein Synthesis Depends on the Correct
      Charging of tRNA by the AA-tRNA Synthetases 521
    Summary: Transfer RNA 523
24–2 Ribosomes and Protein Synthesis.. 524
    Polysomes Consist of Several Ribosomes Attached to the
      Same mRNA 524
    During Protein Synthesis There is a Ribosome-Polysome Cycle...... 524
    The Small Subunit Binds to the mRNA Ribosome Binding Site 526
    Prokaryotic Proteins Start with Formylmethionine....................... 527
    IF_1, IF_2, and IF_3 – Protein Factors That Initiate Protein Synthesis 527
    Eukaryotic Initiation Factors are More Complex Than Those
      in Prokaryotes 529

Eukaryotic IF$_2$ is Phosphorylated by a Cyclic AMP-dependent
Protein Kinase 529
EFTu + EFTs, and EFG are Protein Elongation Factors 529
The Large Ribosomal Subunit is Involved in the Process of
Elongation 530
Chain Termination Involves UAA, UGA, and UAG Codons
and Releasing Factors R$_1$ and R$_2$ 530
Antibiotics and Toxins are Useful in Molecular Biology
and Medicine 531
Secretory Proteins have a Hydrophobic Signal Peptide 531
Summary: Protein Synthesis 533

Chapter 25

GENE REGULATION 536

25–1 Gene Regulation in Prokaryotes 537
Control of Enzyme Induction and Repression—Reprssors
Bind to Operators 537
The Lac Operon Codes for β-Galactosidase, Lac Permease,
and Transacetylase 537
The i Gene Codes for the Repressor Protein, which Binds
to the Inducer 538
The Operator is a DNA Sequence that has Two-fold Symmetry 538
Repressors Bind within Promoters and Prevent Attachment
of RNA Polymerase 539
The Jacob-Monod Model is a Mechanism of Nagative Control 540
Positive Control of Lac Transcription is Regulated by the
CAP-cAMP Complex 540
The Tryptophan Operon—Regulation of Transcription at
Initiation and Termination 541
A Finer Control of Metabolism Occurs at the Level of
Enzyme Activity 543
Summary: Gene Regulation in Prokaryotes 543
25–2 Gene Regulation in Eukaryotes 544
Heterochromatin is Not Transcribed 545
Repetitive DNA Sequences Are Characteristic of the Eukaryotes 545
Satellite DNAs Contain the Most Highly Repetitive Sequences 546
Moderately Repetitive Genes—rDNA, 5S DNA, and
Histone Genes 547
Summary: Gene Regulation in Eukaryotes and Repetitive DNA 547
25–3 Regulation at the Chromosome Level—Giant Chromosomes 548
Polytene Chromosomes have a Thousand DNA Molecules
Aligned Side by Side 548
The Bands Represent Chromomeres Aligned Side by Side 550
Puffs in Polytenes are the Sites of Gene Transcription 550
Puffs Can be Induced by Ecdysone and by Heat Shock 552
Lampbrush Chromosomes in Oöcytes at the Diplotene Stage
of Meiosis 552
The Lateral Loops are Sites of Intense RNA Synthesis 553
Summary: Regulation in Eukaryotes at the Chromosome Level 556

Chapter 26

CELL DIFFERENTIATION 560

26–1 General Characteristics of Cell Differentiation 562
The Differentiated State is Stable 562
Determination can Precede Morphological Differentiation 562
26–2 Nucleocytoplasmic Interactions 562

Acetabularia is Well Suited for the Study of Nucleocyto-
plasmic Interactions 563
Red Blood Cell Nuclei can be Reactivated by Cell Fusion.............. 563
Cytoplasts and Karyoplasts can be Prepared by Cytochalasin
B Enucleation 565
Cell Fusion Yields Pure Antibodies of Medical Importance 566
Gene Expression by Somatic Nuclei is Reprogrammed in
Xenopus Oöcytes.. 567
Summary: Nucleocytoplasmic Interactions 570
26–3 Molecular Mechanisms of Cell Differentiation 571
The Genome Remains Constant During Cell Differentiation—
Nuclear Transplantation... 571
Gene Amplification Cannot Explain Cell Differentiation................ 571
Cell Differentiation Cannot be Explained by Translational
Control.. 572
Cell Differentiation is Probably Controlled at the Level
of Transcription.. 574
Cytoplasmic Determinants Localized in Egg Cytoplasm
are Important in Early Development 575
Cell Differentiation is Intimately Related to Cellular Interactions...... 578
Summary: Mechanisms of Cell Differentiation............................. 579

Part 7

CELL AND MOLECULAR BIOLOGY OF SPECIALIZED CELLS

Chapter 27

CELLULAR AND MOLECULAR BIOLOGY OF MUSCLE 585

27–1 Structure of the Striated Muscle Fiber 586
The Myofibril and Sarcomere are Structures Differentiated
for Contraction 586
Thick and Thin Myofilaments are the Macromolecular
Contractile Components... 587
The Z-line Shows a Woven-basket Lattice and Contains
α-Actinin... 588
The Sarcomere I-Band Shortens and the Banding Inverts
During Contraction 589
Smooth Muscles Lack the Z-line... 590
Summary: The Structure of Muscle 590
27–2 Molecular Organization of the Contractile System............................ 591
The Thick Myofilament—Myosin Molecules; The Cross-
bridges—S_1 Subunits... 591
Actin, Tropomyosin, and the Troponins Constitute the
Thin Myofilament 593
Contractile and Regulatory Proteins—Localization by
Fluorescent Antibodies 593
27–3 The Sliding Mechanism of Muscle Contraction 593
Myosin and Actin have A Definite Polarization within
the Sarcomere... 594
Muscle Contraction—A Cyclic Formation and Breakdown
of Actin-myosin Linkages 595
Summary: Molecular Organization and the Sliding Mechanism 596
27–4 Regulation and Energetics of Contraction 597
Molecular Regulation—Displacement of Tropomyosin after
Binding of Ca^{2+} to Troponin-C 597
Energy for Contraction Comes from Oxidative Phosphoryla-
tion and Glycolysis 598
Summary: Regulation and Energetics of Contraction 598

27–5 Excitation-Contraction Coupling 598
    Sarcoplasmic Reticulum – A Longitudinal Component with
    Terminal Cisternae Forms Part of the Triad 599
    The T-system is in Continuity with the Plasma Membrane
    and Conducts Impulses Inward.. 599
    Stimulation Releases Ca^{2+} from the Terminal Cisternae 599
    A Ca^{2+}-activated ATPase is Present in the Sarcoplasmic
    Reticulum and Acts as a Ca^{2+} Pump.................................... 601
    Summary: Excitation-Contraction Coupling and the
    Sarcoplasmic Reticulum 601

Chapter 28

CELLULAR AND MOLECULAR NEUROBIOLOGY... 604

28–1 General Organization and Function of Nerve Fibers.......................... 605
    Axon Structure – Neurofibrils and Neurotubules 606
    The Biosynthetic Functions of the Neuron Occur in the
    Perikaryon 606
    Macromolecules are Transported Through the Axon in an
    Orthograde Direction 609
    Orthograde Axonal Transport may be Fast or Slow 609
    Macromolecules Such as NGF Undergo Retrograde
    Axonal Transport 610
    Tetanus Toxin and Some Neurotropic Viruses May Undergo
    Retrograde and Trans-synaptic Transport 612
    Nerve Conduction Velocity – Related to Diameter, Myelin,
    and Internode Distance.. 612
    The Action Potential – Non-decremental and Propagated as
    a Depolarization Wave.. 613
    In Myelinated Nerve Fibers Conduction is Saltatory 615
    At the Physiologic Receptors and Synapses Potentials are
    Graded and Non-propagated 615
    Propagated Acton Potentials Depend on Na^+ and K^+ Channels
    in the Axon Membrane.. 615
    Summary: General Organization of the Neuron and
    Function of the Nerve Fiber 617
28–2 Synaptic Transmission and Structures of the Synapse 619
    Nerve Impulse Transmission – Possibly Electrical but
    Mediated Mainly by a Chemical Mechanism............................ 619
    Synaptic Transmission may be Excitatory or Inhibitory 620
    Several Thousand Synapses may Impinge on a Single Neuron........ 621
    The Number of Synapses is Related to that of the Dendritic
    Spines.. 621
    The Synapse Ultrastructure Suggests Many Types of
    Functional Contracts.. 622
    Lectin Receptors and Postsynaptic Densities May Play a Role
    in the Formation and Maintenance of Synapses........................ 624
    The Presynaptic Membrane Shows Special Projections at
    Active Zones... 624
    The Postsynaptic Membrane Shows a Complex Macromolecular
    Organization... 625
28–3 Synaptic Vesicles and Quantal Release of Neurotransmitter 626
    Several Types of Synaptic Vesicles may be Recognized by
    Morphology and Cytochemistry 627
    Neuronal Development – The Type of Neurotransmitter and
    Synaptic Vesicle May be Medium-Determined 627
    Synaptosomal Membranes and Synaptic Vesicles may be
    Isolated by Cell Fractionation 629
    Neurotransmitter Synthesis and Metabolism is Exemplified by
    the Acetylcholine System 631

CONTENTS

The Transport of the Neurotransmitter Involves the Synaptic
Vesicles 632
Transmitter Release is Related to the Role of Synaptic
Vesicles in Nerve Transmission 632
Transmitter Release Probably Involves Exocytosis with
Recycling of Synaptic Vesicles 634
The Depolarization Transmitter—Secretion Coupling is
Mediated by Calcium Ions 634
Summary: Synaptic Transmission and Structure of the Synapse...... 635
28–4 Synaptic Receptors and the Physiologic Response............................ 636
Synaptic Receptors are Hydrophobic Proteins Embedded in
the Lipoprotein Framework of the Membrane 636
The Acetylcholine Receptor is Coupled to the Translocation
of Sodium and Potassium Ions.. 637
Acetylcholine-induced Fluctuations may be Observed in a
Reconstituted Cholinergic Receptor 639
An Oligomeric Model of the Cholinergic Receptor has been
Postulated.. 639
Long-lasting Synaptic Functions Involve the Use of a Second
Messenger 640
Cyclic AMP is Involved in the Phosphorylation of Membrane
and Other Proteins... 641
Summary: Receptors and the Physiologic Response 643

INDEX... 647

Part 1

INTRODUCTION TO CELL BIOLOGY — METHODS OF STUDY

The first four chapters of this book may be considered an elementary introduction to the cell as a biological unit and to the main methods of cellular study. In living matter there is an integration of different levels of organization, from which the manifestations of life originate. From the morphologic viewpoint, these levels of organization are related to those aspects which can be resolved with the different means of observation: the human eye (anatomy), the various types of microscopes (histology and cytology), and the other methods which facilitate a deeper probe into molecular biology and the ultrastructure of the cell.

Chapter 1 contains a brief account of the history of cell biology, with particular emphasis on the *Cell Theory* and the correlations of cell biology with genetics, cell physiology, and biochemistry. The modern aspects of cellular ultrastructure and molecular biology are also presented. For the benefit of the student, literary sources for further reading in cell biology are provided.

In the second chapter the main characteristics of prokaryotic and eukaryotic organisms are emphasized. The bacterium *Escherichia coli* is the best known example of a prokaryotic cell. It is important that, from the beginning, the reader recognize the similarities and differences between these two types of organisms, although these points will be taken up again in other chapters.

Chapter 2 also deals with the general structure of cells in the living state and after fixation. The main components of the nucleus and cytoplasm are mentioned. The chapter introduces the concept of the life cycle of the cell in direct relation to the processes of mitotic and meiotic division.

Chapter 2 is prerequisite to the fourth and fifth parts of this book, in which cytoplasmic organelles and the cellular basis of cytogenetics will be studied. To understand better the modern developments in cell biology, it is important, from the very beginning of these studies, that the reader be exposed to the nomenclature and general concepts of the field.

The extraordinary progress in cell biology has resulted from the development of new methods for the study of the cell and of its molecular and macromolecular components. In the following chapters two of the main groups of techniques used are presented.

Chapter 3 includes the methods that employ electromagnetic waves. These may be either visible or ultraviolet radiations, electrons, or x-rays. In studying these methods, it is helpful to review the chapters in physics textbooks that deal with reflection, refraction, interference, and diffraction of electromagnetic waves. This background reading will promote a better understanding of the instruments used for analysis of cellular and subcellular structure. The importance of electron microscopy and x-ray diffraction is emphasized. The images given by the light, phase, interference, polarization, and electron microscopes are discussed in relation to the various physical principles involved. The study of the electron microscope and of the techniques employed to prepare biological material for electron microscopy is of great importance in cell biology.

The discussion of instrumentation is complemented by Chapter 4, which is a brief presentation of the main methods of cytologic and cytochemical analysis used for observation and experimentation. Since the number of techniques by far surpasses the limits of this book, only a few selected examples are mentioned. The methods for study of living cells, the process of fixation for the preparation of cells and tissues, and the mechanism of staining are considered. At present, the cytochemical techniques for the identification and localization of enzymes and substances within cells are of paramount importance. The examples mentioned deal with the mechanism of cytochemical analysis and the problems that must be overcome. Numerous other techniques of biochemical or biophysical nature and those dealing with the molecular aspects of cell biology are mentioned throughout the book.

INTRODUCTION AND HISTORY OF CELL BIOLOGY

1–1　Levels of Organization in Biology　3
　　　Levels of Organization and Instrumental
　　　　Resolving Power
　　　Summary: Levels of Organization

1–2　History of Cell Biology　7
　　　The Cell Theory
　　　Cytogenetics – Cytology and Genetics

Cell Physiology
Cytochemistry – Chemical and Physicochemi-
　cal Cellular Analysis
Ultrastructure and Molecular Biology
Summary: Modern Cell Biology

1–3　Literary Sources in Cell Biology　11

Ancient philosophers and naturalists, particularly Aristotle in Antiquity and Paracelsus in the Renaissance, arrived at the conclusion that "All animals and plants, however complicated, are constituted of a few elements which are repeated in each of them." They were referring to the macroscopic structures of an organism, such as roots, leaves, and flowers common to different plants, or segments and organs that are repeated in the animal kingdom. Many centuries later, owing to the invention of magnifying lenses, the world of microscopic dimensions was discovered. It was found that a single cell can constitute an entire organism as in Protozoa, or it can be one of many cells that are grouped and differentiated into tissues and organs to form a multicellular organism.

The cell is thus a fundamental structural and functional unit of living organisms, just as the atom is the fundamental unit in chemical structures (Fig. 1–1). If by mechanical or other means cellular organization is destroyed, cellular function is likewise altered. Although some vital functions may persist (such as enzymatic activity), the cell becomes disorganized and dies.

The development and refinement of microscopic techniques made it possible to obtain further knowledge of cellular structure, not only as it appears in the cell killed by fixation but also as seen in the living state. Biochemical studies

have demonstrated that the products of living matter, and even the living matter itself, are composed of the same elements that constitute the inorganic world. Biochemists have isolated from the complex mixture of cell constituents not only inorganic components but also much more complex molecules, such as proteins, fats, polysaccharides, and nucleic acids. Such biochemical studies have also demonstrated the "oneness" of the entire living world. Today it is known that from bacterium to man the chemical machinery is essentially the same in both its structure and its function (see Monod, 1971).

1–1　LEVELS OF ORGANIZATION IN BIOLOGY

The advances in our knowledge of the cell have produced a fundamental change in the interpretation of cellular structures. For example, it has been demonstrated that beyond the organization visible with the light microscope are a number of more elementary structures at the macromolecular level that constitute the "ultrastructure" of the cell. We are living in the era of *molecular biology*, that is, the study of the shape, aggregation, and orientation of the molecules that compose the cellular system as a unit.

Modern studies on living matter demonstrate

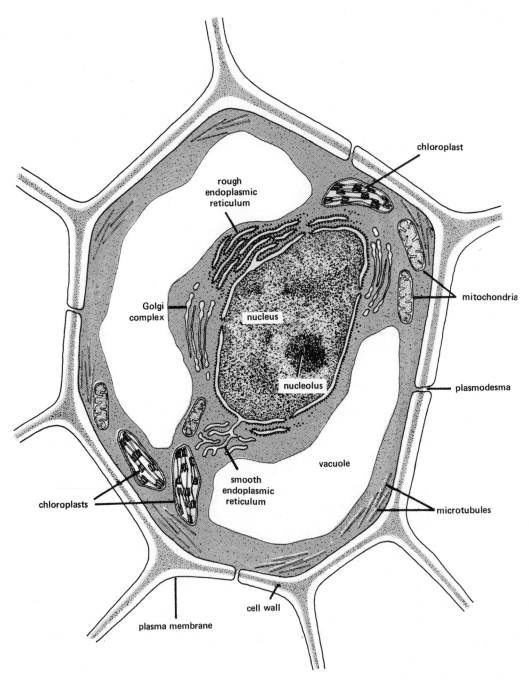

Figure 1–1   Diagram of a plant cell showing the following cell components: the cell wall, the plasmodesmata, chloroplasts, and vacuoles.

that there is a combination of *levels of organization* that are integrated, and that this integration results in the vital manifestations of the organism. The concept of levels of organization as developed by Needham[1] and others implies that in the entire universe — in both the nonliving and living worlds — there are such various levels of different complexity that "The laws or rules that are encountered at one level may not appear at lower levels." This concept can be applied to the different structural constituents of a cell or to the association of numerous cells in a tissue.

In Figure 1–2 the various levels of biological organization are represented by concentric interacting shells — each shell being the environ-

4

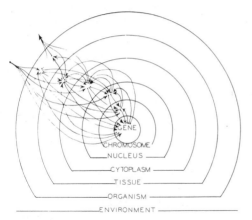

Figure 1–2 The levels of biological organization represented by concentric, interacting shells. (From Weiss, P., *Canad. Cancer Conf.*, 5:243, 1963. © Academic Press, Inc.)

ment for the next nearest inner one. The intricacy of the interrelations between the different levels is indicated by the network of arrows interconnecting them, thus giving an idea of the complexities involved in living matter.

Although both inorganic and living matter are composed of the same atoms, there are fundamental differences between them. According to present concepts, in the nonliving world there is a continuous tendency toward reaching a thermodynamic equilibrium with a random distribution of matter and energy, whereas in the living organism there is a high degree of structure and function that is maintained by energy transformations based on continuous input and output of matter and energy.[2]

Levels of Organization and Instrumental Resolving Power

Table 1–1 shows the limits that separate the study of biological systems at various dimension levels. The boundaries between levels of organization are imposed artificially by the resolving

power of the instruments employed. Note that a great deal of overlapping exists. The human eye cannot resolve (discriminate between) two points separated by less than 0.1 mm (100 μm). Most cells, in general, are much smaller and must be studied under the full resolving power of the light microscope (0.2 μm). However, most cellular components are even smaller and require the resolution of the electron microscope.

From a morphologic point of view, all these fields of biology fall within the discipline of *anatomy* (Gr., *temnein*, to cut; *ana*, up), which is the separation of the different components in such a way as to identify and study them as both isolated parts and integrated parts of the whole organism. Bennett,[3] in a clear interpretation of these concepts, says that "the operational approaches to all branches of anatomy have essential features in common." Whether working in the field of gross, microscopic, or molecular anatomy, one generally proceeds by cutting apart, in some way, the objects of interest. The methodological approach is the same whether a scalpel is used to dissect the cadaver, sections are made for the light or electron micrsocope, or subcellular components are separated by homogenization and centrifugation.

To construct a mental image of the molecular organization of a biological system, one should start with knowledge of the main constituent molecules, particularly those of high molecular weight, such as nucleic acids, proteins and polysaccharides. Lipids, although of smaller molecular size, also play an important role as structural components of cells.

These components must be studied from the point of view of their size, shape, charge, stereochemical characteristics, and main reacting groups. Such studies are difficult when the molecules are isolated or are distributed at random. Frequently, however, the molecules arrange themselves into repeating periodic structures, which can be analyzed with crystallographic techniques. In recent years great advances have been made in the detailed x-ray diffraction anal-

TABLE 1–1 VARIOUS LEVELS OF BIOLOGICAL STRUCTURE

Dimension	Field	Structures	Method
0.1 mm (100 μm) and larger	anatomy	organs	eye and simple lenses
100 μm to 10 μm	histology	tissues	various types of light microscopes,
10 μm to 0.2 μm (200 nm)	cytology	cells, bacteria	x-ray microscopy
200 nm to 1 nm	submicroscopic morphology ultrastructure	cell components, viruses	polarization microscopy, electron microscopy
smaller than 1 nm	molecular and atomic structure	arrangement of atoms	x-ray diffraction

ysis of the molecular configuration of proteins, of nucleic acids and of larger molecular complexes, such as certain viruses. This important field is within the realm of *molecular biology* (Table 1–1).

At a cytologic level, *ultrastructure* or *submicroscopic morphology* is concerned with supramolecular structures that can be analyzed with microscopic techniques. The first technique to be applied, about a century ago, was *polarization microscopy*. German workers, beginning with Nägeli, first recognized ordered structures within biological systems. Later these studies became quantitative and were extended considerably by the work of W. J. Schmidt. This technique makes use of the effect that anisotropic structures have on polarized light (see Chapter 3–1).

The most important tool in the study of submicroscopic morphology is the *electron microscope*. With this instrument, direct information can be obtained about structures ranging from 0.4 to 200 nm, thus bridging the gap between observations with the light microscope and the world of macromolecules. Results obtained by application of electron microscopy have changed the field of cytology so much that a large part of the present book is devoted to discussions of the achievements obtained by this technique.

In Figure 1–3 the sizes of different cells, bacteria, viruses, and molecules are indicated on a logarithmic scale and compared with the wavelengths of various radiations, as well as with the limits of resolution of the eye, the light microscope and the electron microscope. Note that the light microscope (limit of resolution 200 nm) in-

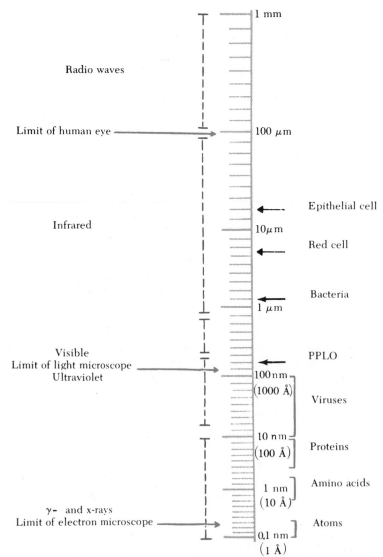

Figure 1–3 Logarithmic scale of microscopic dimensions. Each main division represents a size ten times smaller than the division above. **To the left,** the position of the different wavelengths of the electromagnetic spectrum and the limits of the human eye, the light microscope, and the electron microscope. **To the right,** the sizes of different cells, bacteria, PPLO (the smallest known living organism), viruses, molecules, and atoms. (Modified from M. Bessis.)

TABLE 1–2 RELATIONSHIPS BETWEEN LINEAR DIMENSIONS AND WEIGHTS IN CYTOCHEMISTRY*

Linear Dimension	Weight	Terminology
1 cm	1 gm	conventional biochemistry
1 mm	1 mg or 10^{-3} gm	microchemistry
100 μm	1 μg or 10^{-6} gm	histochemistry ⎱ ultramicrochemistry
1 μm	1 $\mu\mu$g (or 1 picogram or 10^{-12} gm)	cytochemistry ⎰

*From Engström, A., and Finean, J. B., *Biological Ultrastructure*, New York, Academic Press, Inc., 1958. Copyright 1958, Academic Press, Inc.

troduces a 500-fold increase in resolution over the eye (10^5 nm), and the electron microscope (~0.4 nm) provides a 500-fold increase over the light microscope.

Some cytologic structures, such as mitochondria, centrioles, chromosomes, and nucleoli, can be resolved with the optical microscope; but many more, such as ribosomes, the plasma membrane, myofilaments, chromosomal microfibrils, microtubules, and synaptic vesicles can be resolved only with the electron microscope (Fig. 1–1).

In several chapters of this book examples are given of the different levels of organization of biological structures that can be discerned with different magnifying instruments. From the very start, however, the reader must become aware of the importance of these concepts and be able to visualize the proper level of organization that is being considered, i.e., anatomic, histologic, cytologic, ultrastructural, or molecular (Table 1–1).

Table 1–2 shows the general relationships between some of the linear dimensions and weights used in different fields of chemical analysis of living matter. Familiarity with these relationships is essential to the study of cell and molecular biology. The weight of the important components of the cell is expressed in picograms (1 pg = 1 $\mu\mu$g or 10^{-12} gm), that of molecules in *daltons*. The dalton is the unit of molecular weight (MW); one dalton equals the weight of a hydrogen atom. For example, a water molecule weighs 18 daltons; and a molecule of hemoglobin weighs 64,500 daltons.

SUMMARY

Levels of Organization

The cell is the fundamental structural and functional unit of living organisms. In living matter there are levels of organization that are interrelated in a complex manner and are maintained by energy transformations. The boundaries between these levels of organization are imposed by the resolving power of the eye (0.1 mm), the light microscope (0.2 μm) and the electron microscope (~0.4 nm). These limits bring about the differentiation between the fields of anatomy, cytology, ultrastructure, and molecular biology. The electron microscope provides direct information about structures ranging in size from 0.4 to 200 nm. It is important to correlate these boundaries with the spectrum of electromagnetic waves and the actual dimensions of cells, bacteria, viruses, and proteins (Fig. 1–3). The student should have a firm understanding of the relationships between linear dimensions and the weights used in cytochemistry. For example, the weight of cell components, such as nucleic acids, is expressed in picograms (10^{-12} gm).

1–2 HISTORY OF CELL BIOLOGY

Cell biology is one of the youngest branches of the life sciences. It was recognized as a separate discipline by the end of the last century. Its early history is intimately bound to the development of optical lenses and to the combination of lenses in the construction of the compound microscope (Gr., *mikros*, small + *skopein*, to see, to look).

The term *cell* (Gr., *kytos*, cell; L., *cella*, hollow space) was first used by Robert Hooke (1665) to describe his investigations on "the texture of cork by means of magnifying lenses." In these observations, repeated by Grew and Malpighi on different plants, only the cavities ("utricles" or "vesicles") of the cellulose wall were recognized. In the same century and at the beginning of the next, Leeuwenhoek (1674) discovered free cells, as opposed to the "walled in" cells, of Hooke and Grew. Leeuwenhoek observed some organization within cells, particularly the nucleus in some erythrocytes. For more than a century afterward, these observations were all that was known about the cell.

The Cell Theory

More directly related to the origin of cell biology was the establishment of the *cell theory*, one of the broadest and most fundamental of all biological generalizations. It states in its present form that all living beings — animals, plants, or protozoa — are composed of cells and cell products. This theory resulted from numerous investigations that started at the beginning of the 19th century (Mirbel, 1802; Oken, 1805; Lamarck, 1809; Dutrochet, 1824; Turpin, 1826), and finally led to the studies of the botanist Schleiden (1838) and the zoologist Schwann (1839), who established the theory in a definite form.

The cell theory has had wide-ranging effects in all the fields of biological research. It was established immediately that every cell is formed by the division of another cell. Much later, with the progress of biochemistry, it was shown that there are fundamental similarities in the chemical composition and metabolic activities of all cells. The function of the organism as a whole was also recognized to be a result of the sum of the activities and interactions of the cell units.[4]

The cell theory was soon applied to pathology by Virchow, and Kölliker extended it to embryology after it was demonstrated that the organism develops from the fusion of two cells — the spermatozoon and the ovum.

A more general conclusion was reached at the same time by investigators such as Brown (1831), who established that the nucleus is a fundamental and constant component of the cell. Others (Dujardin, Schultze, Purkinje, von Mohl) concentrated on the description of the cell contents, termed the *protoplasm*.

Thus the primitive idea of the *cell* was transformed into the more sophisticated concept of a mass of protoplasm limited in space by a cell membrane and possessing a nucleus. The protoplasm surrounding the nucleus became known as the *cytoplasm* to distinguish it from the *karyoplasm*, the protoplasm of the nucleus. The extraordinary changes produced in the nucleus at each cell division attracted the attention of a great number of investigators. For example, the phenomena of *amitosis*, or direct division (Remak), and of indirect division were discovered by Flemming (in animals) and by Strasburger (in plants). Indirect division was also called *karyokinesis* (Schleicher, 1878) or *mitosis* (Flemming, 1880). It was proved that fundamental to mitosis is the formation of nuclear filaments, or *chromosomes* (Waldeyer, 1890), and their equal division between the nuclei (daughter cells). Other discoveries of importance were the fertilization of the ovum and the fusion of the two pronuclei (O. Hertwig, 1875). In the cytoplasm, the aster (van Beneden, Boveri), the mitochondrion (Altmann, Benda), and the reticular apparatus (Golgi) were discovered. In 1892 O. Hertwig published his monograph *Die Zelle und das Gewebe (The Cell and the Tissue)* in which he attempted to achieve a general synthesis of biological phenomena based on the characteristics of the cell, its structure and function. He thus created cytology as a separate branch of biology.[4]

If one follows the development of cell biology in the present century it is evident that there were two main reasons for the advance of cytologic knowledge: (1) the increased resolving power of instrumental analysis, due essentially to the introduction of electron microscopy and x-ray diffraction techniques, and (2) the convergence of cytology with other fields of biological research, especially genetics, physiology, and biochemistry.

Cytogenetics – Cytology and Genetics

By the middle of the 19th century the universality of cell division as the central phenomenon in the reproduction of organisms was established, and Virchow expressed it in the famous aphorism *"Omnis cellula e cellula."* From this time on, the study of cells and of heredity and evolution converged.

Observations on the germ cells made by van Beneden, Flemming, Strasburger, Boveri, and others gave support to the theory of the continuity of germ plasm proposed by Weissmann in 1883. This theory stated that the transference of hereditary factors from one generation to the next takes place through the continuity of what he called *germ plasm,* located in the sex elements (spermatozoon and ovum), and not through somatic cells.

The discovery of fertilization led to the theory that the cell nucleus is the bearer of the physical basis of heredity. Furthermore, Roux postulated that chromatin, the substance of the nucleus that constitutes the chromosomes, must have a linear organization.

The fundamental laws of heredity were discovered by Gregor Mendel in 1865, but at that time the cytologic changes produced in the sex cells were not sufficiently known to permit an interpretation of the independent segregation of hereditary characters (see Chapter 19–1). For this and other reasons, little attention was paid to Mendel's work until the botanists Correns, Tschermack, and De Vries in 1901 independently rediscovered Mendel's laws. At that time cytology had advanced enough that the mechanism of distribution of the hereditary units postulated by Mendel could be understood and explained. It was known that the somatic cells have a double, or *diploid,* hereditary constitution, whereas in the reproductive cells or gametes this constitution is single, or *haploid.* In addition, cytologists had observed that the cycle the chromosomes undergo in *meiosis* of germ cells was related to hereditary phenomena.

McClung (1901–1902) suggested that sex determination was related to some special chromosomes; this theory was later corroborated by Stevens and Wilson (1905). The chromosome theory of heredity was finally established by Morgan and his collaborators, Sturtevant and Bridges, who assigned to the *genes* (Johanssen), or hereditary units, definite loci within the chromosomes. Thereafter experimental research on heredity and evolution became a separate branch of biology, which Bateson in 1906 called *genetics.* Almost from the beginning, however, the science of genetics maintained a close relationship with cytology. From the convergence of cytology and genetics *cytogenetics* originated (see Chapter 20). In the past decade the study of genetics has become linked to biochemistry and has reached the molecular level. Thus the new fields of *biochemical* and *molecular genetics* were established.

Cell Physiology

Most early cytologic knowledge was based on observations of fixed and stained cells and tissues. By 1899 interest shifted toward the study of living cells, principally as a result of the work of Fischer and Hardy, who showed that several of the structures observed in fixed cells could be reproduced by the action of fixatives on colloidal models. Various types of movement, such as cyclosis (cytoplasmic streaming), ameboid motion, and ciliary, flagellar, and muscular contraction, were studied at a cellular level.

At the end of the 19th century, Overton advanced the theory that the cell membrane was a lipoidal film. Michaelis made membrane models to study the passage of substances and did the first vital staining of mitochondria. The basic concepts of cell irritability and nerve function were established by the middle of the 19th century by Du Bois-Reymond; physiological techniques were then developed to aid in the study of these cells and to measure action potentials and nerve currents.

An important avenue to the study of the living cell was opened in 1909 by Harrison, who demonstrated that nerve cells from an embryo could grow and differentiate in vitro. This gave rise to the technique of tissue culture (Carrel), which had a great impact on cytology.

Another method for the study of the physicochemical properties of the living cell — *microsurgery* — came from the field of bacteriology. Schouten and Barber used fine micropipets, moved by precision instruments, to isolate and culture single bacteria. In 1911 Kite adapted this method to cytology. Levi, Peterfi, Chambers, and others have perfected techniques of intracellular operations and have obtained data on the viscosity, hydrogen concentration, and other physicochemical properties of the cell.

Among the important phenomena studied in cell physiology are: the nature of the cell membrane and of active transport across membranes, the reaction of cells to changes in environment, the mechanisms of cell excitability and contraction, and cell nutrition, growth, secretion, and other manifestations of cellular activity.

Cytochemistry – Chemical and Physicochemical Cellular Analysis

Another modern branch of cell biology is *cytochemistry,* the result of the convergence of methods and sciences devoted to the chemical and physicochemical analysis of living matter. Among many outstanding biochemical studies were those of Fischer and Hofmeister in 1902, who independently recognized that the protein molecule consists of a small number of amino acids united by peptide bonds. Of similar importance to cell biology were the earlier investigations of Miescher (1869) and Kossel (1891), who, by the analysis of pus cells, spermatozoa, hemolyzed erythrocytes of birds, and other cell types, isolated the nucleic acids, whose basic role in heredity and protein synthesis has been recognized only recently.

Another great advance was the introduction into biological thinking by Ostwald of the concept of catalytic activity and the discovery that enzymes are the molecular entities used by the cell to produce the various types of energy transformations necessary for maintenance of living activities. The main types of cellular oxidations were discovered by Wieland (1903) and by Warburg (1908), but the final mechanism was discovered much later by Keilin (1934).

Because of cytology's emphasis on morphology, cytologists were at first very slow in grasping the importance of the biochemical approach; at the same time, biochemists, because of their field's emphasis on organic chemistry, had no interest in cell structure and were concerned principally with isolating chemical components and studying elementary enzyme reactions. The point of convergence of cytology and biochemistry can be traced back to 1934, when Bensley and Hoerr isolated mitochondria from cells in large enough quantities to permit analysis by chemical and physicochemical methods.

Research in this direction was pursued with great success by Claude, Hogeboom, and others and led to the conclusion that mitochondria are centers of cellular oxidations.

Advances in cell fractionation have been of the greatest importance to biology and biochemistry, especially with the development of radioactive tracer techniques that permit a dynamic approach to the study of cell metabolism. Another great advance was the use of the electron microscope for the observation and characterization of cell fractions.

Modern cytochemistry has also developed along the lines of microchemical and ultramicrochemical analysis by means of techniques for assay of minute quantities of material and isolation of single cells and even parts of cells (see Table 1–2). Chemical analysis can be carried out by cytophotometry, which facilitates the study of the localization of nucleic acids and proteins within parts of a single cell (see Chapter 4–6).

Another important branch of cytochemistry arose from the application of numerous enzyme reactions that could be observed under the light and electron microscopes. This last approach is of particular interest since it combines cytochemistry and ultrastructure and thus permits study of the localization of enzymes at the level of resolution of the electron microscope. Of similar importance have been the autoradiographic studies on the localization of radioactive tracers in different cellular structures (see Chapter 4–6).

Ultrastructure and Molecular Biology

In studying the limits and dimensions in biology (see Tables 1–1 and 1–2), the impact of instrumental analysis was mentioned and the modern fields of ultrastructure and molecular biology were delineated. These are the most progressive branches of biology, in which the merging of cytology with biochemistry, physicochemistry, and especially macromolecular and colloidal chemistry becomes increasingly intimate. Knowledge of the submicroscopic organization or ultrastructure of the cell is of fundamental importance, because practically all the functional and physicochemical transformations take place within the molecular architecture of the cell and at a supramolecular level.

The following advances are having an extraordinary impact on cell biology: the discovery that in the protein molecule the exact sequence of amino acids and the three-dimensional arrangement of the polypeptide chain have a direct relationship to definite biological properties; the studies of active groups in different enzymes; the molecular model of DNA suggested by Watson and Crick in 1953; as well as all the knowledge on the stereochemistry of macromolecules. Molecular biology is thus providing valuable information to the fields of genetics (through molecular genetics), biochemistry, and even pathology —the latter through the establishment of molecular diseases.

Both in ultrastructure and molecular biology the integration of morphology and physiology becomes so intimate that it is impossible to separate them; the concepts of *form* and *function* fuse into inseparable unity.

SUMMARY: MODERN CELL BIOLOGY

In summary, it can be said that modern cell biology approaches the problems of the cell at all levels of organization — from molecular structure on. Cell biology is therefore the modern science in which genetics, physiology, and biochemistry converge. Cell biologists, without losing sight of the cell as a morphologic and functional unit within the organism, must be prepared to study biological phenomena at all levels and to use all the methods, techniques, and concepts of the other sciences. This is a great challenge, but there is no other way to proceed if the life of the cell and of the organism is to be interpreted mechanistically, i.e., on the basis of combinations and associations of atoms and molecules.

1–3 LITERARY SOURCES IN CELL BIOLOGY

The preceding considerations of the present scope of cell biology demonstrate why the sources of literature are wide and multidisciplinary. Current studies are presented at scientific meetings and are published in specialized periodicals. Of the long list of literary sources that could be made, only a few of the more specific ones are mentioned: *Cell and Tissue Research, Chromosoma, Cytogenetics and Cell Genetics, Developmental Biology, Experimental Cell Research, Heredity, Hereditas, Journal of Cell Biology, Journal of Cell Science, Journal of Cellular Physiology, Journal of General Physiology, Journal of Molecular Biology, Journal of Supramolecular Structure,* and *Journal of Ultrastructure Research.* Papers on cell biology are frequently published in more general periodicals, such as *American Scientist; Comptes Rendus Hebdomadaires des Seances de l'Academie des Sciences, D: Sciences Naturelles; Nature; Naturwissenchaften; Experientia; Proceedings of the National Academy of Sciences (Wash.); Proceedings of the Royal Society* [London, B: Biological Sciences]; and *Science;* or even in such specialized publications as *Biochimica et Biophysica Acta, Biochemical Journal,* and *Journal of Biological Chemistry.*

Reviews of recent advances are found in the: *Advances in Cell and Molecular Biology, Advances in Genetics, Biological Reviews of the Cambridge Philosophical Society, Physiological Reviews, Plant Physiology, Protoplasma, Quarterly Review of Biology, International Review of Cytology, Scientific American, La Recherche,* and others.

Very useful in the compilation of a bibliography are special journals that give titles of papers or abstracts of the literature. These journals include: *Berichte über die Wissenschaftliche Biologie, Biological Abstracts, Chemical Abstracts, Excerpta Medica,* and *Index Medicus.* Journals such as *Current Contents* and *Bulletin Signalétique du Conseil des Recherches* publish titles of all papers which appear in various journals.

In addition there are many monographs, compendia, and textbooks that will be mentioned in later chapters of this book and which cover various specialized subjects of cell biology. The widest coverage is found in *The Cell,* in six volumes, edited by Brachet and Mirsky, and in *Handbook of Molecular Cytology,* edited by A. Lima-de-Faría (1969); however, both are at present outdated.

REFERENCES

1. Needham, J. (1936) *Order and Life.* Yale University Press, New Haven, Conn.
2. Bertalanffy, L. von. (1952) *Problems of Life.* John Wiley & Sons, New York.
3. Bennett, H. S. (1956) *Anat. Rec.,* 125:2.
4. Hughes, A. (1959) *History of Cytology.* Abelard-Schuman, London and New York.

ADDITIONAL READING

Baker, J. R. Five articles evaluating the cell theory. *Quart. J. Micr. Sci.,* 1948, *89*:103; 1949, *90*:87; 1952, *93*:157; 1953, *94*:407; 1955, *96*:449.
Baumeister, W. (1978) Biological horizons in molecular microscopy. *Cytobiologie,* 17:246.

Bernal, J. D., and Synge, A. (1973) The origin of life. In: *Readings in Genetics and Evolution.* Oxford University Press, London.

Bourne, G. H., and Danielli, J. F., eds. (1952–1978) *The International Review of Cytology,* Vols. 1–55, Academic Press, New York.

Brachet, J., and Mirsky, A. E. (1959–1961) *The Cell.* 6 Volumes. Academic Press, Inc., New York.

Cairns, J., Stent, G. S., and Watson, J. D., eds. (1966) Phage and the origins of molecular biology. *Cold Spring Harbor Laboratory of Quantitative Biology,* New York.

Claude, A. (1975) The coming of age of the cell. *Science, 189*:433–435.

Goldstein, L., and Prescott, D. M., eds. (1977) *Cell Biology: A Comprehensive Treatise,* Vol. 1, Academic Press, New York.

Heilbrunn, L. V., and Weber, F., eds. (1953–1959) *Protoplasmatologia, Handbuch der Protoplasmaforschung.* Springer, Vienna.

Hermann, H. (1968) This is the cell biology that is. *Bull. Inst. of Cell. Biol.* (Univ. of Conn., Storrs, Conn.), *10*:1.

Hughes, A. (1959) *History of Cytology.* Abelard-Schuman, London and New York.

Lima-de-Faría, A., ed. (1969) *Handbook of Molecular Cytology.* North-Holland Pub. Co., Amsterdam.

Monod, J. (1971) *Chance and Necessity.* Random House, Inc., New York.

Nordenskiöld, E. (1966) *The History of Biology.* Tudor Pub. Co., New York.

Olby, R. (1974), *Path to the Double Helix.* University of Washington Press.

Reinert, J., and Ursprung, H., eds. (1971) *Origin and Continuity of Cell Organelles.* Springer-Verlag, Berlin.

Stent, G. S. (1968) That was the molecular biology that was. *Science, 160*:390–395.

Watson, J. D. (1975) *Molecular Biology of the Gene.* 3rd Ed. W. A. Benjamin, Inc., New York.

Weiss, P. (1963) Cell interactions. *Canad. Cancer Conf., 5*:241–276. Academic Press, Inc., New York.

Wolfe, S. L. (1972) *Biology of the Cell.* Wadsworth Publishing Co., Inc., Belmont, Calif.

PROKARYOTIC AND EUKARYOTIC CELLS — GENERAL STRUCTURE

2–1 General Structure of Prokaryotic Cells 13
Escherichia coli (E. coli) — *The Most Studied Prokaryote*
Smallest Living Mass — DNA, RNAs, and a Plasma Membrane

2–2 General Structure of Eukaryotic Cells 16
Cell Shape — Specific for Each Cell Type
Cell Size — Generally Microscopic
Living Cell — Only Some Structural Components Visible
Fixed Cell — Complex Structural Organization

Summary: Prokaryotic and Eukaryotic Cells

2–3 The Nucleus, Chromosomes, and the Cell Cycle — General Concepts 21

2–4 Mitosis and Meiosis — Essentials 22
Mitosis — Maintenance of Chromosomal Continuity and Diploid Number
Meiosis — Reduction of Chromosomal Number to Haploid Set
Summary: Essentials About Nucleus and Chromosomes

2–1 GENERAL STRUCTURE OF PROKARYOTIC CELLS

The typical cell, with the nucleus and cytoplasm and all the cellular organelles, which is described in this book, is not the smallest mass of living matter or protoplasm (Gr., *protos*, first + *plasma*, formation); simpler or more primitive units of life exist. Thus, unlike the higher types of cells, which have a true nucleus (eukaryotic cells), prokaryotic cells (Gr., *karyon,* nucleus), which comprise bacteria and blue-green algae, lack a nuclear envelope, and contain a *nucleoid* that is in direct contact with the rest of the protoplasm.

From an evolutionary viewpoint prokaryotes are considered to be ancestors of eukaryotes. In fact prokaryotes alone are found in fossils three billion years old, while eukaryotes probably appeared one billion years ago (see Chapter 7). In spite of the differences between prokaryotes and eukaryotes (see Table 2–1), there are considerable homologies in their molecular organization and function. We shall see for example that all

living organisms employ the same genetic code and a similar machinery for protein synthesis.

In addition to the prokaryotes and the eukaryotes, one should consider the *viruses*, which although not true cells, do share some of their properties. Recognized at first by the property of being able to pass through the pores of porcelain filters and by the pathologic changes they produced in cells, all viruses can now be visualized with the electron microscope. Viruses can be recognized morphologically and their macromolecular organization can be studied. Although they have properties common to living organisms, such as autoreproduction, heredity, and mutation, viruses are dependent on the host's cells and are considered obligatory parasites. (For the structure of viruses, see Chapter 7–2.)

Escherichia coli (E. coli) — *The Most Studied Prokaryote*

Although this book is dedicated to the more complex eukaryotic cells, it is important to know

TABLE 2–1 COMPARISON OF CELL ORGANIZATION IN PROKARYOTES AND EUKARYOTES

	Prokaryotic Cells *Bacteria,* *blue-green algae, and* *mycoplasm*	*Eukaryotic Cells* *Protozoa,* *other algae,* *metaphyta, and* *metazoa*
Nuclear Envelope	Absent	Present
DNA	Naked	Combined with Proteins
Chromosomes	Single	Multiple
Nucleolus	Absent	Present
Division	Amitosis	Mitosis or Meiosis
Ribosomes	70S (50S + 30S)*	80S (60S + 40S)*
Endomembranes	Absent	Present
Mitochondria	Respiratory and Photosynthetic Enzymes in the Plasma Membrane	Present
Chloroplast	Absent	Present in Plant Cells
Cell Wall	Non-cellulosic	Cellulosic, Only in Plants
Exocytosis and Endocytosis	Absent	Present
Locomotion	Single Fibril, Flagellum	Cilia and Flagella

*S refers to the Svedberg sedimentation unit, which is a function of the size and shape of molecules.

that much of the present knowledge of molecular biology stems from the study of viruses and bacteria. A bacterial cell such as *Escherichia coli (E. coli)* is easily cultured in an aqueous solution containing glucose and some inorganic ions. In this medium, at 37° C, the cell mass doubles and divides in about 60 minutes. This time—the *generation time*—can be reduced to 20 minutes if purines and pyrimidine bases (the precursors of nucleic acids), as well as amino acids, are added to the medium.

As shown in Figure 2–1, one cell of *E. coli* is about 2 μm long and 0.8 μm thick. It is surrounded by a rigid *cell wall*, 10 nm or more thick, containing protein, polysaccharide, and lipid molecules. Inside the cell wall is the true *cell* or *plasma membrane*, a lipoprotein structure that constitutes a molecular barrier to the surrounding medium. This plasma membrane, by controlling the entrance and exit of small molecules and ions, contributes to the establishment of a special internal milieu for the protoplasm of the bacterium. It is interesting that enzymes involved in the oxidation of metabolites, and which constitute the *respiratory chain*, are associated with this plasma membrane. In eukaryotic cells these enzymes are confined to special organelles in the cytoplasm, the mitochondria.

Under the electron microscope (Fig. 2–2) it is possible to recognize light *nuclear* regions or *nucleoids*, where the chromosome of the bacteria, formed by a *single circular molecule of deoxyribonucleic acid (DNA)*, is present. It is important to remember that this DNA, which is about 1 mm long (10^6 nm) when uncoiled, contains all the genetic information of the organism. In fact, the DNA molecule in this bacterium contains sufficient genetic information to code 2000 to 3000 different proteins. The DNA molecule is folded and packed within the nuclear region; it lies free in the protoplasm, and is not separated by a nuclear envelope as in the eukaryotic cell. In Figure 2–1 two chromosomes are shown because replication of the DNA has occurred and the cell is ready to divide. An important point shown in this diagram is the attachment of the DNA to the plasma membrane. This anchoring may play a role in the separation of the two nucleoids after DNA replication. In fact, the growth of the intervening membrane may accomplish this separation.

In addition to a chromosome, certain bacteria that develop resistance to antibiotics contain a small extrachromosomal circular DNA called a *plasmid*. As will be mentioned in the chapters on molecular biology, plasmids can be separated and reincorporated; genes (pieces of DNA) can be inserted into plasmids, which are then transplanted into bacteria using the techniques of *genetic engineering* (see Chapter 21).

Surrounding the DNA, in the dark region of the protoplasm (Fig. 2–2), are 20,000 to 30,000 particles, about 25 nm in diameter, called *ribosomes*, which are composed of *ribonucleic acid (RNA)* and proteins. These particles are the sites of protein synthesis. Ribosomes exist in groups called polyribosomes or polysomes and are composed of a large and a small subunit.

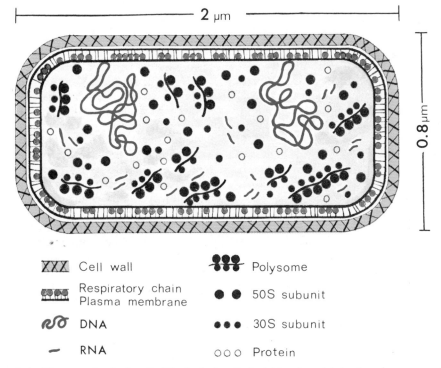

XXX Cell wall	●●● Polysome
Respiratory chain Plasma membrane	● ● 50S subunit
ℛℛ DNA	●●● 30S subunit
— RNA	○○○ Protein

Figure 2–1 Diagram of a single cell of the bacterium *Escherichia coli* containing two chromosomes 1 mm long (10^6 nm) attached to the cell membrane. 50S and 30S refer to ribosomal subunits.

The remainder of the cell is filled with water, various RNAs, protein molecules (including enzymes), and various smaller molecules.

Certain motile bacteria have hair-like processes about 10 nm wide and of variable length, called *flagella*, which are used for locomotion. In contrast with the cilia and flagella of eukaryons, which contain several fibrils, each flagellum in bacteria is made of a single fibril.

Most prokaryotes are *heterotrophic* (see Chapter 6); that is, they feed on organic molecules present in the medium. A few bacteria may oxidize inorganic compounds (such as H_2S to S or H_2 to H_2O), and blue-green algae are autotrophic—i.e., they are able to utilize CO_2 and H_2O with the energy provided by light—photosynthesis. For this purpose blue-green algae contain light-capturing pigments.

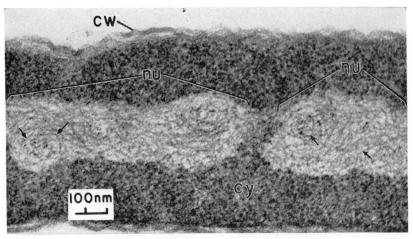

Figure 2–2 Electron micrograph of thin sections of the bacterium *Escherichia coli*. The nucleoid (*nu*) shows the presence of microfibrils of DNA (arrows). Note that the nucleoid lacks a membrane. The cytoplasm (*cy*) is very dense; *cw*, cell wall. ×100,000. (Courtesy of E. Kellenberger.)

Smallest Living Mass — DNA, RNAs, and a Plasma Membrane

Most prokaryotic cells are small, in the range of 1 to 10 μm, but some blue-green algae, such as *Oscillatoria princeps,* may reach 60 μm in diameter. From what has been said about *E. coli,* it is evident that there must be a minimum size limit for a cell. However, the cell must be large enough (1) to have a plasma membrane, (2) to contain the genetic material necessary to provide a code for the various RNAs involved in protein synthesis, and (3) to contain the biosynthetic machinery where this synthesis takes place.

Among agents that have the smallest living mass, the best suited for study are small bacteria, the *mycoplasms* (formerly called pleuropneumonia-like organisms or PPLO, Fig. 1–3), which produce infectious disease in certain animals and man and which can be cultured in vitro like any bacteria. These agents range in diameter from 0.25 μm to 0.1 μm; thus their size corresponds to that of some of the large viruses. These microbes are of general biological interest because each is a living organism a thousand times smaller than the average size bacterium and a million times smaller than a eukaryotic cell.

2–2 GENERAL STRUCTURE OF EUKARYOTIC CELLS

The eukaryotic cell consists of a small mass of protoplasm, the cytoplasm, containing a nucleus and surrounded by the plasma membrane. The cells of a multicellular organism vary in shape and structure and are differentiated according to their specific function in the various tissues and organs. Because of this functional specialization, cells acquire special characteristics, but a common pattern of organization persists in all cells.

Cell Shape — Specific for Each Cell Type

Some cells, such as amebae and leukocytes, change their shape frequently. Other cells always have a typical shape, more or less fixed, which is specific for each cell type; e.g., the spermatozoids, infusoria, erythrocytes, epithelial cells, nerve cells, most plant cells, and others.

The shape of cells depends mainly on functional adaptations and partly on the surface tension and viscosity of the protoplasm, the mechanical action exerted by the adjoining cells, and the rigidity of the cell membrane. In Chapter 9 it will be shown that certain cell organelles, called *microtubules,* also have an important influence on the shape of the cell.

When isolated in a liquid, many cells become spherical, according to the laws of surface tension. For example, leukocytes in the circulating blood are spherical, but in the extravascular milieu they develop pseudopods (thus exhibiting ameboid movement) and become irregular in shape.

Individual cells in a large mass appear to behave like polyhedral solids of minimal surface that are packed without interstices. Although regular polyhedra of four, six, and twelve sides can be packed without interstices, the fourteen-sided polyhedron (tetrakaidecahedron) satisfies most closely the conditions of minimal surface. The study of bubbles in soap foam by Plateau and Lord Kelvin showed that these conditions of minimal surface exist, and that the average bubble has 14 sides (Fig. 2–3, *A* and *B*).

When observing cells under the microscope, one should always think in terms of three dimensions and observe sections of varied orientations. The best way to learn about the actual shape of cells is by making serial sections of known thickness, drawing all of them and making reconstructions in wax — a procedure similar to that used in anatomic reconstructions. Figure 2–3, *C* to *H*, shows some reconstructions of different cell types.

Cell Size — Generally Microscopic

The size of different cells ranges within broad limits. Some plant and animal cells are visible to the naked eye. For example, the eggs of certain birds have a diameter of several centimeters and are composed, at least at first, of a single cell. However, this is an exception; the great majority of cells are visible only with a microscope, since they are only a few micrometers in diameter (Fig. 1–3). The smallest animal cells have a diameter of 4 μm.

In tissues of the human body, with the exception of some nerve cells, the volume of cells varies between 200 μm^3 and 15,000 μm^3. In general, the volume of the cell is fairly constant for a particular cell type and is independent of the size of the organism. For example, kidney or liver cells are about the same size in the bull, horse, and mouse; the difference in the total mass of the organ depends on the number, not the volume, of cells.

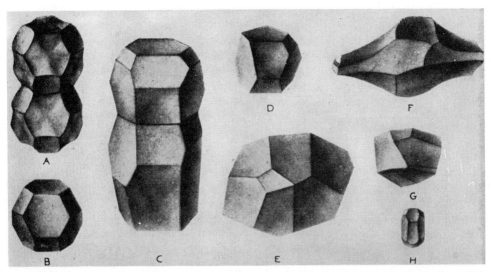

Figure 2–3 Three-dimensional reconstructions of, **A** and **B,** Kelvin's minimal tetrakaidecahedron. **C–H,** wax plate reconstructions of different cell types: **E,** human fat cell; **F, G,** and **H,** outer, middle, and basal cells of stratified epithelium of the mouth of a five-month human embryo. Approximate magnifications: C, × 170; D, × 150; E, × 300; F, G, and H, × 750. (From F. T. Lewis.)

Living Cell — Only Some Structural Components Visible

Living cells can be studied only with light microscopes, since they are, in general, too thick to be studied via electron microscopy. Many animal cells can be observed isolated in an isotonic liquid, such as blood serum, aqueous humor, or physiological salt solutions, or in tissue culture. They appear as irregular, translucent masses of cytoplasm containing a nucleus. In Figure 2–4 most of the cells are in interphase, the nondividing stage, and show a clear nucleus having one or more nucleoli and separated from the cyto-

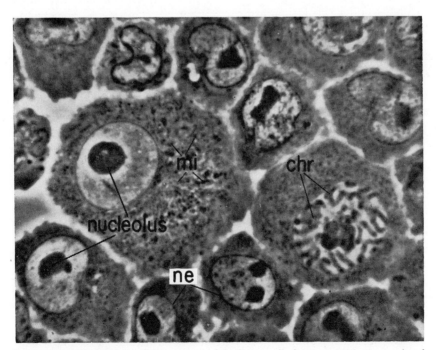

Figure 2–4 Photomicrograph in phase contrast of the living cells from an ascitic tumor. *chr,* chromosomes; *mi,* mitochondria; *ne,* nuclear envelope. (Courtesy of N. Takeda.)

17

plasm by the nuclear envelope. When cells are about to divide, several refractile bodies, the *chromosomes,* appear in the nucleus.

The cytoplasm appears as an amorphous, homogeneous substance, the *ground cytoplasm,* containing refractile particles of various sizes, among which the mitochondria are the most conspicuous (Fig. 2–4). Frequently the peripheral layer of the cytoplasm, the *ectoplasm,* also called the *cortex,* is relatively more rigid and devoid of granules. The internal cytoplasm, the *endoplasm,* which contains different granules, is less viscous than the ectoplasm.

The living cell can be centrifuged, and the effect on the cell components can be observed in a special centrifuge microscope. For example, if a sea urchin egg is subjected to centrifugation, the various components of the endoplasm become stratified in accordance with their densities (Fig. 2–5). The egg elongates and then becomes constricted in the center. The fat droplets accumulate at the centripetal pole. Beneath this is a clear, wide zone, the ground cytoplasm, which contains the nucleus. The mitochondria form the next layer, and the yolk bodies, the next. Pigment granules accumulate at the centrifugal pole. The ectoplasm is not displaced by centrifugation, because of its greater viscosity and rigidity; this property appears to depend on the presence of calcium ions, since the ectoplasm liquefies when eggs are treated with substances that bind Ca^{2+}.

Interesting studies on the colloidal properties of cytoplasm and on the physicochemical forces involved have been made. By increasing the hydrostatic pressure, the cortex can be liquefied, and the cell can no longer change its shape. This effect is reversible within certain limits. The ground cytoplasm behaves, in general, as a reversible sol-gel colloid system. This change can sometimes be produced by mechanical action, a property generally called *thixotropism* (Gr., *thixis,* touch + *trope,* a change).

In addition to mitochondria, other particles observed in the cell, such as highly refractile lipid droplets, yolk bodies, pigment, and secretion granules, are products elaborated by the cell and are found in various amounts. In plant cells, granules called *plastids* can be observed. Among these are the *chloroplasts,* which contain a green pigment — chlorophyll. The function of chlorophyll is *photosynthesis,* a process of immense importance in the biological world (see Chapter 14–5).

In animal and, more commonly, in plant cells, fluid vacuoles surrounded by a membrane may be found.

Mitochondria and chloroplasts are considered cell *organelles,* because of their general presence and their important function in cells. Other cell organelles, such as the *Golgi* apparatus and the *centrioles,* are observed less often in living cells.

Fixed Cell — Complex Structural Organization

Examination of the living cell is based mainly on the differences in refractive index of the different cell components. Sometimes the use of stains that act on the living organism *(vital staining)* facilitates observation of the living cell. However, more important in the morphologic study of the cell are *methods of fixation,* by which

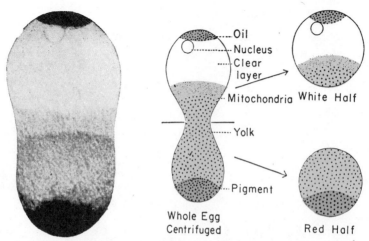

Figure 2–5 **Left,** sea urchin egg (*Arbacia punctulata*) submitted to the action of centrifugal force. The egg has elongated and is being divided into two halves. The cellular materials become stratified (see the description in the text). (Courtesy of Costello.) **Right,** diagram of the stratification of the egg and its division into two halves. (From E. B. Harvey.)

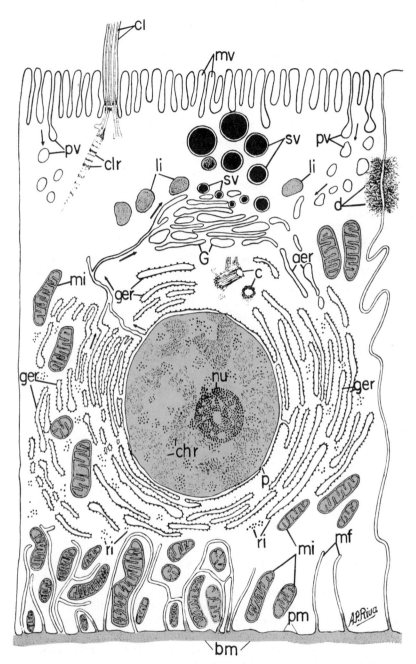

Figure 2–6 General diagram of the ultrastructure of an idealized animal cell. *aer,* agranular endoplasmic reticulum; *bm,* basal membrane; *c,* centriole; *chr,* chromosome; *cl,* cilium; *clr,* cilium root; *d,* desmosome; *G,* Golgi complex; *ger,* granular endoplasmic reticulum; *li,* lysosome; *mf,* membrane fold; *mi,* mitochondria; *mv,* microvilli; *nu,* nucleolus; *p,* pore; *pm,* plasma membrane; *pv,* pinocytic vesicle; *ri,* ribosome; *sv,* secretion vesicle. (From E. De Robertis and A. Pellegrino de Iraldi.)

cell death results in such a way that physiologic structure and chemical composition are preserved as much as possible.

Eukaryotic cells are characterized by a nuclear envelope that divides the cell in two main compartments: nucleus and cytoplasm (Table 2–1). The cytoplasm in turn is limited by the plasma membrane. In a plant cell (Fig. 1–1), the plasma membrane is covered and protected on the outside by a thicker cell wall through which there are tunnels, the plasmodesmata, by which the cell intercommunicates with neighboring cells via fine cell processes. In animal cells (Fig. 2–6), parts of the plasma membrane are covered by a thin layer of material that is generally described as the cell coat. The so-called basement membrane shown in Figure 2–6 corresponds to the cell coat.

The cytoplasmic compartment of the cell has a complex structural organization. When examining this compartment under the electron microscope, one is particularly struck by the prodigious development of membranes. For example, the plasma membrane has numerous infoldings and differentiations (Fig. 2–6). In addition, a basic membranous organization is found in some cell organelles, such as *lysosomes* (Fig. 2–6, *li*) and *mitochondria* (Fig. 2–6, *mi*), in which one and two membranes separate the interior matrix from the surrounding ground cytoplasm. *Chloroplasts* are organelles with a complex organization (see Fig. 1–1). Numerous vesicles, vacuoles, and secretory droplets found in the cytoplasm are also surrounded by membranes.

A complex system of membranes, the *endomembrane system*, pervades the ground cytoplasm, forming numerous compartments and subcompartments. This system is so polymorphic that it is difficult to describe and to encompass within a single denomination. The cytoplasm is separated generally into two parts, one contained within this membrane system and the other, the *cytoplasmic matrix* proper, remaining outside. To the endomembrane system belong the *Golgi apparatus* and the *nuclear envelope;* but the major part is formed by the so-called *endoplasmic reticulum*, which may in turn be differentiated into a *granular* or *rough reticulum*, containing ribosomes, and an *agranular* or *smooth reticulum*. The diagram in Figure 2–6 indicates the possible continuities and functional interconnections of these different portions of the cytoplasmic membrane system.

Other cell organelles, the *centrioles*, are involved in cell division. During cell division two centrioles are contained in a clear, gel-like zone, the so-called *aster*. Centrioles are also related to the differentiation of *cilia* and *flagella*, both motile appendixes of the cell (Fig. 2–6, *cl*).

In spite of this complex structural organization, the most important constituents of the cytoplasm are in the *matrix* (ground cytoplasm), which lies outside the endomembrane system. This matrix constitutes the true internal milieu of the cell and contains the following: the *ribosomes*, the principal structures of the machinery for protein synthesis of the cell; microtubules; microfilaments; structural proteins; and all the components found in a primitive organism, excluding DNA. In Chapter 8, under the name *cytoskeleton*, the microfilaments and microtubules will be specially studied in relation to cell motility and the colloidal changes of the cytoplasm.

SUMMARY:

Prokaryotic and Eukaryotic Cells

Prokaryotic cells such as bacteria and blue-green algae lack a nuclear envelope and contain a single chromosome, formed by a circular DNA molecule, that lies free in the nuclear region (also called nucleoid). A single cell of the bacterium *E. coli* (0.2 × 0.8 μm) is surrounded by a plasma membrane containing the respiratory enzymes. This membrane is surrounded by a more rigid cell wall. The DNA molecule is 1 mm long when uncoiled and is attached at one point to the plasma membrane. It contains information to code some 3000 protein molecules. The protoplasm contains some 20,000 to 30,000 ribosomes, principally as polysomes.

The smallest mass of living matter is represented by the *mycoplasm,* a small microbe 0.2 to 0.1 μm in diameter. At the minimum a living mass should contain (1) a plasma membrane, (2) the genetic material, and (3) the machinery for the synthesis of proteins.

The shape of *eukaryotic* cells is, in general, fixed and depends either on mechanical properties that are intrinsic or on association with other cells. In plant cells the presence of a rigid cell wall is important. Many cells are polyhedral solids and approach the figure of minimal surface area — the *tetrakaidecahedron.* The size of cells varies considerably; the smallest cells may be only a few micrometers in diameter, the largest, several centimeters.

Living cells can be observed only with the light microscope. They show a cytoplasm with an ectoplasm (or cortex) and a more fluid endoplasm, which contains mitochondria, plastids (in plant cells), and various other inclusions. Cell organelles are cell structures that are present in all, or most, cells. Some organelles, such as the Golgi complex or the centrioles, are difficult to see in living cells.

Fixed eukaryotic cells show a complex organization. The cytoplasm is surrounded by the plasma membrane, which has many differentiations. The cell wall in plant cells and the cell coat in animal cells are outside the plasma membrane. The cytoplasm is subdivided into compartments and subcompartments by numerous intracellular membranes that form a large endomembrane system. The main parts of this system are the endoplasmic reticulum (rough and smooth), the nuclear envelope with the nuclear pores, and the Golgi complex. Membrane-bound organelles are lysosomes, mitochondria, secretion vesicles, etc. The ground cytoplasm or matrix lies outside the endomembrane system; it contains the ribosomes, microtubules, microfilaments, soluble proteins, etc. It may have also differentiations such as myofilaments and myofibrils, keratin fibers, and others.

It is important, at this point, to summarize some of the concepts presented in this chapter about pro- and eukaryotic cells (see Table 2–1). These two types of cells differ not only in size but also in complexity of structure. The prokaryon nuclear region lacks the nuclear envelope, contains a single DNA molecule, and divides by binary fission. The nucleus of the eukaryon has a nuclear envelope composed of two membranes with nuclear pores (Fig. 2–6); the DNA is divided into several chromosomes, and one or more nucleoli are present; division is carried out by mitosis or meiosis. The prokaryon cytoplasm lacks membrane-bound organelles, and the respiratory or photosynthetic enzymes are located in the plasma membrane. The eukaryon cytoplasm contains many membranous compartments (i.e., endoplasmic reticulum, Golgi complex) and membrane-bound organelles (i.e., mitochondria, lysosomes, chloroplasts, etc.). The plasma membrane of prokaryons is rather simple, whereas that of eukaryons shows numerous differentiations—i.e., microvilli, infoldings, desmosomes, etc.—see Fig. 2–6). These and other differences are indicated in Table 2–1.

2–3 THE NUCLEUS, CHROMOSOMES, AND THE CELL CYCLE—GENERAL CONCEPTS

The growth and development of every living organism depends on the growth and multiplication of its cells. In unicellular organisms, cell division is the means of reproduction, and by this process two or more new individuals arise from the mother cell. In multicellular organisms, new individuals develop from a single primordial cell, the *zygote;* it is the multiplication of this cell and its descendants that determines the development and growth of the individual.

In many instances cells appear to grow to a certain size before division occurs. This process is repeated in the two daughter cells, so that the total volume eventually becomes four times that of the original cell. The growth of living material progresses rhythmically and according to a geometric progression that has been expressed as follows:

$$\frac{Mn}{Mc}, \frac{2Mn}{2Mc}, \frac{4Mn}{4Mc}, \frac{8Mn}{8Mc}, \text{etc.}$$

where Mn is the *nuclear mass,* and Mc is the *cytoplasmic mass* of the cells. The two masses are in a state of optimum equilibrium, the so-called *nucleoplasmic index* (NP), which is expressed numerically as:

$$NP: \frac{Vn}{Vc - Vn}$$

where Vn is the *nuclear volume,* and Vc is the *cell volume.*

In general, every cell has essentially two periods in its life cycle: *interphase* (nondivision) and *division* (which produces two daughter cells). This cycle is repeated at each cell generation, but the length of the cycle varies considerably in different types of cells. Some cells have a short life cycle, and cell division takes place frequently, whereas others have an interphase that may be as long as the life of the organism (e.g., nerve cells). During cell division the nucleus undergoes a series of complex but remarkably regular and constant changes in which the nuclear envelope and the nucleolus disappear and the chromatin substance becomes condensed into dark-staining bodies — the *chromosomes* (Gr., *chroma*, color + *soma*, body).

Chromosomes are always present in the nucleus. During interphase they are not generally visible because they are dispersed and their macromolecular components are loosely distributed within the nuclear sphere.

In Chapter 16 the cell cycle and the changes occurring in the main chemical components of the cell, i.e., DNA, RNA, and proteins, will be considered in detail. Here it will only be mentioned that the duplication of DNA takes place during a special period of the interphase called the *synthetic* or *S period*. As shown in Figure 2–7 this is preceded and followed by two periods G_1 and G_2 that are intercalated between the S period and mitosis.

The *shape* of the nucleus is sometimes related to that of the cell, but it may be completely irregular. In spheroid, cuboid, or polyhedral cells, the nucleus is generally a spheroid. In cylindrical, prismatic, or fusiform cells, it tends to be an ellipsoid. Examples of irregular nuclei are found in some leukocytes (horseshoe-shaped or multilobate nuclei), in certain *Infusoria* (moniliform nuclei), and in glandular cells of many insects (branched nuclei).

By 1905 Boveri had already noted that, in sea urchin larvae, the size of the nucleus was proportional to the chromosome number. The size of the nucleus is minimal when most of the chromatin is condensed, as in the small lymphocytes. In ovocytes the nucleus (often called the *germinal vesicle*) is very active and may attain a large volume. In general it may be said that each somatic nucleus has a specific size that depends partly on the DNA content and

mainly on the protein content, and that size is related to its functional activity during interphase.

Almost all cells are *mononucleate*, but *binucleate cells* (some liver and cartilage cells) and *polynucleate* cells also exist. The nuclei of polynucleate cells may be numerous (up to 100 per cell in the polykaryocytes of bone marrow — osteoclasts). In the *syncytia*, which are large protoplasmic masses not subdivided into cellular territories, the nuclei may be extremely numerous. Such is the case with striated muscle fiber and certain algae, which may contain several hundred nuclei.

The *position* of the nucleus is variable but is generally characteristic for each type of cell. The nucleus of embryonic cells almost always occupies the geometric center, but it commonly becomes displaced as differentiation advances and as specific parts or reserve substances are formed in the cytoplasm.

In fixed and stained material the structure of the nucleus is distinguished by its complexity and varies according to the type of cell and the fixative used (Fig. 4–4). In general the following structures are recognized in the interphase nucleus: (1) A *nuclear envelope* composed of two membranes and having the *nuclear pores* (Fig. 2–8). (2) The *nucleoplasm* (or *nuclear sap*) that fills most of the nuclear space. This represents uncondensed regions of *chromatin* (i.e., nucleoproteins) where the chromosomes are largely dispersed. These regions correspond to the so-called *euchromatin* (Gr., *eu*, true). (3) The *chromocenters* that along with twisted filaments of chromatin represent parts of the chromosomes that at interphase remain condensed. These condensed regions of so-called *heterochromatin* are frequently found near the nuclear envelope and are also attached to the nucleolus (Fig. 2–8). (4) The *nucleoli*, which are generally spheroid and very large in nerve cells, pancreatic cells, and other cells and are very active in protein synthesis. The nucleoli are either single or multiple, and usually acidophilic, and they contain ribonucleoproteins.

2–4 MITOSIS AND MEIOSIS — ESSENTIALS

It is important to introduce at this point the essentials of mitosis and meiosis, which are studied in detail in Chapters 17 and 18.

All organisms that reproduce sexually develop from a single cell, the *zygote*, produced by the union of two cells, the *germ cells* or *ga-*

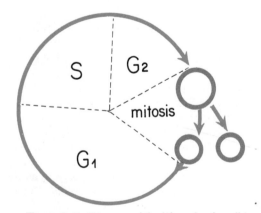

Figure 2–7 Diagram of the life cycle of a cell indicating the mitotic and the interphase periods. Interphase is composed of the G_1, S, and G_2 phases. Duplication of DNA takes place during the synthetic or S phase.

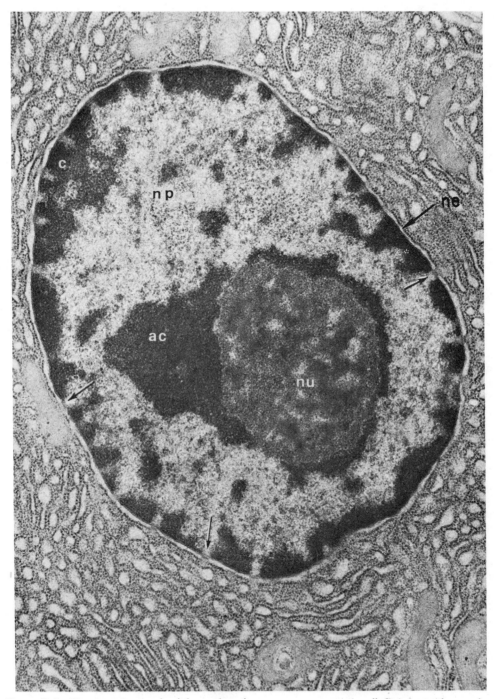

Figure 2–8 Electron micrograph of the nucleus from a mouse pancreatic cell. Staining with uranyl acetate enhances mainly the DNA-containing parts of the cell. Arrows in the interchromatin channels point to nuclear pores; *c,* chromatin; *ac,* chromatin associated with the nucleolus (*nu*); *np,* nucleoplasm; *ne,* nuclear envelope. × 24,000. (Courtesy of J. André.)

metes (a *spermatozoon* from the male, and an *ovum* from the female). The union of an egg and a sperm is called *fertilization*.

Every cell of an individual, with the exception of the *gametes*, contains the same number of chromosomes and hence the same amount of DNA and number of genes. In the somatic cells of a plant or an animal, chromosomes are paired, one member of each pair originally derived from one parent; the other member, from the other parent. Each member of a pair of chromosomes is called a homologue.

Man has 46 chromosomes or 23 pairs; the onion has 8 pairs; the toad, 11 pairs; the mosquito, 3 pairs; and so on (see Table 2–2). Homologues in each pair are alike, but the pairs are generally different from one pair to anoth-

er. The original or *diploid* number of each somatic cell is preserved during successive cell divisions. Only in the gametes is the number reduced to half (the *haploid* number).

Mitosis — Maintenance of Chromosomal Continuity and Diploid Number

The continuity of the chromosomal set is maintained by *cell division*, which is called *mitosis*. At the time of cell division the nucleus becomes completely reorganized, as illustrated in Figure 2–9. Mitosis takes place in a series of consecutive stages known as prophase, metaphase, anaphase, and telophase. In a somatic cell the nucleus divides by mitosis in such a

TABLE 2–2 DIPLOID (2n) NUMBER OF CHROMOSOMES IN SOME PLANTS AND ANIMALS

Plants Common and Scientific Names	Chromosomes	Animals Common and Scientific Names	Chromosomes
Yellow pine, *Pinus ponderosa*	24	Roundworm, *Ascaris megalocephala*, var. univalens	2
Cabbage, *Brassica oleracea*	18	Roman snail, *Helix pomatia*	54
Radish, *Raphanus sativus*	18	Silkworm, *Bombyx mori*	56
Flax, *Linum usitatissimum*	30, 32	Housefly, *Musca domestica*	12
Ombu, *Phytolacca dioica*	36	Fruit fly, *Drosophila melanogaster*	8
Watermelon, *Citrullus vulgaris*	22	Butterfly, *Lysandra atlantica*	446
Cucumber, *Cucumis sativus*	14	Grasshoppers, many Acrididae	24
Papaya, *Cärica papaya*	18	Honeybee, *Apis mellifica*	32, 16
Upland cotton, *Gossypium hirsutum*	52	Mosquito, *Culex pipiens*	6
Cherry, *Prunus cerasus*	32	Frogs, *Rana* spp.	26
Plum, *Prunus domestica*	48	Tree frogs, *Hyla* spp.	24
Pear, *Pyrus communis*	34, 51, 68	Toads, *Bufo* spp.	22
Peanut, *Arachis hypogaea*	40	South African frog, *Xenopus laevis*	36
Ceibo, *Erythrina cristagalli*	42	Chicken, *Gallus domesticus*	ca. 78
Coffee, *Coffea arabica*	44	Turkey, *Meleagris gallipavo*	82
Sunflower, *Helianthus annuus*	34	Pigeon, *Columba livia*	80
Wood rush, *Luzula purpurea*	6	Duck, *Anas platyrhyncha*	80
Potato, *Solanum tuberosum*	48	Opossum, *Didelphys paraguayensis*	22
Tomato, *Lycopersicum solanum*	24	Mouse, *Mus musculus*	40
Tobacco, *Nicotiana tabacum*	48	Rabbit, *Oryctolagus cuniculus*	44
Spiderwort, *Tradescantia virginiana*	24	Albino rat, *Rattus norvegicus*	42
Banana, *Musa paradisiaca*	22, 44, 55, 77, 88	Common rat, *Rattus rattus*	42
Garden pea, *Pisum sativum*	14	Golden hamster, *Mesocricetus auratus*	44
Bean, *Phaseolus vulgaris*	22	Chinese hamster, *Cricetus griseus*	22
Orange, *Citrus sinensis*	18, 27, 36	Guinea pig, *Cavia cobaya*	64
Apple, *Malus silvestris*	34, 51	Mulita, *Dasypus hybridus* S. America	64
Oats, *Avena sativa*	42	Armadillo, *Dasypus novemcinctus* N. America	64
Indian corn, *Zea mays*	20	Dog, *Canis familiaris*	78
Barley, *Hordeum vulgare*	14	Cat, *Felis domestica*	38
Summer wheat, *Triticum dicoccum*	28	Horse, *Equus caballus*	64
Bread wheat, *Triticum vulgare*	42	Donkey, *Equus asinus*	62
Rye, *Secale cereale*	14	Pig, *Sus scrofa*	40
Rice, *Oryza sativa*	24	Sheep, *Ovis aries*	54
Sorghum spp.	10, 20, 40	Goat, *Capra hircus*	60
Black sorghum, *Sorghum almum*	40	Cattle, *Bos taurus*	60
Sugar cane, *Saccarum officinarum*	80	Rhesus monkey, *Macaca mulatta*	42
Broad bean, *Vicia faba*	12	Gorilla, *Gorilla gorilla*	48
Onion, *Allium cepa*	16	Orangutan, *Pongo pygmaeus*	48
Eucalyptus, *Eucalyptus* spp.	22	Chimpanzee, *Pan troglodytes*	48
Passion flower, *Passiflora coerulea*	18	Man, *Homo sapiens*	46
Fern, *Ophioglossum reticulatum* (polyploid)	1260		

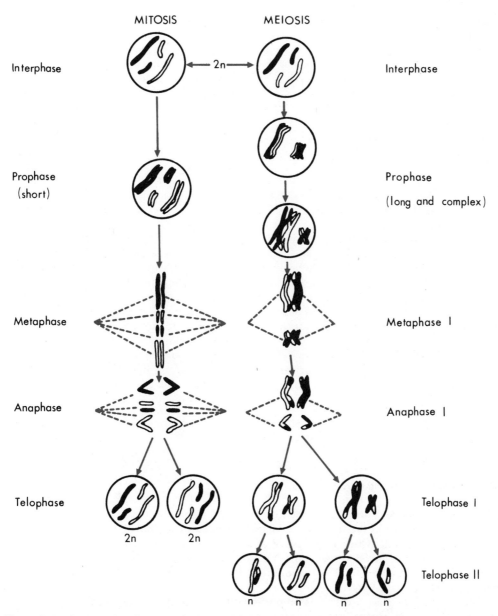

Figure 2–9 Comparative diagram of mitosis and meiosis in idealized cells having four chromosomes (2n). The chromosomes belonging to each progenitor are represented in white and black. In mitosis the division is equational, and in meiosis it is reductional, the two divisions giving rise to four cells having only two chromosomes (n). In meiosis there is, in addition, an interchange of black and white segments of the chromosomes.

fashion that each of the two daughter cells receives exactly the same number and kind of chromosomes that the parent cell had.

Figure 2–9 represents two pairs of homologous chromosomes in a diploid nucleus. Each chromosome duplicates some time during *interphase* before the visible mitotic process begins. At this stage and at early *prophase* chromosomes appear as extended and slender threads. At late prophase chromosomes become short, compact rods by the packing of the nucleoprotein fibers. A spindle arises between the two centrioles and the chromosomes line up across the equatorial plane of the spindle at the *metaphase* plate. At this time each chromosome is made of two filaments called chromatids (Fig. 2–9). At *anaphase* each chromatid separates, forming two daughter chromosomes, which go to opposite poles of the cell. Finally, at *telophase* the daughter chromosomes at each pole again become dispersed, and two daughter nuclei are formed.

In mitosis the original chromosome number is preserved during the successive nuclear divisions. Since the somatic cells are derived from the zygote by mitosis, they all contain the normal double set, or diploid number (*2n*), of chromosomes.

Meiosis — Reduction of Chromosomal Number to Haploid Set

If the gametes (ovum and spermatozoon) were diploid, the resulting zygote would have twice the diploid chromosome number. To avoid this, each gamete undergoes a special type of cell division called *meiosis,* which reduces the normal diploid set of chromosomes to a single (*haploid*) set (*n*). Thus, when the ovum and spermatozoon unite during fertilization, the resulting zygote is diploid. The meiotic process is characteristic of all plants and animals that reproduce sexually, and it takes place in the course of *gametogenesis* (Fig. 2–9).

Meiosis produces the reduction of the chromosome number by means of two nuclear divisions, the *first* and *second meiotic divisions,* that involve only a single division of the chromosomes.

The essential aspects of the process are simple. The homologous chromosomes pair longitudinally, forming a bivalent. Each chromosome is composed of two *chromatids.* The bivalent thus contains four chromatids and is also called a *tetrad.* In the tetrad, one chromatid of the homologue has a pairing partner. Portions of these paired chromatids may be exchanged from one homologue to the other, giving rise to cross-shaped figures, called *chiasmata.* The chiasma is a cytologic manifestation of an underlying genetic phenomenon called *crossing over.*

At metaphase I the bivalents arrange themselves on the spindle, and at anaphase I the homologous chromosomes and their two associated chromatids migrate to opposite poles. Thus, in the first meiotic division the homologous pairs of chromosomes are segregated. After a short interphase the two chromatids of each homologue separate in the second meiotic division, so that the original four chromatids are distributed into each of the four gametes. The result is four nuclei with only a single set (haploid) of chromosomes (Fig. 2–9).

SUMMARY:

Essentials About Nucleus and Chromosomes

Multicellular organisms develop by the division of an initial diploid cell, the *zygote,* which is a product of the union of the haploid gametes (i.e., fertilization). Every cell has a life cycle which is composed of a period of non-division (interphase), and a period of division (generally, mitosis). The interphase period has a synthetic phase, *S,* in which the DNA duplicates. This *S* phase is preceded and followed by the G_1 and G_2 phases.

The size of the interphase nucleus is proportional to the ploidy of chromosomes. By observing a section of the liver, a student may recognize a few larger nuclei corresponding to tetra- and octoploid cells.

The following structures are recognized in the interphase nucleus: (1) the *nuclear envelope;* (2) the *nucleoplasm,* formed by uncondensed regions of chromatin (euchromatin); (3) *chromocenters* and other condensed portions of chromatin (heterochromatin); and (4) nucleoli. The chromosomes are always present in the nucleus, but during interphase they are dispersed, and they are generally not morphologically distinguishable. Each diploid cell has a number of homologous chromosome pairs (23 pairs in human). One chromosome of each pair comes from one parent.

The continuity of the diploid set of chromosomes is maintained by mitosis. Duplication of chromosomes occurs during the *S* phase of interphase; however, by means of *mitosis* (prophase, metaphase, anaphase, and telophase), the daughter chromosomes are distributed equally into the two daughter cells.

Meiosis is a special type of cell division found in germinal cells, by which the number of chromosomes is reduced to the haploid number. Meiosis involves two consecutive divisions without duplication of the chromosomes. The essential feature of meiosis is a long prophase during which the homologous chromosomes pair, forming bivalents. Each bivalent has four chromatids (tetrad). Parts of the paired homologous chromatids interchange at cross-points called *chiasmata*, which are manifestations of an underlying genetic recombination (crossing over). After the second meiotic division, the four resulting cells (gametes) are haploid.

ADDITIONAL READING

Brachet, J., and Mirsky, A. E., eds. (1959–1961) *The Cell,* 6 Volumes, Academic Press, Inc., New York.

Claude, A. (1975) The coming of age of the cell. *Science, 189*:433.

Du Praw, E. J. (1968) *Cell and Molecular Biology.* Academic Press, Inc., New York.

Fawcett, D. W. (1979) *The Cell: Its Organelles and Inclusions,* 2nd Ed. W. B. Saunders Co., Philadelphia.

Hamerton, J. L., ed. (1963) *Chromosomes in Medicine.* Medical advisory committee of the National Spastics Society in association with William Heinemann. Little Club Clinics in Developmental Medicine, No. 5.

Head, J. J., ed. (1973) *Readings in Genetics and Evolution.* Oxford University Press, London.

John, B., and Lewis, K. R. (1965) *The Meiotic System.* Springer-Verlag, Vienna.

Kennedy, D., ed. (1958–1965) *The Living Cell.* W. H. Freeman and Co., San Francisco.

Lima-de-Faría, A., ed. (1969) *Handbook of Molecular Cytology.* North-Holland Publishing Co., Amsterdam.

Macknight, A. D. C., and Leaf, A. (1977) Regulation of cellular volume. *Physiol. Rev., 57*:510.

Margulis, L. (1970) *Origin of Eukaryotic Cells: Evidence and Research.* Yale University Press, New Haven.

Margulis, L. (1971) Symbiosis and evolution. Sci. Am., *225*:48.

Mitchison, J. M. (1971) *The Biology of the Cell Cycle.* Cambridge University Press, New York.

Swanson, C. P., and Webster, P. (1977) *The Cell.* Prentice-Hall, Inc., Englewood Cliffs, N.J.

Schwartz, R., and Dayhoff, M. (1978) Origins of prokaryotes, eukaryotes, mitochondria and chloroplasts. *Science, 199*:395.

White, M. J. D. (1972) *The Chromosomes,* 6th Ed. John Wiley, New York.

White, M. J. D. (1973) *Animal Cytology and Evolution.* 3rd Ed. Cambridge University Press, London.

Woese, C. R., and Fox, G. E. (1977) Phylogenetic structure of the prokaryotic domain: the primary kingdoms. *Proc. Natl. Acad. Sci. (USA), 74*:5088.

3

INSTRUMENTAL ANALYSIS
OF BIOLOGICAL
STRUCTURES

3–1   Various Types of Light Microscopy   28
   Phase Microscopy—Detects Small Differences in Refractive Index
   Interference and Nomarski Microscopy—Detect Continuous Changes in Refractive Index of Cell Structures
   Darkfield Microscopy—Based on Light Scattering at Cell Boundaries
   Polarization Microscopy—Detects Anisotropy with Polarized Light

3–2   Electron Microscopy (EM)   34
   Thin Specimens—Essential for EM Study
   Freeze-Fracturing—Membranes Split on Cleavage Planes

   Preparation of Thin Sections—Epoxy Resins and Ultramicrotomes
   Shadow Casting or Negative Staining—Increased Contrast
   Tracers—Opaque Macromolecules Used
   High Voltage EM—Allows Study of Thicker Specimens
   Scanning EM—Surface View of Cell Structures

3–3   X-ray Diffraction   39
   Summary: *Microscopy*

Before studying this chapter, the student should be familiar with the limits and dimensions in biology (Chapter 1) and with the optical laws and principles on which the ordinary light microscope is based.

Observation of biological structures is difficult because cells are, in general, very small and are transparent to visible light. The search continues for new instruments designed to provide better definition of cell structure down to the molecular level by an increase in *resolving power* and that are also designed to counteract the transparency of the cell by an increase in *contrast.*

The unaided eye detects variations in wavelength (color) and in intensity of visible light. The majority of cell components are essentially transparent, except for some pigments (more frequent in plant cells) that absorb light at certain wavelengths (colored substances). The low light absorption of the living cell is caused largely by its high water content, but even after drying, cell components show little contrast.

One way of overcoming this limitation is to use dyes that selectively stain different cell components and thus introduce contrast by light absorption. In most cases, however, staining tech-

niques cannot be used in the living cell. The tissue must be fixed, dehydrated, embedded, and sectioned prior to staining, and all these procedures may introduce morphologic and chemical changes.

3–1 VARIOUS TYPES OF LIGHT MICROSCOPY

In the *light microscope,* as in any other type of microscope, the resolving power (the capacity of the instrument for showing distinct images of points very close together) depends upon the wavelength (λ) and the numerical aperture (NA) of the objective lens (Fig 3–1). The *limit of resolution,* defined as the minimum distance between two points that allows for their discrimination as two separate points is:

$$\text{Limit of resolution} = \frac{0.61\lambda}{\text{NA}} \qquad (1)$$

The numerical aperture is: NA = $n \times \sin\ \alpha$. Here, n is the refractive index of the medium and sin α is the sine of the semiangle of aperture. Remember that the limit of resolution is

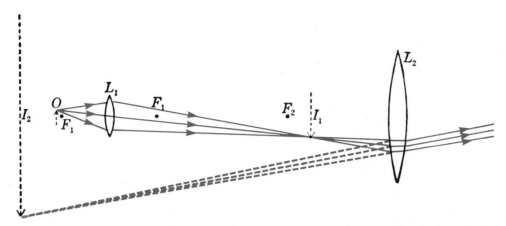

Figure 3–1 Light path in the ordinary light compound microscope. The group of ocular lenses is diagrammatically represented by L_2, the group of objective lenses by L_1. The object (O) on a microscope slide is placed just outside the principal focus of the objective lens (L_1), which has a short focus. This lens produces a real image at I_1, which is formed inside the principal focus of the eyepiece lens (L_2). The eye, looking through the lens L_2, sees a magnified virtual image (I_2) of the image I_1. The eyepiece lens is thus used as a magnifying glass to view the real image (I_1).

inversely related to the resolving power; i.e., the higher the resolving power, the smaller the limit of resolution.

Since sin α cannot exceed 1, and the refractive index of most optical material does not exceed 1.6, the maximal NA of lenses, using oil immersion, is about 1.4. With these parameters it is easy to calculate from formula (**1**) that the limit of resolution of the light microscope cannot surpass 170 nm (0.17 μm) using monochromatic light of $\lambda = 400$ nm (violet). With white light, the resolving power is about 250 nm (0.25 μm). Since in formula (**1**) the NA is limited, it is evident that the only way to increase the resolving power is to use shorter wavelengths. In this case, glass lenses are no longer transparent, and other refractive media should be introduced. For example, with ultraviolet radiation of 200 to 300 nm, quartz lenses or reflecting optical instruments should be used, and the resolution is then increased only by a factor of two, reaching 100 nm (0.1 μm). By similar reasoning, a microscope using infrared radiation of $\lambda = 800$ nm would have a limit of resolution of 0.4 μm.

Phase Microscopy — Detects Small Differences in Refractive Index

In recent years remarkable advances have been made in the study of living cells by the development of special optical techniques, such as *phase contrast* and *interference microscopy*. These two techniques are based on the fact that although biological structures are highly transparent to visible light, they cause phase changes in transmitted radiations.[1, 2]

Figure 3–2 indicates the effects of a nonabsorbent transparent material (A) and an absorbent transparent material (C) on a light ray. In *A*, the wave impinges on a material that has a refractive index different from that of the medium. In passing through the object, the amplitude of the wave is not affected, but the velocity is changed. If the refractive index of the material is higher than that of the medium, there is a *delay*

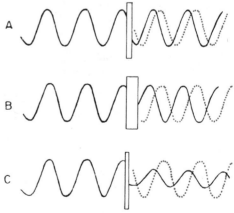

Figure 3–2 Diagram showing: **A,** the effect of a transparent and nonabsorbent material of higher refractive index than the medium, which introduces a phase change (retardation). **B,** the same, but thicker, object. The retardation or phase change is more pronounced. **C,** the effect of a transparent and absorbent object. There is a retardation, but also a decrease, in amplitude (intensity).

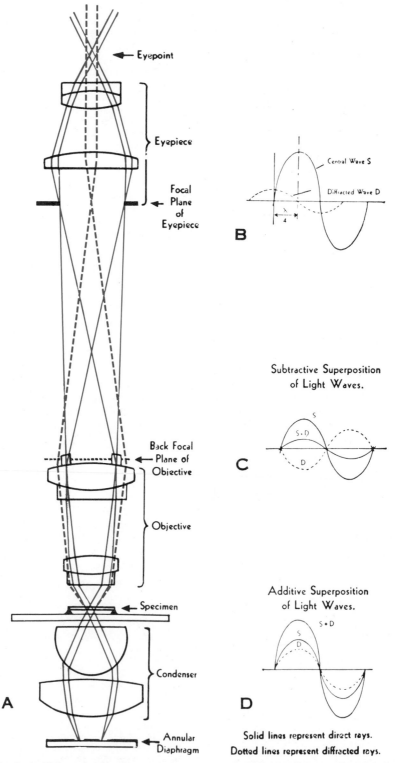

Figure 3–3 **A,** the light path in a phase contrast microscope. **B,** the normal retardation by ¼ wavelength of light diffracted by an object, and its difference in phase from the light passing through the surrounding medium. By phase optics the two waves are superimposed to reinforce each other in bright contrast phase as shown in **D** or to subtract from each other as in dark contrast phase shown in **C**. (From the American Optical Company.)

or *retardation*. After the wave emerges from the object, its original velocity is re-established, but retardation is maintained. This retardation implies a phase change, which can be measured in fractions of a wavelength. The phase change increases in direct proportion to the difference between the refractive indices of the object and the surrounding medium and in direct proportion to the thickness of the object (Fig. 3–2, *B*).

To understand the phase microscope, it is first necessary to analyze the behavior of a ray of light that traverses a thin, transparent particle.

A portion of the light ray that traverses the object (wave *D* in Figure 3–3, *B*) is diffracted and deviates with respect to the rays that do not traverse the object, or with respect to those passing through the exact center (wave *S* in Figure 3–3, *B*).

In biological materials the phase difference between the S and D waves is approximately ¼ wavelength (Fig. 3–3, *B*). These two rays (*S* and *D*) penetrate the objective lens and undergo interference. The resulting ray has a small phase retardation, but this is not detectable with an ordinary light microscope.

In the phase contrast microscope, the small phase differences are intensified. The most lateral light passing through the objective of the microscope is advanced or retarded by an *additional* ¼ wavelength (¼ λ) with respect to the central light passing through the medium around the object, by an annular phase plate that introduces a ¼ wavelength variation in the back focal plane of the objective. In addition, an annular diaphragm is placed in the substage condenser (Fig. 3–3). The phase effect results from the interference between the direct geometric image given by the central part of the objective and the lateral diffracted image, which has been retarded or advanced to a total of ½ wavelength. In *bright*, or *negative*, *contrast* the two sets of rays are added (Fig. 3–3, *D*), and the object appears brighter than the surroundings; in *dark*, or *positive*, *contrast* the two sets of rays are subtracted (Fig. 3–3, *C*), making the image of the object darker than the surroundings (Fig. 2–4). Because of this interference, the minute phase changes within the object are amplified and translated into changes of amplitude (intensity).

A transparent object thus appears in various shades of gray, depending on the product of its thickness and the difference between the refractive indexes of the object and the medium.

Phase microscopy is used routinely to observe living cells and tissues and is particularly valuable for observing cells cultured in vitro during mitosis (Fig. 3–4).

Interference and Nomarski Microscopy — Detect Continuous Changes in Refractive Index of Cell Structures

The *interference microscope* is based on principles similar to those of the phase microscope but has the advantage of giving quantitative data. Interference microscopy permits detection of small, continuous changes in refractive index, whereas the phase microscope reveals only sharp discontinuities. The variations of phase can be transformed into such vivid color changes that a living cell may resemble a stained preparation.[3-5]

A special variation of the interference microscope is the so-called *Nomarski interference-contrast microscope*,[6] in which a single light beam passes through the object and the objective but is then divided into two interfering beams via a special birefringent prism. This microscope also includes polarizer and analyzer filters and a compensating prism above the substage condenser (see Padawer, 1968). The image obtained gives a characteristic relief effect and offers some advantage over ordinary phase contrast optics. It is particularly useful for the study of cells in mitosis, as indicated in Figure 17–5.

Darkfield Microscopy — Based on Light Scattering at Cell Boundaries

Darkfield microscopy, also called *ultramicroscopy*, is based on the fact that light is scattered at boundaries between regions having different refractive indexes. The instrument is a microscope in which the ordinary condenser is replaced by one that illuminates the object obliquely. With this darkfield condenser, no direct light enters the objective; therefore, the object appears bright because of the scattered light, and the background remains dark.

Under the darkfield microscope, objects smaller than those seen with the ordinary light microscope can be detected but not resolved.

The problem of *detectability* also applies to the other types of light microscopes, particularly those using phase contrast optics. Theoretically a fiber of only 5 nm (i.e., about 40 times smaller than the resolving power of a light microscope) could be detected, provided it had enough contrast in comparison to the background.

The detectability of fine structures is important when living cells are observed, as, for example, in following the changes in spindle fibers during the cell cycle (Chapter 17).

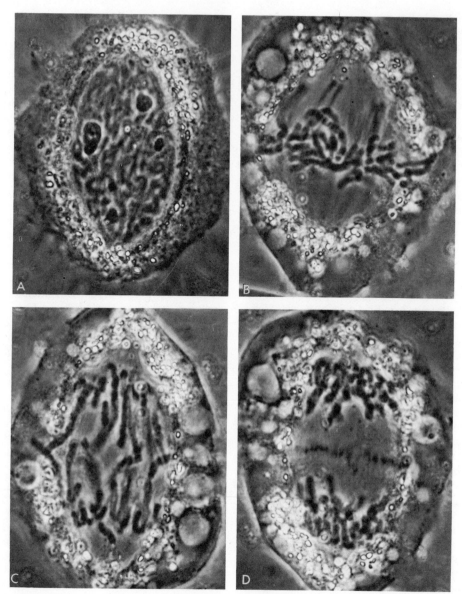

Figure 3–4 Observation by phase contrast microscopy of mitotic cell division in a living cell of endosperm tissue of the plant *Haemanthus*. The same cell has been photographed at the following times: **A,** 10:32 hrs; **B,** 12:48 hrs; **C,** 13:12 hrs; and **D,** 13:21 hrs. **A,** late prophase showing the coiled chromosomes and the nucleoli within the nucleus; **B,** metaphase with chromosomes at the equatorial plane; **C,** anaphase; and **D,** telophase showing the chromosomes at the poles and the formation of the phragmoplast at the equatorial plane. × 700. (Courtesy of A. S. Bajer.)

Polarization Microscopy — Detects Anisotropy with Polarized Light

This method is based on the behavior of certain components of cells and tissues when they are observed with polarized light. If the material is *isotropic,* polarized light is propagated through it with the same velocity, independent of the impinging direction. Such substances or structures are characterized by having the same *index of refraction* in all directions. On the other

hand, in an *anisotropic* material the velocity of propagation of polarized light varies. Such material is also called *birefringent* because it presents two different indexes of refraction corresponding to the respective different velocities of transmission.

Birefringence (B) may be expressed quantitatively as the difference between the two indexes of refraction ($N_e - N_o$) associated with the fast and slow ray. In practice, the retardation (Γ) of the light polarized in one plane is measured relative to that of light polarized in another perpendicular plane with the polarizing microscope. The retardation depends on the thickness of the specimen (t) in this way:

$$B = N_e - N_o = \frac{\Gamma}{t} \qquad (2)$$

Measurement of the retardation is assisted by a form of compensator introduced into the optical system. The measurement is in nm or in fractions of a wavelength (λ).

The *polarizing* microscope differs from the ordinary one in that two polarizing devices have been added: the *polarizer* and the *analyzer*, both of which can be made from a sheet of polaroid film or with Nicol prisms of calcite. The polarizer is mounted below the substage condenser and the analyzer is placed above the objective lens (Fig. 3–5).

In the crossed position, polarized light is not transmitted. Under this condition, if a birefringent specimen is placed on the stage, the plane of polarization will deviate according to the retardation introduced by the object. The usual test with the polarizing microscope consists of rotating the specimen to find the points of maximum and minimum brightness (Fig. 3–5).

In biological fibers, birefringence is *positive* if the index of refraction is greater along the length of the fiber than in the perpendicular plane, and it is *negative* in the opposite case. The sign can be determined by interposing a birefringent material whose slow and fast axes are known.

The main types of birefringence are:

Crystalline (Intrinsic) Birefringence. Crystalline birefringence is found in systems in which molecules or ions have a regular asymmetrical arrangement, and it is independent of the refractive index of the medium. In structures composed of proteins or lipids, a certain degree of crystalline birefringence may appear, which in both cases is positive. On the other hand, fibers of nucleoprotein have a negative birefringence.

Form Birefringence. This is produced when submicroscopic asymmetrical particles are oriented in a medium of a different refractive index. In this instance, the birefringence is changed when the refractive index of the medium varies.

Strain Birefringence. Certain isotropic structures show strain birefringence when subjected to tension or pressure. It occurs in muscle and in embryonic tissues.

Dichroism. This type of birefringence occurs when the absorption of a given wavelength of polarized light changes with the orientation of the object. In dichroism the changes are in amplitude, that is, in the intensity of the transmitted light. Dichroism can be induced in tissues by some staining procedures. For example, organic dyes, such as congo red or thionine, and colloids can produce dichroism in certain structures by the special orientation of the molecules.

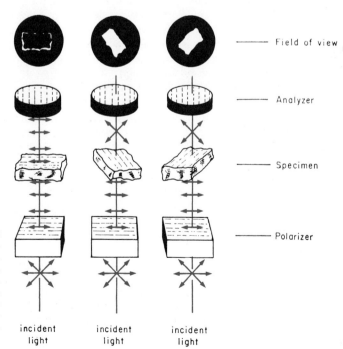

Figure 3–5 Schematic drawing showing variations in darkness and brightness of an anisotropic object when placed between crossed polarizer and analyzer and rotated $\pm 45°$. (Extracted from Wilson, G. B., and Morrison, J. H., *Cytology*, New York, Reinhold Publishing Corporation, 1961, with the permission of Reinhold Publishing Corporation.)

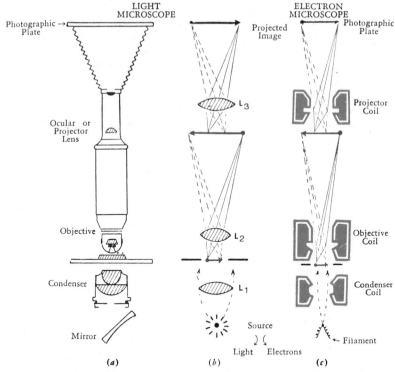

Figure 3–6 Comparison between the optical microscope and the electron microscope. (See the description in the text.) (From G. Thompson.)

3–2 ELECTRON MICROSCOPY (EM)

The electron microscope permits a direct study of biological ultrastructure. Its resolving power is much greater than that of the light microscope. In the electron microscope streams of electrons are deflected by an electrostatic or electromagnetic field in the same way that a beam of light is refracted when it crosses a lens. If a metal filament is placed in a vacuum tube and heated, it emits electrons that can be accelerated by an electrical potential and tend to follow a straight path with properties similar to those of light. Like light, the stream of electrons has a corpuscular and vibratory character, but the wavelength is much shorter (i.e., $\lambda = 0.005$ nm for electrons and 550 nm for light).

The filament or cathode of the electron microscope emits the stream of electrons. By means of a magnetic coil, which acts as a condenser, electrons are focused on the plane of the object and then are deflected by another magnetic coil, which acts as an objective lens and gives a magnified image of the object. This is received by a third magnetic "lens," which acts as an ocular or projection lens and magnifies the image from the objective. The final image can be visualized on a fluorescent screen or recorded on a photographic plate.

In spite of the apparent similarities shown in Figure 3–6, there are great differences between the light and the electron microscope; one of these is the mechanism of image formation. Whereas in the light microscope image formation depends mainly on the degree of light absorption in different zones of the object, image formation in the electron microscope is due principally to electron scattering. Electrons colliding against atomic nuclei in the object are often dispersed, so that they fall outside the aperture of the objective lens. In this case the image on the fluorescent screen results from the absence of those electrons blocked by the aperture. Dispersion may also be the result of multiple collisions, which diminish the energy of the passing electrons.

Electron dispersion, in turn, is a function of the thickness and molecular packing of the object and depends especially on the atomic number of the atoms in the object. The higher the atomic number, the greater is the resultant dispersion. Most of the atoms that constitute biological structures are of low atomic number and contribute little to the image. For this reason, heavy atoms should be added to the molecular structure.

The greatest advantage of the electron microscope is its high resolving power, which depends on the same variants as does the resolving power of the light microscope — see formula (1).

The wavelength of a stream of electrons is a function of the acceleration voltage to which the electrons are subjected; it can be calculated by the formula of De Broglie:

$$\lambda = \frac{12.2}{\sqrt{V}} \, 0.1 \text{ nm} \qquad (3)$$

For example, in current models of the electron microscope, V = 50,000 volts and λ = 0.00535 nm.

Because of the great aberration of the magnetic lenses, the actual numerical aperture of the electron microscope is small and the limit of resolution is theoretically 0.2 nm (Fig. 1–3). In practice, the limit of resolution for biological specimens is larger than that figure.

In the light microscope, magnification is largely determined by the objective, and a maximum magnification of 100 to 120× can be reached. Since the ocular lens can increase this image only 5 to 15 times, a total useful magnification of 500 to 1500× can be achieved.

In the electron microscope the resolving power is so high that the image from the objective can be greatly enlarged. For example, with an initial magnification by the objective of 100×, the image can be magnified 200× with the projector coil, achieving a total magnification of 20,000×.

In the newer instruments a wide range of magnifications can be attained by introducing one or more intermediate lenses. Direct magnifications as high as 1,000,000× may thus be obtained, and the micrographs may be enlarged photographically to 10,000,000× or more, depending on the resolution achieved.

Thin Specimens — Essential for EM Study

Because of its extraordinary resolving power, the electron microscope seems to be an ideal instrument for the study of cellular ultrastructure. Nevertheless, its usefulness is reduced by a number of technical difficulties and limitations.

One limitation is the low penetration power of electrons. In current instruments, if the specimen is more than 500 nm (0.5 μm) thick, it appears almost totally opaque. The specimen is generally deposited on an extremely fine film (7.5 to 15 nm thick) of collodion, carbon, or other substance to support the specimen, and this film must be held up by a fine metal grid. For observation under the electron microscope, the specimen is usually dehydrated and then placed in a vacuum. Techniques for preparing specimens vary considerably. An important method for the study of macromolecules is the so-called monolayer technique of Kleinschmidt,[7] in which the macromolecules are extended on an air-water interface before being collected on a film. This method has given excellent results in the demonstration of DNA and RNA molecules from various sources (Fig. 3–7).

Freeze-Fracturing — Membranes Split on Cleavage Planes

The study of the structure of biological membranes has been greatly improved by the use of techniques that involve the freezing and fracturing of specimens.[8] In general, the specimens are frozen and fractured and then subjected to a certain degree of water sublimation in a vacuum, followed by the deposit of a layer of evaporated carbon and other elements. This layer, called a *replica*, is detached from the object and reveals a natural-looking representation of the surface of the freeze-etched object (Fig. 3–8).

The fracture may disclose either the outer or the inner surface of a membrane or may even split the membrane lengthwise, thus revealing information about components passing through the cell membrane (see Fig. 8–5). The effect of etching is to expose and to render more visible the fine surface details. Freeze-etching can be carried out after treatment of the specimen with enzymes or after labeling with antigens, viruses, or special chemical substances that may yield information about the chemical and biological properties of certain sites in the membrane. The nomenclature used in the freeze-fracture of membranes is given in Chapter 14–3 (see Branton and Kirchanski, 1977; Sleytr and Robards, 1977).

Preparation of Thin Sections — Epoxy Resins and Ultramicrotomes

The morphologic study of cells and tissues is achieved primarily by the use of thin sections.

To satisfy the need for thinner sections, hard embedding media have been utilized. Those most often used are epoxy resins that impregnate the tissue and are then polymerized by proper catalysts. A water-miscible glycol-

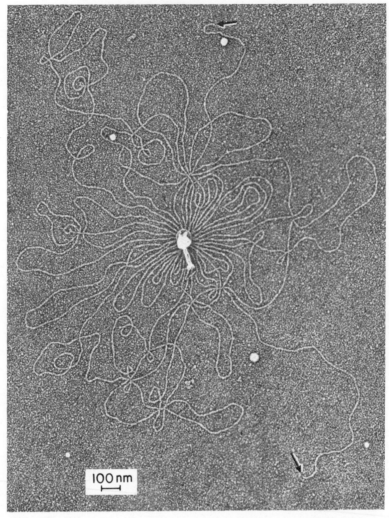

Figure 3–7 Electron micrograph of a bacteriophage (*in the center*) that has undergone an osmotic shock. The DNA molecule that was contained in the "head" of the bacteriophage is now dispersed. Arrows indicate the extremes of the single, unbranched DNA molecule. Preparation shadowcast with platinum. ×76,000 (Courtesy of A. K. Kleinschmidt.)

methacrylate has been developed for cytochemical studies. In this case the section can be submitted to selective extraction and to the action of various enzymes.[9]

To prepare the extremely thin sections, ultramicrotomes with specialized features have been developed. Several microtomes have been designed that have a thermal or a mechanical advancing device. With both types the thinnest sections that can be made are of the order of 20 nm. The limiting factors seem to be proper embedding and the sharpness of the cutting edge of the microtome. Diamond knives are now in general use. Thin sectioning can be performed at low temperature with simple embedding in gelatin.

36

Shadow Casting or Negative Staining — Increased Contrast

One technique, called *"shadow casting,"* consists of placing the specimen in an evacuated chamber and evaporating, at an angle, a heavy metal such as chromium, palladium, platinum, or uranium from a filament of incandescent tungsten. The material is thus deposited on one side of the surface of the elevated particles; on the other side a shadow forms, the length of which permits determination of the height of the particle. Photomicrographs made of such specimens have a three-dimensional appearance that is not found when other techniques are used (Fig. 3–9).

One of the most important techniques in the study of viruses and macromolecules is *"negative staining."* The specimen is embedded in a droplet of a dense material, such as phosphotungstate, which penetrates into all the empty spaces between the macromolecules[10] (Fig. 12–6). These spaces appear well defined in negative contrast. With this technique the numbers of protein molecules (capsomeres) of different viruses have been determined and interesting observations on cellular structures have been made (Fig. 7–3).

A positive increase in contrast in biological structures has been obtained by the use of substances containing heavy atoms, such as osmium tetroxide, uranyl, and lead ions, which, under certain conditions, act as *"electron stains."* These electron stains are comparable to histologic stains, in that they combine with certain regions of the specimen.

Tracers — Opaque Macromolecules Used

Several biological processes may be studied by the use of appropriate tracers that are detected by their electron opacity. For example, the uptake of macromolecules into the cells by pino-

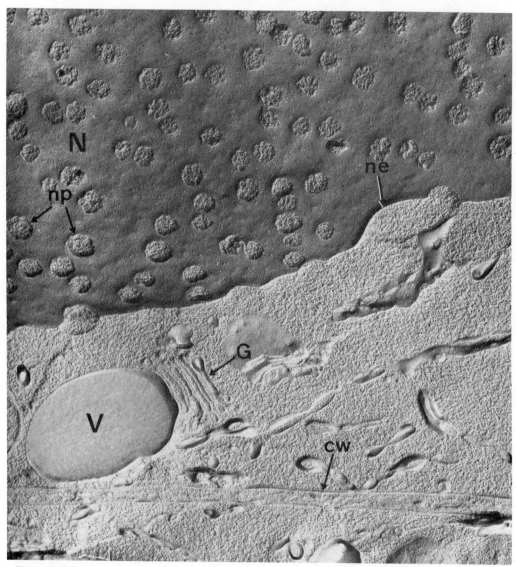

Figure 3–8 Onion root cell which has been submitted to the freeze-etching technique. The upper part of the figure corresponds to the nucleus (N) and shows the nuclear pore complexes (*np*) and the nuclear envelope (*ne*). In the cytoplasm, a Golgi complex (G) and a large vacuole (V) are observed. *cw*, cell wall. × 75,000. (Courtesy of D. Branton.)

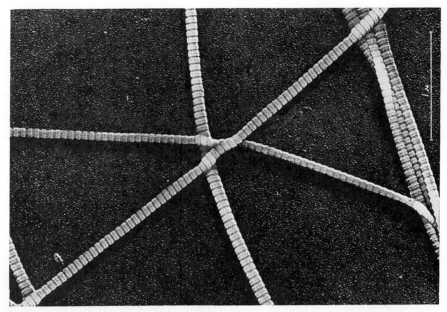

Figure 3–9 Electron micrograph of collagen fibers, shadowed with chromium, from human skin. Bands with a period of 64 nm are shown. ×28,000. (Courtesy of J. Gross.)

cytosis or phagocytosis (see Chapter 13–2) or the transport of molecules across cellular barriers can be studied with various opaque particles.

The ideal tracer should be: (a) nontoxic and physiologically inert, (b) composed of uniform particles of known size, (c) preserved in situ during the processing of the tissue, and (d) composed of particles of a small size. Several colloidal substances, ranging in size from 1.0 to 10 nm, such as gold, mercuric sulfide, iron oxide, thorium dioxide, and colloidal lanthanum have been used.[11] Ferritin, a large protein containing iron, as well as dextrans of various sizes and other polymers, are now being widely used.[12] A variety of tracers are represented by enzymes that are able to produce a reaction that greatly enhances the contrast of biological specimens. The presence of various *peroxidases* is demonstrated by their reaction with peroxide and 3,'3'-diaminobenzidine, for example (see Chapter 4–5).[13] One of the smallest tracers of this kind is the so-called *microperoxidase*, which has a molecular weight of only 1900 daltons.[14]

High Voltage EM — Allows Study of Thicker Specimens

While most electron microscopes use accelerating voltages between 50 and 100 kilovolts, there are instruments now in operation that greatly surpass this voltage and reach 500 to 3000 kV. The design of these newer instruments is essentially similar to the older models, but the construction is much more massive to permit higher acceleration, greater magnetic excitation for the lenses, and the shielding needed to protect against x-radiation. While its main application is in metallurgy, the high-voltage electron microscope is being used increasingly in studies of biological material to examine thick sections (up to 5 μm) and whole cells, with less radiation damage resulting from ionization and temperature effects. With this instrument, theoretically, there is the possibility of examining living cells. The practice of obtaining stereomicrographs by tilting the specimen also gives a greater amount of three-dimensional information (see Fig. 9–3).[15-17]

Scanning EM — Surface View of Cell Structures

A surface view of a specimen may be obtained by using the secondary electron emission that is ejected after the primary electron beam has interacted with the surface of a thick specimen (Fig. 3–10).[18] In the scanning electron microscope a thin beam of electrons moves back and forth across the specimen in the same way that the electron beam moves in a television tube. The secondary electrons are then collected by a photomultiplier tube, and an image is displayed on a television tube.

In some cases, to increase the scattering power

of the surface structures, infiltration with electron-contrasting chemicals or surface coating may be used.

The scanning electron microscope usually has a lower resolving power than the transmission electron microscope; however, with special electron emitters (i.e., lanthanum hexaboride) it is possible to obtain resolutions of the order of 3.0 nm. These brighter and more concentrated electron sources permit x-ray microanalysis of the secondary emission by which qualitative and quantitative estimates of the distribution of certain elements in the specimen are possible.[19, 20]

3–3 X-RAY DIFFRACTION

This technique is based on the diffraction of radiations when they encounter small obstacles. If a ray of white light (wavelength averaging 0.5 μm) impinges upon a diffraction grating that has 1000 lines per millimeter (1 μm spacing), it will be diffracted and will show the various bands of the spectrum. If the wavelength of the light is known, the spacing can be calculated from the diffracted angles, and vice versa. This type of grating would be too wide for x-rays, and no diffraction would be produced.

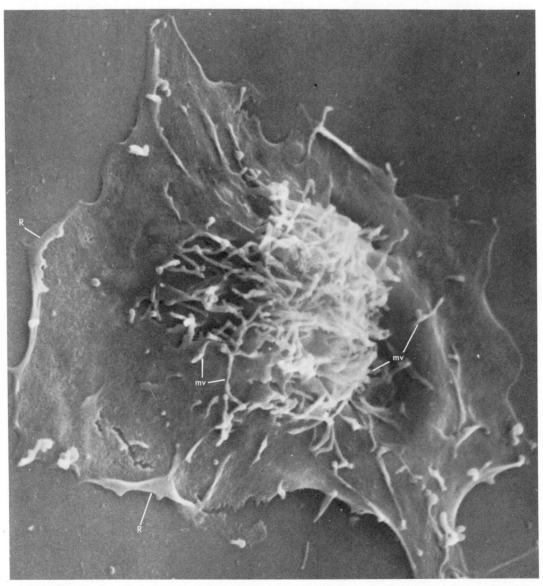

Figure 3–10 Scanning electron micrograph of a cultured cell that is in the process of respreading after cytokinesis. Notice the ruffling edge (R) and the numerous microvilli (mv). ×3260. (Courtesy of R. D. Goldman.)

Laue suggested that gratings of much smaller dimensions, such as those found in natural crystals, would be necessary for the diffraction of x-rays. The atoms, ions, or molecules in crystals constitute a true lattice of molecular dimensions capable of diffracting radiations of this wavelength (see Figure 1–3). This technique has its widest application in the study of inorganic and organic crystals, in which it is possible to determine the precise spatial relationships between the constituent atoms. An analysis of the structure of complex organic molecules, such as proteins and nucleic acids, is much more difficult because of the great number of atoms involved in a single molecule and the irregularities in three-dimensional architecture that most of these large and complex molecules have (Fig. 3–11) However, as shown in Chapter 5, this configuration of molecules is so vitally important to the understanding of biological function that a great deal of work, by Pauling, Perutz, Kendrew, Wilkins, and others, has been carried out to elucidate it. The study of the structure of molecules such as hemoglobin, myoglobin, DNA, and collagen has been of fundamental importance in the development of molecular biology.

In essence, the technique of x-ray diffraction employs a beam of collimated x-rays that traverse the material to be analyzed; a photographic plate is placed beyond this to record the diffraction pattern.

A series of concentric spots or bands, caused by interference between the different diffracted rays, may appear on the plate. The distance between these spots and the center of the pattern depends upon the spaces between the regularly repeating units, or *periods of identity*, in the specimen that produced the diffraction — the smaller the angle of diffraction, the greater the distance between the repeating units; the sharper the spots, the more regular the spacing (Fig. 3–12).

A crystalline structure can be considered a three-dimensional lattice in which the atoms are regularly spaced along the three principal axes. The so-called *unit cell* is a solid parallelepiped (a solid with six faces, each a parallelogram) that represents the minimal repeating unit within the crystal. In practice, it is simpler to think of the crystal as composed of sets of superimposed lattice planes such as indicated in Figure 3–13). In this diagram, d is the spacing of the diffracting planes, and θ, the angle of incidence.

According to Bragg's law, d can be calculated

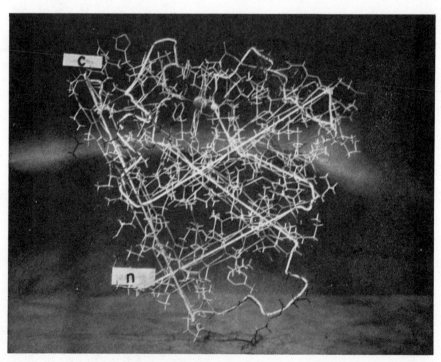

Figure 3–11 Model of the myoglobin molecule. The white cord represents the course of the polypeptide chain. The iron molecule is indicated by a grey sphere. The two terminals of the protein molecule are indicated by c and n. (From Kendrew, J. C., *Science, 139*:1259–1266, 1963, Copyright 1963 by the American Association for the Advancement of Science.)

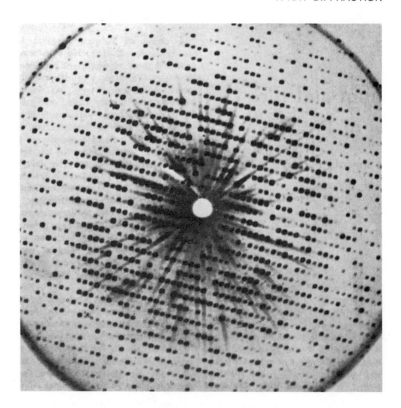

Figure 3-12 X-ray diffraction pattern of a myoglobin crystal. (From Kendrew, J. C., *Science*, 139:1259–1266, 1963, Copyright 1963 by the American Association for the Advancement of Science.)

as follows (*n* is an integer corresponding to the diffraction order):

$$n\lambda = 2d \sin \theta \qquad (4)$$

If the wavelength and the angle of incidence of a definite spot in the diffraction pattern are known, the spacing producing the diffraction can be calculated.

In the more refined methods of x-ray analysis of macromolecules, such as myoglobin, hemoglobin, and DNA, not only is the distance within the unit cell calculated but also the *scattering power* of the individual atoms. This is sim-

ilar to electron dispersion in that the scattering power is related to the atomic number. By introducing heavy atoms (such as mercury) into known points of the organic molecule, it is possible to increase their scattering power. The heavy atom serves as a landmark for the reconstruction of the molecule. This is accomplished by a very complex process that involves plotting the electron density and the Fourier synthesis of all the component waves or diffraction orders. From this mathematical synthesis a three-dimensional representation of the object can be constructed, and models of the entire molecule made (Fig. 3–12).

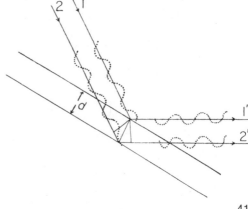

Figure 3-13 Diagram showing the effect of an x-ray beam incident upon two parallel planes in a crystal lattice that are separated by a distance, *d*. Incident rays *1* and *2* form an angle with these planes, and two secondary or diffracted rays are produced (*1'* and *2'*). By simple geometrical considerations, Bragg's law ($n\lambda = 2d \sin \theta$) can be deduced and the distance (*d*) calculated.

SUMMARY

Microscopy

Observation of biological structures is difficult because of their small size and lack of contrast. Optical instruments are specially designed to overcome both difficulties. In the *light microscope* the limit of resolution, $Lm = \dfrac{0.61\lambda}{NA}$, depends on the wavelength of the light and the numerical aperture. The resolving power is the inverse of the limit of resolution. The limit of resolution is, in general, 0.25 μm; with ultraviolet light it can be reduced to 0.1 μm.

Phase microscopy is used for the study of living cells, which are, in general, transparent to light. The principle on which the phase microscope is based is that the light passing through an object undergoes a retardation, or phase change, which normally is not detected. In this instrument, however, the phase difference is advanced or retarded one fourth of the wavelength (λ), and the small variations in phase produced by the various structures are thereby made visible. In phase microscopy the phase changes are translated into changes in light intensity.

Interference microscopy is based on similar principles.

The *Nomarski interference contrast* microscope, in particular, gives extraordinary images of living cells, with a relief effect.

Darkfield microscopy, also called ultramicroscopy, is based on the scattering of light and uses a darkfield condenser. With this microscope, as well as with phase contrast optics, objects smaller than the wavelength of light can be *detected,* although not *resolved.* Theoretically, a fiber of only 5.0 nm could be detected, provided it had enough contrast.

Polarization microscopy uses polarized light. A birefringent material (also called *anisotropic)* has two different indexes of refraction in two perpendicular directions.

Birefringence is related to retardation and to the thickness of the object: $B = N_e - N = \dfrac{\Gamma}{t}$.

In the polarizing microscope there is a polarizer and an analyzer (prisms or polaroid films) that are placed perpendicular to the specimen. The object is rotated 360 degrees, and if it is birefringent, it shows positions of maximum or minimum brightness. Most biological fibers show birefringence along the axis; a fiber of nucleic acid has negative birefringence. Crystalline, or intrinsic birefringence, is independent of the refractive index of the medium, whereas form birefringence changes when the refractive index of the medium varies. In general, both types are present. *Dichroism* is a special type of birefringence in which there is a change in absorption of polarized light with a change in the orientation of the object.

Electron microscopy is the best method for studying biological ultrastructure. Electrons are emitted and accelerated in a vacuum tube, and the electron beam is deflected by electromagnetic coils acting as condenser, objective, and projector lenses. The image formed on the fluorescent screen depends on the dispersion of electrons (electron scattering) by the atomic nuclei present in the object. This dispersion depends on the thickness of the object, the molecular packing, and, in particular, on the atomic number. Most atoms in biological objects do not scatter the electrons; heavy atoms are used as "electron stains" to increase the contrast. The resolving power depends on the wavelength ($\lambda = 0.005$ nm) and numerical aperture, as in the light microscope. The resolution reached is 0.3 to 0.5 nm, and the final magnification can be 10^6 times, or more.

The preparative techniques are of fundamental importance in observation of biological material. Macromolecules such as DNA and RNA can be studied by the monolayer technique. Thick specimens can be studied via the technique of *thin sectioning,* in which the material is embedded in plastic and is cut with glass or diamond knives. The structure of membranes may be observed by freeze-fracturing and freeze-etching. The contrast can be increased by shadow casting, negative staining, or electron staining. Osmium tetroxide is used both as a fixative and an electron stain. Uranyl acetate and lead hydroxide are widely used for staining. Electron opaque substances, such as colloids and certain enzymes that give an opaque reaction, are used as tracers in electron microscopy for the study of several biological processes (e.g., pinocytosis). Thick sections and even living cells may be observed with *high-voltage electron microscopy.*

Scanning electron microscopy gives a surface view of structures by forming an image with the secondary electrons reflected by those structures. Special techniques and microscopes are used for this purpose.

X-ray diffraction is a technique used in molecular biology, especially for the study of nucleic acid and protein structure. The method is based on the fact that a lattice of molecular dimensions (i.e., a crystal) produces a diffraction of x-rays. The most sophisticated techniques permit the determination of the complete three-dimensional structure of proteins such as myoglobin, hemoglobin, and others.

REFERENCES

1. Zernike, F. (1955) *Science, 121*:345.
2. Barer, R. (1965) *Physical Techniques in Biological Research.* Vol. 3, p. 30. (Oster, G., and Pollister, A. W., eds.) Academic Press, Inc., New York.
3. Engström, A., and Finean, J. B. (1958) *Biological Ultrastructure.* Academic Press, Inc., New York.
4. Hale, A. J. (1958) *The Interference Microscope in Biological Research.* The Williams & Wilkins Co., Baltimore.
5. Mellors, R. C., ed. (1959) *Analytical Cytology.* 2nd Ed. McGraw-Hill Book Co., New York.
6. Nomarski,' G. (1955) *J. Phys. Radium (Paris), 16*:95.
7. Kleinschmidt, A. E., Lang, D., Yacherts, D., and Zahn, R. K. (1962) *Biophys. Biochim. Acta, 61*:857.
8. Steere, R. L. (1957) *J. Biophys. Biochem. Cytol., 3*:45.
9. Leduc, E. H., and Bernhard, W. J. (1967) *J. Ultrastruct. Res., 19*:196.
10. Brenner, S., and Horne, R. W. (1959) *Biochim. Biophys. Acta, 34*:103.
11. Revel, J. P., and Karnovsky, M. J. (1967) *J. Cell Biol., 33*:C7.
12. Simionescu, N., Simionescu, M., and Palade, G. E. (1972) *J. Cell Biol., 53*:365.
13. Graham, R. C., and Karnovsky, M. J. (1965) *J. Histochem. Cytochem., 14*:291.
14. Feder, N. J. (1971) *J. Cell Biol., 51*:339.
15. Cosslett, V. E. (1969) *J. Rev. Biophys., 2*:95.
16. Favard, P., Ovtracht, L., and Carasso, N. (1971) *J. Microsc. (Paris), 12*:301.
17. Buckley, I. K., and Porter, K. R. (1975) *J. Microsc. (Oxf.), 104*:107.
18. Carr, K. E. (1971) *Int. Rev. Cytol., 30*:183.
19. Catley, C. W. (1972) *The Scanning Electron Microscope.* Cambridge University Press, New York.
20. Crewe, A. V. (1971) High resolution scanning microscopy of biological specimens. *Philos. Trans. R. Soc. London* [Biol. Sci.], *261*:61.

ADDITIONAL READING

Cosslett, V. E. (1978) Radiation damage in the high resolution electron microscopy of biological materials: a review. *J. Microsc. (Oxf.), 113*:113.

Branton, D., and Kirchanski, S. (1977) Interpreting the results of freeze-etching. *J. Microsc. (Paris), 111*:117.

Buckley, I. K., and Porter, K. R. (1975) Electron microscopy of critical point dried whole cultured cells. *J. Microsc. (Oxf.), 104*:107.

Du Pouy, G. (1973) Three-megavolt electron microscopy. *Endeavour* (London), *32*:66.

Frey-Wyssling, A. (1974) Ultrastructure research in biology before the introduction of the electron microscope. *J. Microsc. (Oxf.), 100*:21.

Engström, A., and Finean, J. B. (1958) *Biological Ultrastructure.* Academic Press, Inc., New York.

Heywood, V. H. (1971) *Scanning Electron Microscopy: Systematic and Evolutionary Applications.* Systematics Assoc., Special Vol., No. 4. Academic Press, Inc., New York.

Hollenberg, M. J., and Erickson, A. M. (1973) The scanning electron microscope: potential usefulness to biologists. *J. Histochem. Cytochem., 21*:107.

Horne, R. W. (1978) Special specimen preparation methods for image processing in transmission electron microscopy: a review. *J. Microsc. (Oxf.), 113*:241.

Kellenperg, E. (1978) Future trend: high resolution electron microscopy. *Trends Biochem. Sci.,* 3:N.135.

Kendrew, J. C. (1963) Myoglobin and the structure of proteins. *Science, 139*:1259.

Muhlethaler, K. (1971) Studies on freeze-etching of cell membranes. *Int. Rev. Cytol., 31*:1.

O'Keefe, M. A., Buseck, P. R. and Jijma, S. (1978) Computed crystal structure images for high resolution electron microscopy. *Nature, 274*:322.

Oster, G. (1956) X-ray diffraction and scattering. In: *Physical Techniques in Biological Research.* Vol. 2, p. 441. (Oster, G., and Pollister, A. W., eds.) Academic Press, Inc., New York.

Padawer, J. (1968) The Nomarski interference-contrast microscope. *J. Roy. Micros. Soc., 88*:305.

Park, R. B. (1972) Freeze-etching, a classical view. *Ann. N. Y. Acad. Sci., 195*:262.

Sjöstrand, F. S. (1967) *Electron Microscopy of Cells and Tissues.* Academic Press, Inc., New York.

Sleytr, U. B., and Robards, A. W. (1977) Freeze-fracturing: a review of methods and results. *J. Microsc. (Paris), 111*:77.

Wischnitzer, S. (1970) *Introduction to Electron Microscopy,* 2nd. Ed. Pergamon Press, Inc., New York.

CYTOLOGIC AND CYTOCHEMICAL METHODS

4–1 Cell Culture and Microsurgery 45
4–2 Fixation 46
    Osmium Tetroxide and Electron Microscopy
    Freeze-Drying and Freeze-Substitution
    Microtomes and Embedding
4–3 Chemical Basis of Staining 49
    Metachromasia — A Change in Original Dye Color
    Summary: Observation of Living and Fixed Cells
4–4 Cytochemistry 51
    Cell Fractionation — Separation of Subcellular Fractions
    Differential or Gradient Centrifugation — Separation of Cell Particles and Macromolecules
    Micro- and Ultramicromethods — Detection of Minute Quantities

4–5 Cytochemical and Histochemical Staining Methods 55
    Schiff's Reagent — Detection of Aldehydes
    Lipids — Detection by Lipid-Soluble Stains
    Enzymes — Detection by Incubation with Substrates
4–6 Cytochemical Methods Using Physical Techniques 59
    Cytophotometric Methods
    Fluorescence Microscopy — Autofluorescence and Fluorochrome Dyes
    Immunocytochemistry — Detection of Antigens with Labeled Antibodies
    Radioautography — Interaction of Radioisotopes with Photographic Emulsions
    Summary: Cytochemistry

In cell biology many different types of specimens and techniques are used in the analysis of cytologic and chemical organization. In general, one or a few types of cells are best suited to each particular problem. Sometimes an entire branch of cytology has developed from the choice of a special material or the development of a certain technique.

As mentioned in the previous chapter, the two main analytic procedures are: (1) direct observation of living cells, and (2) observation of killed cells that have been subjected to procedures that preserve morphology and composition, i.e., *fixation*.

4–1 CELL CULTURE AND MICROSURGERY

The culture of animal and plant cells outside the organism permits the observation of living cells under favorable conditions. Furthermore, a cell culture represents a much simpler experimental system than a whole animal and provides a system that can be studied under carefully controlled conditions.

Since 1912, when Carrel first succeeded in growing tissue explants for many cell generations, considerable progress has been made in the techniques of cell culture. At the present time these techniques represent some of the most powerful methods for the study of fundamental problems in cell biology. In early days of tissue culturing the technique consisted of explanting small portions of different tissues (preferably embryonic) in a medium consisting of blood serum and embryo extract, plus saline solution. This system was extremely complex from the chemical viewpoint, and it was only after 1955 that the first chemically defined culture media became available.[1] At present, the nutritional requirements of eukaryotic cells are well known, and most cells can be grown, with the addition of a small percentage of serum, in synthetic media.

Three main types of cultures can be distinguished: *primary, secondary,* and those using *established cell lines. Primary cultures* are those obtained directly from animal tissue. The organ is aseptically removed, cut into small fragments, and treated with trypsin. This proteolytic enzyme has the property of dissociating the cell aggregates into a suspension of single cells, without affecting viability. Thereafter, the cells are plated in sterile Petri dishes and grown in the appropriate culture medium. This culture can be trypsinized and replated in a fresh medium, resulting in a *secondary culture.*

The other major type of culture utilizes *established cell lines,* which have been adapted to prolonged growth in vitro. Among the best known cell lines are the HeLa cells, obtained from a human carcinoma, the L and 3T3 cells from mouse embryo, the BHK cells from baby hamster kidney, and the CHO cells from Chinese hamster ovary.

Normal mammalian cells do not survive indefinitely in culture and, after a variable time in vitro, they fail to divide and eventually die.[2] Occasionally, some cells will survive and will grow permanently in culture. These established cell lines differ from normal cells in many respects: they grow more tightly packed; they have lower serum requirements; and they are usually *aneuploids;* i.e., their chromosome number varies from one cell to another. Despite these abnormalities, established cell lines are very useful as model systems for the study of cancer (see Tooze, 1973).

One of the major advances in cell culture was the obtaining of a *clone,* i.e., a population of cells derived from a single parent cell. After several days of incubation of cells plated at high dilutions, rounded colonies grow and adhere to the Petri dish. All the cells of this clone are derived from a single cell. If this colony is carefully trypsinized and replated, large numbers of cells may be obtained.[3, 4]

Cell culture techniques have a wide application in cell biology (see Pollack, 1973). In Chapter 8–5 some of the problems involved in the control of cell growth in tissue culture will be considered, particularly the so-called *contact inhibition* of movement and cell division that occurs when normal cells come into contact. It will be shown that in many cases normal cells are able to "communicate" with one another via special cell contacts. Most permanent cell lines do not display contact inhibition, and when they are injected into an animal they develop a cancer-like growth pattern. In several of these cell lines the cells do not communicate with one another.

Microsurgery is another method that has contributed considerably to the knowledge of the living cell. Instruments such as micropipets, microneedles, microelectrodes, and microthermocouples are introduced into cells with the aid of a special apparatus that controls the movement of these instruments under the field of the microscope. Examples of microsurgical procedures are the dissection and extraction of parts of cells or tissues, the injection of substances, the measurement of electrical variables, and the grafting of parts from cell to cell. Laser beams are also used to damage special regions of the cell.

Figure 15–4 shows an example of the application of microsurgery to the study of electrical potentials at the plasma and nuclear membranes.

4–2 FIXATION

Fixation brings about the death of the cell in such a way that the structure of the living cell is preserved with the addition of a minimum number of artifacts. Some fixation methods, at the same time, are useful in maintaining the chemical composition of the cell as intact as possible.

The choice of a suitable fixative is dictated by the type of analysis desired. For example, for studying the nucleus and chromosomes, *acid fixatives* are frequently used. Acetone, formaldehyde, and glutaraldehyde, which produce minimal denaturation and preserve some enzyme systems, are used for the study of enzyme activity.[5, 6]

Some fixing agents produce cross linkages between protein molecules. For example, aldehydes react with the amino, carboxyl, and indole groups of a protein and then produce methylene bridges with other protein molecules. The two-step reaction is shown below. Glutaraldehyde has two aldehyde groups ($HOC-CH_2-CH_2-CH_2-COH$) that can react with amino groups in two adjacent protein monomers.

Chromium salts (e.g., potassium dichromate) produce oxidation and chromium linkages between proteins. They also bind the phospholipids. Mercuric chloride acts on sulfhydryl, carboxyl, and amino groups of proteins, producing mercury linkages between molecules.

When a piece of tissue is immersed in a fixing liquid, cellular death does not occur instantaneously, and "postmortem" alterations due to anoxia, changes in the concentration of hydrogen ions, and enzymatic action (autolysis) may occur. The fixative penetrates the tissue by diffusion in such a way that the external cells are fixed more rapidly and with fewer artifacts than the central cells. For this reason, every fixed tissue has a *gradient of fixation,* which depends upon the *penetrability* of the fixative and its progressive *dilution* with the liquid of the cells. The rate of fixative penetration also depends on the protein barrier of precipitate produced at the periphery of the tissue. For example, with osmium tetroxide the precipitate is very fine. For this reason only very thin pieces (0.5 to 1.0 mm thick) are fixed in osmic liquids.

Diffusion currents that displace the soluble components, such as glycogen, may be observed (Fig. 4–1). Fixatives may also extract soluble substances, such as electrolytes, soluble carbohydrates, and even some lipids.

$$-N-H \quad + \ HCHO \ \rightarrow \quad ----NH \cdot CH_2OH$$
$$\quad \ \ |$$
$$\quad \ \ H$$

amino group formaldehyde methylol

$$-NH \cdot CH_2OH + -N-H \ \rightarrow \ -NH-CH_2-HN + H_2O$$
$$\qquad\qquad\qquad\quad |$$
$$\qquad\qquad\qquad\quad H$$

methylol amino group methylene bridge

The preservation of a structure by fixation depends, to a great extent, on the degree of organization at the macromolecular level. In a well-organized structure, such as a chromosome, a mitochondrion, or a chloroplast, a great number of interacting forces hold the molecules together, and the action of the fixative is insufficient to break structural relationships. However, less organized regions of the cell, such as the cytoplasmic matrix, are more difficult to preserve, and the production of fixation artifacts is more likely to occur.

Osmium Tetroxide and Electron Microscopy

Osmium tetroxide (OsO_4) is one of the most frequently used fixatives for investigation of cell structure under the electron microscope. The reaction that this fixative has with lipids is probably due to double bonds that form unstable osmium esters, which decompose to deposit osmium oxides or hydroxides. The fixative causes proteins to gel initially, presenting a homogeneous structure under the electron microscope. This initial gelation may then be followed by further oxidation and solubilization of some products.

Osmium fixation has been improved by introducing buffer solutions at physiologic pH that maintain osmotic pressure, adding calcium ions, and maintaining a temperature of about 0° C.

 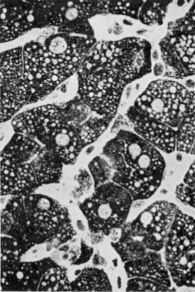

Figure 4–1 **Left,** liver cells of *Ambystoma* fixed in Zenker-formol. The diffusion current produced by the chemical fixative (from the lower to the upper part of the figure) displaces the glycogen of the cell. **Right,** liver cells of *Ambystoma* fixed by freezing-drying. The glycogen appears to be distributed homogeneously in the cytoplasm. Spheroid nuclei and lipid droplets are distinguishable. Stain: Best's carmine. (Courtesy of I. Gersh.)

Freeze-drying and Freeze-substitution

This method consists of rapid freezing of the tissues, followed by dehydration in a vacuum at a low temperature. The initial freezing is generally accomplished by plunging small pieces of tissue in a bath cooled with liquid nitrogen to a temperature of -160 to $-190°$ C. Fixation in liquid helium near absolute $0°$ (Kelvin) has also been used. The tissues are dried in a vacuum at $-30°$ to $-40°$ C. Under these conditions the ice in the tissues is changed directly into a gas, and dehydration is achieved.

The advantages of this method are obvious. The tissue does not shrink; fixation is homogeneous throughout; soluble substances are not extracted; the chemical composition is maintained practically without change; and the structure, in general, is preserved with very few modifications (Fig. 4–1). In addition, fixation takes place so rapidly that cell function can be arrested at critical moments, such as when kidney cells are excreting colored material.

The freeze-drying technique should be considered as intermediary between the examination of fresh and fixed tissues, since many of the cellular components are preserved in the same soluble form as in the living state. Since some cells can resist rapid freezing, this procedure is commonly used to keep them alive (e.g., frozen spermatozoa).

In the freeze-substitution method, the tissue is rapidly frozen and then kept frozen at a low temperature (-20 to $-60°$ C.) in a reagent that dissolves the ice crystals (e.g., ethanol, methanol, or acetone). The advantages of this method are somewhat similar to those just mentioned for fixation by freeze-drying.

Microtomes and Embedding

Tissues should be conveniently sectioned before they are observed under the microscope. For this purpose, *freezing microtomes,* cooled with liquid carbon dioxide, are frequently used.

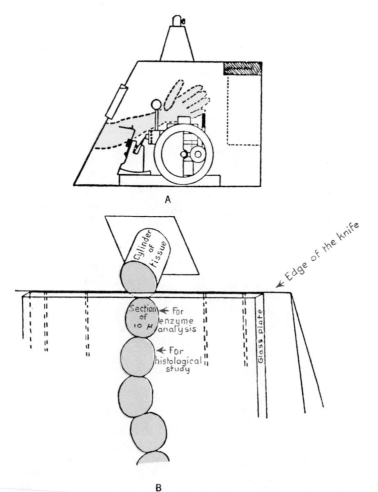

Figure 4–2 A, microtome for frozen sectioning. The apparatus is kept in a refrigerated container that keeps the tissue and the sections frozen. **B,** scheme of sectioning with the freezing microtome. The cylinder of tissue is sectioned and the sections are collected so that each alternate piece is used for enzymatic analysis and the others for histologic control. (Modified from Linderstrom-Lang.)

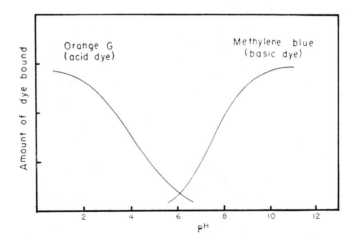

Figure 4–3 Curves indicating the amount of stain fixed by a protein at different pH values. Both the acid and the basic staining show a minimum fixation at the isoelectric point of the protein. (From Singer, M., and Morrison, P. R., *J. Biol. Chem.*, 175:1, 133, 1948.)

Instruments consisting of a microtome enclosed in a chamber at low temperature — the so-called *cryostat* — can make sections of fixed or fresh tissue for cytochemical purposes (Fig. 4–2).

For the most frequently used sectioning techniques the tissue is *embedded* in a material that imparts the proper consistency to the section. For sections to be observed under the light microscope, *paraffin* or *celloidin* is generally used. The fixed tissue is dehydrated and then penetrated by the embedding material. This requires a proper intermediary solvent (e.g., xylene or toluene for paraffin; ethanol-ether for celloidin). (For a discussion of embedding methods required for electron microscopy see Chapter 3–2.)

4–3 CHEMICAL BASIS OF STAINING

Most cytologic stains are solutions of organic aromatic dyes. Since the time of the pioneer work of Ehrlich, two types of dyes have been recognized: basic and acid. In a basic dye the *chromophoric group*, which imparts the color, is basic (cationic). For example, methylene blue is a chlorhydrate of tetramethylthionine, in which the basic part carries the blue color. Eosin is generally used as potassium eosinate, in which the base is colorless. Sometimes the two components of the salt are chromophoric, e.g., eosinate of methylene blue. The most frequently used chromophores for acid dyes contain nitro ($-NO_2$) and quinoid ($O=\langle\quad\rangle=O$) groups. Basic chromophores contain azo ($-N\equiv N-$) and indamin ($-N\equiv$) groups. For example, picric acid has three nitro groups (chromo-

phores) and one OH group, also called *auxochrome*, by which the dye combines with the tissue:

$$\begin{array}{c} OH \\ NO_2 \underset{NO_2}{\overset{}{\bigcirc}} NO_2 \end{array}$$

The properties that enable proteins, certain polysaccharides, and nucleic acids to ionize either as bases or acids should be noted (Chapter 5). Acid ionization may be produced by carboxyl ($-COOH$), hydroxyl ($-OH$), sulfuric ($-HSO_4$), or phosphoric ($-H_2PO_4$) groups. Basic ionization results from amino ($-NH_2$) and other basic groups in the protein. At pH values above the isoelectric point, acid groups become ionized; below the isoelectric point, basic groups dissociate (see Chapter 5). Because of this property, at a pH above the isoelectric point, proteins will react with basic dyes (e.g., methylene blue, crystal violet, or basic fuchsin) and below it, with acid dyes (e.g., orange G, eosin, or aniline blue). The intensity of staining with basic or acid dyes depends on the degree of acidity or alkalinity of the medium (Fig. 4–3). By measuring the amount of dye bound as a function of the pH of the medium, curves can be obtained that are typical of various proteins, nucleic acids, and mucopolysaccharides.

The net charge of nucleic acids is determined primarily by the dissociation of the phosphoric acid groups, and the isoelectric point is very low (pH 2 or less). For this reason, staining with basic dyes (e.g., toluidine blue or azure B) at low pH values is selective for nucleic acids. Toluidine blue is frequently used to stain ribonucleic

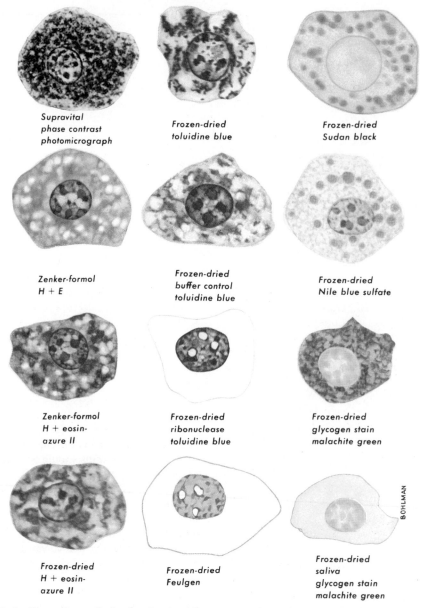

Supravital phase contrast photomicrograph

Frozen-dried toluidine blue

Frozen-dried Sudan black

Zenker-formol H + E

Frozen-dried buffer control toluidine blue

Frozen-dried Nile blue sulfate

Zenker-formol H + eosin- azure II

Frozen-dried ribonuclease toluidine blue

Frozen-dried glycogen stain malachite green

Frozen-dried H + eosin- azure II

Frozen-dried Feulgen

Frozen-dried saliva glycogen stain malachite green

Figure 4–4 Mouse liver cells fixed and stained by a variety of cytochemical procedures to show distribution of deoxyribonucleic acid, ribonucleic acid, glycogen, and lipid droplets. Since these tests were all done on material fixed by freezing and drying, some of the sections are compared with similarly stained sections fixed by Zenker-formol. For further orientation, the fixed cell stained by hematoxylin and eosin and the unfixed cell photographed by phase contrast are also shown. ×1500. (Courtesy of I. Gersh, *in* Bloom, W., and Fawcett, D. W., *Textbook of Histology,* 10th ed., Philadelphia, W. B. Saunders Co., 1975.)

acid, and its specificity can be demonstrated by previous hydrolysis with ribonuclease (Fig. 4–4).

Some histochemical methods based on these staining properties of proteins are now in use. One of the best known is the fast green method for histones.

Metachromasia—A Change in Original Dye Color

Some basic dyes of the thiazine group, particularly thionine, azure A, and toluidine blue, stain certain cell components a different color than the original color of the dye. This property, called *metachromasia,* has interesting histochemical and physicochemical implications. The reaction occurs in mucopolysaccharides and, to a lesser extent, in nucleic acids and some acid lipids. This reaction is strong in cells that contain sulfate groups (such as chondroitin sulfate), e.g., in the cells of cartilage and connective tissue.

Some investigators believe that metachromasia depends on the formation of dimeric and polymeric molecular aggregates of dye on these high molecular weight compounds.

SUMMARY

Observation of Living and Fixed Cells

Observation of cells can be made directly on living specimens or after cells have been fixed. *Tissue culture* consists of explanting a piece of tissue in a suitable medium (e.g., plasma, embryonic extract, or synthetic). The cells spread and divide, forming a zone of growth. Pure strains (i.e., clones) of cells may be obtained after isolation of cells by trypsin. Organs, as well as tissues, can also be cultured. *Microsurgery* is a technique that may give information about the physicochemical properties of living cells. Through this method, transplants of subcellular parts—including the nucleus—can be made.

Fixation is brought about by chemicals which preserve cell structure. For nuclei and chromosomes, acid fixatives are preferred. To preserve the activity of certain enzymes, acetone, formaldehyde, and glutaraldehyde are used. (Aldehydes react with amino groups of proteins, forming methylene bridges.) Glutaraldehyde has two aldehyde groups that can react with adjacent protein monomers. Fixatives penetrate by diffusion and produce a gradient of fixation. The preservation of cell structure depends on the tightness of macromolecular organization. Osmium tetroxide and glutaraldehyde are widely used in fixation for electron microscopy.

Freeze-dying consists of the rapid freezing of tissue using liquid nitrogen (-160 to $-190°$ C), followed by dehydration at -30 to $-40°$ C. The water is sublimed in a vacuum. In *freeze-substitution* the water of the frozen tissue is dissolved with alcohols or acetone. *Embedding* in paraffin or celloidin is used for light microscopy; special plastics are used for electron microscopy. Microtomes and ultramicrotomes are then used for *sectioning* the embedded tissues.

Cytologic stains may be basic (e.g., methylene blue) or acid (e.g., eosin). Chromophores for acid dyes are nitro and quinoid groups; for basic dyes they are azo and indamin groups. The mechanism of staining is based on the ionization of acid groups (carboxyl, hydroxyl, sulfuric, phosphoric) or basic groups (amino) in proteins, polysaccharides, and nucleic acids. The staining of a protein is minimal at the isoelectric point; it stains with acid dyes below, and basic dyes above, the isoelectric point (Fig. 4–3). Nucleic acids are stained selectively at low pH because their isoelectric point is at pH 2.

Metachromasia, the property by which a basic dye has a different color in tissue than in solution, is found in mucopolysaccharides, particularly those containing sulfate groups. This particular characteristic of the dye is probably due to dimerization or polymerization.

4–4 CYTOCHEMISTRY

The immediate goal of *cytochemistry* is the identification and localization of the chemical components of the cell. As R. R. Bensley, one of the founders of modern cytology, once said: The aim of cytochemistry is the outlining "within the exiguous confines of the cell of that elusive and mysterious chemical pattern which is the basis of life." This aim is quantitative as well as qualitative; and once achieved, the next step is to study the dynamic changes in cytochemical organization taking place in different functional stages. In this way it is possible to discover the role of different cellular components in the metabolic processes of the cell.

Cytochemistry is included within the more general subject of *histochemistry*, which deals with the chemical characterization and localization of enzymes, substances, or groups of substances in the cells and intercellular materials of a tissue.[7, 8]

Modern cytochemistry has followed two main methodological approaches. Of these, only one

51

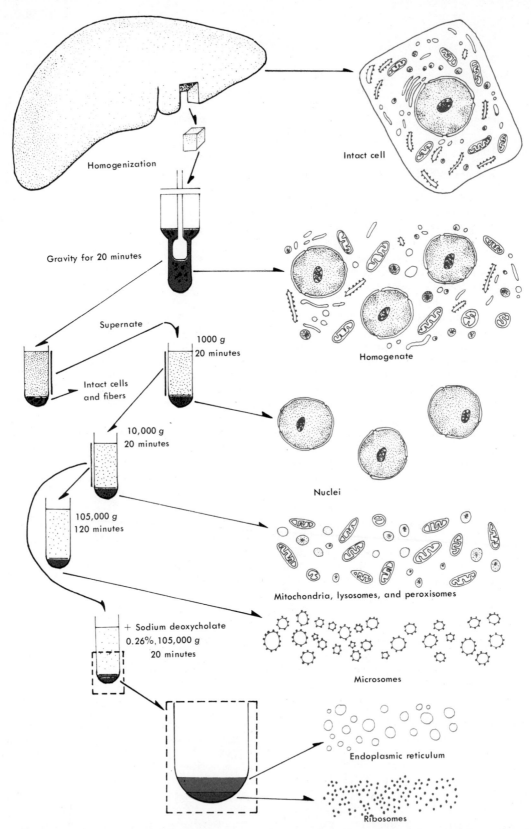

Figure 4–5 Diagram showing the various steps used in the technique of differential centrifugation. A piece of tissue from liver is homogenized and then subjected to a series of centrifugations of increasing centrifugal force, as indicated on the left side of the figure. On the right side are diagrams of the various subfractions as they appear under the electron microscope. (Modified from Bloom, W., and Fawcett, D. W., *A Textbook of Histology*, 10th ed., Philadelphia, W. B. Saunders Co., 1975.)

can be considered strictly *microscopic*, because it comprises a series of chemical and physical methods used to detect or measure different chemical components within the cell. The other method relies on *biochemical* techniques for the isolation and investigation of subcellular fractions, or it applies the techniques of *microchemistry* and *ultramicrochemistry* to the study of minute quantities of material.

Cell Fractionation – Separation of Subcellular Fractions

Cell fractionation methods involve, essentially, the homogenization or destruction of cell boundaries by different mechanical or chemical procedures, followed by the separation of the subcellular fractions according to mass, surface, and specific gravity.

Many different methods of cell fractionation are in use. Most of them are based on the homogenization of the cell in aqueous media —usually sucrose solutions in various concentrations.

A standard cell fractionation procedure is shown diagrammatically in Figure 4–5. The liver of an animal is first perfused with an ice-cold saline solution, followed by cold 0.25 M sucrose. The tissue is then forced through a perforated steel disk and homogenized in 0.25 M sucrose. This classic type of cell fractionation is directed toward the subdivision of the cell components into four morphologically distinct fractions (nuclear, mitochondrial, microsomal, and soluble). In some glandular tissue a fifth fraction containing secretory granules may be obtained.

Note carefully that it is necessary to differentiate between *cell fractions* and the parts of the cell (*organelles*) contained in the fraction. For example the mitochondrial fraction of the liver is composed principally of mitochondria, but the "mitochondrial fraction" of the brain is very heterogeneous and contains nerve endings and myelin in addition to free mitochondria.[9] "Microsomes" do not exist, as such, in the cell; the "microsomal" fraction is composed mainly of broken parts of the endoplasmic reticulum, including the ribosomes, the Golgi complex, and other membranes (see Chapter 10–4).

Differential or Gradient Centrifugation – Separation of Cell Particles and Macromolecules

In the example just described, the method used to separate the subcellular particles is called *differential centrifugation*. Depending on the strength of the centrifugal field needed, *standard centrifuges or preparative ultracentrifuges* are used. The effect of the centrifugal field on particles of different sizes is indicated in Figure 4–6.

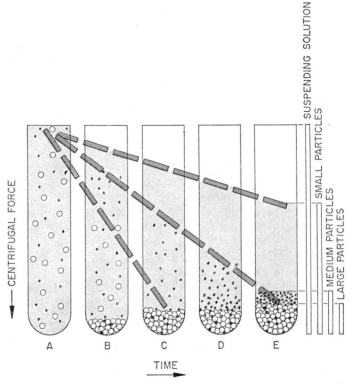

Figure 4–6 Diagram of the effect of a centrifugal field on particles of different sizes. (See the description in the text.) (From Anderson, N. G., *Natl. Cancer Inst. Monogr.* 21, June, 1966.)

Initially, all the particles are distributed homogeneously (A); as centrifugation proceeds, (B)→ (E), the particles settle according to their respective sedimentation ratios. Complete sedimentation of the larger particles is achieved in (C), while in (E), the medium-sized particles have settled.

A similar principle may be used for separation of much smaller particles, such as viruses or macromolecules (i.e., nucleic acids and proteins), with the *analytical ultracentrifuge*. This device has a transparent window in the tube, and with suitable optical and electronic techniques the moving particle boundaries can be visualized and the *sedimentation coefficient* can be determined. This coefficient, expressed in Svedberg (S) units, is related to the molecular weight of the particle (for example, transfer RNA, with 4S, has a MW of 25,000 daltons).

Improvement in the technique of differential centrifugation may be achieved by using a density gradient, which may be either *discontinuous* or *continuous*. If it is discontinuous, the centrifuge tube is loaded with layers of varying densities (for example, with sucrose varying in molarity from 1.6 to 0.5 from the bottom to the top of the tube). However, mixing of two concentrations of sucrose produces a continuous gradient. Once the gradient is formed, the material is layered on the top and centrifuged until the particles reach equilibrium with the gradient. For this reason, this type of separation is also called *isopyknic* (equal density) centrifugation.[10]

Improvements in this type of fractionation technique include the use of heavy water, cesium chloride, and media with different partition coefficients.[11] To avoid drastic changes in osmotic pressure, macromolecular media such as glycogen Ficoll and Percoll are used.

With the use of the so-called *zonal rotors* the density gradient is formed while the rotor is spinning; then the sample is layered and centrifuged until the isopyknic zonal layering of the particles is reached.[10]

Buoyant Density. Isopyknic centrifugation in preparative or zonal rotors permits the determination of the buoyant density of a macromolecule, i.e., the density at which it will reach an equilibrium with the suspending medium. This is important in studies of the molecular biology of nucleic acids (Fig. 4–7).

Micro- and Ultramicromethods – Detection of Minute Quantities

In recent years many ingenious methods have been devised for the quantitative analysis of extremely small quantities of substances.[12-14] For example, if freshly frozen tissue is sectioned in a *cryostat* (Fig. 4–2), some sections can be weighed on a balance made of a fine *quartz fiber*, and enzymatic determinations can be carried out using ultramicropipets and burets and microcolorimetric or microspectrophotometric methods. Of even higher sensitivity are *microfluorimetric* methods, which can be used to determine different enzymes and coenzymes.

Also of considerable interest are *micromanometric* methods. One of these employs the Cartesian diver microrespirometer, which is 1000 times more sensitive than the classic Warburg manometer. With this instrument the oxygen consumption of a single sea urchin egg can be measured during short intervals. Based on similar principles is the *Cartesian diver balance* of Zeuthen, by which a single ameba can be weighed with great accuracy.

By *microchromatographic* and *microelectro-*

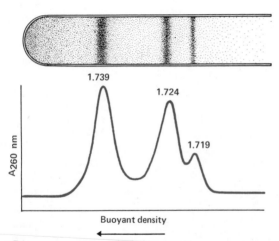

Figure 4–7 Diagram of a continuous gradient of cesium chloride (CsCl) showing the position of three bands with the corresponding buoyant density expressed in grams per cm^2. Since these bands correspond to nucleic acid molecules the concentration of the molecules is measured by absorbancy at 260 nm (A_{260nm}).

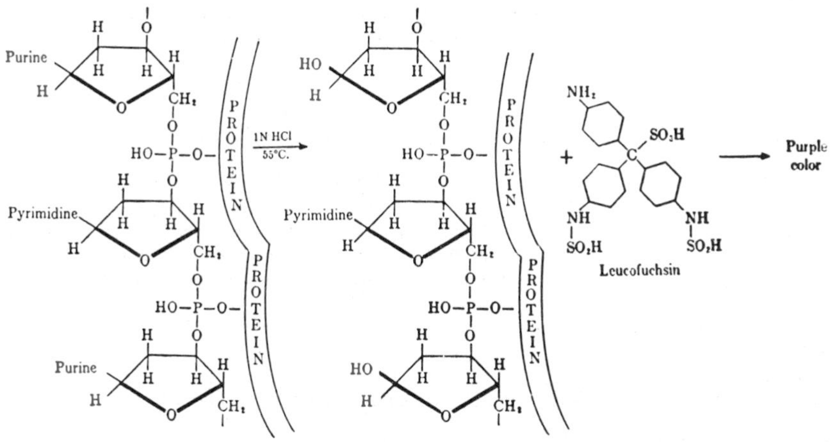

Figure 4–8 Chemistry of the Feulgen reaction. Acid hydrolysis removes the purines and liberates the aldehyde groups, which react with leukofuchsin (Schiff's reagent), resulting in a purple color. In the diagram the size of deoxypentose is greatly exaggerated in relation to the protein. (From Lessler, M. A., *Internat. Rev. Cytol.*, 2:231, 1953.)

phoretic methods the ribonucleic acid content of a single nerve cell has been determined.

4–5 CYTOCHEMICAL AND HISTOCHEMICAL STAINING METHODS

For the cytochemical and histochemical determination of a substance certain conditions must be fulfilled: (A) The substance must be immobilized at its original location. (B) The substance must be identified by a procedure that is specific for it, or for the chemical group to which it belongs. This identification can be made by: (1) chemical reactions similar to those used in analytical chemistry, but adapted to tissues, (2) reactions that are specific for certain groups of substances, and (3) physical methods.

To demonstrate proteins, nucleic acids, polysaccharides, and lipids, some chromogenic agents that bind selectively to some specific groups of these substances may be used. Only a few cytochemical stainings that are widely used are mentioned here:

Detection of Proteins. In the *Millon reaction* a nitrous-mercuric reagent applied to the tissue reacts with the tyrosine groups present in the side-chains of the protein, forming a red precipitate. In the *diazonium reaction* a diazonium hydroxide reacts with tyrosine, tryptophan, and histidine groups, forming a colored complex. *SH groups* in proteins may be detected by agents that make a mercaptide covalent linkage. A red sulfhydryl reagent, 1-(4-chloromercuri-phenyl-azo)-naphthol-2, was first used for this purpose. The —SH content of the cell can be measured quantitatively by photometric analysis of tissues

stained by this technique.[15] Other methods for detection of —SH groups are also widely used.[16, 17]

Schiff's Reagent – Detection of Aldehydes

Deoxyribonucleic acid, certain carbohydrates, and lipids can be demonstrated with a single reagent for aldehyde groups, the so-called *Schiff's reagent*. This reagent is made by treating basic fuchsin, which contains parafuchsin (triaminotriphenyl-methane chloride), with sulfurous acid. Parafuchsin is converted into the colorless compound bis-N aminosulfonic acid (Schiff's reagent), which is then "recolored" by the aldehyde groups present in the tissue (Fig. 4–8).

In the histochemical tests involving Schiff's reagent, three types of aldehydes may be involved: (1) *free aldehydes*, which are naturally present in the tissue, such as those giving the plasmal reaction; (2) *aldehydes produced by selective oxidation* (which give the PAS reaction); and (3) *aldehydes produced by selective hydrolysis* (which give the Feulgen reaction).

Cytochemical staining methods for nucleic acids depend on the properties of the three components of the nucleotide (phosphoric acid, carbohydrate, and purine and pyrimidine bases [Chap. 5]).

Both DNA and RNA absorb ultraviolet light at 260 nm, because of the presence of nitrogenous bases. The deoxyribose present in DNA is responsible for the Feulgen reaction, which is specific for this type of nucleic acid (Table 4–1).

The phosphoric acid residue is responsible for the basophilic properties of both DNA and

TABLE 4–1. SOME SPECIFIC REACTIONS USED IN CYTOPHOTOMETRIC ANALYSIS*

Substance Tested For	Reaction or Test	Maximum Absorption (wavelength in nm)
Total nucleotides	Natural absorption of purines and pyrimidines	260
Soluble nucleotides	Natural absorption of purines and pyrimidines	260
Ribonucleic acid (RNA)	Natural absorption of purines and pyrimidines	260
Deoxyribonucleic acid (DNA)	Natural absorption of purines and pyrimidines	260
Deoxyribonucleic acid (DNA)	Feulgen nucleal reaction for deoxyribose	550 to 575
Deoxyribonucleic acid (DNA)	Methyl green	645
Nucleic acids (phosphoric acid groups)	Azure A	590 to 625
Protein (free basic groups)	Fast green	630
Protein (tyrosine)	Millon reaction	355
Polysaccharides with 1,2-glycol groupings	Periodic acid–Schiff reaction (PAS)	~550

*From Moses, M. J. (1952) *Exp. Cell Res.*, suppl. 2:76.

RNA. Among the basic stains, azure B gives a specific reaction with DNA and RNA. Another stain for DNA, based on the use of methyl green, also depends on the phosphoric acid residues.

Feulgen Reaction. DNA can be studied by means of the *nucleal reaction*, a technique developed in 1924 by Feulgen and Rossenbeck. Sections of fixed tissue are first submitted to a mild acid hydrolysis and then treated with Schiff's aldehyde reagent. This hydrolysis is sufficient to remove RNA, but not DNA. The reaction takes place in the following stages: (1) The acid hydrolysis removes the purines at the level of the purine-deoxyribose glucosidic bond of DNA, thus unmasking the aldehyde groups of deoxyribose. (2) The free aldehyde groups react with Schiff's reagent (Fig. 4–8).[18] The specificity of the reaction can be confirmed by treating the sections with deoxyribonuclease, which removes DNA.[19]

The Feulgen reaction is positive in the nucleus and negative in the cytoplasm (Fig. 4–4). In the nucleus, the masses of condensed chromatin (i.e., heterochromatin) are intensely positive; the nucleolus is Feulgen-negative (Fig. 4–9).

Periodic Acid–Schiff (PAS) Reaction. McManus[20] devised a reaction based on the oxidation with periodic acid of the "1,2-glycol" group of polysaccharides with liberation of aldehyde groups, which give a positive Schiff reaction (Fig. 4–10). This test is done on plant cells for starch, cellulose, hemicellulose, and pectins; and on animal cells for mucin, mucoproteins, hyaluronic acid, and chitin.

Since the PAS reaction is given by a number of substances, different tests can be applied to improve its specificity. For example, an enzyme, such as amylase, can be used to remove glycogen (Fig. 4–4), or hyaluronidase can be used to remove hyaluronic acid.

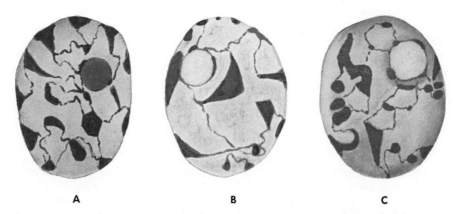

A   **B**   **C**

Figure 4–9 Interphase nuclei of pancreatic cells fixed by freeze-drying. **A,** Azan staining: the nucleolus (in red), the chromonemic filaments with their enlarged portions (chromocenters), and the nuclear sap are visible. **B,** Feulgen reaction: the nucleolus gives a negative reaction; in the nuclear sap the reaction is slightly positive. **C,** action of ribonuclease and staining with Azan. The nucleolus does not stain, owing to the digestion of the ribonucleic acid. (From De Robertis, Montes de Oca, and Raffaele, *Rev. Soc. Arg. Anat. Normal Patol.,* 1945.)

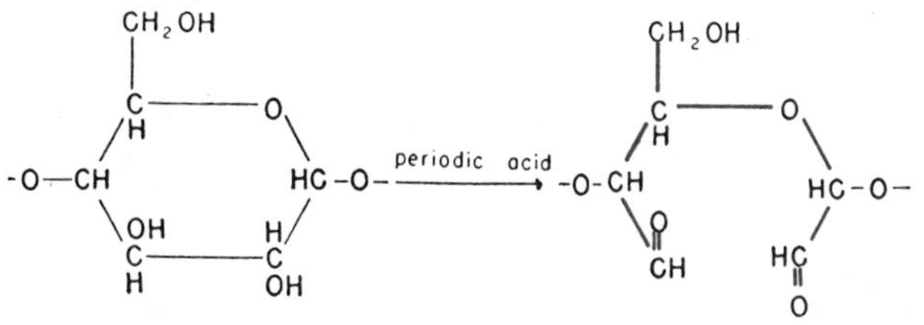

Figure 4–10 Chemical diagram of a polysaccharide, showing the action site of periodic acid in the PAS reaction of McManus. The resulting aldehydes react with Schiff's reagent.

Lipids — Detection by Lipid-soluble Stains

Fat droplets can be demonstrated with osmium tetroxide, which stains them black by reacting with unsaturated fatty acids. Staining with Sudan III or Sudan IV (scarlet red) has a greater histochemical value. These stains act by a simple process of diffusion and solubility and accumulate in the interior of the lipid droplets. Sudan black B has the advantage of being dissolved also in phospholipids and cholesterol, and of producing greater contrast (Fig. 4–4).

Plasmal Reaction. Long-chain aliphatic aldehydes occurring in plasmalogens give the so-called *plasmal reaction* upon direct treatment of the tissue with Schiff's reagent. Since the substances giving the plasmal reaction are soluble in organic solvents, the tissue is not embedded in the usual way but is studied in frozen sections. The compounds are free aldehydes such as *palmitaldehyde*, $CH_3 (CH_2)_{14} CHO$, and *stearaldehyde*, $CH_3(CH_2)_{16} CHO$, corresponding to palmitic and stearic acids, respectively, which together constitute the so-called *plasmal*.

Enzymes — Detection by Incubation with Substrates

Because of the inactivating action of most fixatives, special preparation of tissues for enzyme chemistry is necessary. To detect some enzymes, unfixed frozen sections are made in a cryostat; in other cases, the enzyme can withstand a brief fixation in cold acetone, formaldehyde, glutaraldehyde, or another dialdehyde.[5]

Techniques for identifying and localizing enzymes are based on the incubation of the tissue sections with an appropriate substrate. For example, in the Gomori method for detecting alkaline phosphatase, phosphoric esters of glycerol are used as the substrate.[21] The phosphate ion liberated by hydrolysis is converted into an insoluble metal salt (generally in the presence of Ca^{2+}), and the metal, in turn, is visualized by conversion into metallic silver, lead sulfide, cobalt sulfide, or other colored compounds. In an-

other method a phosphoric ester of β-naphthol is used as the substrate. The hydrolysis liberates β-naphthol, which, in the presence of a diazonium salt, couples immediately, giving a colored azo component at the site of enzymatic activity (Fig. 4–11). Other hydrolytic enzymes, such as esterase, lipase, acid phosphatase, sulfatase, and β-glucuronidase, can be detected with this method by changing the conditions and substrate.

Phosphatases. Phosphatases are enzymes that liberate phosphoric acid from many different substrates. A number of phosphatases are known, and they differ with respect to substrate specificity, optimum pH, and the action of inactivators and inhibitors. The best known are the *phosphomonoesterases*, which hydrolyze simple esters held by P—O bonds, and the *phosphamidases* which hydrolyze P—N bonds. Table 4–2 indicates some of the most common enzymes studied cytochemically and some of the substrates used. An example of the alkaline phosphatase reaction is shown in Figure 4–12.

TABLE 4–2. SOME PHOSPHATASES STUDIED CYTOCHEMICALLY

Type	Substrate
Phosphomonoesterases	
Alkaline phosphatase	α- or β-glycerophosphate
Acid phosphatase	Naphthylphosphate
Adenosine triphosphatase (ATPase)	Adenosine triphosphate
5-Nucleotidase	5-Adenylic acid
Phosphamidase	Phosphocreatine
	Naphthyl phosphoric acid diamines
Glucose-6-phosphatase	Glucose-6-phosphate
Thiamine pyrophosphatase	Thiamine pyrophosphate
Pyrophosphatase	Sodium pyrophosphate
	Dinaphthyl pyrophosphate
Phosphodiesterases	
Ribonuclease	Ribonucleic acid (RNA)
Deoxyribonuclease	Deoxyribonucleic acid (DNA)

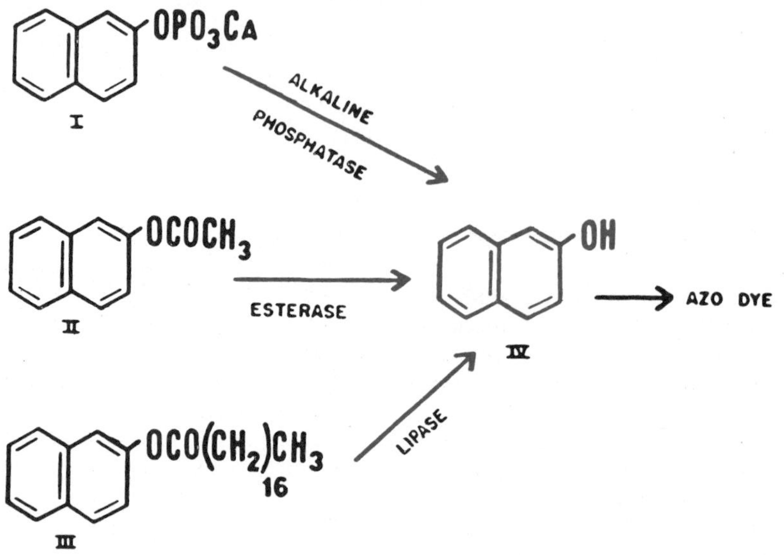

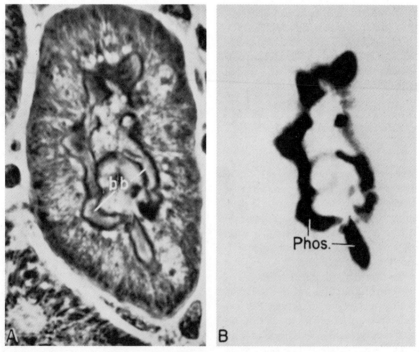

Figure 4–11 Diagram showing the cytochemical steps in the methods used to demonstrate hydrolytic enzymes. (See the description in the text.) (From Nachlas, M. M., and Seligman, A. M., *J. Natl. Cancer Inst.,* 9:415, 1949.)

Figure 4–12 Proximal convoluted tubule from a mouse kidney after freezing-substitution. **A,** observation in phase contrast with a medium of refractive index n = 1460; *bb,* brush border. **B,** observation with transmitted light; *Phos.,* alkaline phosphatase reaction. (Courtesy of B. J. Davies, and L. Ornstein.)

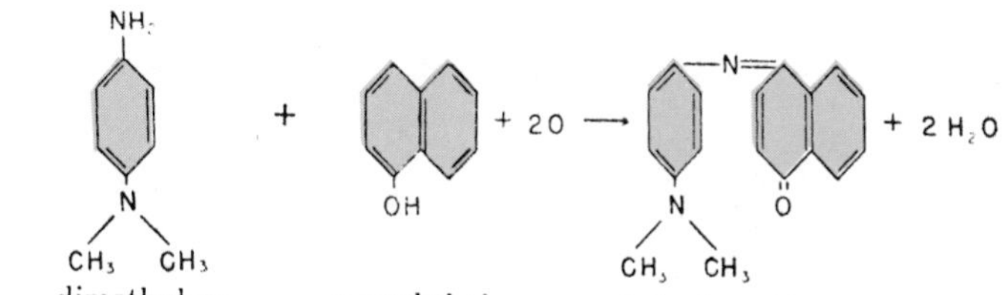

Figure 4–13 Nadi reaction for cytochrome oxidase.

dimethyl-*p*-phenylene diamine   α-naphthol   indophenol blue

Esterases. Esterases are enzymes that catalyze the following reversible reaction:

$$— COOR + HOH \leftrightarrows COOH + R'OH$$

Esterases may be divided into *simple esterases (aliesterases)*, which hydrolyze short chain aliphatic esters; *lipases*, which attack esters with long carbon chains; and *cholinesterases*, which act on esters of choline.

Other *hydrolytic enzymes* studied cytochemically are β-D-glucuronidase, β-D-galactosidase, aryl sulfatase, and aminopeptidase.

Oxidases. Oxidases are enzymes that catalyze the transfer of electrons from a donor substrate to oxygen (Chap. 6). They usually contain iron, e.g., peroxidase and catalase, or copper, e.g., tyrosinase and polyphenol oxidase. Another enzyme in this series is monoamine oxidase, which is involved in the metabolism of indole and catecholamines.

Colorless substrates, such as benzidine, are used to detect peroxidases. These substrates are transformed into stained dyes by H_2O_2 in the presence of the enzyme. *Cytochrome oxidase* gives the so-called *Nadi* reaction. It oxidizes the Nadi reagent, a mixture of α-naphthol and dimethyl paraphenylene diamine (Fig. 4–13). The reagent 3',3'-diaminobenzidine (DAB) has allowed the study of peroxidase[22] and cytochrome oxidase at the electron microscopic level.

Dehydrogenases. The pyridine nucleotide-linked dehydrogenases require the coenzymes NAD^+ or $NADP^+$. Among the best known NAD^+ enzymes are lactic acid dehydrogenase, which converts lactic acid into pyruvic acid, and malic acid dehydrogenase, which converts malic acid

into oxaloacetic acid. Among the $NADP^+$ enzymes are isocitric acid dehydrogenase and the malic enzyme (malate $\rightarrow$ pyruvate $+ CO_2$).

Figure 4–14 represents the mechanism of these histochemical reactions.

4–6 CYTOCHEMICAL METHODS USING PHYSICAL TECHNIQUES

Cytophotometric Methods

Several cell components display a specificity in the way in which they absorb ultraviolet light. For example, the absorption range of nucleic acids is about 260 nm, whereas that of proteins is 280 nm. Some histochemical staining reactions give specific absorption in the visible spectrum and can be analyzed quantitatively with instruments called *cytophotometers*.

A typical apparatus for absorption cytophotometry is represented in Figure 4–15. By changing the light source and the optical system, this instrument can be used for either the ultraviolet or the visible spectrum. The absorption is measured directly by means of a photomultiplier or by densitometry on calibrated photographic plates. Table 4–1 indicates some of the histochemical reactions that can be analyzed by cytophotometric methods.

The specific ultraviolet absorption of nucleic acids is due to the presence of purine and pyrimidine bases, and, for this reason, is the same in DNA, RNA, and nucleotides (Table 4–1). By ultraviolet cytophotometry the two types of nucleic acids can be localized, but not distinguished. On the other hand, the Feulgen

Figure 4–14 Schematic representation of the transfer of electrons to tetrazolium salt. *FAD,* flavin adenine dinucleotide; *FMN,* flavin mononucleotide; NAD^+, nicotinamide adenine dinucleotide; $NADP^+$, nicotinamide adenine dinucleotide phosphate.

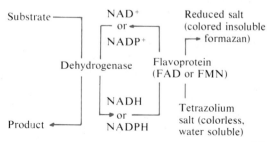

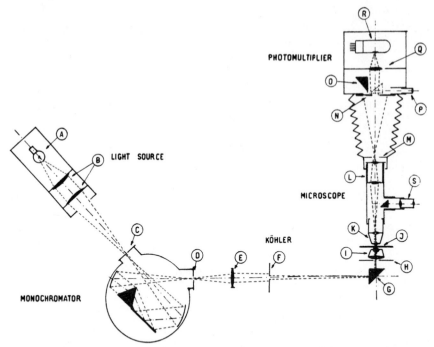

Figure 4–15 Cytophotometer used at the Institute of Cell Biology. *A*, light source of a tungsten light; *B*, condenser lens; *C* and *D*, entrance and exit slits of the monochromator; *E*, lens, *F*, diaphragm; *G*, prism; *H*, diaphragm; *I*, condenser; *J*, slide; *K*, objective; *L*, ocular; *M* and *N*, diaphragms; *O*, prism for observation of the final image; *P*, to displace prism; *Q*, lens; *R*, photomultiplier; *S*, lateral view. (Courtesy of A. O. Pogo and J. Cordero Funes.)

reaction shows the presence of DNA (Fig. 4–4). This reaction can be adapted to quantitative determinations of DNA in tissue sections. The Millon reaction can be used to determine the protein content (Table 4–1).

Fluorescence Microscopy — Autofluorescence and Fluorochrome Dyes

In this method tissue sections are examined under ultraviolet light, near the visible spectrum, and the components are recognized by the fluorescence they emit in the visible spectrum. Two types of fluorescence may be studied: natural fluorescence (*autofluorescence*), which is produced by substances normally present in the tissue, and *secondary fluorescence*, which is induced by staining with fluorescent dyes called *fluorochromes*.

Certain proteins can be tagged with fluorescent dyes, such as fluorescein isocyanate or rhodamine, without denaturing the molecule. These fluorescent proteins may then be injected into the animal and localized in sections within the cell or in the extracellular space.[24]

The most important advantage of fluorescence microscopy is its great sensitivity. Fluorescence often yields specific cytochemical information

because some of the normal components of the tissue have a typical fluorescent emission. Thus vitamin A, thiamine, riboflavin, and other substances can be detected. The cytochemical value of the method is increased considerably by spectrographic analysis of the radiation. Sometimes certain substances incorporated in cells, e.g., sulfonamides, can be localized.

The most common pattern of autofluorescence is a weak, diffuse, bluish fluorescence of the cytoplasm, with a yellow and stronger fluorescence of the granules; usually the nucleus is not fluorescent. Mitochondria of the liver and kidney give a strong fluorescence, calcium deposits appear yellow-white, and free porphyrins have a strong red fluorescence. An important application concerns the so-called lipogenic pigments, which are found in a great number of cells and which increase in number as the cell ages. It is thought that these pigments represent different degrees of oxidation and polymerization of unsaturated fatty acids. With fluorescence two types of pigments, the so-called *lipofuscin* and the *ceroid*, can be determined.[25]

An important development has been the use of paraformaldehyde fixation, which, in freeze-dried tissues, produces condensation with catecholamines and indolamines, emitting a green and a yellow fluorescence, respectively.[26] This fluorescent reaction has also been studied by microspectrographic methods.[27] In Chapter 19 the importance of certain fluorochromes in the study of human chromosomes will be mentioned (Fig. 19–2).

Immunocytochemistry — Detection of Antigens with Labeled Antibodies

Cytochemical techniques have been developed to localize antigens at the light and electron microscopic levels. Antibodies are produced by plasmocytes against most macromolecules (antigens) and also against small molecules, provided they are bound to larger molecular species. Antibodies are present in the γ-globulin fraction of the serum and generally have a sedimentation constant of 6S. Figure 4–16 shows the general principle of some of the cytochemical techniques involving the use of labeled antibodies.

In step (I) of Figure 4–16 a tissue antigen is represented, as well as the corresponding antibody (rabbit anti-X) produced by the injection of the antigen into a rabbit. If the antibody (i.e., γ-globulin) is marked with a fluorescent dye or with an enzyme, which, by reacting with a substrate, gives an opaque deposit, the sites of the

antigen can be observed in the tissue. Antibodies have also been coupled to ferritin, an iron-containing protein that is very opaque to electrons, and thus the reaction can be detected with the electron microscope.[29] (Although the antibody is shown unlabeled, step (I) in Figure 4–16 corresponds to the so-called direct method.)

In the widely used Coons's technique, the antibodies are coupled with fluorescein isocyanate. The tissues are frozen and sectioned in a cryostat, and then the sections are stained with the coupled antiserum. This *direct method* has been widely used to localize viruses and bacterial antigens. The pituitary hormone ACTH was localized in the basophilic cells of the pituitary gland, and several of the enzymes produced by the pancreas have also been localized by this technique.

At present the most frequently used immunocytochemical approach is the *indirect method* in which the primary antigen-antibody reaction is amplified by the introduction of a second antibody. In step (II) of Figure 4–16 it may be observed that goat anti-rabbit γ-globulin (IgG) labeled with a fluorescent dye or with an enzyme can be used. In step (III) the result of this second interaction is observed either by fluorescence microscopy or by light and electron mi-

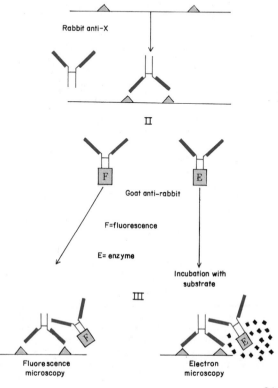

Figure 4–16 Diagram showing the steps involved in the indirect method of immunocytochemistry. In step I, the rabbit anti-X reacts with the tissue antigen. In step II, goat anti-rabbit antibody is used labeled either with a fluorescent dye or with an enzyme (e.g., peroxidase). In step III, the result is observed either by fluorescence microscopy or by electron microscopy. (Courtesy of A. C. Cuello.)

croscopy. The enzyme most often coupled to the anti-goat γ-globulin is *peroxidase*, which is reacted with 3',3'-diaminobenzidine (DAB) in the presence of H_2O_2. This reaction produces an electron-dense deposit that can be detected with the light microscope, as well as with the electron microscope.[30-32]

A large number of proteins have been localized in the cells by immunocytochemistry. In Chapter 9 and in other chapters we will mention studies using anti-tubulin, anti-actin, anti-myosin, and other fluorescent antibodies (see Figs. 9–6 and 9–14).

Radioautography – Interaction of Radioisotopes with Photographic Emulsions

Radioautography is based on the capacity of radioisotopes to act on the silver bromide crystals of the photographic emulsion. The tissue section is put in contact with the emulsion for a certain period; then the radio-autograph is developed like an ordinary photograph. By comparing the radio-autograph with the cells in the tissues seen under a microscope, the radioisotope can be localized fairly accurately.

Radioisotopes used in radioautography may emit one or more of three types of radiation: α- and β-particles and γ-rays. α-Particles are positively charged helium nuclei that produce straight tracks in the emulsion; they can easily be traced back to the point of origin. In biologi-

cal work they have limited use because they are produced primarily by heavy metals. β-Particles are electrons, which have different energy levels. Their tracks are tortuous and may vary in length from a few microns to a millimeter, depending on their energy. γ-Rays are not important in radioautography. Most of the isotopes used are β-emitters.[33] Of the various radioautographic techniques the one most used is that employing *liquid emulsions*. In this case the photographic emulsion, containing the silver halide grains, is in a gel that liquefies at 45° C. Essentially, the method consists of the following steps: (1) The tissue section, mounted on a glass slide or a grid, is immersed in the liquid emulsion at 45° C. (2) The section is removed, so that a film of gelatin covers it, and it is then left at room temperature (Fig. 4–17). (3) The specimens are kept in lightproof boxes for a period of days or weeks to allow the radiation to act on the film. (4) The photographic emulsion is developed. (5) The tissue is stained and observed. The silver grains stand out as black dots on the parts of the specimen where the isotope is localized (Fig. 4–17). For the electron microscope a resolution of about 0.1 μm (100 nm) may be achieved.[34-36]

Quantitative results are obtained by determining the density of the particles in the radioautographs by various optical methods, or by counting the grains.

Substances marked with the β-emitter ^{14}C have been widely used in radioautography.

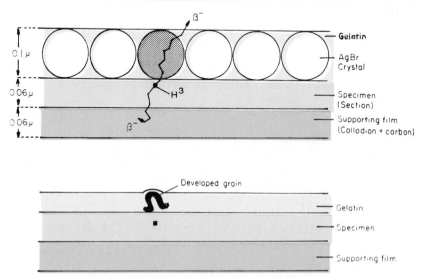

Figure 4–17 Diagrammatic representation of an electron microscope autoradiograph preparation. **Top,** *during exposure:* The silver halide crystals, embedded in a gelatin matrix, cover the section. A beta particle, from a tritium point source in the specimen, has hit a crystal (cross-hatched), causing the appearance of a latent image on the surface. **Bottom,** *during examination and after processing:* The exposed crystal has been developed into a filament of silver; the nonexposed crystals have been dissolved. The total thickness has decreased because the silver halide occupied approximately half the volume of the emulsion. (Courtesy of L. G. Caro.)

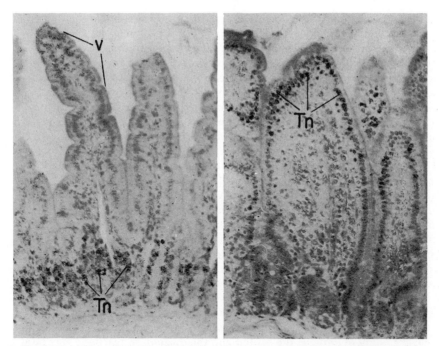

Figure 4–18 Section of intestine of a mouse injected with tritiated thymidine. **Left,** animal killed eight hours after injection; **right,** 36 hours after injection. *Tn,* tagged nuclei; *V,* villus. (See the description in the text.) (Courtesy of C. P. Leblond.)

With such an emitter, *track radioautography* can also be used — a thick liquid emulsion is applied, and it is possible to follow and count the single tracks of β-particles coming out from a definite area, thus providing a quantitative estimation. The resolution of this method is of the order of 1 to 2 μm.

Substances labeled with tritium (^{3}H), a weak β-emitter, are most widely used in radioautography. For the study of deoxyribonucleic acid (DNA) metabolism of the cell, tritiated thymidine is used and is specific for this type of nucleic acid. To study ribonucleic acid (RNA) metabolism, ^{3}H-uridine is used; and for examining the synthesis of proteins, various tritiated amino acids are used. For the polysaccharides and glycoproteins, tritiated monosaccharides, such as ^{3}H-mannose and ^{3}H-fucose, are employed.

Important investigations of the mechanism of DNA replication and RNA metabolism have been made using tritium labeling. In the example shown in Figure 4–18, the nuclei tagged with tritiated thymidine and initially present at the bottom of the intestinal crypts are found after 36 hours near the tip of the villus. This illustrates most graphically the life cycle of the cell — after a division at the bottom of the crypt, the cell ascends along the epithelium until, after a few hours, it is destroyed at the tip of the villus.

A two-emulsion radioautographic technique has been developed that can distinguish between the β-particles emitted from ^{14}C and those from ^{3}H atoms. With this method a tritiated precursor of DNA and a ^{14}C-labeled precursor of RNA or of a protein can be employed simultaneously.

SUMMARY

Cytochemistry

Cytochemistry consists of the identification and localization of chemical components of the cell. Cytochemical (and histochemical) studies may be based on four main analytical techniques: (a) separation of cell fractions by conventional biochemical techniques; (b) isolation of minute amounts of tissues, and even single cells, by micro- and ultramicromethods; (c) direct detection of cell components in the cell by chemical staining; and (d) use of measurement of physical parameters.

Cell fractionation consists of the homogenization of tissue and the separation of the various components of the cell according to mass and specific gravity. In a standard *differential centrifugation,* by the use of increasing centrifugal fields, four main fractions are obtained: nuclear, mitochondrial, microsomal, and soluble. Standard centrifuges and preparative ultracentrifuges are used. The time required for each particle to settle depends on its size and density. Cell fractionation can be improved through the use of discontinuous or continuous *density gradients.* In this case centrifugation is continued until the various particles reach an equilibrium with the density of the gradient (i.e., isopyknic centrifugation). In *zonal rotors* the density gradient is formed and the sample is layered while the rotor is spinning. Small particles, such as ribosomes or macromolecules (DNA, RNA, proteins), may be studied with the *analytical ultracentrifuge* and at the same time the sedimentation constant, S, which is related to the molecular weight, can be determined. Isopyknic ultracentrifugation in gradients of cesium chloride permits demonstration of the existence of DNAs of different *buoyant densities. Microchemical* and *ultramicrochemical* methods are used in the quantitative analysis of tissue sections or isolated cells. Micromanometric methods permit the measurement of the oxygen consumption and of the weight of a single microorganism, such as an ameba.

Cytochemical staining is the method most often used in cytochemical studies. The substance to be investigated must be immobilized and then identified by chemical reactions. For the study of *proteins* the Millon and the diazonium reactions may be used. SH groups in proteins may be determined by certain reagents.

Schiff's reagent, which is the leukobase of basic fuchsin, is recolored by *aldehyde* groups. This reagent is used in the *Feulgen reaction* for DNA. A prior mild acid hydrolysis removes RNA and unmasks aldehyde groups from deoxyribose linked to purine bases. The *periodic acid–Schiff (PAS) reaction* occurs in polysaccharides after oxidation of the 1,2-glycol groups by periodic acid. The *plasmal reaction* occurs in free aldehydes present in certain plasmalogens.

Detection of enzymes is accomplished by means of frozen sections made in a cryostat, or after light fixation with aldehydes. In general, the section is incubated with specific substrates, and the product is converted into a metal precipitate or a colored compound. Various phosphatases, esterases, cholinesterases, and other hydrolytic enzymes may be studied cytochemically. *Oxidases* and *dehydrogenases* are detected by special reagents, such as the Nadi reagent for cytochrome oxidase. The DAB reagent (3,3'-diaminobenzidine) is widely used for peroxidases and other oxidases. Tetrazolium salts produce insoluble formazan dyes with some mitochondrial enzymes.

Several cytochemical methods are based on *physical determinations.* Ultraviolet absorption at 260 nm is characteristic of nucleic acids and results from the presence of purine and pyrimidine bases. Cytophotometric methods can be used in the ultraviolet or the visible spectrum to make quantitative determinations in cells and tissues. Thus, the Feulgen reaction permits the determination of the DNA content of the nucleus and even of single chromosomes.

Fluorescence microscopy, with ultraviolet light near the visible spectrum, is used to study the natural fluorescence *(autofluorescence)* of cell components such as vitamin A, thiamine, riboflavin, calcium, and porphyrins. More important is the use of fluorescent dyes *(secondary fluorescence)* in several cytochemical methods. Catechol and indolamines produce a fluorescent reaction with aldehydes.

Immunocytochemistry uses antibodies for the localization of macromolecules (antigens) in cells. (Antibodies are present in the γ-globulin fraction of serum.) In the direct reaction the antibody may be coupled with a fluorescent dye or an opaque molecule (e.g., ferritin) for the electron microscope. In the *indirect* method the unlabeled antibody-antigen complex reacts with a labeled anti-γ-globulin antibody. This technique permits the study of antibody formation in plasmocytes and the localization of various proteins in different cells.

Radioautography uses substances labeled with radioisotopes. These radioisotopes are then detected using a photographic emulsion that is developed like an ordinary photograph. Most radioisotopes used are β-emitters and contain ^{14}C or tritium. Several techniques, such as the *track radioautography* in a thick liquid emulsion and the thin liquid emulsion for *electron microscope radioautography,* are employed. In most cases ^{3}H-thymidine is used for the study of DNA. ^{3}H-leucine, or other tritiated amino acids, is used for protein synthesis. Radioautography is one of the most frequently used techniques for following within the cell structure the mechanisms of DNA replication, DNA-RNA transcription, and their translation into protein.

REFERENCES

1. Eagle, H. (1955) *Science, 122*:501.
2. Hayflick, L., and Moorehead, P. (1961) *Exp. Cell Res., 25*:585.
3. Puck, T. T., and Marcus, P. I. (1955) *Proc. Nat. Acad. Sci. USA, 41*:432.
4. Thompson, L. H., and Baker, R. M. (1973) In: *Methods in Cell Biology,* Vol. 6, Chapter 7. (Prescott, D. M., ed.) Academic Press, Inc., New York.
5. Sabatini, D. D., Bensch, K. G., and Barrnett, R. J. (1963) *J. Cell Biol., 17*:19.
6. Gersh, I. (1959) Fixation and staining. In: *The Cell,* Vol. 1, p. 21. (Brachet, J., and Mirsky, A. E., eds.) Academic Press, Inc., New York.
7. Pearse, A. G. E. (1968) *Histochemistry, Theoretical and Applied,* 3rd Ed. J. & A. Churchill, London.
8. Glick, D. (1959) Quantitative microchemical techniques of histo- and cytochemistry. In: *The Cell,* Vol. 1, p. 139. (Brachet, J., and Mirsky, A. E., eds.) Academic Press, Inc., New York.
9. DeRobertis, E., Pellegrino de Iraldi, A., Rodriguez de Lores Arnaiz, G., and Salganicoff, L. (1962) *J. Neurochem., 9*:23.
10. Anderson, N. G. (1956) Techniques for the mass isolation of cellular components. In: *Physical Techniques in Biological Research,* Vol. 3, p. 300. (Oster, G., and Pollister, A. W., eds.) Academic Press, Inc., New York.
11. Albertsson, P. (1960) *Partition of Cell Particles and Macromolecules.* John Wiley & Sons, New York.
12. Kirk, P. L. (1950) *Quantitative Ultramicroanalysis.* John Wiley & Sons, New York.
13. Eränkö, Q. (1955) *Quantitative Methods in Histology and Microscopic Histochemistry.* Little, Brown and Co., Boston.
14. Lowry, O. H. (1957) Micromethods for the assay of enzymes. In: *Methods in Enzymology,* Vol. 4. (Colowick, S. P., and Kaplan, N. O., eds.) Academic Press, Inc., New York.
15. Bennett, H. S., and Watts, R. M. (1958) In: *General Cytochemical Methods,* Vol. 1, p. 317. (Danielli, J. F., ed.) Academic Press, Inc., New York.
16. Barrnett, R. J., and Seligman, A. M. (1955) *J. Histochem. Cytochem., 3*:406.
17. Cafruny, E. J., Di Stefano, H. S., and Farah, A. (1955) *J. Histochem. Cytochem., 3*:354.
18. Lessler, M. A. (1953) *Internat. Rev. Cytol., 2*:231.
19. Brachet, J. (1957) *Biochemical Cytology.* Academic Press, Inc., New York.
20. McManus, F. A. (1946) *Nature (London), 158*:202.
21. Gomori, G. (1952) *Microscopic Histochemistry.* University of Chicago Press, Chicago.
22. Graham, R. C., and Karnovsky, M. L. (1966) *J. Histochem. Cytochem., 14*:291.
23. Seligman, A. M., Karnovsky, M. J., Wasserknig, H. L., and Hanker, J. S. (1968) *J. Cell Biol., 38*:1.
24. Mancini, R. E. (1963) *Internat. Rev. Cytol., 14*:193.
25. Price, G., and Schwartz, S. (1956) Fluorescence microscopy. In: *Physical Techniques in Biological Research,* Vol. 3, p. 91. (Oster, G., and Pollister, A. W., eds.) Academic Press, Inc., New York.
26. Carlsson, A., Falck, B., and Hillarp, N. (1962) *Acta Physiol. Scand., 56* (suppl.): 196.
27. Ritzen, M. (1967) *Exp. Cell Res., 44*:250, and *45*:178.
28. Coons, A. H. (1956) *Internat. Rev. Cytol., 5*:1.
29. Rifkind, R. A., Osserman, E. F., Hsu, K. C., and Morgan, C. (1962) *J. Exp. Med., 116*:423.
30. Leduc, E. H., Avrameas, S., and Bouteille, M. (1968) *J. Exp. Med., 127*:109.
31. Stengerger, L. A. (1967) Electron microscopic immunocytochemistry. *J. Histochem. Cytochem., 15*:139.
32. Nakane, P. K. (1970) *J. Histochem. Cytochem., 18*:9.
33. Taylor, J. H. (1956) Autoradiography at the cellular level. In: *Physical Techniques in Biological Research,* Vol. 3, p. 546. (Oster, G., and Pollister, A. W., eds.) Academic Press, Inc., New York.
34. Caro, L. G. (1964) In: *Methods in Cell Physiology,* Vol. 1, p. 327. (Prescott, D. M., ed.) Academic Press, Inc., New York.
35. Saltpeter, M. M. (1966) General area of autoradiography at the electron miscroscope level. In: *Methods in Cell Physiology,* Vol. 2. (Prescott, D. M., ed.) Academic Press, Inc., New York.
36. Stevens, A. R. (1966) *High Resolution Autoradiography,* Vol. 2, p. 255. (Prescott, D. M., ed.) Academic Press, Inc., New York.

ADDITIONAL READING

Allfrey, V. G. (1959) The isolation of subcellular components. In: *The Cell,* Vol. 1, p. 193. (Brachet, J., and Mirsky, A. E., eds.) Academic Press, Inc., New York.

Anderson, N. G., ed. (1966) The development of zonal centrifuges. *Natl. Cancer Inst. Monogr., 21.* Bethesda, Md.

Avrameas, S. (1972) Enzyme markers: their linkage with proteins and use in immuno-histochemistry. *Histochem. J. (London),* 4:321.

Baserga, R., and Malamud, D. (1969) *Autoradiography: Techniques and Application.* Harper and Row, Publishers, New York.

Burstone, M. S. (1962) *Enzyme Histochemistry and Its Application in the Study of Neoplasms.* Academic Press, Inc., New York.

Coons, A. H. (1956) Histochemistry with labeled antibody. *Internat. Rev. Cytol.,* 5:1.

DeDuve, C. (1975) Exploring cells with a centrifuge. *Science, 189:*186.

Engström, A. (1956) Historadiography. In: *Physical Techniques in Biological Research,* Vol. 3, p. 489. (Oster, G., and Pollister, A. W., eds.) Academic Press, Inc., New York.

Ficq, A. (1959) Autoradiography. In: *The Cell,* Vol. 1, p. 67. (Brachet, J., and Mirsky, A. E., eds.) Academic Press, Inc., New York.

Fujiwara, K., and Pollard, T. D. (1978) Simultaneous localization of myosin and tubulin in human tissue culture cells by double antibody staining. *J. Cell Biol., 77:*182.

Gabe, M. (1976) *Histological Techniques.* Springer-Verlag, Berlin.

Glick, D. (1959) Quantitative microchemical techniques of histo- and cytochemistry. In: *The Cell,* Vol. 1, p. 139. (Brachet, J., and Mirsky, A. E., eds.) Academic Press, Inc., New York.

Gomori, G. (1952) *Microscopic Histochemistry.* University of Chicago Press, Chicago.

Hale, A. J. (1957) The histochemistry of polysaccharides. *Internat. Rev. Cytol.,* 6:194.

Harris, M. (1964) *Cell Culture and Somatic Variation.* Holt, Rinehart & Winston, Inc., New York.

Heidelberg, M. (1967) Some contributions of immunochemistry to biochemistry and biology. *Annu. Rev. Biochem., 36:*part 1, 1.

Hopwood, D. (1972) Theoretical and practical aspects of glutaraldehyde fixation. *Histochem. (London),* 4:267.

Lison, L. (1960) *Histochimie et cytochimie animale,* 3rd. Ed. Gauthier-Villars, Paris.

Mazurkiewicz, J. E., and Nakane, P. K. (1972) Light and electron microscopic localization of antigens in tissues embedded in polyethylene glycol with a peroxidase-labeled antibody method. *J. Histochem. Cytochem.,* 20:969.

Miller, H. R. P. (1972) Fixation and tissue preservation for antibody studies. *Histochemical J.,* 4:305.

Pearse, A. G. E. (1968) *Histochemistry, Theoretical and Applied,* 3rd Ed. J. and A. Churchill, London.

Price, G., and Schwartz, S. (1956) Fluorescence microscopy. *Physical Techniques in Biological Research,* Vol. 3, p. 91. (Oster, G., and Pollister, A. W., eds.) Academic Press, Inc., New York.

Paul, J. (1970) *Cell and Tissue Culture,* 4th Ed. The Williams & Wilkins Co., Baltimore, Md.

Pollack, R. (1973) *Readings in Mammalian Cell Culture.* Cold Spring Harbor Laboratory, Cold Spring Harbor, New York.

Puck, T. T. (1972) *The Mammalian Cell as a Microorganism.* Holden-Day Inc., San Francisco.

Roddyn, D. B. ed. (1978) *Subcellular Biochemistry.* Plenum Press, New York.

Tooze, J., ed. (1973) *The Molecular Biology of Tumor Viruses,* Chapter 2. Cold Spring Harbor Laboratory, Cold Spring Harbor, New York.

Waymouth, C. H., ed. (1970) *Advances in Tissue Culture.* The Williams & Wilkins Co., Baltimore, Md.

Part 2

MOLECULAR COMPONENTS AND METABOLISM OF THE CELL

The general organization of the cell presented in the introductory chapters may best be interpreted on chemical and physicochemical evidence. Readers should review their previous studies of organic chemistry, especially those on proteins, carbohydrates, lipids, and nucleic acids — the main molecular components of the cell. Part 2 of this book is dedicated to an elementary and abbreviated survey of these molecular components. In Chapter 5 special emphasis is given to some of the stereochemical characteristics of these components, such as the primary, secondary, tertiary, and quaternary structure of proteins, and to the Watson-Crick model of DNA, which explains how the DNA molecule duplicates and functions as the primary storehouse of genetic information.

Chapter 6 briefly introduces the concept of enzymes as molecular machines used by the cell to produce all chemical transformations. The notion of the enzyme active site and of its possible interpretation at the molecular level is of particular importance. Enzyme kinetics and the various factors and mechanisms by which an enzymatic reaction may be inhibited or activated are presented. The main metabolic pathways utilized by the cell to obtain chemical energy from different foodstuffs are also introduced. The energy that is produced is then used to synthesize new products which either increase the mass of the cell or are eliminated to the environment as secretions. This suggests the idea of high energy bonds and the concept of bioenergetics, which, in turn, introduces the concept of entropy and its importance in biology. The reader is also introduced to the concepts of anaerobic glycolysis, the Krebs cycle, and oxidative phosphorylation; the last two together embody cell respiration. The ideas presented in Part 2 are fundamental to understanding Chapters 12 and 14 which deal with mitochondria and the photosynthetic mechanism of chloroplasts, respectively.

BIOCHEMISTRY OF THE CELL

5–1 Chemical Components of the Cell   69
  Water Is The Most Abundant Cell Component
  Salts and Ions are Essential
  Macromolecules – Polymers of Repeating Monomers
5–2 Proteins   71
  Proteins are Chains of Amino Acids Linked by Peptide Bonds
  Primary Structure – Amino Acid Sequence
  Secondary Structure – α-Helix or Pleated Sheet Conformation
  Tertiary Structure – Three-Dimensional Folding
  Quaternary Structure – Protein Subunits
  Weak Interactions are Essential for Protein Structure
  Electric Charges of Proteins and the Isoelectric Point
  Separation of Cell Proteins – Isoelectric Focusing and SDS Electrophoresis
5–3 Carbohydrates   78
  Complex Polysaccharides
  Glycoproteins – Two-Step Carbohydrate Addition

5–4 Lipids   80
  Triglycerides – Three Fatty Acids Bound to Glycerol
  Compound Lipids – Phospholipids and Biological Membranes
  Summary:
  Molecular Components of the Cell
5–5 Nucleic Acids   83
  Nucleic Acids – A Pentose, Phosphate, and Four Bases
  Regularities in DNA Base Composition – A = T and G = C
  DNA is a Double Helix
  Denaturation and Annealing of DNA
  Circular DNA – Supercoiled Conformation
  RNA Structure – Ribose and Uracil Instead of Deoxyribose and Thymine
  The Simplest Infectious Agent – A Circular RNA Chain
  Summary:
  Nucleic Acids

To understand the organization of biological systems, one should first become familiar with the main constituent molecules, particularly those of high molecular weight, such as proteins, nucleic acids, polysaccharides, and lipids.

5–1 CHEMICAL COMPONENTS OF THE CELL

An early approach to the study of the chemical composition of the cell was the biochemical analysis of whole tissues, such as the liver, brain, skin, or plant meristem. This method had limited cytologic value, because the material analyzed was generally a mixture of different cell types and, in addition, contained extracellular material. In recent years the development of cell fractionation methods and various micromethods has led to the isolation of different subcellular particles and thus to more important and precise information about the molecular architecture of the cell (Chapter 4).

The chemical components of the cell can be classified as *inorganic* (water and mineral ions) and *organic* (proteins, carbohydrates, nucleic acids, lipids, and so forth). Some organic components, such as enzymes, coenzymes, and hormones that have specific activities are mentioned in Chapter 6.

The protoplasm of a plant or animal cell contains 75 to 85 per cent water, 10 to 20 per cent protein, 2 to 3 per cent lipid, 1 per cent carbohydrates, and 1 per cent inorganic material.

Water is the Most Abundant Cell Component

With few exceptions, such as bone and enamel, water is the most abundant component of cells. It serves as a natural solvent for mineral ions and other substances and also as a dispersion medium of the colloid system of protoplasm. For instance, from microinjection experiments it is known that water is readily miscible with protoplasm. Furthermore, water is indispensable for metabolic activity, since physiologic processes occur exclusively in aqueous media. Water molecules also participate in many enzymatic reactions in the cell and can be formed as a result of metabolic processes.

Water exists in the cell in two forms: *free* and *bound*. *Free water* represents 95 per cent of the total cellular water and is the principal part used as a solvent for solutes and as a colloid dispersion medium. *Bound water,* which represents only 4 to 5 per cent of the total cellular water, is loosely held to the proteins by hydrogen bonds and other forces. Because of the asymmetrical distribution of charges a water molecule acts as a *dipole,* as shown in the following diagram.

Because of this polarity, water can bind electrostatically to both positively and negatively charged groups in the protein. Thus, each amino group in a protein molecule is capable of binding 2.6 molecules of water.

Water is also used to eliminate substances from the cell and to absorb heat — by virtue of its high specific heat coefficient — thus preventing drastic temperature changes in the cell.

The water content of an organism is related to the organism's age and metabolic activity. For example, it is highest in the embryo (90 to 95 per cent) and decreases progressively in the adult and in the aged.

Salts and Ions are Essential

Salts dissociated into anions (e.g., Cl^-) and cations (e.g., Na^+ and K^+) are important in maintaining *osmotic pressure* and the *acid-base equilibrium* of the cell. Retention of ions produces an increase in osmotic pressure and thus the entrance of water. Some of the inorganic ions, such as magnesium, are indispensable as cofactors in enzymatic activities; others, such as inor-

ganic phosphate, form adenosine triphosphate (ATP), the chief supplier of chemical energy for the living processes of the cell, through oxidative phosphorylation.

The concentration of various ions in the intracellular fluid differs from that in the interstitial fluid (see Table 00–0). For example, the cell has a high concentration of K^+ and Mg^{2+}, while Na^+ and Cl^- are localized mainly in the interstitial fluid. The dominant anion in cells is phosphate; some bicarbonate is also present.

Calcium ions are found in the circulating blood and in cells. In bone they combine with phosphate and carbonate ions to form a crystalline arrangement.

Phosphate occurs in the blood and tissue fluids as a free ion, but much of the phosphate of the body is bound in the form of phospholipids, nucleotides, phosphoproteins, and phosphorylated sugars. As primary phosphate ($H_2PO_4^-$) and secondary phosphate (HPO_4^{2-}), phosphate contributes to the acid-base equilibrium, thereby buffering the pH of the blood and tissue fluids.

Other ions found in tissues are sulfate, carbonate, bicarbonate, magnesium, and amino acids.

Certain *mineral components* are also found as part of larger macromolecules. For example, *iron,* bound by metal-carbon linkages, is found in hemoglobin, ferritin, the cytochromes, and some enzymes (such as catalase and cytochrome oxidase). Traces of *manganese, copper, cobalt, iodine, selenium, nickel, molybdenum,* and *zinc* are indispensable for maintenance of normal cellular activities.

Macromolecules – Polymers of Repeating Monomers

Structural and other properties of the cell are intimately related to large molecules made of repeating units linked by covalent bonds. These units are called *monomers,* and the resulting macromolecule is called a *polymer.* Molecules having an increasing number of monomers possess widely different characteristics. For example, among the hydrocarbons, methane and ethane are gases, while butane and octane are liquids; further polymerization (20 or more monomers) produces oils, and finally solids, such as paraffins.

The three main examples of polymers in living organisms are as follows: (1) *Nucleic acids* result from the repetition of four different units called *nucleotides.* The repetition of the four nucleotides in the DNA molecule is the primary source of genetic information. (2) *Polysaccharides* can be polymers of monosaccharides, forming starch, cellulose, or glycogen, or may also involve the repetition of other molecules, forming more complex polysaccharides. (3) *Proteins* and *polypeptides* consist of the association in various proportions of some 20 different amino acids linked by peptide bonds. The order in which these 20 monomers can be linked gives rise to an astounding number of combinations in various protein molecules. This can determine not only their specificity, but in certain cases, their biological activity.

5-2 PROTEINS

Proteins are Chains of Amino Acids Linked by Peptide Bonds

The building blocks of proteins are the amino acids (Table 5–1). Essentially, an amino acid is an organic acid in which the carbon next to the — COOH group (called an alpha carbon) is bound to an amino group (— NH$_2$). In addition, the alpha carbon is bound to a side-chain (R) which is different in the various amino acids.

$$H_2N—\overset{\overset{\displaystyle H}{|}}{\underset{\underset{\displaystyle R}{|}}{C}}—COOH$$

(side-chain)

The amino acids differ from one another only in the side-chain; for example, the R in alanine has one carbon, while in leucine it has four carbons. Table 5–1 shows that the properties of the various amino acids depend on the chemical composition of their side-chains; for example, lysine and arginine are basic because their side-chains contain an extra amino group, and the acidic amino acids (glutamic and aspartic acids) contain an extra carboxyl group.

Because of the simultaneous presence of acidic (carboxyl) and basic (amino) groups, amino acids can have both positive and negative charges and are therefore *amphoteric molecules* or *zwitterions*.

The ionized form of an amino acid is:

$$^+H_3N—\overset{\overset{\displaystyle H}{|}}{\underset{\underset{\displaystyle R}{|}}{C}}—COO^-$$

TABLE 5-1 THE TWENTY AMINO ACIDS

Type of Amino Acid	3-Letter Symbol	1-Letter Symbol
Hydrophobic		
(Aliphatic Side-Chain)		
Glycine	Gly	G
Alanine	Ala	A
Valine	Val	V
Leucine	Leu	L
Isoleucine	Ile	I
Basic (Diamino)		
Arginine	Arg	R
Lysine	Lys	K
Acid (Dicarboxylic)		
Glutamic acid	Glu	E
Aspartic acid	Asp	D
Amide-Containing		
Glutamine	Gln	Q
Asparagine	Asn	N
Hydroxyl-Containing		
Threonine	Thr	T
Serine	Ser	S
Sulfur-Containing		
Cysteine	Cys	C
Methionine	Met	M
Aromatic		
Phenylalanine	Phe	F
Tyrosine	Tyr	Y
Heterocyclic		
Tryptophan	Trp	W
Proline	Pro	P
Histidine	His	H

The condensation of amino acids to form a protein molecule occurs in such a way that the acidic group of one amino acid combines with the basic group of the adjoining one, with the simultaneous loss of one molecule of water.

The linkage — NH — CO — is known as the *peptide linkage* or *peptide bond*. The formed molecule preserves its amphoteric character, since an acidic group is always at one end and a basic group is at the other, in addition to the lateral residues (radicals) that can be either basic or acidic. A combination of two amino acids is a *dipeptide;* of three, a *tripeptide*. When a few amino acids are linked together, the structure is an *oligopeptide*. A *polypeptide* contains a large number of amino acids.

$$H_2N \quad H \quad CO(OH) \quad C \quad R''$$

peptide bond

TABLE 5–2 MOLECULAR WEIGHTS OF SOME PROTEINS

Insulin	12,000
Cytochrome (horse heart)	12,100
Trypsin	23,800
Pepsin	35,500
Ovalbumin	44,000
Serum Albumin (bovine)	68,000
Hemoglobin (human)	67,000
γ-Globulin (human)	100,000
Catalase	250,000
Thyroglobulin (pig)	650,000

The distance between two peptide links is about 0.35 nm. A protein with a molecular weight of 30,000 consisting of 300 amino acid residues, if fully extended, should have a length of 100 nm, a width of 1.0 nm, and a thickness of 0.46 nm.

Table 5–2 lists the molecular weights of various proteins. The term *protein* (Gr., *proteuo*, I, occupying first place) indicates that all basic functions in living organisms depend on specific proteins. They constitute the enzymes and the contractile machinery of the cell, and are present in the blood, and other intercellular fluids. Some long-chain proteins, such as *collagen* and *elastin*, play an important role in the organization of the extracellular framework of tissues.

For details of the classification of the proteins, the reader is referred to biochemistry textbooks; however, it is important to stress that the properties of proteins vary considerably. For instance, *keratin* and *collagen* are insoluble and fibrous; the *globular proteins* e.g., egg albumin and serum proteins, are soluble in water or salt solutions and are spherical rather than thread-like molecules.

The *conjugated proteins* are attached to a non-protein moiety, the so-called *prosthetic group*. To such a group belong the *nucleoproteins* associated with nucleic acids, the *glycoproteins*, the *lipoproteins* (e.g., blood lipoproteins), and the *chromoproteins*, which have a pigment as the prosthetic group, such as hemoglobin, hemocyanin, and the cytochromes. Hemoglobin and myoglobin (present in muscle) contain the prosthetic group *heme*, an iron-containing organic compound that combines with oxygen.

Primary Structure – Amino Acid Sequence

The amino acid sequence of the polypeptide chain is known as the *primary structure* of the protein molecule. It is the most important and specific structure and determines the so-called *secondary* and *tertiary* structures. Groups of protein subunits containing secondary and tertiary structures constitute the *quaternary* structure; many proteins are formed by several polypeptide subunits.

In the protein molecule, amino acids are arranged like beads on a string (Fig. 5–1), and their sequence is of great biological importance. In the hemoglobin molecule a change in a single amino acid produces profound biological changes (see Table 21–2).

Secondary Structure – α-helix or Pleated Sheet Conformation

In a protein formed by several hundred amino acids, the chain may sometimes be linear, but more frequently it assumes different shapes that constitute what is known as the *secondary structure*. Fibrous proteins are often arranged in an orderly manner that can be analyzed by x-ray diffraction methods (see Chapter 3). This technique has facilitated classification of proteins into three structural types or groups.

The *β-keratin type* has an identity period of about 0.72 nm, and the adjacent chains are disposed in a *pleated sheet structure* as shown in Figure 5–2, in which the side-chains of the amino acid residues stick out perpendicular to the plane of the chain. The individual chains are held together by hydrogen bonds, forming a "peptide grid."

The *α-helix structure* found in the *α-keratin type* is produced when the polypeptide chain forms a helical structure, like a spiral winding around an imaginary cylinder, in such a way that hydrogen bonds are established within the molecule and not with an adjacent molecule (Fig. 5–3, B). For the *collagen group*, a model made of three helical chains has been proposed (see Fig. 7–1).

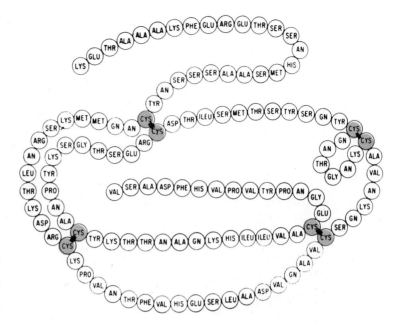

Figure 5–1 The primary structure of bovine pancreatic ribonuclease. Notice the position of the four disulfide bridges between cystine residue. (From C. B. Anfinsen.)

Tertiary Structure – Three-dimensional Folding

In the so-called *globular proteins* the polypeptide chain is held together in a definite way to form a compact structure in three dimensions (Fig. 5–3, C). The arrangement in space of such chains is very complex but may be resolved by x-ray diffraction (Fig. 3–11). In globular proteins the chains are folded in a compact way with the polar groups toward the surface, thus leaving little space in the interior for water molecules (the hydrophobic amino acids are in the center of the molecule). These proteins have regions of α-helix or β-configurations, and other parts exist as a *random coil*, i.e., in a flexible structure that may change at random. The tertiary structure is given by the way in which these helical or

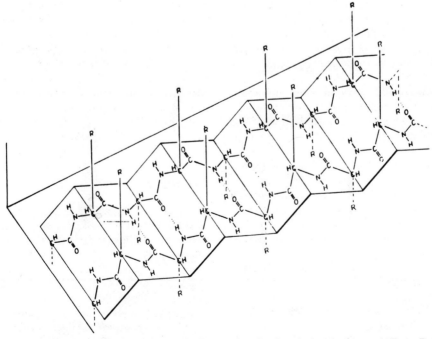

Figure 5–2 Pleated sheet structure of β-protein chains. (See the description in the text.) (From P. Karlson.)

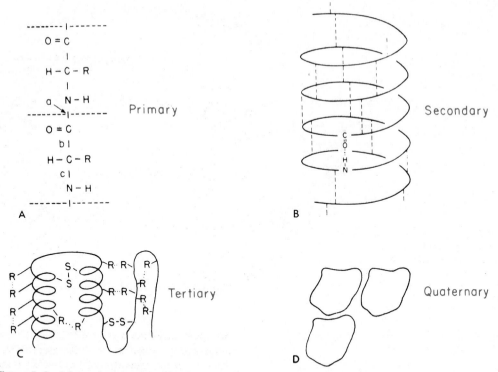

Figure 5–3 Schematic representation of the structural levels in proteins. Amino acid chains are denoted by R, non-covalent interactions. . . . (Courtesy of Prof. C. Nemethy.)

random-coil segments arrange with respect to each other.

The spatial arrangment is predetermined by the sequence of amino acids in the primary structure and by the bonds that can be established among some of the residues. A series of biological properties of proteins, such as enzyme activity and antigenicity, are related to the tertiary structure.

The *denaturation* of a protein is brought about by high temperatures or other nonphysiologic conditions, and consists of a disruption of the tertiary structure. This is usually accompanied by loss of biological activity.

Figure 5–4 shows that in the case of the enzyme ribonuclease, denaturation can be achieved by treatment with β-mercaptoethanol and high concentrations of urea. Mercaptoethanol is a reducing agent which can disrupt—S—S—bridges (disulfide bonds), reducing them to —SH groups; while urea disrupts weak interactions (see later discussion). The tertiary structure of ribonuclease is held together by four disulfide bonds which are established between pairs of cysteines (an amino acid that contains an —SH group), as shown in Figure 5–4. After denaturation, the enzyme can be cor-

rectly refolded into its natural conformation (*renaturation*) by removing the urea and mercaptoethanol gradually by dialysis; after this, the enzymatic activity is recovered (Fig. 5–4). There are 105 possible combinations in which eight cysteines can pair to produce four disulfide bridges, but only the biologically active natural conformation is produced after careful renaturation, because it is thermodynamically the most stable structure. This is a clear demonstration that all the information needed to produce the complex folding of a protein molecule is contained in its amino acid sequence (see Anfinsen, 1973).

Quaternary Structure – Protein Subunits

Unlike the primary, secondary, or tertiary structures, which concern a single polypeptide chain, the quaternary structure involves two or more chains (Fig. 5–3, D). These chains may or may not be identical, but in both cases they are linked by weak bonds (non-covalent). For example, the hemoglobin molecule is composed of four polypeptides or subunits, two designated as α and two as β. Separation and association of the subunits may occur spontaneously. Hemo-

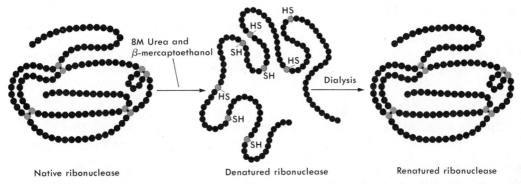

Figure 5–4 Denaturation and renaturation of ribonuclease. This experiment shows that the information for protein folding is contained in the amino acid sequence (primary structure) of proteins. (Cys residues are indicated with shaded beads. See Fig. 5–1.) Sulfhydryl groups, *SH, HS*.

globin may be broken into two half molecules (two α and two β) by urea. When urea is removed, they reassemble, forming complete, functional molecules. This binding is highly specific and takes place only between the half molecules. This is called the *principle of self-assembly*. This principle also applies to the building up of more complex cellular structures, such as the cell membrane, microtubules, and so forth (see Chapter 7). Most enzymes exist as multi-subunit structures.

Weak Interactions are Essential for Protein Structure

Different types of bonds are involved in the structure of proteins. The primary structure is fully determined by *covalent bonds* (peptide bonds).

Disulfide bonds (—S—S—bridges) are also co-valent and are established between the — SH groups of two cysteine residues. Disulfide bonds can be reversibly dissociated by the action of *reducing agents* (such as mercaptoethanol and dithiothreitol — DTT), as seen in Figure 5–4.

In addition, several *weak interactions* are very important in the establishment of the secondary and tertiary structures. All these weak bonds are non-covalent; the main types (illustrated in Fig. 5–5) are:

Ionic or electrostatic bonds, which result from the attractive force between ionized groups of opposite charge (Fig. 5–5, *a*).

Hydrogen bonds result from a H^+ (proton) that is shared between two neighboring electronegative atoms. The H^+ can be shared between nitrogen or oxygen atoms which are close to each other. Hydrogen bonds are very important in biology and are the main force that holds the two strands of DNA together. Hydrogen bonds are essential for the specific pairing between nucleic

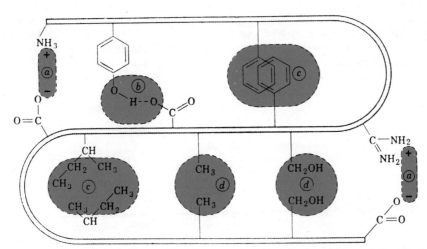

Figure 5–5. Types of non-covalent bonds that stabilize protein structure. *a*, ionic bonds; *b*, hydrogen bonds; *c*, *hydrophobic interactions*; *d*, Van der Waals interactions. (From C. B. Anfinsen.)

acid bases, which is the basis for the coding of genetic information. Figure 5–14 shows hydrogen bonds in DNA, and Figure 5–5, b shows them in proteins.

In hydrophobic interactions (Fig. 5–5, c) water tends to be excluded by non-polar groups, which associate with each other so that they are not in contact with water. In globular proteins the side-chains of the most hydrophobic amino acids (those with long non-polar side-chains, such as leucine, isoleucine, valine, and phenylalanine) tend to aggregate inside the molecule, and the polar (charged) groups protrude from the surface of the tertiary structure. The hydrophobic residues tend to repel the water molecules that surround the protein, thereby causing the globular structure to be more compact.

Van der Waals interactions (Fig. 5–5, d) occur only when two atoms come very close together. The closeness of two molecules can induce charge fluctuations which may produce dipoles and mutual attraction at very short range.

The essential difference between a covalent and a non-covalent bond is in the amount of energy needed to break the bond. For example a hydrogen bond requires only 4.5 kcal mole^{-1}, as compared with 110 kcal mole^{-1} for the covalent O — H bond in water. Although each individual bond is weak, large numbers of them can produce very stable structures, as in the case of double-stranded DNA. In the following chapter, a more detailed discussion will show that covalent bonds are generally broken by the intervention of enzymes, whereas non-covalent bonds are easily dissociated by physicochemical forces.

Electric Charges of Proteins and the Isoelectric Point

In addition to the terminal —NH$_3^+$ and — COO$^-$ charged groups, proteins contain dicarboxylic- and diamino-amino acids (Table 5–1) which dissociate as follows:

1. The acidic groups lose protons and become negatively charged. For example, in aspartic and glutamic acids, the free carboxyl group dissociates into — COO$^-$ + H$^+$.

2. The basic groups, by gaining protons, become positively charged — NH$_2$ + H$^+$ → —NH$_3^+$. This is found in amino acids with two basic groups, such as lysine or arginine. All these so-called ion-producing groups contribute to the acid-base reactions of proteins and to the electrical properties of protein molecules.

The actual charge of a protein molecule is the result of the sum of all single charges. Because dissociation of the different acidic and basic groups takes place at different hydrogen ion concentrations of the medium, pH greatly influences the total charge of the molecule. Figure 5–6 shows that in an acid medium, amino groups capture hydrogen ions and react as bases (—NH$_2$ + H$^+$ → —NH$_3^+$); in an alkaline medium the reverse takes place and carboxylic groups dissociate (— COOH→ COO$^-$ + H$^-$). For every protein there is a definite pH at which the sum of positive and negative charges is zero (Fig. 5–6). This pH is called the isoelectric point (pI). At the isoelectric point, proteins placed in an electric field do not migrate to either of the poles, whereas at a lower pH they migrate to the negative pole (cathode) and at a higher pH, to the positive pole (anode). This migration is called electrophoresis, and it provides a very useful technique for the separation of cellular proteins.

Separation of Cell Proteins — Isoelectric Focusing and SDS Electrophoresis

Each protein has a characteristic isoelectric point, and this property can be used in the separation of proteins. In the technique called isoelectric focusing, proteins are subjected to electrophoresis on a pH gradient. Each protein moves until it reaches a pH equal to its individual isoelectric point. At that moment, migration in the electric field stops because the net charge of the protein is zero.

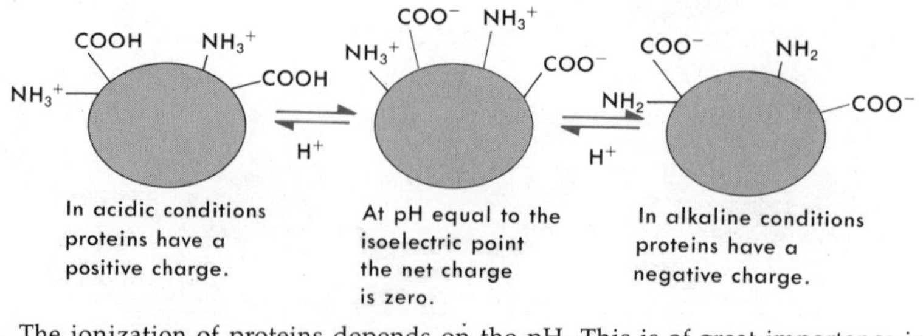

Figure 5–6 The ionization of proteins depends on the pH. This is of great importance in electrophoresis; in acidic conditions proteins migrate to the cathode, in alkaline conditions, to the anode.

Recently the techniques of isoelectric focusing and polyacrylamide gel electrophoresis have been combined to produce *two dimensional* separation of proteins. Figure 5–7 shows how several hundred cellular proteins can be resolved from one another. This technique is increasingly used in Cell Biology, and its great resolving power is due to the use of two *independent properties* of proteins. The proteins are first separated by isoelectric focusing (this is the first dimension), which separates proteins according to their *charge* (isoelectric point). The proteins are subsequently separated by electrophoresis (this is the second dimension) in polyacrylamide gels containing SDS, which separates proteins according to their *size* (molecular weight). This technique results in a series of spots distributed throughout the polyacrylamide gel (if the same property of proteins had been used in both di-

mensions, the spots would be distributed along a diagonal).

When the detergent sodium dodecyl sulfate (SDS) is used with electrophoresis, the proteins are separated mainly according to their molecular weight. This is because SDS binds to the proteins, giving them large numbers of negative charges due to the sulfate. Thus, most of the protein charges will come from the SDS, minimizing the role of charge differences between individual proteins (differences which would otherwise affect electrophoretic mobility), and all the proteins migrate according to their size. The larger proteins move more slowly than the smaller ones because they encounter more resistance when traversing the molecular pores within the polyacrylamide gel used for electrophoresis. SDS electrophoresis is widely used as a method for determining molecular weights of proteins.

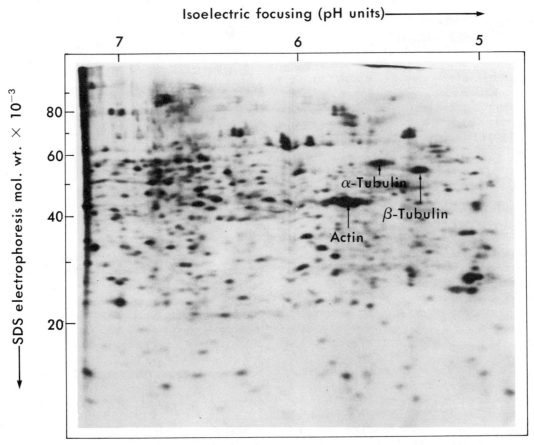

Figure 5–7 Two-dimensional electrophoresis of the proteins of *Xenopus borealis* oocytes. The frog cells were labeled with S^{35}-methionine, homogenized, and the total proteins were separated by isoelectric focusing in the first dimension and by SDS (sodium dodecyl sulphate) electrophoresis in the second. The polyacrylamide gel was exposed against photographic film (autoradiography). Several hundred radioactive proteins can be seen. As in most other cell types, the most abundant protein is actin (3 to 5 per cent of the total cell protein), followed by tubulin (which has α and β subunits, both of which form part of cell microtubules). Notice that the distance migrated by the proteins in the second dimension is proportional to the logarithm of the molecular weight. (From E. M. De Robertis).

5–3 CARBOHYDRATES

Carbohydrates, composed of carbon, hydrogen, and oxygen, are sources of energy for animal and plant cells; in many plants they also form important constituents of cell walls and serve as supporting elements. Animal tissues have fewer carbohydrates; among the most important are glucose, galactose, glycogen, and amino sugars and their polymers.

Carbohydrates of biological importance are classified as monosaccharides, disaccharides, and polysaccharides. The first two, commonly referred to as *sugars*, are readily soluble in water, can be crystallized, and easily pass through dialyzing membranes. Polysaccharides, on the other hand, neither crystallize nor pass through membranes.

MONOSACCHARIDES

Monosaccharides are simple sugars having the empirical formula $C_n(H_2O)_n$. They are classified in accordance with the number of carbon atoms, e.g., trioses, pentoses, and hexoses. The pentoses *ribose* and *deoxyribose* are found in the molecules of nucleic acids, and the pentose *ribulose* is important in photosynthesis. Glucose, a hexose is the primary source of energy for the cell. Other important hexoses are *galactose* and *fructose (levulose)*.

DISACCHARIDES

Disaccharides are sugars formed by the condensation of two monomers of monosaccharides with the loss of one molecule of water. Their empirical formula is therefore $C_{12}H_{22}O_{11}$. The most important of this group are *sucrose* (formed by glucose and fructose) and *maltose* (formed by two glucose monomers) in plants, and *lactose* (formed by galactose and glucose) in animals.

POLYSACCHARIDES

Polysaccharides result from the condensation of many molecules of monosaccharides, with a corresponding loss of water molecules. Their empirical formula is $(C_6H_{10}O_5)_n$. Upon hydrolysis they yield molecules of simple sugars. The most important polysaccharides in living organisms are *starch* and *glycogen*, which are reserve substances in cells of plants and animals, respectively, and *cellulose*, the most important structural element of the plant cell.

Starch is a combination of two long polymer molecules: *amylose*, which is linear, and *amylopectin*, which is branched. Glycogen may be considered to be the starch of animal cells. Figure 5–8 shows the structure of this branched polymer, which is composed of many molecules of glucose. It is found in numerous tissues and organs, but the greatest proportion is contained in liver cells and muscle fibers.

Complex Polysaccharides

In addition to the polysaccharides made of hexose monomers mentioned in the preceding section, there are many more complex long molecules that contain amino nitrogen (e.g., glucosamine) that can, in addition, be acetylated (e.g., acetylglucosamine) or substituted with sulfuric or phosphoric acid. All these polymers are important in molecular organization, particularly as intercellular substances. These polysaccharides may exist either freely or combined with proteins or lipids. The most important are:

Neutral polysaccharides, which contain only acetylglucosamine. The main example is *chitin*, a supporting substance found in the exoskeletons of insects and crustaceans.

Acidic mucopolysaccharides, which contain sulfuric or other acids in the molecule. These molecules are strongly basophilic. To this group belong *heparin*, an anticoagulant substance; *chondroitin sulfate*, present in the cartilage, skin, cornea, and umbilical cord; and *hyaluronic acid*, in skin and other animal tissues. The latter is hydrolyzed by *hyaluronidase*.

Glycoproteins — Two-step Carbohydrate Addition

Glycoproteins are complexes composed of a protein and a prosthetic group of carbohydrates. Several monosaccharides, such as galactose, mannose, and fucose, as well as N-acetyl-D-glucosamine and sialic acid, are found. Glycoproteins can be divided into two major categories: cellular and secretory. Cellular glycoproteins are present mainly in the cell membrane, with the carbohydrate moiety protruding to the outside of the cell, and they have important functions in membrane interaction and recognition (see Chapter 8).

Most secreted proteins are glycoproteins and are produced by various cells: serum glycoproteins (i.e., seroalbumins), secreted by the liver; thyroglobulin, produced in the thyroid gland, immunoglobulins, secreted by plasma cells; ovalbumin, secreted by the hen oviduct; and ribonuclease and deoxyribonuclease, produced by the pancreas. In most glycoproteins the protein is linked to the carbohydrate moiety by a

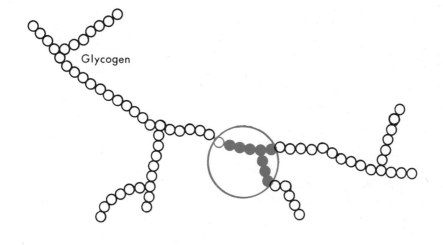

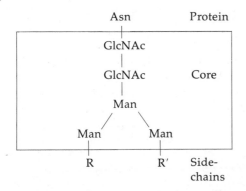

Figure 5–8 Glycogen is a branched polymer formed by as many as 30,000 glucose units. The glycosidic bonds are established between carbons 1 and 4 of glucose, except at the branching points, which involve linkages between carbons 1 and 6. The **top** part of the figure shows a low resolution diagram; the circled area is enlarged on the **bottom** part.

bond between asparagine (Asn) and N-acetyl-D-glucosamine (GlcNAc).

```
            Asn        Protein
             |
           GlcNAc
             |
           GlcNAc      Core
             |
            Man
           /    \
        Man      Man
          \      /
          R      R'     Side-
                        chains
```

This diagram shows that a constant feature of asparagine-containing glycoproteins is the presence of a common pentasaccharide "core" consisting of N-acetylglucosamine and mannose,

which is then continued by two side-chains (R and R'). The side-chains differ in length and composition in the various glycoproteins and may contain galactose, mannose, fucose, and sialic acid. The "core" is added inside the endoplasmic reticulum as the nascent peptides are being synthesized, while the side-chains are added later in the Golgi complex by enzymes called *glycosyltransferases*.

Evidence obtained by Leloir's group at Buenos Aires indicates that the oligosaccharide core of glycoproteins (formed mainly by N-acetyl-D-glucosamine and mannose) is first assembled on a lipid carrier and is then transferred to the protein molecule. The function of the lipid carrier (a derivative of *dolichol)* is to enable the hydrophilic oligosaccharide to traverse the endoplasmic reticulum membrane.

Several exocrine glands, such as the salivary

glands and the mucous glands of the digestive tract, secrete mucoproteins in which the linkage is between N-acetylglucosamine and serine or threonine in the protein. Sialic acid and L-fucose are the terminal sugars in these compounds.

The carbohydrate moiety of glycoproteins is important in determining certain physicochemical properties (the viscosity of mucins may be important in protecting and lubricating the mucous membrane of the digestive tract, for example); in molecule-membrane interactions (for the determination of blood groups and other immunological properties at the surface of erythrocytes); and in membrane-membrane interactions (see Chapter 8–5).

5–4 LIPIDS

The compounds in this large group are characterized by their relative insolubility in water and solubility in organic solvents. The cause of this general property of lipids and related compounds is the predominance of long aliphatic hydrocarbon chains or benzene rings. These structures are non-polar and hydrophobic. In many lipids these chains may be attached at one end to a polar group, rendering it capable of binding water by hydrogen bonds. The following is a classification of lipids:

Triglycerides — Three Fatty Acids Bound to Glycerol

Simple lipids are alcohol esters of fatty acids. Among these are: (1) *Neutral fats* (glycerides), often called triglycerides; they are triesters of fatty acids and glycerol. Neutral fats accumulate in adipose tissue. (2) *Waxes*, which have a higher melting point than natural fats; they are esters of fatty acids with alcohols other than glycerol, such as in beeswax.

Fatty acids have long hydrocarbon chains with the general formula:

$$\begin{array}{c} COOH \\ | \\ (CH_2)_n \\ | \\ CH_3 \end{array}$$

Fatty acids always have an even number of carbons (because they are synthesized from two-carbon acetyl units). For example *palmitic acid* has 16 carbons, and *stearic acid* has 18 carbons. Sometimes the hydrocarbon chain has double bonds ($-C{=}C-$), and in this case the fatty acid is

said to be non-saturated; for example, *oleic acid* has 18 carbons and one double (unsaturated) bond.

The carboxyl groups of fatty acids react with the alcohol groups of glycerol in the following way:

The resulting triglycerides are used by organisms to store spare energy conveniently: The oxidation state of the long hydrocarbon chains is very low, and they therefore liberate large amounts of energy (about twice as many calories per gram as carbohydrates and proteins) when oxidized to form CO_2 and water.

Compound Lipids — Phospholipids and Biological Membranes

Upon hydrolysis these lipids yield other compounds in addition to alcohol and acids. They serve mainly as structural components of the cell, particularly in cell membranes. The following are classified as compound lipids:

Phosphatides (phospholipids) are diesters of phosphoric acid that can be esterified with either glycerol or sphingosine. This group includes the lecithins, cephalins, inositides, and plasmalogens (acetyl phosphatides) (see Table 5–3).

A phospholipid has only two fatty acids attached to the glycerol molecule. The third hydroxyl group of glycerol is esterified to *phosphoric acid* instead of to a fatty acid. This phosphate is also bound to a second alcohol molecule, which can be *choline, ethanolamine, inositol,* or *serine,* depending on the type of phospholipid:

choline
ethanolamine
inositol
serine

Alcohol

$^-O{-}P{=}O$

Fatty acid   Fatty acid

As shown in Figure 5–9, phospholipids have two long *hydrophobic* fatty-acid "tails" and a *hydrophilic* (polar) phosphate-containing "head." Phospholipids are *amphipathic* molecules (molecules with a hydrophilic and a hydrophobic region), and they are the basic constituents of biological membranes. Biological membranes are bilayers of phospholipids with the hydrophilic heads (phosphate-containing regions) positioned at the water interface and the long hydrophobic tails avoiding water and arranged in the interior of this membrane structure (Chapter 8).

Glycolipids and *sphingolipids* are characterized by the fact that glycerol is replaced by the amino alcohol *sphingosine.* To these groups belong the *sphingomyelins,* mainly in the myelin sheath of nerves; the *cerebrosides,* which are characterized by the presence of galactose or glucose in the molecule; the *sulfatides,* which contain sulfuric acid esterified to galactose; and the *gangliosides* (Table 5–3).

The gangliosides deserve special mention because of their presence in cell membranes, their possible role as receptors of virus particles, and their influence on ion transport across membranes. A ganglioside is a complex molecule containing sphingosine, fatty acids, carbohydrates (lactose + galactosamine), and neuraminic acid. This is a long and highly polar molecule.

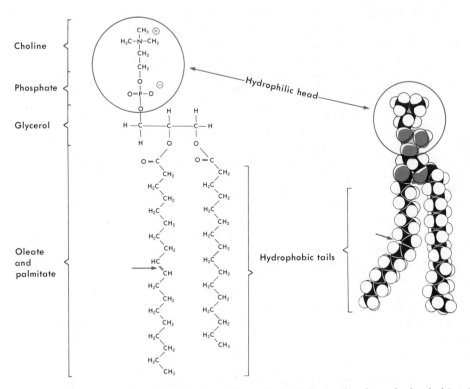

Figure 5–9 A membrane phospholipid molecule has a hydrophilic head and two hydrophobic tails. The phospholipid represented here is palmitoyl-oleoyl-phosphatidyl-choline. Note that the double bond in the oleic acid produces a bend in the hydrocarbon chain (indicated by an arrow). Double bonds in the fatty acids increase cell membrane fluidity because unsaturated chains are more flexible (the rotation of the carbon-carbon single bonds on either side of the arrow is enhanced).

TABLE 5–3   CLASSIFICATION OF PHOSPHATIDES AND GLYCOLIPIDS*

Name	Main Alcohol Component	Other Alcohol Components	P:N Ratio
I. Glycerophosphatides			
1. Phosphatidic acids	Diglyceride (= glycerol diester)		1:0
2. Lecithins	Diglyceride (= glycerol diester)	Choline	1:1
3. Cephalins	Diglyceride (= glycerol diester)	Ethanolamine, serine	1:1
4. Inositides	Diglyceride (= glycerol diester)	Inositol	1:0
5. Plasmalogens ("acetyl phosphatides")	Glycerol ester and enol ether	Ethanolamine, choline	1:1
II. Sphingolipids			
1. Sphingomyelins	N-Acylsphingosine	Choline	1:2
2. Cerebrosides	N-Acylsphingosine	Galactose,† glucose†	0:1
3. Sulfatides	N-Acylsphingosine	Galactose†	(1 H_2SO_4)
4. Gangliosides	N-Acylsphingosine	Hexoses,† hexosamine,† neuraminic acid†	no P

*From Karlson, P., *Introduction to Modern Biochemistry.* 2nd ed., New York, Academic Press, Inc. 1967.
†These components are present as glycosidic linkage, and thus are called glycolipids.

SUMMARY:

Molecular Components of the Cell

This chapter is an elementary survey of the molecular components of the cell. The protoplasm of a plant or animal cell contains 75 to 85 per cent water, 10 to 20 per cent protein, 2 to 3 per cent lipid, and 1 per cent carbohydrate. *Water* is the natural solvent of mineral ions and other small molecules and is a dispersion medium for the colloid system of the cell. About 95 per cent of the water is *free,* the rest is *bound* or immobilized within the structure. To each amino group of a protein 2.6 molecules of water are bound. Anions (e.g., Cl^-) and cations (e.g., K^+, Na^+) help to maintain the osmotic pressure and pH of the cell. Inorganic phosphate bound in ATP is the main supplier of chemical energy. The interior of the cell is rich in K^+ and Mg^{2+}, while the intercellular fluid contains Na^+ and Cl^-.

Macromolecules are long *polymers* composed of *monomers.* Proteins are made of about 20 different monomers — the *amino acids.* These are *amphoteric* molecules because they carry an acidic (— COOH) and a basic group (— NH_2). Amino acids are linked by the *peptide bonds,* forming polypeptides. All basic functions of the cell depend on specific proteins. These proteins are *globular* (e.g., egg albumin, serum proteins) or *fibrous.* Conjugated proteins contain a *prosthetic group* (e.g., nucleoproteins, lipoproteins, chromoproteins).

The *primary structure* of the protein is the amino acid sequence. A change in a single amino acid may produce a profound change in the molecule (e.g., hemoglobins). The secondary structure of a protein may be of the *β-keratin,* the *α-helix* keratin, or the random coil type. The *tertiary* structure of globular protein is very complex and can be determined by x-ray diffraction. These proteins may contain parts with *α*-helix and *β*-configurations, or a random coil. The disruption of the tertiary structure by high temperatures or other agents is called *denaturation.* The *quaternary* structure is characteristic of proteins having more than one subunit. Hemoglobin is a tetramer composed of two *α*- and two *β*-chains.

In proteins, non-covalent bonds, i.e., ionic, hydrogen, and hydrophobic interactions play an important role in determining the secondary and tertiary structure. The electric charge of a protein is determined by the *ion-producing groups* (i.e., acidic or basic) and their degree of dissociation at different pHs. At the *isoelectric point* (pI) the net charge is 0, and the molecule will not migrate in an electric field. This is the basis for *isoelectric focusing,* a technique used to separate proteins.

Carbohydrates serve as sources of energy or play a structural role (e.g., in cell walls). They are classified as mono-, di-, and polysaccharides. The main monosaccharides are *pentoses* (5 carbons) and *hexoses* (6 carbons). The main polysaccharides, which are composed of glucose monomers, are starch (plants) and glycogen (liver and muscle). Complex polysaccharides contain amino sugars (i.e., glucosamine), and sulfuric acid or phosphoric acid. The acidic mucopolysaccharides are strongly basophilic (i.e., heparin, chondroitin sulfate, hyaluronic acid). *Glycoproteins* are either *cellular,* present on the outside of cell membranes, or are *secretory* (e.g., seroalbumins, thyroglobulin, immunoglobulins, ribonuclease, etc.). In most glycoproteins the carbohydrate is linked to the amino acid asparagine by way of N-acetyl-D-glucosamine (G1cNAc). The carbohydrate moiety has a constant pentaoligosaccharide core of mannose and G1cNAc, which is added to the nascent peptides in the endoplasmic reticulum, and two variable side-chains, which are added later in the Golgi complex.

Lipids comprise a large group of different compounds characterized by their solubility in organic solvents. There are *simple* lipids (glycerides), *steroids* (i.e., sex hormones, cholesterol), and *conjugated lipids* (i.e., phosphatides, glycolipids, cerebrosides, sulfatides, and gangliosides). Phospholipids are the main components of biological membranes. They have a hydrophilic (phosphate-containing) region and two hydrophobic (fatty-acid) tails.

5-5 NUCLEIC ACIDS

Nucleic acids are macromolecules of the utmost biological importance. All living organisms contain nucleic acids in the form of deoxyribonucleic acid (DNA) and ribonucleic acid (RNA). Some viruses may contain only RNA, while others have only DNA.

DNA is the major store of genetic information. This information is transmitted by *transcription* into RNA molecules, which are utilized in the synthesis of proteins. In fact, the central dogma of modern biology is:

DNA → RNA → PROTEIN

The biological role of nucleic acids is discussed in detail in the chapters dealing with gene expression. Only those features of their chemical structure relevant to the understanding of their function will be considered here.

In higher cells DNA is localized mainly in the nucleus as part of the chromosomes. A small amount of DNA is present in the cytoplasm and contained within mitochondria and chloroplasts. RNA is found both in the nucleus, where it is synthesized, and in the cy-

TABLE 5-4   DNA AND RNA: STRUCTURE, REACTIONS, AND ROLE IN THE CELL

	Deoxyribonucleic Acid	*Ribonucleic Acid*
Localization	Primarily in nucleus; also in mitochondria and chloroplasts	In cytoplasm, nucleolus, and chromosomes
Pyrimidine bases	Cytosine Thymine	Cytosine Uracil
Purine bases	Adenine Guanine	Adenine Guanine
Pentose	Deoxyribose	Ribose
Cytochemical Reaction	Feulgen	Basophilic dyes with ribonuclease treatment
Hydrolyzing enzyme	Deoxyribonuclease (DNase)	Ribonuclease (RNase)
Role in cell	Genetic information	Synthesis of proteins

toplasm, where the synthesis of proteins occurs (Table 5–4).

Nucleic Acids — A Pentose, Phosphate, and Four Bases

Nucleic acids are formed of sugar moiety (pentose), nitrogenous bases (purines and pyrimidines), and phosphoric acid. After a mild hydrolysis the nucleic acids are decomposed into nucleotides.

Nucleotides are the monomeric units of the nucleic acid macromolecule. Figure 5–10 shows that nucleotides result from the covalent bonding of a phosphate and a heterocyclic base to the pentose. Within the nucleotide, the combination of a base with the pentose constitutes a *nucleoside*. For example, *adenine* is a purine base; *adenosine* (adenine + ribose) is the corresponding nucleoside and *adenosine monophosphate* (AMP), *adenosine diphosphate* (ADP), and *adenosine triphosphate* (ATP) are nucleotides (Fig. 5–10). In addition to constituting the building blocks of nucleic acids, nucleotides are important because they are used to store and transfer chemical energy. Figure 5–10

shows that the two terminal phosphate bonds of ATP contain high energy; when these bonds are cleaved, the energy released can be used to drive a variety of cell functions (Chapter 6–3).

Nucleic acids are linear polymers in which the nucleotides are linked together by means of phosphate-diester bridges with the pentose moiety. These bonds link the 3′ carbon in one nucleotide to the 5′ carbon in the pentose of the adjacent nucleotide. The backbone of nucleic acids consists, therefore, of alternating phosphates and pentoses. The nitrogenous bases are attached to the sugars of this backbone.

As shown in Figure 5–11 the phosphoric acid uses two of its three acid groups in the 3′,5′ diester links. The remaining negative group confers to the polynucleotide its acid properties and enables the molecule to form ionic bonds with basic proteins. In eukaryotic cells, DNA is associated with *histones* (i.e., basic proteins rich in arginine or lysine), forming a *nucleoprotein complex* called *chromatin*. This anionic group also causes nucleic acids to be highly basophilic; i.e., they stain readily with basic dyes.

Pentoses are of two types: *ribose* in RNA, and

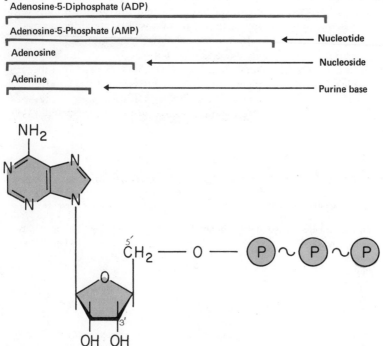

Figure 5–10 Structure of adenosinetriphosphate and its components. Note the presence of two high energy phosphate bonds.

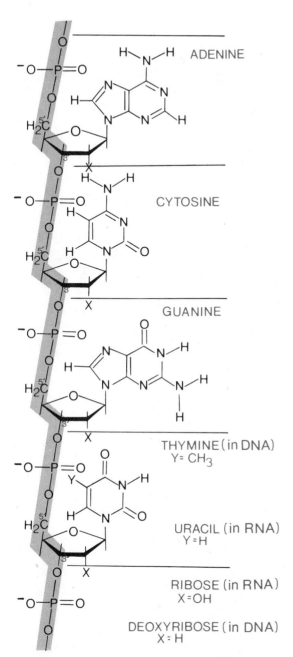

ADENINE

CYTOSINE

GUANINE

THYMINE (in DNA)
Y= CH₃

URACIL (in RNA)
Y = H

RIBOSE (in RNA)
X = OH

DEOXYRIBOSE (in DNA)
X = H

Figure 5–11 A segment of a single hypothetical nucleic acid chain showing the nucleotides and their constituent parts. The pentose-phosphate backbone is indicated.

deoxyribose in DNA. The only difference between these two sugars is that the oxygen in the 2′ carbon is lacking in deoxyribose (Fig. 5–11). A cytochemical reaction that is specific for the deoxyribose moiety (*Feulgen reaction*) can be used to visualize DNA under the microscope (Chapter 4–5).

The bases found in nucleic acids are either *pyrimidines* or *purines*. Pyrimidines have a single heterocyclic ring, whereas purines have two fused rings. In DNA the pyrimidines are *thymine* (T) and *cytosine* (C); the purines are *adenine* (A) and *guanine* (G) (Fig. 5–11). RNA contains *uracil* (U) instead of thymine (Table 5–4). Therefore *between RNA and DNA there are two main differences:* the first one in the pentose moiety (ribose and deoxyribose, respectively) and the second in a pyrimidine base (uracil instead of thymine). This explains why in Cell Biology radioactive thymidine (i.e., the nucleoside) is used to label specifically DNA and radioactive uridine for RNA.

Heterocyclic bases absorb ultraviolet light at 260 nm. A cell photographed at this wavelength shows the nucleolus, the chromatin, and the RNA-containing regions of the cytoplasm absorbing intensely (Fig. 5–12). In fact, the simplest way of determining the concentration of a solution of nucleic acids is to measure its absorbance at 260 nm.

All the genetic information of a living organism is stored in the linear sequence of the four bases. Therefore, a four letter alphabet (A, T, C, G) must code for the primary structure of all proteins (i.e., composed of 20 amino acids). All the excitement in molecular biology, leading to the unraveling of

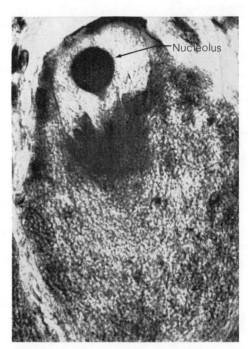

Figure 5–12 Nerve cell photographed with ultraviolet light at 260 nm. The regions in the nucleus and cytoplasm that absorb the ultraviolet light contain nucleic acid. (From H. Hydén.)

the genetic code (Chapter 21), began when the structure of DNA was understood.

Regularities in DNA Base Composition – A = T and G = C

DNA is present in living organisms as linear molecules of extremely high molecular weight. For example, *E. coli* has a single circular DNA molecule with a molecular weight of about 2.7×10^9 daltons and a total length of 1.4 mm.* In higher organisms the amount of DNA may be several thousand times larger; for example, the DNA contained in a single human diploid cell, if fully extended, would have a total length of 1.7 meters.

Between 1949 and 1953 Chargaff studied the base composition of DNA in great detail. It was found that the base composition varies from one species to another, and yet striking regularities were also found. In all cases the amount of adenine was found to be equal to the amount of thymine (i.e., A = T). The number of cytosine and guanine bases were also found to be equal (i.e., G = C). As a consequence, the total quantity of purines equals the total quantity of pyrimidines (i.e., A + G = C + T). On the other hand, there is considerable variation between species regarding the AT/GC ratio. In higher plants and animals in general AT is in excess of GC, whereas in viruses, bacteria, and lower plants the contrary is more common. For example, in man the AT/GC ratio is 1.52; in *E. coli*, it is 0.93.

DNA is a Double Helix

In 1953, based on the x-ray diffraction data of Wilkins and Franklin, Watson and Crick proposed a model for the DNA structure that provided an explanation for the above-mentioned regularities in base composition and for the biological properties of DNA — particularly its duplication in the cell. The structure of DNA is shown in Figure 5–13. It is composed of two right-handed helical polynucleotide chains that form a *double helix* around the same central axis. The two strands are *antiparallel;* i.e., their 3′,5′ phosphodiester links are in opposite directions. Furthermore, the bases are stacked inside the helix in a plane perpendicular to the helical axis.

---

*One base pair of DNA corresponds (on average) to 660 daltons; and 1 μm of double helix corresponds to 2940 base pairs or 1.94×10^6 daltons. The *E. coli* chromosome therefore has about 3,400,000 base pairs.

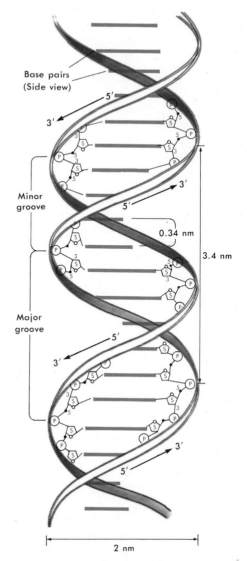

Figure 5–13 The DNA double helix. The phosphate-ribose backbones are indicated as ribbons. The base pairs are flat structures stacked one on top of another perpendicular to the long axis of DNA, and they are therefore represented as horizontal lines in this side view. The base pairs can be seen from a top view in Figure 5–14. Note that the two strands are antiparallel and that the molecule has a minor and a major groove. The double helix gives one complete turn every ten base pairs (3.4 nm). *P*, phosphate group; *S*, sugar.

The two strands are held together by *hydrogen bonds* established between the pairs of bases. Since there is a fixed distance between the two sugar moieties in the opposite strands, only certain base pairs can fit into the structure. As may be seen in Figure 5–14, the only two pairs that are possible are AT and CG. It is important to note that two hydrogen bonds are formed between A and T, and three are formed between C and G, and that therefore a CG pair is more stable than an AT pair. In addition to hydrogen bonds, hydrophobic interactions established between the stacked bases are important in maintaining the double helical structure.

In the Watson and Crick model the distance between the stacked bases is 0.34 nm, which corresponds to the primary period demonstrated by x-ray diffraction. Furthermore, a turn of the double helix is completed in 3.4 nm, a length which corresponds to 10 nucleotide residues (Fig. 5–13).

It is evident from observation of a space-filling model of DNA that the double helix has a mean diameter of 2.0 nm; furthermore, two grooves (a major or deep groove, and a minor or more shallow one) are observed (Fig. 5–15).

The *axial sequence* of bases along one polynucleotide chain may vary considerably, but on the other chain the sequence must be *complementary*, as in the following example:

First chain: 5′ T, G, C, T, G, T, G, G, T,3′
              || ||| ||| || ||| || ||| ||| ||
Second chain: 3′A, C, G, A, C, A, C, C, A,5′

Because of this property, given an order of bases on one chain, the other chain is exactly complementary.

During DNA duplication, the two chains dissociate, and each one serves as a template for the synthesis of two complementary chains. In this way two DNA molecules are produced, each having exactly the same molecular constitution. The mechanism of DNA duplication will be discussed in Chapter 16. The varying sequence of the four bases along the DNA chains forms the basis for genetic information. Four bases can produce thousands of different hereditary characters, because DNA molecules are long polymers along which an immense number of combinations may be produced, as we will see when we discuss the genetic code in Chapter 21.

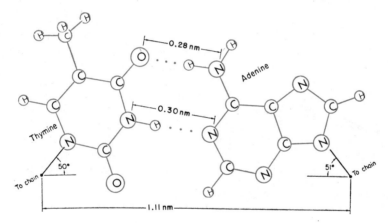

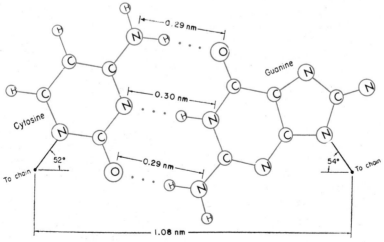

Figure 5–14 The two base pairs in DNA. The complementary bases are thymine and adenine (T—A) and cytosine and guanine (C—G). Observe that between T—A there are two, and between C—G three, hydrogen bonds. (From L. Pauling and R. B. Corey.)

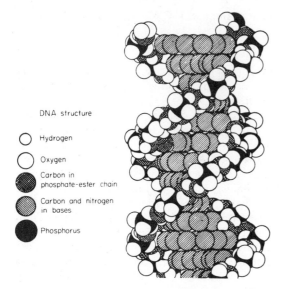

DNA structure

○ Hydrogen

○ Oxygen

◉ Carbon in
phosphate-ester chain

◉ Carbon and nitrogen
in bases

● Phosphorus

Figure 5–15 Space-filling model of a segment of DNA molecule showing the double helix with the phosphate-ribose chain and the bases at right angles. In this model it is evident that the surface of the DNA molecule has a major and a minor groove.

Denaturation and Annealing of DNA

Since the structure of the double helix of DNA is preserved by weak interactions (i.e., hydrogen bonds and hydrophobic interactions established between the stacked bases), it is possible to separate the two strands by heating and other treatments (such as alkaline pH). This separation process, called *melting* or *denaturation of DNA*, can be followed in the spectrophotometer by measuring the absorbance at 260 nm (Fig. 5–16). Since the double helix of DNA absorbs less ultraviolet light than the sum of the two isolated strands, by raising the temperature it may be observed that at a certain point there is a sharp increase in the quantity of ultraviolet light absorbed (sometimes called *hyperchromic shift*). The increase in temperature at this point corresponds to the "melting" (or separation) of the two DNA strands and is variable for DNAs of different base composition. Since the temperature required to break the GC pair (having three hydrogen bonds) is higher than the one needed to break the AT pair (having two hydrogen bonds), the melting point of a given DNA sample will depend on the AT/GC ratio. If after denaturation the DNA is cooled slowly, the complementary strands will base-pair in register, and the *native* (double helical) conformation will be restored (Fig. 5–16). This process is called *renaturation* or *annealing*.

Renaturation of DNA is a very useful tool in

Molecular Biology. Figure 5–17 shows how DNA renaturation can be used to estimate the size (number of nucleotides) of the genome of a given organism. When DNA is renatured under standardized conditions, a large genome (e.g., calf) takes more time to reanneal than a small genome (*E. coli* or bacteriophage T_4). This is because the individual sequences take longer to find the correct partners (the larger the genome, the more chances there are of incorrect molecular collisions).

Renaturation studies have led to the discovery of *repeated sequences* in eukaryotic DNA. When certain copies are repeated many times, the rate of renaturation will be much faster than for sequences present as single copies. Some sequences (called *satellite* DNAs) can be repeated millions of times in the genome (10×10^6 times for mouse satellite).

Single-stranded DNA will also anneal to complementary RNA. Molecular *hybridization* is used extensively in Molecular Biology. Purified RNAs (such as, for example, globin mRNA) can be labeled (the most convenient way is to use the enzyme *reverse transcriptase*, which can transcribe RNA into a radioactive *complementary DNA copy*). These labeled probes can then be used to study the genes from which they arose. For example, from the analysis of the velocity of renaturation, in very much the same way as in Figure 5–17, it can be shown the globin gene is present in only one copy per haploid genome. Furthermore, using the modern genetic engineering methods described in Chapter 21–4, the

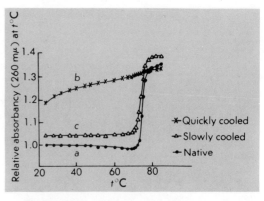

Figure 5–16 Hyperchromic shift in DNA from bacteriophage T2. Observe that the native DNA (*a*), upon reaching a temperature near 80° C, suddenly increases in absorbancy. In *b* and *c*, after heating to 91° C, the DNA was either cooled quickly (*b*) or slowly (*c*). In this last case, the DNA structure is re-formed. (From J. Marmur and P. Doty.)

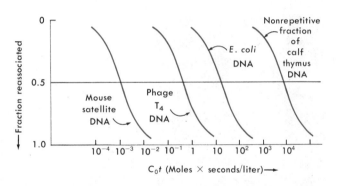

Figure 5–17 Renaturation kinetics of various DNAs C_0t curves). The velocity of renaturation is measured in the C_0t units required to obtain 1/2 renaturation. C_0t (short for Concentration × time) by convention is the initial concentration of DNA (in moles of nucleotides per liter) in the reaction mixture, multiplied by the time (in seconds) during which reannealing was allowed to proceed. A large genome renatures more slowly than a small one. (Redrawn from the classic paper by R. J. Britten and D. E. Kohne. *Science, 161*:530, 1968.)

labeled probes can be used to identify, by hybridization, recombinant plasmids grown in *E. coli* which contain, for example, the globin gene.

Denaturation done under carefully controlled conditions may be used for physical mapping of DNA in a technique known as *partial denaturation mapping*. This technique is based on the fact that the regions rich in AT separate more easily. Under the electron microscope those regions are detected as single-stranded loops, and the distance between the loops and the end of the DNA molecule can be measured (Fig. 5–18).

Circular DNA – Supercoiled Conformation

Many viruses have covalently closed circular DNA molecules that when visualized under the electron microscope show the twisted structure seen in Figure 5–19. This DNA is *supercoiled*, which indicates that the structure is under tension. Supercoiling arises from a relative lack of turns in the DNA double helix.

Let us consider a covalently closed circular DNA molecule with 5000 base pairs. In the *relaxed* conformation (smooth circle in Fig. 5–19), the DNA helix will have 500 helical turns (because

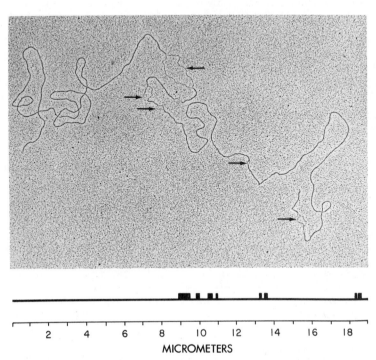

Figure 5–18 Electron micrograph of the lambda phage DNA showing sites in which denaturation (i.e., separation of the strands) has resulted from the action of an alkaline medium (pH 11). A map of this partial denaturation is presented below the electron micrograph, showing at scale the two ends of the DNA molecules and indicating by rectangles the position and length of the denatured sites. (Courtesy of R. B. Inman; see Inman, R. B., and Schnös, M. G., *J. Molec. Biol., 49*:93–8, 1970.)

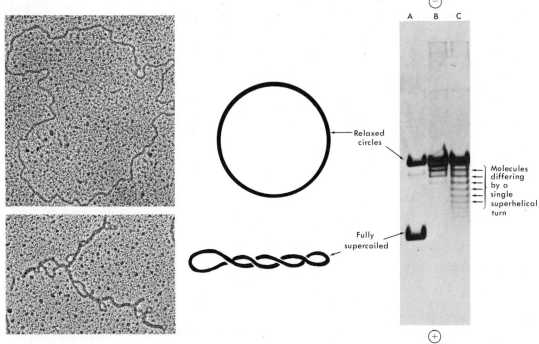

Figure 5–19. Superhelical turns in circular DNA molecules detected by electron microscopy and gel electrophoresis. DNA from the virus SV 40 was electrophoresed in an agarose gel. The structure of the supercoiled molecules is more compact than the relaxed circles, and therefore it migrates faster. (A) SV 40 DNA sample, containing both relaxed and fully supercoiled molecules, (B) the same DNA was treated with *nicking-closing* enzyme that removes superhelical turns, (C) as in B, but treated with less nicking-closing enzyme so that some— but not all—of the superhelical turns are removed. Bands differing between themselves by only one superhelical turn can be clearly seen. (Electron micrographs courtesy of Sue Whytock, gel electrophoresis courtesy of R. A. Laskey.)

the double helix gives one turn every 10 nucleotides). However, in a supercoiled molecule, the helix will have fewer turns (if it has 499, it will have one superhelical coil; if 490, it will have 10 coils). The molecule is under tension and folds upon itself in the supercoiled form. Since the molecule must be covalently closed, supercoils are always present in integral numbers; i.e., a molecule can lack one, two, or three turns, but not fractional amounts, such as 1.1, 1.2, or 1.3 (half-twists are not possible because the two DNA chains have opposite 3′ to 5′ polarities). DNA molecules different by only a single superhelical turn can be seen in Figure 5–19, lane C.

Fully supercoiled DNA has about one superhe-

lical turn every 200 base pairs. It is thought that supercoiling reflects the fact that in eukaryotic cells DNA is normally not free but bound to histone proteins. *In vivo* the DNA is not under tension and adopts the conformation most favorable for chromatin structure. When the histones are removed during purification of the DNA, the tension becomes apparent, and the molecules supercoil.

RNA Structure—Ribose and Uracil Instead of Deoxyribose and Thymine

The primary structure of RNA is similar to that of DNA, except for the presence of ribose and

TABLE 5–5 MAJOR CLASSES OF RIBONUCLEIC ACIDS IN *E. COLI*

Type	Sedimentation Coefficient	Molecular Weight	Number of Nucleotide Residues	Per Cent of Total Cell RNA
mRNA	6S to 25S	25,000 to 1,000,000	75 to 3,000	~2
tRNA	~4S	23,000 to 30,000	75 to 90	16
rRNA	5S	~35,000	~100	
	16S	~550,000	~1,500	82
	23S	~1,100,000	~3,100	

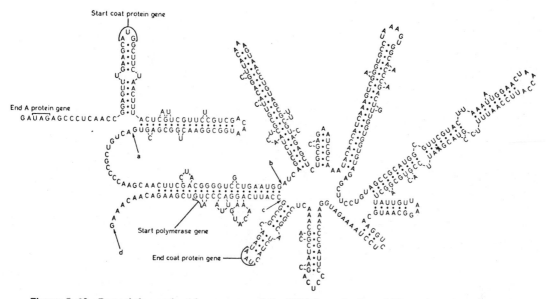

Figure 5–20 Part of the nucleotide sequence of the RNA bacteriophage MS2, a virus that infects *E. coli*. Note that the molecule folds back on itself, forming *hairpin loops*. The starting codon (AUG) for the coat protein is readily accessible to ribosomes. The start for the polymerase protein is blocked by the secondary structure and is accessible only when the structure is opened by the translation of the coat gene by ribosomes. (From Fiers, W., et al., *Nature, 260, 500, 1976*.)

uracil in place of deoxyribose and thymine. The base composition of RNA does not follow Chargaff's rules (see earlier discussion), since RNA molecules are made of only one chain.

RNA is synthesized within the nucleus by using only one strand of DNA as template (i.e., RNA is a faithful copy of the information contained in the DNA). As shown in Table 5–5 there are three major classes of ribonucleic acid: ribosomal RNA (rRNA), transfer RNA (tRNA), and messenger RNA (mRNA). All of them are involved in protein synthesis. Their structure and function will be studied in detail in the chapters on gene expression.

Although each RNA molecule has only a single polynucleotide chain, RNA is not a simple, smooth, linear structure. RNA molecules have extensive regions of complementarity in which hydrogen bonds between AU and GC pairs are formed between different regions of the same molecule. Figure 5–20 shows that as a result the molecule folds upon itself, forming structures called *hairpin loops*. In the base-paired regions the RNA molecule adopts a helical structure which is comparable to that of DNA. The compact structure of RNA molecules folded upon themselves has important biological consequences. For example, a sequence that indicates the starting site for protein synthesis might be in a part of the molecule inaccessible to the ribosomes, and

will only be expressed if some additional factor unfolded the molecule (this does in fact occur in RNA bacteriophages such as MS2 — see Fig. 5–20).

The Simplest Infectious Agent – A Circular RNA Chain

Viruses are small organisms consisting of nucleic acid (either DNA or RNA) which carries the genetic information, packed inside a cover of protein molecules (called a capsid) (see Chapter 7–2). Viruses do not have their own machinery for protein synthesis but rather use the infected cell as a source of ribosomes, energy, and other requirements for successful multiplication.

Viruses are simple enough, but *virioids* are even simpler. Virioids are infectious agents that attack plant cells (some produce economically important plant diseases) and that are made of an RNA molecule only, that is *not* covered with protein. Figure 5–21 shows that a virioid that attacks potatoes has a *circular* RNA molecule only 359 nucleotides long. This naked molecule is able to multiply in plant cells, disperse into the environment, and infect other plants. The compact base-paired structure (Fig. 5–21) probably protects the molecule from ribonucleases, which are very active and ubiquitous enzymes but which in general prefer single-stranded RNA.

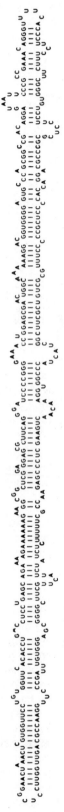

Figure 5–21 Nucleotide sequence and secondary structure of potato spindle tuber virioid. The RNA is circular, and it has only 359 nucleotides and does not contain AUG codons (the signal for start of protein synthesis). Virioids are the simplest infectious agents known and are not covered with protein. (From Gross, J. J., et al. *Nature*, 273:203, 1978.)

SUMMARY:

Nucleic Acids

Nucleic Acids are the repositories of genetic information. In general, deoxyribonucleic acid (DNA) is transcribed into ribonucleic acids (RNA), and these are involved in the translation into proteins.

$$DNA \rightarrow RNA \rightarrow Protein$$

DNA is localized mainly in the nucleus (i.e., in the chromosomes); small amounts are in mitochondria and chloroplasts.

Nucleic acids are composed of a pentose, phosphoric acid, purine bases (adenine, guanine), and pyrimidine bases (thymine, cytosine, uracil). They are linear polymers of *nucleotides* linked by phosphate-diester bonds between the pentoses. Since one acid group is left free, nucleic acids are very acidic and bind to basic proteins. In eukaryotic cells DNA is bound to *histones,* forming a nucleoprotein structure called *chromatin.*

DNA differs from RNA in the pentose (deoxyribose in DNA; ribose in RNA) and in one of the pyrimidine bases (thymine in DNA; uracil in RNA). Radioactive *thymidine* is used to label DNA; *uridine* is used for RNA. Genetic information is stored in the linear sequence of bases; the genetic alphabet consists of four letters: A, T, C, and G. The bases of DNA are in certain molar ratios; the amount of A always equals T, G equals C, and A + G = C + T. The ratio AT/GC varies among the different species.

Watson and Crick established the structure of the DNA molecule in 1953 on the basis of the x-ray diffraction studies of Wilkins and Franklin. The model is formed of two right-handed helical polynucleotide chains that are *antiparallel* and in which the bases are stacked perpendicularly. The two chains are held together by hydrogen bonds between the base pairs. Only two types of base pairs are possible within the double helix: AT and GC. This explains the regularities observed in the molar ratios of bases. AT pairs are held together by two hydrogen bonds, and GC pairs have three; for this reason GC pairs are more stable. The double helix has a major and a minor groove and completes one turn every 10 base pairs.

The axial sequence of bases along one chain is a complement of the other; DNA duplication takes place via a template mechanism, following the unwinding of the two strands. The sequence of bases in a long polymer provides an explanation for the immense number of combinations that carry different genetic information.

By an increase in temperature or other treatments, it is possible to denature (i.e., melt) the DNA. The two strands separate, producing a sharp increase in absorbance at 260 nm. More energy is required to break the G — C pair than is necessary to break the A — T pair.

Following denaturation, if the separated strands are kept at a suitable temperature, they can base-pair in register and restore the double helix (*renaturation* or *annealing*). Renaturation is a very important tool in Molecular Biology, as it allows us to estimate the size of a genome, to detect *repeated sequences* in eukaryotic DNAs, and to *hybridize* RNA to the DNA from which it was transcribed.

RNA has only one chain, and it contains ribose and uracil. In RNA there are regions of secondary structure (hairpin loops) formed by hydrogen bonded AU or GC pairs. The major classes of RNA are ribosomal RNA, transfer RNA, and messenger RNA. The smallest infectious agents are called *virioids* and their only component is a single *circular* RNA molecule.

ADDITIONAL READING

Anfinsen, C. B. (1973) Principles that govern the folding of protein chains. *Science, 181*:223.

Britten, R., and Kohne, D. (1968) Repeated sequences in DNA. *Science, 161*:529.

Cold Spring Harbor Symposium of Quantitative Biology (1973) *Structure and function of proteins at the three-dimensional level,* Cold Spring Harbor Laboratory, Cold Spring Harbor, New York.

Crick, F. H. C. (1957) Nucleic acids. *Sci. Amer., 197*:188.

Davidson, J. N. (1976) *The Biochemistry of the Nucleic Acids.* 8th Ed. (revised by Adams, R. L. P., Burdon, R. H., Campbell, A. M., and Smellie, R. M. S.) Chapman & Hall, London. An excellent advanced manual.

Dayhoff, M. O., ed. (1972) *Atlas of Protein Sequence and Structure.* National Biomedical Research Foundation, Silver Springs, Md.

Fiers, W., et al. (1976) Complete nucleotide sequence of bacteriophage MS2 RNA. *Nature, 260*:500.

Frieden, E. (1972) The chemical elements of life. *Sci. Amer., 227*:52.

Gross, H. J., Domdey, H., Lossow, C., Jank, P., Raba, M., Alberty, H., and Sanger, H. L. (1978) Nucleotide sequence and secondary structure of potato spindle tuber virioid. *Nature, 273*:203.

Karlson, P. (1967) *Introduction to Modern Biochemistry,* 2nd Ed. Academic Press, Inc., New York.

Lehninger, A. L. (1975) *Biochemistry.* Worth Publishers, Inc., New York.

Marmur, J., and Doty, P. (1962) Determination of base composition of DNA from its thermal denaturation temperature. *J. Molec. Biol., 5*:109.

O'Farrell, P. H. (1975) High resolution two-dimensional electrophoresis of proteins. *J. Biol. Chem., 250*:4007.

Perutz, M. (1978) Hemoglobin structure and respiratory transport. *Sci. Am., 239*(6):68.

Phillips, D. C., and North, A. C. T. (1975) *Protein Structure: Oxford Biology Readers,* Vol. 34. Oxford University Press, Oxford.

Sanger, F. (1965) The structure of insulin. In: *Currents in Biochemical Research.* (Green, D. E., ed.) Interscience Publishers, New York.

Staneloni, R. J., and Leloir, L. F. (1979) The biosynthetic pathway of the asparagine-linked oligosaccharides of glycoproteins. *Trends in Biochem. Sci., 4*:65.

Stryer, L. (1975) *Biochemistry.* W. H. Freeman and Co., San Francisco. Highly recommended.

Watson, J. D., and Crick, F. H. C. (1953) Molecular structure of nucleic acid: a structure for deoxyribose nucleic acid. *Nature, 171*:737.

Watson, J. D. (1968) *The Double Helix.* Atheneum Pubs., New York. The story of one of the most important scientific discoveries of this century.

Watson, J. D. (1975) *Molecular Biology of the Gene.* 3rd Ed. W. A. Benjamin, Inc., London.

White, A., Handler, P., and Smith, E. (1973) *Principles of Biochemistry.* McGraw-Hill Book Co., New York.

ENZYMES, BIOENERGETICS, AND CELL RESPIRATION

6–1 Enzymes 95
Enzymes are Proteins
Enzymes are Highly Specific
Some Enzymes Require Cofactors
Substrates Bind to the Active Site
Enzyme Kinetics — Km and Vmax Define Enzyme Behavior
Enzyme Inhibitors can be Very Specific
Isoenzymes
The Cell is Not Simply a Bag Full of Enzymes
Allosteric Enzymes have Multiple Interacting Subunits
6–2 Metabolic Regulation 101
Enzymes are Regulated at the Catalytic and the Genetic Levels
Enzyme Interconversions Also Regulate Metabolism
Cyclic AMP — The Second Messenger in Hormone Action
Summary:
Enzymes in the Cell

6–3 Bioenergetics 106
Entropy is Related to the Degree of Molecular Disorder
Photosynthesis is Essential in the Biological Energy Cycle
Cells Utilize Chemical Energy
ATP has High Energy Bonds
6–4 Cell Respiration 109
Anaerobic Glycolysis Yields Only 2 ATPs per Glucose Molecule
Aerobic Respiration Produces 36 ATPs per Glucose Molecule
The Krebs Cycle — The Common Final Pathway in the Degradation of "Fuel" Molecules
The Respiratory Chain — Stepwise Release of Energy by Electron Pairs
Oxidative Phosphorylation — The Energy Released by Electron Pairs Produces ATP
Summary:
Bioenergetics and Cell Respiration

The cell may be compared to a minute laboratory capable of carrying out the synthesis and breakdown of numerous substances. These processes are carried out by enzymes at normal body temperature, low ionic strength, low pressure, and a narrow range of pH.

The enzymes are not randomly distributed within the cell, but are located in various cell compartments. Frequently they are arranged in an orderly fashion within the macromolecular framework of the cell and cell organelles to form what is called a *multi-enzyme system*. Knowledge of the localization and grouping of enzymes within the cell structure is essential and is emphasized throughout this book.

Metabolism may be defined as the sum of all chemical transformations in the cell. It comprises both the processes of *catabolism*, by which substances are broken down, and *anabolism*, by

which new products are synthesized. Catabolic reactions are mostly *exergonic*, i.e., they liberate energy; anabolic reactions are *endergonic*, i.e., they consume energy. For example, the different substrates taken in by the cell as foodstuffs, such as glucose, amino acids, and lipids, are broken down into smaller molecules, with liberation of energy. This energy may be trapped by substances like adenosine triphosphate (ATP) and, in turn, is utilized by the cell in the synthesis of new and more complex molecules.

6–1 ENZYMES

Enzymes are Proteins

Enzymes are biological catalysts. A *catalyst* is a substance that accelerates chemical reactions but that is not itself modified in the process, so that it can be used again and again. Enzymes are the largest and most specialized class of protein molecules. More than a thousand different enzymes have been identified; many of them have been obtained in pure, and even crystalline, condition. Enzymes represent one of the most important products of the genes contained in the DNA molecule. The complex network of chemical reactions which are involved in cell metabolism is directed by enzymes.

Enzymes *(E)* are proteins with one or more loci, called *active sites,* to which the *substrate (S)* (i.e., the substance upon which the enzyme acts) attaches. The substrate is chemically modified and converted into one or more products *(P).* Since this is generally a reversible reaction, it may be expressed as follows:

$$E + S \rightleftarrows [ES] \rightleftarrows E + P \qquad (1)$$

where [ES] is an intermediary enzyme-substrate complex. Enzymes accelerate the reaction until an equilibrium is reached. They are so efficient that the reaction may proceed from 10^8 to 10^{11} times faster than in the non-catalyzed condition.

Enzymes are Highly Specific

A very important feature of enzyme activity is that it is *substrate-specific;* i.e., a particular enzyme will act only on a certain substrate. Some enzymes have nearly *absolute* specificity for a given substrate and will not act on even very closely related molecules, as, for example, stereoisomers of the same molecule. Other enzymes have *relative* specificity, since they will act upon a variety of related compounds. One example of enzyme specificity for proteinases (i.e., proteolytic enzymes) is shown in Figure 6–1. Each enzyme splits the polypeptide chain at different and very precise sites.

Some Enzymes Require Cofactors

Some enzymes require small non-protein components called *cofactors* for their activity. For example, some enzymes are conjugated proteins having tightly bound *prosthetic groups,* as in the case of the *cytochromes,* which have an iron-porphyrin complex. Iron atoms are essential in many electron-transfer reactions.

Other enzymes cannot function without the addition of small molecules called *coenzymes,* which become bound during the reaction. When joined with a *coenzyme,* these inactive enzymes, also called *apoenzymes,* form active *holoenzymes.* For example, *dehydrogenases* utilize either nicotinamide-adenine dinucleotide (NAD$^+$) or nicotinamide-adenine dinucleotide phosphate (NADP$^+$). These are among the most important coenzymes.

The function of the coenzyme is to accept two electrons and a hydrogen ion from the substrate, thus oxidizing it:

Substrate + NAD$^+$ + ENZYME →
    oxidized substrate + NADH and H$^+$

The two electrons of NADH can then be transferred to a second molecule, which will become reduced (i.e., it gains electrons).

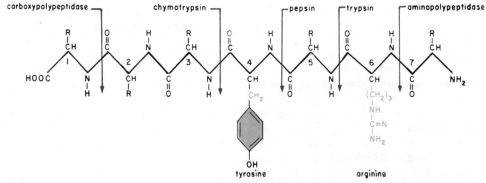

Figure 6–1 A diagram to indicate the specificity of various proteolytic enzymes. The numbers refer to the amino acid residues, of which only two, tyrosine and arginine, are indicated in full detail. The polypeptidases are specific, one to the free carboxyl end (left) of a protein molecule or peptide, the other to the free amino end (right) of such molecules. Pepsin is specific to the amino side of tyrosine (or phenylalanine) residues inside a protein molecule; chymotrypsin is specific to the carboxyl side of such residues; and trypsin is specific to the carboxyl side of arginine or lysine residues. (From Giese, A. C., *Cell Physiology,* 3rd ed., Philadelphia, W. B. Saunders Co., 1968.)

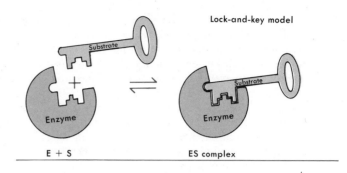

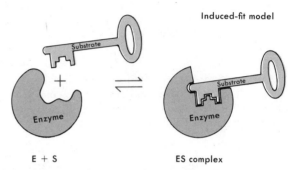

Figure 6-2 Substrates interact with the active site in a precise way. Some enzymes have an *induced-fit:* the shape of the active site is complementary to the substrate only after the substrate is bound.

In the cell the energy-producing catabolic enzymes use NAD as coenzyme; the synthetic processes, however, use NADPH as a hydrogen donor. In many coenzymes, as in NAD^+ and $NADP^+$ (which contain *nicotinamide*), the essential components are vitamins, particularly those of the B group. Some examples are *pantothenic acid* (vitamin B_5), which forms part of the important coenzyme A; *riboflavin* (vitamin B_2), incorporated into the molecules of flavin-adenine dinucleotide (FAD), and *pyridoxal* (vitamin B_6), a cofactor of transaminases and decarboxylases.

Substrates Bind to the Active site

Enzymes have great specificity for their substrates and will frequently not accept related molecules of a slightly different shape. This can be explained by assuming that enzyme and substrate have a *lock-and-key* interaction. As shown in Figure 6–2 the enzyme has an *active site* complementary to the shape of the substrate. If a substrate has a different shape, it will not bind.

Although we can think of enzymes in terms of locks and keys, this does not mean that the active site is a rigid structure. In some enzymes the active site is precisely complementary to the substrate only *after* the substrate is bound, a phenomenon called *induced fit*. As shown in Figure 6–2, the binding of the substrate induces a

conformational change in the protein, and only then will the chemical groups essential for catalysis come in close contact with the substrate. Some enzymes bind preferentially to one out of several possible conformations of the substrate; in this way the flexibility of both enzyme and substrate may contribute to catalysis.

The binding of the substrate to the active site involves forces of a non-covalent nature (ionic and hydrogen bonds, van der Waals forces), which are of very short range. This explains why the enzyme-substrate complex can be formed only if the enzyme has a site that is exactly complementary to the shape of the substrate. The active site is a three-dimensional entity; and because of the folding of the protein chain, the amino acid residues important in the function of the site can be far apart in the linear sequence of amino acids.

In molecular terms one can explain the function of an enzyme in terms of two steps: (1) the formation of the *specific complex* and (2) the *catalytic step proper,* in which the different mechanisms of catalysis, i.e., hydration, dehydration, transfer of groups, and so forth, are produced.

Enzyme Kinetics—Km and Vmax Define Enzyme Behavior

The existence of an enzyme-substrate complex [ES] at the active site was postulated by Michaelis and Menten in 1913 on the basis of kinetic evidence. This concept has been of great importance in the understanding of the mechanism of the enzymatic reactions (see the following discussion). The existence of the [ES] complex was later proven by spectroscopic methods and by the direct isolation of stable covalent derivatives of the complex.

As was mentioned previously, the enzyme-substrate reaction proceeds in two steps (1). The first step can be written as follows:

$$E + S \underset{K_2}{\overset{K_1}{\rightleftarrows}} [ES] \qquad (2)$$

In the second step the [ES] complex breaks down to form the product and the free enzyme, which will now be available for processing a new substrate molecule:

$$[ES] \underset{K_4}{\overset{K_3}{\rightleftarrows}} E + P \qquad (3)$$

(K_1 K_2, K_3, and K_4 are rate constants for the reactions.) All steps are reversible, but in gen-

eral K_4, is negligible, and reaction (3) follows the direction $[ES] \rightarrow E + P$.

As shown in Figure 6–3, the velocity (V) of the reaction depends on the substrate concentration, and the curve describes a hyperbola. At low substrate concentrations, the initial velocity increases rapidly and follows a first order reaction; i.e., the amount of product formed is proportional to the substrate concentration [S]. However, as the [S] increases, the reaction saturates and reaches a point of equilibrium in which the velocity no longer depends on [S]. At this point, because of the great excess of substrate, all the enzyme is in the form of an [ES] complex, and the maximum velocity (Vmax) of the reaction is reached. The equation for this curve is:

$$V = \frac{Vmax \,[S]}{Km + [S]} \qquad (4)$$

This is the Michaelis-Menten equation, which allows for the calculation of the velocity of the reaction for any substrate concentration, provided that Vmax and Km are known. Km is the *Michaelis constant*, which may be defined experimentally as the substrate concentration at which the velocity is half maximal. As shown in

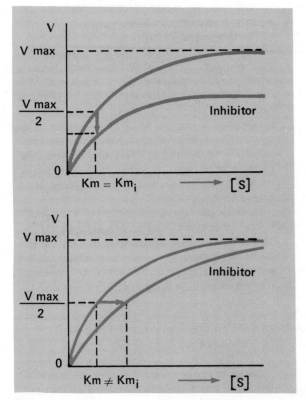

Figure 6–3 Enzyme reaction showing the effect of the substrate concentration [S] on the velocity (V). V max, maximum velocity; Km, Michaelis constant, corresponding to $\frac{V \, max}{2}$. **Upper graph,** the effect of a noncompetitive inhibitor in which $Km = Km_i$. **Lower graph,** the effect of a competitive inhibitor in which $Km \neq Km_i$ (see the description in the text).

Figure 6–3, the Km value can be extrapolated from the point $\dfrac{Vmax}{2}$ on the ordinate. The Km is expressed in moles of substrate per liter. The Km is the concentration of substrate at which half of the enzyme molecules are forming ES complexes. The smaller the value of Km, the greater the *apparent affinity* of the enzyme for the substrate. *The kinetic behavior of the enzyme is defined by the values of Vmax and Km.* From plots such as those of Figure 6–3, however, it is

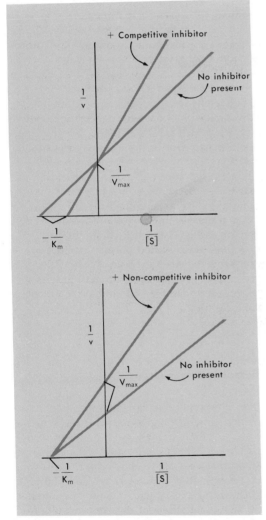

Figure 6–4 Lineweaver-Burk plot of the reactions shown in Figure 4–4. Here the plot is between the reciprocal of the velocity $\dfrac{1}{V}$ and the reciprocal of the substrate concentration $\dfrac{1}{[S]}$. In the presence of a competitive inhibitor V_{max} is unchanged, while with a noncompetitive inhibitor, K_m is unchanged.

difficult to calculate Vmax, since the curve is asymptotic. The calculation is performed more accurately from the double reciprocal, or *Lineweaver-Burk plot,* (i.e., a plot of $\dfrac{1}{V}$ against $\dfrac{1}{[S]}$ [Fig. 6–4]). From this plot the value of $\dfrac{1}{Vmax}$ is obtained at the intercept of the ordinate axis and $-\dfrac{1}{Km}$ at the intercept of the abscissa.

The activity of an enzyme depends on a number of external factors such as temperature, hydrogen ion concentration, and so forth; these factors must be kept constant while the kinetics of the enzyme is studied. With external factors kept constant, and in the presence of an excess of substrate, the reaction catalyzed by an enzyme has a velocity that is proportional to the enzyme concentration.

Enzyme Inhibitors can be Very Specific

Enzyme inhibition may be *reversible* or *irreversible*. There are two major types of reversible inhibition: *competitive* and *noncompetitive*. Competitive inhibition involves a compound similar in structure to the substrate, which forms a complex with the enzyme:

$$E + I \overset{K_1}{\rightleftarrows} [EI] \qquad\qquad (5)$$

where E is the enzyme, I the inhibitor, and K_1 the association constant of the enzyme-inhibitor complex. Unlike the [ES] complex, the [EI] complex does not break down into the products of reaction and the free enzyme. The inhibition of succinic dehydrogenase by malonic acid, the molecular structure of which is very similar to that of succinic acid, serves as an example of competitive inhibition:

$$
\begin{array}{ll}
\text{COOH} & \text{COOH} \\
| & | \\
\text{CH}_2 & \text{CH}_2 \\
| & | \\
\text{CH}_2 & \text{CH}_3 \\
| & \\
\text{COOH} &
\end{array}
$$

Succinic acid   Malonic acid

In this case, both the substrate (succinate) and the inhibitor (malonate) will compete for the active site, and the enzyme activity will be reduced. This inhibition can be reversed, nevertheless, by increasing the substrate concentration, so that the substrate molecules outnumber those of the inhibitor. Therefore, the *Vmax* is not changed by the competitive inhibition (Fig. 6–3) This fact is even more easily observed in the Lineweaver-Burk plot (Fig. 6–4). Note that in competitive inhibition the Km increases, i.e., the apparent affinity of the enzyme for the substrate decreases.

In non-competitive inhibition the inhibitor and the substrate are not structurally related, and the inhibitor binds to a different site than the substrate. Non-competitive inhibition cannot be reversed by high concentrations of the substrate. As shown in Figure 6–4, the Km remains unchanged, but the Vmax is decreased.

Irreversible inhibition involves either the denaturation of the enzyme or the formation of a covalent bond with the enzyme.

Isoenzymes

Isoenzymes are multiple forms of an enzyme which differ by minor variations in amino acid composition and sometimes in regulation. One of the best examples of an isoenzyme is lactic dehydrogenase (LDH), which catalyzes the conversion to pyruvate to lactate. There are five LDH isoenzymes which differ in their electrophoretic mobility in starch gels. LDH is a tetramer which can be formed by two types of subunits: M (predominant in muscle) and H (predominant in the heart). Each subunit is the product of a different gene. The five isoenzymes result from the five possible combinations of these subunits (M_4, M_3H_1, M_2H_2, M_1H_3, and H_4). The relative proportions of the five isoenzymes are characteristic for each tissue and for each stage of embryonic development.

The Cell is not Simply a Bag Full of Enzymes

Enzymes catalyze the thousands of chemical reactions that occur in cells. The enzymes of some pathways are in solution in the cytosol, and the substrates must diffuse freely from one enzyme to the next. However, in other cases the enzymes involved in a chain of reactions are bound to one another and function together as a *multi-enzyme complex*. For example, the seven enzymes that synthesize fatty acids are tightly bound to one another. Similarly, the pyruvate-dehydrogenase complex is formed by three enzymes. Multi-enzyme systems facilitate complex reactions because they limit the distance

through which the substrate molecules must diffuse during the sequence of reactions. The substrate is not released from the complex until all the reactions are completed.

The most complex multi-enzyme systems are associated with biological membranes and the ribosomes. For example, the respiratory enzymes necessary for electron transfer are arranged in a precise way in the inner membrane of mitochondria in eukaryotes, and in prokaryotes they are arranged within the cell membrane. These multi-enzyme systems require this well-defined structure for activity; the enzymes become inactive when removed from the membrane. For this reason, the study of membrane biochemistry is especially difficult.

Enzyme distribution is never random. Some enzymes are packed into lysosomes, and others into secretion granules. Other enzymes, such as the RNA and DNA polymerases, are located in the nucleus and not in the cytoplasm.

Some enzymes are synthesized in an inactive form. This is the case with pancreatic digestive enzymes, which are stored in the secretion granules in inactive forms (for example, trypsinogen and chymotrypsinogen). These *proenzymes* are only activated after excretion into the digestive system. The proenzymes are converted into the active forms (trypsin, chymotrypsin) by the action of proteases that cleave and remove a small segment of the polypeptide chain, thus exposing the active site. This mechanism is useful because it prevents the enzymes from digesting the pancreatic cells. A similar activation of proenzymes by proteolytic cleavage is found in the process of blood clotting.

Allosteric Enzymes have Multiple Interacting Subunits

Not all enzymes display the simple hyperbolic kinetics shown in Figure 6–3, and the V vs. [S] curves are sigmoidal (Fig. 6–5). Enzymes having this kinetic behavior are called *regulatory or allosteric enzymes*. They are *oligomers* containing two *(dimer)*, four *(tetramer)*, or more subunits which are able to intereact with one another. The sigmoidal shape of the curve results from the fact that the binding of the first substrate molecule enhances the affinity for binding the second substrate molecule, and so forth. By observing Figure 6–5 the reader may recognize that there is a region in the curve in which a small increase in [S] causes a very large increase in enzyme activity. Allosteric enzymes are of great regulatory value, since large changes in activity can be obtained by small changes in [S].

Two main models have been proposed to ex-

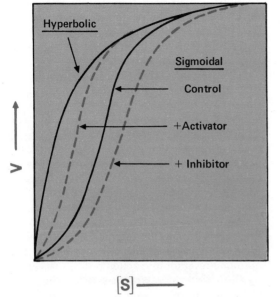

Figure 6-5 Sigmoidal curve characteristic of a regulatory or allosteric enzyme. Observe the difference with an hyperbolic type of curve. Under the action of an activator the curve becomes more hyperbolic, while an inhibitor makes it more sigmoidal.

plain the complex behavior of allosteric enzymes. In the Monod, Wyman, and Changeux model, the oligomeric enzyme can exist in two different conformational states: R and T, in which all the subunits change their conformation in a *concerted* way.

Both states are in equilibrium. In the absence of substrate most of the enzyme moles will be in the T state. Since the substrate binds preferentially to the R state, the addition of substrate shifts the equilibrium toward the R state. In this way the binding of the first molecules of substrate enhances the binding of those that follow, since more enzyme molecules will be in the more active R state.

In the Koshland, Nemethy, and Filmer model the binding of a ligand to one of the subunits induces a sequential *change in conformation*. Since the various monomers interact, the change in conformation of one can modify the behavior of the adjacent one in such a way that the binding of the substrate is enhanced. This allosteric change in a tetrameric enzyme can be represented as shown at the bottom of this page.

Regulatory enzymes are sensitive to the so-called *allosteric modifiers* (i.e., modulators) which act as *inhibitors* or *activators*. Such modifiers do not bind to the active site but to a different site — the allosteric one. The binding of the modifier induces a conformational change in the active site that results in an increased or decreased affinity of the enzyme for the substrate.

Figure 6-5 shows that in the presence of an activator the curve tends toward the hyperbola, while under the influence of an allosteric inhibitor, it is more sigmoidal. As will be shown in the following section, allosteric enzymes are of paramount importance in the regulation of cell metabolism.

6-2 METABOLIC REGULATION

Enzymes are Regulated at the Catalytic and the Genetic Levels

The living cell seldom wastes energy synthesizing or degrading more material than necessary. Therefore, the thousands of chemical reactions that occur inside the cell must be carefully controlled. The regulation of enzyme activity is produced by two major mechanisms: *genetic control* and *control of catalysis* (Fig. 6-6).

Genetic control implies a change in the total amount of enzyme molecules. Best known examples of this form of regulation are *enzyme induction* and *repression* in microorganisms in which the synthesis of the enzyme is regulated at the gene level by the indirect action of certain metabolites. In Chapter 25 we will analyze in detail the cases of the *lactose operon* (induction) and of the *tryptophan operon* (repression) of *E. coli*. We will see that gene expression is regulated by proteins called *repressors* which bind to the DNA, and that small metabolites can affect gene activity by binding to these repressors. As shown in Figure 6-6, in enzyme *induction* the availability of a substrate (e.g., lactose) induces the synthesis of the enzymes that degrade it, while in enzyme repression the accumulation of the end product of a metabolic chain (e.g., tryptophan) turns off production of the enzymes involved in this synthesis. In both cases, the net result is that enzymes are synthesized only when required.

Control of catalysis involves a change in enzyme activity without a change in the total amount of enzyme synthesized. This is frequently produced in regulatory or *allosteric enzymes* by the action of allosteric activators or inhibitors. *Enzyme interconversions* also act on

catalysis. Two important mechanisms of control are *feedback inhibition* and *precursor activation* (Fig. 6–6).

In *feedback inhibition* the end-product of a metabolic pathway acts as an *allosteric inhibitor* of the first enzyme of this metabolic chain. Thus, when enough product is synthesized, the entire chain can be shut off, and useless accumulation of metabolites is avoided. An example is provided by aspartate transcarbamylase, the first enzyme in pyrimidine biosynthesis, which is inhibited by CTP, the end-product of this biosynthetic pathway.

In *precursor activation* the first metabolite of a biosynthetic pathway acts as the *allosteric activator* of the last enzyme of the sequence (Fig. 6–6). For example, glycogen-synthetase is activated by glucose-6-phosphate, a precursor of glycogen.

Genetic control (induction, repression) is generally considered to be a coarse and relatively slow type of regulation; and feedback inhibition, a finer, almost instantaneous way of ensuring that enzyme activity is adequate for cellular requirements.

Enzyme Interconversions Also Regulate Metabolism

Some enzymes may exist in two forms of different activity which are interconvertible. Frequently, the mechanism of interconversion consists of *phosphorylation,* i.e., the covalent binding of a phosphate group that is provided by ATP. One of the best known cases is *glycogen phosphorylase,* the enzyme that degrades glycogen into its glucose units, and which exists in two forms: *a* and *b*. Phosphorylase *a* is a tetrameric protein having the higher activity; *phosphorylase b* is a less active dimer. The *b* form can be converted into the *a* form by the convalent binding of four phosphate groups; the reverse phenomenon takes place by *dephosphorylation.* By this mechanism the cell regulates the rate of glycogen utilization according to the requirements of the organism.

Cyclic AMP —The Second Messenger in Hormone Action

Hormones are molecules that transfer information from one group of cells to another distant tissue. For example, the cells of the anterior lobe of the hypophysis produce hormones that are carried by the bloodstream and control the thyroid gland, the gonads, the adrenal gland, and the growth of cartilage. The molecular bases of

hormonal action were very poorly understood until 1956, when E. W. Sutherland discovered *cyclic adenosine monophosphate* (cAMP), a cyclic nucleotide that has been found to regulate a large number of metabolic processes.

In 3'-5' cyclic AMP the phosphate group is covalently bound to the 3' and 5' carbons of the ribose ring. (Normally in AMP only the 5' carbon is bound to the phosphate; see Fig. 5–10.) This cyclic nucleotide is synthesized by adenylate cyclase, an enzyme that is tightly bound to the cell membrane. Many hormones modify the activity of adenylate cyclase and therefore produce a change in the intracellular level of cyclic AMP. In Sutherland's model of hormonal action (Fig. 6–7) the hormone is regarded as a *first messenger* that interacts with specific receptor sites located in the outer surface of the cell membrane. This hormone-receptor interaction results in a change in the activity of adenylate cyclase. The active site of this enzyme is on the inner surface and using ATP as substrate, it produces cyclic AMP in the cytoplasm. This nucleotide is considered as a *second messenger* that carries the information to the metabolic machinery of the cell. The effect of cyclic AMP depends on the target organ; for example, an increase in cyclic AMP will produce glycogen degradation in the liver and steroid production in the adrenal cortex.

Many hormones are known to act by way of specific receptors that stimulate adenylate cyclase. Among these hormones are: epinephrine (adrenalin), norepinephrine, glucagon, adrenocorticotropic hormone (ACTH), thyroid-stimulating hormone (TSH), melanocyte-stimulating hormone (MSH), parathyroid hormone, luteinizing hormone (LH), vasopressin, and thyroxine.

Each hormone listed has a particular receptor protein. However, all the receptors can interact with the same adenylate cyclase enzyme. Some cells may have more than one receptor (for example, liver cells respond both to adrenalin and to glucagon). As we will see in Chapter 8–2 the receptors are mobile and not permanently bound to the adenylate cyclase; they can diffuse on the lipid bilayer, thus enabling the enzyme to interact with several receptors.

The main way in which cAMP affects metabolism is by stimulating the activity of *protein kinases,* a group of enzymes that catalyze the phosphorylation of proteins:

$$\boxed{\text{protein}} + \text{ATP} \longrightarrow \boxed{\text{protein}} - \textcircled{P} + \text{ADP}$$

(6)

Figure 6–8 shows the way in which cAMP activates protein kinase. The enzyme has two

102

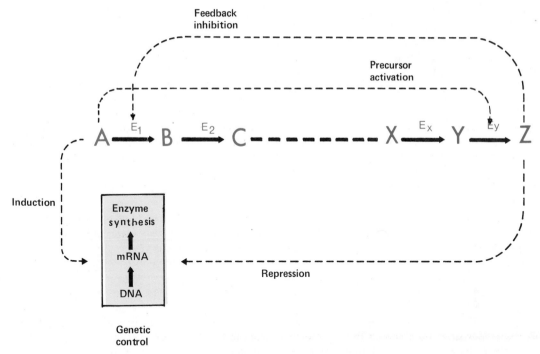

Figure 6–6 Enzymes are regulated at the level of catalysis and at the gene level (enzyme synthesis).

In *feedback inhibition* the end product (Z) of a metabolic chain acts as an allosteric inhibitor of the *first* enzyme of the pathway. In *precursor activation* the first metabolite (A) of a pathway is an allosteric activator of the final enzyme. The *genetic control* mechanisms modify enzyme synthesis according to cellular requirements. In *induction* (e.g., the lactose operon) the presence of a substrate (A) stimulates the synthesis of the enzymes that degrade it, while in *repression* (e.g., the tryptophan operon) the accumulation of the end product (Z) switches off enzyme production.

subunits, one catalytic, and one regulatory which can bind cAMP. In the absence of cAMP the regulatory and catalytic subunits form a complex that is enzymatically inactive. When cAMP binds to the regulatory subunit it causes a conformational change, and the complex dissociates. The catalytic subunit thus freed is enzymatically active and will phosphorylate other enzymes.

The classic example of regulation produced by

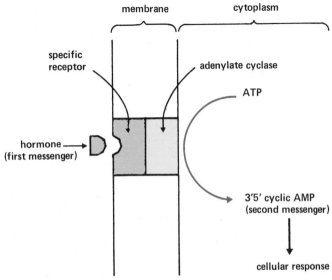

Figure 6–7 Diagram showing the effect of a hormone (first messenger) upon a specific receptor of the cell membrane and its effect on the enzyme adenylate cyclase.

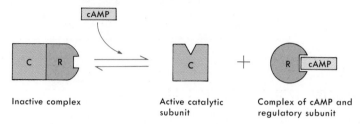

Figure 6-8 Activation of a protein kinase by cyclic AMP. The enzyme has a catalytic (C) and a regulatory (R) subunit, which form an inactive complex. Cyclic AMP binds to the regulatory subunit and induces a conformational change. The free catalytic subunit is the active form, which phosphorylates proteins in the reaction:

cyclic AMP is the degradation and synthesis of glycogen. As shown in Figure 6–9 epinephrine (adrenalin) and glucagon induce glycogenolysis in the liver. These hormones, acting on the corresponding receptor, stimulate adenylate cyclase and raise the cyclic AMP level. Cyclic AMP activates *protein kinase* by the mechanism shown in Figure 6–8. Protein kinase phosphorylates *phosphorylase b kinase,* thus converting it into its active form. This kinase is a specific enzyme that will, in turn, phosphorylate *phosphorylase a. Phosphorylase a* is the active form of the enzyme that degrades glycogen into its glucose units. By this cascade mechanism, the cell amplifies considerably the initial signal given by the hormones at the membrane level (Fig. 6–9).

Glycogen synthesis is also affected by the rise in cAMP. In this case, however, the phosphorylated form of *glycogen synthetase* is the *inactive* form of the enzyme (Fig. 6–9). As a result, an increase in cAMP increases glycogen degradation and coordinately decreases glycogen synthesis in the liver. This is the molecular mechanism by which adrenalin (which is liberated in conditions of alarm or stress) increases the amount of glucose available to the bloodstream.

Cyclic AMP is also involved in certain pathologic conditions that are important in medicine. For example, a toxin of *Vibrio cholerae*, the bacterial agent in cholera, activates adenylate cyclase in the intestine. This stimulates salt and water secretion that may lead to a lethal diarrhea. Another aspect that is being actively investigated

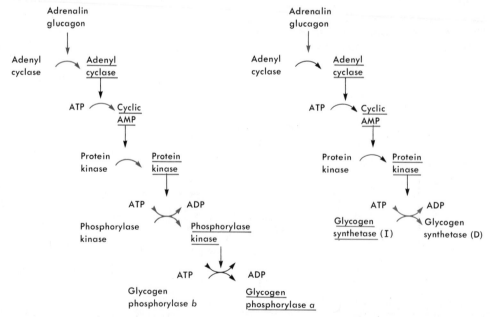

Figure 6-9 Hormones regulate glycogen metabolism by inducing a cascade of reactions mediated by cyclic AMP. The active forms are underlined. Note that phosphorylation activates glycogen phosphorylase (thus increasing glycogen degradation), but inhibits glycogen synthetase. The cascade mechanism greatly amplifies the hormonal signal.

relates to the area of cancer research. It is known that many cancer cells have low levels of cyclic AMP. Furthermore, it is known that certain abnormal features of cancer cells growing in vitro may be restored to normality by the addition of this cyclic nucleotide to the culture medium.

Cyclic GMP is a second type of cyclic nucleotide found in cells. Sometimes (although not always) the levels of GMP are in an inverse relationship to those of cAMP (when cAMP increases, cGMP decreases, and vice versa). The precise function of cGMP is not yet understood.

SUMMARY:

Enzymes in The Cell

Enzymes are proteins that act as biological catalysts, accelerating chemical reactions. They contain a so-called *active site* to which the substrate attaches, forming a temporary [ES] complex; then the substrate is converted into one or more products, and the enzyme becomes free again ($E + S \rightleftarrows [ES] \rightarrow E + P$). The specificity for the substrate is considerable, and frequently analogues of a slightly different shape will not work at all.

The *lock-and-key theory* provides one explanation for the specificity of the active site, which is dependent on the primary, secondary, and even the tertiary structure of the protein. The shape of the active site is complementary to that of the substrate. According to the *induced-fit theory,* the enzyme-substrate interaction may cause a conformational change in the protein. The binding of the substrate to the active site is by short range, non-covalent forces. After the ES complex is formed, the catalytic step, in which the substrate can undergo hydration, dehydration, oxidation, reduction, or transfer of chemical groups (among other processes), proceeds.

Some enzymes require *cofactors* for their activity. For example, the prosthetic group in cytochromes consists of a metalloporphyrin complex. Other enzymes use small non-protein molecules, i.e., *coenzymes,* which become bound during the reaction to activate the enzyme (apoenzyme + coenzyme → holoenzyme). Important coenzymes are NAD^+ and NADP. Vitamins of the B group, such as nicotinamide, pantothenic acid, riboflavin, and pyridoxal, are parts of enzymes or act as cofactors.

In the case of many enzyme-catalyzed reactions, the velocity of the reaction depends on the substrate concentration. The characteristic curve described in these reactions is a hyperbola that reaches a maximum velocity (Vmax) when all the enzyme active sites are saturated. The Km (Michaelis constant) is the substrate concentration at which the velocity is $\dfrac{Vmax}{2}$ (Fig. 6–3). The smaller Km is, the greater is the apparent affinity for the substrate.

Enzyme inhibition may be reversible or irreversible. Reversible inhibition may be *competitive* or *non-competitive.* In competitive inhibition the inhibitor has a molecular structure similar to that of the substrate. In this case, Vmax is not changed, but the Km increases (Fig. 6-3). In non-competitive inhibition the inhibitor is not structurally related to the substrate; Vmax is decreased and Km remains unchanged (Fig. 6-3).

Isoenzymes are multiple molecular forms of the same enzyme that differ in their electrophoretic mobility. Lactic dehydrogenase is a tetramer that has five isoenzymes derived from all the combinations possible between two types of subunits (M_4, M_3H_1, M_2H_2, M_1H_3, and H_4).

In some metabolic pathways the enzymes involved in a chain of reactions are bound to one another in a *multi-enzyme complex.* The substrate passes from one enzyme to the next, without needing to diffuse freely. The most complex multi-enzyme systems are associated with biological membranes, forming a precise two-dimensional array of enzymes.

Allosteric or regulatory enzymes have a sigmoidal V/[S] curve. They are oligomers composed of two (dimer), four (tetramer), or more protein subunits. The binding of a substate molecule to one subunit enhances the affinity for binding a second S molecule, and so forth. These enzymes have a great regulatory value since their activity can be changed by small modifications in the concentration of the substrate [S]. Two models have been suggested to explain this behavior. In one of them a conformational change of the monomers is postulated; in the other, the oligomeric enzyme is thought to exist in two different conformational states (i.e., the subunits can change their shape in a *sequential* or a *concerted* way).

Regulatory enzymes are sensitive to *modifiers* or *modulators* that bind to a separate *allosteric site* which, in turn, influences the active site. In the case of aspartate transcarbamylase, the allosteric inhibitor cytidine triphosphate binds to a different subunit than the substrate.

The *regulation of enzyme activity* is by two major mechanisms: *genetic control* and *control of the catalytic activity*. In genetic control there is a change in the amount of enzyme as, for example, in *enzyme induction* and *repression* (see Chapter 25). Control of catalysis consists of a change in the activity of the enzyme. This control may be achieved by *feedback inhibition, precursor activation,* and *enzyme interconversion* (see Fig. 6–16).

In feedback inhibition the *end-product* of a metabolic pathway acts as an allosteric inhibitor of the *first* enzyme of the pathway, thus ensuring that the cells do not produce more metabolites than necessary.

Some enzymes are interconverted from a less active into a more active form. Frequently, this is done by phosphorylation of the enzyme. For example, *glycogen phosphorylase* (the enzyme that degrades glycogen) is more active when phosphorylated, while *glycogen synthetase* becomes less active when phosphorylated. Proteins are phosphorylated by enzymes, called *protein kinases,* that catalyze the reaction:

$$\text{Protein} + \text{ATP} \longrightarrow \text{Protein} - \textcircled{P} + \text{ADP} \qquad (6)$$

The activity of some protein kinases is stimulated by *cyclic AMP* (Fig. 6–8). The level of cyclic AMP in cells is regulated by *hormones,* which act on specific *receptors* on the outer cell surface, which in turn modify the activity of *adenylate cyclase,* the enzyme that synthesizes cyclic AMP. The cyclic nucleotide is the *second messenger* in the action of many hormones, such as adrenalin, glucagon, ACTH, TSH, MSH, LH, thyroxine, and so forth. Figure 6–9 shows the way in which hormones affect glycogen metabolism, i.e., in a cascade of reactions that involve an increase in cyclic AMP and the phosphorylation of enzymes.

6–3 BIOENERGETICS

The energy which the cell has at its disposal exists as chemical energy primarily locked in high energy bonds. The cell uses only part of the *total energy* (H), also called *enthalpy*, contained in a chemical compound. This portion of the total energy, the *free energy* (G), does not dissipate as heat. Expressed as an energy change:

$$\Delta H = \Delta G + T\Delta S$$

The equation shows that the change in total energy (ΔH) is equal to the change in free or available energy (ΔG) plus the unavailable energy ($T\Delta S$), which dissipates as heat. (In this

equation, T is the temperature in degrees Kelvin, and S is the *entropy* of the system.)

Entropy is Related to the Degree of Molecular Disorder

It is important to have a clear idea of the role of entropy in biological systems. In the preceding equation ΔS is a measure of the irreversibility of a reaction. As the entropy increases, more energy ($T\Delta S$) becomes unavailable, and the process becomes less reversible.

According to the Second Law of Thermodynamics, the entropy of an isolated system of reactions tends to increase to a maximum, at which point an equilibrium is reached and the reaction stops. The concept of entropy is related to the ideas of "order" and "randomness." When there is an orderly arrangement of atoms in a molecule, the entropy is low. The entropy of the system increases when, during a chemical reaction, there is a tendency toward molecular disorder. Thermodynamically, it is well established that the flow of energy proceeds from a higher to a lower level, a phenomenon accompanied by increased entropy. For example, in a cool (low energy) system, the slower moving molecules are better organized in a statistical sense; however, when heat flows from a hotter system and warms up the low energy system, the molecules begin to move faster and the system becomes more disordered. The construction and destruction of a house serves as an illustration of entropy. A great deal of energy (in form of workers' efforts, heat energy, electrical energy, and so forth), is required over a long period of time to build the house, but much less effort is needed to destroy it. The difference between the large amount of energy required to build the house and the much smaller amount needed to destroy it shows that a great deal of energy was lost: this lost energy is entropy.

In any protein molecule (Chapter 5) the sequence of amino acids is very precisely determined. Therefore, the molecule shows a high degree of order and low entropy. The synthesis of such a molecule from the individual amino acids requires considerable amounts of energy or "work" (an *endergonic reaction*). On the other hand, the breakdown of specific proteins into amino acids or into carbon dioxide and water is a highly irreversible process that gives up considerable energy (an *exergonic reaction*).

When energy is forced to flow in a reverse direction (from a lower to a higher level), the entropy decreases. Such processes are thermodynamically unlikely unless they are connected with another system in which the entropy increases accordingly, thus compensating for the decrease. In the plant cell, synthesis of glucose from carbon dioxide and water, simultaneously locking energy derived from the sun into the molecule, is accompanied by a decrease in entropy. On the other hand, in the oxidation of glucose in the animal cell there is a considerable increase in entropy. The interaction of the two systems thus satisfies the second law of thermodynamics, i.e., that entropy must always increase.

These concepts are of great importance in biological systems, since cells are characterized by a high degree of order expressed in their molecular and subcellular structure. When a cell dies, disintegration begins, and entropy increases.

Photosynthesis is Essential in the Biological Energy Cycle

The ultimate source of energy in living organisms comes from the sun. The energy carried by photons of light is trapped by the pigment *chlorophyll*, present in chloroplasts of green plants, and accumulates as chemical energy within the different foodstuffs (see Chapter 14–5). Without the sun, there should be no life on this planet, but interestingly enough, it has been estimated that all life on earth is driven by only 0.24 per cent of the total energy reaching the earth's surface.

All cells and organisms can be grouped into two main classes, differing in the mechanism of extracting energy for their own metabolism. In the first class, called *autotrophs* (i.e., green plants), CO_2 and H_2O are transformed by the process of *photosynthesis* into the elementary organic molecule of *glucose* from which the more complex molecules are made (Fig. 6–10).

The second class of cells, called *heterotrophs* (i.e., animal cells), obtain energy from the different foodstuffs (i.e., carbohydrates, fats, and proteins) that were synthesized by autotrophic organisms. The energy contained in these organic molecules is released mainly by combustion with O_2 from the atmosphere (i.e., oxidation) in a process called *aerobic respiration*. The release of H_2O and CO_2 by heterotrophic organisms completes this cycle of energy (Fig. 6–10).

There is a small group of bacteria that is able to obtain energy from inorganic molecules. For example, the bacteria of the genus *Nitrobacter* oxidize nitrites to nitrates ($NO_2^- + \frac{1}{2} O_2 \rightarrow NO_3^-$). Other bacteria transform ferrous into ferric oxides, and some oxidize SH_2 to sulfate.

Cells Utilize Chemical Energy

The *chemical or potential* energy of foodstuffs is locked in the covalent bonds between the atoms of a molecule. For example, during hydrolysis of a chemical bond (such as a peptide or an ester bond), about 3000 calories per mole is liberated. In *glucose*, between the atoms of C, H, and O, there is an amount of potential energy of about 686,000 calories per mole (i.e., per 180 grams of glucose) which can be liberat-

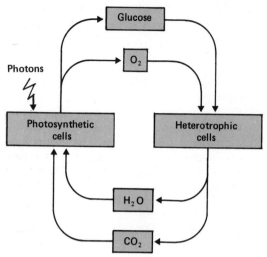

Figure 6–10 Simple diagram of the energy cycle and of the interaction between photosynthetic and heterotrophic cells. (After A. L. Lehninger.)

ed by combustion, as in the following reaction:

$$C_6H_{12}O_6 + 6 O_2 \rightarrow$$
$$6 H_2O + 6 CO_2 + 686{,}000 \text{ calories} \qquad (7)$$

Within the living cell this enormous amount of energy is not released suddenly, as in combustion by a flame. It proceeds in a stepwise and controlled manner, requiring a great number of enzymes that finally convert the fuel into CO_2 and H_2O.

In the engine of a moving car there are great changes in temperature; within the cell this does not occur. Only a part of the energy liberated from foodstuffs is dissipated as heat; the rest is recovered as new *chemical energy*. The energy liberated in the exergonic reactions resulting from the oxidation of foodstuffs is used in the various cellular functions. Thus, energy may be used: (1) to synthesize new molecules (i.e., proteins, carbohydrates, and lipids) by means of *endergonic* reactions (these molecules can then be used to replace others or for the natural growth of the cell); (2) to perform mechanical work, such as cell division, cyclosis (cytoplasmic streaming), or muscle contraction; (3) to carry out *active transport* against an osmotic, or ion gradient; (4) to maintain membrane potentials, as in nerve conduction and transmission, or to produce electric discharges (e.g., in electric fish); (5) in cell secretion, or (6) to produce radiant energy, as in bioluminescence. Only in the reactions of group (1) is the energy provided by the foodstuff transformed into chemical bond energy. In all the

other reactions chemical energy is transformed into other forms of energy.

ATP has High Energy Bonds

Between all these transformations there is a common link, namely, the molecule of *adenosine triphosphate* (ATP). This is a compound found in all cells. Its main characteristic is two terminal bonds with a potential energy much higher than all the other chemical bonds. As shown in Figure 5–10, ATP is composed of the purine base *adenine*, of *ribose*, and of three molecules of *phosphoric acid*. Adenine plus ribose form the nucleoside *adenosine*; this nucleoside, with the first phosphate, forms adenosine monophosphate, a nucleotide. The most important compounds in energy transformation are adenosine diphosphate (ADP) and adenosine triphosphate (ATP). If phosphate is represented by P, the simplified formula of ATP and its transformation into ADP or AMP is as follows:

| Adenosine | ———Ⓟ~Ⓟ~Ⓟ $\rightleftharpoons$ | Adenosine |

$$———Ⓟ~Ⓟ + Pi + 7300 \text{ calories} \qquad (8)$$

or:

| Adenosine | ———Ⓟ~Ⓟ~Ⓟ $\rightleftharpoons$ | Adenosine |

$$———Ⓟ + PPi + 7300 \text{ calories} \qquad (9)$$

Note that the release of any one of the two terminal phosphates of ATP produces about 7300 calories, instead of the 3000 calories from common chemical bonds. The reaction ATP $\rightleftharpoons$ ADP plays a central role between the exergonic processes that liberate energy and those that store or transform energy in the various cellular functions (Fig. 6–11).

The high energy $\sim$ P bond enables the cell to accumulate a great quantity of energy in a very small space and keep it ready for use as soon as it is needed.

Other nucleotides having high energy bonds, such as cytosine triphosphate (CTP), uridine triphosphate (UTP), and guanosine triphosphate (GTP), are involved in biosynthetic reactions. However, the energy source for these nucleoside triphosphates is ultimately derived from ATP. The energy obtained by the transfer of the terminal phosphate of ATP is channeled into the various synthetic processes by the uridine, guanosine, and cytosine triphosphates. Figure 6–11 indicates the nucleoside triphosphates of the ribose and deoxyri-

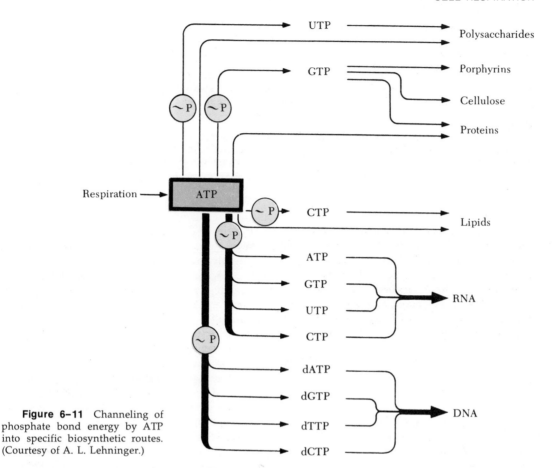

Figure 6-11 Channeling of phosphate bond energy by ATP into specific biosynthetic routes. (Courtesy of A. L. Lehninger.)

bose type (i.e., dATP) that are used as energy sources for the synthesis of important biological compounds. High energy phosphate bonds are also found in phosphocreatine and phosphoenolpyruvic acid.

6-4 CELL RESPIRATION

In the following section we will consider the mechanisms by which organic substances are degraded and the way in which part of this released energy is stored in ATP as high energy bonds. The most common fuel used by the cell is *glucose*; and the way in which it is metabolized will depend on the availability of oxygen. *Anaerobic glycolysis* (fermentation) does not require oxygen, but as a result of this process only a small fraction of the chemical energy of glucose is recovered. On the other hand, in the presence of oxygen, by the process of aerobic respiration, glucose is oxidized to CO_2 and H_2O, with a much higher yield in ATP. Some

of the main differences between aerobic and anaerobic respiration are summarized in Table 6-1.

Anaerobic Glycolysis Yields Only 2 ATPs per Glucose Molecule

Anaerobic glycolysis will degrade the 6-carbon glucose molecule into two 3-carbon lactic acid molecules:

$$C_6H_{12}O_6 \longrightarrow 2\ C_3H_6O_3 \qquad \textbf{(10)}$$
$$\text{glucose} \qquad \text{lactic acid}$$

This process is achieved in 11 successive steps, each one catalyzed by a different enzyme. Glycolytic enzymes are soluble in the cytoplasmic matrix. As shown in Figure 6-12, in this chain of reactions the product of one enzyme serves as substrate for the next reaction. Glucose is first phosphorylated by ATP, with the production of glucose-6-phosphate which is converted into fructose-6-phosphate;

TABLE 6-1 SOME DIFFERENCES BETWEEN AEROBIC AND ANAEROBIC RESPIRATION

Aerobic Respiration (Oxidative Phosphorylation)	Anaerobic Respiration (Fermentation)
Uses molecular O_2	Does not use O_2 as an electron acceptor
Degrades glucose to CO_2 and H_2O	Degrades glucose to trioses and other complex organic compounds
Exergonic	Exergonic
Recovers almost 50 per cent of chemical energy	Recovers less chemical energy
Enzymes localized in mitochondria	Enzymes localized in the cytoplasmic matrix
Yields 36 ATP per glucose molecule	Yields 2 ATPs per glucose molecule

this is then phosphorylated again by ATP (giving fructose diphosphate). After several additional steps, pyruvate is formed. In Figure 6–12 it may be observed that during this part of the chain reaction 4 ATP molecules are produced. Since 2 ATPs were previously used, however, the yield is only two ATP molecules. The general reaction may be written:

$$C_6H_{12}O_6 + 2\,Pi + 2\,ADP \longrightarrow \qquad (11)$$

glucose $\qquad 2\,C_3H_6O_3 + 2\,ATP + 2\,H_2O$

$\qquad$ lactic acid

The fate of *pyruvate*, a key product in glycolysis, depends on whether oxygen is available. In the case of anaerobic conditions, it will be used as a hydrogen acceptor for the 2 NADH generated during glycolysis, and it is converted into *lactate*. Under aerobic conditions pyruvate is converted into *acetyl-coenzyme A* and CO_2 is released. At this point a direct connection with the Krebs cycle is made (Fig. 6–12).

Aerobic Respiration Produces 36 ATPs per Glucose Molecule

The term *aerobic respiration* refers to the series of reactions by which organic substances are broken down to CO_2 and H_2O in the presence of molecular oxygen. The general reaction for the degradation of glucose is:

$$C_6H_{12}O_6 + 6\,O_2 \longrightarrow 6\,CO_2 + 6\,H_2O \qquad (12)$$

glucose

This process releases 686,000 calories of the chemical energy contained in glucose. During anaerobic glycolysis, however, less than 10 per cent of this amount (i.e., 58,000 calories) is released because its final product, lactic acid, is far more complex than CO_2 and H_2O and,

therefore, contains more energy. In the first stage of anaerobic respiration all foodstuffs (carbohydrates, amino acids, and fatty acids) are degraded into acetyl groups (Fig. 6–13). Acetyl groups are transferred bound to *coenzyme A* (A stands for *acetylation* coenzyme), forming a compound called *acetyl-CoA*, which has a central role in metabolism. As do many other coenzymes, CoA has a vitamin as part of its structure, in this case pantothenic acid. During the second stage of aerobic respiration, which takes place in mitochondria, the acetyl groups enter the Krebs cycle, from which CO_2 and hydrogen atoms are produced. Finally, in a third stage the energy contained in the hydrogen is taken up and transferred by the *electron transport or respiratory chain*, and, 36 ATPs per glucose molecule are produced in the process. The hydrogen atoms ultimately combine with molecular oxygen to yield water.

The Krebs Cycle — The Common Final Pathway in the Degradation of Fuel Molecules

The *Krebs cycle*, also called the *tricarboxylic acid cycle*, is the common final pathway of cellular catabolism, in which all "fuel" molecules undergo a final oxidative process. This cycle degrades the acetyl group contained in acetyl coenzyme A to CO_2 and hydrogen atoms:

$$C_2H_4O_2 + 2\,H_2O \longrightarrow 2\,CO_2 + 8\,H \qquad (13)$$

As shown in Figure 6–13, the first step of the Krebs cycle consists of the condensation of the acetyl group (2 carbons) with oxaloacetate (4 carbons) to form citrate, a 6-carbon compound. In the following steps of the cycle, two molecules of CO_2 will be released and oxaloacetate is ultimately formed and used again. The eight hydrogen atoms produced will be used to gen-

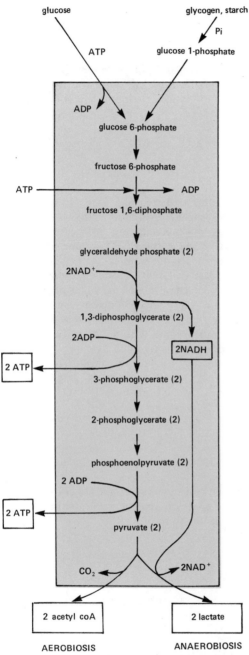

Figure 6-12 Diagram of the degradation of glucose or glycogen by anaerobic glycolysis.

erate ATP molecules by a series of oxidation-reduction reactions in the respiratory chain. This series of reactions illustrates one of the classic examples of nature's economy; the seven enzymes of the Krebs cycle perform their function for *any* type of food which the organism may ingest (Fig. 6–13).

The enzymes of the Krebs cycle are located in the mitochondrial matrix.

The Respiratory Chain—Stepwise Release of Energy by Electron Pairs

The series of reactions comprising the respiratory chain also occur within the mito-

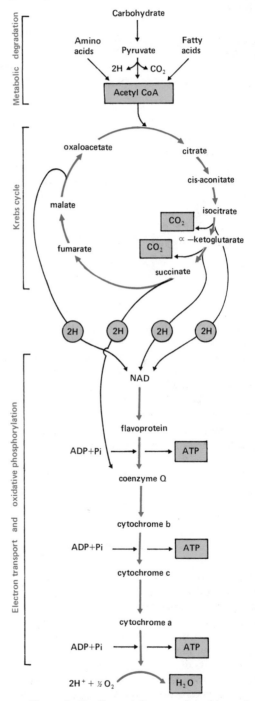

Figure 6-13 General diagram of aerobic respiration showing the Krebs cycle, the respiratory chain, and its coupling with oxidative phosphorylation. (From Lehninger, A. L., *Biochemistry*, New York, Worth Publishers, Inc., 1975.)

chondria and will be described in detail in relation to the molecular structure of the mitochondria in Chapter 12–4.

In the Krebs cycle, specific *dehydrogenases* transfer hydrogen pairs to nicotine adenine dinucleotide (NAD+), thus generating NADH + H+, a key coenzyme in catabolic processes. NADH transfers the hydrogen atoms to the *respiratory chain*, where electrons are transported in a series of oxidation-reduction steps to react, ultimately, with molecular oxygen (Fig. 6–13).

In oxidation-reduction reactions there is transfer of electrons from an electron donor to an electron acceptor. *Oxidation* consists of the loss of electrons; *reduction* consists of the gain of electrons. In many biological oxidations electrons are transferred via hydrogen atoms.

Several of the cytochromes of the respiratory chain are iron-containing molecules. During an electron transfer the iron passes from the ferrous to the ferric state, releasing one electron:

$$Fe^{2+} \rightleftarrows Fe^{3+} + e^-$$

This reaction is the basis of all oxidation-reduction processes. When a pair of electrons is transferred from NADH + H+ to molecular oxygen, a large amount of energy is released. In the respiratory chain this transfer is done in a stepwise fashion in which the electron pairs are passed from one acceptor to another, thus delivering energy more gradually.

The respiratory chain enzymes are located in the inner mitochondrial membrane.

Oxidative Phosphorylation—The Energy Released by Electron Pairs Produces ATP

Coupled with the respiratory chain is the process of *oxidative phosphorylation*. As shown in Figure 6–13, ATP is formed at three steps of the electron transport chain. The equation for this process can be expressed as follows:

$$NADH + H^+ + 3\ ADP + 3\ Pi + \frac{1}{2} O_2 \longrightarrow$$

$$NAD^+ + 4\ H_2O + 3\ ATP \qquad (14)$$

Normally, oxidative phosphorylation is coupled with the respiratory chain; however, these two mechanisms can be separated by the use of *uncoupling agents*. For example, in the presence of 2,4-dinitrophenol, oxygen is consumed, but no ADP is phosphorylated. The molecular mechanisms of oxidative phosphorylation take place in conjunction with the respiratory chain (in the inner membrane of the mitochondrion) and will be described in detail in Chapter 12.

The energy balance of aerobic respiration shows that 36 ATP molecules are produced from each glucose molecule. The overall equation can be written:

$$C_6H_{12}O_6 + 6O_2 + 36\ Pi + 36\ ADP \longrightarrow$$

$$6\ CO_2 + 36\ ATP + 42\ H_2O \qquad (15)$$

The cell will store more than 40 per cent of the chemical energy liberated by the combustion of glucose in the form of ATP.

SUMMARY:

Bioenergetics and Cell Respiration

Bioenergetics deals with the mechanisms by which the cell utilizes chemical energy. A portion of the total energy (H), or *enthalpy*, is dissipated as heat (TΔS), and another portion is retained as free energy (ΔG). Thus, $\Delta H = \Delta G + [T\Delta S]$. In this equation ΔS is a measure of the *entropy* of the system. When entropy is low, there is relative order in the system; when it increases, the system has a tendency to move toward molecular disorder. Flow of energy is related to an increase in entropy. For the synthesis of a protein, molecular energy is used *(endergonic reaction)* and is then released during the catabolism of the protein *(exergonic reaction)*.

The cycle of biochemical energy starts with the *photons* of light trapped by *chlorophyll*. In *autotrophic* organisms (i.e., plants) CO_2 and H_2O are transformed by the process of *photosynthesis* into glucose and other complex molecules. In *heterotrophic* organisms (i.e., animals) the energy is obtained from foodstuffs by combustion with O_2 *(aerobic respiration)* (Fig. 6–10). One mole of glucose (180 grams) contains 686,000 calories. This energy is released in a stepwise fashion and is used for the many functions of the cell. Part of the chemical energy is stored in the ATP molecule, which has two terminal high energy bonds (these bonds contain 7300 calories). The reaction, ATP $\rightleftharpoons$ ADP + Pi, releases the energy contained in the $\sim$ P bond. Other nucleoside triphosphates (i.e., CTP, UTP, and GTP) derive their energy from ATP.

Cell respiration is the series of chemical reactions by which organic substances are degraded and energy is released. In *anaerobic glycolysis* the 6-carbon glucose is degraded to two molecules of lactic acid. This process occurs without O_2 via 11 successive steps that take place in the cytoplasmic matrix. The net result of anaerobic glycolysis is the production of two molecules of ATP. The final product under anaerobic conditions, pyruvate, will be transformed into lactate by acting as the hydrogen acceptor for 2 NADH + H$^+$. In the presence of O_2, pyruvate is converted to CO_2 and acetyl-coenzyme A, which enters the Krebs cycle. Anaerobic glycolysis liberates less than 10 per cent of the energy contained in glucose (58,000 calories).

The *Krebs* or *tricarboxylic cycle*, by a complex series of reactions involving seven enzymes, degrades the acetyl group into CO_2 molecules and H atoms which will be used to generate ATP molecules in the respiratory chain. The Krebs cycle takes place in the mitochondrial matrix.

The *respiratory chain* or *electron transport system* contains a series of oxidation-reduction systems, involving several cytochromes, in which electrons are transferred according to the reaction $Fe^{2+} \rightleftharpoons Fe^{3+} + e^-$. The final cytochrome (cytochrome *a* or cytochrome oxidase) transfers the H atoms to O_2 to produce H_2O. The respiratory chain is coupled with the process of *oxidative phosphorylation* in the inner membrane of the mitochondrion.

The final energy balance of aerobic respiration is as follows:

$$C_6H_{12}O_2 + 6\ O_2 + 36\ Pi + 36\ ADP \longrightarrow 6\ CO_2 + 36\ ATP + 42\ H_2O$$

Thus, in aerobic respiration, 36 ATP molecules are produced from one molecule of glucose. This energy production is equivalent to 40 per cent of the total energy contained in this molecule.

ADDITIONAL READING

Atkinson, D. E. (1966) Regulation of enzyme activity. *Annu. Rev. Biochem, 35*:85.

Baldwin, E. (1967) *Dynamic Aspects of Biochemistry,* 5th Ed. Cambridge University Press, London.

Boyer, P. D. (1970–1975) *The Enzymes,* 3rd Ed., 14 Volumes. Academic Press, Inc., New York.

Boyer, P. D., Chance, B., Ernster, L., Mitchell, P., Racker, E., and Slater, E. C. (1977) Oxidative phosphorylation and photophosphorylation. *Annu. Rev. Biochem., 46*:955.

Dixon, M., and Webb, E. C. (1965) *Enzymes,* 2nd Ed. Academic Press, Inc., New York.

Fersht, A. (1977) *Enzymes, Structure and Mechanism.* W. H. Freeman & Ço., Pubs., San Francisco.

Goldberg, N. D., and Haddox, M. K. (1977) Cyclic GMP metabolism and involvement in biological regulation. *Annu. Rev. Biochem., 46*:823.

Holloway, M. R. (1976) *The Mechanism of Enzyme Action.* Oxford Biology Readers, Vol. 45. Oxford University Press, Oxford.

Koshland, D. E., Jr., Nemethy, G., and Filmer, D. (1966) Comparison of experimental binding data and theoretical models in proteins containing subunits. *Biochemistry, 5*:365. The sequential model of allosteric transitions.

Koshland, D. E., Jr. (1973) Protein shape and biological control. *Sci. Am.* 229(4):52.

Krebs, H. A. (1950) The tricarboxylic acid cycle. *Harvey Lect.,* Ser. 44 (1948–49), p. 165.

Lehninger, A. L. (1971) *Bioenergetics.* W. A. Benjamin, Inc., Menlo Park, Calif.

Lehninger, A. L. (1975) *Biochemistry.* Worth Publishers, Inc., New York.

Monod, J., Changeux, J. P., and Jacob, F. (1963) Allosteric proteins and cellular control systems. *J. Mol. Biol., 6*:306.

Monod, J., Wyman, J., and Changeux, J. P. (1965) On the nature of allosteric transitions. *J. Mol. Biol., 12*:88. The concerted model of allosteric interactions.

Pastan, I. (1972) Cyclic AMP. *Sci. Am , 227*:97.

Perutz, M. (1978) Hemoglobin structure and respiratory transport. *Sci. Am., 239*(6):68. Hemoglobin is the best understood allosteric protein.

Phillips, D. C. (1966) The three-dimensional structure of an enzyme molecule. *Sci. Am., 215*(5):78.

Robinson, A. G., Butcher, W., and Sutherland, E. W. (1971) *Cyclic AMP*. Academic Press, Inc., New York.

Stadtman, E. R. (1970) Mechanisms of enzyme regulation in metabolism. In: *The Enzymes,* 3rd Ed. (Boyer, P. D., ed.) Vol. 1, p. 397.

Stroud, R. M. (1974) A family of protein-cutting proteins. *Sci. Am., 231*(1):24.

Umbarger, H. E. (1978) Amino acid biosynthesis and its regulation. *Annu. Rev. Biochem., 47*:533.

Part 3

SUPRAMOLECULAR STRUCTURE AND THE CELL SURFACE

In the following two chapters we will study how the molecular components presented in Part 2 are organized to form more complex structures.

Chapter 7 deals with important concepts regarding supramolecular structures and the mechanisms involved in their assembly. Several examples, such as those of viruses, collagen fibers, and membranous structures, will be presented to illustrate the principles of molecular assembly that give rise to the biological structures observed with the electron microscope. For this purpose the use of molecular models, such as lipid monolayers and bilayers, and the myelin figures, have been of great importance. A section of this chapter will be dedicated to speculation about the origin of prokaryotic and eukaryotic cells, a problem directly related to the *origin of life* on our planet. Emphasis will be put on the principles of molecular assembly, which might have been at work during the chemical and biological evolution that finally gave rise to the first living cell.

In Chapter 8 the cell membrane and other components of the cell surface will be considered together, as they represent one of the best known examples of supramolecular organization. The surface of the cell will be studied from the point of view of its chemical compositon and the way in which the various molecular species are organized. The important concepts of the fluidity and the molecular asymmetry of the cell membrane will be introduced. The various differentiations of the membrane will be studied, and the important of certain junctions in the coupling of cells will be emphasized. It will be shown that several of the properties of the cell surface are altered in *cancerous cells.* The molecular structure of the cell membrane will be related to its function in the various types of cell permeability. The macromolecules forming the cell coat will be studied from the viewpoint of important properties of the cell surface, such as molecular recognition and cell-to-cell interaction.

SUPRAMOLECULAR ORGANIZATION AND THE ORIGIN OF CELLS

7–1 The Shape of Protein Molecules 118
7–2 The Assembly of Macromolecules 118
   Assembly of Viruses — Nucleic Acids and Proteins
   Collagen Fibers — Supramolecular Assemblies of Tropocollagen
   Fibrinogen and Thrombin in Blood Clotting
   Glycogen Particles — Three Levels of Organization
7–3 Elementary Membranous Structures 123
   Lipids Tend to Form Monolayer Films
   Artificial Lipid Bilayers — Important Model Systems
   Bulk Phospholipid-Water Systems Form Hexagonal and Lamellar Structures

   Liposomes and Phospholipid Vesicles — Possible Applications in Biology and Medicine
   Summary:
   Supramolecular Organization
7–4 The Origin of Cells 127
   Chemical Evolution Produced Carbon-Containing Molecules
   Mechanisms of Assembly were at Work to Form Primitive Proteinoids
   Prokaryotic Cells Preceded the Eukaryotic Cells
   Summary:
   The Origin of Cells

Some of the molecular constituents of the cell, described in Chapter 5, can interact among themselves and become organized into supramolecular units. These units, in turn, are parts of structures recognizable within the cells by means of the electron microscope.

When the molecules are associated linearly, the elementary units are primarily *unilinear* (fibrous); when the molecules are extended in two dimensions forming thin membranes, the units are *two-dimensional;* and when they are crystalline or amorphous particles, the units are *three-dimensional*.

Many of the molecular components are polymers in themselves (e.g., proteins are polymers of amino acids) but, in turn, are monomers of larger units and can polymerize end to end or can interact laterally to form the fibrous, membranous, or crystalline structures.

In certain systems supramolecular structures may aggregate to form higher types of organization visible under the light microscope and even to the naked eye. In animal and plant tissues there are several series of components with this type of organization. They can be classified into three categories: *subcellular*, which comprise parts of cells, such as membranes, cilia, and chromosomes; *extracellular*, such as collagenous and elastic fibers, membranes of cellulose, or chitin situated outside the cells; and *supracellular*, which are macroscopic structures, such as hair, bone, and muscle, with a more complex supramolecular organization.

The function of these supramolecular structures in biological systems will be mentioned throughout this book. Several of these molecular systems are involved in *mechanical functions;* e.g., collagen fibers form tendons; fibrin fibers are used in blood clotting to prevent bleeding; and muscle proteins interact to produce shortening during contraction. Several of these supramolecular structures have *enzymatic properties* and may constitute *multi-enzymatic complexes*. The storing (coding) and transmitting of *genetic information* by nucleic acids is one important function of these complexes. Research has shown that most of the fundamental functions of biological systems, such as osmotic work, association of cells, permeability, and oxidations, are intimately related to these supramolecular structures.

7-1 THE SHAPE OF PROTEIN MOLECULES

The shape of a small molecule is determined by the distribution of the covalent bonds; however, in long polymeric molecules such as proteins and nucleic acids, the secondary bonds (i.e., ionic, hydrogen, and van der Waals bonds) are also important in determining the three-dimensional organization. For example, in a linear molecule internal secondary bonds cause the molecule to bend back upon itself resulting in a compact globular shape. On the other hand, if the bonds are external, as in the α-helix (see Chapter 5), the molecule tends to be extended or fibrous.

Since proteins are most often involved in the formation of the supramolecular structures associated with biological systems, it is important to study the size and shape of these molecules (Table 7-1). In the case of the soluble proteins (see Chapter 5), which are important components of biological tissues, it will be noted that the α-helix content of the polypeptide chain varies from about 100 to 30 per cent. The asymmetry of the molecule is, in general terms, proportional to the α-helix content. For example, the muscle proteins (i.e., tropomyosin, light meromyosin, paramyosin, myosin, and heavy meromyosin) have an α-helix content above 50 per cent and are elongated molecules. On the other hand, the so-called globular proteins have a lower α-helix content and a basically spherical shape.[1]

Both the globular and the elongated protein molecules may associate to form elementary structures of various degrees of complexity. For example, the contraction of a muscle depends on the formation of complexes in which several of the proteins indicated in Table 7-1 are integrated in an elaborate marcromolecular machinery (see Chapter 27).

7-2 THE ASSEMBLY OF MACROMOLECULES

In Chapter 3 the concept of *self-assembly*, by which several protein subunits may form more complex arrangements, was mentioned in conjunction with the quaternary structure of proteins. In hemoglobin, for example two α- and β-chains interact to form the complete molecule (i.e., a tetramer).

Other examples include the case of multi-enzyme complexes, some of which are found free in the cytoplasm (e.g., the huge *pyruvate dehydrogenase* complex of *Escherichia coli,* which contains three groups of enzymes and a total of 88 protein subunits). Other multi-enzyme complexes are embedded in membranes, such as in the case of the respiratory chain in mitochondria (see Chapter 12-4).

In self-assembly, the protein subunits contain the necessary information to produce the larger complex by means of secondary bonds. Complex macromolecules, and even subcellular structures, may be formed in the cell by the principle of self-assembly. In addition to this simple case

TABLE 7-1 MOLECULAR STRUCTURE OF α-PROTEINS*

	Helix Content (%)	Mol. Wt.	Approximate Length
Tropomyosin	>90	53,000	40 nm
Light meromyosin fr. 1	>90	135,000	80 nm
Paramyosin	>90	200,000	140 nm
Myosin	65	530,000	140 nm
Heavy meromyosin	50	350,000	40 nm
Fibrinogen	30	340,000	46 nm
Prekeratin	~40	640,000	
Flagellins	~40	20–40,000	3 to 4 nm
	Globular Proteins		
Myoglobin	70	17,000	3 nm
Bovine serum albumin	45	68,000	5 nm

*(From Cohen, C., *in* Wolstenholme, G. E. W., and O'Connor, M. (eds.), *Ciba Foundation Symposium*, p. 101, London, J. & A. Churchill, 1966.)

of self-assembly, in which no other component is involved, there is also the principle of *aided assembly*, in which certain enzymes may prepare the macromolecules for assembly (e.g., fibrin). Finally, there is the case of *directed assembly*, in which a previous structure is needed for the organization of the new macromolecule. Note that in the duplication of DNA and the transcription of RNA (see Chapters 22 and 24) a template is needed to direct the assembly of the other macromolecule.

Assembly of Viruses – Nucleic Acids and Proteins

We mentioned in Chapter 2–1 that viruses are not cellular organisms and that they are obligatory parasites of either prokaryote or eukaryote cells. When outside the cell, viruses are metabolically inert and may even be crystallized. When they enter the host cell, they use their own genomes to program the replication of new virus particles, but they use the biosynthetic machinery (e.g., ribosomes) of the host cell to express the information that those viral genomes carry. They represent a heterogenous group ranging in size from 30 to 300 nm and contain either DNA or RNA as genetic material and a protein coat or *capsid.*

Viruses represent a beautiful example of structures in which the principles of macromolecular assembly are in action.

The tobacco mosaic virus (TMV), for example, is a particle 40×10^6 daltons in mass, with the form of a cylinder of 16×300 nm. This cylinder contains a single-stranded molecule of RNA consisting of 6500 nucleotides, forming a helix with a radius of 4.0 nm, and having a cylindrical cavity of 2.0 nm. Associated with this RNA helix and forming the protein coat are 2130 identical protein subunits of 18,000 daltons (Fig. 7–1). It has been found possible to dissociate the RNA and the protein subunits and later to reassociate them in order to reconstitute active virus particles. The RNA molecule appears to influence the assembly of the protein subunits.[2]

Many viruses have been observed to display *icosahedral symmetry*. This symmetry depends on the fact that the assembly of the protein subunits (i.e., the *capsomeres)* causes the *capsid* of the virus to be at a state of minimum energy.[3] Such icosahedral symmetry has been found in a virus as small as the ϕX174, which has only 12 capsomeres, and in one as large as the adenovirus, which has 252 capsomeres.

The *bacteriophages* are viruses which attach to

0   10 nm

Figure 7–1 Diagram of the molecular organization of the tobacco mosaic virus. In the center there is a spiral of RNA which is associated with protein subunits. There is one protein monomer for every three bases in the RNA chain. (From Caspar, D. L., and Klug, A., *Cold Spring Harbor Sympos. Quant. Biol.*, 27:1, 1962.)

bacteria and infect them by injecting their own DNA content. An interesting example of a complex virus is the small bacteriophage ϕ 29 of *Bacillus subtilis,* which has the elaborated macromolecular structure shown in Figure 7–2. This phage contains seven major structural proteins.[4] The head is a prolate icosahedron composed of two types of protein subunits with a pentameric and hexameric arrangement, and containing a double-stranded DNA molecule 5.7 μm long. The head is covered by thin fibers composed of another protein. Attached to the head of the phage is a neck with three types of subunits and a tail with a single subunit. A total of 172 protein molecules, of which 145 are in the head capsid, are found in a single phage. By the use of ethylenediaminetetraacetate (EDTA) or dimethyl-sulfoxide it has been possible to produce a progressive disruption of the various parts of the phage. Thus, the head fibers, the tail, and the neck pieces can be detached in a sequential manner. DNA can also be removed, thus leaving the empty capsids.

These various disrupted structures can then be characterized by electron microscopy (Fig. 7–3), and the proteins can be separated by poly-

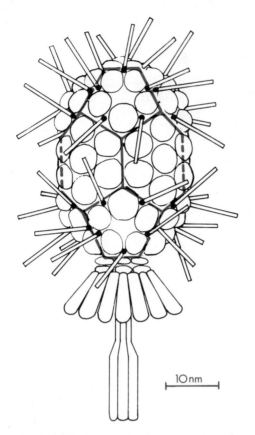

| 10 nm |

Figure 7–2. Diagram of the small bacteriophage ϕ29 of *B. subtilis* having a total of 172 protein molecules of which 145 integrate the capsid. (Courtesy of C. Vásquez.)

acrylamide gel electrophoresis. These various processes constitute a complete biophysical and biochemical characterization of the structural components of this phage.[5]

Collagen Fibers – Supramolecular Assemblies of Tropocollagen

Collagen may be used to illustrate the principles involved in the formation of large molecular complexes. One of the most abundant proteins in the animal kingdom, collagen is synthesized primarily by the fibroblasts and is an important part of major fibrous components of the body, such as skin, tendon, cartilage, and bone. The large aggregates are visible to the naked eye and under the light microscope, but the intimate structure at the molecular level can be studied only by combining electron microscopy, x-ray diffraction, chemical analysis, and other techniques. It has been found, for example, that col-

lagen fibers can be dissociated into smaller and smaller units by the action of acids and then reassembled.

The basic collagen molecule has a molecular weight of 360,000, a length of about 280 nm, and a width of 1.4 nm. It consists of three chains coiled together in a helical fashion as shown schematically in Figure 7–4.[6] It is interesting to recall that collagen has a rather simple amino acid constitution: about one third is glycine, another third is proline and hydroxyproline, and the rest is other amino acids.

The molecular unit of collagen — also called "tropocollagen" — can be considered a macromolecular monomer,[7] because it is capable, by interaction, of forming different collagen structures. The tropocollagen molecule is polarized in the sense of having a definite linear sequence of the amino acid residues in the intramolecular strands. In fact, in relation to its interaction, the tropocollagen molecule behaves as if it had a "head" and a "tail."

The study of native *collagen fibers* with x-ray diffraction and electron microscopy has shown that they are composed of *fibrils* that have a repeating period of 70 nm, which is reduced to 64 nm after drying (Fig. 3–10). The relationship of the tropocollagen molecule of 280 nm to this period of the fibrils has been clarified by reconstituting collagen in the presence of some glycoproteins or ATP (Fig. 7–5). The most probable explanation is that the collagen fibrils — with a period of 70 nm — result from the lateral association of tropocollagen molecules, which overlap at intervals of one fourth their length. It is assumed that in this instance the molecules are longitudinally associated "heads" with "tails" (Fig. 7–5).

In the case of fibrous collagen with long spacing, there is no lateral overlapping, and the tropocollagen molecules are assembled side by side and randomly linked in a linear direction. In the segments with long spacing it is supposed that the tropocollagen molecules do not overlap laterally, and, because they are all in phase, they cannot link longitudinally (Fig. 7–5).[8]

Fibrinogen and Thrombin in Blood Clotting

The important process of *blood clotting* involves fibrinogen and thrombin. The fibrinogen molecule is asymmetrical and has a molecular weight of 340,000.

Under the electron microscope it appears to be composed of three beads, each about 6.5 nm, connected by a very thin strand of 1.5 nm.[9]

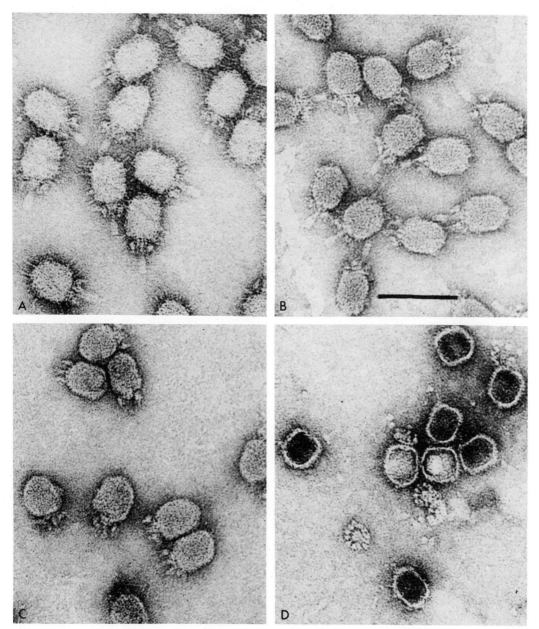

Figure 7-3. Electron micrographs of ϕ29 bacteriophage of *B. subtilis* showing a progressive disruption of the various parts. **A,** normal phages; **B,** phages that have lost the head fibers; **C,** phages that have lost the tail; and **D,** phages that have lost the neck piece. In this last case some capsids have lost their content of DNA. Scale in B = 50 nm. (Courtesy of C. Vásquez.)

Under the action of thrombin, which splits off a small peptide from fibrinogen, fibrinogen is activated and starts to interact with other monomers. The end-to-end association forms long fibrin fibrils, but apparently there is also some lateral staggering and cross-linking with other fibers to form a network. As clotting progresses, aided by the blood platelets,[10] fibrin retracts, squeezing out the serum, and the blood clot is completed.

The Nature of Bonds. The nature of the physicochemical forces involved in these macromolecular interactions varies considerably. For example, the fact that collagen fibrils are soluble in weak organic acids implies that salt linkages and hydrogen bonds are involved.

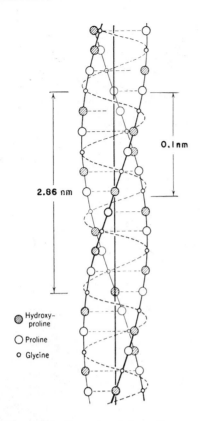

2.86 nm

0.1 nm

● Hydroxy-
 proline

○ Proline

o Glycine

Figure 7–4 Structure of the collagen molecule with the three-stranded helix. (From Rich, A., *Biophysical Science,* New York, John Wiley & Sons, Inc., 1959.)

Stronger bonds, such as $-S-S-$ linkages, are involved in other proteins, such as those forming the various types of keratin fibers. Within the cell, loose and reversible aggregations of corpuscular proteins may occur. These globular-fibrous transformations take place in some processes involving displacement of parts of the cell matrix, such as ameboid motion, cyclosis, or the formation of the mitotic apparatus. The formation of microtubules and microfilaments is generally involved in these transformations (Chap. 9).

Glycogen Particles – Three Levels of Organization

Another interesting example of molecular interaction is observed in the glycogen deposits found in liver cells, muscle, and in many other tissues. The branched structure of the polysaccharides amylopectin and glycogen, is based on 1,6-α-glycosidic bonds, as mentioned in Chapter 5. Electron microscopy has revealed that glycogen particles have three structural levels of organization, each with a characteristic size and morphology.[11, 12] The largest units — called α-particles — are spheroid and measure 50 to 200 nm, with a mean of 150 nm. These particles are composed of smaller units — the β-particles — which are ovoid or polyhedral and measure 30 nm in diameter (Fig. 7–6). Finally, within the

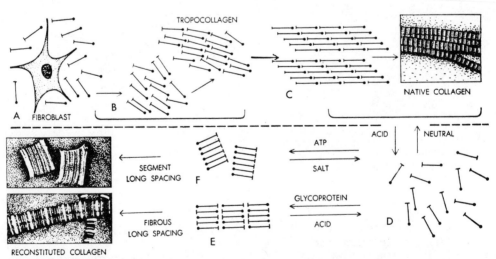

Figure 7–5 Diagram of the formation and reconstitution of collagen. A fibroblast (**A**), manufactures *tropocollagen* molecule (**B**), which form *native collagen* (**C**). Collagen fibrils are solubilized in acid (**D**) and the resulting tropocollagen, in the presence of glycoprotein, produces *fibrous long spacing* (**E**) and, with addition of ATP, *segment long spacing* collagen (**F**). The long spacing of 280 nm results from the lateral aggregation of the tropocollagen molecules without overlapping. The 70 nm spacing of native collagen fibrils is due to the overlapping of the tropocollagen molecule. (Courtesy of J. Gross, 1961.)

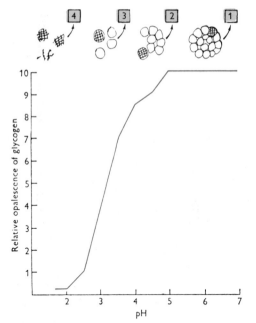

Figure 7–6 The structure of particulate glycogen. At decreasing pH values the glycogen particle dissociates progressively. (See the description in the text.) (From Drochmans, P., *Biochem. Soc. Symp., 23*:127, 1963.)

β-particles a finer structure — the γ-particles — composed of rods of 3 × 20 nm can be observed. These three different units can be demonstrated by acidic treatment of particulate glycogen (Fig. 7–6). Glycogen synthesis is achieved in two successive steps which can be followed under the electron microscope.

7–3 ELEMENTARY MEMBRANOUS STRUCTURES

Biological membranes are known to result from interaction between lipids and proteins, but the molecular arrangement of these two components is very complex. The use of models and artificial monomolecular films has increased our understanding of the natural structures.

Lipids Tend to Form Monolayer Films

The structural importance of certain lipids was mentioned in Chapter 5. Fatty acids, phospholipids, cholesterol, and cholesterol esters can be packed in single layers of constant thickness, and the orientation of the lipids within this structure depends on the dipolar constitution of a polar group and a non-polar hydrocarbon chain (Fig. 7–7, A). These properties of the lipids

can be studied by forming films on the surface of water.

The technique of making *monolayer (monomolecular) films* is of considerable biological importance. The *film balance,* devised by Langmuir in 1917, is still the principal instrument used to study these films. Essentially, it is a shallow trough filled with water on which the substance is spread. A bar or barrier can be pushed across the trough to compress the film. The surface pressure exerted by the film is measured by a sensitive floating, suspended balance. For example, if stearic acid is dissolved in a volatile solvent and deposited on the water, the molecules will spread until they reach an equilibrium. Upon evaporation of the solvent, a film one molecule thick is formed. Because the molecule is bipolar, the polar group ($-COOH$) is attracted by the water molecule, and the non-polar hydrocarbon chain tends to stand straight on the surface. At first, some molecules are not well aligned because of the ample space, but as the barrier is pushed across the trough and the surface area is reduced, the molecules are compressed until they form a packed film (Fig. 7–7, B and C). Under these conditions, molecules exert a pressure that can be measured with the film balance. When the number of molecules and the total surface occupied by the film at the maximum compression are known, the average area of each molecule can be calculated. For example, the stearic acid molecule occupies about 0.2 square nanometer units (Fig. 7–7, A).

By this method the thickness of the monolayer can also be measured (2.5 nm for stearic acid). Thickness depends on the number of carbons in the hydrocarbon chain. The monomolecular film can be deposited on the surface of a glass slide dipped into the water. As shown in Figure 7–7, D, by successive dippings bimolecular or multimolecular layers are built up. Multilayered systems can be obtained that give coherent x-ray diffractions from which the distance or period between the layers can be measured.

Artificial Lipid Bilayers — Important Model Systems

Because of the many experimental applications the so-called lipid bilayers are even of greater interest than the lipid monolayers. In this case an artificial lipid membrane is produced across a small hole separating two aqueous solutions (Fig. 7–8). A droplet of a lipid dissolved in an organic solvent is applied within this hole. After a few minutes, a thin film having

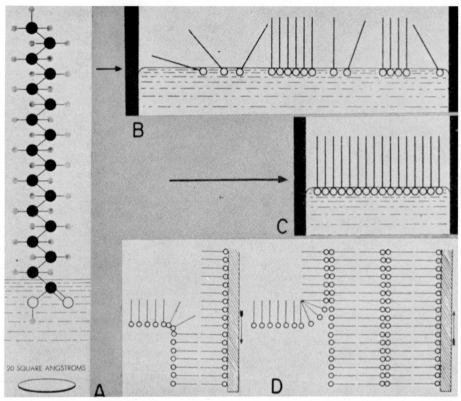

Figure 7–7 Diagram of the technique of making monolayer films. **A,** a molecule of stearic acid with the polar group dipped in water. **B,** at low compression, molecules are oriented at different angles or form packed aggregates. **C,** at high compression, molecules are tightly packed and are vertical. Circles represent polar groups and straight lines the nonpolar hydrocarbon chains. **D,** method of building up molecular films at an air-water interface. *Left,* a glass slide previously coated with a monomolecular film of barium stearate (notice the polar groups attached to the glass surface) is dipped in water that has a monomolecular film at the interface. The second monomolecular layer attaches to the first by the nonpolar ends. *Right,* several bimolecular layers of barium stearate have been deposited on the glass slide by successive dips into the water. (A, B, C, from H. E. Ries, Jr.; D, courtesy of D. Waugh.)

no interference colors (i.e., a black film) is produced.[13]

The apparatus indicated in Figure 7–8, or a similar one, permits the study of the electrical properties of the lipid bilayer as well as its fixation and removal for study under the electron microscope.[14] These artificial membranes are 6 to 9 nm in thickness and have a trilaminar structure, suggesting that they are composed of two layers of lipids, with the hydrophilic groups toward the water interphase and a hydrophobic region in the middle (see below). Lipid bilayers have biophysical properties that are comparable to some biological membranes. They differ, however, in their higher electrical resistance and in the fact that they do not show selectivity for the passage of various ions.[15]

The incorporation of certain polypeptides and proteins (i.e., ionophores) may alter the properties of these membranes considerably. With the addition of these molecules, the membrane may acquire selectivity, electrical excitability, and even chemical receptor properties.[16]

Bulk Phospholipid-Water Systems Form Hexagonal and Lamellar Structures

By mixing phospholipids with water, different bulk systems can be produced. This type of association depends on the concentration of the components and on the temperature.[17] If the phospholipid is anhydrous or the amount of water is low, the lipid molecules tend to form crystals. In a partially hydrated phospholipid a *hexagonal* phase may be formed. This consists of a two-dimensional hexagonal lattice in which cylinders filled with water are embedded in a lipid matrix (Fig. 7–9, *A*).[18] If a hexagonal phase, fixed with osmium tetroxide, is examined under the electron microscope, dense dots corresponding to the water-filled cylinders are observed.

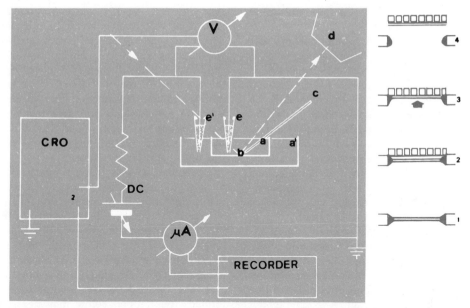

Figure 7–8 **Left,** diagram of the apparatus used to study the electrical properties of artificial membranes. The membrane is formed in a 1 mm hole (*b*) at the bottom of a teflon cup (*a*) immersed in a dish (*a'*). The electrical measurements are done via calomel electrodes (*e, e'*). The membrane is polarized at a constant voltage by means of a DC source. Measurements are made by a microammeter (*μA*) and a voltmeter (*V*). An oscilloscope (*CRO*) and a recorder may be used to register the conductance changes. The pipette (*c*) may be used to study the effect of drugs; (*d*) stereomicroscope. **Right,** diagram of the technique used to remove the artificial membrane and bring it to observation under the electron microscope. *1,* artificial membrane; *2,* placing of the grid; *3,* change in pressure to stick the membrane to the grid; *4,* removal of the grid with the membrane. (From Parisi, Reader, Vásquez, and De Robertis, unpublished.)

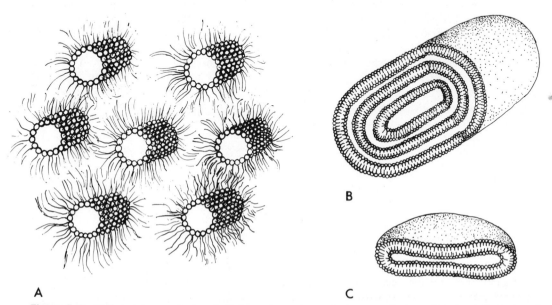

Figure 7–9 Different types of phospholipid-water systems. A, hexagonal phase in a partially hydrated phospholipid. Observe that the water is contained within the cylinders which are limited by polar groups; B, Multilayered smetic mesophase corresponding to one liposome. Water is contained between lipid bilayers; C, Phospholipid vesicle, single bilayered structure of 2 nm. (Courtesy of A. D. Bangham and D. A. Haydon. From Bangham, A. D., and Haydon, D. A., *Br. Med. Bull.,* 24:124, 1968.)

This indicates that the osmium is bound to the polar groups of the lipids.

In the presence of increasing amounts of water (i.e., with more than 50 per cent), *lamellar phases,* generally called *myelin figures,* are formed. Such lamellar phases rearrange in what is called a *smetic mesophase,* which denotes that a substance in a liquid-crystalline state may orient its molecules in parallel planes. Both the hexagonal and lamellar systems, when studied at different temperatures, exhibit special transitions that can be detected by X-ray diffraction, as well as by other physical methods. At these transitions the so-called *thermotropic mesomorphism* occurs. This phenomenon is apparently due to the change of the hydrocarbon chain from a liquid to a solid state. With a temperature above the transition point the mobility of the lipid molecules is increased. In biological membranes it is considered that the phospholipids are in a liquid state at body temperature. In Chapter 8 it will be mentioned that the "fluidity" of the membrane depends on the composition of the membrane, particularly the chain length and degree of saturation of the fatty acids.[19, 20]

Liposomes and Phospholipid Vesicles — Possible Applications in Biology and Medicine

In 1965 Bangham[21] defined a *liposome* as a special type of lamellar phase in which the water is self-contained. If phospholipids extracted from brain or other tissues are mixed with water, then wormlike, concentric, semiliquid structures, which flow from the lipid phase, appear. These structures have a strong birefringence with a radially oriented axis. The lipid molecules are disposed in bimolecular layers attached by their non-polar interfaces.

With time and some mechanical agitation such myelin figures become broken into spheroidal structures of about 2 μm in diameter which are made of several layers separated by water spaces (Fig. 7–9, B). The liposomes may be studied with x-ray diffraction or observed under the electron microscope after fixation with osmium tetroxide. The electron micrographs show alternate parallel light and dark bands, which repeat at approximately 4 nm. The dark bands should be attributed to the osmium deposits in this multi-lamellar structure.

Liposomes possess many properties akin to those of biological membranes and are easy to manipulate; their lipid composition may be varied at will, and many substances may be trapped inside the interlamellar spaces. This can be understood easily by remembering that a dried phospholipid can be swollen by suspension in an aqueous solution containing a molecule that is to be trapped. Once the liposomes have formed, the trapped solute may be separated from the non-trapped one by dialysis or gel filtration.

By sonicating a liposome suspension it is possible to obtain, under certain special conditions, single bilayer vesicles measuring about 2 nm (Fig. 7–9, C). Liposomes and phospholipid vesicles have proven to be useful models for studying some permeability mechanisms, as well as for gaining better information on the mode of interaction between membrane proteins and lipids.[22]

In recent years important applications for liposomes have emerged in the medical field. Liposomes constitute excellent vehicles for carrying different molecules, which are protected within the lipid membrane, into cells and tissues. For example, a missing enzyme can be delivered into an organism that has a hereditary deficiency of that enzyme. Drugs for cancer chemotherapy may be delivered more easily to the sites of attack, and insulin may be given orally with liposomes[23] (see Papahadjopoulos, 1978).

SUMMARY:

Supramolecular Organization

The molecular constituents of cells can be organized into one-dimensional (fibers), two-dimensional (membranes), and three-dimensional (crystals) structures. These structures may be subcellular, extracellular, or supracellular and may have different functions, i.e., *mechanical* (collagen, fibrin), *enzymatic* (multi-enzyme complexes), or *informational* (nucleic acids). Proteins are most often involved in supramolecular structures. There are globular and elongated proteins, and the asymmetry of a molecule is related to the $\propto$-helix content (Table 7–1).

The assembly of macromolecules may be brought about by simple *self-assembly,* in which no other component is involved; by *aided assembly,* in which an enzymatic process may prepare macromolecular parts for assembly, or by *directed assembly,* in which a template directs the formation. Self-assembly is found in some multi-enzyme complexes and in certain virus particles.

In the assembly of macromolecules, the size and shape of the protein subunits may play an important role. In some instances, such as with TMV (tobacco mosaic virus) and with ribosomes, the interaction between RNA and protein may be fundamental to guiding the assembly of the final structure.

Many viruses have an icosahedral symmetry that depends on the assembly of the capsomeres that form the capsid; there are RNA viruses, such as the tobacco mosaic, and DNA viruses, such as the bacteriophages.

Collagen is the result of the assembly of tropocollagen units that are 280 nm in length. The characteristic period of 70 nm of collagen is due to the overlap of one fourth of the tropocollagen molecule. Other examples of macromolecular assemblies are blood clots and glycogen. The bonds involved in supramolecular structures are of various kinds, i.e., hydrophobic interactions, hydrogen bonds, salt linkages, and —S—S— bonds.

Several models of lipid-water and lipid-protein systems may be used to clarify our understanding of natural biological membranes. Lipid monolayers and bilayers, as well as myelin figures or cystalline phospholipid-water phases with a hexagonal configuration, may be produced and analyzed with polarization microscopy, x-ray diffraction, or electron microscopy. Such studies may give information about the relative position of polar and non-polar groups and their association with proteins. Some inferences, based on studies with the electron microscope, may be made about the electron density image.

Some special model systems of a lamellar type are provided by multi-layered liposomes and the single, bilayered phospholipid vesicles. These systems may have important applications in Biology and Medicine.

7–4 THE ORIGIN OF CELLS

In his *Chance and Necessity* Monod considers, under "Molecular Ontogenesis" many of the concepts that have been discussed in this chapter. He explains that the way in which structures of higher and higher order are built in a cell is determined by the genetic information contained in DNA. This determines the primary sequence of the polypeptide chain in the protein, which in turn determines the secondary and tertiary structures, and finally, the formation of the oligomeric complexes. The interaction of different proteins with lipids and nucleic acids, on the other hand, results in the formation of molecular complexes and structures of a higher order of complexity.

A fundamental problem is to determine by which mechanisms supramolecular organization originated on our planet and gave rise to the prokaryotic and eukaryotic cells. Any discussion of this problem should be considered highly speculative since it is directly related to the *origin of life.*

Although we do not know how cells were first formed, from fossil records it has been possible to establish that prokaryotic organisms preceded the eukaryotes and appeared between 3.5 and 3.0 × 10^9 years ago. Before that time there must have been a long period of *chemical evolution* in which carbon-containing molecules and macromolecular precursors, such as amino acids, sugar, and nucleic acid bases, appeared. Then, by polymerization, macromolecules and structures of a higher order of complexity were formed. It is possible that during this period the mechanisms of assembly of macromolecules discussed earlier were at work to give rise to the first self-reproducing supramolecular structures (Table 7–2).

Chemical Evolution Produced Carbon-containing Molecules

In *prebiotic times,* that is, before the origin of life, the atmosphere of the earth was devoid of molecular oxygen, and, as in the case of the major planets Jupiter and Saturn at present, it

TABLE 7–2 EVOLUTIONARY STEPS IN THE ORIGIN OF CELLS

		———————— 4.6×10^9 years ago —
	Biogenic molecules:	Formation of solar system
Chemical	water, ammonia,	
	formaldehyde,	
	hydrogen cyanide,	Electrical discharge,
Evolution	acetonitrile, etc.	ultraviolet light,
		heat, pressure
	Amino acids, sugars,	
	nucleic acid bases	
	↓	
	Proteins	
	Polysaccharides │ Nucleic acids	
	↘ Proteinoids ╱	
Biological		
	↓ Genetic	
Evolution	╱ code	
	First prokaryote ←	between 3.5 and 3.0 $\times 10^9$ years ago
	↓	
	First eukaryote	0.9×10^9 years ago

contained largely hydrogen, nitrogen, ammonia (NH_3), methane (CH_4), carbon monoxide (CO), and carbon dioxide (CO_2). There was also water, as vapor and as liquid covering parts of the earth surface. Although normally these molecules do not react, they could have interacted then because of the energy provided by ultraviolet radiation, heat, and electrical discharges (lightning). This gave rise to highly reactive intermediary molecules such as acetaldehyde, hydrogen cyanide, formaldehyde, and others, from which final products could have been synthesized (e.g., acetic acid, a simple fatty acid; or alanine, a simple amino acid).

By 1920 Oparin in Russia and Haldane in Britain postulated that these molecules, by polymerization, could have given origin to the proteins, nucleic acids, and carbohydrates found in living organisms. In 1953 Stanley Miller performed a fundamental experiment using conditions imitating those of the atmosphere of prebiotic times. He produced electrical discharges in a flask into which water vapor, H_2, CH_4, and NH_3 had been injected. In the condensed water he found that several amino acids, such as glycine, alanine, and glutamic and aspartic acids, had formed. So far 17 amino acids (of the 20 present in proteins) have been obtained in experiments simulating prebiotic conditions. Other compounds that have been formed are: sugars, fatty acids, and the bases that form part of the nucleic acids. For example, adenine, which is present in DNA, RNA, and ATP, has been produced in high yields. It is interesting that carbon-containing molecules have also been found in terrestrial lava, in meteorites, and in moon rock samples.

Mechanisms of Assembly Were at Work to Form Primitive Proteinoids

The next step, *biological evolution*, was probably the polymerization of amino acids to form proteins. This could have been initiated by the catalytic action of clays. All the following prebiotic evolutionary processes probably occurred in a water medium (i.e., in ponds) in which the organic molecules were concentrated, forming a kind of "soup" in which molecular interactions were favored. Once the first protein had been formed, the mechanisms of assembly described earlier in this chapter could have operated. The primary sequence of amino acids could have originated the secondary, tertiary, and — with the formation of oligomers — even the quaternary structure of proteins. In this way the enzymatic functions could have arisen. In the primitive soup the macromolecules probably formed larger complexes called *coacervates* or *"proteinoid" droplets*, which have a membrane-like outer surface and a fluid interior.

Primitive proteinoids could have exhibited enzymatic and transport activities, as in the case of the artificial membranes described above. These proteinoids, without nucleic acids to act as informational molecules, could not have had genetic continuity; thus, it is possible that extensive trial and error occurred, leading to short-lived structures.

Prokaryotic Cells Preceded the Eukaryotic Cells

Only after the origin of the genetic code, determined by the sequence of bases in nucleic acids, could a self-perpetuating organism have arisen in which the laws of natural selection had begun to operate. At that point, a first prokaryote with the minimum living mass (see Chapters 2–1) was originated and life emerged on Earth. It is possible that the first prokaryotes were *heterotrophic* (i.e., they drew their nutrients from the organic molecules) and *anaerobic,* since there was no oxygen in the atmosphere. At a later time *autotrophic* prokaryotes, such as the blue-green algae, that have photosynthetic pigments appeared. Because of photosynthesis, oxygen was produced, and it accumulated in the atmosphere, allowing the formation of *aerobic* prokaryotic cells.

It is only after the appearance of the autotrophic prokaryotes that the eukaryotic cell could have originated. From fossil records it is surmised that eukaryotic organisms appeared 0.9×10^9 years ago. The possible evolution of prokaryotes to eukaryotic cells is studied in the symbiont hypothesis of mitochondria and chloroplasts, which is discussed in Chapters 12 and 14. (Further readings on the origin of cells can be found in Bernal and Synge, 1973; and Fox and Dose, 1972.)

SUMMARY:

The Origin of Cells

From fossil records it has been determined that prokaryotic cells preceded eukaryotic and appeared on our planet between 3.5 and 3.0×10^9 years ago. In prebiotic times there was probably a long period of *chemical evolution* in which H_2, NH_3, CH_4, and CO_2, contained in the atmosphere interacted to form carbon-containing compounds. Experiments using conditions imitating those of the atmosphere at that time have yielded sugars, fatty acids, nucleic acid bases (e.g., adenine), and several amino acids. This step, chemical evolution, was probably followed by another, *biological evolution,* in which proteins were formed by polymerization, and aggregates, called coacervates or proteinoid droplets, appeared in the water medium. With the origin of the genetic code (nucleic acids), self-perpetuating cells could have arisen. It is possible that heterotrophic anaerobic prokaryotes preceded the autotrophic ones, such as blue-green algae. The production of oxygen by photosynthesis allowed the reproduction of aerobic cells. Eukaryotic cells probably first appeared 0.9×10^9 years ago.

REFERENCES

1. Cohen, C. (1966) *Ciba Foundation Symposium,* p. 101. (Wolstenholme, G. E. W., and O'Connor, M., eds.) J. & A. Chuchill, London.
2. Klug, A. (1972) *Fed. Proc., 31*:30.
3. Caspar, D. L., and Klug, A. (1962) *Cold Spring Harbor Symp. Quant. Biol., 27*:1.
4. Mendex, E., Ramirez, G., Salas, M., and Viñuela, E. (1971) *Virology, 45*:567.
5. Salas, M., Vásquez, C., Mendez, E., and Viñuela, E. (1972) *Virology, 50*:180.
6. Crick, F. H. C., and Rich, A. (1957): In: *Recent Advances in Gelatine and Glue Research,* p. 20. Pergamon Press, London.
7. Gross, J. (1961) *Sci. Am., 204*:120.
8. Hodge, A. J., and Schmitt, F. O. (1958) *Proc. Natl. Acad. Sci. USA, 44*:418.
9. Hall, C. E. (1963) *Lab. Invest., 12*:998.
10. DeRobertis, E., Paseyro, P., and Reissig, M. (1953) *Blood J. Hematol., 8*:7.
11. Drochmans, P. (1963) In: Methods of separation of subcellular structural components. *Biochem. Soc. Symp., 23*:127.
12. Drochmans, P. (1968) *Excerpta Medica Internat. Cong. Ser., 166*:49.
13. Mueller, P., Rudin, D. O., Ti Tien, H., and Wescott, W. C. (1963) *J. Phys. Chem., 67*:534.
14. Vásquez, C., Parisi, M., and DeRobertis, E. (1971) *J. Membr. Biol., 6*:353.
15. Henn, F. A., and Thompson, T. E. (1969) *Annu. Rev. Biochem., 38*:341.
16. Schlieper, P., and DeRobertis, E. (1977) *Biochem. Biophys. Res. Commun., 75*:886.

17. Luzzati, V., Reiss-Husson, F., and Saludjian, P. (1966) *Ciba Foundation Symposium,* p. 69. (Wolstenholme, G. E. W., and O'Connor, M., eds.) J. & A. Churchill, London.
18. Luzzati, V. (1968) *Biological Membranes* (Chapman Edition), p. 71. Academic Press, Inc., New York.
19. Chapman, D. (1969) *Lipid Res., 4*:251.
20. Engelman, D. (1970) *J. Molec. Biol., 47*:115.
21. Bangham, A. D., Standish, M. M., and Watkins, J. C. (1965) *J. Molec. Biol., 13*:238.
22. Bangham, A. D., Hill, M. W., and Miller, N. G. M. (1974) In: *Methods in membrane biology.* (Korn, E., ed.) Plenum Press, London, *1*:1.
23. Gregoriadis, G. (1977) *Nature, 265*:407.

ADDITIONAL READING

Bernal, J. D., and Synge, A. (1973) The origin of life. In: *Readings in Genetics and Evolution.* The Clarendon Press, Oxford.

Cavalier-Smith, T. (1975) The origin of nuclei and eukaryotic cells. *Nature, 256*:463.

De Terra, N. (1978) Some regulatory interactions between cell structures at the supramolecular level. *Biol. Rev., 53*:427.

Eigen, M. (1971) Molecular self-organization and the early stages of evolution. *Q. Rev. Biophys., 4*:149.

Fenner, F. (1968) *The Biology of Animal Viruses.* Academic Press, Inc., New York.

Fox, S., and Dose, K. (1972) *Molecular Evolution and the Origin of Life.* W. H. Freeman, San Francisco.

Gregor, H. P., and Gregor, C. D. (1978) Synthetic membrane technology. *Sci. Am., 239*:112.

Klug, A. (1972) Assembly of tobacco mosaic virus. *Fed. Proc., 31*:30.

Miller, A., and Parry, D. A. D. (1973) Structure and packing of microfibrils in collagen. *J. Molec. Biol., 75*:441.

Monod, J. (1971) *Chance and Necessity.* Random House, Inc., New York.

Oparin, A. I., ed. (1969) *Proceedings of the First International Symposium on the Origin of Life on the Earth.* Pergamon Press, Oxford.

Oparin, A. I. (1974) Evolution of the concepts on the origin of life: Seminar on the origin of life. Moscow, 1974.

Oparin, A. I. (1978) The origin of life. *Scientia, 113*:7.

Papahadjopoulos, D. (1978) Liposomes and their uses in biology and medicine. Ann. N. Y. Acad. Sci., *308.* New York.

Schopf, W. (1978) Chemical evolution and the origin of life. *Sci. Am., 239*:70.

Spiegelman, S. (1971) An approach to the experimental analysis of precellular evolution. *Q. Rev. Biophys., 4*:213.

Stoeckenius, W., and Engelman, D. M. (1969) Current models for the structure of biological membranes. *J. Cell Biol., 42*:613.

Tanford, C. (1973) The Hydrophobic Effect: formation of micelles and biological membranes. Wiley-Interscience, New York.

Wolstenholme, G. E. W., and O'Connor, M., eds. (1966) *Principles of Biomolecular Organization. Ciba Foundation Symposium.* J. & A. Churchill, London.

THE CELL MEMBRANE AND PERMEABILITY; INTERCELLULAR INTERACTIONS

8-1 Molecular Organization of the Cell
Membrane 132
    The Cell Membrane—Composed of Proteins, Lipids, and Carbohydrates
    Lipids are Asymmetrically Distributed within the Bilayer
    Carbohydrates—In the Form of Glycolipids and Glycoproteins
    Membrane Proteins—Peripheral or Integral Polypeptides of the Red Cell Membrane
    Every Protein of the Cell Membrane is Distributed Asymmetrically
    Major Polypeptides of the Red Cell Membrane are Well Characterized
    Asymmetrical Distribution of Enzymes
    Summary: *Molecular Organization of the Cell Membrane*

8-2 Molecular Models of the Cell Membrane
139
    Unit Membrane Model—Re-evaluation of the EM Image
    The Fluid Mosaic Model is Now Generally Accepted
    Membrane Fluidity—Studied with Physical and Biological Techniques
    Membrane Fluidity and Coupling of Receptors to Adenylate Cyclase
    The Myelin Sheath and the Photoreceptors—Special Multi-layered Membranes
    Summary: *Membrane Molecular Models*

8-3 Cell Permeability 148
    Different Ionic Concentrations across the Membrane Create Electrical Potentials
    Passive Permeability—Dependent on the Concentration Gradient and the Partition Coefficient
    Passive Ionic Diffusion—Dependent on Concentration and Electrical Gradients
    Active Transport of Ions Uses Energy
    A "Sodium Pump" is Postulated in the Active Efflux of Na^+

    Ionic Transport through Charged Pores in the Membrane
    Anion Transport in Erythrocytes Involves the Special Band-3 Polypeptide
    The Vectorial Function of Na^+ K^+ ATPase— The Carrier Hypothesis
    Various Substances are Transported by a Carrier Mechanism
    Selectivity of Transport—Dependent on Permease Systems
    Penetration of Large Molecules—Various Mechanisms
    Summary: *Cell Permeability*

8-4 Differentiations at the Cell Surface and
Intercellular Communications 158
    Microvilli—A Greatly Increased Cell Membrane Surface Area
    Desmosomes, Intermediary and Tight Junctions—Intercellular Attachments
    Gap Junctions (Nexus) and Intercellular Communications
    Electrical Coupling between Cells Depends on Gap Junctions
    Gap Junctions—Channels Permeable to Ions and Small Molecules
    Coupling between Cells Enables Metabolic Cooperation
    Altered Coupling in Cancer Cells
    Summary: *Differentiations of the Cell Membrane and Intercellular Communications*

8-5 Coats of the Cell Membrane and Cell
Recognition 166
    Numerous Functions are Attributed to the Cell Coat
    Cell-Cell Recognition—Specific Cell Adhesion and Contact Inhibition
    Cancer Cells—Many Changes in the Cell Surface Properties
    Transformation of Cells—Produced by Certain Viruses
    Summary: *Cell Coats and Cell Recognition*

The cell has a different internal milieu from that of its external environment. For example, the ionic content of animal cells is quite dissimilar from that of the circulating blood. This difference is maintained throughout the life of the cell by the thin surface membrane, the *cell* or *plasma membrane*, which controls the entrance and exit of molecules and ions. The function of the plasma membrane of regulating this exchange between the cell and the medium is called *permeability*.

This membrane is so thin that it cannot be resolved with the light microscope, but in some cells it is covered by thicker protective layers that are within the limits of microscopic resolution. For example, most plant cells have a thick cellulose wall that covers and protects the true plasma membrane (Fig. 1–1). Some animal cells are surrounded by cement-like substances that constitute visible cell walls. Such layers, also called *cell coats*, generally play no role in permeability but do have other important functions.

In this chapter we will consider first how the cell membrane is organized *at the molecular level*. This discussion will present the *"fluid mosaic" model*, which dominates present views of membrane organization.[1] With this structure as a base we will consider various aspects of *cell permeability*. Then, in relation to certain *differentiations of the cell membrane* we shall examine the important problem of *intercellular communications*. Finally, together with the study of the *cell coat*, we will analyze some of the mechanisms by which cells *recognize* their partners in a tissue. It will be mentioned that changes in intercellular communication and *molecular recognition* are of paramount importance in the *cancerous transformation of cells*. The cell membrane is becoming one of the fields most fundamental to the cell biology of both the normal and the neoplastic cell.

8–1 MOLECULAR ORGANIZATION OF THE CELL MEMBRANE

In the study of the molecular organization of the cell membrane, the first step is to isolate it from the rest of the cytoplasm in the purest form possible. The isolated membrane is then studied by biochemical and biophysical methods.

Several methods have been used to isolate plasma membranes from a variety of cells. In most cases the purity of the fraction has been controlled by electron microscopy, enzyme analysis, the study of surface antigens, and other criteria.[2]

Plasma membranes are more easily obtained from erythrocytes subjected to hemolysis. The cells are treated with hypotonic solutions that produce swelling and then loss of the hemoglobin content (i.e., *hemolysis*). The resulting membrane is generally called a *red cell ghost*. Two main types of ghosts may be produced: *resealed ghosts* and *white ghosts*. The so-called resealed ghosts are produced when hemolysis is milder; the ghosts can be treated with substances that produce restoration of the permeability functions (i.e., resealing). White ghosts are formed if hemolysis is more drastic. There is complete removal of the hemoglobin, and the ghosts can no longer be resealed. These ghosts can be used for biochemical, but not physiological, studies.

Figure 8–1 is meant to introduce some of the concepts about the molecular organization of the red cell membrane which will be dealt with in more detail later on. Free-floating erythrocytes and a red cell ghost cut in half are represented as seen under the light microscope; there is also a three-dimensional view of the membrane at the molecular level, in which the "fluid mosaic" model is represented. It is important to observe (1) that the membrane is formed by a rather continuous bilayer of lipids into which protein complexes are embedded in a kind of "mosaic" arrangement, (2) that there are other proteins which are peripheral to the bilayer and disposed on the inner surface (from (1) and (2) it can be concluded that the membrane is highly asymmetrical), and (3) that the molecular asymmetry is further emphasized by the oligosaccharide chains that protrude only at the outer surface of the membrane.

The Cell Membrane — Composed of Proteins, Lipids, and Carbohydrates

More is known about the chemical composition of the plasma membrane of the human red cell than about that of any other cell. Protein represents approximately 52 per cent of its mass, *lipids* 40 per cent, and *carbohydrates* 8 per cent. Oligosaccharides are bound to lipids (i.e., *glycolipids*) and, mainly, to proteins (i.e., *glycoproteins*).

From the data shown in Table 8–1 it is evident that there is a wide variation in the lipid-protein ratio between different cell membranes. Myelin is an exception, in the sense that the lipid predominates; in the other cell membranes there is higher protein/lipid ratio. In Table 8–1 it may be observed that in myelin the area occupied by the protein is insufficient to cover that of the lipids, whereas in a red cell ghost the opposite situation is found.

Lipids are Asymmetrically Distributed within the Bilayer

The main lipid components of the plasma membrane are phospholipids, cholesterol, and galactolipids; their proportion varies in different cell membranes.

The major proportion of membrane phospho-

lipids is represented by phosphatidylcholine, phosphatidylethanolamine, and sphingomyelin, all of which have no net charge at neutral pH (i.e., *neutral phospholipids*) and tend to pack tightly in the bilayer (see Chapter 7). (This property is also shared by cholesterol.) Five to 20 per cent of the phospholipids are acidic, including: phosphatidylinositol, phosphatidylserine, car-

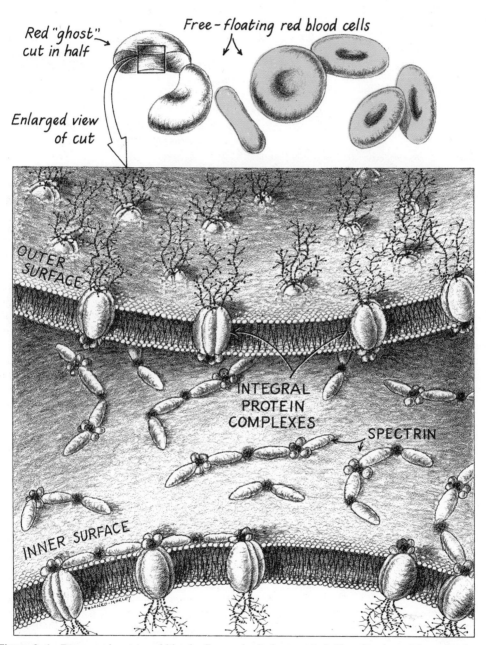

Figure 8–1 Diagram showing red blood cells, a red cell ghost cut in half, and a view of the molecular organization of the membrane according to the "fluid mosaic" model. Integral protein complexes are represented with the oligosaccharide chains sticking out on the outer surface. At the inner surface the peripheral protein spectrin is represented. (Courtesy of G. L. Nicolson, 1978.)

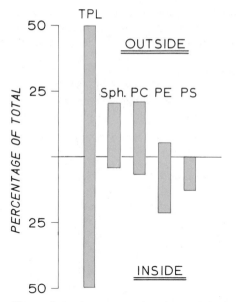

Figure 8–2 Asymmetric distribution of phospholipids between the inner and outer layer of the human erythrocyte membrane. *TPL,* total phospholipids; *Sph,* sphingomyelin; *PC,* phosphatidyl choline; *PE,* phosphatidyl ethanolamine; *PS,* phosphatidyl serine. (From L. van Deenen et al., 1974.)

cytoplasm, and the outer or *external surface* (E_s) in contact with the surrounding fluid medium (Fig. 8–1).

Using non-permeant reagents[3] and various *phospholipases* (i.e., enzymes that hydrolyze different parts of the phospholipid molecule), it has been demonstrated that the distribution of the *phospholipids* is highly asymmetrical (Fig. 8–2). While the *outer layer* consists mainly of lecithin and sphingomyelin, the *inner layer* is composed mainly of phosphatidylethanolamine and phosphatidylserine. Furthermore, the *glycolipids* are mainly in the outer half of the bilayer. It is assumed that this asymmetry is rather stable and that there is no exchange of lipids across the bilayer.[3, 4]

Carbohydrates — In the Form of Glycolipids and Glycoproteins

The distribution of the oligosaccharides is also highly asymmetrical. In both *glycolipids* and *glycoproteins* they are confined exclusively to the external membrane surface. In erythrocyte membranes, hexose, hexosamine, fucose, and sialic acid are bound mainly to proteins. In fact all the proteins present at the outer surface are glycosylated. (see Steck and Hainfeld, 1977).

Because of the presence of sialic acid residues, as well as carboxyl and phosphate groups, the outer surface of the membrane is negatively charged; consequently, positively charged proteins may be bound by electrostatic interactions to the plasma membrane. Only a small amount of sialic acid exists in the form of *gangliosides* (i.e., glycolipids) in the plasma membrane of

diolipin, phosphatidylglycerol, and sulfolipids. *Acidic phospholipids* are negatively charged and in the membrane are associated principally with proteins by way of lipid-protein interactions.[2]

One of the main characteristics of the molecular organization of the plasma membrane is the asymmetry of all of its chemical components. This refers to their non-uniform distribution between the two surfaces, i.e., the inner or *protoplasmic surface* (P_s), in contact with the ground

TABLE 8–1 LIPID AND PROTEIN RATIOS IN SOME CELL MEMBRANES*

	Species and Tissue	Protein (%)	Lipid (%)
Human	CNS myelin	20	79
Bovine	PNS myelin	23	76
Rat	Muscle (skeletal)	65	35
Rat	Liver	60	40
Human	Erythrocyte	60	40
Rat	Liver mitochondrion	70	27–29

		Molar ratio			Area ratio
	Amino Acid	Phospholipid	Cholesterol		Protein: Lipid
Myelin	264	111	75		0.43
Erythrocyte	500	31	31		2.0

*Triggle, D. J., *Neurotransmitter-Receptor Interactions,* New York, Academic Press, Inc., 1971, p. 122.

TABLE 8–2 CRITERIA FOR DISTINGUISHING PERIPHERAL AND INTEGRAL MEMBRANE PROTEINS*

Property	Peripheral Protein	Integral Protein
Requirements for dissociation from membrane	Mild treatments sufficient: high ionic strength, metal ion chelating agents	Hydrophobic bond-breaking agents required: detergents, organic solvents, chaotropic agents
Association with lipids when solubilized	Usually soluble — free of lipids	Usually associated with lipids when solubilized
Solubility after dissociation from membrane	Soluble and molecularly dispersed in neutral aqueous buffers	Usually insoluble or aggregated in neutral aqueous buffers
Examples	Cytochrome *c* of mitochondria; Spectrin of erythrocytes	Most membrane-bound enzymes; histocompatibility antigens; drug and hormone receptors

*From Singer, S. J., "The Molecular Organization of Biological Membranes," *in* Rothfield, L. I., (Ed.). *The Structure and Function of Biological Membranes*, New York, Academic Press, Inc., 1971, p. 145.

liver. However, gangliosides are important constituents of the neuronal surface and are probably involved in ion transfers.[5]

Membrane Proteins — Peripheral or Integral

Proteins represent the main component of most biological membranes (Table 8–1). They play an important role, not only in the mechanical structure of the membrane, but also as carriers or channels, serving for transport; they may also be involved in regulatory or ligand-recognition properties. In addition, numerous enzymes, antigens, and various kinds of receptor molecules are present in plasma membranes.

Membrane proteins have been classified as *integral (intrinsic)* or *peripheral (extrinsic)* according to the degree of their association with the membrane and the methods by which they can be solubilized[1, 6] (Table 8–2).

Peripheral proteins are separated by mild treatment, are soluble in aqueous solutions, and are usually free of lipids. Examples include: the above-mentioned *spectrin*, which may be removed from red cell ghosts by chelating agents;[7] cytochrome *c*, found in mitochondria; and acetylcholinesterase, in electroplax membranes, which are easily removed in high salt solutions.[8]

Integral proteins represent more than 70 per cent of the two protein types and require drastic procedures for isolation. Usually they are insoluble in water solutions and need the presence of detergents to be maintained in a non-aggregated form. The study of integral proteins from different membranes has shown that they are rather heterogeneous in relation to molecular weight. These proteins may be attached to oligosaccharides, thus forming glycoproteins.

Polypeptides of the Red Cell Membrane

The usual method used to separate the membrane polypeptides is to dissolve the erythrocyte ghosts in sodium dodecyl sulfate (SDS) (an ionic detergent) and then to use polyacrylamide gel electrophoresis (PAGE). The gel can be stained for proteins (with Coomassie Blue) or for carbohydrates (with PAS reagent, see Chapter 4–5). Figure 8–3 shows a densitometric scan of the PAGE, which, by the position of the bands, permits estimation of the molecular weight of the polypeptides and, by the surface of the profile, enables determination of its relative mass (see Table 8–3). For example, *polypeptides 1* and *2* corresponding to *spectrin* (i.e., myosin), represent about 30 per cent of the total protein. *Polypeptide 3* is the major intrinsic protein and represents about 25 per cent.[9]

Every Protein of the Cell Membrane is Distributed Asymmetrically

The molecular organization of the proteins is highly asymmetrical, and this can be demonstrated by the use of reagents that are unable to

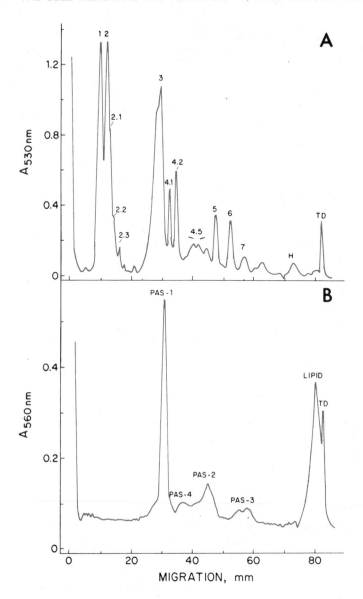

Figure 8–3 Proteins and glycoproteins present in the erythrocyte membrane. **A**, densitometric scan of a gel stained for protein with Coomassie Blue. **B**, the same, stained with PAS reagent for carbohydrates. The nomenclature of the bands, the molecular weight of the proteins, and the nature of some of them are indicated in Table 8–3.

TABLE 8–3   THE MAJOR ERYTHROCYTE MEMBRANE POLYPEPTIDES AND GLYCOPROTEINS*

Component	Mol. Wt.	Percent Stained Protein	Polypeptides per Ghost	Other Designations
Polypeptides				
1	240,000	15.1	216,000	Spectrin
2	215,000	14.7	235,000	Tektin A
3	88,000	24.1	940,000	Myosin-like polypeptide
4.1	78,000	4.2	180,000	
4.2	72,000	5.0	238,000	
5	43,000	4.5	359,000	Actin-like polypeptide
6	35,000	5.5	540,000	G_3PD
7	29,000	3.4	403,000	
Glycoproteins				
PAS-1	55,000	6.7	500,000	Glycophorin
PAS-2				Sialoglycoprotein

*From Steck, L. (1974) *J. Cell Biol.*, *62*:1.

cross the membrane. The reagent is first applied to the *intact erythrocyte* and then to the *white ghost* (see earlier discussion); the difference between the two preparations may give information about the position of a particular protein with respect to the outer or inner surface. Some specific labels can be detected on either side of the membrane by using cytochemical and electron microscopic techniques. For example, using antibodies against spectrin or actin (polypeptide 5) it is possible to show that each is in the cytoplasmic surface of the red cell membrane. On the other hand, acetylcholinesterase can be shown to be in the outer surface by its inactivation with proteolytic enzymes (see Steck, 1974).

It can be said with certainty that *every protein constituent is asymmetrically distributed in the red cell membrane*. All readily soluble polypeptides are localized at the cytoplasmic surface, and those that are at the outer surface are tightly bound to the lipid structure of the membrane. A tentative model representing the position of the main polypeptides in the red cell membrane is shown in Fig. 8–4.[10]

Major Polypeptides of the Red Cell Membrane are Well Characterized

On the cytoplasmic side of the red cell membrane there are *polypeptides 1 and 2* (the myosin-like *spectrin*) and *polypeptide 5 (actin)*. These proteins are associated into supramolecular structures forming *microfilaments* (see Chapter 9), which can be identified by electron microscopy, especially by scanning electron microscopy (Steck and Hainfeld, 1977). Under the red cell membrane these structures form a filamentous network that gives stability to the membrane by providing a kind of skeletal support to the fluid lipid bilayer, with its intrinsic proteins.[11] This system of microfilaments may also control the characteristic shape of the erythrocyte (see Ralston, 1978). Thus, a possible alteration of the microfilaments has been implicated in a hereditary abnormality of the erythrocytes in which they are spheroidal instead of biconcave discs (*hereditary spherocytosis*).[12]

Other polypeptides found on the cytoplasmic surface are in *bands 4.1, 4.2, 6, and 7* (Fig. 8–4). Both 4.2 and 6 are in the membrane as tetramers (i.e., composed of four subunits), band 6 is the monomer of the enzyme *glyceraldehyde -3-P dehydrogenase* (G3PD).

The major intrinsic protein (in *band 3*), with a molecular weight of 90,000 daltons, spans the thickness of the membrane and has a small amount of carbohydrate on the pole at the outer surface (Fig. 8–4). This polypeptide is present in the membrane as a dimer held together by S-S bonds. There are between 500,000 and 600,000 such dimers per cell, enough to account for the 8 nm particles observed in freeze-fractured membranes (see later discussion and Fig. 8–5). This

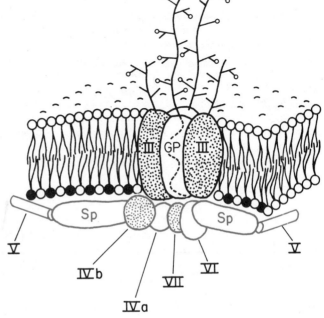

Figure 8–4 Hypothetical model of a red cell membrane showing the lipid bilayer (the black lipid head groups indicate the asymmetric distribution of phosphatidyl serine and phosphatidyl glycerol). *GP*, glycophorin (bands PS-1 and PS-2); *III*, band 3; *Sp*, spectrin (band 1); *IV a* and *IV b*, components 4.1 and 4.2; *V*, actin (band 5); *VI*, G3PD (band 6); *VII*, band 7. The nomenclature of the bands is as in Table 8–3. Peripheral proteins are in red. (Courtesy of G. L. Nicolson, 1977.)

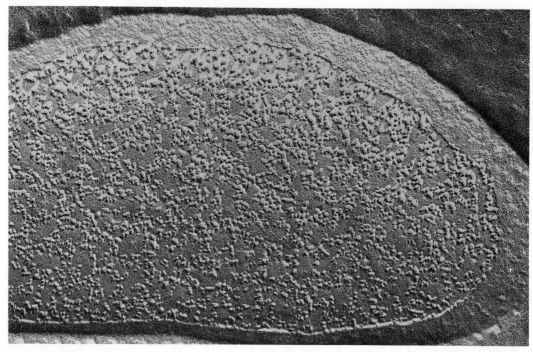

Figure 8–5 Electron micrograph of a freeze-fractured and etched red cell ghost. Most of the surface shows particles that are intercalated in the plane of cleavage of the membrane. ×88,000. (Courtesy of D. Branton.)

dimeric protein appears to be involved in the facilitated *diffusion of anions* (i.e., chloride, bicarbonate) across the membrane.[13]

In the model shown in Figure 8–4 several glycoproteins, including band 3, are represented; however, the major one that spans the membrane is the so-called *glycophorin* (i.e., PAS-1 and PAS-2, Table 8–3). This protein has a molecular weight of 55,000, of which 60 per cent is carbohydrate. Near the COOH end of the molecule there is a region that is very hydrophobic and which interacts with the lipids of the membrane. The COOH end is probably exposed to the interior of the red cell. The NH_2 end is more hydrophilic, is exposed to the external environment, and has the attached oligosaccharides that are at the outer surface of the membrane. At this surface glycoproteins contain protein-bound antigens of the ABO blood groups and other antigens, such as: the MN groups reacting with rabbit antisera, the influenza virus, phytohemagglutinin, and wheat germ agglutinin. It has been calculated that there are some 700,000 copies of glycophorin per human red cell. This protein accounts for 80 per cent of the carbohydrate and 90 per cent of the negatively charged sialic acid present in the cell surface.

Among the most hydrophobic integral proteins of the membrane, the so-called *proteolipids* are characterized by their strong association with lipids and the fact that they are soluble in organic solvents. First isolated by Folch and Lees[14] from myelin, proteolipids are found in practically all cell membranes, and in many of them they represent receptor proteins for synaptic transmitters or form channels across those membranes (see De Robertis, 1975).

Asymmetrical Distribution of Enzymes

More than 30 enzymes have been detected in isolated plasma membranes. Those most constantly found are 5'-nucleotidase, Mg^{2+} ATPase, $Na^+ — K^+$ activated - Mg^{2+} ATPase, alkaline phosphatase, adenyl cyclase, acid phosphomonoesterase, and RNAse.

Some enzymes have a preferential localization; for example, alkaline phosphatase and ATPase are more abundant at the bile capillaries, while disaccharidases are present in microvilli of the intestine. A specific localization with a mosaic arrangement has been postulated for some of these enzymes. Disaccharidase forms 5 to 6 nm globular units coating the membrane of the microvilli. The plasma membrane lacks the respiratory chain and glycolytic activity.

Of all the membrane-associated enzymes, $Na^+ — K^+$ activated - Mg^{2+} ATPase is one of the most important because of its role in ion transfer across the plasma membrane. This enzyme is dependent on the presence of lipids and is inactivated when all lipids are extracted.

Enzymes also show an *asymmetrical distribution*. For example, in the outer surface of erythro-cytes there are acetylcholinesterase, nicotinamide-adenine dinucleotidase, and the ouabain binding site of the $Na^+ K^+$ ATPase. In the inner surface there is NADH-diaphorase, G3PD, adenylate cyclase, protein kinase, and ATPase.[9]

SUMMARY:

Molecular Organization of the Cell Membrane

The first step in the study of the molecular organization of the membrane consists of the isolation of the cell membrane. The easiest to isolate is the red cell membrane obtained after hemolysis. The red cell ghost contains 52 per cent protein, 40 per cent lipid, and 8 per cent carbohydrate that is bound to lipid (glycolipids) or protein (glycoproteins). The lipids form a rather continuous bilayer in which the outer layer contains mainly lecithin and sphingomyelin, and the inner one, phosphatidyl ethanolamine and phosphatidylserine (lipid asymmetry). The oligosaccharides, present in glycolipids and glycoproteins, are located exclusively on the external surface.

Proteins play various roles: mechanical, transport, receptor, antigenic, and enzymatic. Classified by the degree of their association to the membrane, proteins are *peripheral* (extrinsic) or *integral* (intrinsic). The various proteins may be isolated by polyacrylamide gel electrophoresis (PAGE). Every protein in the red cell membrane is asymmetrically distributed. The peripheral proteins *spectrin* (polypeptides 1 and 2) and *actin* (polypeptide 5) form microfilaments associated with the inner surface of the membrane. Other peripheral proteins are glyceraldehyde-3-P dehydrogenase (G3PD) (polypeptide 6) and the protein of band 4.2 (Fig. 8–4).

The major intrinsic protein (band 3), with 90,000 daltons, spans the membrane and is present as a dimer. The major glycoprotein is *glycophorin,* with a MW of 55,000, which also spans the membrane and has several antigenic sites. There are some 30 enzymes in the cell membrane, and these have an asymmetrical distribution.

8–2 MOLECULAR MODELS OF THE CELL MEMBRANE

Before the isolation of plasma membranes, theories on the molecular structure of the membrane were generally based on indirect information. Since substances soluble in lipid solvents penetrate the plasma membrane easily, Overton postulated in 1902 that the plasma membrane is composed of a thin layer of lipid. In 1926 Gorter and Grendell found that the lipid content of hemolyzed erythrocytes was sufficient to form a double layer of lipid molecules over the entire cell surface. This theory was also supported by electrical measurements that indicated a high impedance at the plasma membrane. The high impedance is due to the fact that it is difficult for ions to penetrate a lipid layer.

Other indirect information came from the study of the interfacial tension of different cells. Tension at a water-oil interface is about 10 to 15 dynes per centimeter, whereas surface tension of cells is almost nil. It has been pos-tulated that the low tension is due to the presence of protein layers on the lipid components. In fact, when a very small amount of protein is added to a model lipid-water system, the surface tension is lowered comparably.

To explain all these properties in 1935 Danielli and Davson proposed that the plasma membrane contained a *lipid bilayer*, with protein adhering to both lipid-aqueous interfaces.

Figure 8–6 *a* to *f*, illustrates different membrane models based on the concept of a lipid bilayer, but which differ in the type of protein (i.e., globular, α or β helix) or the penetration of proteins within the bilayer.[15]

Other models containing globular lipid micelles, globular proteins, or a combination of both have been proposed (Fig. 8–6 *g* to *i*); a globular-bilayer transition has also been postulated (Fig. 8–6, *j*). These globular models do not account satisfactorily for the high electrical impedance.

Electron microscopy has thrown some light on the fine structure of the plasma membrane and has revealed the numerous structural dif-

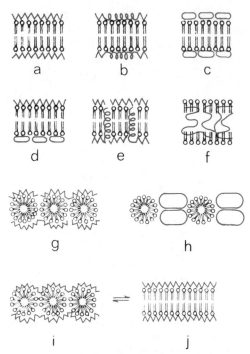

Figure 8–6 A variety of molecular models proposed for the plasma membrane: **a-f**, models based on a lipid bilayer structure; **g-j**, models based on globular arrangements. **a,** protein in β-form; **b,** α-helix; **c,** globular protein; **d,** asymmetry in the protein; **e,** partial penetration with protein channels or pores; **f,** protein within the lipid bilayer; **g,** lipid micelles with β protein; **h,** lipid micelles with globular protein; **i** and **j,** globular-bilayer transformation. (Courtesy of A. L. Lehninger.)

ferentiations that this membrane and the underlying cell cytoplasm have in different cell types. To resolve the structure of the plasma membrane, extremely thin sections (~ 20 nm) must be used, otherwise one would observe different orientations of the plasma membrane with respect to the plane of the section. The membrane appears most thin when it is exactly perpendicular to the plane of the section. Membranes of 6 to 10 mm have been observed at the surface of all cells. The membranes of two cells that are in close contact appear as dense lines separated by a space of 11 to 15 nm, which is strikingly uniform and contains a material of low electron density (Fig. 8–7, A). This intercellular component can be considered as a kind of cementing substance.[16]

The Unit Membrane Model — Re-evaluation of the EM Image

The observation that the cell membrane generally appears to have three layers with two

outer dense layers of 2.0 nm and a middle clear one of about 3.5 nm led Robertson[16] to postulate the so-called *unit membrane* model. The electron microscopic image was interpreted along the lines of the Danielli-Davson model, so that the clear layer corresponded to the hydrocarbon chains of the lipids, and the dense layers, to the proteins. This unit membrane model was also extended to all kinds of intracellular membranes. That this concept is an oversimplification which does not account for the many proteins traversing the membrane can be readily seen by observing some high resolution micrographs in which fine bridges cross the unit membrane (Fig. 8–7, C).

Recently the unit membrane image has been re-evaluated on the basis of experiments using methods designed to avoid the removal of proteins from membranes during the preparation for conventional electron microscopy or freeze-fracturing. With both methods a granular appearance of the membrane is obtained, which is more in keeping with the fact that a high proportion of the proteins traverse the membrane. According to these findings the unit membrane image is, to a great extent, artifactual. (see Luftig et al., 1977).

The Fluid Mosaic Model is Now Generally Accepted

Present knowledge about the molecular organization of biological membranes comes mainly from an integration of the data on chemical analysis and from the application of several biophysical techniques.[1]

The important concepts that have emerged are summarized in the so-called *fluid mosaic model* of membrane structure. This postulates: (1) that the lipid and integral proteins are disposed in a kind of mosaic arrangement; and (2) that biological membranes are quasi-fluid structures in which both the lipids and the integral proteins are able to perform translational movements within the overall bilayer. The concept of fluidity implies that the main components of the membrane are held in place only by means of non-covalent interactions.[17]

To better understand the molecular organization of the membrane, it is necessary to remember that, not only the lipids, but also many of the intrinsic proteins and glycoproteins of the membrane, are amphipathic molecules. The term *amphipathy*, coined by Hartley in 1936, refers to the presence, within the same molecule, of hydrophilic and hydrophobic

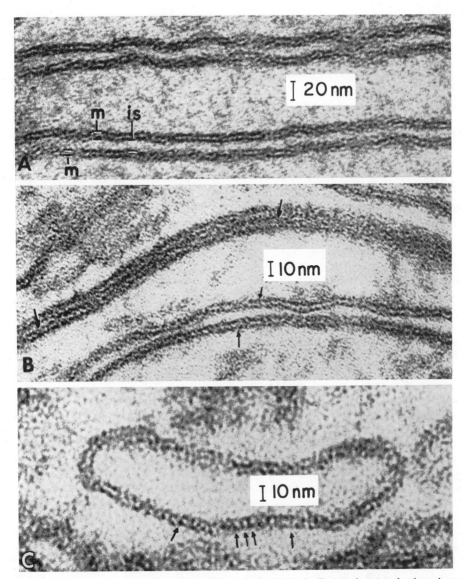

Figure 8–7 **A,** electron micrograph of cell membranes of intestinal cells (*m*), showing the three-layered structure (unit membrane). *is,* intercellular space. ×240,000. **B,** cell membranes in the rat hypothalamus showing the unit membrane structure and, with arrows, some finer details across the membrane. The upper arrows indicate a region in which the two cell membranes are adherent (*tight junction*) and the intercellular space has disappeared. ×360,000. **C,** the same as **B,** showing fine bridges (arrows) across the unit membrane. ×380,000. (From E. De Robertis.)

groups. As was mentioned in Chapter 7, these amphipathic molecules constitute liquid crystalline aggregates in which the polar groups are directed toward the water phase and the non-polar groups are situated inside the bilayer (Fig. 7–9).

In the fluid mosaic model represented in Figure 8–1 the integral proteins of the membrane are intercalated to a greater or lesser extent into a rather continuous lipid bilayer. This arrangement is based on the fact that these integral proteins are also amphipathic, with

polar regions protruding from the surface and non-polar regions embedded in the hydrophobic interior of the membrane.

This arrangement may explain why different enzymes and antigenic glycoproteins may have their active sites exposed to the outer surface of the membrane. It is well recognized that a protein of appropriate size or a cluster of protein subunits may pass across the entire membrane (transmembrane proteins). Such traversing proteins could be in contact with the aqueous solvent on both sides of the membrane. A similar model has been postulated for the cholinergic receptor.[18] This model also reflects the generally accepted view that the major proportion of the phospholipids in the membrane are arranged in a bilayer form.

One of the major supports for this mosaic model of the membrane, with intercalating proteins, comes from the use of freeze-fracturing techniques in erythrocytes and other cell membranes (see Chapter 3–2). The red cell ghosts show a large number of particles, about 8 nm in diameter, that have been interpreted as representing proteins embedded within the plane of cleavage which passes through the middle of the lipid bilayer (Fig. 8–5).

There are between 500,000 and 600,000 such particles per cell, and more are attached to the inner half of the membrane (PF) than to the outer half (EF). (The nomenclature of freeze-fracturing is explained in Chapter 14–3 together with the structure of chloroplasts.) We mentioned earlier that these particles represent dimers of the polypeptide found in *band 3*, and that they may also correspond to anionic permeation channels.[10] The degree of dispersion or aggregation of these particles within the membrane appears to be influenced by the *spectrin-actin system* (see earlier discussion). This suggests that some link between this system and the dimers of band 3 could exist in the living membrane.[9, 10, 19, 20]

The mosaic model allows for the characteristic *asymmetry* of the membrane, in which specific components predominate in the outer half or the inner half of the membrane (see earlier discussion). This implies that there are constraints on the transmembrane rotation of those macromolecules that are well oriented for the carrying of information across the cell membrane.

The fluidity of the membrane in this model also implies that both the lipids and the proteins have considerable freedom of movement within the bilayer. The fluidity of the lipids depends (1) on the degree of saturation of the hydrocarbon chains and (2) on the ambient temperature. A considerable proportion of the lipids in the membrane are unsaturated, so that the melting point of the bilayer is below body temperature..

Membrane Fluidity — Studied with Physical and Biological Techniques

The fluidity of the membrane can be studied with a series of techniques that can be classified as physical or biological.

The *physical techniques* are of two main types: (1) those that involve a minimal perturbation of the membrane, such as *X-ray diffraction* and *nuclear magnetic resonance (NMR) spectroscopy*, and (2) those that use certain added molecules to monitor specific sites of the membrane. Into this second class fall *fluorescence microscopy*, which uses fluorescent probes (see Chapter 4–6), and *electron spin resonance (ESR)* spectroscopy. (Consideration of these techniques is beyond the scope of this book. For references see Nicolson, et al., 1977.) In practice, these physical techniques can be subdivided on the basis of their ability to examine mainly either the lipid or the protein components, or both. ESR spectroscopy uses paramagnetic probes (e.g., nitroxide-containing amphipathic molecules) which are introduced into the lipid bilayer. This can be done with natural and artificial membranes (see Chapter 7–3). The data obtained provide information on lipid-protein packing, lateral diffusion of lipids, lipid-protein interactions, the fluidity of the membrane, and the rate of transmembrane rotation (so-called flip-flop) of molecules across the bilayer.[21, 22]

The *biological techniques* involve light and fluorescence microscopy and electron microscopy, including freeze-fracturing and radioisotope labeling methods. One of the simplest methods consists of binding gold or carbon particles to the cell surface and observing under the light microscope the movement of those particles on the surface. More information has come from studies in which different ligands, such as antibodies and plant lectins, interact with cell surface receptors (see Edidin, 1974 and Nicolson, et al., 1977). If these ligands are labeled with fluorescent dyes their movement can be followed by fluorescence microscopy.

For example, in a classical experiment by Frye and Edidin[23] mouse L cells and human transformed cells which have different surface antigens, were marked with the corresponding antibodies labeled with two distinct fluorescent dyes. These cells were then induced to fuse under the influence of Sendai viruses. While at the beginning both cell surfaces could be recognized by their differing labels, after 40 minutes considerable intermixing of the antigens had occurred, so that the two labels could no longer be recognized. This intermixing was retarded by temperatures below 20° C.

In surface antigens of cultured muscle cells, a diffusion rate of 1×10^{-9} cm per second was calculated.[24] Also, the so-called clustering or "capping" effect observed in lymphocytes may result in the displacement of antigenic molecules within the surface membrane. In this case, upon treatment with a labeled antibody, the antigens are first randomly distributed, but after some time they become clustered, forming patches, and agglomerate at one pole of the cell (i.e., capping) where pinocytosis of the antigen-antibody complexes takes place (Fig. 8–8). This process is also inhibited by temperatures that produce solidification of the lipid bilayer (see Raff, 1976).

A number of histochemical techniques using the electron microscope may give information about the fluidity of the membrane. For example, antigens labeled with ferritin or peroxidase can be used (see Chapter 4–6). Another extremely valuable tool is freeze-fracturing (see Chapter 3–2) with which it is possible to demonstrate the movement of the particles in response to different experimental conditions that produce changes in the molecular structure of the membrane. (For further details, see Nicolson, et al., 1977.)

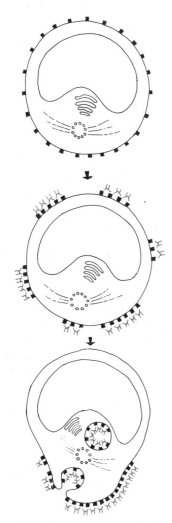

Figure 8–8 Diagram representing the phenomenon of "capping" in lymphocytes. **Top,** corresponds to the normal distribution of antigens on the cell surface. **Middle,** the antigens are now clustered into patches after being cross-linked with a bivalent antibody. **Bottom,** a cap is formed by the active transport of the patches toward the pole that contains the centrosomal-Golgi area. Observe that the membrane is being internalized by endocytosis. Antibodies are in red. (From S. De Petris and M. C. Raff.)

Membrane Fluidity and Coupling of Receptors to Adenylate Cyclase

The fluid mosaic model has been used to explain the fact that many receptors present in the cell membrane are coupled to enzymes such as adenylate and guanylate cyclase. While the problem of membrane receptors will be more extensively addressed in Chapter 28, here we can say that they are macromolecules, in most cases of a protein nature, that act at the interface between the cell surface and the environment.

A receptor has the dual function of recognizing a chemical signal (i.e., one of the many regulatory agents, such as peptide hormones, neurotransmitters, prostaglandins, antigens, plant lectins, and some bacterial toxins) and of initiating a biological response. The function of recognition is determined by the presence of a site that binds specifically to the ligand (i.e., to the binding portion of a particular regulatory agent). This interaction produces a conformational change that may induce the translocation of ions or the activation of membrane enzymes, such as adenylate cyclase, thus producing cyclic AMP (cAMP) (Fig. 28–24).

In a cell there may be several kinds of receptors (at least eight in a fat cell) coupled to the same adenylate cyclase. The ligands do not compete for each specific receptor when binding, and the effect of multiple ligand-receptor bindings on adenylate cyclase is not additive. These findings have been interpreted as indicating that each ligand reacts with a unique receptor and that several receptors may be coupled to a certain number of enzyme molecules in the membrane.

According to the *mobile hypothesis*[25] both the receptors and the enzymes can diffuse independently within the plane of the membrane (see Jacobs and Cuatrecasas, 1977). The greater degree of freedom that this model includes within the lipid realm of the membrane would allow different receptors to couple to a single enzyme.

Experimental support of this hypothesis has been obtained by experiments similar to those of Frey and Edidin[23] mentioned earlier. Fusing cells in which the adenylate cyclase was inactivated but the receptor remained intact, with other cells which contain only the enzyme, resulted in receptor-enzyme coupling. In other words, the receptor of one cell (e.g., the β-adrenergic receptor of an erythrocyte) was coupled to the adenylate cyclase of the other cell (e.g., a tumor cell lacking the receptor). The coupling is manifested by a great increase in cAMP after receptor interaction with the specific ligand (in this case isoproterenol) (see Schramm, et al., 1977). The diagram shown below (from Jacobs and Cuatrecasas, 1977) interprets in a highly simplified manner the fusion of two different cells that results in coupling between a receptor from one cell and adenylate cyclase from the other.

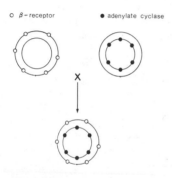

The Myelin Sheath and the Photoreceptors — Special Multi-layered Membranes

Important information about the molecular structure of the cell membrane is derived from studies of natural multi-layered lipoprotein systems, such as the myelin sheath and the photoreceptors.

The Myelin Sheath. The myelin sheath is a lipoprotein membrane that surrounds the axon of the nerve fiber. In peripheral nerves this sheath is formed by the Schwann cells. In central nerves the myelin sheath is produced by the activity of the oligodendroglial cells.

The myelin sheath is a very special membrane structure in which the lipids are more abundant than the protein (Table 8–1). In central myelin there are two main protein moieties: the proteolipid and the basic protein.

It has been known for over a century that the myelin sheath has a strong birefringence, which indicates a high degree of organization

at a submicroscopic level. Studies with x-ray diffraction have revealed a spacing of 17 nm in amphibian and 18.0 to 18.5 nm in mammalian peripheral nerves. Within this period, the proportion corresponding to the lipid, protein, and water content has been estimated.

In the molecular model shown in Figure 8–9 the existence of lecithin-cholesterol and sphingomyelin-cholesterol complexes is postulated. The first type can be accommodated within the thickness of the lipid layer (L), but in the second, the longer sphingomyelin molecules must interdigitate in order to fill the same space.[26] This model also accounts for the localization of protein (HP) and water (HL) and is in accord with the view that each x-ray diffraction period of 18 nm corresponds to two unit membranes, i.e., to two lipid bilayers separated by a watery space (H), that originally corresponds to extracellular space.

Electron microscopic studies have confirmed that myelin has a multi-layered membranous structure. In most cases, however, the x-ray diffraction period is reduced to 10 to 12 nm, thereby introducing a great many artifacts. Recently, with a highly polar embedding medium based on polymerized glutaraldehyde-urea, the periodicity obtained was of the same order as by x-ray diffraction. With this technique the lipids are not extracted; this explains the lack of collapse of the structure.[27]

The intraperiod space H in Figure 8–9 appears in the electron micrograph as a relatively thick layer that is darkly stained in Figure 8–10. It is possible that this space corresponds to hydrated carbohydrate that covers the outer leaflet of the two bilayers. The less dense lines of Figure 8–10 correspond to the hydrocarbon chains of the lipids, and the fine lines (mp) correspond to the main lines of the period.

In nerve conduction the myelin sheath seems to function as an insulator, preventing the dissipation of energy into the surrounding medium. It might act not only as a dielectric (insulating) material but also as a kind of resonant conductor in which the energy waves resonating in the lipid layers between the protein membranes could pass with maximum speed and minimum loss of energy (see Boggs and Moscarello, 1978).

Photoreceptors. The retinal rods and cones are highly differentiated cells that have at their outermost segment a lipoprotein structure that is specialized for photo-reception. Studies with the polarization microscope suggest a submicroscopic organization consisting of transversely oriented protein layers alternating with lipid

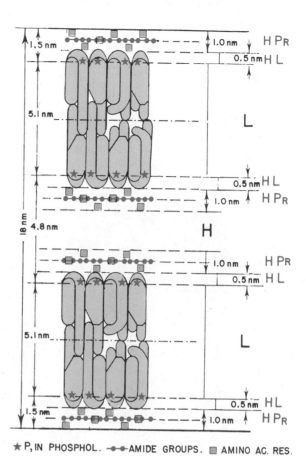

Figure 8–9 Molecular model of the myelin sheath. *HPr,* protein layer represented by a chain backbone; *HL,* water layer; *L,* lipid bilayer made of lecithin-cholesterol and sphingo-myelin-cholesterol complexes (these are interdigitating); *H,* intraperiod water space. (See the description in the text.) (From Vandenheuvel, F. A., Structural studies of biological membranes: the structure of myelin. *Ann. N.Y. Acad. Sci., 122:70,* 1965. © The New York Academy of Sciences, 1965; reprinted by permission.)

★ P, IN PHOSPHOL. ●—● AMIDE GROUPS. ▪ AMINO AC. RES.

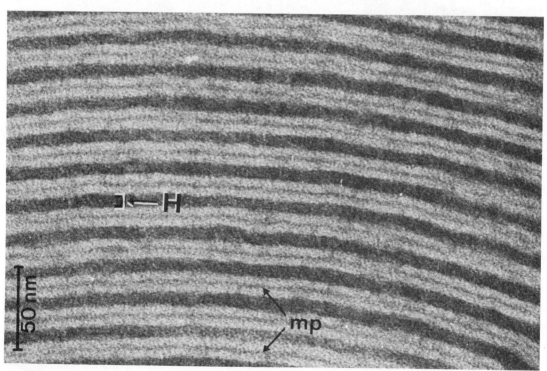

Figure 8–10 Electron micrograph of a section of the myelin sheath using the glutaraldehyde-urea method of embedding. *H,* the intraperiod space stained with silico-tungstic acid; *mp,* the main lines of the period. (See description in the text.) ×440,000. (Courtesy of R. G. Peterson and D. C. Pease.)

molecules arranged longitudinally along the axis of the photoreceptor. This type of layered organization has been demonstrated by electron microscopy in fragmented rod outer segments and in thin sections of the retina. These observations indicate that the rod consists of a stack of superimposed disks (several hundred) along the axis. These disks are really flattened sacs made of two membranes, which surround a thin space of 3 nm and become continuous at the edges (Fig. 8–11). The space between the rod sacs is 5 to 12 nm. The cone outer segments, with minor differences, have a similar structure.[28]

Rod sacs are highly sensitive to osmotic change. In hypotonic solutions they swell con-

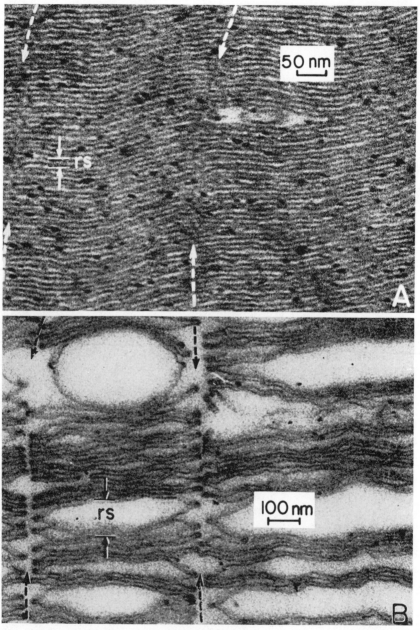

Figure 8–11   Electron micrographs of the outer segment of the retinal rods of a toad. **A,** fixation with isotonic osmium tetroxide maintains the regular organization of the retinal sacs, ×190,000; **B,** fixation with hypotonic osmium tetroxide (Palade's method) produces great swelling of the sacs and separation of the membranes. (From De Robertis, E., and Lasansky, A., "Ultrastructure and chemical organization of photoreceptors," *in* Smelser, G. K., ed., *The Structure of the Eye,* New York, Academic Press, Inc., 1961.)

siderably, and the inner space between the membranes becomes very large (Fig. 8–11, B).

In the frog the rod outer segments can easily be isolated and be submitted to chemical analysis. The protein composition is simpler than in other membranes. The photopigment protein, *rhodopsin*, represents 40 per cent of the total mass of the rod sacs; another 40 per cent is lipid. Rhodopsin is an integral type of protein requiring detergents for extraction. In the plane of the membrane the rhodopsin molecules have a liquid-like distribution.[29] X-ray diffraction studies have led to the interesting conclusion that this molecule may change its position within the lipid layer. In the dark (non-stimulated condition) the molecule is about one third embedded in the lipid and two thirds exposed to the water surface. Upon activation by light (bleached rhodopsin) the molecule becomes more deeply embedded within the lipid bilayer. This change is presumably caused by changes in surface charges of the molecule.

Photoreceptors transform light energy into another type of energy that can be conducted as nerve impulses. This process is based on a cycle of chemical reactions which involve the visual pigments present in the protein membranes of the rod and cone sacs. This multilayered structure is a very effective system that facilitates the maximum absorption and utilization of light by the chromophoric groups present in the visual pigments (retinenes). The acute sensitivity of the photoreceptors, which can react to a single photon, can be explained by the fact that the possibility of striking a sensitive molecule is increased by a factor of hundreds or thousands by the molecular organization of the photoreceptor.

It is thought that the interaction with photons of light produces liberation of Ca^{2+} ions which, in turn, inhibit the sodium current that is characteristic of the photoreceptor in the dark condition. In this way photoreceptors act as energy transducers, transforming light energy into electrical signals.

SUMMARY:

Membrane Molecular Models

To explain various properties of the cell membrane (i.e., permeability of lipid solvents, low surface tension, and so forth) Danielli and Davson (1935) proposed a model in which a lipid bilayer had protein adhering to both lipid-aqueous interfaces.

The observation of thin sections by electron microscopy has led to the concept of a trilayered plasma membrane, also called the "unit membrane." This is interpreted as indicating that the electron-dense outer layers correspond to the protein, and the less dense middle layer corresponds to the hydrocarbon chains of the lipids. The unit membrane concept, however, is certainly an oversimplification; numerous fine details suggest that the molecular organization of the membrane is much more complex. Recent observations lead to the conclusion that the unit membrane image is mainly artifactual and does not account for the many proteins traversing the membrane.

The most favored model for the plasma membrane is the so-called fluid mosaic structure. According to this model, there is a rather continuous lipid bilayer into which the integral proteins of the membrane are intercalated; both these components are capable of translational diffusion within the overall bilayer. The mosaic model of the membrane is supported by the results of freeze-etching techniques in which protein particles are shown at the plane of cleavage of the bilayer.

There are between 500,000 and 600,000 particles per red cell, with the majority of these attached to the inner half of the membrane. Such particles represent the protein of band 3, which may form anionic permeation channels (see later discussion). The mosaic arrangement implies (1) that the macromolecules have a characteristic *asymmetry*, (2) that they are oriented for carrying information across the bilayer, and (3) that they have considerable freedom of movement within the bilayer (*fluidity*). The fluidity of the lipids depends on the temperature and degree of saturation. Most lipids of the membrane are fluid at body temperature. With freeze-fracturing it is possible to demonstrate the movement of protein particles in different experimental conditions.

The fluidity of the lipids is supported by many indirect studies based on x-ray diffraction, differential thermal analysis, and electron spin techniques. The fluidity of the integral proteins is supported by experiments on cell fusion and on those of clustering and "capping" of surface antigens.

Membrane fluidity is used in the *mobile hypothesis* to explain the coupling between several membrane receptors and a single adenylate cyclase. This hypothesis is supported by experiments using cell fusion.

The myelin sheath and the photoreceptors are special multi-layered systems in which a great deal of information about chemical organization and function of membranes has been obtained.

8–3 CELL PERMEABILITY

Permeability is fundamental to the functioning of the living cell and to the maintenance of satisfactory intracellular physiologic conditions. This function determines which substances can enter the cell, many of which may be necessary to maintain its vital processes and the synthesis of living substances. It also regulates the outflow of excretory material and water from the cell.

The presence of a membrane establishes a net difference between the *intracellular* fluid and the *extracellular* fluid in which the cell is bathed. This may be fresh or salt water in unicellular organisms grown in ponds or the sea, but in multicellular organisms the internal fluid, i.e., the blood, the lymph, and especially the *interstitial* fluid, is in contact with the outer surface of the cell membrane.

One of the functions of the cell membrane is to maintain a balance between the osmotic pressure of the intracellular fluid and that of the interstitial fluid.

In higher organisms, the osmotic pressure of the body as a whole is regulated principally by the kidneys, and the osmotic pressure of the interstitial fluid is about the same as that of the intracellular fluid.

In plants, the intracellular fluid has a higher osmotic pressure than the extracellular fluid. The cell is protected from bursting by a rigid cellulose wall.

Animal cells generally lack the turgidity that characterizes plant cells. The unfertilized eggs of some marine animals, such as the sea urchin, behave like genuine osmometers. Since they are spheroid, one can, by measuring the diameter, determine the volume and the changes that the egg undergoes with changes in the osmotic pressure of the medium.

In many unicellular organisms the osmotic equilibrium is maintained by means of a contractile vacuole. This "organelle" extracts water from the protoplasm, releasing its contents into the external medium.

Different Ionic Concentrations across the Membrane Create Electrical Potentials

In all cells there is a difference in ionic concentration with the extracellular medium and an electrical potential across the membrane. These two properties are intimately related, since the electrical potential depends on an unequal distribution of the ions on both sides of the membrane.

As shown in Table 8–4, the interstitial fluid has a high concentration of Na^+ and Cl^-; and the intracellular fluid, a high concentration of K^+ and of larger organic anions (A^-).

Using fine microelectrodes with a tip of 1 μm or less, investigators are able to penetrate through the membrane into a cell and also into the cell nucleus and to detect an *electrical potential* (also called the *resting*, or *steady*, potential), which is always negative inside. The values of

TABLE 8–4 IONIC CONCENTRATION[†] AND STEADY POTENTIAL IN MUSCLE[*]

		Interstitial Fluid	Intracellular Fluid
Cations	Na^+	145	12
	K^+	4	155
Anions	Cl^-	120	3.8
	HCO_3^-	27	8
	A^- and others	7	155
Potential		0	−90 mv

*Modified from Woodbury, J. W., "The Cell Membrane: Ionic and Potential Gradients and Active Transport," *in* Ruch, T. C., Patton, H. D., Woodbury, J. W., and Towe, A. L. (Eds.), *Neurophysiology*, Philadelphia, W. B. Saunders Co., 1961.

†Ionic concentration in mEq.

the membrane potential vary in different tissues between −20 and −100 millivolts (mv).

Passive Permeability — Dependent on the Concentration Gradient and the Partition Coefficient

The passage of molecules and ions across membranes may be the result of two principal mechanisms. Permeability may be *passive*, if it occurs only because of physical laws such as *diffusion* or *active* if it requires energy for transport. There are several kinds of permeability which will be considered in the following sections.

In the absence of an intervening membrane, when two solutions of different concentration are mixed a process of intermixing called *diffusion* occurs. For example, if a concentrated solution of sugar is placed in contact with water, there will be a net movement (i.e., flux) of the solute from the region of higher concentration to that of a lower concentration. In this case the higher the difference in concentration between the two solutions (i.e., the *concentration gradient*), the more rapid the rate of diffusion.

The presence of a lipoprotein membrane, such as the plasma membrane, greatly modifies this diffusion or passive permeability.

We can consider that the passage of solute

from one side of the membrane to the other occurs in three stages: (1) the solute leaves the aqueous phase and enters the hydrophobic region of the membrane, (2) it traverses the lipid bilayer, and (3) it leaves the lipid phase to enter another aqueous phase. For most polar molecules stages 1 and 3 represent a considerable barrier.

At the end of the last century, Overton demonstrated that substances that dissolve in lipids pass more easily into the cell, and Collander and Bärlund, in their classic experiments with the cells of the plant *Chara*, demonstrated that the rate at which substances penetrate depends on their solubility in lipids and the size of the molecule. The more soluble they are, the more rapidly they penetrate, and with equal solubility in lipids the smaller molecules penetrate at a faster rate (Fig. 8–12).

The permeability (P) of molecules across the membrane is:

$$P = \frac{KD}{t} \qquad (1)$$

with K, the partition coefficient; D, the diffusion coefficient, which depends on the molecular weight; and t, the thickness of the membrane. The partition coefficient in most cell membranes is similar to that of olive oil and water.

The *partition coefficient* can be measured by

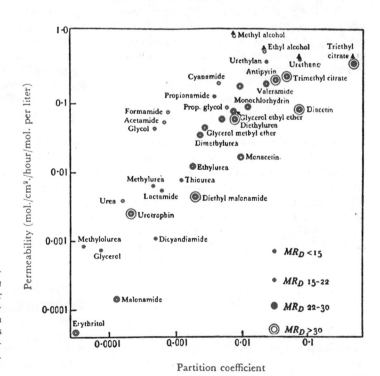

Figure 8–12 Rate of penetration (permeability) in cells of *Chara ceratophylla* in relation to molecular volume (measured by molecular refraction, MR_D), and to the partition coefficient of the different molecules between oil and water. (After Collander, R., and Bärlund, H., *Acta Bot. Fenn.*, 11:1, 1933.)

mixing the solute with an oil-water mixture and then waiting until the phases are separated. The coefficient is the concentration of the solute in the oil, divided by the concentration in the aqueous phase. Today, the *diffusion coefficient* can be easily determined by using radioactive solutes and measuring their rate of entry into the cytoplasm at various external concentrations. From the experiments shown in Figure 8–12 it is evident that many molecules may enter the cell by simple diffusion if the external concentration is higher than the internal. On the other hand, waste products accumulated in the cytoplasm may be eliminated by diffusion.

Passive Ionic Diffusion – Dependent on the Concentration and Electrical Gradients

The diffusion of ions across membranes is even more difficult because it depends not only on the *concentration gradient* but also on the *electrical gradient* present in the system.

From Table 8–4 it is evident that within the cell there is a large concentration of non-diffusible anions (A^-) and that an electrical gradient across the membrane is established.

In 1911, Donnan predicted that if a theoretical cell, having a non-diffusible negative charge inside, is put in a solution of KC1, K^+ will be driven into the cell by both the concentration and the electrical gradients; $C1^-$, on the other hand, will be driven inside by the concentration gradient, but will be repelled by the electrical gradient. As shown by Donnan, the equilibrium concentrations will be exactly reciprocal:

$$\frac{[K^+_{in}]}{[K^+_{out}]} = \frac{[Cl^-_{out}]}{[Cl^-_{in}]} \qquad (2)$$

A Donnan equilibrium involving only physical forces (i.e., without expenditure of energy by the membrane) was confirmed by the demonstration that the membrane potential was negative on the inside and was accompanied by a high K^+ and a low Cl^- concentration (Table 8–1). The relationship between the concentration gradient and the resting membrane potential is given by the Nernst equation:

$$E = RT \log \frac{C_1}{C_2} \qquad (3)$$

where E is given in millivolts, R is the universal gas constant, and T is the absolute temperature.

150

From (2) and (3) the Donnan equilibrium for KCl can now be expressed as follows:

$$E = RT \log \frac{[K^+_{in}]}{[K^+_{out}]} = RT \log \frac{[Cl^-_{out}]}{[Cl^-_{in}]} \qquad (4)$$

According to (4) any increase in the membrane potential will cause an increase in the ion asymmetry across the membrane, and vice versa. While the first measurements of membrane potentials and ion concentration seemed to confirm this type of *passive* or *diffusion equilibrium*, more precise determinations in different cell types demonstrated that this was not the case. As mentioned in the next section, this discrepancy may be explained by the involvement of the active transport of ions.

Active Transport of Ions Uses Energy

In addition to the diffusion or passive movement of neutral molecules and ions across membranes, cell permeability includes a series of mechanisms that require energy. These mechanisms are generally described as *active transport*. Adenosine triphosphate (ATP), which is produced mainly by oxidative phosphorylation in mitochondria, is generally used as the source of energy. For this reason, active transport is generally related to, or coupled with cell respiration.

Active transport against a concentration gradient is explained in Figure 8–13 by analogy with a hydrostatic example in which water has to be moved upstream (i.e., against gravity). The osmotic work to be done is expressed by the Nernst equation. A charged molecule crossing through an electrochemical gradient also may imply expenditure of energy. For example, to maintain a low intracellular concentration (Na^+, the cell must extrude sodium against a gradient (i.e., higher Na^+ concentration outside). In addition, it must do this against an electrochemical barrier since the membrane is negative inside and positive outside (Fig. 8–14, *A* and *B*).

A "Sodium Pump" is Postulated in the Active Efflux of Na⁺

When an ion is transported against an electrochemical gradient, an extra consumption of oxygen is required. It is calculated that 10 per cent of the resting metabolism of a frog muscle is used for transport of sodium ions. This consumption may increase to 50 per cent in some

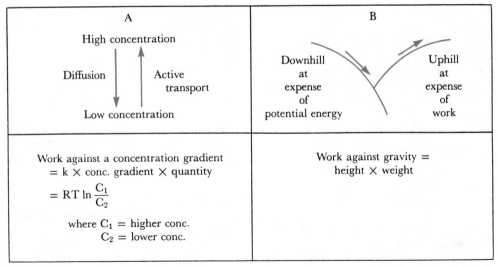

Figure 8–13 Analogy between concentration gradient (left half of **A**) and potential gradient (left half of **B**) and between movement against a concentration gradient (right half of **A**) and work done in moving uphill (right half of **B**). Equations for work against a concentration gradient (osmotic work) and for work in lifting a weight up a height are given at the bottom of the figure. (From Giese, A. C., *Cell Physiology,* 3rd ed., Philadelphia, W. B. Saunders Co., 1968.)

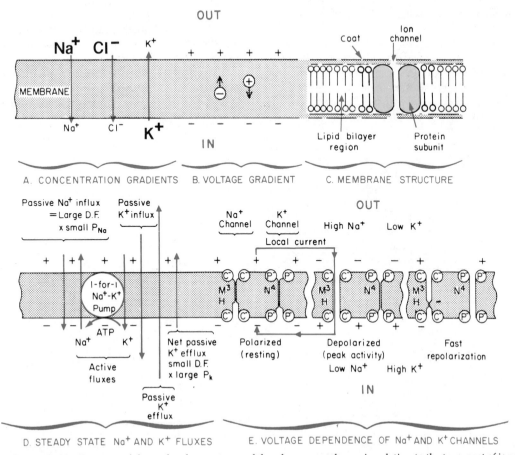

Figure 8–14 Diagram of the molecular structure of the plasma membrane in relation to the transport of ions. (Courtesy of J. W. Woodbury.)

experimental conditions in which the muscle is stimulated.

That the resting membrane potential is due to active transport may be demonstrated in plant and animal cells that have been metabolically blocked by anoxia or specific poisons. In this case leakage of K^+ occurs and the potential may decrease to zero. This is clearly observed when anoxia is combined with the poisoning of glycolysis. As suggested by Krogh in 1946, the membrane potential is not really at equilibrium but in a "steady state" involving the constant expenditure of energy.

An interesting example of active transport has been provided by experiments with isolated frog skin. The epithelium is specialized to transport Na^+ from the pond water to the interstitial fluids, and by this mechanism the frog can trap this essential ion for use in different tissues. The isolated skin can be kept alive for many hours and used as a wall between two chambers in which the ionic concentration and other factors are changed experimentally.[30] By means of this preparation, a difference in potential across the skin has been demonstrated: the inside surface is positive with respect to the outside. The sodium ions are transported from the outer toward the inner surface, and the current produced is due to the flux of sodium. It was also observed that the antidiuretic hormone of the neurohypophysis stimulates the transport of sodium and water. Similar findings have been observed in the isolated toad bladder[31] and the kidney tubules of *Necturus*.

Ionic transport is intense in the salivary and sweat-producing glands, and even more so in the glands of the stomach that produce H^+ and Cl^- that must be replaced by the blood.

The active transport of ions is fundamental to maintenance of the osmotic equilibrium of the cell, the required concentration of anions and cations and the special ions needed for the functioning of the cell. Together with the extrusion of Na^+, which is continuously pumped out by the cell, there is an exit of water molecules. In this way the cell keeps its osmotic pressure constant.

Potassium ions, which are concentrated inside the cell, must pass against a concentration gradient. This can be achieved by a "pumping" mechanism at the expense of energy. As explained before, Na^+ also is transported by an active process, which is sometimes called the "sodium pump."

The diagram in Figure 8–15 summarizes the relationship existing between the transfer of K^+ and Na^+ (i.e., ionic fluxes) by passive and active mechanisms and the resulting steady state potential. The passive (downhill) fluxes are distinguished from the active (uphill) fluxes. Notice that the active pumping out of Na^+ is the main mechanism for maintaining a negative potential inside the membrane of -50 mv. This diagram demonstrates that the distribution of ions across the membrane depends on the summation of two distinct processes: (1) simple electrochemical diffusion forces which tend to establish a Donnan equilibrium (i.e., passive transport)

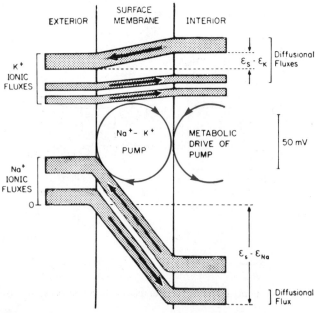

Figure 8–15 Active and passive Na^+ and K^+ fluxes through the membrane in the steady state. The ordinate is the electrochemical potential of the ion ($\epsilon_s - \epsilon_K$ for K^+, $\epsilon_s - \epsilon_{Na}$ for Na^+). The abscissa is the distance in the vicinity of the membrane. The width of the band indicates the size of that particular one-way flux. Passive efflux of Na^+ is negligible and is not shown. (After Eccles, J. C., *Physiology of Nerve Cells*, Baltimore, Johns Hopkins Press, 1957.)

and (2) energy-dependent ion transport processes (i.e., active transport).

In Figure 8–14, *D*, it may be observed that the $Na^+ - K^+$ pump drives ions extruding Na^+ and taking in K^+ (active fluxes). At the same time the passive Na^+ influx depends on a large driving force (D.F.), resulting from the concentration and voltage gradients, and only a slight permeability of the membrane to Na^+ (P_{Na}). Similarly, the net passive K^+ efflux results not from a small driving force, but from a greater permeability to K^+ (P_K).

Ionic Transport through Charge Pores in the Membrane

It can now be ascertained that the molecular machinery involved in ionic transport is located within the cell membrane. This has been demonstrated in two key materials. For example, if red blood cells are hemolyzed so that only the cell membrane remains, they can be filled again with appropriate solutions containing ions and ATP, and Na^+ is transported and K^+ is taken up as in a normal cell (i.e., a resealed ghost).

The giant axon of the squid, which has a diameter of about 0.5 mm, can be emptied of the axoplasm and then refilled with solutions of different electrolytes. The transport of ions against a concentration gradient, steady potentials, and even action potentials with the conduction of impulses can be obtained in this preparation in which most of the axoplasm is lacking and the excitable membrane left alone.[32]

The use of radioisotopes demonstrated that ions can enter into the cell rapidly without obvious osmotic effects. It was then suggested that an ionic interchange across the membrane could take place through electrically charged pores.

Knowledge about the diameter of the different ions in the hydrated state is particularly pertinent. In this respect it is interesting to remember that the sodium ion, although smaller than K^+ and Cl^- in weight, is large in the hydrated condition and enters with more difficulty into the cell (see Fig. 8–16).

A possible molecular interpretation of the membrane pores is shown in Figure 8–14. The pore could be envisioned as the interstice between four adjacent protein subunits which could form a hydrophilic channel across the membrane; two such subunits are shown in Figure 8–14, *C*.

In Figure 8–14 *E*, the ion channels for Na^+ and K^+ are represented in the polarized or resting condition, or in the state of depolarization that

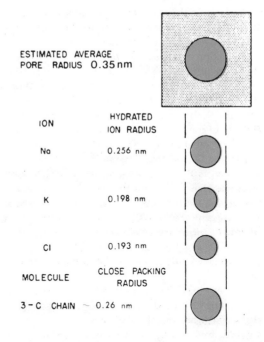

Figure 8–16 Schematic representation of the red cell pore. Notice that the hydrated ion radius is larger for Na^+ than for K^+. (From Solomon, A. K., *J. Gen. Physiol.*, 43:5, part 2, suppl. 1, p. 1, 1960.)

occurs during the action potential. It is postulated that the Na^+ and K^+ channels are closed in the polarized resting membrane.[33]

The rest of the diagram shows the opening of the Na^+ pores at the peak of depolarization of the action potential and the opening of the K^+ pores during repolarization (see Chapter 28–1).

The total area of the pores in the red blood cell has been estimated to be on the order of 0.06 per cent of the surface area. This means that a 0.7 nm pore would be surrounded by a nonporous square 20×20 nm. These findings indicate that the cell uses only a minute fraction of its surface area for ionic interchange.[34]

Anion Transport in Erythrocytes Involves the Special Band-3 Polypeptide

The red cell membrane is endowed with an extremely active anion permeation mechanism to transport CO_2 from the tissues to the lung. This system can exchange chloride and bicarbonate ions across the membrane. Using several chemical probes that inhibit anion permeability, knowledge has been gained about the molecular localization of this mechanism. Some of the probes can be transported by the anion system and, under certain conditions, can be fixed covalently to it. In this way it is possible to determine not only the number of permeation sites but also the polypeptide in which they are localized.

This approach has permitted the identification of the *polypeptide of band 3* (see Fig. 8–4) with the site involved in the transport. This protein spans the membrane and may exist as a dimer (or tetramer). (We have seen earlier that it also corresponds to the 8 nm particles observed in freeze-fractured membranes, (Fig. 8–5).

The model for anion transport that has been proposed is that of a continuous proteinaceous aqueous channel across the membrane having, near the outer surface, one anion binding site with three positive charges, and a hydrophobic barrier to limit the free diffusion of anions (Fig. 8–17). Passage through the barrier could occur only as a consequence of the binding of the anion.

In this model it is assumed that the segment containing the binding site could exist in two conformations, one facing outward, the other inward. In this way the segment could act as a gate in the proteinaceous channel, swinging between the two positions; this would permit the binding site to interact with anions coming from the outside or from the interior of the cell (see Rothstein, et al., 1976). (In this model it is ob-served that the anion binding site is located in a 65,000 daltons segment of the protein which remains in the membrane after a 35,000 daltons segment has been separated by a proteolytic enzyme.)

The Vectorial Function of $N^+ K^+$ ATPase — The Carrier Hypothesis

We have seen previously that the so-called *sodium pump* is a mechanism of active transport by which Na^+ is eliminated from the cell. This mechanism, discovered by Hodkin and Keynes in 1955, was soon associated by Skou with the $Na^+ K^+$ ATPase. This enzyme is able to couple the hydrolysis of ATP with the removal of Na^+ from the cytoplasm against an unfavorable electrochemical gradient.

Figure 8–18 shows an idealized diagram of the Na^+K^+ATPase within the red cell membrane. It may be observed that the hydrolysis of one ATP provides the energy for the linked transport of 2 K^+ ions toward the inside, and 3 Na^+ ions toward the outside of the cell. This diagram shows the vectorial characteristics of the enzyme which is sensitive to ATP on the inside of the membrane, but not on the outside. This ATPase

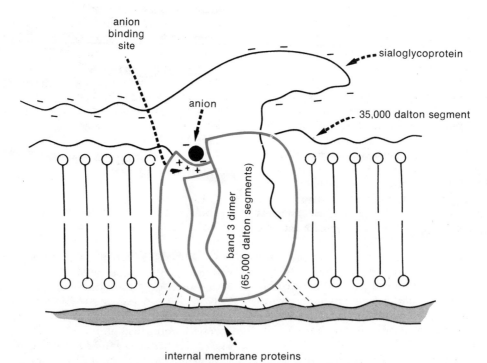

Figure 8–17 Schematic model of the disposition of the band 3 protein and of the anion transport site in the erythrocyte membrane. (From Rothstein, A., Cabantchik, Z. I., and Knauf, P., *Fed. Proc.*, 35:3, 1976.)

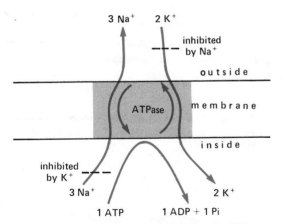

3 Na⁺ 2 K⁺

inhibited by Na⁺

outside

ATPase membrane

inside

inhibited by K⁺

3 Na⁺ 2 K⁺

1 ATP 1 ADP + 1 Pi

Figure 8–18 Diagram of the Na⁺-K⁺ ATPase in the cell membrane. Observe that for each molecule of ATP hydrolyzed at the inner part of the membrane 3 Na⁺ ions are transported outside and 2 K⁺ ions are transported inside. (See description in the text.)

is stimulated by a mixture of both Na^+ and K^+ and is inhibited specifically by the cardiotonic glycoside *ouabain*. Figure 8–18 also shows that the Na^+ and K^+ sites are independent and are competitively inhibited by K^+ and Na^+, respectively. The Na^+ K^+ ATPase has been found to be concentrated in membranes of nerves, brain, and kidney and is particularly rich in the electric organ of the eel and the salt gland of certain marine birds (e.g., albatross) in which the transport of ions is extremely active.

The vectorial properties of the enzyme can be demonstrated cytochemically; thus, by electron microscopy the liberation of phosphate has been shown to occur exclusively on the inner side of the red cell ghost.[35] On the other hand, ouabain binds exclusively to the outer surface of the membrane. The vectorial function of Na^+K^+ ATPase can be best explained by the so-called "carrier" hypothesis. In its most simple form a carrier consists of: (1) a membrane protein having a specific site to combine with the ligand to be translocated, (2) a mechanism of translocation which moves the ligand from one side of the membrane to the other, and (3) a mechanism for the release of the ligand. Most carrier models involve some kind of conformational change of the protein by which the affinity of the binding site for the ligand changes during the translocation. According to the diagram of Figure 8–18, in the case of Na^+ K^+ ATPase it is possible to envision that the enzyme has binding sites for Na^+ and K^+ and a complex carrier mechanism to deliver and release these ions at opposite sides of the membrane. The coupling of the barrier

mechanism to the hydrolysis of ATP generates the energy needed to move the ions against the concentration gradients.

The transport reaction takes place in two major steps. Step I consists of the formation of a *covalent phosphoenzyme intermediate* (E ~ P) on the inner side of the membrane and in the presence of Na^+. This first reaction is inhibited by Ca^{2+}, but not by ouabain. In Step II the [Na^+ E ~ P] complex is hydrolyzed to form free enzyme and Pi. This reaction requires K^+, added to the outside of the membrane, and is inhibited by ouabain, which competes with K^+. Steps I and II can be represented as follows:

Step I $\quad Na^+_{in} + ATP + E \longrightarrow$ (5)
$$[Na^+ \cdot E \sim P] + ADP$$

Step II $[Na^+ \cdot E \sim P] + K^+_{out} \longrightarrow$ (6)
$$Na^+_{out} + K^+_{in} + Pi + E$$

It may be observed that Na^+ and K^+ are ultimately translocated in the opposite directions from which they were bound.

The Na^+ K^+ ATPase is tightly bound to the cell membrane, from which it may be released by the use of detergents. The molecular weight of the enzyme has been estimated to be about 670,000, and it probably contains several polypeptides. Certain lipids are needed for the activity of this enzyme. In a red cell there are about 5000 enzyme molecules, each of which may extrude 20 Na^+ ions per second. (For a review of this field, see Schwartz et al., 1972.)

Various Substances are Transported by a Carrier Mechanism

Two types of Na^+ pumps can be distinguished: (1) The so-called *Na^+ K^+ exchange pump* or *coupled neutral pump* in which, as described before, the transport of K^+ inward is coupled with that of Na^+ outward and in which electrical neutrality is thus achieved. (2) The so-called *electrogenic pump* in which the exit of Na^+ is not compensated by the entrance of K^+. In this case an electric gradient generates the potential which may provide the driving force for transporting other solutes.

It is now well known that the transport of many substances, such as glucose and various amino acids, into the cell may be coupled to the Na^+ pump. For example, the entrance of glucose into the intestine cells is activated by Na^+ and inhibited by K^+ and ouabain. To explain these

155

findings a carrier mechanism has been postulated. In such a model the carrier should have a binding site for Na^+ and another for glucose and should exist in two conformational states, one with low affinity for glucose and the other with high affinity.

A carrier model can be postulated in which the glucose site is first in a high affinity state and binds the glucose molecule at the surface of the cell. In a second step the carrier is translocated, and the glucose site changes to a low affinity and releases the ligand into the cell. The two affinities are modulated by the concentration of Na^+ outside and inside the cell.

The carrier model is based on the existence of a protein that is able to translocate a soluble molecule across the thickness of the membrane. Several hypotheses for the molecular mechanism that is involved have been put forth. In one, the solute binds on the outside surface of the membrane to the carrier, and this then rotates and delivers the solute to the interior of the cell. In the second, it is supposed that the carrier is fixed in place but is able to undergo some kind of conformational change by which the solute is translocated. From the viewpoint of energetics this *fixed-pore mechanism* is more probable, since it would require less expenditure of energy.

Fixed-pore mechanisms, such as that illustrated in Figure 8–17, are also more directly related to the structure of the membrane, with the integral proteins spanning its entire thickness. It is postulated that along the channels there are hydrophilic groups which can translocate the soluble molecules. A model of this type is depicted in Figure 28–28 for the Na^+-K^+ channel of the cholinergic receptor.

Selectivity of Transport — Dependent on Permease Systems

The transport of many different molecules across the membrane shows a high degree of specificity. In other words, the permeability of a molecule is related to its chemical structure. For example, while one molecule may readily enter the cell, another, with the same size but a slightly different molecular structure, may be completely excluded. This type of selectivity is similar to that of the *enzymes* (Chap. 6) and to that of the *receptors*, in which a *binding site* is able to "recognize" a definite *substrate* or *ligand*. Such selective transport has been attributed to certain proteins of the membrane which are called *permeases*.

Permeases *accelerate transport*, provide for *special selectivity*, and are *recycled*; this means that they remain unchanged after having assisted in the entry or exit of a molecule. Some permeases can transport only if there is a favorable concentration gradient, while others may also do so against an unfavorable one. The first type of transport, being driven by a passive mechanism, is sometimes called *facilitated diffusion*.

When the permease transport is against the gradient, it represents another case of *active transport*. One of the best known permease systems is the *lac permease* of *E. coli*. As will be shown in Chapter 25–1 lac permease is coded within the *lac operon* and is a single unit protein that acts as a carrier for the entrance of lactose.

Permeases are genetically determined; for example, in *E. coli* there are mutants which are incapable of transporting lactose. Lac-permease has been isolated and purified.[36] In *E. coli* there may be many permease systems specialized for the transport of various molecules. Permeases are found in the cell membrane of eukaryotic cells and also in the inner mitochondrial membrane (see Chapter 12–5).

Penetration by Large Molecules — Various Mechanisms

What has been said demonstrates that the general term "cell permeability" comprises a variety of different mechanisms. In addition to certain foreign substances that can penetrate the cell because they are lipid-soluble (e.g., anesthetics), generally ions penetrate through charged pores and other molecules penetrate by permease systems.

There is no doubt that under certain conditions large molecules, such as certain proteins, penetrate the cell. This is the case with ribonuclease, an enzyme that penetrates living plant cells readily and also eggs, flagellates, ascitic tumors, and so forth.

Basic proteins of the protamine and histone types have been reported to enter into living cells. In Chapter 21 it will be mentioned that DNA penetrates certain bacteria and produces a genetic change known as *transformation*.

In Chapter 13–2 *phagocytosis* and *pinocytosis*, by which solid or fluid material in bulk can be ingested by the cell, will be studied.

SUMMARY:

Cell Permeability

The study of permeability requires knowledge of the chemical and molecular organization of the cell membrane. This membrane regulates the inflow of substances to the cell and the outflow of water ions and other materials. The cell membrane maintains a balance between the osmotic pressure of the intracellular and the interstitial fluid. Solutions may be isotonic, hypotonic, or hypertonic with respect to the intracellular fluid.

The *ionic concentration* in the intracellular fluid differs from that in the interstitial fluid. In the latter, Na^+ and Cl^- are high, and K^+ is low. Inside the cell, K^+ is high, Na^+ and Cl^- are low, and there is a large pool of organic anions (A^-) that do not cross the cell membrane. A *membrane potential* of -20 to -100 millivolts is detected in all cells. Cell membranes are polarized, being negatively charged inside and positively charged outside.

Movement of substances across the cell membrane may be by *passive diffusion* or by several mechanisms of *active transport*. *Diffusion* occurs when there is a concentration gradient and takes place from a high to a low concentration. *Diffusion of molecules* across membranes depends on molecular volume and the lipid solubility of the substance. *Diffusion of ions* depends on both the concentration and the electric gradients across the membrane.

The distribution of K^+ and Cl^- generally follows a Donnan equilibrium that depends on the presence of non-diffusible anions inside the cell. The relationship between concentration gradient $\left(\dfrac{C_1}{C_2}\right)$ and membrane potential (E) is given by the Nernst equation:

$$E = RT \log \frac{C_1}{C_2}$$

The *active transport* of neutral molecules and ions requires energy (ATP) and is generally coupled to the energy-yielding mechanisms of the cell. Active transport is blocked by cooling or by certain metabolic poisons of respiration and glycolysis. With complete energy block the membrane potential may decrease to zero and there is leakage of K^+. Active transport may be demonstrated by simple experiments using kidney tubules, frog skin, and toad bladder. Ionic transport is particularly intense in electric tissues of the eel and in salt-secreting glands. By active transport Na^+ and water are pumped out of the cell and K^+ penetrates the cell. Both electrochemical diffusion (i.e., passive transport) and energy-dependent ion transport determine the distribution of ions and the membrane potential across the membrane. The passage of ions is thought to be across charged pores in the membrane. The size of the hydrated ion is important in the transport through the pores.

In a red blood cell 0.06 per cent of the area may be occupied by pores. This membrane has an active anion permeation mechanism to transport chloride and bicarbonate. This mechanism has been localized to the polypeptide of band 3 that spans the membrane and to the 8 nm particles seen in *freeze-fractured membranes*. A channel-model for the transport of anions has been postulated.

$Na^+ K^+$ ATP*ase* is related to the active transport of Na^+ and K^+ in red blood cells, nerves, brain membranes, electric tissues, and others. The hydrolysis of 1 ATP $\rightarrow$ ADP + Pi in the inner surface of the membrane is coupled with the transport of 3 Na^+ from the inside toward the outside of the membrane and with the transport of 2 K^+ in the opposite direction.

All these properties indicate that the enzyme has a vectorial orientation across the membrane. The $Na^+ K^+$ ATPase is specifically inactivated with *ouabain*. The function of this enzyme is explained by a complex carrier mechanism which comprises: (1)

binding sites for Na^+ and K^+, (2) a mechanism for translocation, and (3) a mechanism for the release of the ligand.

The production of a covalent phosphoenzyme intermediate has led to the concept that the transport is carried out in two steps as follows:

Step I $\quad Na^+_{in} + ATP + E \longrightarrow [Na^+ \cdot E \sim P] + ADP$

Step II $\quad Na^+ \cdot E \sim P + K^+_{out} \longrightarrow Na^+_{out} + K^+_{in} + Pi + E$

in which $[Na^+ \cdot E \sim P]$ is the phosphorylated complex of the enzyme.

Glucose and various amino acids are transported coupled to the Na^+ pump. The transport of many molecules shows a high degree of specificity and is attributed to proteins of the membrane called *permeases*. Permease transport may be driven by a passive mechanism (*facilitated diffusion*) or by active transport. One of the best known permeases is the lac permease of *E. coli*, which is coded within the lac operon and acts in the transport of lactose.

8–4 DIFFERENTIATIONS AT THE CELL SURFACE AND INTERCELLULAR COMMUNICATIONS

Regions of the cell surface of certain cells are related to absorption, secretion, fluid transport, and other physiologic processes. Topographically they are referred to as specializations of the cell surface.

Figure 8–19 is a diagram of an idealized columnar cell in which various types of differentiations of the cell membrane may be observed. The apical surface is projected into slender processes called *microvilli*. On the surface corresponding to the edge of the cell in contact with an adjacent cell there are several differentiations (see later discussion). At the base of the cell the plasma membrane is covered by a thick basement membrane of extracellular material where infoldings of the plasma membrane may be observed.

Microvilli — A Greatly Increased Cell Membrane Surface Area

In the intestinal epithelium microvilli are very prominent and form a compact structure that appears under the light microscope as a *striated border*. These microvilli, which are 0.6 to 0.8 μm long and 0.1 μm in diameter, represent cytoplasmic processes covered by the plasma membrane. Within the cytoplasmic core fine microfilaments are observed which in the subjacent cytoplasm form a terminal web. As will be mentioned in Chapter 9–4, these filaments contain actin and are attached to the tips of the microvilli by α-actinin; their function is to produce contraction of the microvilli.

The outer surface of the microvilli is covered by a coat of filamentous material (fuzzy coat) composed of glycoprotein macromolecules. Microvilli increase the effective surface of absorption. For example, a single cell may have as many as 3000 microvilli, and in a square millimeter of intestine there may be 200,000,000. The narrow spaces between the microvilli form a kind of sieve through which substances must pass during absorption. Numerous other cells have microvilli, although they are fewer in number. They have been found in mesothelial cells, in the epithelial cells of the gall-bladder, uterus, and yolk sac, in hepatic cells, and so forth.

The *brush border* of the kidney tubule is similar to the striated border, although it is of larger dimensions. An amorphous substance between the microvilli gives a periodic acid-Schiff reaction for polysaccharides. Between the microvilli the cell membrane invaginates into the apical cytoplasm. These invaginations are apparently pathways by which large quantities of fluid enter by a process of pinocytosis. (Other specializations of the cell surface, such as cilia and flagella, are described in Chapter 9).

Desmosomes and Intermediary and Tight Junctions — Intercellular Attachments

Essentially four types of differentiations are present at the lateral surfaces of epithelial cells: (1) the *macula adherens* or *desmosome*; (2) the *zonula adherens*, also called *intermediary junction* or *terminal bar*; (3) the *zonula occludens* or *tight junction*; and (4) the *gap junction* or *nexus*. The first three types may be observed in the electron micrograph of Figure 8–20.

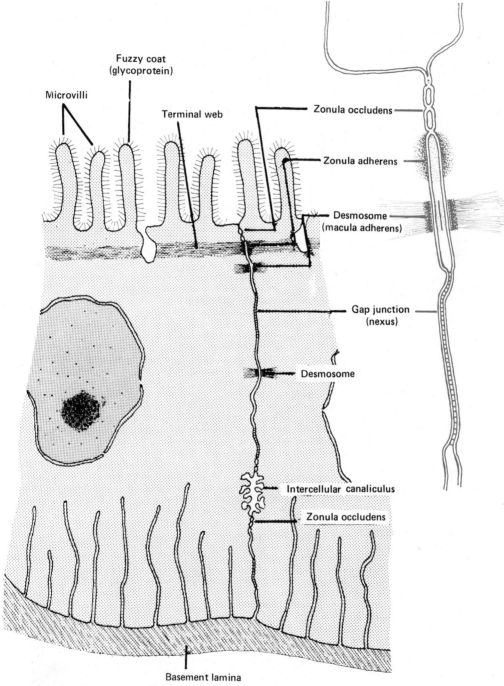

Figure 8–19 Diagram of an idealized columnar epithelial cell showing the main differentiations of the cell membrane. **To the right,** at a higher magnification, the series of differentiations found between two epithelial cells are indicated. (See the description in the text.)

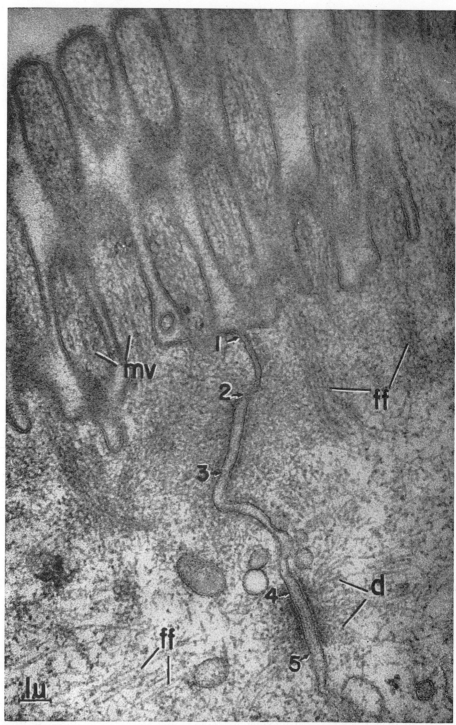

Figure 8-20 Electron micrograph showing the apical region of contact between two intestinal cells. *1-2,* tight junction; *2-3,* intermediary junction; *4-5* desmosome. (See the description in the text.) *d,* desmosome; *ff,* fine filaments in the matrix; *mv,* microvilli. ×96,000. (From Farquhar, M., and Palade, G. E., *J. Cell Biol., 17:*375, 1963.)

The so-called *desmosomes,* found in a number of epithelial cells, appear under the light microscope as darkly stained bodies. These structures are formed by a circular area, about 0.5 μm in diameter, of the plasma membranes of two adjacent cells that are separated by a distance of 30 to 50 nm. Under the membrane there is a dense intracellular plaque toward which numerous *tonofilaments* converge. These filaments describe a kind of loop in a wide arc and course back into the cell. Within the intercellular gap a coating material may be observed which sometimes forms a discontinuous middle dense line. This extracellular material contains acid mucopolysaccharides and proteins. In fact, desmosomes are broken by trypsin, collagenase, and hyaluronidase and are sensitive to agents that chelate calcium. The dense intracellular attachment plaques are digested by proteolytic enzymes.[37]

While the tonofilaments provide the intracellular mechanical support, cellular adhesion at the desmosome depends on the extracellular coating material. Frequently there are regions of looser contact between the desmosomes and even intercellular spaces for free circulation of fluids.

Along the basal surface of some epithelial cells *hemidesmosomes* may be observed. These are similar to desmosomes in fine structure, but represent only half of them, the outer sides frequently being substituted with collagen fibrils.[38]

The zonula adherens or *terminal bar* also called *intermediary junction* is generally found at the interface between columnar cells just below the free surface. Under the electron microscope the terminal bar appears somewhat similar to the desmosome. The membrane is thickened and the adjacent material is dense, but filaments are generally lacking (Fig. 8–20).

Within the *zonula adherens* there are parts of the two adjacent membranes that come into close contact at certain points, forming a kind of anastomosing network that extends around the apical region of the cell. Such a network is especially visible in freeze-etched preparations.[39]

Cell contacts may be specially differentiated to create a barrier or "seal" to diffusion.[40] In the *tight junction (zonula occludens)* the adjacent cell membranes adhere, and therefore there is no intercellular space for a variable distance (Fig. 8–20, 1–2). The tight junction is situated just below the apical border, and at this point the outer leaflets of the unit membranes adhere, forming a single intermediary line. Experiments have demonstrated that macromolecules that are put into the lumen of an epithelial cavity cannot penetrate the intercellular space. Tight junctions may also play an important role in brain permeability at the level of the blood-brain barrier and the synaptic barrier.

Gap Junctions (Nexus) and Intercellular Communications

The concept of the cell as the basic unit of all living matter should not lead us to disregard the fact that multicellular organisms are made of populations of cells that interact among themselves. Such *cellular interaction* is essential for the coordination of activities, and furthermore, the propagation between cells of signals for growth and differentiation is indispensable for development. It is now known that most cells in an organized tissue are interconnected by *junctional channels* and that they share a common pool of many small metabolites and ions that pass freely from one cell to another. However, their individuality is maintained by macromolecules which are not exchanged between cells.

Electrical Coupling between Cells Depends on Gap Junctions

One of the manifestations of cellular interaction is electrical coupling between cells. By introducing intracellular microelectrodes into adjacent cells of a tissue, investigators have been able to demonstrate that in many animal cells there are intercellular communications.[41] In this case the cells are electrically coupled and have regions of low resistance in the membrane through which there is a rather free flow of electrical current carried by ions. The other parts of the cell membrane, which are not coupled, show a much higher resistance. This type of coupling, called *junctional communication,* is found extensively in embryonic cells. In adult tissue it is usually found in epithelia, cardiac cells, and liver cells. Skeletal muscle and most neurons do not show electrical coupling.

Electrical coupling seems to be related to the so-called *gap junctions* or *nexus* that are shown in the diagram of Figure 8–19. An excellent material for the study of gap junctions is the myocardial tissue, in which the action potential is transmitted from cell to cell by an electrical coupling.

In a thin section the gap junction appears as a plaque-like contact in which the plasma membranes of adjacent cells are in close apposition, separated by a space of only 2 to 4 nm. This gap can be filled with electron opaque materials such as lanthanum and ruthenium red.[42] In tangential sections gap junctions show a hexagonal array of 8- to 9-nm particles (Fig.

8–21, 2a). The electron-dense material is able to penetrate in between the particles, thus delineating their polygonal arrangement, and also into the central region of each particle. This central region is 1.5 to 2 nm in diameter and corresponds to the probably location of the channel.

With freeze-fracturing it is possible to split

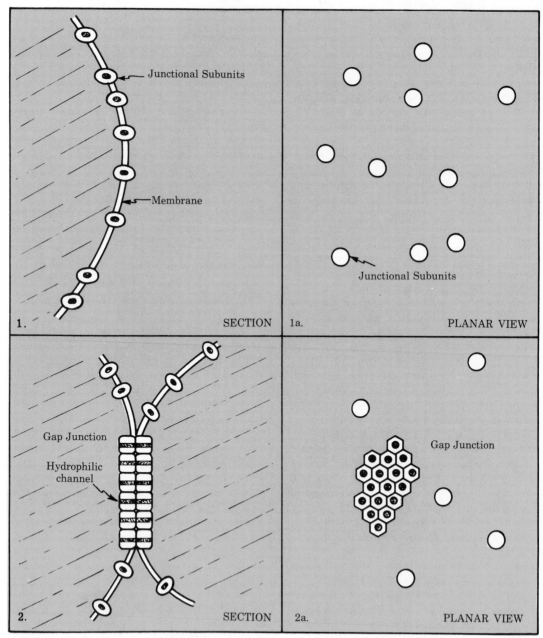

Figure 8–21 Diagram of the formation of gap junctions in transverse section and planar view. Larger precursor particles appear first (**1** and **1a**), the space between cells is reduced, and the particles are confluent with the plaques that make the channels (**2** and **2a**). In a planar view the gap junctions show a hexagonal array. (Courtesy of J. D. Pitts.)

the junctions and to better define their internal structure. Upon the inner half-membrane of the fracture (PF face) the 8-to-9-nm particles are observed, while in the external half (EF face) a complementary array of pits or depressions appears.[39] Gap junctions are resistant to mechanical disruption, proteolysis, and removal of Ca^{2+} and can be isolated from the cells (see Gilula, 1977). After isolation they can be shown by electron microscopy and x-ray diffraction to have the same polygonal lattice structure, with 8-to-9 — nm particles, as seen in the intact junction (see Caspar, et al., 1977). The unit of the gap junction has been named the "*connexon*," and it appears to span the bilayer of each of the two connected cell membranes, as well as the gap in between, thus providing for intercellular communication (Fig. 8–22).

The development of gap junctions has been observed in various cell types. The process consists of the appearance of plaques between cells, with a reduction in the intercellular space, the appearance of larger particles (probably precursors), and finally the arrangement of the particles in polygonal arrays (Fig. 8–21) (see Pitts, 1977).

When two cells are brought into direct contact, junctional communications may be formed in a matter of seconds, and this does not require new protein synthesis. It is therefore thought that the 8-nm protein particles are already present on the cell surface and that when cells come in contact these particles diffuse and interact with similar particles on the other membrane (Fig. 8–21).

From the physiologic viewpoint we can envision the gap junction as resulting from the apposition of two channels, one present in each membrane. Such channels not only traverse each membrane but also project into the intercellular gap and are joined there, thus creating the direct intercellular communication of the connexon (Fig. 8–22, *bottom*). The essential components of such coupling are represented in the diagram of Figure 8–22, *top*. It can be observed that the channels represent a region of high permeability that is separated from the intercellular fluid by a kind of junctional seal of low permeability.

Gap Junctions – Channels Permeable to Ions and Small Molecules

The size of the channels present in gap junctions has been estimated by (1) electron mi-

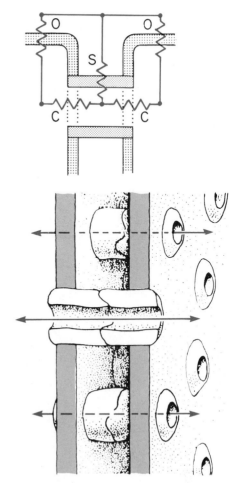

Figure 8–22 **Top,** diagram of the gap junction channel as defined by electrical measurements and intracellular tracer diffusion. *O,* non-junctional membrane; *C,* junctional membrane; *S,* junctional insulation. **Bottom,** model of the gap junction in which two protein channels, one in each membrane, are in continuity. (Courtesy of W. R. Loewenstein.)

croscopy (see earlier discussion), (2) by electrical measurements, and (3) by the diffusion of tracer substances.

By determining the conductance of single channels and the resistance of the cytoplasm and by assuming a cylindrical channel 20 nm long, the diameter can be calculated to be of the order of 1 nm, a size which is in agreement with morphological data (see Loewenstein, 1976).

The first tracer used was the fluorescent dye *fluorescein* (300 daltons), which when injected into a cell was shown to diffuse into adjacent cells. Recently, with the use of fluorescent peptides in cells of *Chironomus* salivary glands; a weight limitation of 1300 to 1900 daltons was reached, which corresponds to a channel pore of about 1.4 nm.[43]

The permeability of the channel is regulated by Ca^{2+}. Normally the Ca^{2+} content of the cell is low ($10^{-7}M$), and the permeability of the channel is high. If the intracellular Ca^{2+} concentration increases; the permeability is reduced. The extracellular concentration of Ca^{2+} is about $10^{-3}M$, so that any damage to the cell membrane or the junctional seals may increase the Ca^{2+} inside the cells. When the Ca^{2+} level reaches a certain limit the channels close,[44] and they open again if the cell eliminates the excess Ca^{2+}.

It has been found that the junctional communications depend on the energy provided by oxidative phosphorylation (i.e., ATP). Treatments that inhibit cell metabolism, such as cooling at 8° C, dinitrophenol, cyanide, and oligomycin, produce uncoupling between the cells; this may be reversed by injection of ATP. It is thought that the uncoupling may be due to the influx of Ca^{2+}, either from the medium or from mitochondria, during the action of the inhibitors.

Recently, the intracellular Ca^{2+} concentration has been monitored by the injection of *aequorin*, a luminescent protein in which the light emission is proportional to the Ca^{2+} concentration.[45] With such a monitoring system and with the use of fluorescent peptides of various sizes, it has been possible to demonstrate that the permeability changes caused by Ca^{2+} are *graded*. In other words, tracers of different molecular sizes can cross the channel according to the intracellular Ca^{2+} concentra-

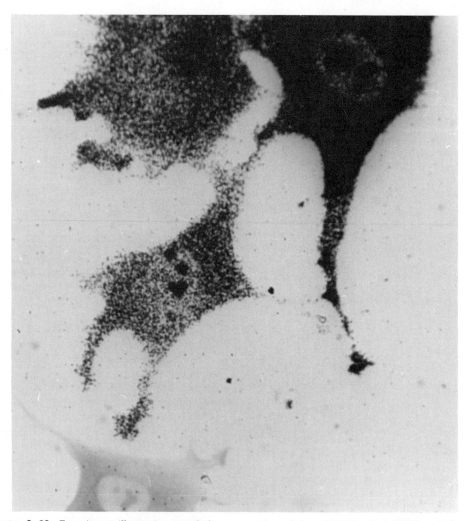

Figure 8–23 Experiment illustrating metabolic cooperation. Cultured BHK cells labeled with ³H-uridine were added to similar but unlabeled cells. (In this radioautograph the highly labeled cell on the top right was added to the culture.) It may be observed that the label was transferred to the two cells that make direct contact but not to another which is not in contact (bottom). ×1160. (Courtesy of J. D. Pitts.)

tion. As the Ca^{2+} is elevated from 10^{-7} to $10^{-5}M$, the limit of permeation decreased gradually, and at $5 \times 10^{-5}M$ the channel is closed. It is thought that this control by Ca^{2+} may provide a powerful mechanism for regulation of intercellular communications (see Loewenstein, 1977).

Coupling between Cells Enables Metabolic Cooperation

Several functions may be ascribed to junctional communications. In cardiac muscle and in electrical synapses between certain neurons, the junctions are related to communication of electrical signals between these cells. Junctional interconnections are particularly widespread during embryonic development at stages in which cell differentiation takes place. Because diffusion of molecules can readily take place through them, these junctions are well suited for the dissemination of signals controlling cellular growth and differentiation at close range.

The coupling between cells is not only electrical but also *metabolic*. In addition to permitting free passage of electric current, junctional communications allow the passage of ions and small molecules such as nucleotides, sugars, vitamins, and other metabolites. This gives rise to the phenomenon of *metabolic cooperation* between cells. Figure 8–23 shows the transfer of nucleotides between cells in culture. A few cells prelabeled with a nucleotide were added to a culture of unlabeled cells; it can be seen that the nucleotide is transferred only between those cells that have direct contacts. Other experiments have shown that while nucleotides are easily transferred, macromolecules such as enzymes and RNA are not transferred between cells.

Metabolic cooperation was first demonstrated in cell cultures of fibroblasts in which a mutation made the cells unable to incorporate exogenous ^3H-hypoxanthine into their nucleic acids. If such cells were grown in contact with wild-type fibroblasts (i.e., non-mutants), there was incorporation in both types of cells. These findings indicate that a metabolite (the nucleotide inosine monophosphate) from the wild-type fibroblast had penetrated the mutant fibroblast through intercellular channels.[46]

Altered Coupling in Cancer Cells

Another important finding was that certain cancer cells show no intercellular coupling and

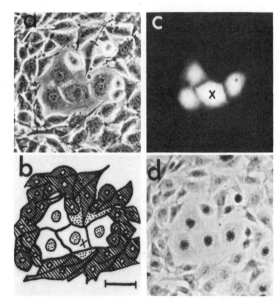

Figure 8–24 Experiment that demonstrates the lack of coupling between normal cells and cancerous cells. **a,** phase contrast micrograph of a culture having four normal liver cells surrounded by cancerous cells; **b,** tracing of micrograph in **a; c,** the cell marked with X was injected with fluorescein; it is observed that the stain has diffused into the three other normal cells, but not into the cancerous ones; **d,** since the fluorescein is labeled with tritium, in the radioautograph only the four normal cells are ^3H-labeled. (From Azarnia, R., and Loewenstein, W. R., *J. Membr. Biol.,* 6:368, 1971.)

that they fail to communicate with normal cells[47] (Fig. 8–24). It is assumed that these cancer cells have a genetic defect which has interrupted the passage of growth-controlling molecules between them. Recent experiments in which heterokaryons have been produced between a cancer cell and a normal cell (i.e., cell fusion) have shown that it is possible to obtain a noncancerous hybrid.[48] In this hybrid the intercellular communications, which were absent in the original cancerous cell, were now established between the hybrid cells.[49]

Hybrid cells were made between human normal fibroblasts and mouse cancer cells that showed no intercellular communication. These hybrids were observed for several generations by studying the type of growth (i.e., cancerous or normal) and the presence or absence of channels. It has been found that a human factor probably linked to one chromosome can correct both for the growth and the channel defect.[50]

These experiments and others suggest that coupling and permanent uncoupling could be genetically determined. They also point to the possible regulatory role of molecules of small molecular weight — probably nucleotides — between cells controlling growth and differentiation.[46] Such regulatory agents are not transferred to the cancerous cells having no intercellular communication (see Pitts, 1977). It is evident that in most cells, as in a society, the exchange of information is essential to maintain normal health.

SUMMARY:

Differentiations of the Cell Membrane and Intercellular Communications

The cell membrane may present regional differentiations that are related to specialized functions: absorption, fluid transport, electrical coupling, and so forth. *Microvilli* are found at the apical surface of the intestinal epithelium and form the *brush border* of kidney tubules. They increase the absorption surface and are covered by a coat of glycoproteins. *Desmosomes* are zones of attachment between epithelial cells, and they have a mechanical function. The *zonula adherens* (intermediary junction) between adjacent cells also has a mechanical function. *Tight junctions (zonulae occludens)* are regions in which the two adjacent membranes fuse, forming a kind of seal; at these points the intercellular space disappears.

The so-called *gap junctions (nexus)* are essential in intercellular communications. They represent regions in which there are junctional channels through which ions and molecules can pass from one cell to another. Cells having gap junctions are electrically coupled; i.e., there is a free flow of electrical current carried by ions. At the gap junction the membranes are separated by a space of only 2 to 4 nm, and there is a hexagonal array of 8-to-9–nm particles. At the center of each particle there is a channel 1.5 to 2 nm in diameter.

Gap junctions provide direct intercellular communication by allowing the passage of molecules up to a limiting weight of 1300 to 1900 daltons (in *Chironomus* salivary glands). The permeability is regulated by Ca^{2+}; if the intracellular Ca^{2+} level increases the permeability is reduced or abolished. Through the gap junction, metabolites (i.e., labeled nucleotides) can pass from one cell to another. In several strains of cancer cells there is no coupling as seen in normal cultured cells. Coupling is genetically determined, and probably genes linked to one chromosome can correct the cancerous growth and the channel defect. Junctional communication may convey electrical signals between certain neurons (i.e., electrical synapses) and between cardiac cells; however, most neurons and skeletal muscle lack electrical coupling. Gap junctions are also used in the transfer of substances that control growth and differentiation in cells.

8–5 COATS OF THE CELL MEMBRANE AND CELL RECOGNITION

At the beginning of this chapter the *coats* surrounding cell membranes were mentioned. These are very conspicuous in eggs of marine animals and in amphibia. A glycoprotein-like substance called *mucin* is the main constituent. Mucins also cover and protect the cell surface lining the gastrointestinal tract. Polysaccharides constitute the *pectin* and *cellulose* of plant cell walls and the *chitin* of crustacea. (The cell wall of plant cells is discussed in Chapter 14–1.)

Glycoproteins and polysaccharides, in the form of hyaluronic acid, are found at the base of most epithelial cells, in capillaries, and also in many intercellular spaces. The name *glycocalyx*[51] has been coined to designate the glycoprotein and polysaccharide covering that surrounds many cells.

The glycocalyx contains the oligosaccharide side-chains of the glycolipids and glycoproteins which are exposed to the outer surface of the membrane. The cell coat has negatively charged sialic acid termini, on both the glycoproteins and gangliosides, which may bind Ca^{2+} and Na^+ ions. When the membranes are treated with *neuraminidase*, an enzyme that removes sialic acid, there is a reduction in the negative charge of the membrane.

The use of PAS and Alcian blue staining may render the surface of many cell types more visible. Staining with ruthenium red or with lanthanum may reveal the thin cell coat present in biliary capillaries (Fig. 8–25).

Visualization of the carbohydrates may be made more specific by the use of *lectins*. These are proteins, derived from plants, that bind to the cell surface and cause agglutination. *Concanavalin A* isolated from Jack beans is specific for glucose and mannose residues; and *germ agglutinin*, for N-acetylglucosamine. These lectins can be labeled with fluorescent dyes or with electron-dense materials for observation under the electron microscope.

The cell coat generally appears as a layer 10 to 20 nm thick which is in direct contact with the outer leaflet of the plasma membrane. In some cases, as in the ameba, the coat is formed by fine filaments 5 to 8 nm thick and 100 to 200 nm long. The strength and durability of the coat varies from cell to cell. For example, the coat of the intestinal epithelium resists vigorous mechanical and chemical attacks; in other cells the coat is labile and is depleted by washing or exposure to enzymes. (see Luft, 1976).

In Chapter 11 it will be mentioned that the biosynthesis of the glycoproteins forming the glycocalyx takes place in the ribosomes of the endoplasmic reticulum and that the final assembly with the oligosaccharide moiety is accomplished in the Golgi complex. The cell coat can be considered to be a secretion product of the cell which is incorporated into the cell surface and undergoes continuous renewal.

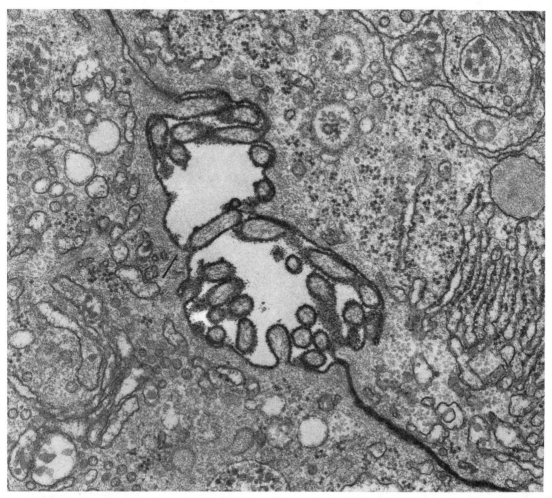

Figure 8–25 Electron micrograph of a biliary capillary stained with lanthanum nitrate to enhance the electron density of the surface coat of cells. Note the dark material between the liver cells and around the microvilli. ×38,400. (Courtesy of D. Ferreira.)

The components of the cell undergo active turnover, and some of them are shed into the surrounding medium. This is particularly evident in cultured cells, which leave a kind of "footprint" on the glass to which they adhere. (see Doljanski and Kapeller, 1976).

Numerous Functions are Attributed to the Cell Coat

In addition to the protection of the cell membrane numerous functions have been attributed to the cell coat. Although it is not absolutely necessary for the integrity of the cell and for the permeability of the plasma membrane, it nevertheless performs functions that are of great significance to studies in cell biology.

Filtration. The extracellular coats that surround many vertebrate capillaries, especially the kidney glomerulus, act as a filter and regulate the passage of molecules according to size. Hyaluronate in connective tissue may control diffusion.

Microenvironment. The glycocalyx may change the concentration of different substances at the cell surface by acting as a diffusion barrier. Because of its charge, the glycocalyx may act as a kind of exchange resin and change the cationic environment of the cell. For example, a muscle cell with its excitable plasma membrane is surrounded by a glycocalyx that can trap sodium ions. Certain components, such as hyaluronate, can drastically change the electrical charge and pH at the cell surface. Because of this, enzymes present at the plasma membrane may change their activity while they are kept in the microenvironment of the cell.

Enzymes. In the intestinal cells the apical coat is remarkably stable and cannot be separated from the underlying striated border made of microvilli. Histochemical techniques have demonstrated alkaline phosphatase in the coat as well as on the surface of the microvilli. When these structures are isolated, practically all the enzymes involved in the terminal digestion of carbohydrates and proteins are found in them.

Molecular Recognition. The oligosaccharides present in the cell coat may constitute a kind of molecular code for the cell surface. The number of permutations of their individual components (i.e., galactose, hexosamine, mannose, fucose, sialic acid, and so forth) makes for each cell type a special kind of *fin-gerprint* by which molecular recognition between cells could be established. It is also well known that hormonal receptors and various antigenic properties of the membrane are present in the cell coat.

Blood Group Antigens. The classical division of the ABO blood groups is based on the occurence of antigens on the red cell surface and of specific antibodies in the serum. The specific ABO antigens are found not only on the erythrocytes but in other tissues and also in secretions. Each one of the groups has a specific structure for the terminal carbohydrates (see Harrison and Lunt, 1975).

Other Antigens. Among these antigens are those of histocompatibility, which permit the recognition of the cells of one organism and the rejection of other cells that are alien to it (e.g., the rejection of grafts from another organism). This function, essentially, is related to molecular recognition.

The major sialoglycoprotein of the red cell membrane carries the M and N antigens that appear, infrequently, in man. It also contains the receptor sites for the influenza virus and various lectins.

Molecular recognition reaches its maximum expression in the nervous system, where a neuron can make synaptic contacts with numerous other neurons, thereby forming specific neuronal circuits of immense complexity. This property of the cell membrane (i.e., molecular recognition) is dependent on the expression of genes located within the nucleus. Certainly, molecular recognition is one of the fields of cell biology and neurobiology which must be strongly developed in the future (see Singer and Rothfield, 1973).

Cell-cell Recognition — Specific Cell Adhesion and Contact Inhibition

Other properties of the cell coat are related to the capacity of cells to "recognize" similar cells in a tissue, to adhere to one another, to dissociate and reassociate, and to produce an inhibition in the neighboring cells by the so-called *contact inhibition*.

Cell Adhesion. There are three main types of contact between cells which lead to their aggregation in a tissue.

1. *The aggregation may form by the inclusion of the cell in a common matrix.* In most cases this extracellular matrix consists of the *coat* of

the cell. The substances of the coat (cellulose, hyaluronic acid, and so forth) accumulate after the cells have made contact with each other.

2. *The aggregation of cells may have little intercellular material.* Electron microscopy shows the gap between cells to be frequently on the order of 15 nm (Fig. 8–7). However, even in these small intercellular spaces there is some substance, possibly a mucoprotein, that accounts for the specificity of cell association. Furthermore, the enzymes that are more effective in separating these cells are proteases and mucases.

3. *The aggregation may imply the presence of intercellular channels.* Cells in a tissue use these channels to interchange more or less freely. In plant cells, cytoplasmic bridges or plasmodesmata (Fig. 1–1) have been recognized for a long time. These provide narrow connections between the cells and across the cellulose walls and permit the free passage of ions and probably macromolecules.

In 1908 Wilson described how living sponges forced through a fine silk mesh disaggregate into isolated motile cells, and then, upon standing, reaggregate to form fresh sponges. Much later it was found that embryonic tissues treated with trypsin dissociate into individual cells and then reaggregate to form the specific patterns of the original tissue.[52]

The process of reaggregation depends on the motility of cells. If all the cells of a chick embryo are dissociated with trypsin and allowed to stand, they will reaggregate. When similar cells attach to each other, they form aggregates that are characteristic for a given cell population, e.g., retinal cell, kidney, or bone.

The process of reaggregation is not species specific. If cells of chick and mouse embryos are mixed, they reaggregate according to the cell population rather than the species. Reaggregation does not take place in the absence of calcium ions or if the treatment used to disaggregate the tissue selectively blocks certain carbohydrates and glycoproteins.

The mechanism by which a cell can "recognize" and associate with another of similar kind is a property of the cell surface in which the coat is directly involved. The membrane carbohydrates have been implicated in this mechanism. For example, the morphology and type of growth of a cultured cell may be changed by the addition of certain monosaccharides to the medium.

An interesting phenomenon related to cell-cell recognition is the so-called *homing of the lymphocytes*, in which lymphocytes leave the bloodstream and enter the lymphatic tissues. This process is dependent on specific carbohydrates present on the lymphocyte membrane. The course of these cells has been followed by labeling lymphocytes with ^{32}P. Lymphocytes taken from the thoracic ducts of rats and injected into other rats went into the spleen and lymph nodes. Treatment of lymphocytes with neuraminidase or with trypsin caused the lymphocytes to accumulate in the liver. These and other experiments suggest the involvement of glycoproteins in this cell-cell recognition mechanism.

Cancer Cells — Many Changes in the Cell Surface Properties

Cancer involves *uncontrolled cell growth*, invasion of other tissues, and *dissemination* to other sites of the organism by way of *metastasis*. All these characteristics indicate that cancer cells have in some way escaped from the controls that regulate normal growth. Many of the changes observed in such cells can be related to alterations in the plasma membrane and especially, in the cell coat.

Contact Inhibition of Movement. Normal cells growing in tissue culture tend to establish cell contacts by adhesion to neighboring cells. The morphology at the points of adhesion changes; some kind of electron-dense "plaque" is formed in both contacting cells. At the same time there is a slowing down of the ameboid processes which results in *contact inhibition of movement* (Fig. 8–26). In contrast cancer cells are unable to form adhesive junctions and do not show this type of contact inhibition. Such cells tend to have more ruffles and blebs on the cell surface than normal cells (Fig. 8–26). As will be mentioned in Chapter 9, the differences in the movement of normal and cancerous cells are accompanied by changes in the various microfilaments that bring about cell movement.

Contact Inhibition of Mitosis. Normal cultured cells generally divide every 24 hours, as long as they float freely in the medium. However, when they come in close contact in a monolayer, the rate of mitosis slows down and there is inhibition of cell division. The inhibition depends on some unknown signal between cells in contact, and not on a diffusible substance acting at a distance.

In cancer cells, *the mitotic rate is not inhibited,*

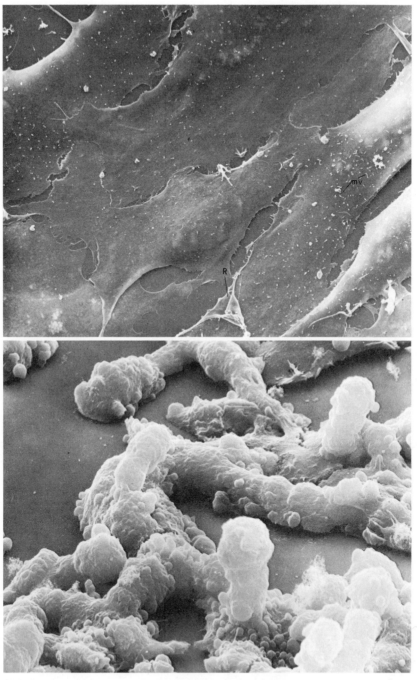

Figure 8–26 Scanning electron micrographs of cultured cells. **Above,** well-spread normal hamster embryo cells. A few microvilli (*mv*) and ruffles (*R*) are observed. ×1620. **Below,** the same cell type after transformation by human adenovirus. Observe that the cells have blebs on the surface and tend to make several layers. ×3145. (Courtesy of R. D. Goldman.)

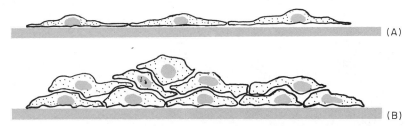

Figure 8–27   Schematic representation of the growth pattern of normal (**A**) and transformed (**B**) cells cultured on a solid substratum. Normal cells are inhibited by contact and stop multiplying, forming a monolayer. Transformed cells continue to multiply, forming multilayers. (From Ringertz, N. R., and Savage, R. E., in *Cell Hybrids,* Academic Press, Inc., New York, 1976.)

and in cultures the cells tend to pile up, forming irregular masses several layers deep. Figure 8–27, *A* shows that normal cultured cells form a monolayer and stop cell division by contact inhibition and that transformed (cancer) cells continue to multiply, giving rise to multilayers (Fig. 8–27, *B*).Cancerous cells show less adhesion to a solid support or among themselves and motility is more pronounced. These properties may explain why neoplasms invade other tissues and follow an uncontrolled growth. The *loss of contact inhibition* is easily studies in normal cultured cells that are "transformed" into cancerous cells by oncogenic viruses (i.e., viruses capable of inducing cancer) such as the polyoma or the SV40 (i.e., simian virus 40). The changes occurring in cultured hamster cells after transformation by human adenovirus are remarkable when observed by scanning electron microscopy (Fig. 8–26).

Surface Properties.   Changes in the surface properties of cancer cells have been observed. For example, using specific staining procedures for mucopolysaccharides (i.e., Hale's reaction or ruthenium red) a considerable increase of this substance was discovered in cells infected with polyoma virus.[53] These cells also showed an increased electrophoretic mobility.

It was observed under the electron microscope that the gap junctions tend to disappear in cells transformed by oncogenic viruses. This may explain the lack of electrical coupling and contact inhibition between cancer cells mentioned earlier.

Another difference is demonstrated by the use of lectins (cell agglutinating proteins) that bind to glycoprotein receptors at the cell surface. Apparently in cancer cells these receptors tend to diffuse more easily within the lipid bilayer than in normal cells and tend to form patches that agglutinate the cells. This greater diffusibility of receptors, which is also found in normal cells during mitosis, may be due to a disorganization of the microfilaments that attach to the plasma membrane.

Molecular Changes of the Cell Membrane.   Differences in the glycolipids and glycoproteins of normal and cancer cells have been observed. A reduction in gangliosides and in the enzymes of synthesis has been observed. In transformed cells there is a disappearance of certain proteins of the membrane.

Membrane Transport.   In virally transformed cells a great increase in sugar uptake has been found. Tumor cells frequently release intracellular enzymes to the medium, indicating leakage from cell membranes.

A major protease secreted by cancer cells is called *cell factor* (MW 40,000). This enzyme is able to act on *plasminogen* and activate it to *plasmin*. This is a proteolytic enzyme that dissolves blood clots and also removes exposed protein groups at the cell surface of cancer cells. If plasminogen is removed from the medium the morphology of cancer cells returns to normal. Thus, it is possible that this enzymatic digestion could activate the cancer cell into division.

Membrane Antigens.   Frequently tumor cells carry new antigens not present in normal cells. For this reason they can induce an immunologic response that in certain favorable conditions may eliminate the cancer cells. Some of the many changes of the cell surfaces of transformed cells are shown in Figure 8–28. (For further details see Nicolson et al., 1977.)

171

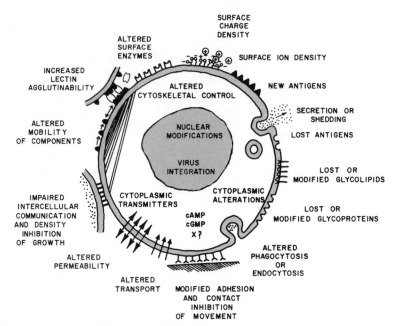

Figure 8–28 Highly schematic diagram of the surface changes that a cell may undergo after neoplastic transformation by an oncogenic virus. (Courtesy of G. L. Nicolson, 1977, ed.). From Nicolson, G. L., et al., *Modifications in transformed and malignant tumor cells, in International Cell Biology,* p. 138. (Brinkley, B. R., and Porter, K. R., eds.) New York, Rockefeller University Press, 1977.

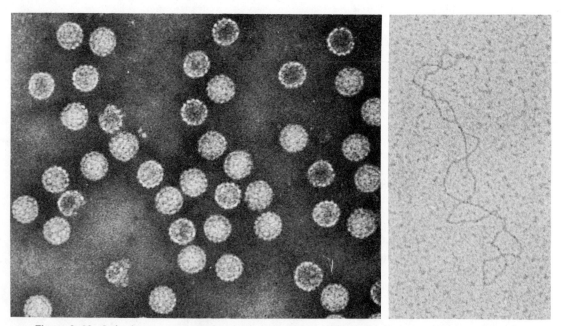

Figure 8–29 **Left,** electron micrographs of polyoma virus. The virus has a diameter of 45 nm with 72 subunits in the capsid; **right,** polyoma DNA, which appears as a twisted filament. This DNA has a length of 1.6 μm, a sedimentation constant of 20S, and a molecular weight of 3×10^6 daltons. In both micrographs the magnification is approximately ×150,000. (Courtesy of C. Vásquez.)

Transformation of Cells — Produced by Certain Viruses

In recent years considerable progress has been made on the mechanisms by which a normal cell may be transformed into a cancerous cell that shows all the characteristics just mentioned. More than 50 years ago Peyton Rous demonstrated that a chicken sarcoma could be reproduced by inoculation of a free cell extract. Viruses that produce leukemia and other tumors in mice and birds belong to the so-called *myxoviruses* which have a single-stranded chain of RNA in their genome.

The small viruses polyoma and SV40 are the most often used in cell transformation. Polyoma virus has a circular double-stranded DNA genome only 1.6 μm in length (Fig. 8–29) and a capsid with 72 capsomeres. When added to a culture of normal cells, the virus multiplies and kills only a small proportion of the cells. Most of the cells become transformed into cancerous cells and the virus apparently disappears. There are many indications that the DNA of the virus becomes integrated into the genome of the cell in regions in which there is some genetic homology. This can be demonstrated by studies of hybridization between the DNAs of the two genomes.

With cell transformation, the host cell shows some of the changes that were just mentioned, i.e., lack of control of cell division and of cell motility, and changes in cell contact and cell surface. Another change that may be demonstrated by fluorescent antibodies is the appearance of the two antigens that are specific to the virus — one in the nucleus, the other in the cell membrane.

The use of oncogenic viruses in cancer research is of fundamental importance in reaching a better understanding of the molecular biology of cancer — an understanding that is prerequisite to definite progress in its treatment (see Watson, 1975).

SUMMARY:

Cell Coats and Cell Recognition

Most cell membranes have a coat, sometimes referred to as the *glycocalyx*, made of glycoproteins or polysaccharides. The cell coat is negatively charged and may bind Na^+ and Ca^{2+}. Several cytochemical techniques are used to reveal the cell coat (e.g., PAS and ruthenium red). The oligosaccharides may be visualized by the use of *lectins*. The cell coat is a kind of secretion product that undergoes an active turnover.

Many functions are attributed to the cell coat. It may act mechanically, protecting the membrane, and participate in the *filtration* and *diffusion* processes The cell coat makes a kind of *microenvironment* for the cell. It contains enzymes involved in the digestion of carbohydrates and proteins.

Molecular recognition between cells may depend on a molecular code made up of the individual monosaccharides (i.e., galactose, hexosamine, mannose, fucose, sialic acid and others). The classical ABO blood groups are based on specific antigens of the red cell coat which are specified by their terminal carbohydrates. Several other antigens are found on the cell coat. Molecular recognition reaches maximum expression in the nervous tissue. Cell adhesion and cell dissociation and reassociation are dependent on the coat. Cells are able to recognize similar cells in a tissue. This is dramatically shown by Wilson's experiments with sponges. The *homing of lymphocytes* also depends on the cell coat.

In cancer cells many properties of the cell membrane and cell coat are altered; i.e., such cells have lack of contact inhibition, loss of electrical coupling and gap junctions, increased transport of sugars, and new antigens. These changes are observed in cultured cells after transformation (into cancerous cells) by oncogenic viruses (e.g., polyoma and SV40 viruses).

The DNA of polyoma is a double-stranded circular DNA of 1.6 μm containing genetic information to code for a few proteins. The viral genome becomes integrated in certain regions of the lost genome that have some genetic homology. The viral genome expresses itself in the many changes that are characteristic of a cancerous cell.

REFERENCES

1. Singer, S. J., and Nicolson, G. L. (1972) *Science,,* 175:720.
2. Maddy, A. H. (1966) *Int. Rev. Cytol.,* 20:1
3. Bretcher, M. S. (1973) *Science, 181:622.*
4. Van Deenen, L. L. M., et al. (1973) In: *24th Int. Congress of Pure and Applied Chemistry,* Vol. 2. Pergamon, Elmsford, N. Y.
5. Lapetina, E. G., Soto, E. F., and De Robertis, E. (1967) *Biochim. Biophys. Acta,* 35:33.
6. Capaldi, R. A., and Vanderkooi, G. (1972) *Proc. Natl. Acad. Sci. USA,* 69:930.
7. Marchesi, S. L., Steers, E., Marchesi, V. T., and Tillack, T. W., (1970) *Biochemistry,* 9:50.
8. De Robertis, E., and Fiszer de Plazas, S. (1970) *Biochim. Biophys. Acta,* 219:388.
9. Steck, L. (1974) *J. Cell Biol.* 62:1.
10. Nicolson, G. L. (1976) *Biochim. Biophys. Acta,* 457:57.
11. Nicolson, G. L., Marchesi, V. T., and Singer, S. J. (1971) *J. Cell Biol.,* 51:437.
12. Jacob, H., Amsden, T., and While, J. (1972) *Proc. Natl. Acad. Sci. (USA)* 69:471.
13. Rothstein, A., Cabanchik, Z. I., and Knauf, P. *Fed. Proc.,* 35:3.
14. Folch, J., and Lees, M. (1951) *J. Biol. Chem.,* 191:807.
15. Lehninger, A. L. (1968) *Proc. Natl. Acad. Sci. USA,* 60:1069.
16. Robertson, J. D. (1959) *Biochem. Soc. Symp.,* 16:3.
17. Gitler, C. (1972) *Annu. Rev. Biophys. Bioeng.,* 1:51.
18. De Robertis, E. (1971) *Science, 1971:963.*
19. Yu, I., and Branton, D. (1976) *Proc. Natl. Acad. Sci. USA,* 73:3891.
20. Marchesi, V. T., Furtmayr, H., and Tomita, M. (1976) *Annu. Rev. Biochem.,* 45:667.
21. Hubbell, W. L., and McConnel, H. M. (1971) *J. Am. Chem. Soc.,* 93:314.
22. Deveaux, P., and McConnell, H. M. (1972) *J. Am. Chem. Soc.,* 94:4475.
23. Frye, L. D., and Edidin, M. (1970) *J. Cell Sci.,* 7:319.
24. Edidin, M., and Fambrough, D. (1973) *J. Cell Biol.,* 57:27.
25. Cuatrecasas, P., and Hollenberg, M. D. (1976) *Adv. Protein Chem.,* 30:251.
26. Vanderheuvel, F. A. (1965) *Ann. N. Y. Acad. Sci.,* 122:57.
27. Peterson, R. G., and Pease, D. C. (1972) *J. Ultrastruct. Res.,* 41:115.
28. De Robertis, E., and Lasansky, A. (1958) *J. Biophys. Biochem. Cytol.,* 4:743.
29. Blasie, J. K. (1972) *Biophys. J.,* 12:191.
30. Ussing, H. H. (1960) *J. Gen. Physiol.,* 43:5, part 2, suppl. 1, p. 135.
31. Leaf, A. (1960) *J. Gen Physiol.,* 43:5, part 2, suppl. 1, p. 175.
32. Backer, P. F., Hodgkin, A. L., and Shaw, T. I. (1962) *J. Physiol.* (London), 164:330 and 335.
33. Woodbury, J. W. (1969) In: *Basic Mechanism of Epilepsies.* (Jasper, H.H., Ward, A. A., and Pope, A., eds.) Little, Brown & Co., Boston.
34. Solomon, A. K. (1960) *J. Gen. Physiol.,* 43:5, part 2, suppl. 1, p. 1.
35. Marchesi, V. T., and Palade, G. E. (1967) *J. Cell. Biol.,* 35:385.
36. Kennedy, E. P., Fred Fos, C., and Carter, J. R. (1966) *J. Gen. Physiol.,* 49:347.
37. Douglas, W. H. J., Ripley, R. C., and Ellis, R. A. (1970) *J. Cell Biol.,* 44:211.
38. Kelly, D. E. (1960) *J. Cell Biol.,* 28:51.
39. Goodenough, D. A., and Revel, J. P. (1970) *J. Cell Biol.,* 45:272.
40. Farquhar, M., and Palade, G. E. (1963) *J. Cell Biol.,* 17:375.
41. Loewenstein, W. R., and Kanno, Y. (1964) *J. Cell Biol.,* 22:565.
42. Revel, J. P., and Karnovsky, M. J. (1967) *J. Cell Biol.,* 33:C7.
43. Simpson, I., Rose, B., and Loewenstein, W. R. (1977) *Science,* 195:294.
44. Oliveira-Castro, G. M., and Loewenstein, W. R. (1971) *J. Mebr. Biol.,* 5:51.
45. Rose, B., and Loewenstein, W. R. (1975) *Science,* 190:1204.
46. Subak-Sharpe, H., Buck, P., and Pitts, T. D. (1969) *J. Cell Sci.* 4:353.
47. Azarnia, R., and Loewenstein, W. R. (1971) *J. Membr. Biol.,* 6:368.
48. Harris, H. (1971) *Proc. R. Soc. London [Biol]* 179:1.
49. Azarnia, R., and Loewenstein, W. R. (1973) *Nature, 241:455.*
50. Azarnia, R., and Loewenstein, W. R. (1977) *J. Membr. Biol.,* 34:1.
51. Bennet, H. S. (1963) *J. Histochem. Cytochem.,* 11:14.
52. Moscona, M., and Moscona, A. (1963) *Science,* 142:1070.
53. Martinez-Palomo, A., and Brailowsky, C. (1968) *Virology, 34:379.*

ADDITIONAL READING

Bell, G. I. (1978) Models for the specific adhesion of cell to cell. *Science, 200:618.*

Boogs, J. M., and Moscarello, M. A. (1978) Structural organization of the human myelin membrane. *Biochem. Biophys. Acta, 515:1.*

Branton, D. (1971) Freeze-etching studies of membrane structure. *Philos. Trans. R. Soc. London [Biol. Sci.], 261:33.*

Bretcher, A., and Weber, K. (1978) Purification of microvilli and analysis of protein components of microfilament core bundle. *Exp. Cell Res., 116:397.*

Bretscher, M. S. (1973) Membrane structure: some general principles. *Science, 181:622.*

Bretscher, M. S., and Raff, M. C. (1975) Mammalian plasma membranes. *Nature* (London), 258:43.

Capaldi, R. A. (1974) A dynamic model of cell membranes. *Sci. Am., 230:26.*

Caspar, D. L. D., Goodenough, D. A., Mankowski, L., and Phillips, W. C. (1977) Gap junction structures. I. Correlated electron microscopy and x-ray diffraction. *J. Cell Biol.*, 74:605.

De Pierre, J. W., and Karnovsky, M. L. (1973) Plasma membranes of mammalian cells. *J. Cell Biol.*, 56:275.

De Robertis, E. (1975) *Synaptic Receptors: Isolation and Molecular Biology*. Marcel Dekker, New York.

Doljanski, F., and Kapeller, M. (1976) Cell surface shedding—the phenomenon and its possible significance. *J. Theor. Biol.*, 62:253.

Edelman, M. (1976) Surface modulation in cell recognition and cell growth. *Science*, 192:218.

Edidin, M. (1974) Rotational and translational diffusion in membranes. *Annu. Rev. Biolphys. Bioeng.*, 3:179.

Ellory, J. C., and Lew, V. L. (1977) *Membrane Transport in Red Cells*. Academic Press, New York.

Gilula, N. B. (1977) Gap junctions and cell communication. In: *International Cell Biology*, p. 61. (Brinkley, B. R., and Porter, K. R., eds.) The Rockefeller University Press, New York.

Glaser, L. (1976) Cell-cell recognition. *Trends in Biochem. Sci.*, 1:84.

Guidotti, G. (1976) The structure of membrane transport systems. *Trends in Biochem. Sci.*, 1:11.

Harrison, R., and Lunt, G. G. (1975) *Biological Membranes*. Blackie & Son, Glasgow.

Jacobs, S., and Cuatrecasas, P. (1977) The motile receptor hypothesis for cell membrane receptor action. *Trends Biochem. Sci.*, 2:280.

Kahane, I., and Gitler, C. (1978) Red cell membrane glycophorin labeling from within the lipid bilayer. *Science*, 201:351.

Kalckar, H. M. (1965) Galactose metabolism and cell sociology. *Science*, 150:305.

Keynes, R. D. (1979) Ion channels in the nerve cell membrane. *Sci. Am.*, 240:126.

Kirkpatrick, F. H. (1976) Spectrin: Current understanding of its physical, biochemical and functional properties. *Life Sci.*, 19:1.

Lerner, R. A., and Bergsma, D., eds. (1978) *Molecular Bases of Cell-cell Interaction*. Alan R. Liss, Inc., New York.

Lodish, H. F., and Rothman, J. E. (1979) The assembly of cell membranes. *Sci. Am.*, 240:48.

Loewenstein, W. R. (1976) Permeable junctions. *Cold Spring Harbor Symp. Quant. Biol.*, 11:49.

Loewenstein, W. R. (1977) Permeability of the junctional membrane channel. *International Cell Biology*, p. 70. (Brinkley, B. R., and Porter, K. R., eds.) The Rockefeller University Press, New York.

Loewenstein, W. R. (1979) Junctional intercellular communication and the control of growth. *Biochem. Biophys. Acta*, 560:1.

Loewenstein, W. R., Kanno, Y. and Socolar, S. J. (1978) Quantum jumps of conductance during formation of membrane channels at cell-cell junctions. *Nature*, 274:133.

Luft, J. H. (1976) The structure and properties of the cell surface coat. *Int. Rev. Cytol.*, 45:291.

Luftig, R. B., Wehrli, E., and McMillan, P. N. (1977) The unit membrane image: a re-evaluation. *Life Sci.*, 21:285.

Marchesi, V. T., Furthmayr, H., and Tomita, M. (1976) The red cell membrane. *Annu. Rev. Biochem.*, 45:667.

Nicolson, G. L. (1976) Transmembrane control of the receptors on normal and tumor cells. *Biochim. Biophys. Acta*, 457:57.

Nicolson, G. L. (1979) Cancer metastasis. *Sci. Am.*, 240:66.

Nicolson, G. L., Giotta, G., Lotan, R., Neri, A., and Poste, G. (1977) Modifications in transformed and malignant tumor cells. *International Cell Biology*, p. 138. (Brinkley, B. R., and Porter, K. R., eds.) The Rockefeller University Press, New York.

Pastan, I., and Willingham, M. (1978) Cellular transformation and the morphologic phenotype of transformed cells. *Nature*, 274:645.

Pitts, J. D. (1977) Direct communication between animal cells. *International Cell Biology*, p. 43. (Brinkley, B. R., and Porter, K. R., eds.) The Rockefeller University Press, New York.

Poste, G., and Nicolson, G. L. (1978) Membrane fusion. *Cell Surf. Rev.*, Vol. 5. North-Holland, Amsterdam.

Quinn, P. J., ed. (1976) *Molecular Biology of Cell Membranes*. University Park Press, Baltimore.

Raff, M. C. (1976) Cell-surface immunology. *Sci. Am.*, 234:30.

Ralston, G. B. (1978) The structure of spectrin and the shape of the red blood cell. *Trends Biochem. Sci.*, 3:195.

Rothfield, L. I., ed. (1971) *Structure and Function of Biological Membranes*. Academic Press, New York.

Rothstein, A., Cabantchik, Z. I., and Knauf, P. (1976) Mechanism of anion transport in red blood cells: role of membrane proteins. *Fed. Proc.* 35:3.

Rothstein, A., Grinstein, S., Ship, S., and Knauf, P. A. (1978) Asymmetry of functional sites of erythrocyte anion transport protein. *Trends Biochem. Sci.*, 3:126.

Schramm, M., Orly, J., Eimerl, S., and Korner, M. (1977) Coupling of hormone receptors to adenylate cyclase of different cells by cell fusion. *Nature*, 268:310.

Schwartz, A., Lindenmayer, G. E., and Allen, J. C. (1972) The Na^+, K^+ ATPase membrane transport system: importance in cellular function. In: *Current Topics in Membranes and Transport*, Vol. 3, p. 2, (Bronner, F., and Kleinzeller, A., eds.) Academic Press, Inc., New York.

Shires, T. K., Pitot, H. C., and Kauffmann, S. A. (1974) The membron: a functional hypothesis for the translational regulation of genetic expression. *Biomembranes, Vol. 5,* p. 81. (Manson, L. A., ed.) Plenum Press, New York.

Singer, S. J. (1974) The molecular organization of membranes. *Annu. Rev. Biochem., 43*:805.

Singer, S. J. (1977) Thermodynamics, the structure of integral membrane protein and transport. *J. Supramol. Struct. 6*:313.

Singer, S. J., and Nicolson, G. L. (1972) The fluid mosaic model of the structure of cell membranes. *Science, 175*:720.

Steck, L. (1974) The organization of proteins in the human red blood cell membrane. *J. Cell Biol., 62*:1.

Steck, T. L. and Hainfeld, J. F. (1977) Protein ensembles in the human red-cell membrane. *International Cell Biology,* p. 6 (Brinkley, B. R., and Porter, K. R., eds.) The Rockefeller University Press, New York.

Watson, J. D. (1976) *Molecular Biology of the Gene.* 3rd Ed. W. A. Benjamins, Inc., London.

Part 4

THE CYTOPLASM AND CYTOPLASMIC ORGANELLES

In the following six chapters the structural, biochemical and physiologic characteristics of the cytoplasm of animal and plant cells, as well as their main organelles, are considered. The discussion is based on the latest studies of electron microscopy, cytochemistry, and structural evolution of biological systems.

Chapter 9 studies the structure and function of the cytoplasmic matrix or cytosol, the true internal milieu of the cell. This part of the cell carries out many fundamental functions, such as glycolysis, synthesis of proteins, and the various forms of cell motility. Within the cytoplasmic matrix there is a fabric of microfilaments and microtubules that constitute a kind of cytoskeleton for the cell. Microtubular structures, such as cilia, flagella, and centrioles are also studied in this chapter.

Chapter 10 initiates the study of the intracytoplasmic membrane system and its main components, the endoplasmic reticulum, the Golgi apparatus, and the nuclear envelope (Chap. 15). This membrane system subdivides the cytoplasm into several compartments that may function independently. It is postulated that this system interchanges, circulates, and segregates the products that are absorbed by the cell or that are synthesized on the ribosomes. Some of these products may be prepared for export (i.e., secretion) out of the cell.

In general terms, most of the metabolic and biosynthetic functions of the cell occur in the cytoplasm. The cytoplasm is differentiated by the activity of the genes contained in the nucleus and becomes adapted to the division of cellular work. In different types of cells, therefore, the cytoplasm may be markedly different, while the nucleus appears to be comparatively uniform.

In this chapter the so-called signal hypothesis for the synthesis of exported proteins will also be discussed.

Chapter 11 studies the morphology, structure, and function of the Golgi apparatus, with emphasis on its role in the glycosidation of lipids (glycolipids) and proteins (glycoproteins), as well as in the packaging and release of secretory products. The entire process of cell secretion, which involves the activity of all parts of the cell, will be considered in relation to this organelle.

Chapter 12 presents the mitochondria as macromolecular machines whose chemical and molecular organization is admirably adapted to their function in cell respiration. Special emphasis will be placed on the ultrastructure and compartmentalization of these organelles and on the studies concerning the separation of these compartments. The coupling of oxidation and phosphorylation will be explained on the basis of the special localization of the enzymes and their peculiar asymmetry on the mitochondrial crests. The findings that both the mitochondrion and the chloroplast contain DNA and special ribosomes and are capable of some local protein synthesis are also discussed in Chapter 12. The symbiont idea is based on the theory that these organelles contain genetic information and have a certain degree of autonomy within the cell.

Chapter 13 studies the lysosomes, organelles that contain numerous hydrolytic enzymes, and the peroxisomes, which contain some oxidases. The lysosomes are initially related to the digestion of substances that enter the cell by endocytosis and of parts of the cytoplasm of the cell itself (autophagy). The fundamental functions of the lysosomes and their importance in pathology and medicine will be emphasized.

Chapter 14 is dedicated to the plant cells and emphasizes some of the differences between them and animal cells, in particular, the presence of rigid cell walls, a type of cell division peculiar to plants, and the presence of special organelles, generally called plastids, which are related to the special metabolic properties of plants. Most of this chapter concerns the ultrastructure and macromolecular organization of the chloroplasts and their fundamental function in photosynthesis.

THE CYTOSKELETON AND CELL MOTILITY: MICROTUBULES AND MICROFILAMENTS

9–1 The Cytosol and the Cytoskeleton 179
The Cytoskeletal Fabric—Microtubules and Microfilaments
Summary: The Cytoskeleton
9–2 Microtubules 183
Tubulin—The Main Protein of Microtubules
Microtubules—Assembled from Tubulin Dimers
Detection of Microtubules in Culture Cells by Anti-Tubulin Antibodies
Functions of Cytoplasmic Microtubules
Summary: Properties of Microtubules
9–3 Microtubular Organelles: Cilia, Flagella, and Centrioles 187
Ciliary and Flagellar Motions—Present in Cells and in Tissues
The Ciliary Apparatus—The Cilium, Basal Bodies, and Ciliary Rootlets
The Axoneme Contains Microtubular Doublets
Basal Bodies (Kinetosomes) and Centrioles Contain Microtubular Triplets
Ciliary Movement—A Sliding of Microtubular Doublets That Involves Dynein
Kartagener's Syndrome—A Mutation Involves a Lack of Dynein

Photoreceptors are Derived from Cilia
Cilia and Flagella Originate from Basal Bodies
Summary: Structure, Motion, and Origin of Cilia and Flagella
9–4 Microfilaments 197
Microtrabecular Lattice in Cytosol—Revealed by High-Voltage EM
Cytochalasin B Impairs Several Cellular Activities Involving Microfilaments
Actin, Myosin, and Other Contractile Proteins—Present in Non-Muscle Cells
Contractile and Regulatory Proteins—Detected by Specific Antibodies; Stress Fibers
Two Recognizable Types of Microfilaments
Microfilaments—Involved in All Motility Events in Non-Muscle Cells
Cytoplasmic Streaming (Cyclosis)—Observed in Large Plant Cells
Ameboid Motion—Characteristic of Amebae and Many Free Cells
Summary: Microfilaments, Cyclosis, and Ameboid Motion

In Chapter 2 we mentioned that there were notable similarities between prokaryotic and eukaryotic cells (Table 2–2) in spite of their differences in structure. The cytosol, the *cytoplasmic matrix* or *ground cytoplasm*, of a eukaryotic cell contains the same components as a bacterium: ribosomes, RNA molecules, globular proteins, enzymes, and so forth (Fig. 2–1). The new components that have evolved in higher cells are the many membranes that constitute the *endomembrane* or *vacuolar system* with its several portions (e.g., nuclear envelope, endoplasmic reticulum, and Golgi apparatus) and the membrane-bound organelles (e.g., mitochondria, chloroplasts, lysosomes, peroxisomes, and vacuoles). As a con-

sequence of the vacuolar system, numerous compartments and subcompartments are formed in the cell.

9–1 THE CYTOSOL AND THE CYTOSKELETON

In spite of the compartmentalization of the cell there is one continuum in the cytoplasm, the cytosol, which fills all the spaces of the cell and constitutes its true *internal milieu*. Many fundamental functions of the cell take place in this part of the cytoplasm, which is particularly rich in

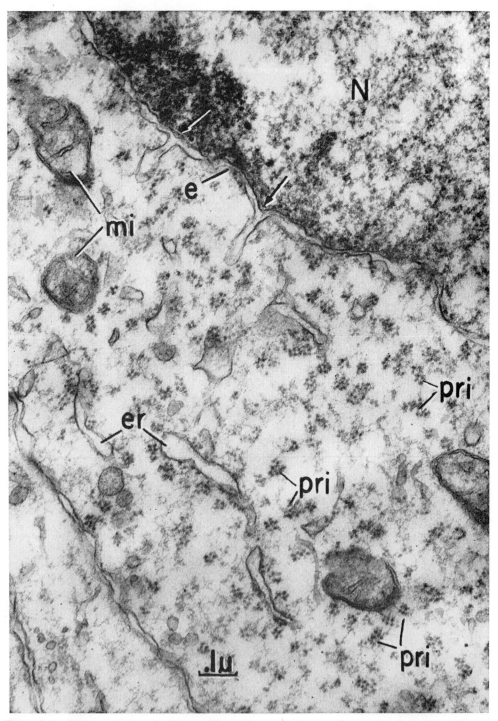

Figure 9-1 Electron micrograph of a neuroblast of the cerebral cortex of a rat embryo, showing the cytoplasm rich in matrix with numerous ribosomes and little development of the vacuolar system. *e,* nuclear envelope sending projections into the cytoplasm (arrows); *er,* endoplasmic reticulum; *mi,* mitochondria; *N,* nucleus; *pri,* polyribosomes (groups of ribosomes). ×45,000. (From E. De Robertis.)

differentiating cells (Fig. 9–1). The colloidal properties of the cell, such as those essential to sol-gel transformations, viscosity changes, intracellular motion (cyclosis), ameboid movement, spindle formation, and cell cleavage, depend, for the most part, on the cytoplasmic matrix.

After the nuclear, mitochondrial, and microsomal fractions have been separated by cell fractionation (see Chapter 4–4), the remaining supernatant, or *soluble fraction* or *cytosol* (Fig. 6–5), contains the soluble proteins and enzymes found in the cytoplasmic matrix. These proteins and enzymes constitute 20 to 25 per cent of the total protein content of the cell. Among the important *soluble enzymes* present in the matrix are those involved in glycolysis and in the activation of amino acids for protein synthesis. The enzymes of many reactions that require ATP are found in the soluble fraction. Soluble (transfer) RNA and all the machinery for protein synthesis, including the ribosomes, are found in this part of the cytoplasm.

While it was initially thought that the ground cytoplasm was essentially amorphous, at the end of the nineteenth century it was discovered that portions of the cytoplasm in certain cells have a differential staining property. Because these areas stained with basic dyes, they were called *basophilic* or *chromidial cytoplasm*. The still common name *ergastoplasm* (Gr., *ergazomai*, to elaborate and transform) was coined by Garnier in 1887 to imply that biosynthesis is the fundamental role of this substance.

The ergastoplasm includes basophilic regions of the ground cytoplasm, such as the Nissl bodies of the nerve cells, the basal cytoplasm of serous cells (e.g., secretory cells of the pancreas and the parotid gland and chief cells of the stomach), and the basophilic clumps of liver cells. Caspersson, Brachet, and others demonstrated that the intense basophilic property of the ergastoplasm is due to the presence of ribonucleic acid (Fig. 9–2).

It was shown that the ergastoplasm loses its staining properties if the cell is treated with ribonuclease, an enzyme that hydrolyzes RNA (Fig. 9–2). RNA is contained principally in the ribosomes, and as a result, a relationship between the ergastoplasm and protein synthesis was postulated.

The Cytoskeletal Fabric — Microtubules and Microfilaments

The name *cytoskeleton* was coined many years ago, but it was abandoned because many of the structures observed in the cytoplasm with the light microscope were considered to be fixation artifacts (see Chapter 1–2). The existence of an organized fibrous array in the structure of the protoplasm was postulated in 1928 by Koltzoff. He concluded that "each cell is a system of liquid components and rigid skeletons which generate the shape, and even though we rarely see the skeletal fibrils in living and fixed cells, that only means that these fibrils are very thin or that they are not distinguished by their refractive index

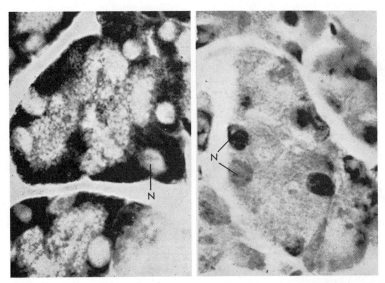

Figure 9–2 **Left,** pancreatic acini frozen and dried and stained with toluidine blue. The basophilic substance appears intensely stained. **Right,** the same tissue, but after digestion with ribonuclease; the basophilic substance has disappeared. *N,* nucleus. (From E. De Robertis.)

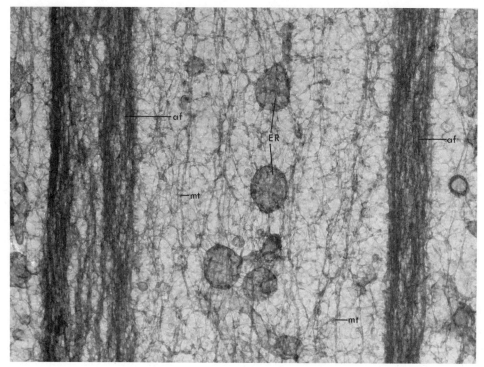

Figure 9–3 Electron micrograph of a thin section of a cultured cell made with the high-voltage electron microscope at 10^6 volts. Two bundles of actin filaments (*af*) run vertically. There are vesicles of the endoplasmic reticulum (*ER*) and a few microtubules (*mt*). Observe the lattice of microtrabeculae that pervades the matrix ×40,000. (Courtesy of K. R. Porter. From Porter, K. R., Introduction: motility in cells, *in Cell Motility*, Vol. 1, p. 1. (Goldman, R. D., et al., eds.) Cold Spring Harbor, New York, Cold Spring Harbor Laboratory, 1976.)

from the surrounding colloidal solution." He thus conceived of a cytoskeleton that determines both the shape of the cell and the changes in its form.

Confirming this assumption, the electron microscope has recently revealed that the cytoplasm of most eukaryotic cells contains a cytoskeletal fabric formed of *microtubules* and of various types of *microfilaments* (see Cohen, 1977). Different morphological and experimental studies have demonstrated that these components play a fundamental role in a variety of cell functions, such as those mentioned above: cyclosis, ameboid movement, mitosis, and cell cleavage. Furthermore, microtubules are the main constituents of certain organelles (such as *cilia* and *flagella*) involved in cell mobility and are also present in *centrioles*.

Great progress has been made in the isolation of proteins that constitute these cytoskeletal structures, such as *tubulin, actin, myosin, tropomyosin,* and related molecules, and in the mechanism of their assembly. In addition by the production of specific antibodies against these proteins it has been possible to examine under the light and the electron microscope the dispo-

sition of the microtubules and the microfilaments. The use of high-voltage electron microscopy on whole cells has also helped to demonstrate that there is a highly structured, three-dimensional lattice in the ground cytoplasm[1,2] (Fig. 9–3).

A fundamental line of thought in the study of *cell motility* is that its various forms result from interaction among the several kinds of microfilaments and microtubules embedded in the ground cytoplasm. The most elaborated system of cell motility is represented by the *myofibril* of skeletal muscle, in which there is a highly differentiated macromolecular *mechanism for contractility.* However, it is now evident that nonmuscle cells use mechanisms essentially similar to those of muscle. While many of the concepts about molecular structure, including the biochemistry of the contractile proteins, apply to the material presented in this chapter, they will be treated in detail in Chapter 27. Here, first we shall describe the microtubules and microfilaments, and then we will try to summarize the role they play in the cytoskeleton and in the various forms of cell motility, including those of cilia and flagella.

SUMMARY:

The Cytoskeleton

Within the cell cytoplasm the *cytosol* or *cytoplasmic matrix* represents a continuum that constitutes the true internal milieu of the cell. In the cytosol many functions related to sol-gel transformations take place (i.e., cyclosis, ameboid movement, spindle formation, and cell cleavage). In the cytosol are contained the glycolytic enzymes and the whole machinery for protein synthesis. Basophilic regions of the cytosol which are rich in RNA (ribosomes) were long ago named the *ergastoplasm*.

The name *cytoskeleton* can be applied to the fabric of microtubules and microfilaments that pervade the cytosol and which are related to the various forms of cell motility. The main proteins that are present in the cytoskeleton are *tubulin* (in the microtubules), actin, myosin, tropomyosin, and others (in the microfilaments) which are also in muscle. Thus, the same proteins are involved in the contraction of non-muscle cells as in that of muscle cells.

9–2 MICROTUBULES

Microtubules are structures universally present in the cytoplasm of eukaryotic cells and are characterized by their tubular appearance and their uniform properties in the different cell types. Most cytoplasmic microtubules are rather labile and do not resist the effects of fixatives such as osmium tetroxide; because of this, intensive studies began only after 1963, when glutaraldehyde fixation was introduced in electron microscopy.

The first observation of these tubular structures, in the axoplasm extruded from myelinated fibers, was made by De Robertis and Franchi in 1953 (see Figure 28–2). Here the so-called *neurotubules* appeared as elongated, unbranched, cylindrical elements 20 to 30 nm in diameter and of indefinite length.[3] Microtubules were observed in a variety of animal cells studied in sections[4,5] (Fig. 9–4).

Cytoplasmic microtubules are uniform in size and are remarkably straight. They are about 25 nm in outer diameter and several micrometers in length. In cross section they show an annular configuration with a dense wall about 6 nm thick and a light center. Each microtubule is surrounded by a zone of low electron density from which ribosomes or other particles are absent.

The wall of the microtubule consists of individual linear or spiraling filamentous structures about 5 nm in diameter, which, in turn, are composed of subunits.[6] In cross section there are about 13 subunits with a center-to-center spacing of 4.5 nm (Fig. 9–5). Application of negative staining techniques has shown that microtubules have a lumen and a subunit structure in the wall.[7] Occasionally, dense dots

or rods have been detected in the center portion of some microtubules.

Although all the microtubules studied show approximately the same morphologic characteristics, it is evident that they differ in other properties. For example, microtubules of cilia and flagella are much more resistant to various treatments. The microtubules forming the spindle fibers and the others present in the cytoplasm are, in general, labile and transitory structures. Cytoplasmic microtubules usually disappear if stored at 0° C or after treatment with colchicine.

Tubulin — The Main Protein of Microtubules

Microtubules are composed of protein subunits that are rather similar, even though they are found in a variety of cell types. The term *tubulin*, used for the principal protein of cilia and flagella, is also used for the protein of cytoplasmic microtubules. Tubulin is a dimer of 110,000 to 120,000 daltons. The monomers of similar size are believed to be composed of 4 nm × 6 nm subunits. It has been shown that two different monomers — tubulin A and B — are present in flagella. In most cases, tubulin is a heterodimer having two monomers of different kinds although they are quite similar in molecular weight (55,000 daltons) (see Stephens, 1974).

The 8-nm spacing along the longitudinal axis of microtubules that is observed by electron microscopy probably reflects the pairing of the two types of tubulin monomers (see Mohri, 1976) (Fig. 9–5). One dimer of tubulin binds to a molecule of ^{3}H-colchicine,[8] and this specific property is used for assay of this protein. Tubulin also binds to the Vinca alkaloid *vinblas-*

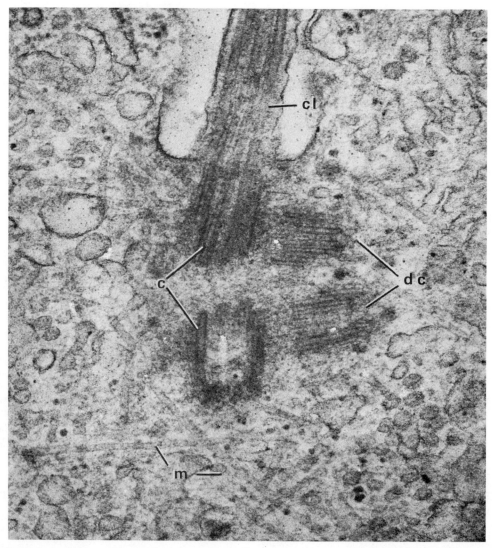

Figure 9–4 Electron micrograph of the pancreas of a chick embryo showing cytoplasmic microtubules, and the replication of centrioles; *c,* the two centrioles; *dc,* daughter centrioles; *cl,* cilium; *m,* microtubules. ×50,000. (Courtesy of J. Andrè.)

tine, but at a site other than for colchicine. Vinblastine tends to produce crystal-like structures of tubulin in the cytoplasm, and in homogenates it produces precipitation of this protein, thus allowing rapid purification. The amino acid composition of tubulin from different sources shows little variation, although some differences are found between the monomers. Some enzymatic activities have been reported for tubulin (e.g., protein kinase activity). Some studies indicate that tubulin, in addition to being present as a pool of free dimers and integrating the microtubules, may form an integral part of some membranes.

Microtubules — Assembled from Tubulin Dimers

The assembly of microtubules from the tubulin dimers is a specifically oriented and programmed process. In the cell there are sites of orientation, i.e., centrioles, basal bodies of cilia, and centromeres, from which the polymerization is directed in some way. It has been found that in a concentrated solution of tubulin containing guanosine triphosphate (GTP) and Mg^{2+}, depolymerization can take place only if the level of Ca^{2+} is kept low.[9] The quantity of polymerized tubulin is high at in-

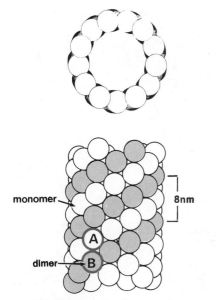

Figure 9–5 Diagram of the molecular structure of a microtubule showing the 13 subunits in cross section and the monomers and heterodimers in longitudinal view. The spacing of 8 nm may correspond to the heterodimer.

terphase (cytoplasmic microtubules) and metaphase (spindle microtubules), but low at prophase and anaphase.

Within the cell, microtubules are in equilibrium with free tubulin. Phosphorylation of the tubulin monomers by a cyclic AMP-dependent kinase favors the polymerization. In cultured epithelial cells cyclic AMP promotes the formation of microtubules, and the cell becomes elongated, with a fibroblastic appearance. Thus there is a relationship between cell shape, the number and direction of microtubules, and cyclic AMP.

Microtubular Associated Proteins. One of the best materials in which to study the biochemistry of tubulin assembly is brain protein extract. It was found that microtubules assemble at 37° C and disassemble on cooling to 5° C.[9] However, the microtubules purified by several cycles of assembly-disassembly do not consist of tubulin only but contain about 5 per cent other proteins which have been called, generically, *microtubular associated proteins* (MAPs). A number of MAPs ranging in size between 300,000 and 55,000 daltons have been isolated. Although the particular role of each of the proteins is unclear, it is evident that MAPs are involved in microtubular assembly. In fact, when very purified tubulin is used there is no

microtubular formation. These accessory factors appear to be involved stoichiometrically rather than catalytically in stimulating tubulin assembly (see Amos, 1977; Sherline and Schiavone, 1978).

Detection of Microtubules in Culture Cells by Anti-tubulin Antibodies

The isolation of tubulin has permitted specific antibodies against this protein to be obtained and an *immunofluorescent probe* to be produced for localizing microtubules in the cytoplasm of a wide variety of cultured cells.[10, 11] Similar studies with antibodies against microfilament proteins (anti-actin antibodies and others) have permitted differentiation of the microtubules from the microfilaments in the same cell type (Fig. 9–6).

In a cultured cell, cytoplasmic microtubules often extend radially from the nucleus and appear as straight or curved filaments that seem to terminate near the cell surface. These filaments disappear when treated with *Colcemid*, a derivative of colchicine, or when cooled, and they reappear if the conditions are reversed. Such studies reveal that microtubules arise near the nucleus, from one or two focal points corresponding to the centrosomal region. (i.e., the centrosphere).

In electron micrographs microtubules are often observed irradiating from the region containing the centrioles (see Fig. 9–4). These observations suggest that the centrosphere is the main microtubule organizing center. In cells entering mitosis the cytoplasmic microtubules disappear and are replaced by those integrating the spindle and asters.

It has been postulated that the reversible changes in microtubular structure may be modulated by the concentration of free Ca^{2+} in the cytosol. It has been found that Ca^{2+} inhibits biochemically the polymerization of tubulin in vitro.

Cells transformed by a virus or by various chemicals (i.e., cells that are becoming cancerous) show disorganization and even disappearance of the microtubular filaments, with only a diffuse fluorescence after immunofluorescent treatment.[12] These changes are morphologically similar to those observed in actin microfilaments during cell transformation.[13] (Further considerations about the structure and function of microtubules in mitosis will be found in Chapter 17).

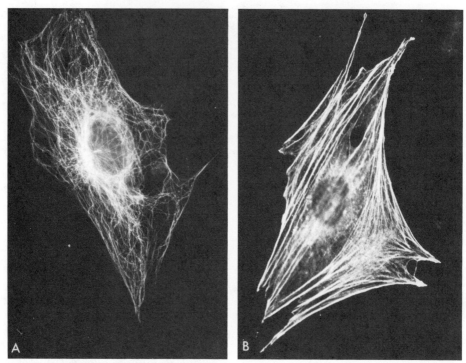

Figure 9–6 Cultured 3T3 cells stained by immunofluorescence with antibodies against tubulin (**A**) and actin (**B**) ×600. (Courtesy of M. Osborn and K. Weber, 1976.)

Functions of Cytoplasmic Microtubules

A rather long list of functions has been attributed to microtubules:

Mechanical Function. The shape of some cell processes or protuberances has been correlated to the orientation and distribution of microtubules. They are considered to be a framework which determines the shaping of the cell and redistributes its content. The integrity of microtubules is necessary for the individual shape of many cells and the rigidity of elongated processes. This is particularly evident in axons (Fig. 28–2) and dendrites.

Morphogenesis. Related to their mechanical function is the role that microtubules play in the shaping of the cell during *cell differentiation*. For example, the elongation of the cells during the induction of the lens placode in the eye is accompanied by the appearance of numerous microtubules. The morphogenetic changes that occur during spermiogenesis provide another interesting example. The enormous elongation that takes place in the nucleus of the spermatid is accompanied by the production of an orderly array of microtubules that are wrapped around the nucleus in a direction perpendicular to the

nuclear axis and that have a double helical arrangement.

Cellular Polarity and Motility. The determination of the intrinsic polarity of certain cells is also related to the mechanical function. Treatment of various culture cells with Colcemid results in a change in motion. The following forms of movement have been observed to persist: (1) membrane ruffling, (2) endocytosis, (3) attachment to the surface, and (4) extension of microvilli. Only a saltatory movement of particles was found to be inhibited after the destruction of the microtubules. However, the directional gliding of the cell is replaced by a random movement.[14]

Culture cells treated in vitro with dibutyryl cyclic AMP become elongated and the number of microtubules increases. All these observations suggest that microtubules are responsible for maintenance of the shape of the cell, the polarity of movement, and the distribution of certain organelles.

Circulation and Transport. Microtubules may also be involved in the *transport* of macromolecules in the cell's interior; to this end, they probably form channels in the cytoplasm. A very interesting example is the protozoan *Actinosphaerium* (Heliozoa) which sends out long, thin pseudopodia within which cytoplasmic particles migrate back and forth. These pseudopodia contains as many as 500 microtubules disposed in a helical configuration. When these protozoans are exposed to cold,[15] or high pressure, the pseudopodia are withdrawn and the microtubules depolymerize.

Another example of an association between microbules and transport of particulate material is the melanocyte, in which the melanin granules move centrifugally and centripetally with different stimuli. The granules have been ob-served moving between channels created by the microtubules in the cytoplasmic matrix.

In the erythrophores found in fish scales the pigment granules may move at a speed of 25 to 30 μm per second between the microtubules. The possible role of microtubules in axoplasmic transport will be discussed in Chapter 28–1.

Sensory Transduction. Regularly arranged bundles of microtubules are common in sensory receptors, and a possible function in transduction of different incident energy has been attributed to them (see Wiederhold, 1976).

SUMMARY:

Properties of Microtubules

Microtubules are found in all eukaryotic cells — either free in the cytoplasm or forming part of centrioles, cilia, and flagella. They are tubules 25 nm in diameter, several micrometers long, and with a wall 6 nm thick with 13 subunits. The stability of different microtubules varies. Cytoplasmic and spindle microtubules are rather labile, whereas those of cilia and flagella are more resistant to various treatments. The main component is a protein called *tubulin*. This heterodimer of 110,000 to 120,000 daltons, is formed of two different monomers (tubulin A and B) of the same molecular weight (i.e., 55,000). The monomers are 4 nm $\times$ 6 nm and correspond to the subunit lattice structure seen in the tubular wall. Tubulin binds colchicine and vinblastin at different binding sites.

The assembly of tubulin in the formation of microtubules is a specifically oriented and programmed process. Centrioles, basal bodies, and centromeres are sites of orientation for this assembly. Calcium may be a regulating factor in the in vivo polymerization of tubulin. The level of polymerized tubulin is high at interphase and metaphase (i.e., spindle microtubules) and lower at prophase and anaphase.

Specific anti-tubulin antibodies have permitted the localization of microtubules in cultured cells and study of the changes that occur during mitosis and after treatment with colchicine. Cell transformation (i.e., cancer) produces a disorganization of microtubules.

Several functions, some of which are related to the primitive forms of cell motility described in this chapter, have been attributed to microtubules. They play a *mechanical* function and the *shape* of the cell and cell processes is dependent on microtubules. This function is particulary apparent during *cell differentiation* of certain placodes, in the nerve cells, and in spermiogenesis. The *polarity* and directional gliding of cultured cells depend on microtubules. These structures are associated with *transport* of molecules, granules, and vesicles within the cell. They play a role in the contraction of the spindle and movement of chromosomes and centrioles, as well as in ciliary and flagellar *motion*. A possible role in *sensory transduction* has been postulated.

9–3 MICROTUBULAR ORGANELLES: CILIA, FLAGELLA, AND CENTRIOLES

Several cell organelles can be considered to be derived from special assemblies of micro-tubules. In fact, cilia, flagella, basal bodies, and centrioles have, as one of their main components, groups of microtubules arranged in a special fashion. For this reason and in spite of the complexity of their structure, these organelles will be considered at this point, along with some of their physiological aspects.

Ciliary and Flagellar Motions — Present in Cells and in Tissues

Ciliary and flagellar cell motility is adapted to liquid media and is executed by minute, specially differentiated appendices that vary in size and number. They are called *flagella* if they are few and long, and *cilia*, if short and numerous. In Protozoa, especially the Infusoria, each cell has hundreds or thousands of minute cilia, and their movement permits a rapid progression of the organism in the liquid medium. In some special regions of the Infusoria, several cilia fuse and form larger conical appendices, *the cirri*, or membranes known as *undulating membranes*.

One entire class of Protozoa, the Flagellata, is characterized by the presence of flagella. The spermatozoa of metazoans move as isolated cells by means of flagella. On the other hand, epithelial cells that possess vibratile cilia and constitute true ciliated sheets are relatively common. These may cover large areas of the external surface of the body and determine the motion of the animal. Such is the case with some Platyhelminthes and Nemertea and also with larvae of Echinodermata, Mollusca, and Annelida. More often, the ciliated epithelial sheets line cavities or internal tubes, such as the air passages of the respiratory system or various parts of the genital tract. In these organs all the cilia move simultaneously in the same direction, and fluid currents are thus produced. In some cases, the currents serve to eliminate solid particles in suspension (e.g., in the respiratory system). The eggs of amphibians and mammals are driven along the oviduct with the aid of vibratile cilia.

The Ciliary Apparatus — The Cilium, Basal Bodies, and Ciliary Rootlets

The essential components of the ciliary apparatus are: (1) the *cilium*, which is the slender cylindroid process, limited by the plasmalemma (cell membrane), that projects from the free surface of the cell, (2) the *basal body*, or granule, the intracellular organelle similar to the centriole from which it originates, and (3) in some cells fine fibrils — called *ciliary rootlets* — that arise from the basal granule and converge into a conical bundle, the pointed extremity of which ends at one side of the nucleus.

The basal bodies are embedded in the ectoplasmic layer beneath the cell surface. In general, they are spaced uniformly and in parallel rows.

Cilia and flagella are extremely delicate filaments whose thickness is often at the limit of the resolving power of the light microscope.

Various epithelia have appendices similar in shape to cilia, but immobile; these are called *stereocilia*. Examples are the processes of the epithelial cells of the epididymis. In the macula and crista of the inner ear, there are stereocilia in addition to motile cilia, or *kynocilia*. Stereocilia do not contain microtubules.

The Axoneme Contains Microtubular Doublets

Axoneme is the term applied to the axial basic microtubular structure of cilia and flagella, and is it thought to be the essential motile element. In various flagella *accessory structures* made of fibers, microtubules, crystalline bodies, or mitochondrial derivatives may be found.

The axoneme of a cilium or flagellum may range from a few microns to 1 to 2 millimeters in length, but the outside diameter is about 0.21 μm. It is generally surrounded by an *outer ciliary membrane*, which is continuous with the plasma membrane. All the components of the axoneme are embedded within the *ciliary matrix* (Fig. 9–7).

Figure 9–8 shows a general diagram of the axoneme, with the fundamental 9 + 2 microtubular pattern and the other essential structural components of the axoneme. A plane perpendicular to the line joining the two central tubules divides the axoneme into a right and a left symmetrical half. It is generally admitted that the plane of the ciliary beat is perpendicular to this plane of symmetry.

The pairs of peripheral microtubules have an ellipsoidal profile, whereas the central ones are circular. The diameter of the peripheral tubules varies from 18 to 25 nm with the central tubules being slightly larger. The two microtubules of each peripheral pair can be clearly distinguished by several morphologic features. The doublets are skewed at about 10 degrees, so that one tubule, designated subfiber A, lies closer to the axis than the other (i.e., subfiber B). The microtubule of subfiber A is smaller, but complete, whereas that of subfiber B is larger and incomplete, since it lacks the wall adjacent to A. In fact, while A has 13 tubulin subunits, B has only 11. Furthermore, subfiber

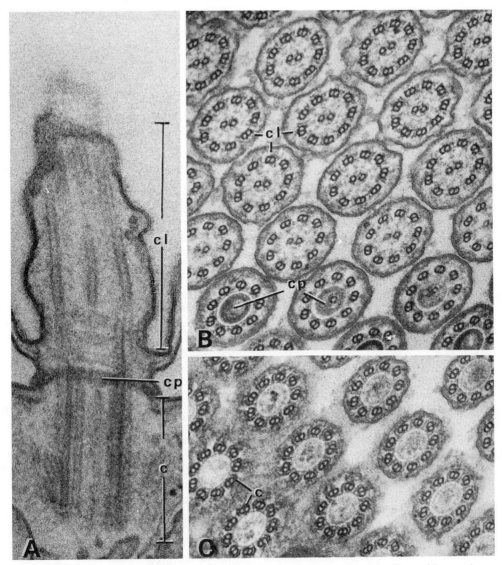

Figure 9-7 Electron micrographs of cilia in longitudinal and cross sections. **A**, cilium of *Paramecium aurelia* showing the centriole or basal body (*c*), the ciliary plate (*cp*), and the cilium (*cl*) proper. **B** and **C**, cross sections through cilia of *Euplotes eurystomes*. **B**, the section passes through the cilium proper showing the typical structure and the ciliary plate (*cp*). **C**, the section passes through the centriole or basal body (*c*). Notice the absence of central tubules and triple number of peripheral tubules. A, ×110,000; B and C, ×72,000. (Courtesy of J. Andrè and E. Fauret-Fremiet.)

Figure 9-8 Diagram of a cross section of a cilium showing the axoneme with the "9 + 2" microtubular structure. The axoneme is viewed from base-to-tip, with the arms directed clockwise. The inset shows doublet 2 (d$_2$) at high magnification with indication of the tubulin subunits. (See the description in the text.) Modified from Warner, F. D., Macromolecular organization of eukaryotic cilia and flagella, *in* DuPraw, E. J. (Ed.) *Advances in Cell and Molecular Biology*, Vol. 2, p. 193, New York, Academic Press, Inc., 1972.)

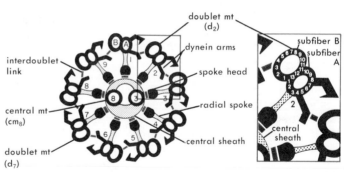

A has processes—the so-called *dynein arms*—that are oriented in the same direction in all microtubules. This orientation is clockwise when the axoneme is viewed from base to tip. In Figure 9–8 the nine pairs are numbered, starting from the one lying in the plane of symmetry, and continuing in the clockwise direction.

Dynein Arms. The arms of subfiber A are generally called *dynein arms* because they contain dynein, a high molecular weight ATPase (300,000 to 400,000 daltons). This is a Mg^{2+}- and Ca^{2+}-activated enzyme, which after solubilization can recombine at the same position on the A microtubules. Two isoenzymatic forms of the ATPase, dynein I and II, have been separated by electrophoresis.[16] Dynein I constitutes the largest proportion of the axonemal ATPase and is present in the arms of subfiber A. As will be mentioned later, the interaction between tubulin and dynein is thought to underlie the basic mechanism of ciliary and flagellar motion (see Mohri, 1976, Gibbons, 1977).

Nexin Links. As shown in Figure 9–8 the doublets are linked by *interdoublet* or *nexin links*. This name is derived from the fact that another protein called *nexin* has been isolated from these links. This protein has a molecular weight of 150,000 to 160,000 daltons.[17, 18] The function of these links is unknown, but they may serve as structures that maintain the geometric integrity of the axoneme during the sliding motion (see later discussion).

Radial Spokes. In Figure 9–8 it may be observed that there are radial bridges or links between the A subfiber and the sheath containing the central microtubules. These spokes terminate in a dense knob or head, which may have a forklike structure. Certain observations that the spokes are attached perpendicularly to the ciliary axis where it is straight and that they are relatively detached in bent or tilted regions of the axis has led to the hypothesis that they may be active in the conversion of active sliding between the outer doublets into local axial bending.[19]

Basal Bodies (Kinetosomes) and Centrioles Contain Microtubular Triplets

Since the classic works of Henneguy and Lenhossek in 1897, it has been suggested that basal bodies (or kinetosomes) of cilia and flagella are homologous with the centrioles found in mitotic spindles (Fig. 17–1). In some cells it was discovered that a centriole engaged in mitosis could carry a cilium at the same time. This homology was fully confirmed with the introduction of electron microscopy.

Centrioles are cylinders that measure, on the average, 0.2 μm × 0.5 μm; at times they may be as long as 2 μm. This cylinder is open on both ends, unless it carries a cilium. In the latter case, it is separated from the cilium by a *ciliary plate* (Fig. 9–7, A).

In freeze-fractured cilia it has been possible to observe the so-called *ciliary necklace*.[20, 21] This consists of two to six rows of membrane particles present just below the ciliary plate where the central pair of microtubules end.

The wall of the centriole has nine groups of microtubules arranged in a circle. Each group is a triplet formed of three tubules (rather than two, as in cilia) that are skewed toward the center (Fig. 9–9). Since the angle made by the set of tubules with the tangent to the cylinder varies from 70 degrees in the proximal end to 20 to 30 degrees in the distal end, the tubules twist from one end to the other or describe a helical course. As shown in Figure 9–9, the tubules are designated *subfiber A, B,* and *C,* from the center toward the periphery. Both subfibers A and B cross the ciliary plate and are continuous with the corresponding subfibers in the axonemene; subfiber C terminates near the *ciliary or basal plate.* There are no central microtubules in the centrioles and no special arms; however, they are linked by connectives. The proximal portion of the centriole has a cartwheel appearance with spokes that radiate from a central hub and connect to subfiber A of each triplet. The presence of the cartwheel structure at the proximal end provides the centriole with a structural and functional polarity. The growth of the centriole is from the distal end, and in the case of kinetosomes, it is from this end that the cilium is formed. Furthermore, the *procentrioles,* which are formed at right angles to the centriole, are located near the proximal end (see Wolfe, 1972).

Both centrioles may give rise to a cilium, but in most cases only one does, and the sister centriole remains at a right angle to the kinetosome (Fig. 9–4). In most multiciliated cells there are single basal bodies. It will be shown later that in sensory cells having nonmotile ciliary derivatives and a 9 + 0 configuration, there are two basal bodies (see Fig. 9–11). The two centrioles of a pair are always located at a distance from each other of at least 0.8 μm and are perpendicular to each other.

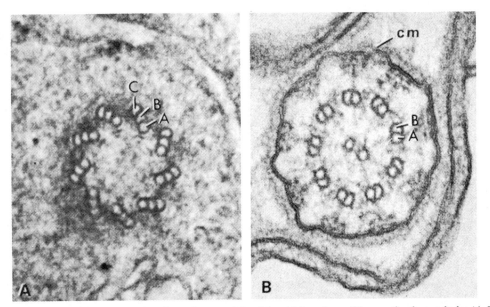

Figure 9-9 **A,** transverse section through a centriole of the chick embryo. Observe the three tubules (A, B, C) present in each of nine groups at the periphery of the centriole. Also note the absence of central tubules and the density of the centriolar wall. **B,** transverse section through a cilium (compare with the centriole structure); *cm,* ciliary membrane. ×150,000. (See Figure 22–1 for further details.) (Courtesy of J. Andrè.)

Ciliary Rootlets. In some cells, ciliary rootlets originate from the basal body. Most rootlets are striated, having a regular crossbanding with a repeating period of 55 to 70 nm. Five intraperiodic sub-bands have been observed in the cross-striated rootlets.

The striated fibers are composed of parallel microfilaments, 3 to 7 nm in diameter, which in turn are formed of globular subunits. These fibers and filaments may serve a structural role such as anchoring the kinetosomes. By analogy with other microfilaments (see below), a contractile role has been postulated for the ciliary rootlets. Furthermore, ATPase has been found associated with the cross-bands of certain rootlets.

Basal Feet and Satellites. Basal feet are dense processes that are arranged perpendicularly to the basal body in a particular direction and originate from two to three of the triplets. These processes impose a structural asymmetry on the basal body that has been related to the direction of the ciliary beat. The basal foot, which is composed of microfilaments that terminate in a dense bar, may be a focal point for the convergence of microtubules. *Satellites* or *pericentriolar bodies* are electron-dense structures lying near the centriole that probably are nucleating sites for microtubules.[22]

Ciliary Movement — A Sliding of Microtubular Doublets That Involves Dynein

Ciliary movement can be analyzed easily by scraping the pharyngeal epithelium of a frog or toad with a spatula and placing the scrapings in a drop of physiologic salt solution between a slide and a coverglass. On the free surface of the epithelial cell, the rapid motion of the vibratile cilia can be seen. If a row of cilia is observed, the contraction is *metachronic* in the plane of the direction of motion; that is, it starts before or after the contraction of the next cilium. In this way true waves of contraction are formed.

On the other hand, in a plane perpendicular to the direction of motion, the contraction is *isochronic*; all the cilia are observed in the same phase of contraction at a given time. The rhythmic contraction of cilia has been interpreted in different ways. A two-step process that involves *intraciliary* excitation followed by *interciliary* conduction has been proposed to explain the metachronic rhythm of the ciliary beat. This mechanism evidently does not depend on the nervous system, since it persists after the epithelium has been separated from the rest of the organism. However, cytoplasmic continuity is indispensable to its maintenance, for if a cut is made in the row of cilia the waves of contraction of the two isolated pieces becomes uncoordinated.

The direction of the effective ciliary beat also depends on the underlying cytoplasm. If a piece of epithelium is removed from the pharynx of a frog and implanted with a reversed orientation, the movement is maintained but in the opposite direction.

Ciliary contractions are generally rapid (10 to 17 per second in the pharynx of the frog). Analysis of the motion has been facilitated greatly by stroboscopic and ultrarapid microcinematography.

Ciliary movement may be pendulous, unciform (hooklike), infundibuliform, or undulant. The first two are carried out in a single plane. In the pendulous movement, typical of the ciliated Protozoa, the cilium is rigid and the motion is carried out by a flexion at its base. On the other hand, in the unciform movement, the most common type in the Metazoa, the cilium is doubled upon contraction and takes the shape of a hook. In the infundibuliform movement, the cilium or flagellum rotates, describing a conical or funnel-shaped figure. In the undulant motion, characteristic of flagella, contraction waves proceed from the site of implantation and pass to the free border. It was shown above that the plane of the beat of a cilium is perpendicular to that passing through the central microtubules. In Figure 9–8, this plane passes through doublet 1 and between doublets 5 and 6. However, there are many indications that there is a three-dimensional movement during the beat. Although the effective stroke is a single plane, the cilium moves away from the original plane during the recovery phase This movement, which can be beautifully demonstrated by scanning electron microscopy,[23] implies that either the whole cilium rotates at its base or that the central microtubules rotate.

Since the last century, ciliary and flagellar motion have been compared to that of muscle. However, the complex structure of the axoneme suggests that this motion may be more complex than the planar sliding of myofilaments in muscle (see Chapter 27).

The key process appears to be the sliding of the microtubular doublets over each other, with the associated making and breaking of the cross-bridges between adjacent doublets, represented by the dynein arms. The sliding of doublets during the bending of the cilium or flagellum has actually been observed and measured. This displacement of the doublets has been determined because the central microtubules extend to the very tip, whereas subfibers B terminate at a distance from the tip of 1 to 2 μm, and subfibers A terminate slightly beyond subfibers B. This arrangement has also made it possible to demonstrate that the two subfibers do not move relative to one another within the doublet. In the diagram of Figure 9–10 the amount of sliding displacement is shown to depend on the distance separating the tubules and on the bend angle (see Sleigh, 1971).

Most of the experimental work on ciliary motion has been aimed at demonstrating the involvement of ATP; it was mentioned above that the ATPase activity is associated with the protein *dynein* in the arms of subfibers A. Dynein constitutes about 8 per cent of the total ciliary protein. In glycerin-extracted cilia, flagella, and spermatozoan tails, the addition of ATP produces rhythmic activity that may persist for a few minutes or even hours. Interaction between tubulin and dynein seems to be essential in this type of cell motility. When studying myofilaments in muscle, we shall see that the actin-myosin system is the other process involved in contraction.

Many approaches are currently being used to study the molecular mechanisms in ciliary motion (see Gibbons, 1977). One of them has been the use of sperm flagella in which the membranes have been removed by detergents. These sperm can be reactivated with ATP. Extraction of dynein arms by KC1 leads to a reduction in the frequency of beating, which can be restored to the original level by the addition of *dynein*.[24] An anti-dynein antiserum inhibits the movement of spermatozoa without membranes, and ATP induces, it.[25]

Another important observation is that, unlike muscle, cilia and flagella do not require Ca^{2+} for activation. In many cases, however, Ca^{2+} has a regulating effect on the wave form of the beating.

Kartagener's Syndrome — A Mutation Involves a Lack of Dynein

A recent approach to the study of ciliary motion is the use of mutants of *Chlamydomonas* in which various deficiencies in axoneme structure have been observed. From the medical point of view, this research is of great importance in the so-called *Kartagener's syndrome*, in which there is infertility with non-motile sperm, chronic bronchitis, and sinusitis. It has

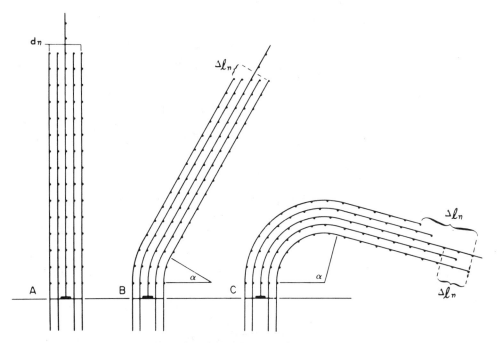

Figure 9–10 Diagram of the sliding model of the axoneme. The amount of displacement ($\triangle_{ln}$) is a function of the distance separating the microtubules (d_n) and the bend angle (α). The central microtubules are represented by the longer lines. (From Sleigh, M. A., The physiology and biochemistry of cilia and flagella, *in* Lima-De-Faría, A. (Ed.), *Frontiers of Biology: Handbook of Molecular Cytology*. Vol. 15, p. 1243. Amsterdam, North-Holland Publishing Co., 1970.)

been demonstrated that this syndrome is due to a mutation involving the dynein arms.[26, 27] In these patients *cilia and sperm are non-motile because they lack dynein arms, which contain ATPase.*

Photoreceptors are Derived from Cilia

Studies on the structure of retinal rods and cones have shown that the short, fibrous connection found between the outer and inner segments is of a ciliary nature. Cross sections of this so-called connecting cilium have revealed nine pairs of filaments similar to those found in cilia (Fig. 9–11).[28] In this structure, however, there are no central microtubules.

The outer segment of this cilium is composed of numerous double membrane rod sacs arranged like a stack of coins. The first stage in the development of a rod is a primitive cilium projecting from the bulge of cytoplasm that constitutes the primordium of the inner segment (Fig. 9–12). This cilium contains the nine pairs of filaments and the two basal centrioles.[28] The apical end is filled with a vesicular material. In the second stage the apical region of the primitive cilium enlarges greatly, owing to the rapid building up of the vesicles and cisternae that constitute the primi-

tive rod sacs. The proximal part of the primitive cilium remains undifferentiated and constitutes the connecting cilium of the adult (Fig. 9–12).

Other structures have also been recognized as ciliary derivatives. For example, the so-called crown cells of the saccus vasculosus found in the third ventricle of fishes are modified cilia with swollen ends that are filled with vesicles.[29] Also, the primitive sensory cells of the pineal eye found in certain lizards have a ciliary structure.[30, 31]

Cilia and Flagella Originate from Basal Bodies

The origin of centrioles and basal bodies, or kinetosomes, is viewed from the standpoint that these are possibly semiautonomous cell organelles and in this way similar to mitochondria and chloroplasts, as discussed in Chapters 12 and 14. Isolation of kinetosomes from *Tetrahymena* has verified the presence of RNA and DNA, and this suggests that they are capable of some protein synthesis. More direct evidence of DNA in centrioles was obtained by use of the fluorescent dye acridine orange and by ^{3}H-thymidine incorporation followed by treatment with DNAse. However, most of these studies refer to the pellicle, and

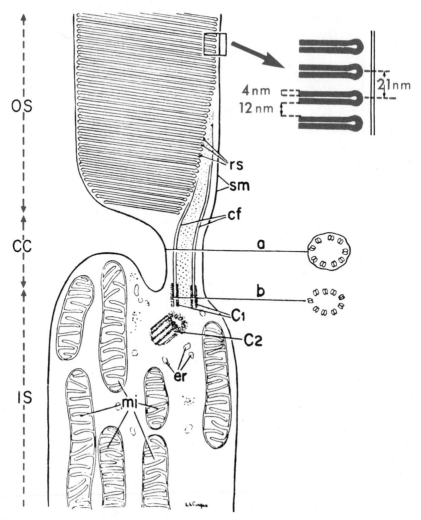

Figure 9-11 Diagram of a retinal rod cell in the rabbit: the outer segment (*OS*) with the rod sacs (*rs*); the connecting cilium (*CC*) and the inner segment (*IS*) are shown. *a* and *b* correspond to a cross section through the connecting cilium and the centriole (C_1). C_1 and C_2, centrioles; *cf*, ciliary filaments; *er*, endoplasmic reticulum; *mi*, mitochondria; *sm*, surface membrane.

the exact localization of DNA has not been determined. Furthermore, these findings do not resolve the question of whether the synthesis of ciliary proteins depends on the DNA-RNA components of the centriole (see Rattner and Phillips, 1973).

Early cytologists realized that the development of cilia and flagella was directly related to the presence of centrioles. Flagella may be experimentally removed and their regeneration followed; thus, in *Chlamydomonas* it was found that the growth rate of a flagellum is about 0.2 μm per minute.

Studies on the origin of centrioles and kinetosomes were hampered by the small size of these structures. Light microscopists thought that cen-

trioles originate by division from pre-existing centrioles and that they possessed a genetic continuity comparable to that of chromosomes. The electron microscope has failed to confirm that such a division process takes place.

There are now several indications that centrioles may be generated by two different mechanisms.[32, 33] Centrioles destined to form mitotic spindles of single cilia arise directly from the wall of the pre-existing centriole. The *daughter centrioles* appear first as annular structures (*procentrioles*) (Fig. 9-4), which lengthen into cylinders. The groups of three tubules originate from single and double groups that first appear at the base of the procentriole. When they are half grown the daughter centrioles are released into the cyto-

194

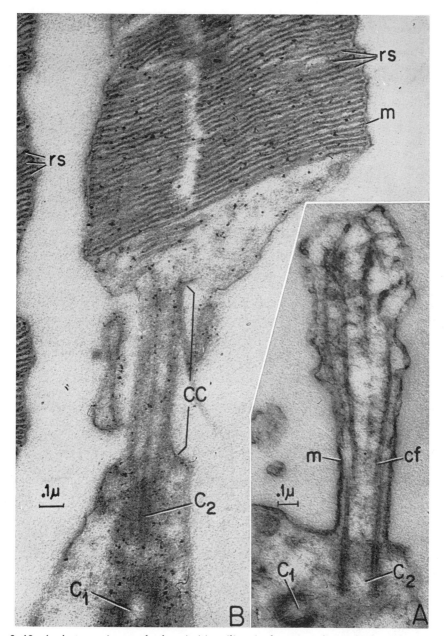

Figure 9-12   **A,** electron micrograph of a primitive cilium in the retina of an eight day old mouse. C_1 and C_2, centrioles; *cf,* ciliary fibrils; *m,* ciliary membrane. ×72,000. **B,** electron micrograph of an adult rod, showing the outer segment with the rod sacs (*rs*) and the outer membrane (*m*); the connecting cilium (*CC*) and the basal centrioles (C_1 and C_2). ×62,000. (From E. De Robertis and A. Lasansky.)

plasm to complete their maturation. Usually, although not always, daughter centrioles are formed one at a time.

The other mechanism in the formation of centrioles is found in those destined to become kinetosomes, as in a ciliated epithelium. The centrioles are assembled progressively from a precursor fibrogranular material located in the apical cytoplasm. The newly formed centrioles become aligned in rows beneath the apical plasma membrane, and each centriole may then produce satellites from the side, a root from its base and a cilium from its apex.

Development of the cilium begins with the appearance of a vesicle that becomes attached to the distal end of the centriole. The growing ciliary shaft invaginates the vesicular wall, which forms a temporary ciliary sheath until the permanent

195

one is formed. In many epithelial cells short cilia enclosed within vesicles may be observed. (For a review of ciliogenesis, see Brenner and Anderson, 1973.)

Electron microscopic studies of centrioles have revealed that they have a specific tubular struc-ture, that they do not divide, and that they are found in association with other tubular struc-tures. This relationship suggests that the func-tion of the centriole may be to regulate the syn-thesis and organization of microtubules within the cytoplasm (see Wolfe, 1972).

SUMMARY:

Structure, Motion, and Origin of Cilia and Flagella

Cilia and *flagella* are motile processes found in protozoa and in many animal cells; in plants, only the antherozoids have flagella. *Kynocilia* (i.e., motile cilia) should be differentiated from *stereocilia,* which are non-motile and lack a microtubular struc-ture. The ciliary apparatus comprises: (1) the *cilium,* (2) the *basal body,* and (3) the *ciliary rootlets.*

The *axoneme,* which is the motile element, is the basic microtubular structure of cilia and flagella. It is surrounded by the *ciliary membrane* and is embedded in the *ciliary matrix.* The fundamental structure is 9 + 2 i.e., 9 pairs of peripheral micro-tubules and 2 single and central microtubules. The peripheral doublets can be num-bered from 1 to 9, starting from the doublet cut by the plane perpendicular to the line joining the two central tubules (Fig. 9–8). This plane coincides with the plane of the ciliary beat. In the doublet, *subfiber A* and *subfiber B* are distinguished by several morphologic features. Subfiber A has two arms that are oriented in the same direc-tion. These so called *dynein* arms contain dynein, a high molecular weight ATPase. The interaction between tubulin and dynein is thought to be the basic mechanism of contraction in cilia and flagella. *Nexin* is another protein that links the microtubular doublets. *Radial spokes* connect each subfiber A with the central sheath that contains the central microtubules. These radial spokes, in longitudinal sections, are seen as periodically spaced processes.

Basal bodies or *kinetosomes* have the same structure as centrioles. A centriole is a cylinder (0.2 μm $\times$ 0.5 μm) opened on both ends; the basal body at the distal end has a *ciliary plate* that separates it from the cilium. The centriolar wall contains triplets of microtubules. From the center to the periphery these tubules are designated A, B, and C. Tubules A and B traverse the ciliary plate and are continuous with the corresponding tubules in the cilium, whereas tubule C terminates near the plate. Centrioles lack the central microtubules. Centrioles are polarized structures; at the proximal end they have a cartwheel structure which con-nects with the triplets. Procentrioles are formed from the proximal end, whereas a cilium or flagellum may be formed from the distal end. Centrioles are generally in pairs and at right angles.

Basal bodies may be related to cross-banded fibers that constitute the *ciliary roots* and which serve a structural role. Other accessory structures are the basal *foot* and *satellites* or *pericentriolar bodies.*

Ciliary movement may be analyzed from a scraping of frog pharyngeal epithelium. The contraction of cilia is *metachronic* in the plane of motion and *isochronic* in a plane perpendicular to the direction of motion. The macromolecular mechanisms of ciliary motion are probably more complex than those taking place in muscle myofila-ments. To account for the movement in cilia, a sliding mechanism is favored. Since the central microtubules end at the very tip of the cilium, and the peripheral ones end at a distance, it has been possible to measure the actual sliding of the doublets and to demonstrate that it increases with the angle of bend. Sperm flagella in which the membranes have been removed by detergents are reactivated by ATP. Extraction of dynein stops contraction, and this is restored by the addition of this protein. A human syndrome in which cilia and sperm are non-motile (they lack dynein) is ge-netically determined.

The outer segments of the retinal rods and cones are connected to the inner segment by a *connecting cilium* that has a 9 + 0 pattern. The photoreceptor develops from a primitive cilium. Other ciliary derivatives are in the crown cells of the *saccus vasculosus* and in the pineal eye of certain lizards.

The *centrioles do not divide*. Those which later form mitotic spindles or single cilia arise directly from the wall of pre-existing centrioles. *Procentrioles* and *daughter centrioles* are formed at right angles. In the case of multiple centrioles, there is a fibro-granular material located in the apical cytoplasm from which the rows of basal bodies are formed. Cilia are formed from the distal end of the centriole.

9–4   MICROFILAMENTS

We mentioned earlier that the ground cytoplasm or cytosol contains a kind of *cytoskeleton* formed by microtubules and various types of microfilaments. From what was said about the function of cytoplasmic microtubules the reader may have gathered the impression that they play mainly a mechanical function, constituting a kind of *passive portion* or *framework of the cytoskeleton*. This concept may not be completely accurate, since these microtubules could store energy by distortion, and during disassembly they could release elastic forces that might transport cytoplasmic structures (see Porter, 1976). Furthermore, we have seen before that the interaction between the tubulin of microtubules and dynein is the main mechanism of motion of cilia and flagella.

In recent years the *active* or *motile function of the cytoskeleton* has been attributed to some of the microfilaments that are observed by electron microscopy as thin filaments between 6 and 10 nm thick. Various studies have suggested that such filaments represent contractile systems which may be involved in cytoplasmic streaming, ameboid motion, and other types of cell activities in which a motive force is involved. These studies have recently led to very important conclusions with the demonstration that the contractile proteins of muscle (i.e., *actin* and *myosin*) are also widely found in nonmuscular systems and that they may be chemical components of some of the microfilaments.

Microtrabecular Lattice in Cytosol — Revealed by High-Voltage

Because of its greater penetration, high-voltage electron microscopy has helped in identifying microfilaments in whole culture cells and in obtaining a three-dimensional view of the so-called *microtrabecular lattice*.[1, 2]

As shown in Figure 9–3, just beneath the membrane there are bundles of filaments that can be seen to be made of actin (see later discussion) and that are in continuity with a lattice of similar filaments that pervades the cytosol. The channels or spaces found in between this *trabecular lattice* are of the order of 50 to 100 nm and in the living cell may provide for the rapid diffusion of fluids and metabolites through all parts of the cytosol. The filaments of the lattice are in contact with vesicles of the endoplasmic reticulum, with microtubules, and with polysomes, all of which seem to be contained or supported within the lattice (Fig. 9–13). On the other hand, mitochondria appear to be free of the lattice, and this may explain the fact that they move rather freely in the cytosol (see Chapter 12–1). This image of the cytoskeleton explains the sol-gel transformation that occurs in the cytosol. The polymerization and depolymerization of this lattice may also provide for small movements of the contained organelles.[2]

Similar observations have been made on thin sections of various cells. Networks of microfilaments have been observed in the growth cones of neuroblasts and in the undulating membranes of cultured cells. A thin sheath of microfilaments forming a network is observed below the plasma membrane of fibroblast and glial cells. In nerve endings a three-dimensional network of actin filaments has been observed.[34]

Cytochalasin B Impairs Several Cellular Activities Involving Microfilaments

While in the sutdy of microtubules the depolymerizing effect of colchicine has been of paramount importance, in the case of microfilaments the action of the drug *cytochalasin B* has had a decisive influence (see Wessels et al., 1971; Allison, 1973). Although the mechanism of action is still unknown, cytochalasin B has been found to impair numerous cell activities in which some types of microfilaments are involved. For example, it inhibits smooth muscle contraction, beat of heart cells, migration of cells, cytokinesis, endocytosis, exocytosis, and other processes.

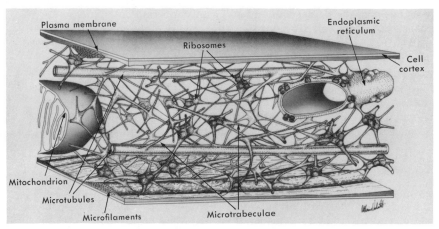

Figure 9–13 A model of the structure of the cytoskeleton of a cell showing the various components contained in the cytoplasmic matrix. Below the plasma membrane there are bundles of actin microfilaments. **A,** microtrabecular lattice pervades the cytosol and is in contact with the endoplasmic reticulum, the microtubules, and ribosomes. ×90,000. (Courtesy of K. R. Porter. From Porter, K. R., Introduction: motility in cells, *in Cell Motility,* Vol. 1, p. 1. (Goldman, R. D., et al., eds.) Cold Spring Harbor, New York, Cold Spring Harbor Laboratory, 1976.)

When applied to the alga *Nitella,* cytochalasin B produces rapid cessation of cyclosis, which is reinitiated after removal of the drug. Cytochalasin B disrupts the regular arrangement of microfilaments associated with some of these functions. For example, the contractile ring of microfilaments observed during cell cleavage (Fig. 17–11).[35] is altered by this drug. Not all microfilaments, however, are sensitive to the drug. These and other studies have led to the general conclusion that the cytopchalasin-B–sensitive microfilaments are the contractile machinery of non-muscle cells.

Actin, Myosin, and other Contractile Proteins — Present in Non-muscle Cells

The finding that the contractile proteins *actin* and *myosin* are present in non-muscle cells has, for the first time, suggested a molecular mechanism for cellular movements. It is now known that all eukaryotic cells contain high concentrations of *actin* and low concentrations of *myosin,* the two *force-generating proteins.* In addition, there are other proteins, such as *tropomyosin* and *troponin* that, as in muscle, play a regulatory function (see Chapter 27). Finally, there are *actin-binding* proteins that can cross-link action filaments, and *α-actinin,* a protein found in the Z-line of muscle and which may participate in actin-filament–membrane interactions (see Tilney, 1977; Pollard, 1977).

Actin. Actin comprises a large proportion of the cytoplasmic proteins of many cells; in developing nerve cells, it may constitute up to 20 per cent of these proteins. Actin is present principally in its globular form (G-actin) with a molecular weight of 45,000, and it may quickly polymerize to form the microfilaments of fibrous actin (F-actin). This G- to F-actin transition seems to represent the basis of the classical sol-gel transition in the cytoplasm of moving cells.

Cytoplasmic actin is very similar in structure to muscle actin and forms identical 6-nm wide microfilaments consisting of a double helical array of globular actin molecules (Fig. 27–6). Actin filaments can be identified under the electron microscope by virtue of the fact that they become *decorated* with myosin (heavy meromyosin) to form arrow-like complexes (see Goldman, 1975) (see Fig. 27–9). *Myosin* has been found in cells having ameboid motion, such as amebae and slime molds, and also in platelets.[36]

Myosin. Compared with actin, myosin is found in a low concentration and is rather heterogenous in size and composition (see Pollard, 1977). In most cases it has a molecular weight of 460,000 daltons, similar to that of muscle myosin, but the two molecules contain different subunits. In *Acanthamoeba* there is a "minimyosin" with a MW of 180,000 daltons. In spite of these differences, all myosins bind reversibly to actin filaments and contain a Ca^{2+}-activated ATPase. As in muscle, both these functions are localized in the globular heads of the molecule (see Chapter 27).

Identification of myosin filaments with the electron microscope, at present, is less satisfactory. However, filaments that are 13 to 22 nm thick and 0.7 μm long, which are of probable myosin nature, have been observed.

Contractile and Regulatory Proteins — Detected by Specific Antibodies Stress Fibers

As in the case of the microtubules, which can be stained with anti-tubulin antibodies,[11] the microfilaments can be detected by fluorescent antibodies against *actin,*[37] *myosin,*[38] *tropomyosin,* and *α-actinin.*[39] In culture cells actin has a typical pattern of straight and often parallel fibers that are predominantly localized on the lower, contacting side of the cell and that are in close contact with the plasma membrane (Fig. 9–14, *A*). This pattern is completely disorganized by the action of cytochalasin B,[40] which does not affect the microfilaments.

Staining of culture cells with anti-myosin antibodies produces a somewhat similar image to that of actin; however, the majority of the filaments show interruptions or striations[41] (Fig. 9–14, *B*). Staining with anti–α-actinin antibodies also reveals a similar pattern, with a periodic arrangement along the filament. These results suggest that α-actinin (and also *tropomyosin*) may be involved in the organization of the actin bundles of microfilaments and in the attachment of them to the plasma membrane, especially in the areas in which there is contact with a solid substrate or with other cells.[39] In these contacting regions actin microfilaments are arranged in parallel, forming bundles that in a living spreading cell may be visible as *stress fibers.*

A very interesting case is that of the microvillus of the intestine, which contains a bundle of actin filaments that is attached to the plasma membrane by way of a dense material that stains with antibodies against α-actinin. This actin bundle has a polarization similar to that found in a muscle sarcomere (see Fig. 27–7), except that in this case the Z-line is replaced by the α-actinin at the tip of the microvillus (see Tilney, 1977).

Two Recognizable Types of Microfilaments

In addition to microtubules (20 to 30 nm) and actin microfilaments (5 to 7 nm), in a variety of cells there is a third cytoskeletal component composed of filaments of an intermediate size (7 to 12 nm). These have been identified in fibroblasts, muscle cells, and glial cells. Other examples include the *neurofilaments* of nerves, the *tonofilaments* of epithelia, and the *prekeratin filaments* of cells in which keratinization occurs.

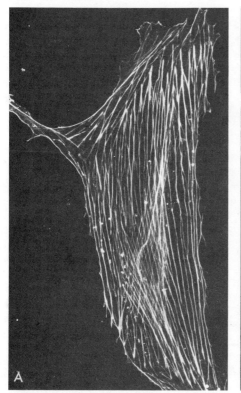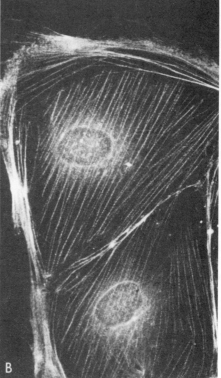

Figure 9–14 **A,** Cultured mammary rat cell stained with anti-actin antibodies. **B,** Cultured rat cell stained with anti-myosin antibody. ×600. (**A,** courtesy of K. Weber, 1976; **B,** courtesy of K. Weber and U. Groeschel-Stewart, 1974.)

Studies with fluorescent antibodies in culture cells have shown the presence of a system of wavy fibers, some of which lie near the cell boundary. This system of fibers is affected neither by cytochalasin, which disorganizes actin filaments, nor by Colcemid, which depolymerizes microtubules. This *intermediate system* of microfilaments appears to be related to the anchorage and maintenance of the epithelioid cell shape;[42, 43] in nerve axons the neurofilaments could be involved in axoplasmic transport. Isolation of intermediate filaments is facilitated in cell cultures of baby hamster kidney treated with colchicine. In these cells the 10-nm filaments aggregate to form juxtanuclear caps, which can be isolated. Two proteins of 54,000 and 55,000 daltons were resolved by SDS gel electrophoresis of these intermediary filaments (see Starger and Goldman, 1977).

Microfilaments — Involved in All Motility in Non-muscle Cells

The basis of contraction in non-muscular systems is still hypothetical, but the scheme that is evolving is in some ways similar to that of muscle (see Chapter 27). In non-muscular systems it is based on the interaction of actin and myosin filaments, with the consequent production of a shearing force. The differences between the mechanism here and in muscle are in the more random distribution of these proteins and the much smaller concentration of myosin. These factors may help to account for the much slower concentration in these primitive contractile systems. The attachment of active filaments to the plasma membrane, however, is favorable to the production of shearing forces acting at the level of the membrane.

Many motile events in non-muscle cells, e.g., cell spreading and shape changes, locomotion, membrane ruffling, cytokinesis, phagocytosis, and formation of pseudopodia, require a tight coupling between the cell surface and the contractile apparatus, and this could be achieved by α-actinin molecules that bind the actin microfilaments to the plasma membrane.

In contrast to muscle cells, in which actin filaments are stable, most non-muscle cells have transitory filaments that may appear or disappear within one minute. Thus the cell must have some mechanism that controls the assembly and disassembly of the actin monomers and the formation of microfilament bundles (i.e., stress fibers). Such mechanisms are al-

tered in cells undergoing transformation by oncogenic viruses (see Chapter 8–5), some of which may show a reduced number of bundles. (For further discussion on this important subject, see Goldman et al., 1976; Tilney, 1977; and Pollard, 1977.)

We shall now consider two classic types of cell motility in which microfilaments and actin-myosin interactions are involved. One of them, called *cytoplasmic streaming* or *cyclosis*, is most remarkable in plant cells. The other, *ameboid motion*, is found mainly in certain protozoa and in animal cells.

Cytoplasmic Streaming (Cyclosis) — Observed in Large Plant Cells

Cytoplasmic streaming, or *cyclosis*, is easily observed in plant cells, in which the cytoplasm is generally reduced to a layer next to the cellulose wall and to fine trabeculae crossing the large central vacuole. Continuous currents can be seen that displace chloroplasts and other cytoplasmic granules.

In some plant cells the protoplasmic current can be initiated by chemicals (*chemodynesis*) or by light (*photodynesis*). Cyclosis is modified by temperature, by the action of ions, or by changes in pH. Cyclosis is stopped by mechanical injuries, electric shock, or some anesthetics. Some auxins (plant growth hormones) increase the rate of cyclosis. In general, all the factors that decrease cell viscosity increase the speed of protoplasmic current and vice versa. Cyclosis decreases progressively in cells subjected to increased hydrostatic pressure at the same time that the protoplasm becomes more liquid.

The classic experimental work on cyclosis has utilized the cylindroid cells of *Nitella*, which have a thin protoplasmic layer of about 15 μm surrounding a central vacuole of 0.5 mm by 10 cm. This protoplasmic layer is divided into a cortical region of structured cytoplasm in which are embedded non-motile chloroplasts and a layer of isolated cytoplasm (the endoplasm) where cyclosis takes place. The whole system is surrounded by the plasma membrane, under which is a single layer of microtubules. Bundles of actin filaments of about 0.2 μm are situated just beneath the chloroplasts in the cortical region and in a direction that is parallel to that of cyclosis. (Each bundle is composed of about 100 microfilaments of 5 to 6 nm.) It is assumed that the separation of the ectoplasm and endoplasm is labile and that actin filaments may be incorporated (or polymerized) into the endoplasm.[44] There is evidence that plant cytoplasm also contains a myosin-like protein.[45] Thus, the actual motive force may be provided by a mechanism of actin-myosin interaction that takes place in the region between the stationary ectoplasm and the moving endoplasm. Furthermore, it has been found that cytoplasmic streaming is stopped when the cortex of the *Nitella* is treated with cytochalasin B.[46]

Ameboid Motion — Characteristic of Amebae and Many Free Cells

In ameboid motion the cell changes shape actively, sending forth cytoplasmic projections called *pseudopodia,* into which the protoplasm flows. Although this special form of locomotion can be observed easily in amebae, it also occurs in numerous other types of cells. One need only to place a drop of blood between a slide and coverglass to see that the leukocytes, at first spheroidal, change their shape, emit pseudopodia, and move about. In tissue cultures, cells move out actively, forming the zone of migration. These changes also occur in vivo. For example, in epithelial repair, the cells free themselves and slide along actively toward the depth of the wound. In an inflammatory proc-

ess, leukocytes wander out of the blood vessels (*diapedesis*) by active ameboid motion and progress toward the focus of infection.

Some amebae are predominantly *monopodial* (one pseudopodium), but others may be temporarily or permanently *polypodial.* The shape of pseudopodia varies between a stout, almost cylindrical *lobopodium* and a fine filamentous or branching *filopodium* (Fig. 9–15). Sometimes these fine processes may be anastomosing (*reticulopodia*), as in Foraminifera.

As shown by the classic studies of Mast, the protoplasm of the ameba has a clear ectoplasm, which expands considerably toward the end of the pseudopodium. As shown in Figure 9–16, the axial endoplasm is surrounded by a "shear zone" where particles move more freely. At the advancing end is the hyaline cap, and just posterior to it, the "fountain zone," where the axial endoplasm appears to contract actively and flows below the ectoplasmic tube. At the opposite end is the tail process, also called the *uroid,* and near it, the *recruitment zone,* where the endoplasm is recruited from the walls of the ectoplasm in the posterior third of the cell.[47]

The rate of progression varies between 0.5 and 4.6 μm per second among different amebae. In the neutrophilic leukocytes it is approximately 0.58 μm per second. This rate is modified by

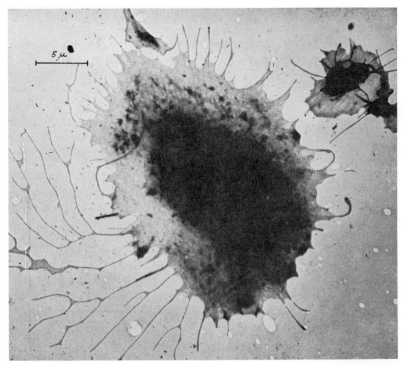

Figure 9–15 Electron micrograph of a polymorphonuclear leukocyte showing fine filopodia. ×3000. (From E. De Robertis.)

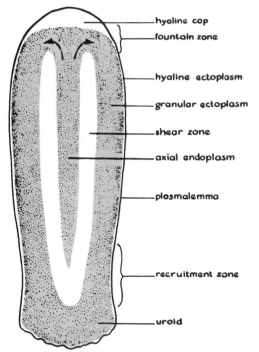

Figure 9–16 A new schema for ameboid structure and movement based on studies of cytoplasmic flow. New terminology is proposed for various regions of the cytoplasm of the ameba. (From Allen, R. D., "Ameboid Movement," *in* Brachet, J., and Mirsky, A. E. (Eds.), *The Cell*, Vol. 2, p. 35. New York, Academic Press, Inc., 1961.)

temperature and other environmental factors. Insufficient oxygen supply does not stop the movement but does slow it. Calcium is required for this type of locomotion. Severe mechanical injury, electric shock, or ultraviolet radiation causes retraction of pseudopodia.

An important factor in ameboid motion is *adhesion to a solid support*. An ameba that floats freely in the liquid medium can emit pseudopo-

dia but does not progress; only when it adheres to a solid surface does it commence this type of locomotion (Fig. 9–17). In tissue cultures the fibers of the coagulum serve as support; in connective tissue, the collagenous or reticular fibers may serve this purpose.

There are substances that influence the motion by attracting or repelling the cells. This property, which is called *chemotaxis*, has great importance in defense mechanisms, especially during inflammation.

All that has been said thus far about microfilaments and cell motility applies to ameboid motion. Actin filaments and myosin aggregates, together forming thicker filaments, have been found in amebae, and there seems no doubt that the actin-myosin interaction provides the actual motive force for ameboid motion. However, the views are divided regarding the most likely site of contraction. Whereas some investigators regard the posterior region of the ectoplasmic tube as more active, others give more importance to the "fountain zone" at the advancing end of the ameba.[47]

The diagram of Figure 9–17 emphasizes this last viewpoint. An ameba is shown in a vertical section showing the zone of attachment to a surface. From the velocity profiles indicated (v), it is evident that the streaming of the axial endoplasm increases toward the front, where the main region of contraction is indicated. The various portions of the cytoplasm are in states of contraction, relaxation, transition, or stabilized equilibrium which can be recognized by the viscoelastic properties and the variations in birefringence revealed by the polarizing microscope. As the endoplasm streams forward, it becomes more rigid, assuming the viscosity of the ectoplasmic tube. On the other hand, the contraction is propagated in a direction opposite to that of the streaming. In

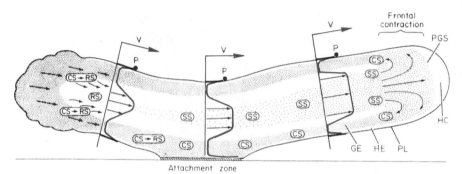

Figure 9–17 A schematic diagram corresponding to a frontal contraction model for ameboid motion. *SS,* stabilized regions; *CS,* contracted and *RS* relaxed regions; *V,* velocity profile; *HC,* hyaline cap; *PGS,* plasma gel sheet; *HE,* hyaline ectoplasm; *GE,* granular ectoplasm; *PL,* plasmalemma. (From D. L. Taylor et al., 1973.)

this model the sol-gel cycles described in the older literature are interpreted as contraction-relaxation cycles, and the recruitment zone corresponds to one of transition between the contracted and relaxed states. From the results of experiments in which Ca^{2+} was injected into different parts of the ameba, it is evident that free calcium ions induce contraction and may regulate the overall ameboid movement, which involves the interaction of actin and myosin. (For further details on the mechanism of ameboid movement, see Allen, 1974.)

SUMMARY:

Microfilaments, Cyclosis, and Ameboid Motion

Microfilaments ranging between 6 and 10 nm represent the active or motile part of the cytoskeleton and appear to play the major role in cyclosis and ameboid motion. With high-voltage electron microscopy a three-dimensionsional view of microfilaments has been obtained (i.e., an image of *microtrabecular lattice*).

Some microfilaments are sensitive to cytochalasin B, an alkaloid that also impairs many cell activities such as the beat of heart cells, cell migration, cytokinesis, endocytosis, and exocytosis, among others. Applied to the alga *Nitella*, cytochalasin B produces a rapid halt in cyclosis. A network of microfilaments is often seen attached to the plasma membrane of moving cultured cells; this is altered by cytochalasin B. It is generally assumed that the cytochalasin-B—sensitive microfilaments are the contractile machinery of non-muscle cells.

The contractile proteins *actin* and *myosin*, as well as tropomyosin, troponin-C, α-actinin, and other proteins found in muscle, are present in microfilaments. These proteins can be localized with specific antibodies. Actin filaments are 6 nm wide and are made of globular actin molecules (*G-actin*). In the presence of myosin they become "decorated," as can be seen with the electron microscope.

Actin is one of the most abundant proteins in many cells. The globular (G-actin) fibrillar transition is the basis of the classical sol-gel transition in the cytoplasm of moving cells.

Myosin is found in amebae, blood platelets, and slime molds but in a much lower concentration than actin. It contains a Ca^{2+}-activated ATPase as is found in muscle myosin. Identification of myosin filaments is more difficult.

An intermediary system of microfilaments of 7 to 10 nm is represented by the neurofilaments of nerves, the tonofilaments of epithelia, and the prekeratin filaments.

The *mechanism of contraction* in non-muscle tissue, which includes cyclosis and ameboid motion, is thought to involve the interaction of actin and myosin filaments (as in muscle) and the production of a shearing force. The random distribution of these filaments and the lower content of myosin may explain the slowness of this contraction, which requires ATP and the Ca^{2+}-activated ATPase.

Cytoplasmic streaming or *cyclosis* is found in all kinds of cells and produces the displacement of various organelles. The alga *Nitella* is the organism most often used for this study. Cyclosis may be started by chemicals or by light. Various injuries and *cytochalasin* B stop cyclosis. Microtubules may provide the framework for cyclosis, but the actual motive force is provided by microfilaments.

Ameboid motion is observed in amebae, leukocytes, culture cells, and in healing wounds. The number and shape of pseudopodia vary (e.g., lobopodial, filopodial, and reticulopodial). In amebae there is an axial endoplasm separated by a shear zone from the peripheral endoplasm. The ectoplasm forms a hyaline cap at the advancing end. At the tail there are the *uroid* and the *recruitment zone*. In slime molds there is a pulsating movement that is inhibited by high pressure and is increased by ATP. Other important factors in ameboid motion are the levels of Ca^{2+} and the adhesion to a solid support.

REFERENCES

1. Buckley, I. K., and Porter, K. R. (1975) *J. Microsc.* (Oxf.), *104*:107.
2. Porter, K. R. (1976) In: *Cell Motility*, Vol. 1, p. 1. (Goldman, R. D., et al., eds.) Cold Spring Harbor Laboratory, Cold Spring Harbor, New York.
3. De Robertis, E., and Franchi, C. M. (1953) *J. Exp. Med.*, *98*:269.
4. Slautterback, M. C. (1963) *J. Cell Biol.*, *18*:367.
5. Slautterback, M. C., and Porter, K. R. (1963) *J. Cell Biol.*, *19*:239.
6. Ledbetter, M. C., and Porter, K. R. (1964) *Science*, *144*:872.
7. Gall, J. G. (1966) *J. Cell Biol.*, *31*:639.
8. Weisenberg, R. E., Borisy, G. G., and Taylor, E. W. (1968) *Biochemistry*, *7*:4466.
9. Weisenberg, R. C. (1972) *Science*, *177*:1104.
10. Osborn, M., and Weber, K. (1976) *Proc. Natl. Acad. Sci. USA*, *73*:867.
11. Weber, K. (1976) In: *Cell Motility*, Vol. 1, p. 403. (Goldman, R. D., et al., eds.) Cold Spring Harbor Laboratory, Cold Spring Harbor, New York.
12. Brinkley, B. R., Fuller, E. M., and Highfield, D. P. (1976) *Cell Motility*, Vol. 1, p. 435. (Goldman, R. D., et al., eds.) Cold Spring Harbor Laboratory, Cold Spring Harbor, New York.
13. Pollack, R., Osborn, M., and Weber, K. (1975) *Proc. Natl. Acad. Sci. USA*, *72*:994.
14. Freed, J. J., Bhisly, A. N., and Libowitz, M. M. (1968) *J. Cell Biol.*, *39*:46a.
15. Tilney, L. G., and Porter, K. R. (1967) *J. Cell Biol.*, *34*:327.
16. Ogawa, K., and Gibbons, R. (1976) *J. Biol. Chem.*, *251*:5793.
17. Stephens, R. E. (1970) *Biol. Bull.*, *139*:438.
18. Stephens, R. E. (1974) In: *Cilia and Flagella*, p. 39. (Sleigh, M. A., ed.) Academic Press, Inc., New York.
19. Warner, F. D., and Satir, P. (1974) *J. Cell Biol.*, *63*:35.
20. Gilula, N. B., and Satir, P. (1972) *J. Cell Biol.*, *53*:494.
21. Matsuasaka, T. (1974) *J. Ultrastruct. Res.*, *48*:405.
22. Tilney, L. G., and Goddard, J. (1970) *J. Cell Biol.*, *34*:345.
23. Tamm, S. L., and Horridge, G. A. (1970) *Proc. R. Soc. London* [Biol.], *175*:219.
24. Gibbons, B. H., and Gibbons, I. R. (1976) *Biochem. Biophys. Res. Commun.*, *73*:1.
25. Okuno, M., Ogawa, K., and Mohri, H. (1976) *Biochem. Biophys. Res. Commun.* *68*:901.
26. Pedersen, H., and Rebbe, H. (1975) *Biol. Reprod.*, *12*:541.
27. Afzelius, B. A. (1976) *Science*, *193*:317.
28. De Robertis, E. (1956) *J. Biophys. Biochem. Cytol.*, *2*:319.
29. Porter, K. R. (1957) *Harvey Lect.*, *51*:175.
30. Steyn, W. (1969) *Nature* (London), *183*:764.
31. Eakin, R. M. (1961) *Proc. Natl. Acad. Sci. USA*, *47*:1084.
32. Mizukanni, I., and Gall, J. G. (1966) *J. Cell Biol.*, *29*:97.
33. Sorokin, S. P. (1968) *J. Cell Sci.*, *3*:207.
34. Le Beux, Y. J., and Willemont, J. (1975) *Cell Tissue Res.*, *160*:37.
35. Schroeder, T. E. (1972) *J. Cell Biol.*, *53*:419.
36. Pollard, T. D., and Korn, E. D. (1973) *J. Biol. Chem.*, *248*:4682 and 4691.
37. Lazarides, E., and Weber, K. (1974) *Proc. Natl. Acad. Sci. USA*, *71*:2268.
38. Pollard, T. D. (1977) In: *International Cell Biology*, p. 378. (Brinkley, B. R., and Porter, K. R., eds.) The Rockefeller University Press, New York.
39. Lazarides, E. (1975) *Cell*, *6*:289.
40. Weber, K., Rathke, P. C., Osborn, M., and Franke, W. W. (1976) *Exp. Cell Res.*, *102*:285.
41. Weber, K., and Groeschel-Stewart, U. (1974) *Proc. Natl. Acad. Sci. USA*, *71*:4561.
42. Osborn, M., Werner, W. F., and Weber, K. (1977) *Proc. Natl. Acad. Sci. USA*, *74*:2490.
43. Lazarides, E. (1978) *Exp. Cell Res.*, *112*:265.
44. Allen, R. D. (1974) *Symp. Soc. Exp. Biol.*, *28*:15.
45. Kersey, Y. M., Hepler, P. K., Palevitz, A. and Wessels, N. K. (1976) *Proc. Natl. Acad. Sci. USA*, *73*:165.
46. Kamiya, N. (1977) In: *International Cell Biology*, p. 361. (Brinkley, B. R., and Porter, K. R., eds.) The Rockefeller University Press, New York.
47. Allen, R. D. (1961) Ameboid movement. In: *The Cell* Vol. 2, p. 135. (Brachet, J., and Mirsky, A. E., eds.) Academic Press, Inc., New York.

ADDITIONAL READING

Allen, R. D. (1974) Some new insights concerning cytoplasmic transport. *Symp. Soc. Exp. Biol.*, 28:15.

Allison, A. C. (1973) The role of microfilaments and microtubules in cell movement, endocytosis, and exocytosis. In: "*Locomotion of Tissue Cells*," *Ciba Found. Symp.* 14:109. Elsevier Publishing Co. Amsterdam.

Amos, L. A. (1977) Tubulin assembly. *Nature*, 270:98.

Brenner, R. M., and Anderson, R. G. W. (1973) Endocrine control of ciliogenesis in the primate oviduct. In: *Handbook of Physiology, Section 7: Endocrinology, Vol. 2, p. 123.* The Williams & Wilkins Co., Baltimore, Md.

Cohen, C. (1977) Protein assemblies and cell form. *Trends in Biochem. Sci.* 2:51.

Fawcett, D. W. (1961) Cilia and flagella. In: *The Cell*, Vol. 2, p. 217. (Brachet, J., and Mirsky, A., eds.) Academic Press, Inc., New York.

Forer, A. (1974) *Cell Cycle Controls*. (Padilla, G. M., Cameron, I. L., and Zimmerman, A. M., eds.) p. 319. Academic Press, Inc., New York.

Gaffiani, G., ed. (1979) Cytoskeleton in normal and pathologic processes. In: *Cell Biology.* S. Kurper, A. G., Basel.

Geiger, B., and Singer, S. L. (1979) The participation of α-actinin in the capping of cell membrane components. *Cell, 16*:213.

Gibbons, I. R. (1977) Structure and function of flagellar microtubules. *International Cell Biology,* p. 348. (Brinkley, B. R., and Porter, K. R., eds.) The Rockefeller University Press, New York.

Goldman, R. D. (1975) The use of heavy meromyosin binding as an ultrastructural cytochemical method for localizing and determining the possible functions of actin-like microfilaments in non-muscle cells. *J. Histochem. Cytochem., 23*:529.

Herman, I. M., and Pollard, T. D. (1979) Comparison of purified anti-actin and fluorescent heavy meromyosin staining pattern in dividing cells. *J. Cell Biol., 80*:509.

Hitchcock, S. E. (1977) Regulation of motility in non-muscle cells. *J. Cell Biol., 74*:15.

Kirschner, M. W. (1978) Microtubules: Assembly and nucleation. Int. Rev. Cytol., *54*:1.

Korn, E. D. (1978) Biochemistry of actomyosin dependent cell motility. *Proc. Natl. Acad. Sci. USA, 75*:588.

Mohri, H. (1976) The function of tubulin in motile systems. *Biochem. Biophys. Acta, 456*:85.

Ogawa, K. O., Mohri, T. and Mohri, H. (1977) Identification of dynein as the outer arms of sea urchin sperm axonemes. *Proc. Natl. Acad. Sci. USA, 74*:5006.

Pollard, T. D. (1977) Cytoplasmic contractile proteins. *International Cell Biology,* p. 378. (Brinkley, B. R., and Porter, K. R., eds.) The Rockefeller University Press, New York.

Porter, K. R. (1976) Introduction: motility in cells. In *Cell Motility,* Vol. 1, p. 1. (Goldman, R. D., et al., eds.) *Cold Spring Harbor Laboratory,* Cold Spring Harbor, New York.

Rieder, C. L. (1979) Ribonucleoprotein staining of centrioles and kinetochores in newt lung cell spindles. *J. Cell Biol., 80*:1.

Rott, H. D. (1979) Kartagener's syndrome and the syndrome of immotile cilia. *Human Genet., 46*:249.

Samson, F. E. (1976) Pharmacology of drugs that affect intracellular movement. *Annu. Rev. Pharmacol. Toxicol., 16*:143.

Sherline, P., and Schiavone, K. (1978) High molecular weight MAPs are part of the mitotic spindle. *J. Cell Biol., 77*:R9.

Sleigh, M. A. (1971) Cilia. *Endeavour, 30*:11.

Sleigh, M. A. (1970) The physiology and biochemistry of cilia and flagella. In: *Frontiers of Biology: Handbook of Molecular Cytology,* Vol. 15, p. 1243. (Lima-de-Faría, A., ed.) North-Holland Publishing Co., Amsterdam.

Sleigh, M. A. (1979) Contractility of the roots of flagella and cilia. *Nature, 277*:263.

Small, J. V., and Celis, J. E. (1978) Direct visualization of the 10-nm filament network in whole and enucleated cultured cells. *J. Cell Sci., 31*:393.

Spencer, M. (1978) Sliding microtubules. *Nature, 273*:595.

Starger, J. M., Brown, W. E., Goldman, A. E., and Goldman, R. D. (1978) Biochemical and immunological analysis of rapidly purified 10-nm filaments from baby hamster kidney cells. *J. Cell Biol., 78*:93.

Stephens, R. E. (1974) *Cilia and Flagella,* p. 39. (Sleigh, M. A., ed.) Academic Press, Inc., New York.

Stephens, R. E. (1978) Primary structural differences among tubulin subunits from flagella, Cilia, and the Cytoplasm. *Biochemistry, 17*:2882.

Summer, K. (1975) The role of flagellar structures in motility. *Biochim. Biophys. Acta, 416*:153.

Taylor, D. L. (1977) Dynamics of cytoplasmic structure and contractility. *International Cell Biology,* p. 367. (Brinkley, B. R., and Porter, K. R., eds.) The Rockefeller University Press, New York.

Tilney, L. G. (1977) Actin: its association with membranes and the regulation of its polymerization. *International Cell Biology.* p. 388. (Brinkley, B. R., and Porter, K. R., eds.) The Rockefeller University Press, New York.

Vandekerckhove, J., Weber, K. (1978) Mammalian cytoplasmic actins are products of at least two genes *Proc. Natl. Acad. Sci. USA, 75*:1106.

Warner, F. D. (1972) Macromolecular organization of eukaryotic cilia and flagella. *Advances in Cell and Molecular Biology,* Vol. 2, p. 193. Academic Press, Inc., New York.

Weber, K. (1976) Visualization of tubulin-containing structures by immunofluorescence microscopy of cytoplasmic microtubules, mitotic figures, and vinblastine-induced paracrystals. In: *Cell Motility,* Vol. 1, p. 403. (Goldman, R. D., et al., eds.) Cold Spring Harbor Laboratory, Cold Spring Harbor, New York.

Wiederhold, M. L. (1976) Mechanosensory transduction in "sensory" and "motile" cilia. *Annu. Rev. Biophys. Bioeng., 5*:39.

Wolfe, J. (1972) Basal body fine structure and chemistry. In: *Advances in Cell and Mollecular Biology,* Vol. 2, p. 151. Academic Press, Inc., New York.

10

THE ENDOPLASMIC RETICULUM, AND CELL SECRETION I

10–1   General Morphology of the Endomembrane System   208
    The Rough ER — Ribosomes and Protein Synthesis
    Ribosomal Binding to the ER — 60S Subunit and Ribophorins Involved
    The Smooth ER Lacks Ribosomes
    Summary: *The Endoplasmic Reticulum*
10–2   Microsomes — Biochemical Studies   213
    Microsomal Membranes — A Complex Lipid and Protein Composition
    Two Microsomal Electron Transport Systems — Flavoproteins and Cytochromes b$_5$ and P-450
    Microsomal Enzymes — Glycosidation and Hydroxylation of Amino Acids
    Microsomal Enzymes — Asymmetry Across the Membrane
    Summary: *Microsomes*
10–3   Functions of the Endoplasmic Reticulum   218

    Membrane Biogenesis Involves a Multi-Step Mechanism
    ER-Membrane Fluidity and Flow Through the Cytoplasm
    Ions and Small Molecules — Transport Across ER Membranes
    Special Functions of Smooth ER — Detoxification, Lipid Synthesis, and Glycogenolysis
    Summary: *Functions of the Endoplasmic Reticulum*
10–4   The Endoplasmic Reticulum and Synthesis of Exportable Proteins   221
    Special Initial Codons for Signal Peptides — In RNA of ER-Bound Polysomes
    The Signal Peptide is Removed by a Signal Peptidase
    Membrane Proteins are Made and Assembled in Different Compartments
    Summary: *Synthesis of Exportable Proteins — The Signal Hypothesis*

In Chapter 2 we mentioned that the eukaryotic cell has a high degree of ultrastructural organization. In both plant (Fig. 1–1) and animal (Fig. 2–6) cells the cytoplasm is permeated by a complex system of membrane-bound tubules, vesicles, and flattened sacs (cisternae) that, at many points, are intercommunicating. This *endomembrane system* may be interpreted three-dimensionally as a vast network that subdivides the cytoplasm into two main compartments, one enclosed within the membranes, the other situated outside and constituting the *cytoplasmic matrix* or *cytosol*.

The existence of the endoplasmic reticulum was discovered in 1945 after the introduction of electron microscopy applied to cultured cells.[1] The first micrographs showed a lacelike arrangement of tubules that did not reach the periphery of the cell, hence the term *"endoplasmic"* (Fig.

10–1). The recent use of high-voltage electron microscopy on such cells has rendered a clearer three-dimensional view of the endoplasmic reticulum (Fig. 9–3).[2] A more detailed analysis of the system became possible with the introduction of thin sectioning and freeze-fracturing of cells.

Other advances were made by cell fractionation methods followed by biochemical analysis and the use of cytochemical techniques for the study of specific components — particularly enzymes. As in many other areas of cell biology, rapid progress has resulted from the convergence of various technical and scientific approaches.

A eukaryotic cell can no longer be considered as a bag containing enzymes, ribonucleic acid (RNA), deoxyribonucleic acid (DNA), and solutes surrounded by an outer membrane, as in

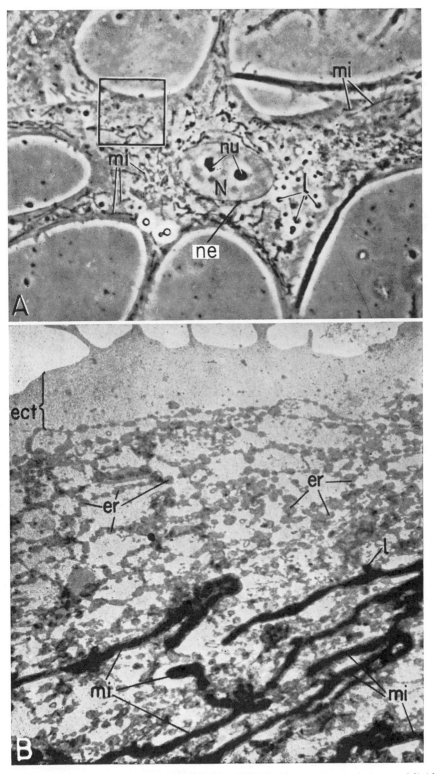

Figure 10–1 **A,** living cell of a tissue culture observed under the phase contrast microscope: *l,* lipid; *mi,* mitochondria; *ne,* nuclear envelope; *nu,* nucleoli. The region indicated in the inset is similar to B. (Courtesy of D. W. Fawcett.) **B,** electron micrograph of the marginal region of a mouse fibrocyte in tissue culture: *er,* endoplasmic reticulum; *mi,* filamentous mitochondria; *l,* lipid. The peripheral region *(ect)* is homogeneous. ×7000. (Courtesy of K. R. Porter.)

the most primitive bacterium (Fig. 2–1). Numerous membrane-bound compartments are responsible for vital cellular functions, among which are the separation and association of enzyme systems, the creation of diffusion barriers, the regulation of membrane potentials, ionic gradients, different intracellular pH values, and other manifestations of cellular heterogeneity. Furthermore, there is evidence that enzymes are spatially organized, forming multi-enzyme systems within the insoluble membranous framework of the cell.

10–1 GENERAL MORPHOLOGY OF THE ENDOMEMBRANE SYSTEM

The main components of the *endomembrane system*, also called the *cytoplasmic vacuolar system*, are the following: (1) The *nuclear envelope*, consisting of two non-identical membranes, one apposed to the nuclear chromatin and the other separated from the first membrane by perinuclear cisternae, the two being in contact at the *nuclear pores* (Figs. 1–1 and 2–6). (2) The *endoplasmic reticulum* (ER), which is in general the most developed portion of the endomembrane system. (3) The *Golgi complex*, which is a specialized region of the system, mainly related to some of the terminal processes of cell secretion (see Chapter 11).

The endoplasmic reticulum comprises two parts differentiated by the presence or absence of ribosomes on the outer or cytoplasmic surface. The portion that has ribosomes is generally referred to as *"rough"* or *"granular"* *endoplasmic reticulum* (RER) and the one lacking ribosomes is the *"smooth"* or *"agranular"* *endoplasmic reticulum* (SER) (Fig. 10–2).

The fact that the entire system of endomembranes represents a kind of barrier separating cytoplasmic compartments is clearly indicated in Figure 10–3. This diagram specially emphasizes the two faces of each membrane, (1) the so-called *cytoplasmic face* or protoplasmic face and (2) the *luminal face*, also called the endoplasmic or extracellular face.[3]

The cytoplasmic face is directly opposed to the cytosol and in this sense is equivalent to the inner face of the plasma membrane and to the surfaces of the outer mitochondrial membranes and of the other intracellular organelles, such as the Golgi complex, the lysosomes, the peroxisomes, and the secretory granules (Fig. 10–3).

The luminal face borders the perinuclear cisternae, the cavities of the rough and smooth endoplasmic reticulum, and the Golgi elements. This surface also corresponds to the interior of the secretory granules, the lysosomes, and the peroxisomes and to the faces of the mitochondria confronting the outer chamber. It is important to remember that after freeze-fracturing, the

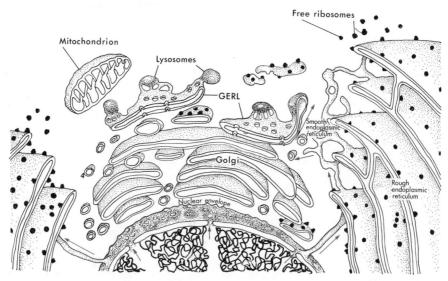

Figure 10–2 Three-dimensional diagram of the endomembrane system of the cell. The nucleus with its chromosomal fibrils shows interchromatin channels (arrows) leading to nuclear pores. Note the double-membrane organization of the nuclear envelope. Cisternae of rough endoplasmic reticulum are interconnected and have ribosomes attached to the outer surface. Some of these cisternae are extended by tubules of smooth endoplasmic reticulum. The Golgi apparatus shows the GERL region. The large arrows indicate the probable dynamic relationship of the portions of the endomembrane system.

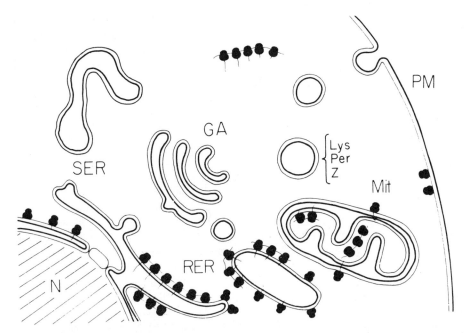

Figure 10-3 Diagram of cellular membranes and their relationship to compartments containing the ribosomes. In each membrane the luminal faces are shown in thick lines, while the cytoplasmic faces are depicted by thin lines. Ribosomes are always on the cytoplasmic (or matrix) side. *N*, nucleus; *GA*, Golgi apparatus; *Lys*, lysosome; *Mit*, mitochondria; *RER*, rough endoplasmic reticulum; *Per*, peroxisome; *PM*, plasma membrane; *SER*, smooth endoplasmic reticulum; *Z*, zymogen granule. (Courtesy of D. D. Sabatini and G. Kreibich. From Sabatini, D. D., and Kreibich, G., Functional specialization of membrane-bound ribosomes in eukaryotic cells, *in The Enzymes of Biological Membranes*, Vol. 2, p. 432. Martonosi, A., ed. New York, Plenum Press, 1976.)

membrane halves containing the cytoplasmic (A or P) faces show many more intramembranous particles than those consisting of the luminal (B or E) faces.[4]

As in the case of the plasma membrane, we may consider that in the endomembrane system there is a lipid bilayer with peripheral and integral proteins, some of which may be exposed only toward either the cytoplasmic or the luminal face, and in addition this bilayer may have integral proteins that have a transmembrane disposition (see Fig. 8–1). A very important problem to be considered later in this chapter is a possible mechanism of synthesis of these various kinds of membrane proteins.

The development of the endoplasmic reticulum varies considerably in different cell types and is related to their functions. It is often small and relatively undeveloped in eggs and in embryonic or undifferentiated cells but increases in size and complexity with differentiation. In spermatocytes only a few vacuoles can be observed. A simple smooth endoplasmic reticulum is found in cells engaged in lipid metabolism, such as adipose, brown fat, and adrenocortical cells. In the reticulocytes, which produce only proteins to be retained in the cytosol (hemoglobin), the endoplasmic reticulum is poorly developed or non-existent, although the cell may contain a large number of ribosomes.

The Rough ER — Ribosomes and Protein Synthesis

The rough endoplasmic reticulum is especially well developed in cells actively engaged in protein synthesis, such as the enzyme-producing cells. In general, it occupies the regions of the cytoplasm that appear to be basophilic (corresponding to the *ergastoplasm*) when viewed under the light microscope. This property is due to the presence of the RNA-containing ribosomes.

In rapidly dividing cells, such as those of the intestinal crypts and of plant and animal embryos, as well as of cancers the cytoplasm may be strongly basophilic, but the endoplasmic reticulum is poorly developed. In cells engaged in the production of large amounts of proteins for export (enzymes), such as the cells of the pancreas, the rough endoplasmic reticulum is highly developed and consists of parallel stacks of large flattened cisternae occupying the cytoplasm of the base and the lateral regions of the cell.

Table 10–1 indicates the relative volumes and

TABLE 10–1 RELATIVE CYTOPLASMIC VOLUMES AND MEMBRANE SURFACE AREAS OF SECRETORY COMPARTMENTS IN RESTING GUINEA PIG PANCREATIC EXOCRINE CELLS*

Compartment	Relative Cytoplasmic Volume Per Cent	Membrane Surface Area $\mu m^2/Cell$
RER	~20	~8000
Golgi complex	~8	~1300
Condensing vacuoles	~2	~150
Secretory granules	~20	~900
Apical plasmalemma		~30
Basolateral plasmalemma		~600

*Data originally from Bolender, 1974; and Amsterdam and Jamieson, 1974; (from Jamieson, 1977).

membrane areas of the main secretory compartments of the pancreatic cell and shows the large structural amplification of the rough endoplasmic reticulum in relation to the areas of other membranes (see Jamieson, 1977). In liver cells the rough endoplasmic reticulum occurs as groups of cisternae interspersed with regions of smooth endoplasmic reticulum that is mixed with glycogen particles (Fig. 10–4).

The cavity of the rough endoplasmic reticulum is sometimes very narrow, with the two membranes closely apposed; but more frequently there is a true space between the membranes that may be filled with a material of varying opacity. This space is much distended in certain cells actively engaged in protein synthesis, such as the plasma cells and goblet cells. In these cases a dense macromolecular material can be observed inside the cisternae. In the pancreas, intercisternal secretion granules, smaller than the zymogen granules, may be observed.

The membrane of the endoplasmic reticulum is about 5 to 6 nm thick. Although it is thinner than the plasma membrane, it does exhibit a "unit membrane" structure, i.e., two dense layers separated by a lighter one. The total surface of the endoplasmic reticulum contained in 1 ml of liver tissue has been calculated to be about 11 square meters, two thirds being of tha granular, or rough type.[5]

Ribosomal Binding to the ER — 60S Subunit and Ribophorins Involved

As was mentioned before, the main characteristic of the rough endoplasmic reticulum is the presence of attached ribosomes on the outer surface. These are present as polysomes held together by mRNA (see Chapter 23–1) and are often arranged in typical "rosettes" or spirals (Fig. 10–5).

Another characteristic, which is shown in Figures 10–2 and 10–6, is that the ribosomes are always attached to the membrane by the large 60S subunit. It is now generally thought that in the membranes of the endoplasmic reticulum are specific sites for the binding of ribosomes and that the binding is rather complex. It involves electrostatic interactions which can be disrupted in media of high salt content, but they are also attached by the nascent polypeptide chain that grows from the ribosome and penetrates across the membrane into the cavity of the endoplasmic reticulum (Fig. 10–7).

Recent studies suggest that ribosomes could be bound by way of two transmembrane glycoproteins — the so-called *ribophorins I and II* of 65,000 and 63,000 daltons respectively which are present in rough membranes and absent in smooth.[6] It has been found that rough membranes stripped of ribosomes have a high-affinity binding to these two proteins, while smooth membranes lack this property. Furthermore, the number of proteins is stoichiometrically related to that of bound ribosomes. It is also thought that the ribophorins have an extended configuration forming a kind of supporting network that may confer the characteristic shape to the cisternae and at the same time prevent the segregation of the ribosomes to other regions of the endomembrane system.

The Smooth ER Lacks Ribosomes

Although the smooth endoplasmic reticulum forms a continuous system with the rough portion, it has a different morphology. In the liver cell it is made of tubular elements that apparently start at the edge of the cisternae and make a tubular network that pervades large regions of the cytoplasmic matrix. These fine tubules, with some fenestrated smooth membrane segments, are present in regions rich in glycogen and can be observed as dense particles in the matrix (Fig. 10–4). Another characteristic observed in liver cells is the relationship of the smooth endoplasmic reticulum to *peroxisomes* (i.e., organelles containing peroxidase, catalase, and other oxidases) and with the Golgi complexes.

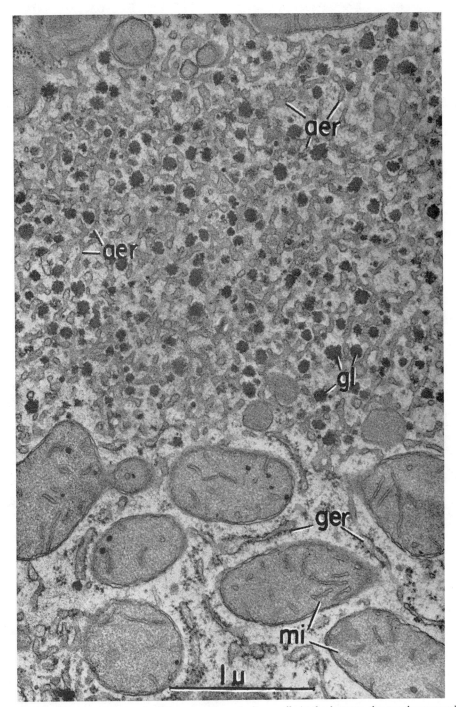

Figure 10–4 Electron micrograph of the cytoplasm of a liver cell. At the bottom, the rough or granular endoplasmic reticulum (*ger*) and mitochondria (*mi*); at the top, the smooth or agranular endoplasmic reticulum (*aer*) mixed with glycogen particles (*gl*). ×45,000. (Courtesy of G. E. Palade.)

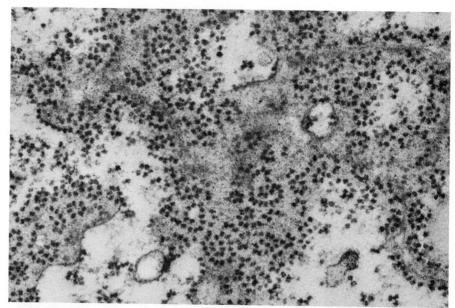

Figure 10–5 Electron micrograph of a root hair from an epidermal cell of the radish. The tangential section through the membrane of the endoplasmic reticulum shows groups of ribosomes (i.e., polyribosomes) disposed in recurrent patterns. ×57,000. (Courtesy of H. T. Bonnett, Jr., and E. H. Newcomb.)

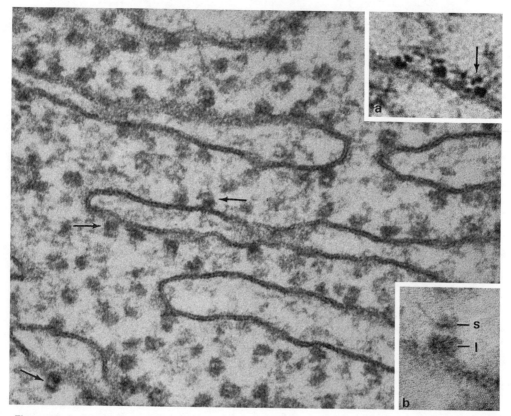

Figure 10–6 Electron micrograph showing ribosomes attached to the membranes of the endoplasmic reticulum. Observe the "unit membrane" structure. Arrows indicate ribosomes where the attachment of the large (60S) subunit is best observed. ×208,000. (Courtesy of G. E. Palade.) Inset **a,** attachment of the large and small subunits forming a "cap," which is subdivided into two portions (arrow). Insert **b,** at higher magnification, the small (s) and large (l) subunits appear to be separated by a clear cleft. **a,** ×200,000; **b,** ×410,000. (Both insets courtesy of N. T. Florendo.)

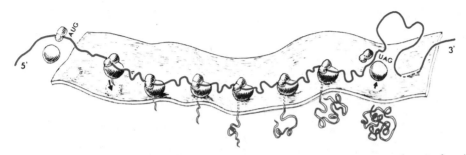

Figure 10–7 Three-dimensional diagram showing the assembly of the ribosomes on the cytoplasmic side of the membrane and the penetration of the polypeptide chain through the membrane into the lumen. The messenger RNA extends between the initial codon AUG at the 5° end, and the terminal codon UAG at the 3° end. See that the mRNA might also be attached to the membrane by a segment near the 3° end. (Courtesy of D. D. Sabatini and G. Kreibich. From Sabatini, D. D., and Kreibich, G., Functional specialization of membrane-bound ribosomes in eukaryotic cells, *in The Enzymes of Biological Membranes,* Vol. 2, p. 431. Martonosi, A., ed. New York, Plenum Press, 1976.)

SUMMARY:

Endoplasmic Reticulum

The cytoplasm of animal and plant cells is traversed by a complex system of membranes that form closed compartments in structural continuity. The main components of the system are the endoplasmic reticulum, with its three portions (the nuclear envelope, the rough and the smooth endoplasmic reticulum, and the Golgi complex). The rough endoplasmic reticulum has ribosomes attached to its outer surface and is particularly well developed in the basophilic regions of the cytoplasm (i.e., the ergastoplasm). The tubular and cisternal cavities of the endoplasmic reticulum may be closed, but more often contain material that has been synthesized on the ribosomes (e.g., protein) or by the enzymes present in the membranes (e.g., lipids, oligosaccharides). The smooth endoplasmic reticulum is devoid of ribosomes. Frequently it forms a tubular network, and in the liver it is related to glycogen deposits and peroxisomes.

The structure of the endoplasmic reticulum membrane is similar to that of the cell membrane in the sense that there is a lipid bilayer with peripheral and integral proteins. The ribosomes are bound by their 60S (large) subunits. In the rough endoplasmic reticulum are two proteins (ribophorins I and II) that are absent in the smooth membranes. These are transmembrane glycoproteins that may correspond to the ribosomal binding sites.

10–2 MICROSOMES — BIOCHEMICAL STUDIES

In Chapter 4 the concept of the "microsomal fraction" that can be isolated by differential centrifugation following the separation of the nuclear and mitochondrial fractions was introduced (Fig. 4–5). Electron microscopic studies have revealed that such a fraction is rather complex. In addition to fragments of the plasma membrane, microsomes include various parts of the vacuolar system, i.e., the rough and smooth endoplasmic reticulum and the Golgi complex (Fig. 10–8). Thus, the concept of microsome-at-large must not be confused with that of a particular membrane system of the cell. Microsomes, as discrete entities, are not found in the intact cell, but are the result of

the fragmentation of most of the cytoplasmic membranous components.

During cell disruption the cisternae of the endoplasmic reticulum break but immediately reseal, maintaining intact the topological relationships between the membranes, the lumen, and the ribosomes (Fig. 10–9, *A*). However, on account of the difference in density, the smooth and the rough microsomes may be separated. It is also possible to subfractionate the microsomes into their components. Thus, the membranes may be stripped of the ribosomes by high salts and treatment with puromycin, which stops the growth of the polypeptide chain.

The ribosomes may be separated by treatment with desoxycholate, a surface-active agent that solubilizes the membrane (Figs.

213

Endomembrane System

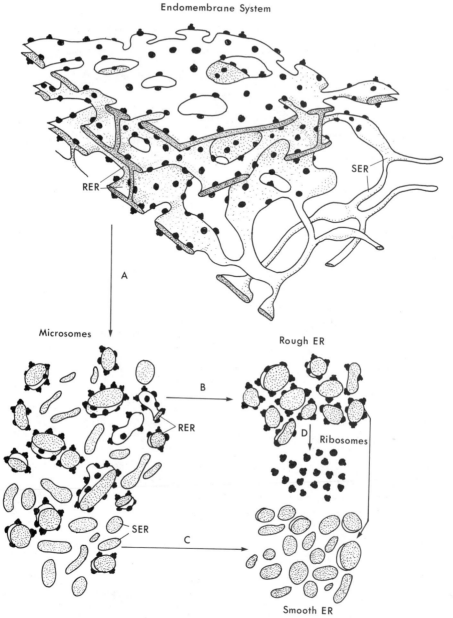

Figure 10-8 Three-dimensional diagram of the endomembrane system. **A,** isolation of microsomes by homogenization and differential centrifugation; **B** and **C,** separation of rough and smooth vesicles of the endoplasmic reticulum; D separation of ribosomes from the RER. *RER,* rough endoplasmic reticulum; *SER,* smooth endoplasmic reticulum.

10–8 and 10–9, *B*). Finally, it is possible to extract the content of the lumen, consisting mainly of precursors of secretory proteins (albumin and other serum proteins in the case of the liver), by the use of sonication or very low concentrations of detergents that make holes in the membranes, enabling the exit of the luminal content.[7]

Microsomal Membranes — A Complex Lipid and Protein Composition

Microsomes constitute about 15 to 20 per cent of the total mass of the liver cell. They contain 50 to 60 per cent of the cellular liver RNA which is due to the presence of ribosomes. Microsomal membranes have a high

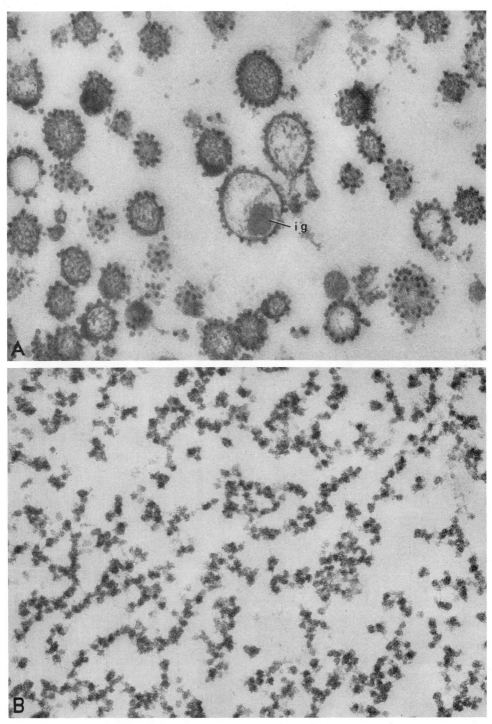

Figure 10–9 **A,** microsomes from the pancreas of a guinea pig. The vesicles show the ribosomes attached to the outer surface. *ig,* intracisternal granule. ×91,200. **B,** ribosomes from the pancreas after solubilization of the membranes. ×136,000. (Courtesy of G. E. Palade.)

lipid content which includes: phospholipids, neutral lipids, phosphatidylinositol, plasmalogens, and some gangliosides. There is more lipid in relation to proteins in the smooth endoplasmic reticulum and Golgi membranes than in the rough endoplasmic reticulum. The latter also contains less sphingomyelin and cholesterol.[8] Microsomal membranes also have some special proteins — particularly of the enzymatic type —which serve as markers for the recognition of this particular fraction and to differentiate it from Golgi membranes, secretory granules, or plasma membranes.[9]

The membrane proteins of the endoplasmic reticulum have been separated by polyacrylamide gel electrophoresis with sodium dodecylsulfate, a technique that separates polypeptides by molecular weight (Fig. 8–3). As many as 30 polypeptide bands ranging from 15,000 to 150,000 daltons have been identified in the rough endoplasmic reticulum of the pancreas. These bands differ from the bands (which are few in number) found in the Golgi membranes and from those corresponding to zymogen granules.

We mentioned earlier the existence of two protein bands — the so-called ribophorins I and II — present in the rough endoplasmic reticulum and absent in the smooth, and the possible role of these components in the binding of the ribosomes.

Two Microsomal Electron Transport Systems — Flavoproteins and Cytochromes b_5 and P-450 Involved

The endoplasmic reticulum contains many of the enzymes utilized in the synthesis of triglycerides, phospholipids, and cholesterol. These lipids are not only incorporated into the reticular membrane itself but are transferred to other organelles by way of specific carrier proteins which can shuttle through the cytosol.

Table 10–2 indicates some of enzymatic activities of microsomes. Many of the metabolic functions are carried out by the use of two electron transport chains present in these membranes. Microsomes contain as electron transfer components at least two flavoproteins (*NADH-cytochrome-c-reductase* and *NADH-cytochrome-b5-reductase*) and two hemoproteins (*cytochrome b_5* and *cytochrome P-450*).

Study of this system has centered especially on the role of cytochrome P-450, which is characterized by a MW of 50,000 daltons and absorption at 450 nm.[10] Of the two hemoproteins,

this one has the concentration which is higher in the liver and which changes quantitatively in animals treated with various "inducing agents," such as barbiturates and polycyclic hydrocarbons.

Cytochrome P-450 functions as a terminal oxidase and in the liver is used to detoxify or inactivate pesticides and many drugs by oxidation, and to hydroxylate steroid hormones (see Table 10–2). Paradoxically, the same system may activate carcinogens (i.e., it may transform a drug into a carcinogen by oxidation). The administration of barbiturates to a rat produces a rapid increase in smooth endoplasmic reticulum and a great increase in cytochrome P-450, while the other electron carriers do not change. It has been calculated that in these conditions cytochrome P-450 may represent 10 per cent of the total protein of the microsome and that one molecule of NADPH-cytochrome-c-reductase should interact with 20 to 30 molecules of cytochrome P-450.

Special cluster arrangements of the electron transport carriers occurring as discrete complexes within the membrane have been postulated.[11] In this model the flavoprotein is located at the core of the complex, surrounded by molecules of cytochrome b_5 and cytochrome P-450.

TABLE 10–2 SOME MICROSOMAL ENZYME ACTIVITIES*

Synthesis of glycerides:
　Triglycerides
　Phosphatides
　Glycolipids and plasmalogens
Metabolism of plasmalogens
Fatty acid synthesis
Steroid biosynthesis:
　Cholesterol biosynthesis
　Steroid hydrogenation of unsaturated bonds
$NADPH_2 + O_2$-requiring steroid transformations:
　Aromatization
　Hydroxylation
$NADPH_2 + O_2$-requiring drug detoxification:
　Aromatic hydroxylations
　Side-chain oxidation
　Deamination
　Thio-ether oxidation
　Desulfuration
L-Ascorbic acid synthesis
UDP-uronic acid metabolism
UDP-glucose dephosphorylation
Aryl- and steroid-sulfatase

*Modified from Rothschild, J., The isolation of microsomal membranes. *In The Structure and Function of the Membranes and Surfaces of Cells. Biochem. Soc. Symp.* 22:4, 1963, New York, Cambridge University Press.

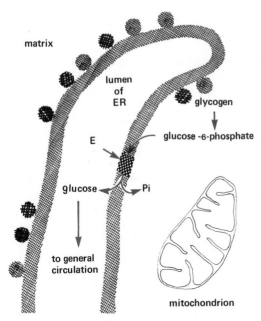

Figure 10–10 Diagram of the intervention of the smooth endoplasmic reticulum in glucogenolysis with the consequent release of glucose. The enzyme (E), glucose-6-phosphatase, is present in the membrane and has a vectorial disposition by which it receives the glucose-6-phosphate from the matrix surface. The product—glucose—penetrates the lumen of the endoplasmic reticulum.

The second electron transport chain in microsomes is composed of *cytochrome* b_5, *NADH-cytochrome-b_5-reductase*, and *fatty acyl CoA desaturase*. This system uses electrons from NADH to desaturate fatty acids.

Microsomal Enzymes — Glycosidation and Hydroxylation of Amino Acids

In the luminal face of the endoplasmic reticulum are enzymes which are able to modify compounds released into the cavity. For example, *peptidases* may cut out portions of protein precursors (see later discussion), and other enzymes may modify the nascent polypeptides by *glycosidation* or *hydroxylation* of amino acid residues.

Other important enzymes are Mg^{2+}-activated ATPase and glucose-6-phosphatase. In the liver this last enzyme has the important function of splitting off the phosphate from glucose-6-phosphate, thus allowing glucose to be liberated. As shown in Figure 10–10 glucose-6-phosphate originates from the degradation of glycogen in the cytoplasmic matrix, and the enzyme which is embedded in the membrane, probably acts in a vectorial manner sending glucose into the lumen of the endoplasmic reticulum.

Studies done in newly born animals reveal that during differentiation of the liver cell glucose-6-phosphatase and NADPH-cytochrome-c-reductase appear first in the rough, and then in the smooth, portions of the endoplasmic reticulum.[12]

A special chemical differentiation is found in the *sarcoplasmic reticulum* of muscle. A Ca^{2+} *activated ATPase* carries out the major function, pumping calcium into the lumen, thus reducing the calcium concentration in the cytosol and causing muscle relaxation. The stimulus for contraction produces the opposite effect, causing the release of calcium ions from the sarcoplasmic reticulum (see Chapter 27–5).

Microsomal Enzymes – Asymmetry Across the Membrane

As in other membranes the main enzymes of the endoplasmic reticulum show a definite asymmetry across the membrane. When this organelle is broken up into microsomes, their outer surface corresponds to the cytoplasmic surface of the endoplasmic reticulum, and thus any protein present in that surface is quite vulnerable to attack by a protease or other agent. Among the proteins present on the cytoplasmic surface are cytochrome b_5, NADH-cytochrome-b_5-reductase, NADH-cytochrome-c-reductase, cytochrome P-450, and 5'-nucleotidase. Among those in the luminal surface are glucose-6-phosphatase, nucleoside diphosphatase, and β-glucuronidase. Several of the enzymes of the former group, particularly cytochrome b_5, are bound to the membrane by a hydrophobic tail that penetrates into the lipid bilayer.

SUMMARY:

Microsomes

By differential centrifugation most of the components of the endomembrane system can be isolated in the so-called microsomal fraction. Further separation of the rough and smooth endoplasmic reticulum and Golgi complexes may be achieved by gradient centrifugation; the membranes, ribosomes, and luminal contents may also be separated.

The endoplasmic reticulum contains enzymes for the synthesis of triglycerides, phospholipids, and cholesterol. There are two electron transport systems which contain two flavoproteins (NADH-cytochrome-c-reductase and NADH-cytochrome-b_5-reductase) and two hemoproteins (cytochrome b_5 and cytochrome P-450) which absorbs at 450 nm) has the higher concentration in the liver, and this increases with inducing agents (i.e., barbiturates, polycyclic hydrocarbons, and others). It is used to detoxify certain drugs by oxidation, but it may also transform a drug into a carcinogen. A cluster arrangement within the membrane is postulated to explain the relation of one molecule of NADH-cytochrome-c-reductase to 20 to 30 molecules of cytochrome P-450. The second electron transport chain contains NADH; cytochrome-b_5-reductase, cytochrome b_5, and fatty acid acyl CoA desaturase. Other enzymes of the endoplasmic reticulum are peptidases, glycosyl transferases, and hydroxylases; these modify the nascent polypeptides. Glucose-6-phosphatase produces the degradation of glycogen in the smooth endoplasmic reticulum. The various enzymes have different topologies with respect to the luminal or cytoplasmic surfaces of the endoplasmic reticululum.

10–3 FUNCTIONS OF THE ENDOPLASMIC RETICULUM

The polymorphism of the endoplasmic reticulum in a variety of cells and its changes with differentiation and cellular activity have made very difficult an analysis of its multiple functions. Several of these are suggested by its morphology; others, more certain, are based on data derived from cell fractionation or experimental evidence. The great complexity of this membrane system stems from the fact that a large portion of it is associated with the ribosomes, and thus it plays a fundamental role in the storage and processing of proteins which are destined for export from the cell. This is done, in many cases, through an elaborate series of steps that lead to *cell secretion* (see Chapter 11). Other functions are more related to the fact that the endoplasmic reticulum represents a fundamental part of the endomembrane system, which divides the cytoplasm into two definite compartments. Furthermore, these membranes contain many enzyme systems that are able to carry out various functions, including the biogenesis of some of the membrane components.

Membrane Biogenesis Involves a Multi-Step Mechanism

The origin of the endoplasmic membrane is not definitely known. Some electron microscopic observations of differentiating cells suggest that it may develop by evagination from the nuclear envelope (Fig. 9–1). At telophase,

however, the nuclear envelope is re-formed by vesicles of the endoplasmic reticulum. The close relationship between these two portions of the system is also suggested by cytochemical studies.

The relationship between rough and smooth endoplasmic reticulum may be studied in differentiating cells. In rat liver cells before birth there is a preferential increase of the rough type, whereas after birth the growth is mainly of the smooth type. Studies using the protein or lipid precursors ^{14}C-leucine and ^{14}C-glycerol have shown that in the period of rapid growth of the endoplasmic reticulum, the incorporation into proteins and lipids is greater in the rough than in the smooth type. This finding suggests that the synthesis of membranes follows the direction rough → smooth endoplasmic reticulum.

In addition to the *structural* and *functional continuity* between the various membrane compartments of the vacuolar system there is also a *temporal continuity*. This refers to the fact that one cell receives a full set of membranes from its ancestor cell. On the basis of present experimental evidence, there is no *de novo* synthesis of membranes; they grow by expansion of pre-existing membranes.[13, 14]

Current concepts of membrane biogenesis generally assume a multi-step mechanism involving, first, the synthesis of a basic membrane of lipids and intrinsic proteins and, thereafter, the addition, in a sequential manner, of other constituents such as enzymes, specific sugars, or lipids. The process by which a membrane is modified chemically and structurally may be regarded as *membrane differentiation*.

The distribution of various membrane biosynthetic enzymes (see microsomes) leads to the conclusion that the bulk of the intracellular membranes are synthetized in the rough endoplasmic reticulum and then transformed, through a progressive series of changes, into the membranes of the smooth portion and the Golgi complex. The process of membrane biogenesis, however, can not be understood without taking into consideration the fluidity of the membrane, which allows for the circulation, assortment, and segregation of its components in the plane of the membrane.

ER-Membrane Fluidity and Flow through the Cytoplasm

The endoplasmic reticulum may act as a kind of *circulatory system* for intracellular circulation of various substances. Membrane flow may also be an important mechanism for carrying particles, molecules, and ions into and out of the cells. The continuities observed in some cases between the endoplasmic reticulum and the nuclear envelope suggest that the membrane flow may also be active at this point. This flow would provide one of the several mechanisms for export of RNA and nucleoproteins from the nucleus to the cytoplasm. More will be said later about this possible mechanism of membrane flow, particularly in the chapters concerned with pinocytosis, phagocytosis, and secretion.

We may consider the membrane of the endoplasmic reticulum at body temperature to be highly dynamic and also fluid. There is evidence from freeze-fracturing studies that the bound ribosomes are mobile on the membrane and that their mobility is controlled by its fluidity.[15] The concept of flow refers not only to the content of the endoplasmic reticulum but to the membrane proper. Different studies suggest that the transfer of secretory proteins is accompanied by a flow of newly synthesized membrane proteins that are incorporated into the rough endoplasmic reticulum. However, these membrane proteins have a slower rate of synthesis than the secretory ones.[16-18]

Ions and Small Molecules — Transport Across ER Membranes

The endoplasmic reticulum, along with the cytoplasmic matrix, may participate in many of the mechanical functions of the cell. By dividing the fluid content of the cell into compartments the endoplasmic reticulum provides supplementary *mechanical support* for the colloidal structure of the cytoplasm.

The enormous internal surface of the endoplasmic reticulum (about 11 m^2/ml in liver cells) gives an idea of the importance of the exchanges taking place between the matrix and the inner compartment.

It is known that in the cell the vacuolar system has *osmotic properties*. After isolation, microsomes expand or shrink according to the osmotic pressure of the fluid. Diffusion and active transport may take place across the membranes of the vacuolar system. As in the plasma membrane, the presence of *carriers* and *permeases* that are involved in active transport across the membrane has been postulated. The existence of a vacuolar system separating the cytoplasm into two compartments makes possible the existence of ionic gradients and electrical potentials across these intracellular membranes. This concept has been applied especially to the *sarcoplasmic reticulum*, a specialized form of endoplasmic reticulum found in striated muscle fibers, which is considered as an intracellular conducting system.

Role of the Endoplasmic Reticulum in Plant Cells. Plant cells contain an endoplasmic reticulum with morphology and function similar to those in animal cells. However, some special differences regarding the biosynthesis and transport of certain enzymes, polysaccharides, and lipids, as well as the formation of protein bodies, glyoxysomes, lysosomes, and vacuoles, will be studied in Chapter 14.

Special Functions of the Smooth ER — Detoxification, Lipid Synthesis, and Glycogenolysis

In addition to the functions just mentioned that apply to both parts of the endoplasmic reticulum, the following are some which predominate in the smooth portion:

Detoxification. As mentioned earlier large amounts of drugs, such as phenobarbital, administered to an animal result in increased activity of enzymes related to detoxification, as well as other enzymes, and a considerable hypertrophy of the smooth endoplasmic reticulum.[20] This mechanism for detoxification also applies to endogenous or administered steroid hormones. Carcinogens such as 3-methylcholanthrene and 3,4-benzopyrene are among the most potent inducers of drug-metabolizing enzymes.[21]

The enzymes involved in the detoxification of aromatic hydrocarbons such as 3,4-benzopyrene are *aryl hydroxylases*. It is now known that benzopyrene (found in charcoal-broiled meat) is not carcinogenic but under the action of aryl hydrolases in the liver becomes converted to 5,6-epoxide, which is a powerful carcinogen.

The effect of these drugs is to produce a true induction of the enzymes of the endoplasmic reticulum. In other words, there is an increased synthesis of the enzymes which can be prevented by inhibitors of protein synthesis, such as puromycin. The cytochrome P-450 content of a liver cell, for example, may increase 50 to 100 per cent after prolonged phenobarbital administration.[10]

Synthesis of Lipids. While rough endoplasmic reticulum predominates in cells actively synthesizing proteins, the smooth type is abundant in those involved in the synthesis of lipids. An interrelationship of the membranous components of the vacuolar system has been observed during the synthesis of triglycerides and also during the formation of lipoprotein complexes. This interrelationship seems to be associated mainly with the smooth endoplasmic reticulum and the Golgi complex.[22]

We mentioned earlier that the smooth endoplasmic reticulum is particularly rich in cells engaged in lipid metabolism and in those producing steroid hormones.

Glycogenolysis. In fasted animals it has been observed that the residual glycogen remained associated with the tubules and vesicles of the endoplasmic reticulum. When feeding was resumed, there was an increase in smooth endoplasmic reticulum which maintained its association with the accumulating glycogen. In plant cells the smooth endoplasmic reticulum develops along the surface where the cellulose walls are being formed.

The enzyme UDPG-glycogen transferase,[23] which is directly involved in the synthesis of glycogen by addition of uridine diphosphate glucose (UDPG) to primer glycogen, is bound to the glycogen particle rather than to the membranous component.[24] This suggests that the reticulum is related to glycogenolysis, but not glycogenesis.

In prenatal liver cells, just before birth, the amount of glycogen increases and then decreases simultaneously with an increased amount of glucose-6-phosphatase. This depletion of glycogen is accompanied by an increase in smooth endoplasmic reticulum.

It has been mentioned that the withdrawal of glucose from the glycogen deposits could be mediated by the smooth endoplasmic reticulum by way of the glucose-6-phosphatase (Fig. 10–10).

SUMMARY:

Functions of Endoplasmic Reticulum

Numerous functions are attributed to the endoplasmic reticulum. It contributes to the *mechanical support* of the cytoplasm, has *osmotic properties*, and is involved in intracellular *exchanges* between the matrix and the internal cavity. These exchanges are brought about by diffusion or *active transport*, in which carriers and permeases may be involved. In liver there are about 11 square meters of membrane per milliliter available for exchange. *Ionic gradients* and electrical potential may be generated across these membranes, and *conduction of intracellular impulses* have been postulated in the case of the sarcoplasmic reticulum. The endoplasmic reticulum serves as a *circulatory system* for the transport of various substances. Although slower, the flow of membranes may be effective in various types of intracellular transport. The *synthesis of proteins* for export is one of the main functions of the rough endoplasmic reticulum. The membrane of the endoplasmic reticulum is highly dynamic and fluid (the bound ribosomes are mobile), and the transfer of secretory proteins is accompanied by a flow of newly synthesized membrane proteins. The *synthesis of lipids and lipoproteins* is associated with the rough and smooth endoplasmic reticulum. The *synthesis of glycogen* is accomplished in the cytoplasmic matrix, but the smooth endoplasmic reticulum is involved in *glucogenolysis* through the action of glucose-6-phosphatase. Another important function of the smooth endoplasmic reticulum is in the *detoxification* of many endogenous and exogenous compounds. The prolonged administration of certain drugs produces an increase in the smooth endoplasmic reticulum and the induction of the specific enzymes.

10–4 THE ENDOPLASMIC RETICULUM AND SYNTHESIS OF EXPORTABLE PROTEINS

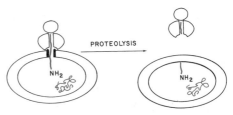

Although the function of ribosomes in protein synthesis will be dealt with in the chapters of Part 6, here we will consider briefly the consequences of the specific association between ribosomes and membranes of the endoplasmic reticulum. We remind the reader that such association occurs by way of the large subunit of the ribosome (60S) and that there are specific binding sites at the outer surface of the rough endoplasmic reticulum, probably associated with special intrinsic proteins, the *ribophorins*, mentioned before.

Although the process of protein synthesis takes place in the cytosol through the interaction of the polysomes, the mRNA, and the tRNAs, which carry the various amino acids, it is now well established that the polypeptide chain grows through a "groove" or "tunnel" that is in the large subunit and is linked directly to a channel in the membrane. In this way, at some stage of its elongation, the polypeptide chain begins to emerge and is finally deposited into the lumen of the endoplasmic reticulum.[25] Knowledge of this mechanism originated mainly from study of the action of proteolytic enzymes on polysomes, which demonstrated the presence of a protected polypeptide chain of about 40 amino acids at the −COOH end of the protein. Furthermore, in bound ribosomes the rest of the chain, with the −NH₂ end, was found to be protected inside the cisternal space where protein is segregated (Fig. 10–11).

Figure 10–11   Diagram showing the location of the nascent polypeptide chains within free and attached ribosomes. By proteolysis, it is possible to demonstrate the existence of a protected segment in both types of ribosomes and also in those attached to the endoplasmic reticulum. (Courtesy of G. Blobel, from D. D. Sabatini and Blobel, 1970.)

Special Initial Codons for Signal Peptides — In mRNA in ER-bound Polysomes

At this point question arose as to the mechanism by which the mRNA that selectively translates a protein for export is able to use bound and not free ribosomes. In other words, what is the difference between the mRNAs that are translated by free ribosomes and the mRNAs that normally use bound ribosomes?

The so-called *signal hypothesis* postulates that the mRNA for secretory proteins contains a set of special signal codons localized immediately after the initiation codon AUG (see Chapter 24–2). The initiation of the synthesis starts on free ribosomes, and it is only when the so-called *signal peptide* emerges from the large subunit that the ribosome becomes attached to

the membrane (Fig. 10–12). The signal peptide is thought to recognize and interact with special receptor proteins on the membrane which then form the tunnel through which the polypeptide penetrates.[26, 27]

The Hydrophobic Signal Peptide is Removed by a Signal Peptidase

The signal hypothesis also postulates that the signal peptide is removed by a peptidase (i.e., a *signal peptidase*) present at the luminal surface of the membrane. The now "processed" polypeptide continues to grow until it is completely segregated into the cisterna. It is proposed that after completion of synthesis, the ribosome is detached from the membrane by a "detachment factor."[28]

According to this hypothesis there is coordination of the translation of the polypeptide, the binding of the ribosome to the membrane, and the removal of the signal peptide. On the other hand, after translation is completed, there is removal of the ribosome and the tunnel with restoration of the original condition (Fig. 10–12).

The signal hypothesis is supported by abundant evidence, only one example of which will be considered here (see Blobel, 1977). Messenger RNAs for several secreted proteins have been translated on free ribosomes (in the absence of microsomal membranes) and shown to produce larger polypeptide chains. These have been called *preproteins* to differentiate them

221

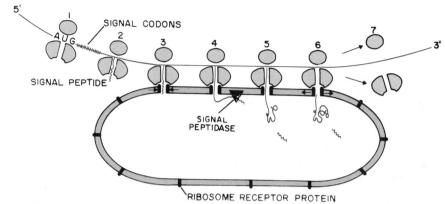

Figure 10–12 Schematic diagram illustrating the signal hypothesis. (Steps are described in the text.) (Courtesy of G. Blobel. From Blobel, G., Synthesis and segregation of secretory proteins, the signal hypothesis, *in International Cell Biology*, p. 318. (Brinkley, B. R., and Porter, K. R. eds.) New York, The Rockefeller University Press, 1977.)

from the *proproteins*, such as proinsulin, proalbumin, proparathyroid hormone, and others, which are larger precursors of the secreted proteins (see Fig. 11–14) and which are present in the endoplasmic reticulum and Golgi complex. For example, a mRNA for albumin[29] first produces a *preproalbumin*; this, after removal of the signal peptide, is converted into *proalbumin*, and later on it is again cleaved at a specific secretory stage to produce the final product, *albumin*. The preproteins contain at the $-NH_2$ terminal the so-called *signal peptide*, which is a sequence of 15 to 30 amino acid residues, most of which are hydrophobic, and it is this sequence that is removed by the *signal peptidase* (see Blobel, 1977).

The function of the signal hydrophobic peptide is to establish the initial association of the ribosome with the membrane, which would then be more firmly bound by the receptor sites probably provided by the ribophorins. Experiments with ribonuclease have detected a site that is protected from the attack of the enzyme at the 3' end of the mRNA that is encoded for exportable proteins. This finding has suggested the hypothesis that the poly A segment of the mRNA could be associated with the membrane of the endoplasmic reticulum[30-32] (Fig. 10–7).

Membrane Proteins are Made and Assembled in Different Compartments

An important problem is raised by the synthesis of the membrane proteins of the endoplasmic reticulum and other portions of the endomembrane system, including the nuclear envelope, the Golgi complex, and the membranes of other cytoplasmic organelles. As was

emphasized in Figure 10–2, each of these membranes has a cytoplasmic and a luminal face and, as does the plasma membrane itself, each contains peripheral and intrinsic proteins that may be exposed to one or the other face (see Chapter 8–1). In addition, these membranes contain other intrinsic proteins that are completely embedded in the lipid bilayer or that traverse the membrane and are exposed on both faces. What might the mechanism of synthesis be for these various types of proteins?

Peripheral and integral proteins of the cytoplasmic face (Fig. 10–13, *f* and *e*) could be synthesized by either free or bound ribosomes which are located in the same cell compartment. In reticulocytes, which contain no endoplasmic reticulum, all proteins of the cytoplasmic face are produced by free ribosomes. On the other hand, the synthesis of a protein by a

cytoplasmic side

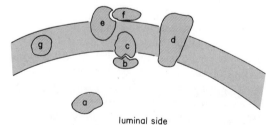

luminal side

Figure 10–13 Diagram of the possible disposition of peripheral and integral membrane proteins. *a*, luminal protein; *b*, terminal peripheral protein; *c*, luminal integral protein; *d*, transmembrane protein; *e*, cytoplasmic integral protein; *f*, cytoplasmic peripheral protein; *g*, intramembranous protein (hypothetical). (Courtesy of D. D. Sabatini and G. Kreibich. From Sabatini, D. D., and Kreibich, G., *in The Enzymes of Biological Membranes*, Vol. 2, p. 431. Martonosi, A., ed. New York, Plenum Press, 1976.)

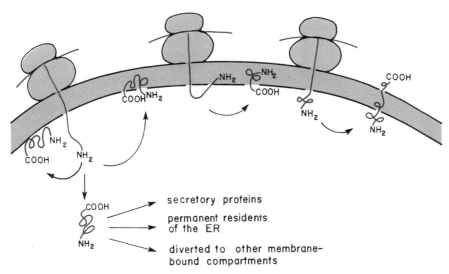

secretory proteins

permanent residents of the ER

diverted to other membrane-bound compartments

Figure 10–14 Diagram of the possible mechanisms by which bound ribosomes may give rise to proteins that are deposited into the lumen of the endoplasmic reticulum or are incorporated either as peripheral or integral proteins at the luminal face, or as transmembrane proteins. (Courtesy of D. D. Sabatini and G. Kreibich. From Sabatini, D. D., and Kreibich, G., Functional specialization of membrane-bound ribosomes, *in Enzymes of Biological Membranes,* Vol. 2, p. 431. Martonosi, A., New York, Plenum Press, 1976.)

bound ribosome may ensure the direct deposition of the protein in the endoplasmic reticulum. This could be true for cytochrome b_5, an amphipathic integral protein that is exposed toward the cytoplasmic face of the membrane.

In the case of integral and peripheral proteins of the luminal face (Fig. 10–13, *b* and *c*) it has been proposed that they are synthesized exclusively by bound ribosomes and are discharged into the lumen of the endoplasmic reticulum. The sites of synthesis of proteins that span the membrane and are exposed on both surfaces and the mechanism of insertion into the membrane present a very special case (Fig. 10–13, *d*). The insertion could be achieved if there is an incomplete vectorial discharge of the polypeptide chain with a more hydrophobic segment that remains embedded in the membrane.

Figure 10–14 summarizes the various mechanisms by which bound ribosomes may give rise to proteins which lie in the lumen or which are incorporated into the membrane as either peripheral, integral, or transmembranal proteins. Observe that in the last type there is a special orientation of the protein, with the

$-NH_2$ end facing the lumen and the $-COOH$ terminal at the cytoplasmic face. This arrangement is reminiscent of the erythrocyte transmembrane protein *glycophorin* (Fig. 8–1), which, however, has its end groups in the opposite orientation. This reversal of polarity could be easily explained if a vesicle carrying the protein fuses with the outer membrane.

The process of membrane biogenesis can not be clearly understood if one does not take into account the fluidity of the membrane and the fact that the various domains of the endomembrane system and the plasma membrane are in spatial and temporal continuity. As shown in other chapters, this continuity is achieved by membrane fusion during secretion, exocytosis, and endocytosis.

In summary, the synthesis of membrane proteins and their assembly in it is a highly compartmentalized process. Proteins of the cytoplasmic face might be synthesized on free ribosomes or could be deposited directly by bound ribosomes. Proteins lining the luminal cavity or traversing the membrane are synthesized by the ribosomes bound to the endoplasmic reticulum.

SUMMARY:

Synthesis of Exportable Proteins — The Signal Hypothesis

The molecular mechanism involved in the synthesis of proteins will be studied in the chapters on molecular biology. Although this process occurs in the cytosol, the polypeptide chain of the exportable protein grows into a "groove" or "tunnel" in the

60S ribosomal subunit and penetrates, through a channel, into the lumen of the endoplasmic reticulum.

The signal hypothesis tries to explain the mechanism by which the messenger RNA (mRNA) is able to recognize free or bound ribosomes. It is postulated that the mRNA for secretory proteins contains a set of *special signal codons* localized after the initial codon AUG. Once the ribosome "recognizes" the signal the ribosome becomes attached to the membrane, and the polypeptide penetrates. It is also postulated that at the luminal surface there is a *signal peptidase* that removes the signal peptide. Thus the mRNA produces a *preprotein* of larger MW than the final protein. The signal peptide has between 15 and 30 amino acids which are generally hydrophobic. Such a signal peptide probably establishes the initial association of the ribosome with the membrane.

Another interesting problem is posed by the probable mechanism of synthesis of the various types of proteins that are present in the membrane. A special case is that of the proteins that traverse the membrane. It is thought that they are synthesized on ribosomes attached to the membrane, with a more hydrophobic segment in the middle that remains embedded in the membrane. the possible mechanisms of synthesis for the various types of proteins are indicated in Figure 10–14.

REFERENCES

1. Porter, K. R., Claude, A., and Fullan, E. F. (1945) *J. Exp. Med., 81*:233.
2. Buckley, I. K., and Porter, K. R. (1975) *J. Microsco.* (Oxf.), *104*:107.
3. Sabatini, D. D., and Kreibich, G. (1976) In: *Enzymes of Biological Membranes*, Vol. 2, p. 531. (Martonosi, A., ed.) Plenum Press, New York.
4. Branton, D., et al. (1975) *Science, 190*:54.
5. Weibel, E. R., Stäbli, W., Gnagi, R., and Hess, E. A. (1969) *J. Cell Biol., 42*:68.
6. Kreibich, G., Ulrich, B. L., and Sabatini, D. D. (1978) *J. Cell Biol., 77*:464.
7. Kreibich, G., and Sabatini, D. D. (1974) *J. Cell Biol., 6*:789.
8. Meldolesi, J., Jamieson, J. D., and Palade, G. E. (1971) *J. Cell Biol., 49*:68.
9. Meldolesi, J., and Cova, D. (1972) *J. Cell Biol., 55*:1.
10. Dehlinger, P. J., and Schimke, R. T. (1971) *J. Biol. Chem., 246*:2574.
11. Estabrook, R. W., et al. (1976) In: *The Structural Basis of Membrane Function*, p. 429 (Hatefi, Y., and Javadi, D., Ohaniance, L., eds.) Academic Press, New York.
12. Dallner, G. P., Siekevitz, P., and Palade, G. E. (1966) *J. Cell Biol., 30*:97.
13. Leskes, A., Siekevitz, P. and Palade, G. E. (1971) *J. Cell Biol, 49*:264.
14. Eytan, G., and Ohad, I. (1972) *J. Biol. Chem., 247*:112.
15. Ojakian, G. K., Kreibich, G., and Sabatini, D. D. (1977) *J. Cell Biol., 72*:530.
16. Franke, W. W., et al. (1971) *Z. Naturforsch, 266*:1031.
17. Morré, D. J., Kennan, T. W., and Mollenhauer, H. H. (1974) In: *Advances in Cytopharmacology*, Vol. 2 p. 107. Raven Press, New York.
18. Meldolesi, J. (1974) In: *Advances in Cytopharmacology*, Vol. 2, p. 71. Raven Press, New York.
19. Palade, G. E. (1975) *Science, 189*:347.
20. Jones, A. L., and Fawcett, D. N. (1966) *J. Histochem. Cytochem., 14*:214.
21. Conney., A. H., Schneidman, K., Jacobson, M., and Kuntzman, R. (1965) *Ann. N. Y. Acad. Sci., 123*:98.
22. Claude, A. (1970) *J. Cell Biol., 47*:745.
23. Leloir, L. F., and Cardini, C. E. (1957) *J. Am. Chem. Soc., 79*:6340.
24. Luck, D. J. L. (1961) *J. Biophys. Biochem. Cytol., 10*:195.
25. Sabatini, D. D., and Blobel, G. (1970) *J. Cell Biol., 45*:146.
26. Blobel, G., and Dobberstein, B. (1975) *J. Cell Biol., 67*:835 and 852.
27. Blobel, G. (1977) In: *International Cell Biology*, p. 318. (Brinkley, B. R., and Porter, K. R. eds.) The Rockefeller University Press, New York.
28. Blobel, G. (1976) *Biochem. Biophys. Res. Commun., 68*:1.
29. Kemper, B., et al. (1976) *Biochemistry, 15*:15.
30. Lande, M., Adesnik, M., Sumida, M., Tashiro, Y., and Sabatini, D. D. (1975) *J. Cell Biol., 65*:513.
31. Milcarek, C., and Penman, S. (1974) *J. Molec. Biol., 89*:327.
32. Adesnik, M., Lovele, M., Martin, T., and Sabatini, D. D. (1976) *J. Cell Biol., 71*:307.

ADDITIONAL READING

Autori, A., Svensson, H., and Dallner, G. (1975) Biogenesis of microsomal membrane glycoproteins in rat liver: I, II, and III. *J. Cell Biol., 67*:687.

Beaufay, H., et al. (1974) Analytical study of microsomes and isolated subcellular membranes. *J. Cell Biol., 61*:213.

Blobel, G. (1977) Synthesis and segregation of secretory proteins, the signal hypothesis. In: *International Cell Biology,* p. 318. (Brinkley, B. R., and Porter, K. R., eds.) The Rockefeller University Press, New York.

Claude, A. (1969) Microsomes, endoplasmic reticulum, and interactions of cytoplasmic membranes. In: *Microsomes and Drug Oxidations,* p. 3. Academic Press, Inc., New York.

De Pierre, J. W. and Dallner, G. P. (1975) Structural aspects of the membrane of endoplasmic reticulum. *Biochem. Biophys. Acta, 415:*411.

Dobberstein, B., and Blobel, G. (1977) Functional interaction of plant ribosomes with animal microsomal membranes. *Biochem. Biophys. Res. Commun., 74:*1675.

Eriksson, L. C., De Pierre, J. W., and Dallner, G. (1978) Preparation and properties of microsomal fractions. *Pharmacology and Therapeutics,* Vol. 2, p. 281.

Haguenau, F. (1958) The ergastoplasm: Its history, ultrastructure, and biochemistry. *Int. Rev. Cytol., 7:*425.

Fleischer, S., Zahler, W. L., and Ozawa, H. (1971) Membrane-associated proteins. In: *Biomembranes,* Vol. 2, p. 105–119 (Manson, L. A., ed.) Plenum Press, New York.

Kruppa, J., and Sabatini, D. D. (1977) Release of poly A (+) mRNA from rat liver rough microsomes. *J. Cell Biol., 74:*414.

Lingappa, V. R., Katz, F. N., Lodish, H. F., and Blobel, G. (1978) A signal sequence for insertion of transmembrane glycoprotein. *J. Biol. Chem., 253:*8867.

Novikoff, A. B. (1976) The endoplasmic reticulum: A cytochemist's view (a review). *Proc. Natl. Acad. Sci. USA, 73:*2781.

Rolleston, F. S. (1974) Membrane-bound and free ribosomes. *Subcell. Biochem., 3:*91.

Sabatini, D. D., and Kreibich, G. (1976) Functional specialization of membrane-bound ribosomes in eukaryotic cells. In: *The Enzymes of Biological Membranes.* Vol. 2, p. 531. (Martonosi, A., ed.) Plenum Press, New York.

11
GOLGI COMPLEX AND CELL SECRETION II

11–1 Morphology of the Golgi Complex (Dictyosomes) 226
    Dictyosomes — A Forming Face, and a Maturing Face Near the GERL
    Polarization of Dictyosomes and Membrane Differentiation
    Summary: *Morphology of the Golgi Complex*
11–2 Cytochemistry of the Golgi Complex 232
    Chemical Composition of the Golgi Complex — Intermediate Between Those of the ER and the Plasma Membrane
    Glycosyl Transferases are Concentrated in the Golgi
    Summary: *Cytochemistry of the Golgi Complex*
11–3 Functions of the Golgi Complex 235

Synthesis of Glycosphingolipids and Glycoproteins — A Major Role of the Golgi
Cancer Cells — Changes in Glycosidation of Lipids and Proteins
Secretion — the Main Function of the Golgi Complex
The Secretory Cycle — Continuous or Discontinuous
The Secretory Process in the Pancreas — Six Consecutive Steps
The GERL Region and Lysosome Formation
Insulin Biosynthesis — A Good Example of the Molecular Processing of Secretion
Summary: *Functions of the Golgi Complex*

In the previous chapter we mentioned that the *Golgi apparatus* or *Golgi complex* is a differentiated portion of the endomembrane system found in both animal and plant cells (Figs. 1–1 and 2–6). This membranous component is spatially and temporally related to the endoplasmic reticulum on one side and, by way of secretory vesicles, may fuse with specific portions of the plasma membrane. Because of the important functions that this structure plays as an intermediary in secretory processes, it will be studied in this chapter together with *cell secretion*.

In 1898 by means of a silver staining method Camillo Golgi discovered a reticular structure in the cytoplasm of nerve cells (Fig. 11–1). Since its refractive index is similar to that of the cytosol, the Golgi complex in the living cell was difficult to observe with the light microscope, and this led to many controversies regarding its true nature. However, the use of electron microscopy provided a distinct image of this organelle, and its structure could thus be studied in detail. Later on, the study of thick sections with the high-

voltage electron microscope,[1] the use of freeze-etching (Fig. 3–9), and the observation of isolated Golgi membranes by negative staining contributed to the study of this structure. Other advances came with the use of autoradiography with special radioactive precursors and cell fractionation, a technique which has enabled isolation of purified fractions and performance of biochemical studies. Through all these advances, the role of this organelle in cellular functions has been, to a great extent, elucidated (see Cook, 1975; Whaley, 1975; and Morré, 1977).

11–1 MORPHOLOGY OF THE GOLGI COMPLEX (DICTYOSOMES)

The Golgi complex is morphologically very similar in both plant and animal cells (Fig. 11–2). It consists of *dictyosome* units formed by stacks of flattened disc-shaped cisternae and associated secretory vesicles.

In cells that have a polarized structure the Golgi

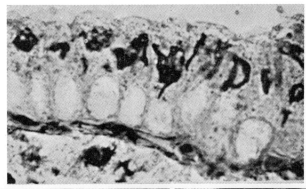

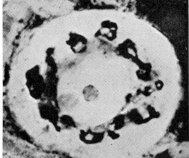

Figure 11-1 **Above,** Golgi complex (Golgi apparatus) in cells of the thyroid gland of the guinea pig, apical position. Osmic impregnation. **Below left,** ganglion cell, perinuclear Golgi apparatus. **Below right,** same structure as at left, optical section tangential with respect to the nucleus. Silver impregnation. (From E. De Robertis.)

complex is, in general, a single large structure and occupies a definite position between the nucleus and the pole of the cell where the release or secretion takes place. This is the case, for example, in thyroid cells (Fig. 11–1), in the exocrine pancreas, or in the mucous cells of the intestinal epithelium. In nerve (Fig. 11–1) and liver cells (Fig. 11–3) and in most plant cells there are many dictyosomes that do not show a special polarity. In liver cells there are some 50 *dictyosomes* per cell. These structures represent some 2 per cent of the total cytoplasmic volume.[2] In plant cells and in invertebrate tissues dictyosomes are dispersed throughout the cytoplasm. Although the localization, size and development of these organelles vary from one cell to another and also with the physiologic state of the cell, they show morphological characteristics that permit their differentiation from the other parts of the endomembrane system.

One such quality is a lack of attached ribosomes; in fact most of the Golgi complex appears to be surrounded by a zone from which most ribosomes, glycogen, and mitochondria are absent; the so-called *zone of exclusion.*[3] However, some free polysomes, present at the periphery of the Golgi complex, have been observed, and the possibility of specific protein synthesis that may lead to membrane modification of the Golgi has been recognized.[4]

In general, three membranous components are recognized under the electron microscope: (1) flattened sacs (i.e., cisternae); (2) clusters of tubules and vesicles of about 60 nm; and (3) larger vacuoles filled with an amorphous or granular content. The Golgi cisternae are arranged in parallel and are separated by a space of 20 to 30 nm which may contain rodlike elements or fibers. Often the cisternae are arranged concentrically with a convex and a concave face. There may be from three to seven of these structures in most animal and plant cells. In certain algae, however, there may be as many as 10 or 20 cisternae.

Dictyosomes — A Forming Face, and a Maturing Face Near the GERL

Each stack of cisternae forming a dictyosome is a polarized structure having a *proximal* or *forming* face generally convex and closer to the nuclear envelope or the endoplasmic reticulum, and a *distal* or *maturing* face of concave shape, which encloses a region containing large secretory vesicles (Fig. 11–4). The forming face is characterized by the presence of small transition vesicles or tubules that converge upon the Golgi cisternae, forming a kind of fenestrated plate (Fig. 11–2). These *transition vesicles* are thought to form as

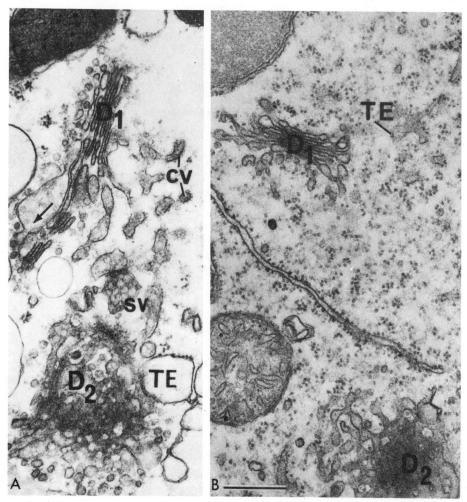

Figure 11–2 Electron micrographs of the Golgi apparatus in: **A,** rat liver and **B,** onion (*Allium cepa*) stem. D_1, dictyosome in cross section showing the stacked cisternae; D_2, dictyosome in tangential section showing a face view of cisternae with a central plate-like region and a fenestrated margin; *TE,* transition element of the endoplasmic reticulum adjacent to the Golgi; *cv,* coated vesicles; *sv,* secretory vesicles. An arrow shows a connection between the smooth endoplasmic reticulum and a secretory vesicle. ×35,000. (Courtesy of D. J. Morré. From Morré, D. J., Membrane differentiation and the control of secretion: A comparison of plant and animal Golgi apparatus, *in International Cell Biology,* p. 293. Brinkley, B. R., and Porter, K. R., eds. New York, Rockefeller University Press, 1977.)

blebs from the endoplasmic reticulum and to migrate to the Golgi where, by coalescence, they form new cisternae. As will be mentioned later, a mechanism of membrane flow is postulated in which new cisternae are formed at the proximal end and thus compensate for the loss at the maturing face that occurs with the release of secretory vesicles (see Beams and Kessel, 1968; Cook, 1973; Morré, 1977).

Associated with the maturing face there is often a saccular structure which is rich in acid phosphatase and has been called the GERL. (Fig. 10–1). This denomination indicates that it has been interpreted as a region of *smooth endoplasmic reticulum, near the Golgi,* and which is involved in the production *of lysosomes.* More re-

cent work relates the GERL to *Golgi condensing vacuoles* or *presecretory granules.*[5]

Polarization of Dictyosomes and Membrane Differentiation

The polarization of the dictyosome is also expressed in terms of what may be called *membrane differentiation.* As shown in Table 11–1 the thickness of the membranes increases progressively from the endoplasmic reticulum to the Golgi and the plasma membrane, and even within the various cisternae of the dictyosome. There are also differences in staining properties; thus the cisternae at the forming face stain more strongly with

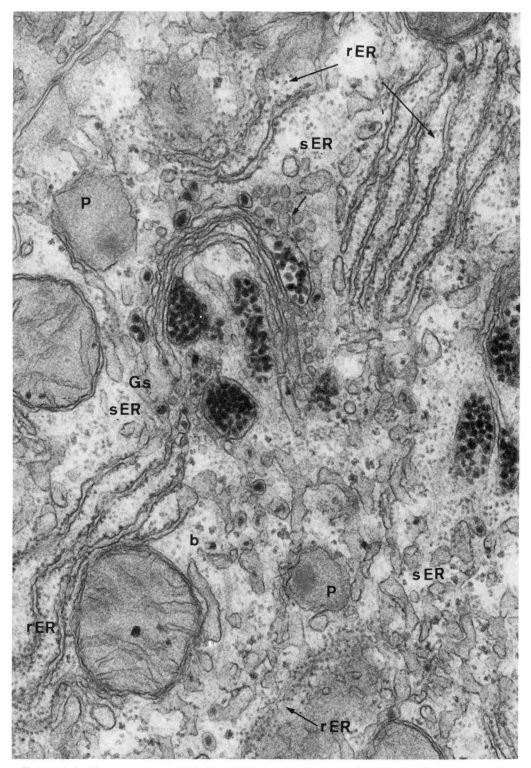

Figure 11–3 Electron micrograph of a liver cell of an animal having a diet rich in fat. The synthesis and transport of the lipoprotein granules is observed. The rough endoplasmic reticulum (*rER*) is observed as two stacks of lamellae converging toward a Golgi complex having Golgi sacs (*Gs*) on its convex or "forming face." There are portions of smooth endoplasmic reticulum (*sER*) connecting both parts of the vacuolar system. Mitochondria and peroxisomes (*P*) are observed. ×56,000. (Courtesy of A. Claude.)

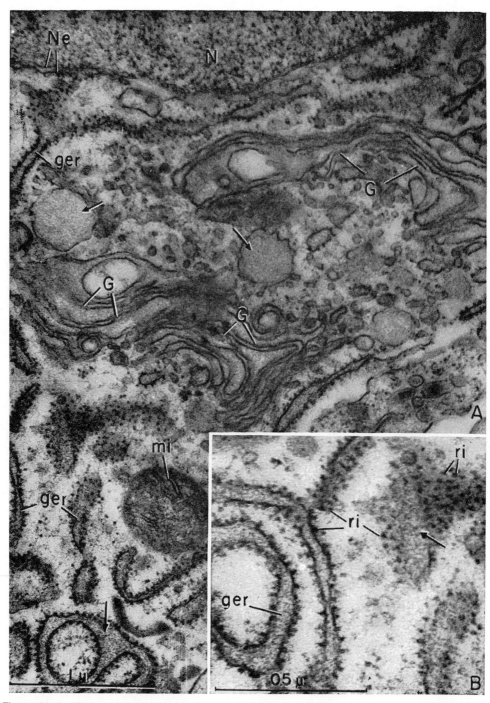

Figure 11–4 Electron micrograph of a plasma cell showing near the nucleus (*N*) a large Golgi complex (*G*) formed of flat cisternae and small and large vesicles. Some of the large vesicles (arrows) are filled with material. Surrounding the Golgi complex is abundant granular endoplasmic reticulum (*ger*) having cisternae filled with amorphous material (arrows). *mi,* mitochondrion; *Ne,* nuclear envelope; *ri,* ribosomes. ×48,000; inset ×100,000. (From E. De Robertis and A. Pellegrino de Iraldi.)

TABLE 11–1   MEMBRANE DIFFERENTIATION IN THE GOLGI COMPLEX OF ANIMALS AND PLANTS*

Membrane Type	Membrane Thickness (nm)			
	Rat Liver	*Rat Mammary Gland*	*Onion Stem*	*Soybean Hypocotyl*
Nuclear envelope	65 ⎱	60	56	56
Endoplasmic reticulum	65 ⎰		53	56
Golgi complex:				
Cisterna 1	65 ⎫		53	56
Cisterna 2	68 ⎬	70	60	38
Cisterna 3	72 ⎪		65	61
Cisternae 4 and 5	80 ⎭		75	69
Secretory vesicle	83	85	88	78
Plasma membrane	85	97	93	88

*From Morré, D. J.: in *International Cell Biology,* p. 293. (Brinkley, B. R., and Porter, K. R., eds.) The Rockefeller University Press, New York, 1977.

osmium, and a stain for glycoproteins, such as periodic acid–methenamine, appears in the maturing cisternae. This is particularly evident in the Golgi complex of certain algae in which scales of cellulose are being produced and secreted (Fig. 11–5) (see Brown and Willison, 1977). The staining gradient seen in the figure, also found in thyroid cells and others, suggests that carbohydrate is being added as the contents of the cisternae pass from the forming to the maturing face of the Golgi complex (see Leblond and Bennett, 1977).

The transitions between the various parts of the Golgi complex are more clearly observed in liver cells under certain experimental conditions that make it possible to trace the synthesis and transport of lipoproteins, which appear as discrete dense granules of about 40 nm. As shown in Figure 11–3 these granules appear first within tubules of the smooth endoplasmic reticulum and then enter the outer fenestrated cisternae of the Golgi. After longer periods they accumulate in the large vacuoles that are formed by dilatation of the edge of the sacs and, finally, they are detached as secretory vesicles. Although not recognizable morphologically, the protein moiety of the lipoprotein is synthesized in the rough endoplasmic reticulum. The granules, however, become visible in the smooth portion because of the addition of triglycerides.[2]

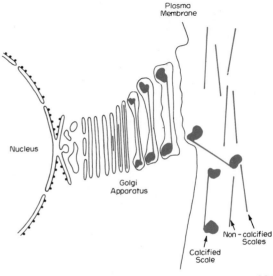

Figure 11–5 Diagram of the process of scale formation in a chrysophycean alga. The scales of cellulose are made progressively from the forming face to the maturing face of the Golgi apparatus. At the scale margin, calcification occurs (*black projections*) prior to discharge to the cell surface. (Adapted from Brown et al., 1970; courtesy of D. J. Morré and H. H. Mollenhauer, 1976.)

SUMMARY:

Morphology of the Golgi Complex

The Golgi apparatus or complex is a differentiated portion of the endomembrane system which is morphologically very similar in animal and plant cells. It is spatially and temporally related to the endoplasmic reticulum on one side, and to secretory vesicles leading to the cell membrane on the other.

The Golgi complex consists of *dictyosomes* formed by stacks of curved cisternae, associated tubules, and secretory vesicles. The cisternae lack ribosomes and are surrounded by a zone in which organelles are excluded. However, some free ribosomes may be at the periphery. Each dictyosome is polarized, with a proximal or *forming face* (convex and close to the nucleus) and a distal or *maturing face* (concave). At the forming face are vesicles and tubules that converge, forming a fenestrated plate. There is a continous mechanism of membrane flow to compensate for the loss of secretory vesicles and also a process of differentiation of membrane revealed by an increase in its thickness and by its staining for glycoproteins. The interrelationship between the various parts of the Golgi and between it and the endoplasmic reticulum and secretory vesicles is clearly seen in the secretion of lipoprotein granules by the liver.

11–2 CYTOCHEMISTRY OF THE GOLGI COMPLEX

Although earlier the Golgi complex had been isolated from cells of the epididymis, in recent years the isolation has been achieved in numerous plant and animal cells.[6, 7] The methods used are based on gentle homogenization, which tends to preserve the stacks of cisternae, thus allowing large fragments of Golgi complexes to be obtained by differential and gradient centrifugation (see Chapter 4–4). The Golgi complexes have a lower specific density than the endoplasmic reticulum or the mito-

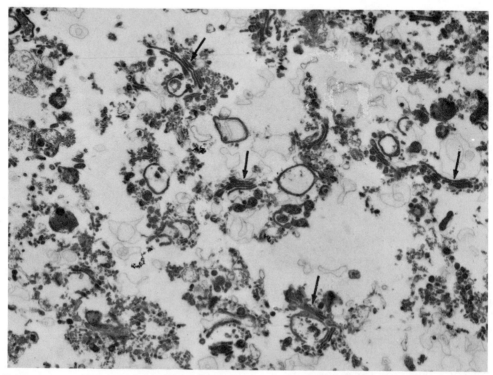

Figure 11–6 Isolated Golgi complexes from liver cells. The complexes which best show the stacks of cisternae are indicated by arrows. (Courtesy of D. J. Morré.)

chondria and they are equilibrated in a band having a density of 1.16. Under the electron microscope the stacked cisternae appear to be bordered by an extensive system of tubules and vesicles (the secretory products remain within these vesicles) (Fig. 11–6). Washing the Golgi complexes in distilled water results in further purification with a consequent loss of the secretory products.

A subfractionation of the Golgi complex has been achieved after unstacking of the cisternae of the dictyosome by the use of proteolytic enzymes. Several subfractions, some richer in secretory vesicles, others containing mainly cisternae plates and other portions of the Golgi have been separated by gradient centrifugation and were characterized by some biochemical parameters.[8]

Chemical Composition of the Golgi Complex — Intermediate Between Those of the ER and the Plasma Membrane

Information about the biochemical makeup of the different endomembranes has come from studies in which highly purified fractions of Golgi complex have been compared with others from the endoplasmic reticulum and plasma membranes isolated in parallel.

The Golgi complex isolated from rat liver consists of about 60 per cent protein and 40 per cent lipid. By the use of gel electrophoresis it was found that the Golgi complex and the endoplasmic reticulum contain some proteins in common, but the former has fewer protein bands. In the plasma membrane fraction, however, fewer proteins were observed. The decreasing complexity in protein band patterns

from endoplasmic reticulum, through Golgi complex, to plasma membranes, is in some ways consistent with the view that membrane proteins are synthesized by ribosomes of the rough endoplasmic reticulum and then transferred to other portions of the endomembrane system.

Studies carried out on isolated Golgi complex from different plant and animal cells show marked differences in protein and enzyme content. The Golgi complex has a phospholipid composition that is intermediate between those of endoplasmic reticulum and plasma membrane. In contrast to rat liver, in which the most abundant phospholipid is phosphatidylcholine, plants have large amounts of phosphatidic acid (30 to 40 per cent) and phosphatidylglycerol (see Morré, 1977).

Plant membranes, including those of the Golgi, lack sialic acid; on the other hand, this sugar is very high in rat liver, and glycosphingolipids are abundant.[9] Both animal and plant cells have some carbohydrate components in common (i.e., glucosamine, galactose, glucose, mannose, and fucose), but plants also have pentoses (i.e., xylose and arabinose) and other special carbohydrates.

Glycosyl Transferases are Concentrated in the Golgi

In the isolation of subcellular organelles it is of great importance to determine the degree of contamination between the various fractions. Table 11–2 shows some enzyme markers used for plasma membranes (5'-nucleotidase), for the endoplasmic reticulum (glucose-6-phosphatase), for mitochondria (cytochrome oxi-

TABLE 11–2 DISTRIBUTION OF ENZYME MARKERS IN THE CORRESPONDING FRACTION AND IN THE GOLGI COMPLEX FROM RAT LIVER*

Enzyme	Fraction	Specific Activity in Corresponding Fraction	Specific Activity in Golgi Apparatus	Ratio
5'-Nucleotidase	Plasma membrane	41.9	5.8	0.14
Glucose-6-phosphatase	Endoplasmic reticulum	11.4	1.1	0.10
Cytochrome oxidase	Mitochondria	67.8	1.8	0.03
Uric acid oxidase	Peroxisomes	50.0	0.5	0.01

*From Morré, D. J., Keenan, T. W., and Mollenhauer, H. H., Golgi Apparatus Function in Membrane Transformations and Product Compartmentalization: Studies with Cell Fractions Isolated from Rat Liver *in Advances in Cytopharmacology*, Vol. 1, First International Symposium on Cell Biology and Cytopharmacology. (Clementi, F., and Ceccarelli, B., eds.) New York, Raven Press, 1971, Table 1.

TABLE 11–3 GLYCOSYL TRANSFERASE ACTIVITY IN GOLGI COMPLEX FRACTION OF RAT LIVER*

Glycosyl Transferase	Specific Activity† in		Per Cent of Total Activity in Golgi Complex
	Total Homogenate	Golgi Complex	
Sialyl	50	422	44
Galactosyl	11	128	42
N-acetylglucosaminyl	24	219	43
Galactosyl-N-acetylglucosamine	6	64	40

*Modified from Morré, D. J., Keenan, T. W., and Mollenhauer, H. H., Golgi Apparatus Function in Membrane Transformations and Product Compartmentalization: Studies with Cell Fractions Isolated from Rat Liver *in Advances in Cytopharmacology*, Vol. 1, First International Symposium on Cell Biology and Cytopharmacology. (Clementi, F., and Ceccarelli, B., eds.) New York, Raven Press, 1971, (Unpublished data of H. Schachter.)

†Specific activity of the four glycosyl transferases is expressed in mμ moles/hrs/mg protein of sugar nucleotide bound to the protein acceptor.

dase), and for peroxisomes (uric acid oxidase). It has been calculated that a fraction of Golgi membranes purified by washing contains small percentages of contamination: plasma membrane, 6 per cent; endoplasmic reticulum, 2 per cent; and mitochondria, 1 per cent.[10] The enzymes that are concentrated in the Golgi fraction are *thiamine-pyrophosphatase* and several *glycosyl transferases*. The Golgi fraction also contains acid phosphatase and other lysosomal enzymes, presumably in relation to primary lysosomes (see Chapter 13).

The most characteristic enzymes of the Golgi fraction are those related to the transfer of oligosaccharides to proteins (i.e., glycosyl transferases) with the resulting formation of glycoproteins (see below). Table 11–3 shows the specific activity of four glycosyl transferases that are involved in the transfer of CMP-neuraminic acid (CMP-NANA), UDP-galactose,

and UDP-acetylglucosamine to protein acceptors. (All of these transferases are highly concentrated in the Golgi fraction of the rat liver.)

We mentioned earlier that the isolated Golgi complex can be subfractionated. In rats to which ethanol is administered before sacrifice, the Golgi vacuoles become filled with low-density lipoproteins. This facilitates the separation of two light fractions (GF_1 and GF_2) containing small and large vacuoles filled with lipoproteins from a heavier fraction (GF_3) corresponding to the cisternae. It was found that glucose-6-phosphatase and thiamine-pyrophosphatase are mainly in GF_1 and GF_2 and also that galactosyltransferase is higher in these fractions.[11] Golgi membranes also contain other enzymes in common with the endoplasmic reticulum, for example, NADH-cytochrome-b_5-reductase, NADH-cytochrome-c-reductase, and 5'-nucelotidase.

SUMMARY:

Cytochemistry of the Golgi Complex

The Golgi complexes of many plant and animal cells can be separated by gentle homogenization, followed by differential and gradient centrifugation. Both the endoplasmic reticulum and mitochondria have a higher specific density than the Golgi complex (density: 1.16). The Golgi from liver contains 60 per cent protein and 40 per cent lipids. (The pattern of protein is less complex than in the endoplasmic reticulum and more complex than in the plasma membrane)

Various subcellular fractions can now be characterized by enzyme markers, and the degree of contamination of the Golgi fraction by mitochondria, plasma membranes, endoplasmic reticulum, and peroxisomes may be determined. The enzymes that appear most concentrated in the Golgi fraction are thiamine-pyrophosphatase and several glycosyl transferases that transfer oligosaccharides to the glycoproteins.

11–3 FUNCTIONS OF THE GOLGI COMPLEX

The major functions of the Golgi complex appear to be related to the fact that it represents a special membranous compartment interposed between the endoplasmic reticulum and the extracellular space. Through this compartment there is a continous traffic of substances which may have been synthesized elsewhere but which are modified and transformed while transported. This process also involves the flow and differentiation of membranes. Within the walls of the Golgi complex a vast diversity of materials may be found, varying from simple fluids to macromolecules and even preformed wall units, as in the case of certain algae. Virtually every major class of macromolecule is transported through the Golgi and secreted, and this implies a continuous and fast turnover of the Golgi membranes. Thus, dictyosomes in algae and liver cells are estimated to turn over every 20 to 40 minutes, and cisternae are released at the rate of 1 every few minutes.[12-14]

Synthesis of Glycosphingolipids and Glycoproteins — A Major Role of the Golgi

The Golgi complex plays a major role in the glycosidation of lipids and proteins to produce *glycosphingolipids* and *glycoproteins* (see Chapter 3). The carbohydrate prosthetic groups of both these complex substances are added in a sequential manner in the Golgi complex by the action of the various glycosyl transferases

(Table 11–3). These enzymes are responsible for transfering a sugar residue (from a sugar nucleotide) either to an amino acid residue of a protein or to another carbohydrate in a glycoprotein or glycolipid.

From numerous studies using radioactive precursors (e.g., galactose, fucose, and sialic acid) the following biosynthetic scheme has emerged for the types of secreted glycoproteins mentioned earlier. The protein backbone of the glycoprotein is synthesized by membrane-bound ribosomes, and then the monosaccharides are added, one by one, as the protein moves through the channels of the endoplasmic reticulum and Golgi complex. In the process the system of glycosyl transferases that are bound to the membranes of the Golgi plays an important role. The enzymes adding the oligosccharide core containing N-acetyl-glucosamine and mannose are, at least in part, assembled in the rough endoplasmic reticulum. The fucosyl- and the sialyl-galactosyl-N-acetyl-glucosamine termini of the prosthetic group, on the other hand, are incorporated in the Golgi complex by the corresponding transferases.[15]

In Table 11–4 the three main patterns of incorporation are indicated. When D-mannose is used as precursor, the rough endoplasmic reticulum is the first to be marked and then the other compartments: Golgi complex → secretory granules → extracellular space or plasma membrane. When sialic acid, L-fucose or D-galactose is used the Golgi complex is the first to be labeled. With D-glucosamine both the rough endoplasmic reticulum and the Golgi complex are simultaneously labeled. A compar-

---

TABLE 11–4 INCORPORATION OF VARIOUS PRECURSORS INTO GLYCOPROTEINS*

Pattern Type	Radioactive Precursor	Interpretation of Time Course Experiments
A	Leucine D-mannose	RER → Golgi → SG, EC, or PM ↑ Label
B	Sialic acid L-fucose D-galactose	RER → Golgi → SG, EC, or PM ↑ Label
C	D-glucosamine	RER → Golgi → SG, EC, or PM ↑ ↑ Label Label

*Abbreviations: *RER*, rough endoplasmic reticulum; *SG*, secretory granule; *EC*, extracellular space; *PM*, plasma membrane. This is a table which presents, in abbreviated form, experiments by various groups of investigators. The experimental work involved the injecting of whole animals, perfused tissues, or tissue slices with the precursors and the fractionating of the tissue at various time intervals. (Modified from Schachter, H., *J. Biol. Chem.*, 248:974–76, 1973.)

ative study of the secretion of glycoproteins and the secretion of albumin by the liver cell revealed differences in the time of transport through the different segments of the vacuolar system. While albumin passes directly into the cisternal cavities, glycoproteins remain attached to the membranes, where they are submitted to glycosidation by the glycosyl transferases.[16]

For cytological studies using electron microscopic radioautography, ^{3}H-fucose is preferred because it does not enter into mucopolysaccharides, and in glycoproteins fucose is present only in the terminal group. In intestinal cells, two minutes after the injection of ^{3}H-fucose the labeling is almost exclusively in the Golgi complex. Then there is migration of the glycoproteins to the sides of the cell and into the apical cell membranes, where it is concentrated four hours after injection. The intracellular migration of the glycoproteins is carried out by small vesicles formed in the Golgi region.[17]

The discovery of various lipid-sugar compounds (i.e., dolichol phosphates) in animal tissues has raised the possibility that these substances could serve as intermediates in the synthesis of glycoproteins.[18] These compounds might be active in the Golgi complex of the cell, serving as carriers of the corresponding sugars.[19] In addition, the Golgi appears to be involved in the addition of sulfate to the carbohydrate moiety of the glycoproteins. In cartilage cells, mucopolysaccharides as well as glycoproteins are synthesized in the Golgi complex.[20, 21]

The diagram of Figure 11–7 is an idealized representation of the synthesis and secretion of glycoproteins and the intervention of the various segments of the vacuolar system: rough and smooth endoplasmic reticulum → Golgi membranes → secretory vesicles and plasma membrane.

In addition to the glycoproteins that are secreted and those that are incorporated into the plasma membrane are others that become incorporated into lysosomes. For example, in liver cells ^{3}H-fucose is first incorporated into the Golgi and later on is found in lysosomes, in the blood plasma, and on the cell membranes (see Leblond and Bennett, 1977).

In summary the polypeptide backbone of the glycoprotein is synthesized on the ribosomes, and side-chains of mannose and N-acetylglucosamine are added in the endoplasmic reticulum, whereas the terminal side-chains of galactose, fucose, and sialic acid are added in the Golgi complex (Fig. 11–7).

236

Cancer Cells — Changes in Glycosidation of Lipids and Proteins

The Golgi complex plays a central role in the biosynthesis of *gangliosides* and other glycosphingolipids. Several transferases involved in the glycosidation of glycosphingolipids have been found to be concentrated in the Golgi complex and to a lesser extent in the endoplasmic reticulum. As in the case of the glycoproteins the synthesis of glycosphingolipids occurs by a stepwise addition of sugars (from sugar nucleotides) to ceramide or to growing glycolipid acceptors. There is evidence that in cancer cells there are surface membrane changes that involve a loss of glycosphingolipids, and these alterations are due, in part, to the reduction of one or more glycosyltransferases present in the Golgi complex. The possibility that changes in surface glycoproteins and glycosphingolipids may be related to the development of the malignant properties of cells is at present receiving wide attention.[22, 23]

Secretion — The Main Function of the Golgi Complex

The study of cell secretion in relation to the endoplasmic reticulum and the Golgi system is justified at this point because these organelles are the ones more directly involved in the synthesis, transport, and release of macromolecules from the cell. The relationship between the Golgi complex and cell secretion was postulated by Cajal in 1914 in his study of goblet cells. Cell secretion is not confined to animal cells alone; in fact plant cells secrete polysaccharides and proteins to make their cell walls. Furthermore, the enzymes present *in lysosomes* and *peroxisomes* of both animal and plant cells are produced by a kind of secretory process.

Generalizing the concept of cell secretion, it may be said that such activity appears even in the prokaryotic cells, in relation to the production of the cell wall of bacteria and to the secretion of a variety of enzymes to the medium. In certain protozoa, vacuoles similar to the Golgi complex are found, which by their contraction expel water into the medium.

The Secretory Cycle — Continuous or Discontinuous

Secretion involves a continuous change that can be best interpreted by studying the cell

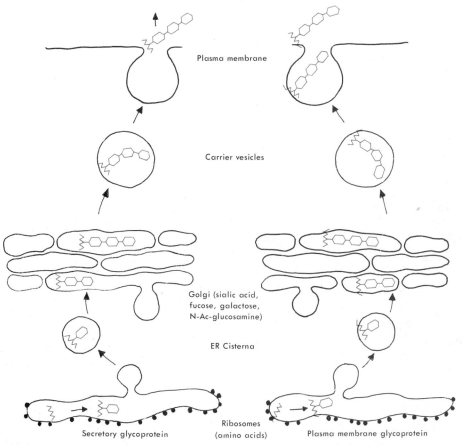

Plasma membrane

Carrier vesicles

Golgi (sialic acid, fucose, galactose, N-Ac-glucosamine)

ER Cisterna

Secretory glycoprotein

Ribosomes (amino acids)

Plasma membrane glycoprotein

Figure 11–7 Diagram comparing possible mechanisms for the elaboration of secretory glycoproteins (**left**) and plasma membrane glycoproteins (**right**). Both proteins are first synthesized by ribosomes. The secretory proteins are released into the lumen, while the plasma membrane glycoproteins remain inserted in the wall of the ER and are transported by membrane flow. The various oligosaccharides are added in a sequential manner in the ER and the Golgi apparatus (see Table 11–4). The glycoproteins are transported from the Golgi by carrier vesicles and released by exocytosis at the plasma membrane. (Modified from Schachter, 1974. Courtesy of C. P. Leblond and G. Bennett. From Leblond, C. P., and Bennett, G., Role of the Golgi apparatus in terminal glycosylation, *in International Cell Biology*, p. 326. Brinkley, B. R., and Porter, K. R., eds. New York, Rockefeller University Press, 1977.)

throughout the different stages of cellular activity. In some secretory cells secretion is continuous; the product is discharged as soon as it is elaborated. In these cells all the phases of the secretory process take place simultaneously.

As was already mentioned, in the secretion of glycoprotein by the liver cell and in the case of the antibodies produced by the plasma cells (Fig. 11–4), the secretion produced is not accumulated into special storage granules, and the release of secretory materials is more or less simultaneous with the synthesis and intracellular transport of these substances.

In other cells the secretory cycle is *discontinuous*: it is specially timed so that the synthesis

and intracellular transport are followed by the accumulation of the secretion product in special storage granules which are finally released to the extracellular space. In these cells the secretory cycle has extremely variable cytologic expressions, but it is generally characterized by products, visible with the microscope, which accumulate in the cell, and then are ultimately eliminated. These products may be dense and refractile granules, vacuoles, droplets, or other structures having a definite location in the cell and, at times, characteristic histochemical reactions. The dense secretory granules containing enzymes, generally in an inactive form (proenzyme), are called the *zymogen granules* (Fig. 11–8, *A*).

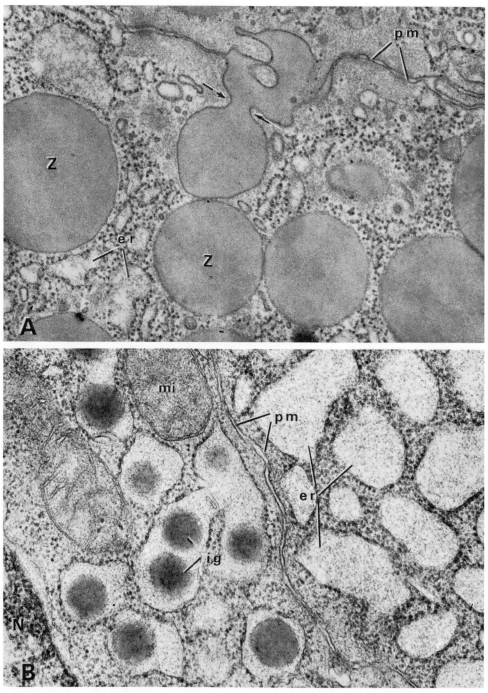

Figure 11–8   **A,** apical region of an acinar cell from the pancreas of a guinea pig showing zymogen granules (Z), one of which is being expelled into the lumen by exocytosis followed by membrane fusion; *er,* granular endoplasmic reticulum; *pm,* plasma membrane. ×30,000. (Courtesy of G. E. Palade.) **B,** the same as above, but from the basal portion showing the enlarged cisternae of the endoplasmic reticulum (*er*), some of which contain intracisternal granules (*ig*); *mi,* mitochondria; *N,* nucleus; *pm,* plasma membrane. ×30,000. (Courtesy of D. Zambrano.)

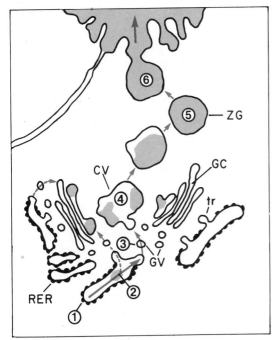

Figure 11–9 Diagram of the processing of secretory proteins in a typical glandular cell. Steps 1 through 6 are described in the text. *RER,* rough endoplasmic reticulum; *tr,* transitional vesicles; *GC,* Golgi cisternae; *CV,* condensing vacuole; *ZG,* zymogen granules. (Courtesy of J. D. Jamieson and G. E. Palade. From Jamieson, J. D., and Palade, G. E., Production of secretory proteins in animal cells, *in International Cell Biology,* p. 308. Brinkley, B. R., and Porter, K. R., eds. New York, Rockefeller University Press, 1977.)

The Secretory Process in the Pancreas — Six Consecutive Steps

In a glandular exocrine cell, such as that of the pancreas, in which the secretory process has been more thoroughly studied, the following six steps that involve the endomembrane system may be recognized (Fig. 11–9) (see Jamieson and Palade, 1977):

Step 1 or The Ribosomal Stage. The initial stage involves the *synthesis of proteins* in direct contact with the polysomes attached to the rough endoplasmic reticulum. This step, which is actually begun in the cytosol, was studied in the previous chapter (for protein synthesis, see also Chapter 24–2).

Step 2 or The Cisternal Stage. This corresponds to the *vectorial transport* of synthesized proteins into the cisternae of the endoplasmic reticulum and was also studied in the previous chapter. The student should bear in mind here the mechanism of the binding of ribosomes to the membrane, the *signal hypothesis* with the signal codons in the mRNA, the signal pep-

tide, the ribosomal receptor proteins (probably the ribophorins), the signal peptidase, and the mechanisms by which the protein is processed and stored within the endoplasmic reticulum. The protein material appears in most cases as a dilute solution of macromolecules, but in certain conditions it may have small *intracisternal granules* (Fig. 11–8, *B*).

Step 3 or Intracellular Transport. In this stage the secretory proteins diffuse through the rough endoplasmic reticulum and enter the so-called *transitional elements,* which are located at the boundary between the rough endoplasmic reticulum and the Golgi complex. These transitional components have ribosomes attached to most of their surface except in the regions facing the forming face of the Golgi, which are smooth and bud off into vesicles similar to the Golgi peripheral vesicles.

According to the studies of Palade and coworkers these vesicles reach the next station of the secretory pathway and fuse with the large condensing vacuoles of the Golgi that are present at the maturing face.

The intracellular transport of the secretory protein has been followed cytochemically by the use of *radiolabeled precursors,* such as ³H-leucine, either injected into the animal or applied to tissue slices. Then, at specific times, the tissues are fixed and processed for radioautography.

The kinetics of the labeling of the secretory protein can be studied by the use of a *short pulse* followed by a so-called *chase*. For example, slices of pancreas are exposed first to a leucine-free medium (i.e., a saline solution containing all amino acids except leucine) which produces depletion of the leucine-endogenous pool. After a short time the tissue is submitted to a medium containing ³H-leucine for a few minutes, and this procedure is then followed by the chase, which consists of washing and incubation with a solution containing excess non-labeled leucine. In the fixed tissue, it is possible to observe that after a few minutes the isotope is localized in the endoplasmic reticulum of the basal region. Later, the newly synthesized protein passes into the Golgi complex. In this region apparently it becomes progressively concentrated into prozymogen granules or condensing vacuoles surrounded by a membrane. After a longer time, the label is found principally in the zymogen granules and in the lumen of the acinus. Figures 11–10 and 11–11 show electron microscopic radioautographs of the sequence of events in sections of pancreas previously pulse labeled with ³H-leucine and followed by a chase at different time intervals.

Quantitative data may be obtained by measur-

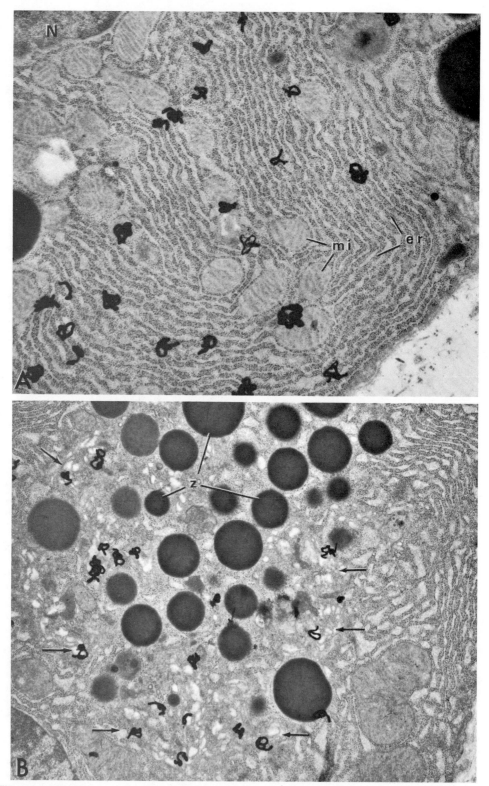

Figure 11-10   Electron microscopic radioautograph of acinar cells from the pancreas of a guinea pig. **A,** three minutes after pulse labeling with ³H-leucine. The radioautographed grains are located almost exclusively on the granular endoplasmic reticulum (*er*); *mi,* mitochondria; *N,* nucleus. ×17,000. **B,** the same as above, but incubated for seven minutes after pulse labeling. The label is now in the region of the Golgi complex (arrows); *z,* zymogen granules. ×17,000. (Courtesy of J. D. Jamieson and G. E. Palade.)

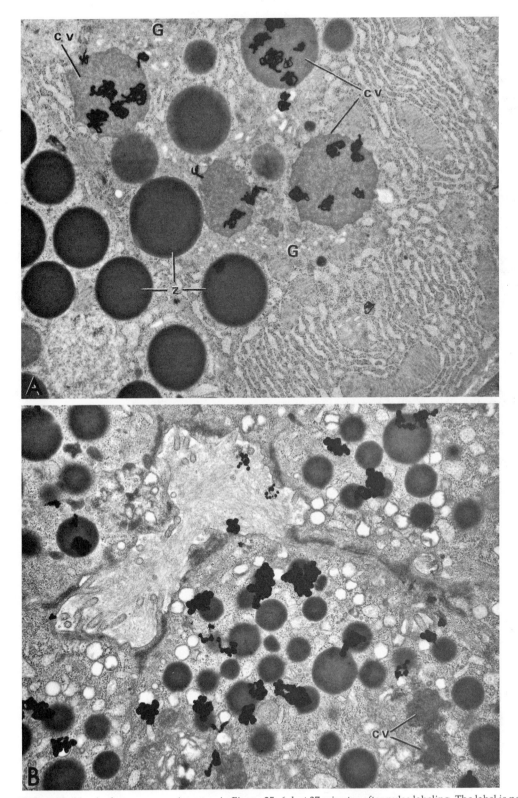

Figure 11–11 **A,** the same experiment as in Figure 25–6, but 37 minutes after pulse labeling. The label is now concentrated in the condensing vacuoles (*cv*) of the Golgi complex (*G*). The zymogen granules (*z*) are unlabeled. ×13,000. **B,** the same as above, but incubated for 117 minutes after pulse labeling. The radioautographed grains are now localized primarily over the zymogen granules, while the condensing vacuoles (*cv*) are devoid of label. Some grains are in the lumen of the acinus, indicating the secretion. ×13,000. (Courtesy of J. D. Jamieson and G. E. Palade.)

ing the number of radioautographic grains over the various cell components. Immediately after the pulse labeling there is a sudden increase in radioactivity in the rough endoplasmic reticulum with a tendency to decline rapidly. The radioactivity then becomes high in the Golgi complex, while it increases more slowly in both the immature and the mature secretory granules. Figure 11–12 readily shows that in the case of the parotid gland, there is a wave-like movement of the pulse-labeled secretory protein through the various intracellular compartments in the following order: rough endoplasmic reticulum → Golgi complex → immature granules (i.e., condensing vacuoles and prozymogen) → mature secretory (zymogen) granules.[24] In the rat pancreas the total life span of a zymogen granule has been estimated at 52.4 minutes.[25]

With small variations, the transport of the secretory protein from the endoplasmic reticulum to the Golgi is the same in all protein-producing cells, such as thyroid and parotid cells and odontoblasts, all of which have been analysed by radioautography.

Since the intracellular transport is made against an apparent concentration gradient, it is of interest to determine what its possible metabolic requirements are. The transport of the newly synthesized polypeptide chain from the ribosome into the cisternae of the endoplasmic reticulum does not require additional energy and seems to be controlled mainly by the structural relationship of the large ribosomal subunit with the membrane of the reticulum (see Chapter 10–

1). When the process of protein synthesis is inhibited by puromycin, the incomplete peptides formed are transported into the vacuolar cavity in the pancreas[26] and also in neurosecretory cells.[27] A similar study performed with another inhibitor of protein synthesis — cycloheximide — also demonstrated that the intracellular transport does not depend on the synthesis of secretory proteins and may continue even in the absence of such a synthesis.[28]

Transport from the endoplasmic reticulum to the condensing vacuole is blocked by inhibitors of cell respiration (nitrogen, cyanide, antimycin A) or of oxidative phosphorylation (dinitrophenol, oligomycin). An important conclusion to be drawn from these studies is that at the periphery of the Golgi complex, in transitional elements between the endoplasmic reticulum and the small vesicles of the Golgi complex, there is a kind of *energy-dependent lock* in the transport. Such a lock may regulate the flow of secretory proteins.

In summary, intracellular transport is arrested by the lack of ATP production. It is possible that energy is required for the processes of membrane fission and fusion that occur as the transitional elements generate transporting vesicles and as these fuse with the condensing vacuoles (see Jamieson and Palade, 1977).

Step 4 or Concentration of the Secretory Protein. During this period the condensing vacuoles are converted into zymogen granules by the progressive filling and concentration of their content, and finally acquire their characteristic electron-opaque content. This conversion is not dependent on a supply of metabolic energy, since it continues after inhibition of glycolysis or respiration. Furthermore, the membranes of the zymogen granules do not contain a Na^+-K^+-ATPase that could participate in an active transport.

These findings provide no explanation for the mechanism by which the secretory protein is condensed. However, the fact that ^{35}S-sulfate is incorporated into the Golgi elements and into the zymogen granules has suggested possible mechanisms.[29, 30] A sulfated peptidoglycan which acts as a large polyanion interacting with the basic secretory proteins has been found in the Golgi complex and condensing vacuoles. It is thought at present that the formation of osmotically inactive aggregates may result in a passive flow of water from the secretion granule to the relatively hyperosmotic cytosol.[32]

Step 5 or Intracellular Storage. The previous step culminates with the storage of the secretory product into *secretory granules*, which are re-

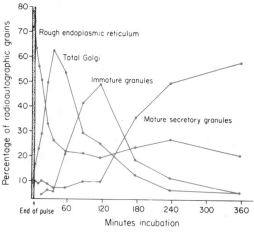

Figure 11–12 Diagram showing the radioactivity of ^3H-leucine, measured as grains under the electron microscope, as a function of time. The curves indicate the variations that occur in the various cell compartments. (See the description in the text.) (From Castle, J. D., Jamieson, J. D., and Palade, G. E., *J. Cell. Biol.*, 53:290, 1972.)

leased when the appropriate stimulus (a hormone or neurotransmitter) acts on the cell. This storage provides a mechanism by which the cell may cope with a demand for secretory product that exceeds its rate of synthesis. This mechanism of storage is utilized not only in the protein-secreting cells but also in others producing smaller molecules such as peptides and amines. For example, the *catecholamine-containing granules* of the adrenal medulla contain several proteins (the so-called chromogranins) and ATP, with which the amines are complexed.[33]

Step 6 or Exocytosis. The discharge of the secretory granule is effected by a process of exocytosis which involves its movement toward the apical region of the cell and fusion between its membrane and the luminal plasma membrane. As a result of the fusion of the membranes and the consequent fission, with elimination of the layers in between, an orifice is made through which the secretory product is discharged (Fig. 11–8, *A*).

The mechanism of exocytosis was first described in the adrenal medulla after electrical stimulation of the splanchnic nerves.[34] As shown in Figure 11–13, it was found that the cathecol-

containing granules first attach themselves to the plasma membrane, then they swell, and finally their content is evacuated, leaving empty membranes.[35] Biochemically, exocytosis requires the intracellular elevation of Ca^{2+} and the production of ATP. Thus, exocytosis is an energy-dependent process and signifies a second energy-requiring step in the entire secretory process. Presumably, part of this energy requirement is related to the process of membrane fusion-fission taking place during exocytosis, but it is possible that energy could be consumed in the propulsion of the granule to the cell apex.

A pancreatic cell in which protein synthesis has been inhibited can go through an entire secretory cycle of transport, concentration, storage, and release. This fact suggests that within the period of time of one cycle (i.e., 60 to 90 min) the synthesis of new membrane proteins, serving as containers of the secretion, is not required. In other words, intracellular membranes have a much longer half-life and are probably re-utilized extensively during the secretory process.

The membrane of the zymogen granules may

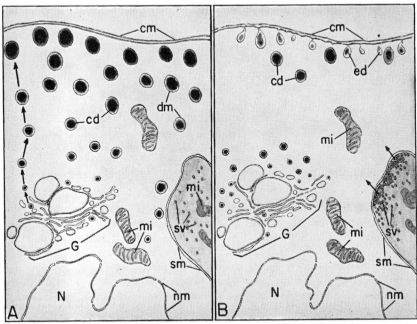

Figure 11–13 Diagrammatic interpretation of the mechanism of secretion in the chromaffin cell. **A,** cell in the resting stage, showing the storage of mature catechol droplets in the outer cytoplasm. Near the nucleus within the Golgi complex new secretion is being formed at a slow rate. At the right, a portion of a nerve terminal, showing the synaptic vesicles (*sv*) and mitochondria (*mi*); *cd*, catechol droplets; *cm*, cell membrane; *dm*, droplet membrane; *ed*, evacuated droplets; *G*, Golgi complex; *N*, nucleus; *nm*, nuclear membrane; *sm*, surface membrane. **B,** cell after strong electrical stimulation by way of the splanchnic nerve. Most of the catechol droplets have disappeared; the few that remain can be seen in different stages of excretion into the intercellular cleft. The Golgi complex is now forming new droplets at a higher rate. The nerve ending shows an increase of synaptic vesicles with accumulation at "active points" on the synaptic membrane. (From E. De Robertis and D. D. Sabatini.)

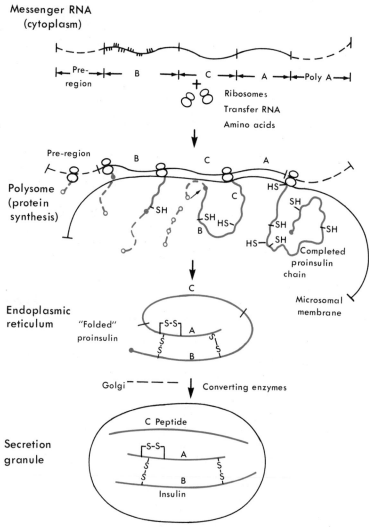

Figure 11–14 Diagram indicating steps in the molecular processing of insulin secretion. Observe that the pre-region that forms part of pre-proinsulin is separated within the endoplasmic reticulum. The converting enzyme present in the Golgi transforms the proinsulin into insulin by removal of the C peptide. (From Chan, S. J., and Steiner, D. F., *Trends Biochem. Sci.,* 2:254, 1977.)

be regarded as a vacuole that shuttles between the Golgi apparatus and the cell surface. In the removal of excess membrane from the apical region of the cell it has been suggested that patches of membranes are invaginated from the surface as small vesicles that move back into the Golgi region, to be re-utilized in the packing of more secretion. Thus, the process of exocytosis is coupled with that of endocytosis. This is supported by the finding of an increase in the number and size of Golgi components after intense in vitro stimulation of the pancreas by carbamylcholine. This increase takes place even under conditions in which protein synthesis has been almost completely inhibited by cycloheximide.

In the parotid gland stimulated by adrenergic

drugs acting on beta receptors, the zymogen granules are discharged by exocytosis in a sequential manner, i.e., in the order in which they are present at the apical region. The excess membrane that is the result of this mechanism is reduced later on by a process of microvesiculation with the formation of small apical vesicles.[36] It has been postulated that cyclic 3'-5' AMP may be implicated in the process of exocytosis. In fact, the secretion can be induced directly by the use of dibutyryl cyclic AMP.[37]

The GERL Region and Lysosome Formation

In addition to serving as the site for packaging the secretory products and providing a limiting membrane to zymogen granules, the Golgi membranes are involved in the formation of the primary lysosomes. This process appears to have the same sequence as that described above, i.e., synthesis, aggregation, transport, and packaging of the enzymes.[38] Since a number of lysosomal enzymes are glycoproteins, the Golgi has also been implicated in their glycosidation.

We mentioned earlier that in the maturing face of the Golgi is the GERL region, in which acid phosphatase, a characteristic lysosomal enzyme, is present. This region has been implicated in the formation of primary lysosomes (see Chapter 13–1). If similar transport mechanisms are used for secretory and lysosomal proteins it seems plausible that some kind of traffic control must exist to direct the various products to their final destination, i.e., to the zymogen granules or the lysosomes. This still unknown traffic regulator should be present in the Golgi complex.[39]

According to Novikoff[5] the GERL of the Golgi complex appears also to be involved in the formation of melanin granules, in the processing and packaging of secretory material in endo- and exocrine cells, and in lipid metabolism.

Insulin Biosynthesis — A Good Example of the Molecular Processing of Secretion

It is well known that many secreted proteins are first synthesized as biologically inactive precursors which are activated later. Activation consists essentially of the removal of a portion of the polypeptide chain and may occur at different sites. For example, the zymogens secreted by the pancreas are activated extracellularly, i.e., after the release of the secretion. Various polypeptide hormones are produced as inactive *prohormones* which are then activated intracellularly by the *converting* (i.e., proteolytic) *enzymes* presumably present in the Golgi apparatus.

Examples of such *proproteins* are: proalbumin, proparathyroid hormone, proglucagon, progastin, and the well-known case of *proinsulin*. The hormone insulin, produced by the beta cells of the pancreatic islands, has a molecular weight of only 12,000 daltons, and two chains: the A chain of 21 amino acids and the B chain of 30 amino acids; the chains are linked by two — S — S bonds.

In Figure 11–14 is represented the insulin mRNA with its B and A regions (cistrons) which code for the B and A chains. In addition the mRNA contains a preregion which codes for the *signal peptide* (see Chapter 10–4), a C segment, and a poly A tail. This mRNA was isolated and found to contain information coding for a polypeptide chain of 330 amino acids. The complete translation of this mRNA on free ribosomes gives rise to *pre-proinsulin*. However, as described in the previous chapter, in the endoplasmic reticulum the preregion (i.e., a signal peptide 23 amino acids long) is removed by the *signal peptidase*.[40] The completed *proinsulin chain* is then activated in the Golgi through the removal of the C peptide by the converting enzyme and is packaged as active hormone within the secretion granule (see Chan and Steiner, 1977).

A possible mechanism linking the processing of the proproteins with exocytosis has been reported for proalbumin.[41] In this case the conversion of proalbumin into albumin depends on the fusion of small Golgi vesicles containing a protease. This process requires Ca^{2+} and is inhibited by the microtubule-disrupting alkaloid colchicine, which also blocks the secretion of several hormones and prevents their conversion to the active form. The general problem of molecular processing of secretory proteins is at present of great importance and is related to the possibility that cleavage products may have different functions. For example, in pituitary cells a single precursor molecule contains the hormone ACTH (adrenocorticotrophin) and β-endorphin, a peptide with opiate-like activity.[42, 43]

SUMMARY:

Functions of the Golgi Complex

The Golgi complex may be considered as an intermediary compartment interposed between the endoplasmic reticulum and the extracellular space, through which there is a continuous traffic of substances (i.e., fluids, macromolecules, cell wall units, and other cell constituents). This traffic also involves the flow and differentiation of membranes and a rapid turnover (every 20 to 40 minutes in dictyosomes of the liver).

One of the major functions of the Golgi is the *glycosidation* of *lipids* and *proteins* to produce glycolipids and glycoproteins. The sugar residues are transferred in a sequential order by the glycosyl transferases. In the case of glycoproteins, after the synthesis of the protein backbone in the endoplasmic reticulum, the oligosaccharides of the core are added and then those of the terminal group (Table 11–4). Glycoproteins may be secreted or incorporated into the cell coat or the lysosomes. The glycosidation of lipids that leads to the synthesis of gangliosides and glycosphingolipids may be altered in cancerous cells.

The main function of the Golgi complex is *cell secretion*, not only of exportable proteins but also of the enzymes present in lysosomes and peroxisomes (Chap. 13). Secretion may be continuous, the secretion products being discharged without storage (i.e., glycoproteins by the liver cells, antibodies by the plasma cells), or it may be discontinuous, with storage in secretory or zymogen granules (e.g., in the pancreas, parotid gland, and others).

The following six steps of secretion can be recognized in the pancreas and other zymogen-secreting glands: (1) *The ribosomal stage*. The synthesis of proteins by polysomes attached to the rough endoplasmic reticulum. (2) *The cisternal stage*. This step corresponds to the vectorial transport of the proteins into the lumen of the endoplasmic reticulum. (The student should remember here the signal hypothesis described in Chapter 10–4). The secreted material may be a dilute solution, but may sometimes have intracisternal granules. (3) *Intracellular transport*. The secreted proteins enter the *transitional* tubules and vesicles that lead to the Golgi complex, where they fuse with the large *condensing vacuoles* present at the maturing face of the Golgi. The transport may be followed by the use of labeled amino acid precursors with subsequent study by radioautography or cell fractionation. In the parotid gland the total life span of a zymogen granule has been estimated to be 52.4 minutes. The transport from the endoplasmic reticulum to the condensing vacuole requires the use of energy (ATP). (4) *Concentration of the secretion*. By a process of concentration the condensing vacuole is converted into the zymogen granule. This conversion does not require energy and is probably due to the formation of osmotically inactive aggregates of sulfated peptidoglycans, with loss of water to the cytosol. (5) *Intracellular Storage*. Formation of the secretory granules. (6) *Exocytosis*. The discharge of the secretory granule involves its movement toward the apical region and fusion of the membranes (of the granule and the cell). Exocytosis requires Ca^{2+} and energy (ATP). The energy is presumably related to the process of fusion-fission of membranes.

At the apical region there is a mechanism of removal and recycling of the excess membrane. Cyclic 3'-5' AMP may be also implicated in the process of exocytosis.

The Golgi apparatus is involved in the formation of primary lysosomes. A number of lysosomal enzymes are glycoproteins and are glycosidated at the Golgi complex.

An excellent example of the molecular processing of secreted proteins is provided by the hormone insulin, which has two precursors: preproinsulin and proinsulin.

REFERENCES

1. Rambourg, A., Clermont, Y., and Marraud, A. (1974) *Am. J. Anat.*, 140:27.
2. Claude, A. (1970) *J. Cell Biol.*, 47:745.
3. Morré, D. J. (1977) In: *International Cell Biology*, p. 293. (Brinkley, B. R., and Porter, K. R., eds.) The Rockefeller University Press, New York.
4. Elder, J. H., and Morré, D. J. (1976) *J. Biol. Chem.*, 251:5054.
5. Novikoff, A. B. (1976) *Proc. Natl. Acad. Sci. USA*, 73:2781.
6. Morré, D. J., Cheetham, R., and Yunghans, W. (1968) *J. Cell Biol.*, 39:961.
7. Fleischer, B. (1969) *Fed. Proc.*, 28:404.
8. Ovtracht, L., Morré, D. J., Cheetham, R. D., and Mollenhauer, H. H. (1973) *J. Micros. (Paris)*, 18:87.
9. Franke, W. W., and Kartenbeck, J. (1976) In *Progress in Differentiation Research*, p. 213. (Müller-Berat M., ed.) American Elsevier Pub. Co., New York.
10. Morré, J., Keenan, T. W., and Mollenhauer, H. H. (1971) *Adv. Cytopharmacol.*, 1:159.
11. Farquhar, M. G., Bergeron, J. J. M., and Palade, G. E. (1974) *J. Cell Biol.*, 60:8.
12. Schnepf, E., and Kock, W. (1966) *Z. Pflanzenphysiol.*, 55:97.
13. Brown, R. M. (1969) *J. Cell Biol.*, 41:109.
14. Neutra, M., and Leblond, C. P. (1966) *J. Cell Biol.*, 30:119, and 137.
15. Schachter, H. (1974) *Biochem. Soc. Symp.*, 40:57
16. Redman, C. M. and Cherian, M. G. (1972) *J. Cell Biol.*, 52:231.
17. Bennett, G., and Leblond, C. P. (1971) *J. Cell Biol.*, 51:875.
18. Leloir, L. F. (1971) *Science*, 172:1299.
19. Lucas, J. J., Wachter, C. J., and Lennarz, W. J. (1975) *J. Biol. Chem.*, 250:1992.

20. Neutra, M., and Leblond, C. P. (1969) *Sci. Am.* 220:100.
21. Young, R. W. (1973) *J. Cell Biol.,* 57:175.
22. Richardson, C. L., Baker, S. R., Morré, D. J., and Keenan, T. W. (1975) *Biochim. Biophys. Acta,* 417:175.
23. Cook, G. M. W. (1975) The Golgi Apparatus, Oxford Biology Reader, Vol. 77, p. 1, Oxford University Press, Oxford.
24. Castle, J. D., Jamieson, J. J., and Palade, G. E. (1972) *J. Cell Biol.,* 53:290.
25. Warshawsky, H., Leblond, C. P., and Droz, B. (1973) *J. Cell Biol.,* 16:1.
26. Redman, C. M., and Sabatini, D. D. (1966) *Proc. Natl. Acad. Sci. USA,* 56:608.
27. Zambrano, D., and De Robertis, E. (1968) *Z. Zellforsch.,* 76:458.
28. Jamieson, J. D.,and Palade, G. E. (1971) *J. Cell Biol.,* 50:135.
29. Berg, N. B., and Young, R. W. (1971) *J. Cell Biol.,* 50:469.
30. Reggio, H., and Palade, G. E. (1976) *J. Cell Biol.,* 70:360.
31. Tartakoff, A. M., Greene, L. J., and Palade, G. E. (1974) *J. Biol. Chem.,* 249:7420.
32. Reggio, H. A., and Palade, G. E. (1978) *J. Cell. Biol.,* 77:288.
33. Blaschko, H., Smith, A. D., Winkler, H., Van der Bosch, H., and Van Deenen, L. M. (1967) *Biochem. J.,* 103:300.
34. De Robertis, E., and Vaz Ferreira, A., (1957) *J. Biophys. Biochem. Cytol.,* 3:611.
35. De Robertis, E., and Sabatini, D. D. (1960) *Fed. Proc.,* 19:70.
36. Herzog, V., and Farquhar, M. G., (1977) *Proc. Natl. Acad. Sci.,* 74:5073.
37. Schramm, M. Selinger, Z., Salomon, Y., Eytan, E., and Batzri, S. (1972) *Nature,* 240:203.
38. Baiton, D. F., Nichols, B. A., and Farqhuar, M. G. (1976) In: *Lysosomes in Biology and Pathology,* Vol. 5, p. 3. (Dingle, J. T., and Dean, R. T., eds.) North-Holland Pub. Co., Amsterdam.
39. Palade, G. E. (1975) *Science,* 189:347.
40. Steiner, D. F. (1977) *Diabetes,* 26:322.
41. Judah, J. D., and Quin, P. S. (1978) *Nature,* 271:384.
42. Mains, R. E., Eipper, B. A., Ling, N. (1977) *Proc. Natl. Acad. Sci.,* 74:3014.
43. Mains, R. E., and Eipper, B. A., (1978) *J. Biol. Chem.,* 253:651.

ADDITIONAL READING

Beams, H. W., and Kessel, R. G. (1968) The Golgi apparatus, structure, and function. *Int. Rev. Cytol.,* 23:209.

Bennett, G., and Leblond, C. P. (1977) Biosynthesis of the glycoprotein present in plasma membrane, lysosomes, and secretory materials, as visualized by radioautography. *Histochem. J.,* 393:417.

Brown, R. M., Jr., and Willison, J. H. M. (1977) Golgi apparatus and plasma membrane involvement in secretion and cell surface deposition. In *Intern. Cell Biol.,* p. 267, (Brinkley, B. R., and Porter, K. R., eds.) The Rockefeller University Press, New York.

Case, R. M. (1978) Synthesis, intracellular transport and discharge of exportable proteins in the pancreatic acinar cell and other cells. *Biol. Rev.,* 53:169.

Chan, S. J., and Steiner, D. F. (1977) Preproinsulin, an new precursor in insulin biosynthesis. *Trends Biochem. Sci.,* 2:254.

Cook, G. M. W. (1975) *The Golgi apparatus. Oxford Biology Reader.,* Vol. 77, p. 1.

Fleischer, B., Zambrano, F., and Fleischer, S. (1974) Biochemical characterization of the Golgi complex of mammalian cells. *J. Supramol. Struct.,* 2:737.

Ito, A., and Palade, G. E. (1978) Presence of NADPH-cytochrome P450 reductase in rat liver Golgi membranes. *J. Cell Biol.,* 79:590.

Jamieson, J. D., and Palade, G. E. (1977) Production of secretory proteins in animal cells. In: *International Cell Biology,* p. 308. (Brinkley, B. R., Porter, K. R., eds.) The Rockefeller University Press, New York.

Kraehenbuhl, J. P., Racine, L., and Jamieson, J. D. (1977) Immunocytochemical localization of secretory proteins in bovine pancreatic exocrine cells. *J. Cell Biol.,* 72:406.

Leblond, C. P., and Bennett, G. (1977) Role of the Golgi apparatus in terminal glycosylation. In: *International Cell Biology,* p. 326. (Brinkley, B. R., Porter, K. R., eds.) The Rockefeller University Press, New York.

Leloir, L. F. (1971) Two decades of research on the biosynthesis of saccharides. *Science,* 172:1299.

Morré, D. J. (1975) Membrane biogenesis. *Annu. Rev. Plant Physiol.,* 26:441.

Morré, D. J. (1977) Membrane differentiation and the control of secretion: A comparison of plant and animal Golgi apparatus. In: *International Cell Biology,* p. 293. (Brinkley, B. R., Porter, K. R., eds.) The Rockefeller University Press, New York.

Morré, D. J. (1977) The Golgi apparatus and membrane biogenesis. In: *Cell Surface Reviews.* Poste, G., and Nicholson, G. L., eds. Elsevier, Amsterdam, in press.

Morré, D. J., and Mollenhauer, H. H. (1974) The endomembrane concept: A functional integration of endoplasmic reticulum and Golgi apparatus. In: *Dynamic Aspects of Plant Ultrastructure,* p. 84. Robards, A. W., ed. McGraw-Hill Book Co., New York.

Neutra, M., and Leblond, C. P.,(1969) The Golgi apparatus. *Sci. Am., 220*:100.

Northcote, D. H. (1971) The Golgi apparatus. *Endeavour, 30*:26.

Northcote, D. H. (1973) The Golgi complex. In *Cell Biology in Medicine.* (Bittar, E. E., ed.) John Wiley & Sons, New York.

Palade, G. E. (1975) Intracellular aspects of the process of protein secretion. *Science, 189*:347.

Parodi, A., and Leloir, L. (1976) Lipid intermediates in protein glycosidation. *Trends Biochem Sci., 1*:58.

Reggio, H. A., and Palade, G. E. (1978) Sulfated compounds in the zymogen granules of the guinea pig pancreas. *J. Cell. Biol., 77*:288.

Schachter, H. (1974) The subcellular sites of glycosylation. *Biochem. Soc. Symp., 40*:57.

Tartakoff, A. M., Greene, L. J., Jamieson, J. D., and Palade, G. E. (1974) Parallelism of processing of pancreatic proteins. In: *Advances in Cytopharmacology,* Vol. 2, p. 177. (Clementi, F., and Ceccarelli, B., eds.) Raven Press, New York.

Whaley, W. G. (1975) The Golgi apparatus. In: *Cell Biology Monographs,* Vol. 2, p. 1. Springer-Verlag, Berlin.

Winkler, H. (1977) The biogenesis of adrenal chromaffin granules. *Neuroscience, 2*:657.

12

MITOCHONDRIA AND OXIDATIVE PHOSPHORYLATION

12–1 Morphology of Mitochondria 250
Summary: *Mitochondrial Morphology*
12–2 Mitochondrial Structure 252
*The Mitochondrial Matrix Contains
Ribosomes and a Circular DNA*
*F_1 Particles—On the M Side of the Inner
Mitochondrial Membrane*
*Mitochondrial Structural Variations in
Different Cell Types*
*Mitochondrial Sensitivity to Cell Injury and
Resultant Degeneration*
Summary: *Structure of Mitochondria.*
12–3 Isolation of Mitochondrial Membranes
259
*Outer and Inner Membranes—Structural
and Chemical Differences*
*Mitochondrial Enzymes are Highly
Compartmentalized*
*The Inner Membrane—Regional Structural
and Enzymatic Differences*
Summary: *Isolation of Mitochondrial
Membranes*
12–4 Molecular Organization and Function
of Mitochondria 263
*Electrons Flow Along a Cytochrome Chain
Having a Gradient of Redox Potential*
*The Respiratory Chain—Four Molecular
Complexes*
*Electron Transport—Coupled to the
Phosphorylation at Three Points*
*Mitochondrial ATPase—Structurally
Complex Proton Pump*

*Topological Organization of the Respiratory
Chain and Phosphorylation System*
*The Chemiosmotic Hypothesis—An Electro-
chemical Link Between Respiration and
Phosphorylation*
*The Chemical-Conformational Hypothesis
Involves Short-Range Interactions*
Summary: *Molecular Organization and
Function*
12–5 Permeability of Mitochondria 271
*ADP, ATP, and Pi—Transport by Specific
Carriers*
*Mitochondrial Conformation—Changes
with Stages of Oxidative Phosphorylation*
*Mitochondrial Swelling and Contraction—
Agent Induced*
*Mitochondrial Accumulation of Ca^{2+} and
Phosphate*
Summary: *Mitochondrial Permeability*
12–6 Biogenesis of Mitochondria 275
*Mitochondria are Semiautonomous
Organelles*
*Mitochondrial DNA—Circular and
Localized in the Matrix*
*Mitochondrial Ribosomes—Smaller than
Cytoplasmic Ribosomes*
*Mitochondria Synthesize Mainly Hydro-
phobic Proteins*
*The Symbiont Hypothesis—Mitochondria
and Chloroplasts are Intracellular Pro-
karyotic Parasites*
Summary: *Mitochondrial Biogenesis*

Mitochondria (Gr., *mito-*, thread + *chondrion*, granule), granular or filamentous organelles present in the cytoplasm of protozoa and animal and plant cells, are characterized by a series of morphologic, biochemical, and functional properties. Among these are their size and shape, visibility in vivo, special staining properties, specific structural organization, lipoprotein composition, and content of a large "battery" of enzymes and coenzymes that interact to produce cellular energy transformations. From the physiologic viewpoint, mitochondria are energy-transducing systems that recover the energy contained in footstuffs (through the Krebs cycle and the respiratory chain) and convert it by phosphorylation into the high energy phosphate bond of adenosine triphosphate (ATP) (see Chapter 6). Thus, mitochondria are the "power plants" that produce the energy necessary for many cellular functions (Fig. 12–1).

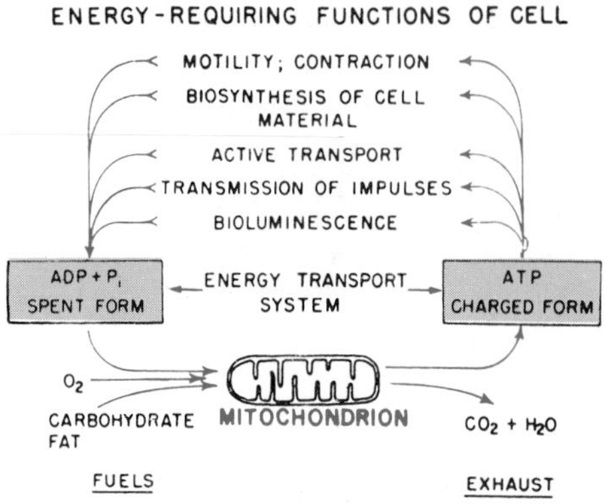

ENERGY-REQUIRING FUNCTIONS OF CELL

MOTILITY; CONTRACTION

BIOSYNTHESIS OF CELL MATERIAL

ACTIVE TRANSPORT

TRANSMISSION OF IMPULSES

BIOLUMINESCENCE

| ADP + P$_i$ SPENT FORM | ENERGY TRANSPORT SYSTEM | ATP CHARGED FORM |

O_2

CARBOHYDRATE FAT   MITOCHONDRION $CO_2 + H_2O$

FUELS   EXHAUST

Figure 12-1 Diagram showing that the mito-chondrion constitutes the central "power plant" of the cell. The adenosine triphosphate (ATP) produced is used in the various functions that are indicated. (From Lehninger, A. L., *Physiol. Rev.*, 42:3, 467, 1962.)

First observed at the end of the nineteenth century and described as "bioblasts" by Alt-mann (1894), these structures were called "mi-tochondria" by Benda (1897). Altmann predicted the relationship between mitochondria and cel-lular oxidation, and Warburg (1913) observed that respiratory enzymes were associated with cytoplasmic particles.

A most important advance was the first isola-tion of liver mitochondria by Bensley and Hoerr in 1934. This established the possibility of a direct study of the organelles by biochemical methods. The final demonstration that the mi-tochondrion was indeed the site of cellular re-spiration was made in 1948 by Hogeboom et al.[1] In recent years important advances in the study of its ultrastructural organization have been made with the aid of the electron microscope. The study of this organelle is particularly thrill-ing because the mitochondrion is one of the best known examples of structural-functional inte-gration within the cell. Lately, more progress has been made, with the demonstration that mi-tochondria contain a specific type of DNA, that they have their own machinery for protein syn-thesis, and that they may participate in inheri-tance and differentiation.

12-1 MORPHOLOGY OF MITOCHONDRIA

Although the examination of mitochondria in living cells is somewhat difficult because of their low refractive index, they can be observed easily in cells cultured in vitro, particularly under darkfield illumination and phase contrast (Fig.

10-1, *A*). This examination has been greatly fa-cilitated by coloration with a dilute solution of *Janus green*. The resultant greenish blue stain is due to the action of the cytochrome oxidase sys-tem present in mitochondria, which maintains the vital dye in its oxidized (colored) form. In the surrounding cytoplasm the dye is reduced to a colorless leukobase.

Fluorescent dyes, which are more sensitive, have been used in isolated mitochondria and recently in intact cultured cells. Such dyes are suitable for metabolic studies of mitochondria in situ.[2]

Micromanipulation has demonstrated that mi-tochondria are relatively stable and can be dis-placed by the microneedle without alterations. The *specific gravity* is greater than that of the cytoplasm. By ultracentrifugation of living cells at 200,000 to 400,000 g, mitochondria are depos-ited intact at the centrifugal pole (Fig. 2–5).

Observation in vivo is of particular interest when supplemented by time-lapse cinematogra-phy. In cultured fibroblasts, continuous and sometimes rhythmic changes in volume, shape, and distribution of mitochondria can be ob-served.

The two main types of mitochondrial motion are *displacement* from one part of the cell to an-other, mainly by cytoplasmic streaming, and *bending* and *stretching cycles*. It has been found that the zones of bending are constant and corre-spond to regions in which the mitochondrial crests are vesicular (see later discussion).[3]

Active changes in volume and shape of mi-tochondria may be caused by chemical, osmotic, and mechanochemical changes. In living cells, low amplitude contraction cycles associated with oxidative phosphorylation have been observed. The property of swelling and contraction is best studied in isolated mitochondria and will be considered again in the discussion of the physi-ology of mitochondria.

Shape. The *shape* of mitochondria is vari-able, but in general these organelles are *filamen-tous* or *granular* (Fig. 12–2). During certain func-tional stages, other derived forms may be seen. For example, a long mitochondrion may swell at one end to assume the form of *a club* or be hollowed out to take the form of a *tennis racket*. At other times mitochondria may become *vesi-cular* by the appearance of a central clear zone. The morphology of mitochondria varies from one cell to another, but it is more or less constant in cells of a similar type or in those performing the same function.

Size. The *size* of mitochondria is also vari-able; however, in most cells the width is rela-

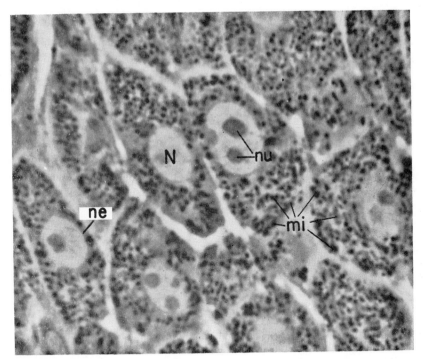

Figure 12–2 Liver cells of the rat fixed at −180° C and dehydrated in acetone at −40° C. Photomicrograph made with phase contrast microscope in a medium of n = 1.460. *mi,* mitochondria; *N,* nucleus; *ne,* nuclear envelope, *nu,* nucleoli. (Courtesy of S. Koulish.)

tively constant (about 0.5 μm), and the length is variable, reaching a maximum of 7μm. Depending on the functional stage of the cell, however, it is possible to find rods that are either thinner or thicker. The size and shape of the fixed mitochondria depend also on the osmotic pressure and the pH of the fixative. Using buffered fixatives at physiologic pH has been shown to be important.

Distribution. Mitochondria are, in general, uniformly distributed throughout the cytoplasm, but there are many exceptions to this rule. In some cases, they accumulate preferentially around the nucleus or in the peripheral cytoplasm. Overloading with inclusions, such as glycogen and fat, displaces these organelles. During mitosis, mitochondria are concentrated near the spindle, and upon division of the cell they are distributed in approximately equal number between the daughter cells.

The distribution of mitochondria within the cytoplasm should be considered in relation to their function as energy suppliers. In some cells they can move freely, carrying ATP where needed, but in others they are located permanently near the region of the cell where presumably more energy is needed. For example, in certain muscle cells (e.g., diaphragm), mitochondria are grouped like rings or braces around the I-band of the myofibril. In the rod and cone cells of the retina all mitochondria are located in a portion of the inner segment. The basal mitochondria of the kidney tubule are intimately related to the infoldings of the plasma membrane in this region of the cell. It is assumed that this close relationship with the membrane is related to the supply of energy for the active transport of water and solutes.

Orientation. Mitochondria may have a more or less definite orientation. For example, in cylindrical cells they are generally oriented in the basal-apical direction, parallel to the main axis. In leukocytes, mitochondria are arranged radially with respect to the centrioles. It has been suggested that these orientations depend upon the direction of the diffusion currents within cells and are related to the organization of the cytoplasmic matrix.

Number. Mitochondria are found in the cytoplasm of all eukaryotic cells but not in prokaryotes. In bacteria the respiratory enzymes are located in the plasma membrane (Fig. 12–1). The mitochondrial content of a cell is difficult to determine, but, in general, it varies with the cell type and functional stage. It is estimated that in liver, mitochondria constitute 30 to 35 per cent of the total protein content of the cell, and in kidney, 20 per cent. In yeast, ultrastructural evidence, based on serial sections, points to the existence of a single branched organelle per cell (see Hoffman and Avers, 1973).

A normal liver cell contains between 1000 and 1600 mitochondria (Table 12–1), but this number diminishes during regeneration and also in cancerous tissue.[4] This last observation may be related to decreased oxidation that accompanies the increase to anaerobic glycolysis in cancer. Another interesting finding is that there is an increase in the number of mitochondria in the muscle after repeated administration of the thyroid hormone, thyroxin. An increased number of mitochondria has also been found in human hyperthyroidism.

Some oöcytes contain as many as 300,000 mitochondria — the largest number recorded for a cell. There are fewer mitochondria in green plant cells than in animal cells, since some of their functions are taken over by chloroplasts.

TABLE 12–1 MEASUREMENTS IN RAT LIVER MITOCHONDRIA*

	Peripheral Cells	Midzonal Cells	Central Cells
Cytoplasmic volume (%)	19.8	19.1	12.9
Number per cell	1060.0	1300.0	1600.0
Diameter (μm)	0.56	0.47	0.32
Length (μm)	3.85	4.32	5.04

*From Loud, A. V., J. Cell Biol., 37:27, 1968.

SUMMARY:

Mitochondrial Morphology

Mitochondria are organelles present in the cytoplasm of all eukaryotic cells. They provide an energy-transducing system by which the chemical energy contained in foodstuffs is converted, by oxidative phosphorylation, into high-energy phosphate bonds (ATP). Mitochondria may be observed in the living cell; visibility is increased by the vital stain, *Janus green*. They display passive and active motion and show changes in volume and shape that are related to their function. Swelling of mitochondria can be induced by Ca^{2+}, various hormones, and certain drugs.

The morphology of mitochondria is best studied after fixation. In general, they are rod-shaped, with a diameter of about 0.5 μm and a variable length that may range up to 7 μm. There are 1000 to 1600 mitochondria in a liver cell and 300,000 in some oöcytes. Green plants contain fewer mitochondria than animal cells. The distribution of mitochondria may be related to their function as suppliers of energy. Their orientation in the cell may be influenced by the organization of the cytoplasmic matrix and vacuolar system.

12–2 MITOCHONDRIAL STRUCTURE

As indicated in Figure 12–3, a mitochondrion consists of two membranes and two compartments, the larger of which contains the *mitochondrial matrix*. An outer limiting membrane, about 6 nm thick, surrounds the mitochondrion. Within this membrane, and separated from it by a space of about 6 to 8 nm, is an inner membrane that projects into the mitochondrial cavity complex infoldings called *mitochondrial crests*. This inner membrane, also about 6 nm thick, divides the mitochondrion into two chambers or compartments: (1) the outer chamber contained between the two membranes and in the core of the crests and (2) the inner chamber, bound by the inner membrane. This inner chamber is filled with a relatively dense proteinaceous material usually called the *mitochondrial matrix*. This is generally homogeneous, but in some cases it may contain a finely filamentous material, or small, highly dense granules (see Figure 12–4). These granules are now considered to be sites for binding divalent cations, particularly Mg^{2+} and Ca^{2+}. The mitochondrial crests that project from the inner membrane are, in general, incomplete septa or ridges that do not interrupt the continuity of the inner chamber; thus, the matrix is continuous within the mitochondrion.

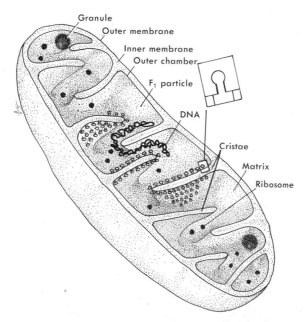

Figure 12–3 Three-dimensional diagram of a mitochondrion cut longitudinally. The main features are shown. Observe that the cristae are folds of the inner membrane and that on their matrix side they have the F_1 particles. The inset shows an F_1 particle with the head piece and stalk.

The Mitochondrial Matrix Contains Ribosomes and a Circular DNA

Within the mitochondrial matrix are small ribosomes (Fig. 12–4) and a circular DNA (see later discussion). As shown in Fig. 12–5 the mitochondrial membranes may show the *unit membrane structure* which was mentioned in our earlier discussion of the plasma membrane (see Chapter 8–2).[5] The outer and inner membranes and the crests can be considered to be fluid molecular films with a compact molecular structure; the matrix is gel-like and contains a high concentration of soluble proteins and smaller molecules. This double (solid-liquid) structure is important in providing an explanation for some of the mechanical properties of mitochondria (e.g., deformation and swelling under physiologic or experimental conditions).

F_1 Particles — On the M Side of the Inner Mitochondrial Membrane

The use of negative staining has enabled recognition of other details of mitochondrial structure. If a mitochondrion is allowed to swell and break in a hypotonic solution and is then immersed in phosphotungstate, the inner membrane in the crest appears covered by particles of 8.5 nm that have a stem linking each with the membrane (Fig. 12–6). These so-called "elementary" or "F_1" particles,[6] are regularly spaced at intervals of 10 nm on the inner surface of these membranes. According to some estimates, there are between 10^4 and 10^5 elementary particles per mitochondrion. These particles correspond to a special ATPase involved in the coupling of oxidation and phosphorylation.[7]

The presence of such F_1 particles on the matrix side (M side) confers to the inner mitochondrial membrane a characteristic asymmetry which is of fundamental importance to its function (see later discussion). It is interesting to recall here that bacterial plasma membranes contain F_1 particles with a similar asymmetric distribution (i.e., toward the bacterial matrix). On the other hand, the opposite orientation is found in chloroplasts (see Chapter 14–4), where there are F_1 particles on the outer or stromal side of the thylakoid membrane. As will be shown later, the asymmetric location of the F_1 particles in the membranes is related to the direction of the proton pump (i.e., the flow of H^+).

253

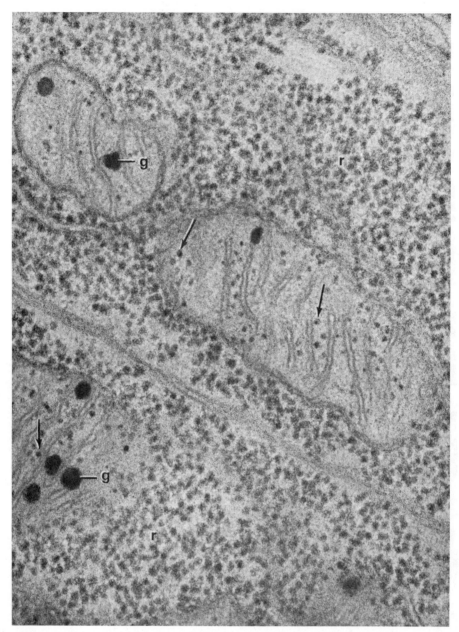

Figure 12–4 Electron micrograph of the intestinal epithelium showing a large accumulation of ribosomes (r) in the cytoplasm. In mitochondria, dense granules (g) and ribosomes (arrows) are observed. ×95,000. (Courtesy of G. E. Palade.)

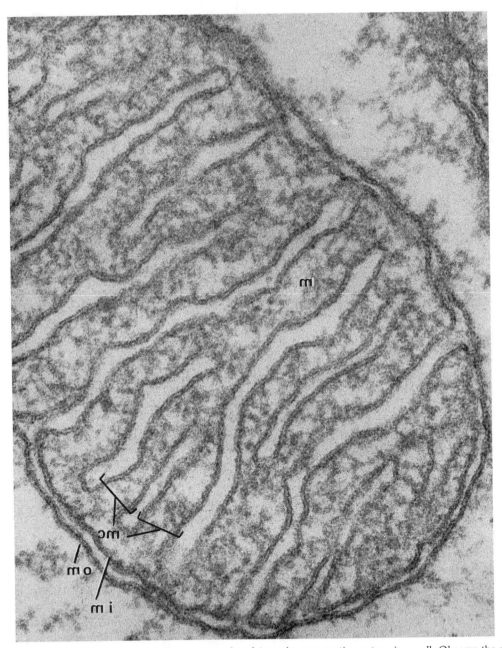

Figure 12–5 Electron micrograph of a mitochondrion of a pancreatic centroacinar cell. Observe the unit membrane structure in the outer membrane (*om*), the inner membrane (*im*), and mitochondrial crests (*mc*). *m*, matrix. ×207,000. (Courtesy of G. E. Palade.)

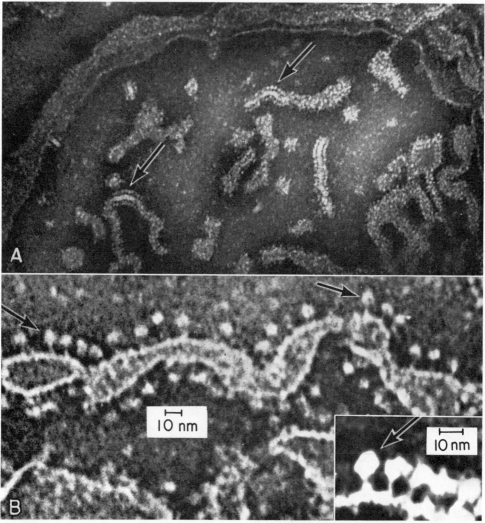

Figure 12–6   Electron micrograph of a mitochondrion swollen in a hypotonic solution and negatively stained with phosphotungstate. **A,** at low power; isolated crests can be observed in the middle of the swollen matrix. Arrows point to some of these crests. **B,** at higher magnification (×500,000), a mitochondrial crest showing the so-called ''elementary particles'' on the surface adjacent to the matrix. Inset at ×650,000, showing the elementary particles with a polygonal shape and the fine attachment to the crest. (Courtesy of H. Fernández-Morán.)

Mitochondrial Structural Variations in Different Cell Types

Some investigators assume that a common pattern of mitochondrial structure developed at an early stage of evolution and was subsequently transmitted, without considerable modifications, from protozoa to mammals and from algae to flowering plants. Detailed structural variations can be observed, however. The crests (Fig. 12–7) may be arranged longitudinally (e.g., in nerve and striated muscle) or they may be simple or branched, forming complex networks. In protozoa, insects, and adrenal cells of the glomerular zone, the infoldings

may be tubular instead of lamellar, and the tubules may be packed in a regular fashion (Fig. 12–7).

The number of crests per unit volume of a mitochondrion is also variable. Mitochondria in liver and germinal cells have few crests and an abundant matrix, whereas those in certain muscle cells have numerous crests and little matrix. In some cases the crests are so numerous that they may have a quasi-crystalline disposition. The greatest concentration of crests is found in the flight muscle of insects. In general, there seems to be a correlation between the number of crests and the oxidative activity of the mitochondrion.

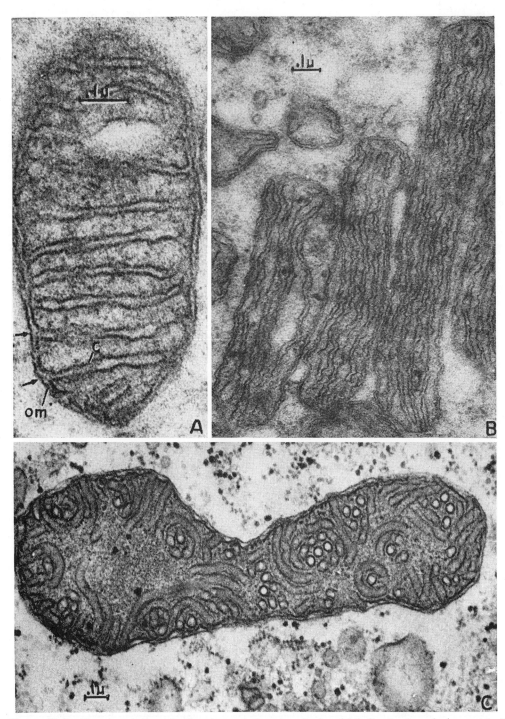

Figure 12–7 Electron micrographs showing variations in mitochondrial ultrastructure. **A,** from rat testicle; *c,* transverse crests; *om,* outer membrane; arrows show origin of crests at the inner membrane. **B,** from ovotestis of *Helix.* Longitudinal crests in mitochondria of spermatocytes. **C,** from *Paramecium,* tubular crests. **A,** ×130,000; **B,** ×68,000; **C,** ×60,000. (Courtesy of J. Andrè.)

A particularly interesting variation in fine structure is observed in cells of the different regions of the adrenal cortex (Fig. 12–8). One characteristic of these mitochondria is the enlargement of the space within the crests or tubules which, according to the diagram in Figure 12–3, corresponds to the outer compartment of the mitochondrion. This is very conspicuous and appears as tubular or vesicular openings, which are much less opaque than the inner matrix. These structures seem to be related to the specific activity of the gland. In this case, mitochondria, in addition to functioning in cell oxidations, are actively engaged in the synthesis of steroid hormones.

Relationship to Lipids. Various authors since the time of Altmann have observed that the *disposition of lipids* may be related to mitochondrial activity. In pancreas and liver cells, for example, after a short period of starvation,

the mitochondria come into contact with lipid droplets. The relationship may be so tight that only the inner mitochondrial membrane can be seen adjacent to the lipid in some regions. The electron microscopic images suggest that an active process of fat utilization takes place under the action of the fatty acid oxidases present in mitochondria.[9]

Intramitochondrial Inclusions. An accumulation of pigment derived from hemoglobin has been observed in mitochondria of amphibians. Ferritin molecules accumulate within mitochondria in subjects suffering from Cooley's hereditary anemia. Another example is the transformation of mitochondria into yolk bodies in eggs of the mollusks. In these, the masses of protein molecules may assume a regular crystalline disposition. In amphibian oöcytes, hexagonal, crystalline yolk bodies also form within mitochondria.

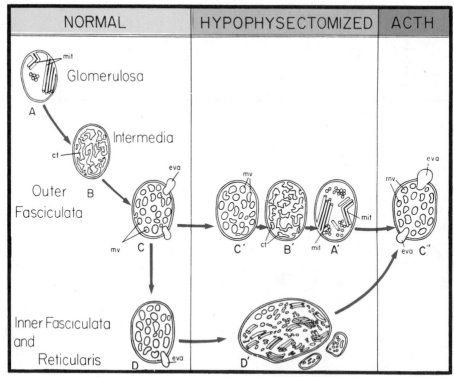

Figure 12–8 General diagram of the mitochondrial changes found in adrenal cortex after hypophysectomy and restorational therapy. *A, B, C,* and *D,* normal mitochondria at the glomerulosa, intermedia, outer fasciculata, inner fasciculata, and reticularis. *C', B',* and *A',* progressive changes of mitochondria of the outer fasciculata, leading to a pattern similar to that of glomerular cells (*A*). (Notice the lack of extruding vacuoles.) *D',* a chondriosphere of the inner fasciculata and reticularis formed by fusion of altered mitochondria. *C",* mitochondria of the fasciculata after the injection of ACTH in a hypophysectomized animal. Notice the restoration of the vesicular pattern and the extruding vacuoles. *eva,* extruding vacuoles; *mit,* mitochondrial tubules; *mv,* mitochondrial vesicle. (From Sabatini, D. D., DeRobertis, E., and Bleichmar, H. B., *Endocrinology, 70:*390, 1962.)

Mitochondrial Sensitivity to Cell Injury and Resultant Degeneration

Mitochondria are one of the most sensitive indicators of injury to the cell. Although mitochondria may be readily altered by the action of various agents, the changes are, within certain limits, reversible. If the alteration reaches a certain critical point, however, it becomes irreversible, and this is generally considered degeneration of the mitochondria. Essentially there are three types of change: (1) fragmentation into granules, followed by lysis and dispersion, (2) intense swelling with transformation into large vacuoles, and (3) a great accumulation of materials with transformation of mitochondria into hyalin granules. This last change is characteristic of the so-called cloudy swelling and hyalin degeneration that frequently results in cellular death.

A relatively frequent observation in an otherwise normal cell is the presence of degenerating mitochondria in foci of autolysis, constituting a type of lysosome (cytolysosome).

Another type of degeneration is the fusion of mitochondria to form large bodies called *chondriospheres*. This degeneration has been found in patients with scurvy and seems to be normal in the adrenal gland of the hamster.[10]

SUMMARY:

Structure of Mitochondria

The mitochondrion contains two compartments — an inner one filled with the mitochondrial matrix and limited by an inner membrane, and an outer one located between the inner and outer membranes. Both membranes have a trilaminar (unit membrane) structure; however, by negative staining the outer membrane is smooth but the inner membrane shows, on its inner surface, particles of 8.5 nm linked to the membrane (F_1 particles). It will be shown later that these particles contain a special ATPase. Complex infoldings of the inner membrane, called mitochondrial crests, project into the matrix. The shape and disposition of these crests vary in different cells and their number is related to the oxidative activity of the mitochondrion. Mitochondria may show a close relationship with lipid droplets. They may accumulate iron-containing pigments or protein molecules, thereby forming yolk bodies. Injury to mitochondria may produce degenerative changes consisting of fragmentation, intense swelling, or accumulation of material. Degenerating mitochondria may be found forming cytolysosomes or large chondriospheres.

12–3 ISOLATION OF MITOCHONDRIAL MEMBRANES

In Chapter 4 the methods of cell fractionation for the separation of the mitochondrial fraction were described (Fig. 4–5). Morphologically homogeneous fractions of mitochondria may be isolated from liver, skeletal muscle, heart, and other tissues. Since the early investigations of Hogeboom et al. (1948)[1] and others it has been demonstrated that the mitochondrion has a lipoprotein compositon — 65 to 70 per cent protein, and 25 to 30 per cent lipid. Most of the lipid content consists of phosphatides (e.g., lecithin and cephalin); cholesterol and other lipids are present in small amounts. Ribonucleic acid was consistently found in about 0.5 per cent of the dry weight.

In recent years the two mitochondrial membranes and the compartments they limit have been separated by density gradient centrifugation.[11, 12] The outer membrane can be separated by causing a swelling, with breakage followed by a contraction of the inner membrane and matrix. Figure 12–9 shows one of the most frequently used procedures in which two detergents, digitonin and lubrol, are used. The outer membrane is much lighter and needs stronger centrifugal forces or a less dense gradient to be separated.

A so-called *mitoplast*, which includes the inner membrane and matrix both intact, has been produced by separating these two elements from the outer membrane with digitonin (Fig. 12–10). The mitoplast has pseudopodic processes and is able to carry out oxidative

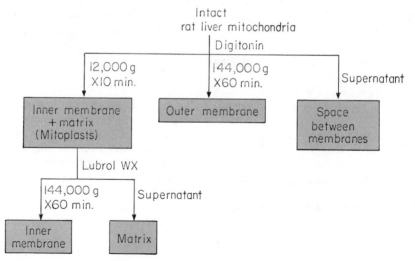

Figure 12–9 Fractionation procedure used to separate the outer and inner membranes of the mitochondria. This method also permits separation of the matrix and provides information about the content of the outer chamber of the mitochondrion. (Courtesy of C. Schnaitman and J. W. Greenawalt.)

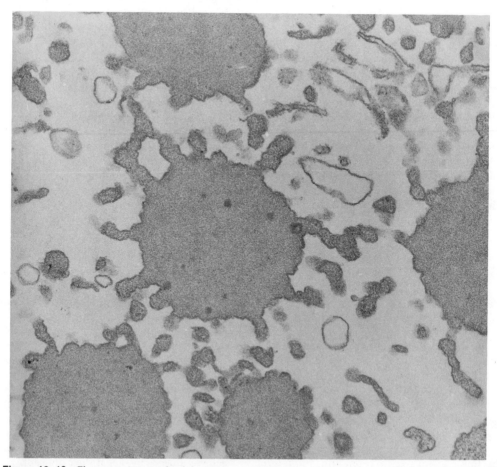

Figure 12–10 Electron micrograph of the inner membrane and matrix (i.e., mitoplast) separated from liver mitochondria. Note the fingerlike processes and the intact appearance of the inner membrane. ×97,500. (Courtesy of C. Schnaitman and J. W. Greenawalt.)

phosphorylation. This separation has provided a clearer definition of the exact localization of the mitochondrial enzyme systems and has revealed interesting differences between the two membranes.

Outer and Inner Membranes — Structural and Chemical Differences

The outer membrane fraction has a 40 per cent lipid content (compared to 20 per cent in the inner membrane), contains more cholesterol, and is higher in phosphatidyl inositol; on the other hand, it is lower in cardiolipin. These figures demonstrate that a fundamental difference between the outer and inner membrane is in the lipid/protein ratio (i.e., about 0.8 in the outer membrane and about 0.3 in the inner membrane). This low lipid/protein ratio indicates that in the inner membrane there is a greater degree of intercalation of the protein within the lipid bilayer (see the fluid model of membrane structure in Chapter 8–2). This fact of structure has been confirmed by electron microscopic observations with freeze-fracturing methods.[13]

From the morphologic viewpoint the outer membrane lacks the elementary particles that are prominent in the inner membrane (Fig. 12–6). As shown in Figure 12–11, the outer membrane has a characteristic "folded bag" appearance in negatively stained preparations.

Mitochondrial Enzymes are Highly Compartmentalized

Table 12–2 lists some of the enzymes present in the various mitochondrial fractions. The ac-

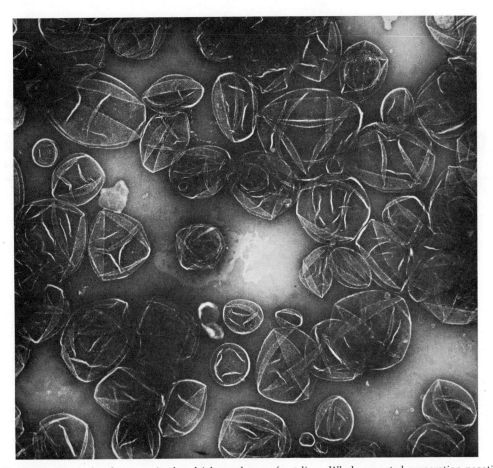

Figure 12–11 Isolated outer mitochondrial membranes from liver. Whole mounted preparation negatively stained. Notice the folded bag appearance of the membrane and the absence of F_1 particles. ×35,000. (Courtesy of D. F. Parsons.)

TABLE 12–2 ENZYME DISTRIBUTION IN MITOCHONDRIA*

Outer membrane
    Monoamine oxidase
    Rotenone-insensitive NADH-cytochrome *c*
      reductase
    Kynurenine hydroxylase
    Fatty acid CoA ligase

Space between outer and inner membranes
    Adenylate kinase
    Nucleoside diphosphokinase

Inner membrane
    Respiratory chain enzymes
    ATP synthetase
    Succinate dehydrogenase
    β-Hydroxybutyrate dehydrogenase
    Carnitine fatty acid acyl transferase

Matrix
    Malate and isocitrate dehydrogenases
    Fumarase and aconitase
    Citrate synthetase
    α-Keto acid dehydrogenases
    β-Oxidation enzymes

*Courtesy of A. L. Lehninger.

tivities of some enzymes that may be considered as markers of the fractions are presented in Table 12–3.

The *outer membrane* contains an NADH-cytochrome-c-reductase system which consists of a flavoprotein and cytochrome b_5. The most specific enzyme system of the outer membrane is *monoamine oxidase*, which may serve as an enzyme marker (Table 12–3).[12] This membrane also contains kynurenine hydroxylase, fatty acid coenzyme A ligase, a phospholipase, and various enzymes of the phospholipid metabolism. The supernatant, obtained after treatment with digitonin (Fig. 12–9), is considered to originate from the outer compartment between the two membranes. In this compartment are found adenylate kinase, nucleoside diphosphokinase, DNAase I, and 5'-endonuclease.

The *mitoplast fraction* (inner membrane plus matrix) contains the components of the respiratory chain and oxidative phosphorylation and the soluble enzymes present in the matrix.[14]

The *inner membrane* carries all the components of the respiratory chain and the oxidative phosphorylation system (see later discussion). It also has glycerol-phosphate dehydrogenase, choline-dehydrogenase, and several *carriers or translocators* for the permeation of phosphate,, glutamate, aspartate, ADP, and ATP. It is calculated that the respiratory chain represents about 20 per cent and the phosphorylating system about 15 per cent of the total protein content of the membrane. These impressive figures give an idea of the importance of these systems in the inner membrane and mitochondrial crests.

The inner membrane contains a remarkably high concentration of cardiolipin (polyglycerophosphatides), which appear to be important in all systems involving electron transport.[15] The *inner compartment* or *mitochondrial matrix* contains all the soluble enzymes of the citric acid or Krebs cycle and those involved in the oxidation of fatty acids. The matrix also contains DNA, ribosomes, and other RNA species and enzymes involved in the synthesis of protein (see later discussion).

TABLE 12–3 ENZYME ACTIVITY OF VARIOUS MITOCHONDRIAL COMPONENTS*

Component	Inner Membrane	Outer Membrane	Matrix	Outer Chamber
Enzyme marker	Cytochrome oxidase	Monoamine oxidase	Malate dehydrogenase	Adenylate kinase
Specific activity of marker in component	9315	551	3895	6690
Specific activity in whole mitochondria	1980	22	2608	421
% Protein in component	21.3	4.0	66.9	6.3

*From Schnaitman, C. A., and Greenawalt, J. W. *J. Cell Biol.*, *38*:158, 1968.

The Inner Membrane — Regional Structural and Enzymatic Differences

Various studies have led to the suggestion that the inner mitochondrial membrane has two major portions, one corresponding to the *mitochondrial crests* and the other to the region that lines the *outer membrane*. For example, it has been reported that the F_1 particles and the coupling factors are present only in the cristae (Fig. 12–3). Furthermore, the activities of several enzymes (i.e., succinate dehydrogenase, β-hydroxybutyrate dehydrogenase, rotenone-sensitive NADH-cytochrome-c-reductase, and ATPase) predominate in the cristae.[16]

Recent studies have focused attention on the fact that mitochondrial membranes have a net electronegative charge due to anionic sites (i.e., the membranes move toward the anode in an electrophoretic field). Using *polycationic ferritin* as an electron microscopic probe for studying the density and distribution of anionic sites, it was found that while the crests have few sites, there are patches of high density in the regions lining the outer membrane. These areas may correspond to those found in the localization of anionic glycoproteins.[17] Another regional difference of the inner membrane involves its association with the outer membrane at the *contact zones*. These are observed especially when the mitochondria are in the condensed state (see Fig. 12–17, *B*); there are about 100 such contacts per liver mitochondrion. At each of these points is a concentration of cytoplasmic ribosomes, which has led to the suggestion that the contact zones could be the sites through which cytoplasmic polypeptides enter the mitochondrion.

SUMMARY:

Isolation of Mitochondrial Membranes

Mitochondria, as well as their membranes and compartments, may be separated by subcellular fractionation. The so-called mitoplast is a mitochondrion from which the outer membrane has been stripped by osmotic action or by digitonin. The outer membrane comes off in a lighter fraction because it has a much higher lipid/protein ratio than the inner membrane. By negative staining it shows a smooth surface and a folded bag appearance.

The mitochondrial enzymes show a definite compartmentation. The outer membrane contains NADH-cytochrome-c-reductase, which consists of a flavoprotein and cytochrome b_5. Monoamine oxidase is the specific enzyme marker of this membrane. The *outer compartment* contains adenylate kinase and other soluble enzymes. The *inner membrane* carries all the components of the respiratory chain and of oxidative phosphorylation. Taken together, these components represent 35 per cent of the protein of the membrane. In addition to other bound enzymes the inner membrane contains several specific carriers or translocation proteins involved in the permeation of metabolites. The *mitochondrial matrix* contains soluble enzymes of the Krebs cycle, DNA, RNA, and other components of the machinery for protein synthesis of the mitochondrion.

12–4 MOLECULAR ORGANIZATION AND FUNCTION OF MITOCHONDRIA

The various functions of mitochondria are so intimately related to their molecular structure that one topic cannot be studied without the other. This section requires some knowledge of the enzymatic mechanisms in which mitochondria are involved. The concepts relating to cell respiration summarized in Chapter 6 should be reviewed.

To express the complexity of the enzyme system in numerical terms, in a mitochondrion more than 70 enzymes and coenzymes and numerous cofactors and metals essential to mitochondrial functions work together in an orderly fashion. Besides oxygen, the only fuel that a mitochondrion needs is phosphate and adenosine diphosphate (ADP); the principal final products are ATP plus CO_2 and H_2O.

Figure 6–13 indicates the final common pathway of biological oxidation, which takes place within the mitochondrion. The three major foodstuffs of the cell (carbohydrate, fat, and protein) are ultimately degraded in the cytoplasm to acetate, a two-carbon unit that is bound to coenzyme A to form acetyl coenzyme A. This penetrates the mitochondrion, and the acetate group condenses with *oxaloacetic acid* to form citric acid, a six-carbon compound. Citric acid is oxidized and loses two carbons as CO_2. In this way the four-carbon compound *succinic acid* is formed, and later on it is oxidized to oxaloacetic acid, starting a new cycle. At each turn of the circle one molecule of acetate penetrates and two CO_2 are released (for more details on the Krebs cycle see textbooks of biochemistry).

An important consequence of the Krebs cycle is that at each turn four pairs of hydrogen atoms are removed from the substrate intermediates by enzymatic dehydrogenation. These hydrogen atoms (or the equivalent pairs of electrons) enter the respiratory chain, being accepted by either NAD^+ or FAD (Fig. 6–13). Three pairs of hydrogens are accepted by NAD^+, reducing it to NADH, and one pair by FAD, reducing it to $FADH_2$ (this last pair comes directly from the succinic dehydrogenase reaction).

Since it takes two turns of the cycle to metabolize the two acetate molecules that come by glycolysis from one molecule of glucose, a total of six molecules of NADH and two of $FADH_2$ are formed at the starting points of the respiratory chain.

Electrons Flow Along a Cytochrome Chain Having a Gradient of Redox Potential

The early work of Warburg and Keilin using spectrophotometric techniques led to the concept that cell oxidations are brought about by electron carriers arranged in a chain of increasing *oxidation-reduction potential*. Keilin employed a simple hand spectroscope to study insect muscle and found that there were special pigments which underwent changes with oxidation and reduction during respiration. Such pigments, which had special absorption bands in the reduced state, were called *cytochromes* and were later on demonstrated to be *iron-containing proteins*. At present *cytochromes b, c, c_1, a,* and a_3 have been identified in mitochondria, and we have seen in Chapter 10 that the endoplasmic reticulum contains *cytochrome b_5*. In each cytochrome the iron may be in either a reduced (Fe^{2+}) or an oxidized state (Fe^{3+}), and the electrons flow along a cytochrome chain in which the redox potentials become more positive (Fig. 12–12). With the **exception of** *cytochrome c,* which is a peripheral protein that can be extracted with salt solutions, these pigments are tightly

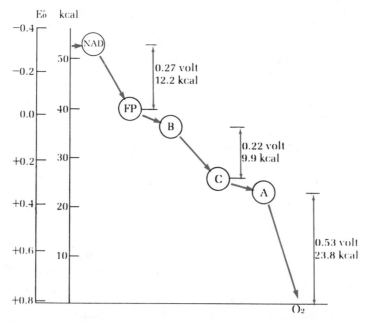

Figure 12–12 Diagram showing the decline of free energy as electron pairs flow down the respiratory chain from NAD to O_2. E'_0, oxidoreduction potential; *kcal,* free energy. At the three points indicated there is enough energy drop to generate a molecule of ATP from ADP and phosphate. (Courtesy of A. L. Lehninger. From Lehninger, A. L., *Biochemistry,* 2nd Ed., p. 516. New York, Worth Publishers, 1975.)

bound to the inner mitochondrial membrane (as intrinsic proteins), and detergents are needed to solubilize and extract them (see Table 8–2).

As shown in Figure 6–13, the electron transport chain does not start with the cytochromes but with the coenzyme *nicotine adenine dinucleotide* (NAD⁺), which is reduced to NADH. As seen in Figure 12–12, the couple NAD⁺/NADH has the most negative redox potential ($E_0 = -0.32$ volts) of the chain and the highest free energy ($\Delta G = -52.7$ kcal/mol), which is released upon oxidation. The reversible NAD⁺/NADH system is followed in the chain by a coenzyme Q_{10} (ubiquinone), a flavoprotein containing FMN and FAD, and then by the series of cytochromes to end with O_2, which has the most positive redox potential ($E_0 = 0.82$) with no free energy to be utilized ($\Delta G = 0$). As shown in Figure 12–12, this sequence of electron transfer reactions is consistent with the redox potentials of the various electron carriers, which become more positive as electrons pass from the substrate to oxygen. Furthermore, sensitive spectrophotometric methods developed by Chance have confirmed the functional nature of this sequence, showing that NAD⁺ is the most reduced member of the chain and that cytochrome a and a_3 (which form part of *cytochrome oxidase*) are the most oxidized.

In recent years it has found that the electron transport chain is extremely complex and that there are additional components. Of particular interest is *coenzyme Q (CoQ)* or *ubiquinone* (because it is ubiquitous in cells), which is a lipid-soluble benzoquinone with a long side-chain of ten isoprenoid units (thus the name CoQ_{10}.) It is thought that CoQ is a kind of shuttling system between the flavoproteins and cytochrome b (see Fig. 6–13). In addition, the mitochondrial respiratory chain contains enzyme-bound copper and iron-sulfur proteins. According to Slater (1973) in the respiratory chain there are about 30 reaction centers able to accept electrons.

The Respiratory Chain — Four Molecular Complexes

All the previously discussed components of the respiratory chain, as well as those involved in the mechanism of phosphorylation, are integrated within the molecular structure of the inner mitochondrial membrane. It is important to remember that this membrane has a high protein content (60 to 70 per cent) and that most of these proteins are of the integral or intrinsic type, which are intercalated into the lipid bi-

layer. In fact it is calculated that one third of this membrane is occupied by phospholipids and the rest by protein.

As in the case of the plasma membrane, there is here a certain degree of *fluidity* that allows for the lateral diffusion of the proteins. This has been demonstrated by freeze-fracturing of the membrane. The study of the distribution of the intramembrane particles under conditions in which there is a slow freezing of the mitochondria shows that this treatment causes aggregation of the particles into patches, while rapid freezing gives a random distribution.[18]

The respiratory chain, together with the phosphorylating system, represents about 35 per cent of the protein of the inner membrane; the rest is represented by the various translocators and structural proteins.

The components of the respiratory chain can be separated into multi-molecular complexes by the use of mild detergents.[19, 20] Such complexes have a considerable amount of lipid which is essential for their activity. In fact, if the lipids are extracted, the proteins denature or aggregate, and thus become unable to transfer electrons. Extraction of phospholipids by acetone inactivates some catalytic functions of the membrane; however, restoration of function can be achieved by addition of the lipids and CoQ_{10}.

Green and associates have recognized four main complexes (I to IV) which, if mixed in correct stoichiometric ratios, can reconstitute to form the electron transport chain. Three of these complexes are indicated in Figure 12–13 (i.e., I, III, and IV). Complex II corresponds to succinate-Q-reductase (i.e., succinate dehydrogenase), which can transfer electrons directly from succinate to CoQ (Fig. 6–14). In Table 12–4 the main characteristics of these complexes are indicated. Listed there are the molecular weight, the number of protein subunits, the various prosthetic groups in the flavoproteins (FMM and FAD), and the iron-sulfur centers and cytochromes (hemes) present in each of the complexes. The table also gives the topology of the active sites in relation to the matrix (M side) or cytosol (C side) (see Papa, 1976; De Pierre and Ernster, 1977).

Complex I (NADH-Q-reductase). This is the largest complex, with a molecular weight well above 500,000 and a structure consisting of 15 subunits.[21] It contains as prosthetic groups *Flavin mononucleotide* (FMN) and six iron-sulfur centers. The NADH reaction site is at the M side; contact between this protein and CoQ is apparently made in the middle of the membrane, in the hydrophobic area. Complex I spans the

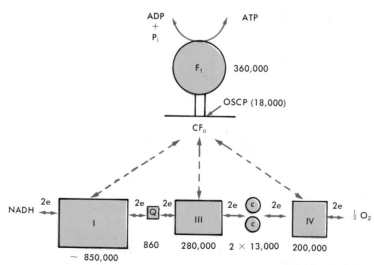

Figure 12–13 Diagram of the electron-transferring complexes (i.e., respiratory chain) and the ATPase complex (F_1) present in the mitochondrial inner membrane; *complex I*, NADH-Q-reductase; *complex III*, QH$_2$-cytochrome-c-reductase; *complex IV*; cytochrome-c-oxidase; *CF$_0$*, F$_0$ coupling factor; *OSCP*, oligomycin sensitive protein; *Q*, ubiquinone; *c*, cytochrome. (See the description in the text.) (Courtesy of E. C. Slater.)

inner mitochondrial membrane and is able to translocate protons across it, from the M side to the C side. This is probably operated through a very hydrophobic subunit of the complex.

Complex II (Succinate-Q-reductase). This complex is composed of two polypeptides with a total molecular weight of 97,000 daltons. It contains *flavin adenine dinucleotide* (FAD) and three iron-sulfur centers. The succinate binding site appears to be on the M side, and the Q site, in the hydrophobic domain of the membrane. In contrast to complex I, succinate-Q-reductase apparently is unable to translocate protons across the membrane.

Complex III (QH$_2$-cytochrome-c-reductase). This complex contains a number of subunits with a total molecular weight of about 280,000 daltons. It contains cytochromes b, cytochrome

TABLE 12–4 COMPOSITION AND TOPOLOGY OF THE MITOCHONDRIAL RESPIRATORY CHAIN*

Enzyme, Complex, MW	Number of Subunits	Prosthetic Groups	Association with Membrane	Topology of Catalytic Sites
NADH-Q-reductase (Complex I) ~ 850,000	16	1 FMN 16–24 FeS (5–6 centers)	Integral	NADH site: M side; Q site: Middle
Succinate-Q-reductase (Complex II) 97,000	2	1 FAD (covalently bound) 8 FeS (3 centers)	Integral	Succinate site: M side Q site: Middle
QH$_2$-cytochrome-c-reductase (Complex III) 280,000	6–8	2 b-hemes 1 c-heme (c$_1$) 2 FeS	Integral	Q site: Middle Cytochrome-c site: C side
Cytochrome c 13,000	1	1 c-heme	Peripheral	Cytochrome-c$_1$ site: C side Cytochrome-a site: C side
Cytochrome-c-oxidase (Complex IV) 200,000	6–7	2 a-hemes (a, a$_3$) 2 Cu	Integral	Cytochrome-c site: C side O$_2$ site: M side (?)

*From De Pierre, J. W., and Ernster, L. *Annu. Rev. Biochem., 46*:201, 1977.

c_1, and iron-sulfur protein. A proton pump linked to one of the b cytochromes has been postulated. In the topology of this complex the Q-site may be in the middle of the membrane in the hydrophobic area, and the cytochrome c site, on the C side.

Complex IV (Cytochrome-c-oxidase). Cytochrome oxidase is a large complex consisting of several polypeptides and having two cytochromes (a and a_3) and two copper atoms. The molecular weight is about 200,000 daltons. This complex is thought to traverse the mitochondrial membrane, protruding on both surfaces. Such a *transmembranal orientation* is associated with the *vectorial transport of protons* across the membrane.

In experiments with yeast mitochondria it has been shown that cytochrome oxidase is made of seven subunits that can be resolved by gel electrophoresis and that range between 40,000 and 5000 daltons. It is of great biological interest that the three larger polypeptides are very hydrophobic and that their synthesis occurs in mitochondrial ribosomes (see later discussion), and that the four small subunits are relatively hydrophilic and originate in the cytoplasm.[22]

The seven subunits are arranged in the membrane in a functional sequence, being in contact with cytochrome c on the C side. The electrons then pass to cytochrome a (also in the C side), then to Cu^{2+} (possibly in the hydrophobic subunits); and finally to cytochrome a_3 and oxygen at the M side (Fig. 12–14).

Using antibodies against cytochrome oxidase, it was found that this complex was inhibited when the antibodies were applied to either side of the inner membrane. Such a finding is in agreement with the above-mentioned traverse topography of complex IV. By the use of freeze-fracturing it has been found that cytochrome oxidase, like other integral proteins, is highly mobile in the lateral plane of the membrane. This lateral diffusion is inhibited by the action of cross-linked antibodies.

Electron Transport — Coupled to the Phosphorylation at Three Points

The electron transport system of mitochondria is *coupled* at three points with the phosphorylating system (Figs. 6–13 and 12–13). The protons

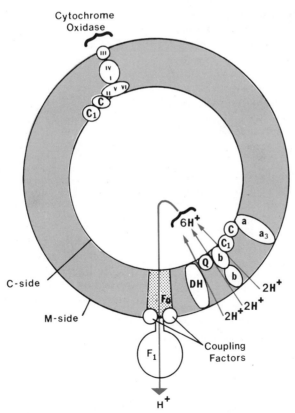

Figure 12–14 The topography of a vesicle originated from the inner membrane of mitochondria, the transmembrane distribution of the various complexes integrating the respiratory chain, and ATPase are indicated. (Courtesy of E. Racker, 1977.)

(H$^+$) originating from electron transfer are translocated by the respiratory chain across the membrane from the M side to the C side. According to the *chemiosmotic hypothesis* of Mitchell (see later discussion), this translocation creates a pH difference and a membrane potential (see Fig. 12–16). Both constitute the proton-motive force that tends to move H$^+$ from the C side back to the M side of the membrane. Since the inner mitochondrial membrane is highly impermeable to H$^+$ ions, these can only reach the M side through the "proton channel" of the ATPase (Fig. 12–14). When H$^+$ moves from the C side to the M side, the F$_1$-ATPase, operating in reverse, catalyzes ATP synthesis. Conversely, when the F$_1$-ATPase hydrolyzes ATP, it functions as a "proton pump" and ejects H$^+$ from the M side to the C side. Figure 12–14 indicates that for each NADH that is oxidized, six protons (H$^+$) are translocated through the inner membrane; these six H$^+$, when returning to the M side through the F$_1$-ATPase, give rise to three molecules of ATP.

Mitochondrial ATPase — A Structurally Complex Proton Pump.

The mitochondrial ATPase molecule is indeed a multi-polypeptide complex. Three parts can be distinguished: (1) The *F$_1$ particle* or soluble ATPase, which is in the head piece of the projection observed on the matrix side of the mitochondrial crests (Fig. 12–6). F$_1$ contains five associated subunits. (2) A complex of very hydrophobic proteins (i.e., proteolipids) localized in the membrane bilayer and probably containing the proton translocating mechanism (this corresponds to Fo in Fig. 12–14). (3) A protein stalk that connects the two. This portion corresponds to the *oligomycin-sensitive conferring protein* (OSCP) and to another protein needed to bind F$_1$ to the membrane (Fig. 12–13). (For further details on this complex system see De Pierre and Ernster, 1977.)

The function of the phosphorylating system is best demonstrated by the elegant experiment of Racker, shown in Fig. 12–15.[25] If isolated mitochondria are subjected to ultrasonic fragmentation, vesicles are formed that show a reverse organization ("inside-out" vesicles), with the F$_1$ particles situated on the outer surface and with the C-side facing the interior of the vesicle. These vesicles are able not only to respire but also to phosphorylate.

After treatment with urea, which removes the F$_1$ coupling factor, the membranes no longer phosphorylate. Reconstitution of the complete

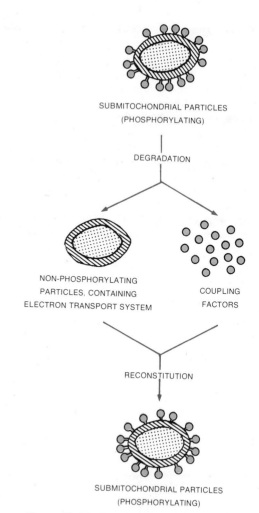

SUBMITOCHONDRIAL PARTICLES
(PHOSPHORYLATING)

DEGRADATION

NON-PHOSPHORYLATING
PARTICLES, CONTAINING
ELECTRON TRANSPORT SYSTEM

COUPLING
FACTORS

RECONSTITUTION

SUBMITOCHONDRIAL PARTICLES
(PHOSPHORYLATING)

Figure 12–15 Diagram showing the experiment by which the submitochondrial particles, corresponding to the inner mitochondrial membranes, are submitted to urea to remove the coupling factor F$_1$, thus leaving non-phosphorylating particles. The lower part of the figure shows the subsequent reconstitution of a phosphorylating submitochondrial particle. (From Racker, E., The membrane of the mitochondrion. *Sci. Am., 218*:32, 1968. Copyright © 1968 by Scientific American, Inc. All rights reserved.)

phosphorylating system can be achieved by mixing the non-phosphorylating membranes with the F$_1$ particles. After it is isolated, the F$_1$ particle causes the hydrolysis of ATP to ADP and phosphate; however, when it is attached to the membrane it functions as a synthesizing enzyme (Fig. 12–13).[26]

Topological Organization of the Respiratory Chain and the Phosphorylation System

All that was just said about the various complexes of the respiratory chain and the phosphorylating system indicates that the inner mitochondrial membrane is highly asymmetrical in its molecular organization. In other words, there is a special "sidedness" of the most important components, which is summarized in Table 12–4 and in the diagram of Figure 12–14. Under the electron microscope the sideness has been demonstrated by the use of antibodies (see earlier discussion) and also by reagents that produce electron-dense deposits in the mitochondrial crests.[27]

By using both a special tetrazole, which is converted to formazine by succinate dehydrogenase (see Fig. 4–14), and 3',3'-diaminobenzidine (DAB), which is oxidized by cytochrome c, it has been shown that the dense products accumulate on the outside surface of the mitochondrial crest.

From experimental evidence it can be demonstrated that the electron transport system is accessible to NADH and succinate only from the inner matrix side, while cytochrome c is reached from the outer surface of the membrane.

In addition to this transverse topology there is also a lateral topology of the components of the respiratory chain and oxidative phosphorylation. The current therory is that the compact assemblies of macromolecules are arranged in a mosaic which is regularly spaced by other proteins and lipids.

The number of assemblies varies according to the tissue and the folding of the membrane. A mitochondrion from liver has about 15,000 assemblies, while one from the flight muscle of an insect may have as many as 100,000. Within this assembly or *unit of phosphorylation* the main enzymes are in equimolecular quantities (see Table 12–5). This finding has resulted from the use of ultra-rapid spectrophotometric studies[28] and from the determination of binding sites of drugs that specifically affect the various components of the system (Fig. 12–13). For example, *aurovertin* binds to the ATPase; *oligomycin* acts at the level of the binding of ATPase to the membrane; *antimycin* is an inhibitor acting on the QH_2 cytochrome-c-reductase; *rotenone* acts on NADH-Q-reductase; and *cyanide* has its well known poisoning effect on cytochrome-c-oxidase.

The minimum unit of oxidative phosphorylation contains numerous proteins with a total molecular weight approaching 5×10^6 daltons. This unit or assembly must, therefore, cover a considerable area of the membrane. From biochemical and electron microscopic studies it is possible to imagine a mosaic organization for the inner membrane of the mitochondrion in which each unit of phosphyorylation should cover a surface area of about 20×20 nm.

TABLE 12–5 STOICHIOMETRY OF ELECTRON-TRANSFER AND ATP-SYNTHESIZING COMPONENTS IN HEART MITOCHONDRIA*

Component†	Stoichiometry	Molecular weight ($\times 10^{-3}$)
I	1	$1 \times 550 = 550$
III	4	$4 \times 230 = 920$
c	8	$8 \times 12 = 96$
IV	8	$8 \times 200 = 1600$
ATPase	4	$4 \times 360 = 1440$
OSCP	4	$4 \times 18 = 72$
	29	4678
CFo	4	
Q	64	

*Modified from Slater, E. C., Electron transfer and energy conservation. *9th International Congress on Biochemistry*, Stockholm, 1973.

†Components: I, Complex I; III, Complex III; c, cytochrome c; IV, Complex IV; OSCP, oligomycin sensitive protein; CF_0, F_0 coupling factor; and Q ubiquinone or coenzyme Q. (See Figure 10–12 for further information.)

The Chemiosmotic Hypothesis — An Electrochemical Link Between Respiration and Phosphorylation

The above-mentioned studies have demonstrated that for each pair of electrons received from the Krebs cycle and transported by the electron transfer system there is one molecule of ATP synthesized by the ATPase. The fundamental problem is that of establishing the nature of the link between these two systems (represented by the dotted lines between the respiratory chain and the F_1 coupling factor in Fig. 12–13).

Several hypotheses have been proposed to explain the mechanism of this link (for a recent review see Boyer et al., 1977). Here we will mention two of the best known views on the subject.

The main feature of the *chemiosmotic coupling* proposed by Mitchell[29] is that the link is essentially electrochemical in nature. This model is based on the fact that the mitochondrial membrane is impermeable to H^+ and OH^- ions, so that pH differences, acting as energy-rich gradients, can be produced across it. In this hypothesis the electron transport system is organized in "redox loops" which translocate protons from the M side onto the C side of the membrane (Fig. 12–16,*A*). This vectorial movement of protons gives rise to a pH difference and an electrical potential across the membrane (Fig. 12–16,*B*).

The chemiosmotic hypothesis postulates that the primary transformations produced by the

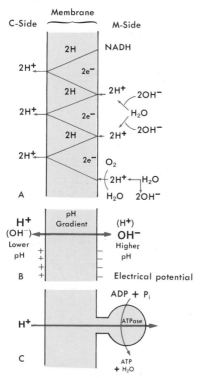

Figure 12–16 Diagram of chemiosmotic coupling according to P. Mitchell. **A,** During electron transport H$^+$ ions are driven to the C-side of the inner mitochondrial membrane. **B,** This process produces a pH gradient and an electrical potential across the membrane. **C,** This gradient drives the proton pump of the ATPase, and ATP is synthesized from ADP and Pi.

respiratory chain guide the *osmotic work* needed to accumulate ions. The energy generated by electron transport is conserved in the energy-rich form of a H$^+$-ion gradient. The pH difference due to the high concentration of OH$^-$ at the M side and of H$^+$ at the C side also maintain the electrical gradient. As indicated in Figure 12–6,C, this gradient, through the action of the proton pump of the ATPase, drives the oxidative

phosphorylation of ADP to form ATP, by which mechanism free energy is conserved:

$$ADP + Pi \rightleftharpoons ATP + H_2O$$

In this reaction H$_2$O is also formed because of the dehydration which leads to the removal of H$^+$ and OH$^-$ ions.

A special feature of this hypothesis is that both the respiratory chain and the ATPase may operate independently. Since coupling by electrical fields can operate at *long-range* distances there is no need for a direct contact between the macromolecules involved. In recent years much experimental evidence has supported the validity of Mitchell's chemiosmotic hypothesis, and the mechanism of the redox proton pump has been elucidated by experiments in which various components of the electron transport and phosphorylating system were reconstituted into vesicular bilayers, the so-called liposomes (see Racker, 1975).

The Chemical-Conformational Hypothesis Involves Short-Range Interactions

An alternate theory — the chemical-conformational hypothesis — postulates that the electron transport and phosphorylating systems are in molecular contact and that information is transmitted by short-range interactions which may, in part, be electrostatic in nature (see Slater, 1973). It is postulated that upon acceptance of the two electrons, each electron transfer protein undergoes a conformational change which is then transmitted to the ATPase by the short-range interactions.

According to some recent views the chemical-conformational and chemiosmotic mechanisms of electron transport linkage to phosphorylation do not represent mutually exclusive alternatives but may indeed be two aspects of the same mechanism (see Ernster, 1977).

SUMMARY:

Molecular Organization and Function

Mitochondria function as energy-transducing organelles into which the major degradation products of cell metabolism penetrate and are converted into chemical energy (ATP) to be used in the various activities of the cell. This entire process requires the entrance of O$_2$, ADP, and phosphate, and it brings about the exit of ATP, H$_2$O, and CO$_2$. The process is based on three coordinated steps: (a) the *Krebs cycle,* carried out by a series of soluble enzymes present in the mitochondrial matrix, which produces CO_2 by decarboxylation and removes electrons from the metabolites (Fig. 6–13); (b) the *respiratory chain* or *electron transport system,* which captures the pairs of electrons and transfers them through a series of electron carriers, which finally leads by combination with activated oxygen to the formation of H_2O; (c) a *phosphorylating system*, tightly coupled to the respiratory chain, which at three points gives rise to ATP molecules.

The electron transport system starts with NAD^+, which is reduced to NADH. NAD^+/NADH has the most negative redox potential ($E_0 = -0.32$ volts) and highest free energy ($\Delta G = -52.7$ kcal/mol). This is followed by a FMN- or a FAD-flavoprotein coenzyme Q_{10}, and a series of cytochromes, ending with O_2 ($E_0 = 0.82$ and $\Delta G = 0$) (Fig. 12–12). The respiratory chain also contains enzyme-bound copper and iron-sulfur proteins. The respiratory chain and the phosphorylating system are in the inner mitochondrial membrane. This is rich in proteins (60 to 70 per cent), most of which are of the integral type.

The components of the respiratory chain may be separated into multi-molecular complexes in which lipids are essential for the activity. Four main complexes have been isolated (Table 12–4):

Complex I or *NADH-Q-reductase* is the largest (MW >500,000) and has 15 subunits. It contains FMN and six iron-sulfur centers. The NADH reaction site is at the M side (matrix).

Complex II or *succinate-Q-reductase* contains FAD and three iron-sulfur centers. This complex transfers electrons from succinate to CoQ.

Complex III or *QH_2-cytochrome-c-reductase* contains cytochromes b_1, c_1, and an iron-sulfur protein. Cytochrome-c is on the C side (cytosol), and CoQ is in the hydrophobic region of the membrane.

Complex IV or *cytochrome-c-oxidase* is a large complex with cytochromes a and a_3 and two copper atoms. As do other complexes, it has a transmembranal orientation that is associated with the vectorial transfer of protons. In yeast mitochondria, complex IV has seven subunits of which the three larger ones are hydrophobic and synthesized on mitoribosomes.

The *phosphorylating* system is represented by the F_1-ATPase. This is a multipeptide complex with three main parts: (1) the F_1 particle or soluble ATPase, (2) a complex of very hydrophobic proteins (i.e., proteolipids) which represent the proton translocating portion (Fo in Fig. 12–14), and (3) a protein stalk that connects the two and contains the coupling factors. By proper methods it is possible to separate the F_1 particle and then to reconstitute the system.

The respiratory chain and phosphorylating system have a fine topology that runs transversely and laterally in a mosaic arrangement. The entire assembly may be 5×10^6 daltons and may cover a surface of 20×20 nm.

The nature of the link between the respiratory chain and the ATPase is unknown, but there are two favored hypotheses: the *chemiosmotic* and the *chemical-conformational*. According to the chemiosmotic hypothesis the ATPase synthesizes ATP under the influence of the vectorial field generated by the transfer of the electrons through the respiratory chain; this effect operates at long range and direct molecular contact is not needed. The chemical-conformational hypothesis postulates that the transfer of electrons originates a conformational change in the proteins of the respiratory chain which, by short-range interaction, transmits the signal to the ATPase that is needed for the synthesis of ATP.

12–5 PERMEABILITY OF MITOCHONDRIA

Since the most important metabolic activities of mitochondria take place within the inner mitochondrial compartment, there should be a rapid and active flow of certain metabolites across the two membranes. The products of extramitochondrial metabolism, for example, must reach the mitochondrial matrix in order to undergo oxidation (Fig. 6-13), and ADP and phosphate must enter to form ATP. Simultaneously, end products such as H_2O, ATP, urea, and ammonia must leave the mitochondrion.

There are important differences in permeability between the mitochondrial membranes. The outer membrane is freely permeable to electrolytes, water, sucrose, and molecules as large as 10,000 daltons. The inner membrane, on the other hand, is normally impermeable to ions, as well as to sucrose.

ADP, ATP, and Pi — Transport by Specific Carriers

The inner mitochondrial membrane uses specific *carriers* or *permeases* for the translocation of various substances. We studied this problem in general in Chapter 8 when discussing cell permeability. We mentioned that carriers are genetically determined intrinsic proteins that are specific for the translocation of metabolites that normally do not cross the membrane.

In mitochondria the existence of specific carriers for ATP (or ADP), phosphate, succinate (or malate), isocitrate, glutamate, aspartate, and bicarbonate has been suggested. A carrier mechanism has been postulated for Ca^{2+}, Mn^{2+}, or Sr^{2+}.[30]

Of all these the most important carriers are those involved in the *transport of ADP, ATP, and Pi* (inorganic phosphate) through the inner mitochondrial membrane. Such mechanisms provide for the entrance of ADP and Pi into the mitochondrial matrix and for the exit of ATP to the cytoplasm. Study of the ADP-ATP translocation is facilitated by the fact that there are specific inhibitors of this transport, the experimental use of which has pointed to the *gated pore* as the most plausible of the models (see Chapter 8–3, for the anion-gated-pore model). Such an ADP-ATP carrier would be asymmetrically disposed through the membrane, with two conformations, one facing the C side, and the other the M side. The same carrier could be used for the reverse translocation of ADP and ATP. Isolation of the carrier protein has been achieved by the use of ^{35}S *carboxyatractylate* (^{35}S CAT), which inhibits transport at the C side, followed by the use of detergents. In this way, a ^{35}S CAT–carrier complex having a molecular weight of 29,000 daltons has been separated from mitochondria. Two of these subunits probably constitute the carrier (see Klingenberg, et al., 1976).

Mitochondrial Conformation — Changes with Stages of Oxidative Phosphorylation

At the beginning of this chapter it was noted that low amplitude contraction cycles that could be associated with stages in oxidative phosphorylation were observed in living cells. Similar observations have been made in isolated mitochondria by using absorbancy or light scattering to study them. Such changes were interpreted as being the result of small variations in volume due to a energy linked swelling-contraction phenomenon. The changes in absorbancy and light scattering, however, may also reflect a rearrangement of the internal structure of the mitochondrion without a concomitant modification of the actual size. By means of a quick sampling method which permits the fixation of isolated mitochondria at different stages of their metabolism, reversible ultrastructural changes were observed.[31]

Mitochondria may alter their internal conformation between the two extreme states shown in Figure 12–17. One is the so-called *orthodox* state that is usually observed in intact tissues. The other corresponds to the *condensed* state, in which there is a dramatic contraction of the inner compartment of the mitochondria accompanied by accumulation of fluid in the outer compartment. In the orthodox conformation the inner membrane shows the characteristic crests; the matrix fills practically the entire volume of the mitochondrion and has a reticular or granular aspect (Fig. 12–17,A). In the condensed conformation the inner membrane is folded and the matrix, now more homogeneous, represents only about 50 per cent of the mitochondrial volume (Fig. 12–17,B). In this state, at certain points, it is possible to see contact zones between the inner and outer membranes of the mitochondrion.

The electron transport system is required for the change from condensed to orthodox conformation to take place. Inhibition of the respiratory chain by cyanide, antimycin A, or Amytal will impair this transformation.[31] The orthodox state is induced when the external ADP becomes low and there is none left to be phosphorylated. If at this time ADP is added, respiration is rapidly enhanced, and the contraction of the inner membranes takes place. It is thought that during the transition from the orthodox to the condensed stage there is a change both in the inner membrane and in the matrix. The inner membrane is believed to contain the contractile elements or "mechano-enzymes,"[32] but the possible role of the matrix in this contraction should also be considered.

Mitochondrial Swelling and Contraction — Agent-Induced

During swelling the mitochondrial volume may increase three to five times its normal value

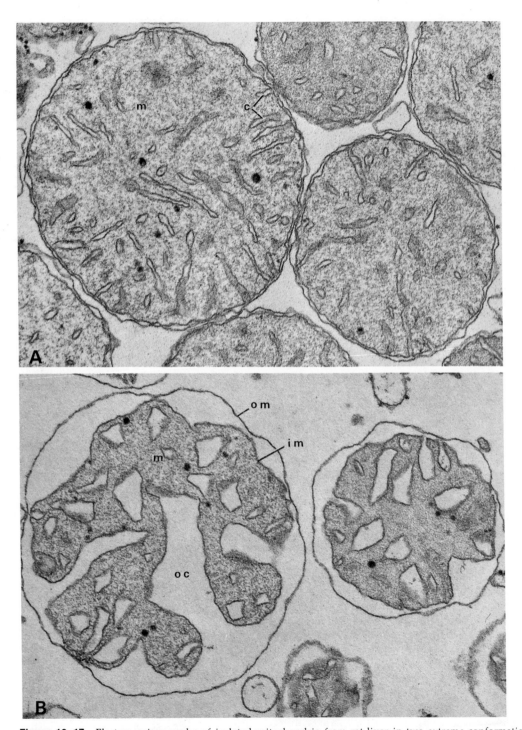

Figure 12–17 Electron micrographs of isolated mitochondria from rat liver in two extreme conformation states. ×110,000. **A,** the *orthodox* conformation. The inner membrane is organized into crests (*c*), and the matrix (*m*) fills the entire mitochondrion. **B,** the *condensed* conformation. Mitochondrial crests are not observed, and the outer chamber (*oc*) represents about 50 per cent of the volume; *om,* outer membrane; *im,* inner membrane. (Courtesy of C. R. Hackenbrock.)

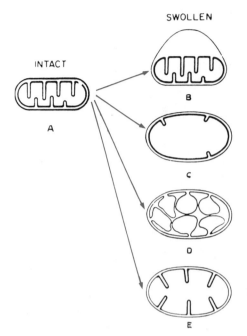

Figure 12-18 Structure of a normal intact mitochondrion (*A*) and of different types of swelling. Entrance of solutes in the outer chamber produces dilation of the intermembranal space (*B*), or of the intracristal spaces (*D*). Penetration in the inner chamber produces dilution of the matrix with (*C*) or without (*E*) unfolding of the crests (From Lehninger, A. L., *Physiol. Rev.*, 42:467, 1962.)

in the absence of ADP. With the addition of ATP, or the restoration of respiration, the mitochondrion may regain its original size. Studies on the phenomenon of swelling demonstrate that mitochondria have an important function in the uptake and extrusion of intracellular fluid.

Agents known to induce swelling are phosphate, Ca^{2+}, reduced glutathione, and, in particular, the thyroid hormone thyroxine. Thyroxine is the most effective swelling agent and is capable of producing swelling at physiologic concentrations.

Since mitochondria have two membranes and two compartments, they may swell in two general ways (Fig. 12–18). Water, K^+, and Na^+ penetrate both membranes very rapidly and may produce the structural changes shown in Figure 12–18,*C* and *E*, with dilution of the matrix. On the other hand, sucrose may produce an "inflation" of the outer chamber or the space within the crest without dilution of the matrix (Fig. 12–18,*B* and *D*).

Figure 12–19 shows the swelling effect of thyroxine as measured by light absorption, and the reversible contraction produced by the addition of ATP. This experiment also shows that the

P:O ratio (i.e., the ratio between the passage of inorganic phosphorus to ADP and the O_2 consumed) is lowered by thyroxine and returned to normal range by ATP. The water uptake is associated with the uncoupling of oxidative phosphorylation, which is restored to normal levels after the water extrusion. Mitochondria can actively squeeze out water and small molecules in the proportion of several hundred per each ATP molecule split.

Mitochondrial Accumulation of Ca²⁺ and Phosphate

Although respiration and oxidative phosphorylation are the most important functions of mitochondria, another related function is the accumulation of cations. Ca^{2+} can be concentrated in isolated mitochondria up to several hundred times normal values. Phosphate enters the mitochondria along with Ca^{2+}. The amounts of Ca^{2+} and phosphate accumulated may be so great that the dry weight may increase by 25 per cent, and microcrystalline, electron-dense deposits may become visible within the mitochondria.[33] Electron microscopy coupled with high resolution microincineration have been used to study these deposits.[34] This process usually occurs in the osteoblasts present in tissues undergoing calcification.

In the presence of Ca^{2+}, mitochondria no longer phosphorylate but, instead, accumulate Ca^{2+} and phosphate. Both oxidative phosphorylation and accumulation of Ca^{2+} depend on the

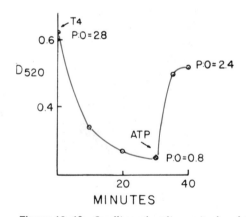

Figure 12–19 Swelling of rat liver mitochondria in the presence of thyroxine and later contraction by ATP. The decrease in optical density corresponds to an increase in water content, and vice versa. Note that the P:O ratio declines during swelling, but is restored again during the contraction stage. *P:O*, ratio between the passage of inorganic phosphorus to ADP and the O_2 consumed. (Courtesy of Lehninger, A. L.)

maintenance of cell respiration (i.e., the Krebs cycle). The accumulation of Ca^{2+} and other cations inside the mitochondrion is accompanied by loss of H^+, so as to maintain an electrical equilibrium. The matrix becomes more alkaline because of OH^- accumulation, while acid is simultaneously released to the outside. Interestingly enough, chloroplasts behave precisely in the reverse direction by pumping out OH^- and accumulating H^+.

SUMMARY:

Mitochondrial Permeability

Since most metabolic processes in mitochondria occur in the inner compartment, an active flow of metabolites across both membranes takes place. The outer membrane is freely permeable, but the inner one is rather impermeable to ions and metabolites and must use specific carriers or translocators to achieve this goal. Carriers for ATP (or ADP), phosphate, succinate or malate, isocitrate, glutamate, aspartate, bicarbonate, and Ca^{2+} have been postulated. In isolated mitochondria cycles of contractions associated with oxidative phosphorylation have been observed. Physiologically, the mitochondrion passes from a conformational stage called *orthodox* into that called *contraction*. In the orthodox conformation, found at low external ADP, the matrix fills the entire volume of the inner compartment, and the crests may be seen clearly. In this stage the organelle does not phosphorylate. If ADP is added, the contraction stage is brought about, the matrix becomes condensed, and about 50 per cent of the water diffuses into the outer compartment.

Mitochondrial swelling is induced by several agents, such as phosphate, Ca^{2+}, and especially thyroxine. This hormone increases the volume of the mitochondrion by water uptake at the same time that it uncouples phosphorylation. With ATP the swelling is reversed; Ca^{2+} and phosphate are accumulated in large amounts. Mitochondria tend to pump out H^+ and to retain OH^-. Chloroplasts act in a reverse manner and tend to pump out OH^- and retain H^+.

12–6 BIOGENESIS OF MITOCHONDRIA

Two main mechanisms for the biogenesis of mitochondria have been postulated: origination by division from parent mitochondria or origination *de novo* from simpler building blocks. Mitochondria are distributed between the daughter cells during mitosis, and their number increases during interphase. It has been observed with time-lapse cinematography that mitochondria gradually elongate and then fragment into smaller mitochondria. This observation has been verified in *Neurospora*.[35] After labeling of a choline deficient mutant of *Neurospora* with radioactive choline, the radioactivity was followed in the mitochondria of the second and third generations. By radioautography it was found that all mitochondria of the original progeny were labeled. The mitochondria of each daughter cell were also labeled but contained about half the radioactivity. This seemed to indicate that mitochondria had divided and grown by the addition of new lecithin molecules to the existing mitochondrial framework.

Yeast cells grown anaerobically lack a complete respiratory chain (cytochromes b and a are absent); under the electron microscope they show no typical mitochondria. When the yeast cells are placed in air, however, the mitochondrial membranes present fuse, unfold, and form true mitochondria that contain the cytochromes. This observation is somewhat in agreement with the *de novo* synthesis theory of the biogenesis of mitochondria.

The differences between the outer and the inner membrane should be considered in relation to the origin of mitochondria. The chemical composition of the outer membrane is similar to that of the endoplasmic reticulum and very different from that of the inner membrane. Furthermore, continuities between the outer membrane of the mitochondrion and the endoplasmic reticulum, and even with the axon membrane, have also been observed.[36]

Mitochondria are Semiautonomous Organelles

In recent years the study of mitochondrial and chloroplast biogenesis became of great interest because it was demonstrated that these organelles contain DNA, as well as ribosomes, and are able to synthesize proteins. The term *semiautonomous organelles* was applied to the two structures in recognition of these findings. However, at the same time, the name indicated that the biogenesis was highly dependent on the nuclear genome and the biosynthetic activity of the ground cytoplasm. It is now known that the mitochondrial mass grows by the integrated activity of both genetic systems, which cooperate in time and space to synthesize the main components. The mitochondrial DNA codes for the mitochondrial, ribosomal, and transfer RNA and for a few proteins of the inner membrane. However, most of the proteins of the mitochondrion result from the activity of the nuclear genes and are synthesized on ribosomes of the cytosol (Fig. 12–20). The study of mitochondrial biogenesis involves knowledge of many concepts that are considered in later chapters dealing with gene expression (Part 6). For this reason at this point we can treat this subject in only a very elementary way (for further readings see Lloyd, 1974; Saccone and Kroon, 1976; Borst, 1977; O'Brien, 1977).

Mitochondrial DNA — Circular and Localized in the Matrix

Although Chevremont had observed that under certain conditions mitochondria of cultured cells give a positive Feulgen reaction, until 1963 the general opinion was that DNA is exclusively confined to the nucleus. In that year M. and S. Nass[37] observed filaments within mitochondria that they interpreted as DNA molecules. This finding was fully confirmed in cell sections, as well as in DNA extracted and studied by the surface spreading technique (Fig. 12–20). The mitochondrial DNA (mtDNA) is localized in the inner compartment or matrix and is probably attached to the inner membrane at the point where DNA duplication starts. This duplication is under nuclear control and the enzymes used (i.e., polymerases) are imported from the cytosol (Fig. 12–20). A single mitochondrion may contain one or more DNA molecules, depending on its size; i.e., the larger the mitochondrion, the more DNA molecules present. Thus, a normal eukaryotic cell has at least as many copies of mtDNA as it has mitochondria. It has been calculated that in an adult man there may be 10^{17} mtDNA molecules, an immense number; it would be interesting to know whether they are similar or different in their nucleotide sequence. So far the results obtained with various techniques suggest that they are identical

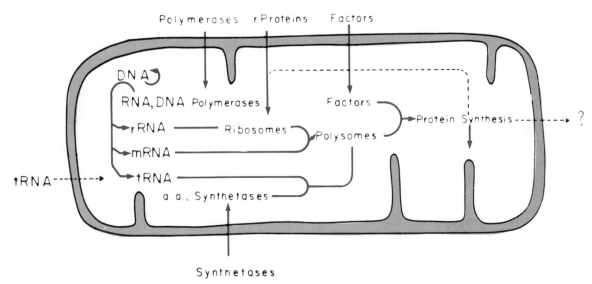

Figure 12–20 Diagram indicating the mitochondrial enzymes, molecules, and structural components involved in mitochondrial transcription and translation. Solid arrows indicate well-established pathways, and dashed lines, less firmly documented ones. *rRNA,* ribosomal RNA; *mRNA,* messenger RNA; *tRNA,* transfer RNA; *rProteins,* ribosomal proteins; *aa,* amino acids. (Courtesy of T. W. O'Brien.)

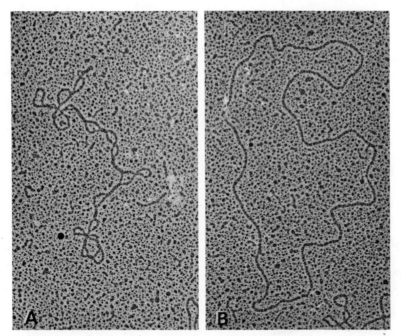

Figure 12–21   DNA extracted from rat liver mitochondria and observed by the spreading technique. **A,** configuration in twisted circle ("super-coiled"); **B,** configuration in open circle. (Courtesy of B. Stevens.)

and that genes are represented in single copies.

In most animals and plants the mtDNA is circular, although it may be highly twisted (Fig. 12–21). In most animal cells mtDNA is only 5.5 μm long, while in yeast it may be longer (i.e., 26 μm).[26] Because of its higher guanine-cytosine content, mtDNA has a higher buoyant density and denatures at a higher temperature than nuclear DNA. It is relatively easy to separate the strands of mtDNA and to localize genes after hybridization with RNA molecules. Genetic maps have been constructed for mtDNAs of yeast and mammalian cells (see Borst, 1977; O'Brien, 1977).

Mitochondrial DNA from human cells is a circular DNA molecule of about 15,000 base pairs. A significant proportion of this DNA is devoted to coding for the protein synthesizing machinery. MtDNA codes for the mitochondrial ribosomal RNAs (12S and 16S), for about 19 tRNAs (probably a full set is encoded, i.e., tRNAs for all the amino acids), and for the mRNAs of about 20 proteins. Most of the proteins for the mitochondrial ribosomes come from the outside and are encoded by the nuclear genes. The mRNAs transcribed from the mitochondrial DNA code mostly for highly hydrophobic proteins. It seems that the reason for having all this biosynthetic machinery inside the mitochondrion is probably that there

is no other way of getting these very hydrophobic proteins inside the organelle.

Mitochondrial DNA duplicates by a mechanism which will be described in Chapter 16. Incorporation of ^{3}H-thymidine into DNA has been observed in mitochondria of *Tetrahymena*. All mitochondria incorporate ^{3}H-thymidine within a population doubling time. In synchronized cultured human cells labeled with ^{3}H thymidine, it was found that mitochondrial DNA was synthesized during a period extending from the G_2 phase to cytokinesis.

Mitochondrial Ribosomes — Smaller than Cytoplasmic Ribosomes

Mitochondria contain ribosomes and polyribosomes (see Chapter 23) which can be demonstrated, in many cases, by electron microscopy (Fig. 12–4). In yeast and *Neurospora*, ribosomes have been assigned to a 70S class similar to that of bacteria; in mammalian cells, however, mitochondrial ribosomes are definitely smaller and have a total sedimentation coefficient of 55S, with subunits of 35 and 25S.[38] Ribosomes in mitochondria appear to be tightly associated with the membrane.

In spite of their low sedimentation coefficient (55S), animal ribosomes are slightly larger

277

than the 70S bacterial ribosomes; this is because the animal structures contain more protein. All the ribosomal proteins are imported from the cytosol (Table 12–6).

Mitochondria Synthesize Mainly Hydrophobic Proteins

We mentioned earlier that mitochondria can synthesize about 20 different proteins which are incorporated into the inner mitochondrial membrane. These proteins are very hydrophobic (i.e., they are proteolipids), which makes their study more difficult. However, it has been recognized that the three largest subunits of cytochrome oxidase and one protein subunit of the cytochrome b-c$_1$ complex are made on mitoribosomes. Furthermore, in yeast several subunits of the ATPase and probably a few hydrophobic proteins of the membrane are also made on *mitoribosomes*.

Table 12–6 shows a list of proteins originating in the cytosol and in the mitochondria and clearly demonstrates the cooperation of the two genetic systems in the synthesis of important mitochondrial protein complexes. One of the best known differences between the two mechanisms of protein synthesis is in the effect of some inhibitors. The mitochondrial system is inhibited by *chloramphenicol* (as in bacteria), while synthesis in the cytosol is not affected. On the other hand, *cycloheximide* has the reverse effect.

The Symbiont Hypothesis — Mitochondria and Chloroplasts are Intracellular Prokaryotic Parasites

At the end of the last century, cytologists like Altmann and Schimper speculated on purely morphological grounds that mitochondria and chloroplasts might be intracellular

parasites that had established a *symbiotic* relationship with the eukaryotic cell. Bacteria were thought to have originated the mitochondria, and blue-green algae, the chloroplasts. In modern cell biology, with the recognition that these organelles have a certain degree of autonomy, this symbiont hypothesis has been reframed. According to the revised theory the original host cell is conceived of as an anaerobic organism deriving its energy from glycolysis, a process that occurs in the cytoplasmic matrix (see Chapter 6), and the parasite contains reactants of the Krebs cycle and the respiratory chain and is able to carry on respiration and oxidative phosphorylation. The symbiont hypothesis is even more plausible in the case of plant cells, since the parasite would be the chloroplast, an autotrophic microorganism able to transform energy from light (see Margulis, 1971; Bogorald, 1975; Mahler and Raff, 1975; Saccone and Quagliarello, 1975; Bucher et al., 1977).

The symbiont hypothesis is based on the many similarities between mitochondria, chloroplasts, and prokaryotes, which when considered from an evolutionary viewpoint, may be more than circumstantial. There are similarities in the *localization of the respiratory chain*. In bacteria this is situated in the plasma membrane, which also has the F$_1$-ATPase projecting into the matrix.

Certain bacteria have membranous projections extending from the plasma membrane, forming the so-called *mesosomes*. Such membranous projections, which are comparable to mitochondrial crests, have been separated and shown to contain the respiratory chain.[39] The inner mitochondrial membrane and the matrix may represent the original symbiont enclosed within a membrane of cellular origin.

The *mitochondrial DNA* is circular, as it frequently is in chromosomes of prokaryons. Mitochondria contain ribosomes that are smaller

TABLE 12–6 BIOSYNTHESIS OF MAJOR MITOCHONDRIAL ENZYME COMPLEXES IN YEAST*

		Number of Subunits	
Enzyme Complex	*Total*	*Made on Cytosol Ribosomes*	*Made on Mitochondrial Ribosomes*
Cytochrome oxidase	7	4	3
Cytochrome b-c$_1$ complex	7	6	1
ATPase (oligomycin-sensitive)	9	5	4
Large ribosomal subunit	30	30	0
Small ribosomal subunit	22	21	1

*From Borst, P., Structure and Function of Mitochondrial DNA, in *International Cell Biology*, p. 237. New York, The Rockefeller University Press, 1977.

than those belonging to the cell-at-large; however, they are generally smaller than those in bacteria.

As just mentioned, protein synthesis in mitochondria and in bacteria is inhibited by chloramphenicol, whereas the extramitochondrial protein synthesis of the higher cell is not affected.

In mitochondria we have also already noted that there is evidence of a DNA-dependent RNA synthesis, which indicates partial autonomy of this organelle. The amount of information carried by the mitochondrial DNA, however, is insufficient for an autonomous biogenesis.

In support of the possible prokaryotic origin of mitochondria and chloroplast is the fact that *intracellular symbiosis* may be found in nature. Thus, *paramecia* may contain certain bacteria, and blue-green algae may occur in simple animals. The endosymbiosis of a photosynthetic prokaryote confers upon the host the ability to capture light energy to synthesize various products. On the other hand, this provides for the prokaryote a constant environment in which to grow and reproduce. In evolutionary terms it is possible to conceive that a symbiotic relationship could have evolved into the present situation, in which the organelles have only a certain degree of autonomy and depend on the nucleus and cytosol of the cell for the synthesis of most of their specific components.

SUMMARY:

Mitochondrial Biogenesis

Mitochondrial and chloroplast biogenesis became of great interest with the demonstration that these organelles contain DNA and ribosomes and are able to synthesize proteins. They are called *semiautonomous* organelles because their biogenesis is the result of the cooperation of two genetic systems (i.e., mitochondrial and nuclear). A mitochondrion has one or more circular DNA molecules about 5.5 μm long localized in the matrix. This DNA is able to code for ribosomal RNA, transfer RNA, and a few proteins of the inner membrane. Mitoribosomes are in general smaller than cytoplasmic ribosome (55S against 80S). The proteins of these ribosomes are imported from the cytosol.

Mitochondria can synthesize some ten proteins which are very hydrophobic (i.e., they are proteolipids). Mitochondrial protein synthesis is inhibited by chloramphenicol (as in bacteria), while cycloheximide inhibits that of the cytosol. From Table 12–6 it is evident that the two genetic systems cooperate to make the mitochondrial proteins.

The symbiont hypothesis postulates that mitochondria (and chloroplasts) originated from the symbiosis of a prokaryotic organism with a host cell which was anaerobic and derived its energy only from glycolysis. Mitochondria could be the result of a bacterium parasite, and chloroplasts, the result of a blue-green algae (having chlorophyll). In evolutionary terms it is possible that a symbiotic relationship could have evolved into the present situation, in which these organelles have only a certain degree of autonomy.

REFERENCES

1. Hogeboom, G. H., Schneider, W., and Palade, G. (1948) *J. Biol. Chem.*, *172*:619.
2. Bereiter-Hahn, J. (1976) *Biochim. Biophys. Acta*, *423*:1.
3. Bereiter-Hahn, J. (1976) *Cytobiologie*, *12*:429.
4. Loud, A. V. (1968) *J. Cell Biol.*, *37*:27.
5. Fleischer, S., Fleischer, B., and Stoeckenius, W. (1967) *J. Cell Biol.*, *32*:193.
6. Fernández-Morán, H. (1963) *Science*, *140*:381.
7. Racker, E. (1967) *Fed. Proc.*, *26*:1335.
8. Sabatini, D. D., De Robertis, E., and Bleichmar, H. (1962) *Endocrinology*, *70*:390.
9. Palade, G. E. (1958) *Anat. Rec.*, *130*:352.
10. De Robertis, E., and Sabatini, D. D. (1958) *J. Biophys. Biochem. Cytol.*, *4*:667.
11. Levy, M., Toury, R., and André, J. (1967) *Biochim. Biophys. Acta*, *135*:599.
12. Schnaitman, C. V., Erwin, V. G., and Greenawalt, J. W. (1967) *J. Cell Biol.*, *32*:719.
13. Melnick, R. L., and Packer, L. (1971) *Biochim. Biophys. Acta*, *253*:503.
14. Sottocasa, G. L., Kuylenstierna, B., Ernster, L., and Berstrand, A. (1967) *J. Cell Biol.*, *32*:415.
15. Parsons, D. F., Williams, G. R., Thompson, W., and Chance, B. (1967) In: *Mitochondrial Structure and Compartmentation*. p. 5. (Quagliariello,

E., et al. eds.). Libreria Adriatica Universitá de Bari.

16. Werner, S., and Neupert, W. (1972) *Eur. J. Biochem.*, 25:369.

17. Hackenbrock, C. R., and Miller, K. J. (1975) *J. Cell Biol.*, 65:615.

18. Höchli, M., and Hackenbrock, C. R. (1976) *Proc. Natl. Acad. Sci. USA*, 73:1636.

19. Green, D. E., and Goldberger, R. F. (1966) In: *Molecular Insights into the Living Process.* Academic Press, Inc., New York.

20. Sottocasa, G. L. (1967) *Biochem. J.*, 105:1.

21. Ragan, C. I. (1975) International Symposium on Electron Transfer Chains and Oxidative Phosphorylation, Fasano, Italy, p. 25. Universitá de Bari.

22. Schatz, G. (1976) In: *The Structural Basis of Membrane Function*, p. 45. (Hatefi, Y., and Javadi-Ohaniance, L. D., Academic Press, Inc., New York.

23. Hackenbrock, C. R., and Hammon, K. M. (1975) *J. Biol. Chem.*, 250:9185.

24. Hackenbrock, C. R. (1977) In: *The Structure of Biological Membranes*, p. 199. (Abrahamsson, S., and Pascher, I., eds.) Nobel Foundation Symposium Ser. No. 34. Plenum Press, New York.

25. Racker, E. (1968) *Sci Am.*, 218:32.

26. Racker, E. (1975) *Biochem. Soc. Trans.*, 3:27.

27. Seligman, A. M., Karnowski, M. J., Wasserbrug, H. L., and Hanker, J. S. (1968) *J. Cell Biol.*, 38:1.

28. Chance, B., and Williams, G. R. (1956) *Adv. Enzymol.*, 17:65.

29. Mitchell, P. (1967) *Fed. Proc.*, 26:137.

30. Reynafarje, B., and Lehninger, A. L. (1969) *J. Biol. Chem.*, 244:584.

31. Hackenbrock, C. R. (1968) *J. Cell Biol.*, 37:345.

32. Lehninger, A. L. (1964) *The Mitochondrion.* W. A. Benjamin, Inc., New York.

33. Greenawalt, J. W., Rossi, C. S., and Lehninger, A. L. (1964) *J. Cell Biol.*, 23:21.

34. Thomas, R. S., and Greenawalt, J. W. (1968) *J. Cell Biol.*, 39:55.

35. Luck, D. J. L. (1965) *J. Cell Biol.*, 24:461.

36. De Robertis, E., and Bleichmar, H. B. (1962) *Z. Zellforsch.*, 57:572.

37. Nass, M. M. K., and Nass, S. (1963) *J. Cell Biol.*, 39:55.

38. Attardi, B., Attardi, G., and Aloni, Y. (1971) *J. Molec. Biol.*, 55:231, 251, and 271.

39. Salton, M. R. J., and Chapman, J. A. (1962) *J. Ultrastruct. Res.*, 6:489.

ADDITIONAL READING

Amzel, L. M., and Pedersen, P. L. (1978) Adenosinetriphosphatase from rat liver mitochondria: Crystallization and X-ray studies of the F_1 component. *J. Biol. Chem.*, 253:2067.

Bandolw, W., Schweyen, R. J., Thomas, D. Y., Wolf, K., and Kaudewitz, F., eds. (1976) *Genetics, Biogenesis and Bioenergetics of Mitochondria.* Gruyter & Co., Berlin.

Brown, W., Shine, J., Goodman, H. (1978) Analysis of 7S DNA from the origin of replication. *Proc. Natl. Acad. Sci. USA*, 75:735.

Bogenragen, D., and Clayton, D. (1977) Mouse mtDNA are selected randomly for replication throughout the cell cycle. *Cell*, 11:719.

Bogorald, L. (1975) Evolution of organelles and eukaryotic genomes. *Science*, 188:891.

Borst, P. (1977) Structure and function of mitochondrial DNA. In: *International Cell Biology*, p. 237. Rockefeller University Press, New York.

Boyer, P. D., Chance, B., Ernster, L., Mitchell, P., Racker, E., and Slater, E. C. (1977) Oxidation phosphorylation and photophosphorylation. *Annu. Rev. Biochem.*, 46:995.

Bücher, T., Neupert, W., Sebald, W., and Werners, S., eds. (1977) *Genetics and Biogenesis of Chloroplasts and Mitochondria.* North-Holland Publishing Co., Amsterdam.

Bygrave, F. L. (1978) Mitochondria and the control of intracellular calcium. *Biol. Rev.*, 53:43.

Chance, B. (1972) The nature of electron transfer and energy coupling reactions. FEBS *Lett.* 23:1.

De Pierre, J. W., and Ernster, L. (1977) Enzyme topology of intracellular membranes. *Annu. Rev. Biochem.*, 46:201.

Erecinska, M., and Wilson, D. F. (1978) Cytochrome-c-oxidase: A synopsis. *Arch. Biochem. Biophys.*, 188:1.

Ernster, L. (1977) Chemical and chemiosmotic aspects of electron transport–linked phosphorylation. *Annu. Rev. Biochem.*, 46:981.

Freedman, J. A., and Chan, S. H. P. (1978) Biosynthesis of mitochondrial membrane proteins. Mol. Cell Biochem., 19:135.

Gillham, N. W. (1978) *Organelle Heredity.* Raven Press, New York.

Gillum, A., and Clayton, D. (1978) D-loop replication initiation. *Proc. Natl. Acad. Sci. USA*, 75:677.

Hoffman, H. F., and Avers, C. (1973) Mitochondria of yeast, ultrastructural evidence for one giant branched organelle. *Science*, 181:749.

Hinkle, P. C., and McCarty, R. E. (1978) How cells make ATP. *Sci. Am.*, 238:104.

Klingenberg, M., Riccio, P., Aguila, H., Buchanan, B. B., and Grebe, K. (1976) Mechanism of carrier transport and ADP, ATP carrier. In: *The Structural Basis of Membrane Function*, p. 293. (Hatefi, Y., and Djavadi-Ohaniance, L., eds.) Academic Press, Inc., New York.

Lehninger, A. L. (1964) *The Mitochondrion.* W. A. Benjamin, Inc., New York.

Lehninger, A. L. (1965) *Bioenergetics.* W. A. Benjamin, Inc., New York.

Lehninger, A. L. (1971) The molecular organization of mitochondrial membranes. *Adv. Cytopharmacol.*, 1:199.

Linnane, A. W., and Nagley, P. (1978) Structural mapping of mitochondrial DNA. *Arch. Biochem. Biophys.*, 187:277.

Lloyd, D., ed. (1974) *The Mitochondria of Microorganisms.* Academic Press, Inc., Ltd., London.

Mahler, H. R., and Raff, R. A. (1975) The evolutionary origin of the mitochondrion, a nonsymbiotic model. *Int. Rev. Cytol,* 43:2.

Margulis, L. (1971) Cell organelles such as mitochondria may have once been free-living organisms. *Sci. Am.,* 225:48.

Mitchell, P. (1977) A commentary on alternative hypotheses of protonic coupling. *FEBS Lett.,* 78:1.

Novikoff, A. B. (1961) Mitochondria (chondriosomes). In: *The Cell,* Vol. 2, p. 299. (Brachet, J., and Mirsky, A. E., eds.) Academic Press, Inc., New York.

O'Brien, T. W. (1977) Transcription and translation in mitochondria. In: *International Cell Biology,* p. 245. Rockefeller University Press, New York.

Papa, S. (1976) Proton translocation reactions in the respiratory chains. *Biochim. Biophys. Acta,* 456:39.

Perlman, P. S., et al. (1977) Localization of genes for variant forms of mt. proteins on mtDNA. *J. Mol. Biol.,* 115:675.

Racker, E. (1968) The membrane of the mitochondrion. *Sci. Am.,* 218:32.

Racker, E. (1970) *Membranes of Mitochondria and Chloroplasts.* Van Nostrand Reinhold Co., New York.

Racker, E. (1975) Reconstitution, Mechanism of Action, and Control of Ion Pumps. *Biochem. Soc. Trans.,* 3:27.

Saccone, C., and Quagliarello, E. (1975) Biochemical studies on mitochondrial transcription and translation. *Int. Rev. Cytol.,* 43:125.

Saccone, C., and Kroon, A. M., ed. (1976) *The Genetic Function of Mitochondrial DNA,* North-Holland Publishing Co., Amsterdam. p. 354.

Saltzgaber, J., Cabral, F., Birchmeier, W., Kohler, C., Frey, T., and Schatz, G. (1977) The assembly of mitochondria. In: *International Cell Biology,* p. 256. Rockefeller University Press, New York.

Sjöstrand, F. S. (1978) The structure of mitochondrial membranes: A new concept. *J. Ultrastruct. Res.,* 64:217.

Slater, E. C. (1972) Mechanism of energy conservation. In: *Mitochondrial Biomembranes,* p. 133. North-Holland Publishing Co., Amsterdam.

Slater, E. C. (1973) Electron transfer and energy conservation. In: *9th International Congress on Biochemistry.* Stockholm.

Tzagoloff, A. (1976) Genetic origin of cytochrome oxidase. *Trends Biochem. Sci.,* 1:139.

13

LYSOSOMES, THE CELL DIGESTIVE SYSTEM, AND PEROXISOMES

13–1 Major Characteristic of Lysosomes 282

    Various Cytochemical Procedures for Identifying Lysosomes

    Lysosomes are Very Polymorphic

    Primary Lysosomes and Three Types of Secondary Lysosomes

    Lysosomal Enzymes — Synthesized in the ER, Packaged in the Golgi

    Summary: *Lysosomal Morphology and Cytochemistry*

13–2 Endocytosis 288

    Phagocytosis — The Process of Cellular Ingestion of Solid Material

    Pinocytosis — The Ingestion of Fluids

    Micropinocytosis and the Ingestion and Transport of Fluids

    Micropinocytosis — Frequently Associated with the Formation of Coated Vesicles

    Endocytosis — An Active Mechanism Involving the Contraction of Microfilaments

    Summary: *Endocytosis and Lysosomes*

13–3 Functions of Lysosomes — Intracellular Digestion 294

    Autophagy by Lysosomes — The Renovation and Turnover of Cell Components

    Lysosomal Removal of Cells and Extracellular Material and Developmental Processes

    Release of Lysosomal Enzymes to the Medium for Extracellular Action

    Lysosomal Enzyme Involved in Thyroid Hormone Release and in Crinophagy

    Leukocyte Granules are of a Lysosomal Nature

    Lysosomes are Important in Germ Cells and Fertilization

    Lysosomal Involvement in Human Diseases and Syndromes

    Storage Diseases — Caused by Mutations That Affect Lysosomal Enzymes

    Lysosomes in Plant Cells and Their Role in Seed Germination

    Summary: *Functions of Lysosomes*

13–4 Peroxisomes 297

    Peroxisomes Contain Nucleoids; Microperoxisomes Lack Them

    Peroxisomal Enzymes — Made in the Rough ER

    Peroxisomes Contain Enzymes Related to the Metabolism of H_2O_2.

    Plant Peroxisomes are Involved in Photorespiration

    Summary: *Peroxisomes*

Every eukaryotic cell has a group of cytoplasmic organelles, *the lysosomes,* of which the main function is intracellular or extracellular digestion. They may be distinguished from other organelles by their morphology and especially by the following functions they perform: (1) digestion of food or various materials taken by *phagocytosis* or *pinocytosis,* (2) digestion of parts of the cell by a process called *autophagy,* and (3) breakdown of extracellular material by the release of enzymes into the surrounding medium.

13–1 MAJOR CHARACTERISTICS OF LYSOSOMES

The concept of the lysosome originated from the development of cell fractionation techniques, by which different subcellular components were isolated. By 1949, a class of particles having centrifugal properties somewhat intermediate between those of mitochondria and microsomes was isolated by De Duve and found to have a high content of acid phosphatase and other hydrolytic enzymes. Because of their enzymatic properties, in 1955 these particles were named lysosomes (Gr., *lysis*, dissolution; *soma*, body).[1]

At present some 50 lysosomal hydrolases are known (Table 13–1), which are able to digest most of the biological substances.[2] Lysosomes have been found both in animal and plant cells and in *Protozoa*. In bacteria there are no lysosomes, but the so-called *periplasmatic space*, sometimes found between the plasma membrane and the cell wall, may play a role similar to that of the lysosomes.[3]

One important property of lysosomes is their stability in the living cell. The enzymes are enclosed by a membrane and are not readily available to the substrate. After isolation by mild methods of homogenization, the amount of enzyme that can be measured is small. The yield increases considerably if the particles are treated with hypotonic solutions or surface-active agents (e.g., triton). This so-called *latency* of the enzymes is due to the presence of the membrane.

The lysosomal membrane is resistant to the enzymes that it encloses, and the entire process of digestion is carried out within the lysosome. The membrane protects the rest of the cell from

TABLE 13–1 Lysosomal enzymes*

Function	Enzyme	Function	Enzyme
		Glycoside hydrolases (Cont.)	α-Acetylglucosaminidase
Transferases transferring phosphorus-containing groups			β-Glucuronidase
Nucleotidyltransferases	Ribonuclease		Hyaluronidase
			β-Xylosidase
Hydrolases acting on ester bonds			α-L-Fucosidase
Carboxylic ester hydrolases	Esterase	*Enzymes acting on peptide bonds (peptide hydrolases)*	
	Lipase	Peptidyl amino acid hydrolases	Carboxypeptidase
	Phospholipase A₁ and A₂		Cathepsin A
Thiolester hydrolases	Hydroxyisobutyryl-CoA-hydrolase		Acetylphenylalanyl-tyrosine hydrolase
	Alanyl-CoA-hydrolase		Acid carboxypeptidase
Phosphoric monoester hydrolases	Acid phosphatase (several types)	Dipeptidehydrolases	Dipeptidase
	Phosphotidate phosphatase	Peptidylpeptide hydrolases	Cathepsin B₁ and B₂
	Phosphoprotein phosphatase		Cathepsin C
			Dipeptidylaminopeptidase
Phosphoric diester hydrolase	Phosphodiesterase (exonuclease)		Cathepsin D
	Phospholipase C		Cathepsin E
	Deoxyribonuclease II		Renin
			Collagenase
Sulphuric ester hydrolases	Arylsulphatase A and B		Neutral proteinase
	Chondrosulphatase		Kininogen activator
			Plasminogen activator
Hydrolases acting on glycosyl compounds		*Enzymes acting on C−N bonds other than peptide bonds in linear amides*	Amino acid naphthylamidases
Glycoside hydrolases	Dextranase [α-(1→6)-glucosidase]		Dipeptide naphthylamidases
	Lysozyme (muramidase)		Aspartylglucosylaminase
	Neuroaminidase		Ceramidase
	α-Glucosidase		
	β-Glucosidase	*Enzymes acting on acid anhydride bonds* in phosphoryl-containing anhydrides	Pyrophosphatase
	α-Galactosidase		Arylphosphatase
	α-Mannosidase		
	α-Acetylglucosaminidase		
	α-Acetylgalactosaminidase	*Enzymes acting on P−N bonds*	Phosphoamidase

*From Allison, A. C., *Lysosomes*, Oxford Biology Readers, Vol. 58. Oxford University Press, 1974.

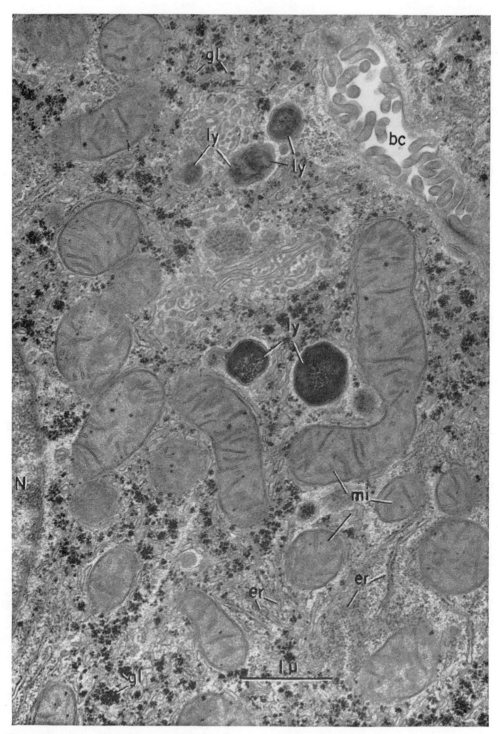

Figure 13–1 Peripheral region of a liver cell, showing a biliary capillary (*bc*) and several bodies interpreted as lysosomes (*ly*). *er,* endoplasmic reticulum; *gl,* glycogen; *mi,* mitochondria; *N,* nucleus. ×31,000. (Courtesy of K. R. Porter.)

the possible destructive effect of the enzymes, and its stability is of fundamental importance to the normal function of the cell. In fact, pathologic conditions are known in which this membrane becomes more labile and permits the exit of the enzymes with catastrophic consequences to the cell. Another interesting point is that most of the lysosomal enzymes act in an acid medium. It is possible that the membrane has a *proton pump* that accumulates H^+ inside the lysosome and acidifies its content (see Reijngoud, 1978).

Various Cytochemical Procedures for Identifying Lysosomes

While the concept of lysosome was, at the beginning, based on purely biochemical grounds, its identification soon became possible with the use of the electron microscope and various cytochemical techniques (Fig. 13–1).[4] The most widely used procedure is the Gomori stain for *acid phosphatase* (see Chapter 4–5) (Fig. 13–2). Cytochemical reactions for *β-glucuronidase, aryl sulfatase, N-acetyl-β-glucosaminidase,* and *5-bromo-4-chloroindoleacetate-esterase* can also stain the lysosomes. Certain substances that are taken up by lysosomes may be used in their identification. Thus, *peroxidase* ingested by the cell may be detected cytochemically. Certain dyes added to living cell may accumulate in lysosomes and then be demonstrated by fluorescence microscopy. For example, neutral red stain, some anti-malarial drugs, and vitamin A are taken up by lysosomes.

Lysosomes are Very Polymorphic

The most remarkable morphological characteristic of the lysosome is its polymorphism, particularly regarding the size of the particle and the irregularities of its internal structure (Fig. 13–1).

Observation of the lysosomal fraction of the liver under the electron microscope led to the recognition of dense bodies about 0.4 μm in diameter, having a single outer membrane and small granules of high electron opacity similar to the ferritin molecules. Rather pure fractions of

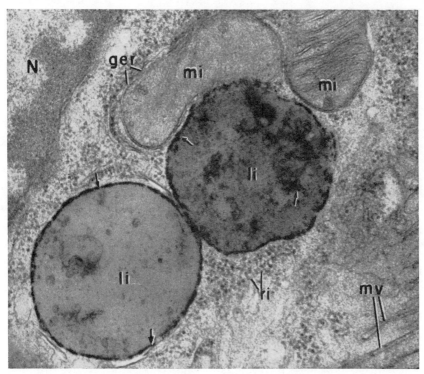

Figure 13–2 Electron micrograph of a proximal convoluted tubule cell of mouse kidney, two hours after injection of crystalline ox hemoglobin. Two absorption droplets (phagosomes, lysosomes) have formed at the apical region, and the acid phosphatase reaction becomes positive at the surface (arrows) and penetrates inside the lysosome. *ger,* granular endoplasmic reticulum; *li,* lysosomes; *mi,* mitochondria; *mv,* microvilli; *N,* nucleus; *ri,* ribosomes. ×60,000. (Courtesy of F. Miller.)

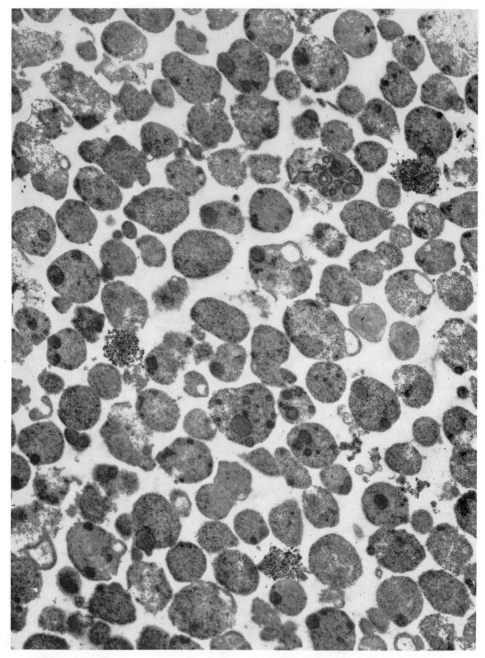

Figure 13–3 Lysosomes isolated by differential centrifugation from rat liver, showing the very dense particles and the variety of other dense material contained within the single membrane of the lysosome. ×60,000. (Courtesy of C. De Duve.)

liver lysosomes have been obtained (Fig. 13–3), and large-scale separation of these particles, as well as of peroxisomes, also from liver, has been achieved.[5] Bodies with similar morphologic characteristics were observed in intact liver cells and named "pericanalicular dense bodies" because of their preferential location along the fine bile canaliculi (Fig. 13–1).

According to the current interpretation, the polymorphism is the result of the association of primary lysosomes with the different materials that are phagocytized by the cell. A summary of these concepts is presented in Figure 13–4.

Primary Lysosomes and Three Types of Secondary Lysosomes

At present four types of lysosomes are recognized, of which only the first is the *primary lysosome;* the other three may be grouped together as *secondary lysosomes.*

(1) The *primary lysosome* (i.e., *storage granule*) is a small body whose enzymatic content is synthesized by the ribosomes and accumulated in the endoplasmic reticulum. From there it penetrates into the Golgi region, where the first acid-phosphatase reaction takes place[4] (Fig. 13–4). The primary lysosome may be charged preferentially with one type of enzyme or another; it is only in the secondary lysosome that the full complement of acid hydrolases is present. The formation of primary lysosomes may be followed in cultures of monocytes, which become transformed into macrophages. In a short time there is considerable synthesis of hydrolytic enzymes, which may be blocked by puromycin. In these activated cells, using ^{3}H-leucine and radioautography at the electron microscopic level, the transfer of protein was observed in the following sequence: endoplasmic reticulum → Golgi complex → lysosomes.

(2) The *heterophagosome* or *digestive vacuole* results from the phagocytosis or pinocytosis of foreign material by the cell. This body, which contains the engulfed material within a membrane, shows a positive phosphatase reaction, which may be due to the association with a primary lysosome[7] (Fig. 13–4). The engulfed material is progressively digested by the hydrolytic enzymes which have been incorporated into the lysosome. The extent of this digestion depends on the amount and chemical nature of the material and the activity and specificity of the lysosomal enzymes. Under ideal conditions digestion leads to products of low molecular weight which pass through the lysosomal membrane and are incorporated into the cell to be used again in many metabolic pathways.

(3) *Residual bodies* are formed if the digestion is incomplete. In some cells, such as ameba and other protozoa, these residual bodies are eliminated by *defecation* (Fig. 13–4). In other cells they may remain for a long time and may be important in the aging process. For example, the pigment inclusions found in nerve cells of old animals may be a result of this type of process.

(4) The *autophagic vacuole, cytolysosome,* or *autophagosome* is a special case, found in normal cells, in which the lysosome contains a part of the cell in the process of digestion (e.g., a mitochondrion or portions of the endoplasmic reticulum) (Fig. 13–4). During starvation the liver cell shows numerous autophagic vacuoles in some of which mitochondrial remnants can be found. This is a mechanism by which the cell can achieve the degradation of its own constituents without irreparable damage (see De Duve,

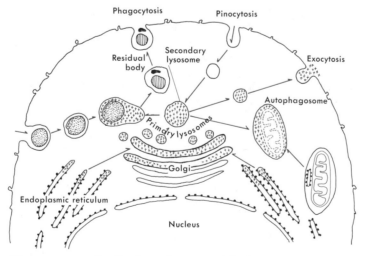

Figure 13–4 Diagram representing the dynamic aspects of the lysosome system. Observe the relationships between the processes of phagocytosis, pinocytosis, exocytosis, and autophagy.

1974; Allison, 1974). In the liver, autophagy may be induced by injection of the pancreatic hormone *glucagon*. This treatment produces a considerable increase in cytolysosomes while the small primary lysosomes diminish in number. This indicates that the pre-existing lysosomes are probably the source of hydrolytic enzymes in the autophagic vacuoles.[8]

Lysosomal Enzymes — Synthesized in the ER, Packaged in the Golgi

According to the current concepts, primary lysosomes are considered as a secretion product of the cell which, like other secretions (see Chapter 11–7), is synthesized by ribosomes, enters the endoplasmic reticulum, and reaches the Golgi region for final packaging. Since by this mechanism a cell may produce different types of lysosomes, and also *peroxisomes* and many other

secretion products, it is likely that there is a kind of *topological specificity* in this endoplasmic reticulum–Golgi system. In other words, the different secretions may be dispersed through different channels of the intracellular membrane system. In Chapter 11–3 we mentioned the relationship between the GERL region of the Golgi complex and the formation of lysosomes.

In polymorphonuclear leukocytes the primary lysosomes become concentrated in the Golgi complex to form specific granules. These are maintained in storage in the cytoplasm until used.[9] On the other hand, in most types of cells the concentration step is omitted, and the primary lysosomes appear as small vesicles with a content of low density and with identification dependent on cytochemical tests.[10] In Chapter 11 it was mentioned that the Golgi complex may be responsible for the glycosidation of proteins that are later conveyed into secretion granules, to the cell coat, and also to lysosomes.

SUMMARY:

Lysosomal Morphology and Cytochemistry

Lysosomes are cytoplasmic organelles that contain numerous (about 50) hydrolytic enzymes and in which the main functions are intracellular and extracellular digestion (Table 13–1). These structures digest materials taken in by endocytosis (phagocytosis and pinocytosis), parts of the cell (by *autophagy*), and extracellular substances. Lysosomes are separated as a fraction that is intermediate between mitochondria and microsomes (see Chapter 4–4). The lysosomal enzymes are enclosed within a membrane (accounting for their *latency*) and generally act at acid pH. Under the electron microscope they can be recognized by using different cytochemical reactions (e.g., acid-phosphatase).

Lysosomes show considerable polymorphism. The *primary lysosomes* (i.e., storage granules) are dense particles of about 0.4 μm surrounded by a single membrane. Their enzymatic content is synthesized by ribosomes in the endoplasmic reticulum and appears in the Golgi region. The formation of primary lysosomes can be blocked by puromycin. *Secondary lysosomes* (i.e., digestive vacuoles) result from the association of primary lysosomes with vacuoles containing phagocytized material. The so-called *phagosome* fuses with lysosomes (i.e., *heterophagosome*) and is digested by hydrolytic enzymes. Sometimes *residual bodies* containing undigested material are formed. These structures may be eliminated, but in most cases they remain in the cell as pigment inclusions and may be related to the aging process. The autophagic vacuole or *cytolysosome* is a special case in which parts of the cell are digested. This normal process is stimulated during starvation and by the pancreatic hormone glucagon.

13–2   ENDOCYTOSIS

The intracellular secretion of primary lysosomes is in some way coupled to another system of extracellular origin that is formed by the

process of *endocytosis*, which is related to the activity of the plasma membrane. Endocytosis includes the processes of *phagocytosis, pinocytosis*, and *micropinocytosis*, by which solid or fluid materials are ingested in bulk by the cell

(Fig. 13–4). (We have seen in Chapter 11 that *exocytosis* is the reverse process, by which membrane-lined products are released at the plasma membrane.)

Phagocytosis — The Process of Cellular Ingestion of Solid Material

Phagocytosis (Gr., *phagein*, to eat), is found in a large number of protozoa and among certain cells of the metazoa. In metazoa, rather than serving for cell nutrition, phagocytosis is, in general, a means of defense, by which particles that are foreign to the organism, such as bacteria, dust, and various colloids, are disposed of. Phagocytosis is highly developed in granular leukocytes (first described by Metschnikoff at the end of the last century), and also in the cells of mesoblastic origin ordinarily grouped under the common term *macrophagic* or *reticuloendothelial system*. The cells belonging to this group include the histiocytes of connective tissue, the reticular cells of the hematopoietic organs, and those endothelial cells lining the capillary sinusoids of the liver, adrenal gland, and hypophysis. All these cells can ingest not only bacteria, protozoa, and cell debris, but also smaller colloidal particles. In this instance phagocytosis is called *ultraphagocytosis*.

In protozoa, phagocytosis is intimately linked to ameboid motion. An ameba ingests large particles, including microorganisms, by surrounding them with pseudopodia to form a food vacuole within which the digestion of food takes place. Macrophages put out hyaline, thin (about 0.25 μm), lamellar pseudopodia that adhere to and extend over the surface of the particle until it is completely surrounded.

Analyzing the process of phagocytosis, one may distinguish two distinct phenomena. First the particle *adheres* (is *adsorbed*) to the mass of the protoplasm, and then the particle actually penetrates the cell. In some cases it has been possible to dissociate these two phases of phagocytosis. For example, at low temperature, bacteria may adhere to the cytoplasm of a leukocyte without being ingested.

Pinocytosis — The Ingestion of Fluids

The uptake of fluid vesicles by the living cell, first observed by Edwards in amebae and by Lewis in cultured cells, has been called pinocytosis (Gr., *pinein*, to drink). As can be read-ily seen in Lewis's motion pictures, the uptake of fluids is accompanied by vigorous cytoplasmic motion at the edge of the cell, as if vesicles of fluid were being surrounded and engulfed by clasping folds of cytoplasm. Vacuoles taken up at the edge of the cell are then transported to other portions of the cell.

Pinocytosis is a mechanism by which proteins and other soluble materials are incorporated into the cell. This may be easily demonstrated with proteins labeled with fluorescent dyes. The presence of the protein seems to act as a stimulus to pinocytosis, and the uptake of protein is surprisingly high. During the "feeding period," the ameba ingests approximately one third its volume of the protein solution. This material is then eliminated in five to six days. The ameba practically "drinks" the protein solution, and with it the organism may absorb other substances that normally do not penetrate. For example, ^{14}C-glucose, if dissolved in the protein solution, can enter the ameba in considerable quantities.

Pinocytosis is induced not only by proteins but also by amino acids and certain ions. In ameba the pinocytotic activity, once started, is kept going for about 30 minutes, during which time some 100 fluid channels are formed.[11] Then the process comes to a stop, and the ameba has to wait for two to three hours before starting another pinocytotic cycle. This has been interpreted as an indication that the surface membrane available for invagination is exhausted in the 30-minute period.

That phagocytosis and pinocytosis are essentially similar phenomena can be demonstrated by allowing the ameba to phagocytize some ciliated cells first and then inducing pinocytosis. The number of channels formed is much lower under these conditions. In the reverse experiment it has been found that an ameba can ingest much fewer ciliates for food.

Pinocytosis is related to the cell coat that covers the plasma membrane. A fluorescent protein may be concentrated 50 times in the cell coat before the ameba starts to engulf it.[12]

The binding to the coat explains why the concentration of the incorporated protein may reach such enormous figures.[13] A kinetic analysis of the uptake by macrophages has been made using radioactive colloidal gold. The interaction involves (1) the reversible adsorption phase, which can be related to the concentration in the extracellular fluid, and (2) the irreversible passage of the surface-bound gold into the cell. The rate of ingestion is proportional to the amount of gold attached to the cell surface

It has been calculated that in these cells between 2 and 20 per cent of the cell surface may be engulfed in a minute![14]

Figure 13–5,A shows the plasmalemma of an ameba with the membrane proper and the cell coat formed by filaments about 6 nm in diameter and 100 to 200 nm in length. After the electron-opaque material is added, it becomes heavily concentrated upon the filaments (Fig. 13–5,B). Since acid mucopolysaccharides carry a strong positive charge, they may bind the inducer by electrostatic attraction. The inducer of pinocytosis has been found to be a negatively charged substance, i.e., charged ions, acid dyes, or proteins with an isoelectric point in the acid range. In the ameba the inducer may cause a 50-fold increase in electrical resistance of the membrane prior to the formation of the typical channels.

Micropinocytosis and the Ingestion and Transport of Fluids

The use of the electron microscope demonstrated that the plasma membrane of numerous cells could invaginate, forming small vesicles of about 65 nm, and that this process could be related to pinocytosis but only in smaller vesicles. These vesicles were first found in endothelial cells lining capillaries in which they concentrate in the region adjacent to the inner and outer membranes. The presence of vesicles opening on both surfaces and of others traversing the cytoplasm suggested a possible transfer of fluid across the cell. A similar component was then observed in Schwann and satellite cells of nerve ganglia and in numerous other cell types, particularly in macrophages, muscle cells, reticular cells, etc.

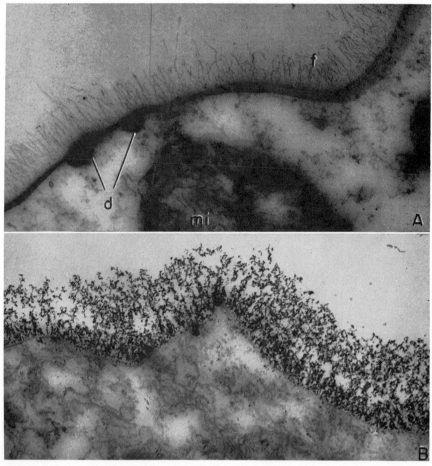

Figure 13–5 **A,** electron micrograph of the cell membrane of the ameba *Chaos chaos,* showing the extraneous coat formed by fine filaments of 5.0 to 8.0 nm in diameter and 100 to 200 nm long. *d,* dense bodies; *f,* filaments; *mi,* mitochondrion. ×55,000. **B,** same as **A,** but after the addition of thorium dioxide particles. These particles are attached to the filaments prior to the formation of channels and penetration of this material into the ameba. ×38,000. (Courtesy of P. W. Brandt and G. D. Pappas.)

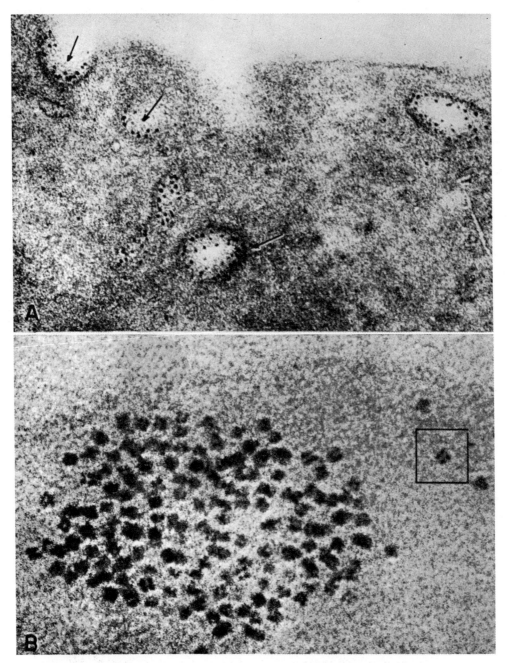

Figure 13–6 A, electron micrograph of the peripheral region of an erythroblast, showing the penetration of ferritin molecules by micropinocytosis. The arrows indicate several phases of the process starting at the surface. ×180,000. **B,** molecules of ferritin inside the cytoplasm of a reticular cell. Inset: one molecule with four dense points of about 1.5 nm, each of which contains about 300 atoms of iron. ×850,000. (Courtesy of M. Bessis.)

The transport of fluid across the capillary endothelium is assumed to occur in quantal amounts; this process implies the invagination of the plasma membrane, the formation of a closed vesicle by membrane fission, the movement of the vesicle across the endothelial cytoplasm, the fusion of the vesicle with the opposite plasma membrane, and the discharge of the vesicular content. The transendothelial passage of fluid has also been studied by the injection of peroxidase.[15, 16]

An interesting example of the physiologic importance of pinocytosis is found in the bone marrow. In the so-called erythroblastic islands, it is possible to observe reticular cells filled with iron-containing macromolecules of ferritin. These molecules leave the reticular cells and enter the erythroblast, to be used in the manufacture of hemoglobin, by the process of pinocytosis (Fig. 13–6). These findings show how cells can utilize again and again the iron resulting from the destruction of old red blood cells to form new erythrocytes.[17]

Phagocytosis and pinocytosis are active mechanisms in the sense that the cell requires energy for their operation. During phagocytosis by leukocytes oxygen consumption, glucose uptake, and glycogen breakdown all increase significantly. Induction of phagocytosis also produces an increased synthesis of phosphatidic acid and phosphatidylinositol.[18] In cultured cells, addition of ATP increases the rate of endocytosis, and this is inhibited by respiratory and other metabolic poisons.

In endocytosis the attachment of the particles to the plasma membrane apparently leads to some kind of change beneath the attachment site, which in turn may be related to the microfilaments of actin inserted in the cytoplasmic side of the membrane. These may be associated with the invagination of the membrane and the formation of pseudopods. By endocytosis, large amounts of plasma membrane may be interiorized (i.e., taken up into the cell).

A macrophage may incorporate almost twice its surface area in an hour, suggesting that there is a continuous recycling of the membrane back to the cell surface.

Micropinocytosis—Frequently Associated with the Formation of Coated Vesicles

The electron microscopic observation of vesicles undergoing endocytosis has shown that in many cases they have tiny regularly spaced bristles that cover the cytoplasmic side. It was postulated that these *coated vesicles* are involved in the uptake of proteins by pinocytosis. The use of lipoproteins conjugated with ferritin has enabled the study of the formation of coated vesicles in fibroblasts.[19] It was observed that these ferritin conjugates attach to coated regions of the membrane and then seem to invaginate at these points. Since they are observed near the Golgi region, coated vesicles may also be involved in other functions. In some cases they may contribute to the forma-

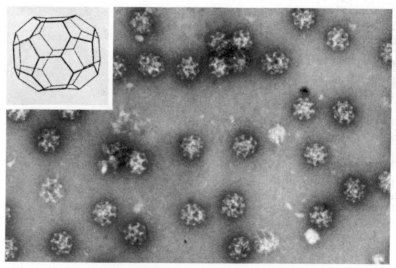

Figure 13–7 Isolated negative-stained coated vesicles. **Inset,** model showing the pentagonal and hexagonal structure of the clathrin network. ×67,500. (Courtesy of B. Pearse.)

tion of new plasma membrane. Recently coated vesicles have been isolated by cell fractionation, and a major protein named *clathrin,* of 180,000 daltons, which appears to form the network of bristles, was identified.[20, 21] The lattice of bristles shows a polyhedral structure with pentagons and hexagons formed by the clathrin molecules (Fig. 13–7). (For more details on the process of endocytosis, see Silverstein et al., 1977.)

Endocytosis — An Active Mechanism Involving the Contraction of Microfilaments

The mechanism of endocytosis which we have just described involves the contraction of microfilaments of actin and myosin present in the peripheral cytoplasm, which causes the plasma membrane to invaginate and form the endocytic vacuole. Involvement of actin microfilaments is demonstrated by the action of the drug *cytochalasin B,* which inhibits endocytosis and disorganizes actin microfilaments (see Chapter 9–4). After entering the cell, the endocytic vacuole or the phagosome moves toward the Golgi region where primary lysosomes attach to it, fuse, and liberate their contents, forming secondary lysosomes (*heterophagosomes*) (Fig. 13–4). With phase-contrast microscopy it has been possible to observe the explosive discharge of lysosomes into the endocytic vacuoles in living cells.

SUMMARY:

Endocytosis and Lysosomes

Primary lysosomes are coupled with an extracellular system that is the result of the activity of the cell membrane (endocytosis).

Bulk ingestion of solid or fluid material by the cell is called, respectively, *phagocytosis* and *pinocytosis.* Both processes show similarities and are included within the more general concept of *endocytosis.*

Phagocytosis is found in many protozoa and in certain cells of metazoa where, principally, it plays a defense role against bacteria and colloidal particles. In the ameba, phagocytosis is related to ameboid motion. The particles are surrounded by pseudopodia and are digested inside food vacuoles. In phagocytosis two distinct phases may be distinguished: adsorption of the particle and actual penetration. The macrophagic system in metazoa accumulates negatively charged vital dyes and colloids, concentrating them into vacuoles. Pinocytosis is easily observed in the ameba as the formation of fluid channels that break up inside the cell into vacuoles. This process is induced by addition of negatively charged proteins, amino acids, and ions. In a fluorescent protein solution the cell coat of an ameba concentrates the protein 50 times or more; then the protein is taken up by invagination of the cell membrane. The relation between pinocytosis and the cell coat is best observed using colloidal gold and the electron microscope. With this instrument it is observed that pinocytosis is widely present in endothelial and other cells as small vesicles of 65 nm (i.e., *micropinocytosis*). These vesicles are engaged in transcellular fluid transport. An important case is found in the erythroblasts, in which the ferritin molecules used to manufacture hemoglobin are transferred into the cell by micropinocytosis. Endocytosis, in general, is an active process that requires energy. Large amounts of plasma membrane may be interiorized by endocytosis. In this process microfilaments of actin are involved. Endocytosis may be inhibited by cytochalasin B. Endocytosis is also related to the presence of *coated vesicles.* These are formed in regions of the cell membrane that have some kind of tiny bristles on the cytoplasmic surface. A major protein (*clathrin*) is the main component of these bristles. Coated vesicles are also seen in the Golgi region. After entering the cell the endocytic vacuole fuses with primary lysosomes to form heterophagosomes.

13–3 FUNCTIONS OF LYSOSOMES — INTRACELLULAR DIGESTION

Within the secondary lysosome the ingested materials or those resulting from autophagy are subjected to the action of the large number of hydrolases already present in the lysosome (Table 13–1). The majority of these enzymes have acid pH optima, and secondary lysosomes may indeed be acidic (pH about 4.0). *Digestion of proteins* usually ends at the level of the dipeptide, which can pass through the membrane and be further degraded into amino acids.

Carbohydrates are usually hydrolyzed to monosaccharides, which are easily released. However, disaccharides or polysaccharides (cellobiose, inulin, or dextran) are not digested and remain within the lysosome.

Sucrose may be taken into macrophages by pinocytosis, but it is not hydrolyzed and remains trapped in secondary lysosomes that become swollen by the increase in osmotic pressure; such cells are intensely vacuolated. It is interesting to note that if at this point the enzyme *sucrase* is given to the cell, it penetrates into the lysosomes and normality is restored. This experiment points out one way in which certain lysosomal diseases could be treated (see later discussion).

Autophagy by Lysosomes — The Renovation and Turnover of Cell Components

Many cellular components, such as mitochondria, are constantly being removed from the cell by the lysosomal system. Cytoplasmic organelles become surrounded by membranes of smooth endoplasmic reticulum, then the lysosomal enzymes are discharged into the autophagic vacuoles, and the organelles are digested. Autophagy is a general property of eukaryotic cells and is related to the normal renovation and turnover of cellular components. Autophagy may be greatly increased in certain conditions; for example, in protozoa deprived of nutrients or in the liver of a starving animal, many autophagic vacuoles appear. This is a mechanism by which parts of the cell are broken down to facilitate survival.

Lysosomal Removal of Cells and Extracellular Material and Developmental Processes

Many developmental processes involve the shedding or remodeling of tissues, with removal of whole cells and even extracellular material.

During metamorphosis of amphibians there is considerable remodeling of tissues with destruction of numerous cells, and this is accomplished by lysosomal enzymes. For example, the degeneration of the tadpole tail is produced by the action of cathepsins (i.e., proteolytic enzymes) contained in the lysosomes. It has been found that with the regression of the tail the concentration of cathepsin increases progressively, while the total amount of enzyme remains constant.

In other organs undergoing regression, such as wolffian ducts in the female embryo and Müller's ducts in the male, there is a considerable increase in lysosomal enzymes. Some tissues that undergo regression after a period of activity may do so by the action of lysosomes. For example, the human uterus just after birth weighs 2 kg and in 9 days returns to its former size, weighing 50 gm! During this process there is a large infiltration of phagocytic cells carrying lysosomes which digest all the debris, extracellular material, and parts of the endometrium. The muscle cells of the uterus also decrease, but in volume, not number. Autophagy is prominent in the mammary gland after lactation and is similarly involved in the uterus during the menstrual cycle.[2]

Release of Lysosomal Enzymes to the Medium for Extracellular Action

We mentioned earlier that the contents of primary lysosomes may be released to the medium by a process of exocytosis somewhat similar to that of cell secretion (Fig. 13–4). The so-called *osteoclasts*, the multi-nucleated cells that remove bone, may do so by the release of lysosomal enzymes that degrade the organic matrix. This process is activated by the parathyroid hormone. In bone rudiments cultured in vitro, lysosomal enzymes may be activated by vitamin A. This causes the degradation of the chondromucoid matrix of the bone rudiments. Vitamin A intoxication may cause spontaneous fractures in animals. Release of lysosomal enzymes also occurs when cells are treated with anti-cellular antibodies in the presence of serum complement. Such a process may occur in rheumatoid arthritis, a human disease in which the cartilage of the joints may be eroded by lysosomes.

Lysosomal Enzyme Involvement in Thyroid Hormone Release and in Crinophagy

Thyroid hormones (thyroxine and triiodothyronine) are released from a large protein molecule (thyroglobulin) stored within the follicles. In 1941 De Robertis found that this release was due to the activation of a proteolytic enzyme.[22] It is now known that this enzymatic mechanism involves lysosomal proteases.

The name *crinophagy* has been applied to a mechanism by which secretory granules produced in excess of physiologic needs may be removed. This mechanism, which is a special case of autophagy, was first observed in the pituitary gland, but it is likely that it is present in many types of endocrine and exocrine glands.

Leukocyte Granules are of a Lysosomal Nature

All leukocytes of vertebrates contain granules of a lysosomal nature. In the *polymorphonuclear leukocytes* are various types of granules, demonstrable by electron microscopy and gradient centrifugation, that contain several of the known lysosomal enzymes (see Bainton et al., 1976). *Monocytes* have few lysosomes, but when they enter the tissues and are transformed into *macrophages*, they gain many lysosomes.

Lysosomes are Important in Germ Cells and Fertilization

The *acrosome* of the spermatozoon, which develops from the Golgi region and covers the anterior end of the nucleus, can be considered as a special lysosome. Indeed, it contains *protease* and *hyaluronidase* (an enzyme that digests cell coats containing hyaluronic acid) and abundant *acid phosphatase*. During fertilization of the oöcyte, hyaluronidase disperses the cells around it *(cumulus oöphorus)* and protease digests the zona pellucida, making a channel through which the sperm nucleus penetrates (see Allison, 1974). In *eggs*, lysosomes play a role in the digestion of the stored reserve materials.

Lysosomal Involvement in Human Diseases and Syndromes

Lysosomes are of particular importance in medicine, since they are involved in many diseases and syndromes. Here we shall mention only a few examples. In certain pathologic conditions, such as the above-mentioned *rheumatoid arthritis, silicosis* and *asbestosis* (diseases produced by the inhalation of silica or asbestos particles), and *gout* (in which crystals of urate accumulate in the joints), there is a release of lysosomal enzymes from the macrophages and an acute inflammation of the tissues that may lead to an increase in collagen synthesis (fibrosis).

An acute release of lysosomal enzymes occurs in states of anoxia, acidosis, and shock, and the enzymes are found to be increased in the blood. A very common example is that of the myocardial infarct. In regard to all these diseases it is important to remember the nature of the membrane of the lysosome. This structure may be made labile by many substances. For example, all the *liposoluble vitamins* (A, K, D, and E) and the steroid sex hormones tend to make the membrane more susceptible to rupture. On the other hand, cortisone, hydrocortisone, and other drugs having an anti-inflammatory action tend to stabilize the lysosomal membrane.

We have seen that lysosomes of *leukocytes* and *macrophages* are essential to the defense of the organism against bacteria and viruses. However, there are diseases, such as leprosy, tuberculosis, and others, in which bacteria are resistant to the lysosomal enzymes (because of special components of the cell wall) and survive within the macrophages. The same is true of certain mycoses and parasitoses, e.g., toxoplasmosis (see De Duve, 1974).

Storage Diseases — Caused by Mutations That Affect Lysosomal Enzymes

Several congenital diseases have been found in which the main alteration consists of the accumulation within the cells of substances such as glycogen or various glycolipids. These are also called *storage diseases* and are produced by a mutation that affects one of the lysosomic enzymes involved in the catabolism of a certain substance. For example, in *glycogenosis* type II, the liver and muscle appear filled with glycogen within membrane-bound organelles. In this disease, *α-glucosidase,* the enzyme that degrades glycogen to glucose, is absent (see Hers and Van Hoof, 1973). Some 20 congenital diseases involving lysosomes are presently known. Most of them involve glycolipids and mucopolysaccharides which accumulate in the tissues. These diseases are also due to the lack of certain lysosomal enzymes, such as β-glucosidase, sulfatidase, and others (see Table 13–2).

Many of the diseases involving the *glycosphingolipids* affect mainly the brain, because these lipids are principal components of the myelin sheath (e.g., Tay-Sachs disease, Niemann-Pick disease, and other disorders). In these patients the synthesis of sphingolipids is normal, but they can not be degraded and accumulate in the lysosomes. (For further details, see Dingle and Fell, 1969; Dingle, 1973; Hers and Van Hoof, 1973.)

Lysosomes in Plant Cells and Their Role in Seed Germination

Membrane-bound organelles containing digestive enzymes have been identified in plant cells. Acid proteases, nucleases, and glycosidases have been found. In corn seedlings the large vacuoles of parenchymatous cells may show some characteristic lysosomal enzymes, e.g., protease, carboxypeptidase, DNAse, RNAse, β-amylase, α-glucosidase, and others. In Chapter 14 we shall mention the so-called protein or aleurone bodies, the starch granules present in seeds, and the changes they undergo during germination. These changes are accomplished by the release of various lysosomal enzymes (e.g., α-amylase) that are able to hydrolyze the stored material to be used by the growing plant. Thus, lysosomes in plants may be involved in intracellular and extracellular digestion and also in the process of development (see Allison, 1974).

TABLE 13–2   STORAGE DISEASES CAUSED BY A LACK OF A LYSOSOMAL ENZYME*

Disease	Major Polysaccharide or Sphingolipid Accumulated	Enzyme Defect
Type II glycogenosis (Pompe's disease)	Glycogen	α-Glucosidase
Gaucher's disease	Ceramide glucoside (glucocerebroside)	β-Glucosidase
Niemann-Pick disease	Sphingomyelin	Sphingomyelinase
Krabbe's disease	Ceramide galactoside (galactocerebroside)	β-Galactosidase
Metachromatic leukodystrophy	Ceramide galactose-3-sulphate (sulphatide)	Sulphatidase
Ceramide lactoside	Ceramide lactoside	β-Galactosidase
Fabry's disease	Ceramide trihexoside	α-Galactosidase
Tay-Sachs disease	Ganglioside GM_2	Hexosaminidase A
Tay-Sachs disease variant	Globoside (plus ganglioside GM_2)	All hexosaminidases
Generalized gangliosidosis	Ganglioside GM_1	β-Galactosidase

*From Allison, A. C., *Lysosomes*, Oxford Biology Readers, Vol. 58, Oxford University Press, 1974.

SUMMARY:

Functions of Lysosomes

Lysosomal enzymes digest proteins into dipeptides and carbohydrates into monosaccharides. Some disaccharides (sucrose) and polysaccharides (inulin, dextran) are not digested and remain in the lysosomes. Through the process of autophagy, lysosomes are involved in the renovation and turnover of cellular components. In development, lysosomes are active in the remodeling of tissues (e.g., removal of the tadpole tail, regression of wolffian and Müller's ducts, and like processes). Digestion of extracellular material involves the release of primary lysosomes by exocytosis (e.g., osteoclasts). Degradation of bone is activated by vitamin A and parathyroid hormone. In rheumatoid arthritis lysosomal enzymes erode cartilage. Crinophagy is a process by which excess secretory granules are removed. The acrosome of the spermatozoon, which develops in the Golgi, is a special lysosome.

The study of lysosomes is particulary important in medicine. For example, they act in rheumatoid arthritis, silicosis, abestosis, and gout. Lysosomes of leukocytes (specific granules) and monocytes are essential in defense against bacteria and viruses. There are about 20 congenital diseases called *storage diseases*, in which there is accumulation of substances (i.e., glycogen, glycolipids) in lysosomes. These diseases are due to the lack of certain lysosomal enzymes (Table 13–2).

Lysosomes are found in plant cells. In seedlings they are involved in the hydrolysis and removal of protein and starch during germination.

13–4 PEROXISOMES

With improved cell fractionation methods a second group of particles, in addition to the lysosomes, has been isolated from liver cells and other sources. These particles, which are rich in the enzymes peroxidase, catalase, D-amino-acid oxidase, and, to a lesser extent, urate oxidase, received the name *peroxisomes*.[23]

In addition to being present in the liver and kidney, *peroxisomes* have been found in protozoa, yeast, and many cell types of higher plants. Although these structures in plant cells show some morphologic similarities to the peroxisomes in animal cells, they have a different enzymatic content, including the enzymes of the glyoxylate cycle, hence their name, *glyoxysomes*.

Peroxisomes Contain Nucleoids; Microperoxisomes Lack Them

Electron microscopic studies have suggested that these particles correspond morphologically to the so-called microbodies found in kidney and liver cells. These microbodies, or peroxisomes, are ovoid granules limited by a single membrane; they contain a fine, granular substance that may condense in the center, forming an opaque and homogeneous core or neucleoid (Fig. 11–3). In a quantitative study on rat liver cells the average diameter of peroxisomes was shown to be 0.6 to 0.7 μm. The number of peroxisomes per cell varied between 70 and 100, whereas 15 to 20 lysosomes were found per liver cell.[24] In many tissues peroxisomes show a crystal-like body made of tubular subunits. The number of organelles having

these bodies is sometimes correlated with the content of urate oxidase.

In contrast to the nucleoid-containing peroxisomes found in liver and kidney are others that are smaller and lack a nucleoid. These *microperoxisomes* are found in all cells and are related to the endoplasmic reticulum. They may be considered as regions of endoplasmic reticulum in which catalase and other enzymes are compartmentalized.

Peroxisomal Enzymes — Made in the Rough ER

Peroxisomes are intimately related to the endoplasmic reticulum. Both in animal and plant cells they appear to be formed as dilatations of this part of the vacuolar system. Regions of endoplasmic reticulum become swollen and filled with an electron-dense substance. Frequently, there are continuities between the endoplasmic reticulum and the membrane of peroxisomes.

The formation of these organelles has been followed in embryonic hepatocytes by using a histochemical reaction that uses the oxidation of 3'-3'-diaminobenzidine (DAB).[25] The enzymes of peroxisomes are synthesized in the ribosomes attached to the endoplasmic reticulum. In fact, catalase was found to be formed in the microsomal fraction of liver. From studies of this type it was assumed that peroxisomes grow slowly and are destroyed after a life span of four to five days.[26] The half-life of catalase was found to be about 36 hours.[27] Studying the catalase content of peroxisomes of different size, investigators found that all of them had the same content of enzyme. It has been postulated that peroxisomes are formed very rapidly and that their transport from the endoplasmic reticulum is completed in about one hour. Then, after four to five days, these organelles are probably destroyed by autophagy.

Peroxisomes Contain Enzymes Related to the Metabolism of H_2O_2

Isolation of liver peroxisomes has demonstrated that these organelles contain four enzymes related to the metabolism of H_2O_2. In fact, three of them — urate oxidase, D-amino oxidase, and α-hydroxylic acid oxidase — produce peroxide (H_2O_2), and catalase destroys it. *Catalase* appears to be in the matrix of liver peroxisomes and represents up to 40 per cent of the total protein. Since H_2O_2 is toxic to the cell, catalase probably plays a protective role. The enzyme *urate oxidase* and two other enzymes present in amphibian and avian peroxisomes are related to the catabolism of purines.

Plant Peroxisomes are Involved in Photorespiration

In green leaves there are peroxisomes that carry out a process called *photorespiration*. In this process, *glycolic acid*, a two-carbon product of photosynthesis that is released from chloroplasts, is oxidized by *glycolic acid oxidase*, an enzyme present in peroxisomes. This oxidation, carried out by oxygen, produces hydrogen peroxide, which is then decomposed by catalase inside the peroxisome. Photorespiration is so called because light induces the synthesis of glycolic acid in chloroplasts. The entire process involves the intervention of two basic organelles: chloroplasts and peroxisomes.

In microbodies isolated from certain fungi zoospores, has been found *catalase*, a typical peroxisomal enzyme, as well as *malate synthetase* and *isocitrate lyase*, two typical enzymes of glyoxysomes (see Powell, 1976). *Glyoxysomes*, which are special plant cell organelles involved in the metabolism of stored lipids, will be studied in Chapter 14, together with the plant cell endoplasmic reticulum from which they originate.

SUMMARY:

Peroxisomes

Peroxisomes are organelles rich in peroxidase, catalase, D-amino-acid oxidase and urate oxidase. They are abundant in the liver, kidney, and in many cell types of animals and plants. Their morphology consists of 0.6 to 0.7 μm granules having a single membrane and a dense matrix. Frequently, a crystal-like condensation is observed. The enzymes of peroxisomes are involved in the metabolism of hydrogen peroxide (H_2O_2). Catalase decomposes peroxide ($2H_2O_2 \rightarrow 2H_2O + O_2$) and has a protective effect on the cell. Urate oxidase and other enzymes are related to purine metabolism.

In plants peroxisomes carry out the process of photorespiration, which involves the cooperation of chloroplasts and peroxisomes. *Glyoxysomes* are special plant organelles involved in the metabolism of stored lipids (see Chapter 14).

REFERENCES

1. De Duve, C. (1963) General properties of lysosomes. In: *Lysosomes*, p. 1. Ciba Foundation Symposium, J. & A. Churchill, London.
2. Allison, A. C. (1974) *Lysosomes, Oxford Biology Readers*, Vol. 58. Oxford Unversity Press.
3. De Duve, C. (1974) Les lysosomes. *La Recherche*, 5:815.
4. Essner, E., and Novikoff, A. B. (1962) *J. Cell Biol.*, 15:289.
5. Leighton, F., Poole, B., Beaufay, H., Baudhuin, P., Coffey, J. W., Fowler, S., and De Duve, C. (1968) *J. Cell Biol.*, 37:207.
6. Cohn, Z. A. (1968) *Excerpta Med. Int. Congr. Ser.*, 166:6.
7. Strauss, W. (1967) *J. Histochem. Cytochem.*, 15:375 and 381.
8. Deter, R. L., Baudhuin, P., and De Duve, C. (1967) *J. Cell Biol.*, 35:11C.
9. Bainton, D. F., Nichols, B. A., and Farquhar, M. G. (1976) In: *Lysosomes in Biology and Pathology*, Vol. 5, p. 3. (Dingle, T., and Dean, R. T., eds.) North-Holland Publishing Co., Amsterdam.
10. Bentfield, M. E., and Baiton, D. F. (1975) *J. Clin. Invest.*, 56:1635.
11. Chapman-Andersen, C., and Holter, H. (1955) *Exp. Cell Res. Suppl.*, 3:5.
12. Brandt, P. W. (1962) Symposium on the plasma membrane. *Circulation*, 26:1075.
13. Chapman-Andersen, C., and Holter, H. (1964) *C. R. Trav. Lab. Carlsberg*, 34:211.

14. Gosselin, R. E. (1967) *Fed. Proc., 26*:987.
15. Karnowsky, M. J. (1967) *J. Cell Biol., 35*:213.
16. Palade, G. E., and Bruni, R. R. (1968) *J. Cell Biol., 37*:633.
17. Bessis, M., and Breton-Gorius, J. (1959) *J. Rev. Hémat., 14*:165.
18. Sastry, P. S., and Hokin, L. E. (1966) *J. Biol. Chem., 241*:3354.
19. Anderson, R. G., Goldstein, J. L., and Brown, M. S. (1976) *Proc. Natl. Acad. Sci. USA, 73*:2434.
20. Crowther, R. A., Finch, J. T., and Pearse, B. M. F. (1976) *J. Molec. Biol., 103*:785.
21. Pearse, B. M. F. (1976) *Proc. Natl. Acad. Sci. USA, 73*:1255.
22. De Robertis, E. (1941) *Anat. Rec., 80*:219.
23. Beaufay, H., and Berther, J. (1963) *Biochem. Soc. Symp., 23*:66.
24. Loud, A. V. (1968) *J. Cell Biol., 37*:27.
25. Essner, E. (1968) *J. Cell Biol., 39*:42a.
26. De Duve, C., and Baudhuin, P. (1966) *Physiol. Rev., 238*:3952.
27. Poole, B., Leighton, F., and De Duve, C. (1970) *J. Cell Biol., 41*:536.

ADDITIONAL READING

Allison, A. C. (1971) Lysosomes and the toxicity of particular pollutants. *Arch. Intern. Med., 128*:131.

Allison, A. C. (1974) *Lysosomes*, Oxford Biology Readers, Vol. 58. Oxford University Press.

Bainton, D. F., Nichols, B. A., and Farquhar, M. G. (1976) Primary lysosomes of blood leukocytes. In: *Lysosomes in Biology and Pathology*, Vol. 5, p. 3–32. (Dingle, J. T., and Dean, R. T., eds.) North-Holland Publishing Co., Amsterdam.

Berlin, R. D., and Oliver, J. M. (1978) Analogous ultrastructure and surface properties during capping and phagocytosis in leukocytes. *J. Cell Biol., 77*:789.

Ciba Foundation Symposium (1963) *Lysosomes*. J. & A. Churchill Ltd., London.

Dean, R. T. (1977) Lysosomes and membrane recycling: A hypothesis. *Biochem. J., 168*:603.

De Duve, C. (1967) Lysosomes and phagosomes. *Protoplasma, 63*:95.

De Duve, C. (1969) The peroxisome: a new cytoplasmic organelle. *Proc. R. Soc. London* [Biol.], *173*:71.

De Duve, C. (1974) Les lysosomes. *La Recherche, 5*:815.

De Duve, C. (1973) Biochemical studies on the occurrence, biogenesis and life history of mammalian peroxisomes. *J. Histochem. Cytochem., 21*:941.

De Duve, C. (1975) Exploring the cell with a centrifuge. *Science, 189*:186.

De Duve, C., Lazarow, P. B., and Poole, B. (1974) Biogenesis and turnover of rat liver peroxisomes. *Adv. Cytopharmacol., 2*:219.

Dingle, J. T., and Fell, H. F. (1969) *Lysosomes in Biology and Pathology*, Vol. I and II. John Wiley & Sons, Inc., New York.

Dingle, J. T., and Dean, R. T. (1976) *Lysosomes in Biology and Pathology*, Vol. 5. North-Holland Publishing Co., Amsterdam.

Geisow, M. (1979) Coated vesicles. *Nature, 277*:90.

Hers, H. G., and Van Hoof, F. (1973) *Lysosomes and Storage Diseases*. Academic Press, New York.

Keen, J. H., Willigham, M. C., and Pastan, I. H. (1979) Clathrin coated vesicles. *Cell, 16*:303.

Leighton, F., Coloma, L., and Koenig, C. (1975) Structure, composition, physical properties and turnover of peroxisomes. *J. Cell Biol., 67*:281.

Masters, C., and Holmes, R. (1977) Peroxisomes: New aspects of cell physiology and biochemistry. *Physiol. Rev., 57*:816.

Novikoff, A. B. (1964) Golgi apparatus and lysosomes *Fed. Proc., 23*:1010.

Novikoff, A. B., and Novikoff, P. M. (1973) Microperoxisomes. *J. Histochem. Cytochem., 21*:963.

Orci, L., et al. (1973) Exocytosis-endocytosis coupling in the pancreatic beta cell. *Science, 181*:561.

Pearse, B. M. F. (1978) On the structural and functional components of coated vesicles. *J. Molec. Biol., 126*:803.

Pfeifer, U., and Scheller, H. (1975) A morphogenetic study of cellular autophagy. *J. Cell Biol., 64*:608.

Powell, M. J. (1976) Ultrastructure and isolation of glyoxysomes (microbodies) in zoospores of the fungus *Entophlyctis sp. Protoplasma, 89*:1.

Reijngoud, D. J. (1978) The pH and transport of protons in lysosomes. *Trends. Biochem. Sci., 3*:178.

Silverstein, S. C., Steinman, R. M., and Cohn, Z. A. (1977) Endocytosis. *Annu. Rev. Biochem., 46*:669.

Stacey, D. W., and Allfrey, V. G. (1977) Evidence for autophagy of microinjected proteins in HeLa cells. *J. Cell Biol., 75*:807.

Vigil, E. L. (1973) Plant microbodies. *J. Histochem. Cytochem., 21*:958.

Walters, M. N. I., and Papadimitrion, J. M. (1978) Phagocytosis: A review. *CRC Crit. Rev. Toxicol., 5*:377.

Woods, J. W., Woodward, M. P., and Roth, T. F. (1978) Common features of coated vesicles from dissimilar tissues: Composition and structure. *J. Cell Sci., 30*:87.

14

THE PLANT CELL AND THE CHLOROPLAST

14–1   The Cell Wall of Plant Cells 301
  The Cell Wall—A Network of Cellulose Microfibrils and a Matrix
  Development of Primary and Secondary Walls and Plant Cell Differentiation
  Cell Wall Components—Synthesis Associated with the Golgi or the Plasma Membrane
  Plasmodesmata Establish Communication Between Adjacent Cells
  Summary: *The Plant Cell Wall*
14–2   Plant Cell Cytoplasm 305
  The Endoplasmic Reticulum of Plants—Formation of Protein Bodies, Glyoxysomes, and Vacuoles
  Seed Development—Cell Division, Cell Expansion, and a Drying Phase
  Seed Germination Involves Synthesis of Hydrolytic Enzymes
  Glyoxysomes—Organelles Related to Triglyceride Metabolism
  Dictyosomes are involved in Several Secretion Processes
  Mitochondria can be Distinguished from Proplastids
  Summary: *Plant Cell Cytoplasm*
14–3   The Chloroplast and Other Plastids 309
  Life on Earth is Dependent on the Function of Chloroplasts in Photosynthesis
  Chloroplast Morphology Varies in Different Cells
  Chloroplasts are Motile Organelles and Undergo Division
  The Envelope, Stroma, and Thylakoids are the Main Components of Chloroplasts

  Freeze-Fracturing Best Reveals the Substructure of the Thylakoid Membrane
  Chloroplast Development—The Formation of Granal and Stromal Thylakoids
14–4   Molecular Organization of Thylakoids 314
  Several Chlorophyll-Protein Complexes in the Thylakoid Membrane
  The Photophosphorylation Coupling Factor and the Photosystems—Vectorially Disposed Across the Thylakoid Membrane
  Summary: *Structure and Molecular Organization of Chloroplasts*
14–5   Photosynthesis 319
  The Primary Reaction of Photosynthesis—The Protochemical Reaction
  Photosynthetic Carbon Reduction—The Major Group of Chemical Reactions in Photosynthesis
  The C_4 Pathway is Found in Some Angiosperms
  Summary: *Photosynthesis and the Chloroplast*
14–6   A Structural-Functional Model of the Chloroplast Membrane 325
  Ion Fluxes and Conformational Changes—Caused by Light and Darkness
  Summary: *A Structural-Functional Model and Conformational Changes*
14–7   Chloroplasts as Semiautonomous Organelles 327
  Summary

Although in many chapters of this book examples taken from the Plant Kingdom are mentioned, most of the emphasis is on the eukaryotic animal cell. Here, we would like to refer to some special characteristics of plant cells, in particular, the thick *cell wall* outside the plasma membrane (Fig. 1–1) and certain organelles — the endoplasmic reticulum, the Golgi apparatus, and especially the *plastids* — that are related to

the synthesis and storage of substances that are found in these cells. Of the plastids, the most important are the *chloroplasts,* which, along with the mitochondria, are biochemical machines that produce energy transformations. In a chloroplast the electromagnetic energy contained in light is trapped and converted into chemical energy by the process of *photosynthesis.*

14–1 THE CELL WALL OF PLANT CELLS

The Cell Wall — A Network of Cellulose Microfibrils and a Matrix

The cell wall constitutes a kind of *exoskeleton* that provides protection and mechanical support for the plant cell. This includes the maintenance of a balance between the osmotic pressure of the intracellular fluid and the tendency of water to penetrate the cell. When plant cells are placed in a solution that has an osmotic pressure similar to that of the intracellular fluid, the cytoplasm remains adherent to the cellulose wall. When the solution of the medium is more concentrated than that of the cell, it loses water, and the cytoplasm retracts from the rigid cell wall. On the other hand, when the solution of the medium is less concentrated than that of the intracellular fluid, the cell swells and eventually bursts.

The *growth* and differentiation of plant cells — such as the formation, from cambial cells, of *xylem vessels* (which become lignified) or of *phloem tissue* (used to transport to the rest of the plant the material that has been photosynthesized in the leaves) — is mainly the result of the special synthesis and assembly of the cell wall.

The cell wall, in some respects, can be compared to a piece of plastic reinforced by glass fibers. The wall consists of a *microfibrillar* network lying in a gel-like *matrix* of interlinked molecules. The microfibrils are mostly *cellulose* (the most abundant biological product on earth) consisting of straight polysaccharide chains made of glucose units linked by 1–4 β-bonds. Figure 14–1 indicates schematically the structural elements of cellulose, from the microscopic to the molecular level. Macrofibrils visible with the light microscope are composed of microfibrils about 25 nm in diameter, each of which is, in turn, composed of about 2000 cellulose chains. About 100 cellulose chains are held together in an elementary fibril. X-ray diffraction reveals a crystalline pattern in cellulose, with a repeating period of 1.03 nm along the fiber axis. This corresponds to a *cellobiose* unit composed of two β-glucose molecules.

In addition to some protein, the matrix contains other polysacchardies and *lignin*. Two

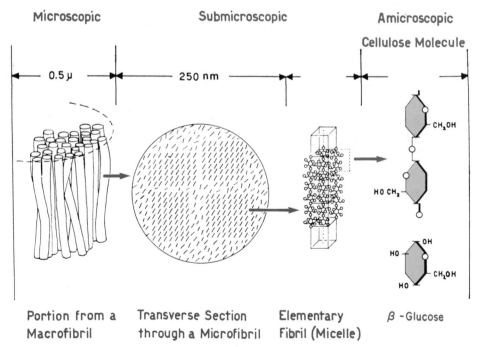

Figure 14–1 Structural elements of cellulose at different levels of organization. (From Mühlethaler, K., "Plant Cell Walls," *in* Brachet, J., and Mirsky, A. E., (Eds.), *The Cell,* Vol. 2, New York, Academic Press, Inc., 1961.)

major polysaccharide fractions are 1) *pectin sub-stances* (containing galactose, arabinose, and ga-lacturonic acid), which are soluble in water, and 2) *hemicelluloses* (composed of glucose, xylose, mannose, and glucuronic acid), which are ex-tracted with alkali.

Lignin is found only in mature cell walls and is made of an insoluble aromatic polymer result-ing from the polymerization of phenolic alco-hols. Certain cell walls may have *cuticular sub-stances* (*cutin waxes*) and *mineral* deposits in the form of calcium and magnesium carbonates and silicates. In fungi and yeasts the cell wall is com-posed of *chitin*, a polymer of glucosamine.

Development of Primary and Secondary Walls and Plant Cell Differentiation

We mentioned that cell walls are complex and highly differentiated in certain plant tissues. In fact, to a great extent the cell wall determines the *shape* of cells and serves in the classification of plant tissues. (Consideration of specialized plant tissues is beyond the scope of this book, see Loewus, 1973; Preston, 1974; Northcote, 1974; Albersheim, 1975.) In general, *primary* and *sec-ondary* walls develop in a special sequence and are differentiated by the disposition of the cellu-lose microfibrils and by the composition of the matrix.

Classification of the layers in a mature wall depends upon the knowledge of its development during and after cell division. As will be de-scribed in Chapter 17–2, the *cell plate*, which is formed mainly from vesicles of the Golgi com-plex that become aligned at the equatorial plane (Fig. 14–2), forms the intercellular layer or *middle lamellae* of the more mature cell wall; this layer contains pectin substances. After cell division each daughter cell deposits other layers that con-stitute the *primary cell wall*, which is composed of pectin, hemicellulose, and a loose network of the cellulose microfibrils, most of which are oriented transverse to the long axis of the cell.

When the cell has enlarged to its mature size, usually with the development of large vac-uoles (Fig. 1–1), the *secondary wall* appears. This consists of material that is added to the inner surface of the primary wall, either as a *homogene-ous thickening*, as in cells forming sieve tubes (*phloem*), or as *localized thickenings*, as in xylem vessels. In both cases this material consists mainly of cellulose and hemicellulose, with little pectin.

In xylem development a further differentiation

consists of the penetration of *lignin* from the outside into the secondary thickenings. This hy-drophobic polymer replaces water and finally encrusts all the microfibrils and the matrix. At this stage the wall is lignified and the cell dies.

Cell Wall Components — Synthesis Associated with the Golgi or the Plasma Membrane

In the biogenesis of cellulose and other cell wall components two main pathways have been described. One involves the *Golgi complex,* and the other appears to be directly associated with the *plasma membrane* (see Brown and Willison, 1977).

Involvement of the Golgi is very evident in some scale-bearing algae. These scales consist of a radial microfibrillar network associated with a spiral one, and an amorphous wall. The synthe-sis of polysaccharides in the Golgi of the algae has been followed by the use of periodic acid and silver methenamine. It is evident that the Golgi membranes with their *glucosyl transferase* content are able to polymerize glucan chains into cellulosic microfibrils. The whole process of syn-thesis, assembly, and release of the scales at the cell surface has been followed under the electron microscope (see Fig. 11–5 and Brown and Ro-manovicz, 1976).

The plasma membrane is the most common site of cellulose synthesis in plant cells. Howev-er, one must not disregard the essential func-tions that are probably carried out by the en-doplasmic reticulum and the Golgi complex (see Chapters 10 and 11). It is possible that the glyco-syl transferases synthesized in the endoplasmic reticulum could be transferred to the Golgi, and from there to the plasma membrane, where they may become active in the synthesis of microfi-brils.[1] Using freeze-fracturing, globular com-plexes situated at the end of growing microfi-brils have been observed in the secondary walls of cotton fibers. These globules have been inter-preted as enzyme complexes involved in the synthesis of the microfibrils. Such globules, which are mobile, may be guided by cytoplas-mic microtubules (see Brown and Willison, 1977).

Plasmodesmata Establish Communication Between Adjacent Cells

A characteristic of most plant cells is the pres-ence of bridges of cytoplasmic material that es-

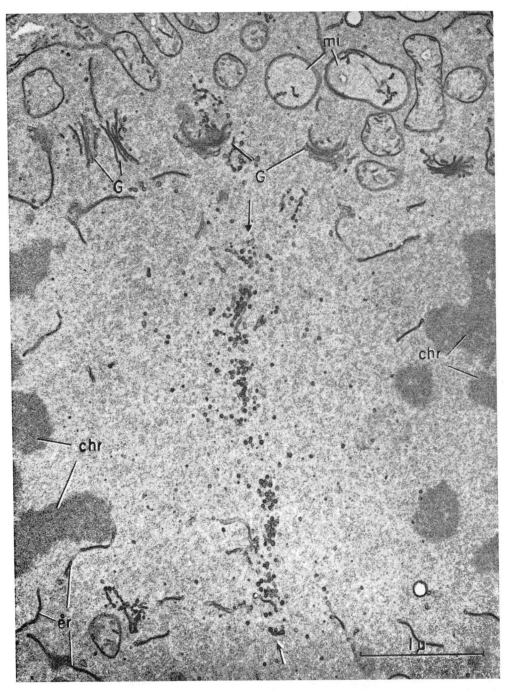

Figure 14-2 Electron micrograph of root cells of *Zea mays* at telophase. This region corresponds to the cell plate. Note at the top the marginal mitochondria (*mi*) and the Golgi complex (*G*) (dictyosomes). Between the arrows the vesicles are aligned to form the first evidence of a cell plate. *chr,* telophase chromosomes in the two daughter cells; *er,* endoplasmic reticulum. ×45,000. (Courtesy of W. Gordon Whaley and H. H. Mollenhauer.)

tablish a continuity between adjacent cells. These bridges, called *plasmodesmata*, pass through the thickness of the pectocellulose membrane. They have a thin plasma membrane and many contain tubules that establish continuities between the endoplastic reticulum of both cells. The presence of plasmodesmata permits the free circulation of fluid, which is essential to the maintenance of plant cell tonicity, and probably also allows passage of solutes and even of macromolecules. According to these concepts, cell walls do not represent complete partitions between cells, but constitute a vast syncytium supported by a skeleton formed by the pectocellulose membranes (Fig. 14–3).

The formation of plasmodesmata is related to the formation of the *cell plate,* mentioned earlier, which appears at the equator of dividing cells during telophase. At this time the cell plate is crossed by vesicles and tubules of the endoplasmic reticulum that determine the location of the plasmodesmata (see Chapter 17–2).[2, 3]

It has been suggested that plasmodesmata may play a role in cell differentiation.[4] In a cell that undergoes elongation the frequency of plasmodesmata is reduced along the axis and remains highest in the transverse walls. Plasmodesmata may be sites from which lytic enzymes attack some of the cell wall constituents, and thus its asymmetrical distribution may be related to changes occurring in later stages of differentiation.

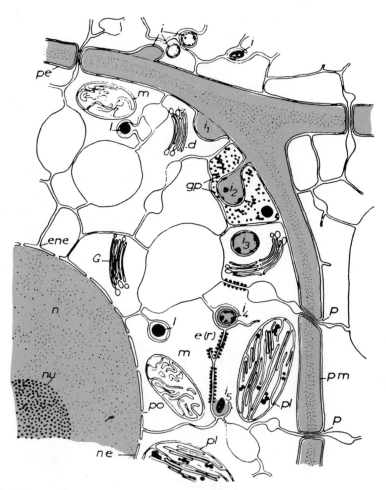

Figure 14–3 Diagram showing the interpretation of plant ultrastructure and its intercellular relationships. G, Golgi complex; *ene,* evagination of nuclear envelope; *er,* endoplasmic reticulum; i_1-i_5, steps in pinocytosis; *l,* lipid; *m,* mitochondrion; *ne,* nuclear envelope; *n,* nucleus; *nu,* nucleolus; *p,* plasmodesmata; *pl,* plastid; *pm,* plasma membrane; *po,* pores. (Courtesy of R. Buvat.)

SUMMARY:

The Plant Cell Wall

The plant cell wall represents a kind of exoskeleton that protects and provides mechanical support, as well as serving to balance the intracellular osmotic pressure with that of the medium. Differentiation of cambial cells into xylem vessels and phloem tissue is also dependent on the cell wall. The wall consists of a microfibrillar network (mainly of cellulose) and a gel-like matrix (containing pectin, hemicellulose, and lignin).

Development of the cell wall starts with the middle lamellae formed by the cell plate after cell division. Then the primary cell wall is made by the daughter cells. After growth of the cell, the secondary wall is added. In lignified tissues the secondary wall becomes encrusted with lignin. Biogenesis of cell wall components either involves the Golgi apparatus or is associated with the cell membrane. The formation of scales in algae occurs in the Golgi. The transfer of glucosyl transferases from the endoplasmic reticulum to the Golgi and the plasma membrane aids in the production of the cell wall.

Plasmodesmata are bridges of cytoplasm that establish communications between cells across the cell wall, and they may play a role in cell differentiation.

14–2 PLANT CELL CYTOPLASM

In Chapter 2 the basic similarities between the cytoplasm of animal and plant cells were mentioned. In meristematic cells the membranes of the cytoplasmic vacuolar system are relatively scanty and are masked by the numerous ribosomes that fill the cytoplasmic matrix. Indeed, in undifferentiated cells most of these particles are not attached to the membranes, but are free in the matrix (Fig. 14–3).

Microtubules are preferentially localized below the plasma membrane and are oriented tangentially (Fig. 1–1). A role in wall deposition has been postulated for these structures. Microtubules are found in the regions of cytoplasm that underlie the points where the microfibrils of the secondary wall are being deposited.[5] As mentioned earlier, they may play a role in guiding the enzymes involved in the biosynthesis of the microfibrils, and may thus determine the pattern (spiral or reticular) in which the microfibrils are deposited (see Newcomb, 1969).

The Endoplasmic Reticulum of Plants — Formation of Protein Bodies, Glyoxysomes, and Vacuoles

As in animal cells, the endoplasmic reticulum in plants has rough portions with attached ribosomes, and smooth sections. The diagram in Figure 14–3 indicates the possible connections of the endoplasmic reticulum with the nuclear envelope and the plasma membrane. It has been postulated that the plasma membrane of the plant cell invaginates and actively takes in fluid (pinocytosis). The same process also occurs in numerous animal cells (see Buvat, 1961). The great development of the endomembrane system during cell differentiation is related to the intense hydration of the cytoplasm. This process may give rise to huge vacuoles that are filled with liquid and may be confluent. As a result, the cytoplasm may become compressed in a thin layer against the cellulose membrane and may show cytoplasmic movements, called *cyclosis* (Chapter 9).

The endoplasmic reticulum plays a role in the biosynthesis of proteins, lipids, and polysaccharides and in the formation of certain organelles, such as *protein bodies, glyoxysomes,* and vacuoles. We mentioned before that the plasmodesmata are crossed by tubules of smooth endoplasmic reticulum. In cells that differentiate to form sieve tubes, pads of *callose,* a polysaccharide composed of glucose units linked by 1–3 β-bonds, is deposited over the endoplasmic reticulum. Thus, this organelle is also related to the formation of sieve pores. (see Northcote, 1974).

Development — Cell Division, Cell Expansion, and a Drying Phase

The seed can be considered as a resting stage of the plant in which the cells are in a highly dehydrated state. In addition to the primordia of the root and shoot, the seed is composed of food stores in the form of starch, fat, and proteins.

Seed development involves the following three main stages: (1) *A division phase,* in which the cells of the embryo and the cotyledons are formed, (2) *A phase of cell expansion,* in which proteins and other reserves are synthesized and stored in cotyledons,[6] and (3) A *drying phase,* in which there is loss of water with further deposition of the reserves.

Many of the proteins (i.e., *zein* in corn endosperm, *vicilin* and *legumin* in legume cotyle-

dons) are localized in special organelles called *protein bodies* or *aleurone grains* (Fig. 14–4,*A* and *B*). During the expansion phase, the cells have an extensive rough endoplasmic reticulum which is involved in the biosynthesis of stored proteins. These are first accumulated in the central vacuole, and later on smaller protein bodies are formed, possibly within cisternae of the endoplasmic reticulum or the Golgi complex. This process has been followed radioautographically by the use of labeled amino acids.[7] *Zein,* the main protein of corn, is synthesized by mRNA that is specifically associated with ribosomes attached to the endoplasmic

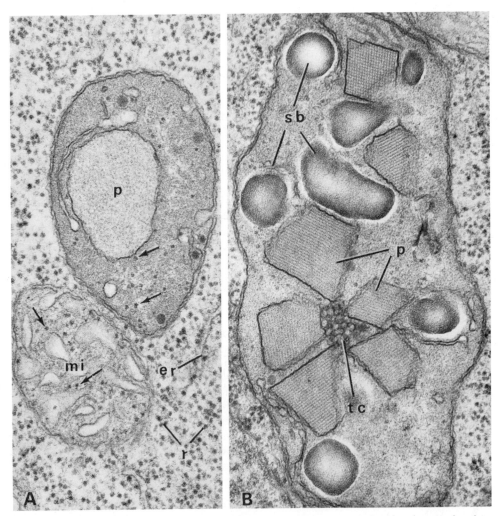

Figure 14–4   Electron micrograph of plastids in bean root tips. **A,** a young plastid and a mitochondrion (*mi*) are observed in a meristematic cell. The plastid contains a protein body (*p*). The stroma is dense and contains granules and ribosomes (arrows). Numerous ribosomes (*r*) are in the cytoplasm. **B,** a plastid containing several crystalline protein bodies (*p*) and starch bodies (*sb*). Near the center it has a system of tubules called the tubular complex (*tc*). A, ×67,000; B, ×77,000. (Courtesy of E. H. Newcomb.)

reticulum.[8] It has been found that protein bodies also have bound ribosomes that may produce zein.[9]

Seed Germination Involves Synthesis of Hydrolytic Enzymes

Seed germination is accompanied by the metabolism of the protein bodies of cotyledons. An *endopeptidase* synthesized in the seedlings penetrates into the protein bodies, causing hydrolysis of the stored proteins.[10] Enzyme synthesis is preceded by proliferation of the rough endoplasmic reticulum.

When treated with the hormone *gibberellic acid,* the aleurone cells of cereal endosperm produce α-amylase and other hydrolytic enzymes, which are released extracellulary and attack the starch reserves. The synthesis of these enzymes is correlated with an extensive proliferation of the rough endoplasmic reticulum in these aleurone cells (see Chrispeels, 1977).

Glyoxysomes – Organelles Related to Triglyceride Metabolism

The germination of the castor bean *(Ricinus)* is accompanied by breakdown of a large amount of fat stored in the endosperm. This process is accompanied by the development of *glyoxysomes,* special organelles involved in triglyceride metabolism. Glyoxysomes consist of an amorphous protein matrix surrounded by a limiting membrane. Both the phospholipid components and the enzymes of these organelles originate from the endoplasmic reticulum.[11, 12]

The enzymes of the glyoxysome are used to transform the fat stores of the seed into carbohydrates by way of the *glyoxylate cycle,* which is a modification of the Krebs cycle. The overall equation of this cycle is as follows: 2 acetyl CoA → succinate + $2H^+$. The difference between the glyoxylate cycle and the Krebs cycle is that the former uses two auxiliary enzymes, *isocitrate lyase* and *malate synthetase,* and requires two molecules of acetyl CoA, instead of one. The other three enzymes of the cycle *(citrate synthetase, aconitase,* and *malate dehydrogenase)* are the same as those present in the Krebs cycle (Fig. 14–5, *Top*).

All the enzymes of the glyoxylate cycle have been found in isolated glyoxysomes of the castor bean endosperm. When seeds of the castor bean germinate, considerable change is produced involving the synthesis of glyoxysomic enzymes. If the endosperm is fractionated at different states of germination, it is possible to observe that the marker enzyme malate synthetase shifts from fractions corresponding to the endoplasmic reticulum to others representing the glyoxysomes (Fig. 14–5, *Bottom*). These findings support the concept that these organelles are made in the endoplasmic reticulum (see Chrispeels, 1977).

Dictyosomes are Involved in Several Secretion Processes

Dictyosomes are dispersed throughout the cytoplasm without definite polarization. At telophase they aggregate at the periphery of the cell plate and form small vesicles, which fuse to form the plate (Fig. 14–2). As in animal cells, the dictyosomes of some plant cells (e.g., root cap cells of maize) are directly related to secretion. The Golgi cisternae become filled with secretion products, which are then concentrated and discharged.

Dictyosomes and their associated vesicles are numerous in cells involved in the synthesis of mucilage, and in those of the outer root cap of the bean. The mucilage is produced at the expense of the starch bodies present in plastids.[13] The dictyosomes of plant cells contain specific enzymes, such as thiamine pyrophosphatase and inosinic diphosphatase. Incorporation of labeled glucose is highest during the formation of the cell plate.[14] The role of the Golgi apparatus in cell wall formation was described above.

Mitochondria can be Distinguished from Proplastids

Mitochondria of plant cells have a structure essentially similar to that of those in animal cells. In meristems, mitochondria have relatively few crests and an abundant matrix. During differentiation this internal structure may vary. In cells engaged in photosynthesis (leaf cells), mitochondria show an increased number of crests; in cells containing starch granules (amyloplasts), mitochondria remain undifferentiated, as in meristems (Fig. 14–4, *A*).

One important point is the relationship between mitochondria and chloroplasts. Guillermond studied this problem in leaf meristems and postulated that in early stages there are two types of organelles. One is typically com-

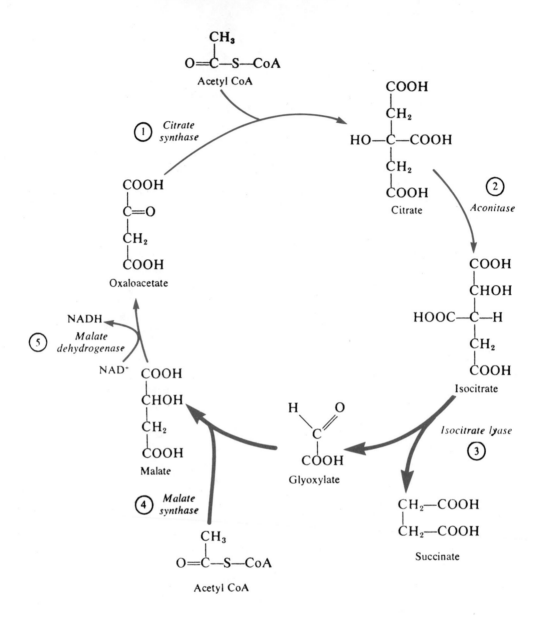

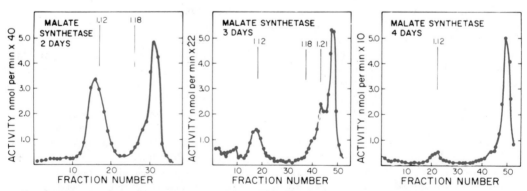

Figure 14-5 The glyoxylate cycle in castor bean endosperm and an experiment showing the localization of enzyme synthesis. **Top,** the glyoxylate cycle, showing the special steps (heavy arrows) that differentiate it from the Krebs cycle of mitochondria. **Bottom,** fractionation of the enzyme malate synthetase during germination of the castor bean endosperm. Germination was carried out for two, three, and four days, and the endosperm homogenate was centrifuged on a continuous gradient. Observe that at two days there are two peaks of enzyme activity corresponding to the endoplasmic reticulum and the glyoxysomes; with time, the first peak is reduced and the second becomes predominant. (From Gonzalez, E. and Beevers, H., *Plant Physiol.,* 57:406, 1976.)

posed of short mitochondria and the other of long filamentous bodies. These elongated organelles increase in thickness and may give rise to vesicles, starch granules, or chloroplasts, whereas the short mitochondria remain unchanged. According to this view, both types of organelles are independent. The introduction of electron microscopy has, in general, confirmed this viewpoint. Although early plastids, also called proplastids, in some ways resemble mitochondria, they are readily distinguishable because of fewer projections of the inner membrane, their large size, and the presence of dense granules in the matrix (Fig. 14–4,*A*).

SUMMARY:

Plant Cell Cytoplasm

Microtubules are preferentially localized under the plasma membrane, and they may play a role in guiding the enzymes involved in cell wall biosynthesis.

The endoplasmic reticulum (rough and smooth) is involved in the formation of protein bodies, glyoxysomes, and vacuoles. A seed can be considered as the plant primordium, in a highly dehydrated state, together with food stores of starch, fat, and proteins. Many of the proteins (zein, vicilin, legumin) are in protein or aleurone bodies. During seed germination an endopeptidase is synthesized which penetrates into the protein bodies and hydrolyzes the proteins. In cereal endosperm treated with gibberellic acid (plant hormone), α-amylase is synthesized and released to attack the starch stores. In *Ricinus* germination results in the development of *glyoxysomes*, which become involved in triglyceride metabolism. Fat stores are transformed into carbohydrates by the glyoxylate cycle.

Two special enzymes of glyoxysomes are *isocitrate lyase* and *malate synthetase* (Fig. 14–5). Golgi complexes in the form of dictyosomes are dispersed throughout the cytoplasm and are involved in secretion (e.g., of mucilage).

Mitochondria of plant cells do not differ from those of animal cells and can be distinguished from early plastids or proplastids.

14–3 THE CHLOROPLAST AND OTHER PLASTIDS

In 1883 Schimper first used the term "plastid" for special cytoplasmic organelles present in eukaryotic plant cells. The most important of all, *the chloroplasts*, are characterized by the presence of pigments such as *chlorophyll* and *carotenoids* and by their fundamental role in *photosynthesis* (see later discussion).

In addition to chloroplasts, other colored plastids may be observed. These are grouped under the name *chromoplasts*. Yellow or orange chromoplasts occur in petals, fruits, and roots of certain higher plants. In general, they have a reduced chlorophyll content and are thus less active photosynthetically. The red color of ripe tomatoes is the result of chromoplasts that contain the red pigment *lycopene*, a member of the carotenoid family. Chromoplasts containing various pigments (e.g., *phycoerythrin* and *phycocyanin*) are found in algae.

Colorless plastids or leukoplasts are found in embryonic and germ cells. During embryonic development leukoplasts in certain differentiated zones of the root produce starch granules, called *amyloplasts* (Fig. 14–4,*B*). They may be seen under the polarization microscope, because of their characteristic birefringence, or they may be distinguished by histochemical reactions for starch. Leukoplasts are also found in meristematic cells and in those regions of the plant not receiving light.

Plastids located in the cotyledon and the primordium of the stem are colorless at first but eventually become filled with chlorophyll and acquire the characteristic green color of chloroplasts.

We mentioned earlier the protein bodies or *proteinoplasts*; others contain starch (*amyloplasts*), fat (*elaioplasts or oleosomes*), or essential oils.[15] It was noted in *Phaseolus vulgaris*, that a single leukoplast may store both starch and protein; i.e., it may qualify as both an amyloplast and a proteinoplast (Fig. 14–4,*B*). The starch granules are deposited in the matrix or stroma of the plastid, while the protein, which sometimes shows a crystalline arrangement, is always accumulated within membrane-bound sacs.

Life on Earth is Dependent on the Function of Chloroplasts in Photosynthesis

Chloroplasts are the most common of the plastids and also those with the greatest biological importance. By the process of *photosynthesis*, they produce oxygen and most of the chemical energy used on our planet by living organisms. Life is esentially maintained by the chloroplasts; without them there would be no plants or animals, since the latter feed on the foodstuffs produced by plants. From a human point of view we may say that each molecule of oxygen used in respiration and each carbon atom in our bodies has at one time passed through a chloroplast. Chloroplasts are localized mainly in the cells of the leaves of higher plants and in algae.

From the evolutionary viewpoint the first living organisms were anaerobic and appeared at least one billion years before photosynthesis. This process appeared first in prokaryotes and then in eukaryotes. With the release of oxygen, a by-product of photosynthesis, the atmosphere favored aerobic organisms, and anaerobic organisms became a small fraction of all living forms.

Photosynthesis is carried out in special membranes in photosynthetic prokaryotes and in the chloroplasts of eukaryotes. These organelles are able to capture light energy and to transform it into chemical energy in the form of ATP, NADP$^+$, and the various foodstuffs that are produced. We shall see in this section that the chloroplast has an extremely elaborate molecular organization adapted to the complex chemical pathways that characterize photosynthesis.

Chloroplast Morphology Varies in Different Cells

The *shape* of chloroplasts may vary in different cells within a species, but these organelles are relatively constant within cells of the same tissue. In leaves of higher plants, each cell contains a large number of spheroid, ovoid, or discoid chloroplasts. Some are club-shaped, having a

thin middle zone and bulging ends filled with chlorophyll. Chloroplasts are frequently vesicular, with a colorless center. The presence of starch granules is detected by the characteristic blue iodine reaction.

The *size* of chloroplasts varies considerably. The average diameter in higher plants is 4 to 6 μm. This is constant for a given cell type, but sexual and genetic differences are found. For instance, chloroplasts in polyploid cells are larger than those in the corresponding diploid cells. In general, chloroplasts of plants grown in the shade are larger and contain more chlorophyll than those of plants grown in sunlight.

The *number* of chloroplasts is relatively constant in the different plants. Algae, such as *chlamydomonas*, often possess a single huge chloroplast (Fig. 14–6). In higher plants there are 20 to 40 chloroplasts per cell. It has been calculated that the leaf of *Ricinus communis* contains about 400,000 chloroplasts per square millimeter of surface area. When the number of chloroplasts is insufficient, it is increased by division; when excessive, it is reduced by degeneration.

The *distribution* of the chloroplasts within the cell may vary with the amount of light energy. Chloroplasts are sometimes distributed homogeneously within the cytoplasm, but are frequently packed near the nucleus, or close to the cell wall.

Chloroplasts are Motile Organelles and Undergo Division

Observation of living epidermal cells from leaves of *Iris* and other genera has shown that chloroplasts are displaced and deformed by the action of cytoplasmic streaming (cyclosis). In addition to this passive *motility*, active movements of an ameboid or contractile type, which are sometimes related to the degree of illumination, have been observed.

Changes in shape and volume caused by the presence of light have been observed in chloroplasts isolated from spinach. The volume decreases considerably after the chloroplasts are struck by light and photophosphorylation is initiated; this effect is reversible. In the dark, contraction of chloroplasts may be induced by addition of ATP.[16] Two proteins having contractile properties, which may account for this phenomenon, have been extracted from isolated chloroplasts.[17] Chloroplasts apparently multiply by division — elongation of the plastid and constriction of the central portion. They have a

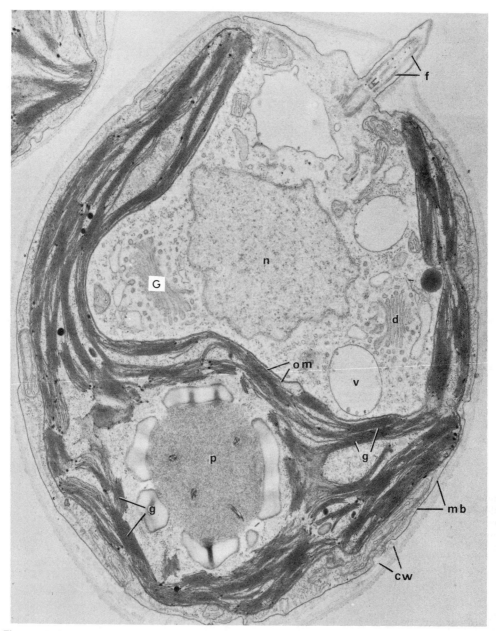

Figure 14–6 Electron micrograph of *Chlamydomonas* showing the huge chloroplast; grana (*g*), the pyrenoid (*p*), Golgi complex (G), flagellum (*f*), membrane (*mb*), cell wall (*cw*), nucleus (*n*), outer membrane of the chloroplast (*om*), vacuole (*v*), ×8000. (Courtesy of G. E. Palade.)

higher density than the cytoplasm and migrate to the centrifugal pole of the cell when submitted to the action of centrifugal force.

The Envelope, Stroma, and Thylakoids are the Main Components of Chloroplasts

Studies using the light microscope had already demonstrated that many chloroplasts have a het-erogeneous structure made up of small granules called *grana*, which are embedded within the stroma, or matrix. The size of the grana varies from 0.3 to 1.7 μm depending on the species (Fig. 14–6). The smallest ones, within the limit of microscopic visibility, are more numerous. They are flat bodies shaped like platelets or disks, which in a lateral view appear as dense bands perpendicular to the chloroplast surface (Fig. 14–7).

The electron microscope has revealed the true

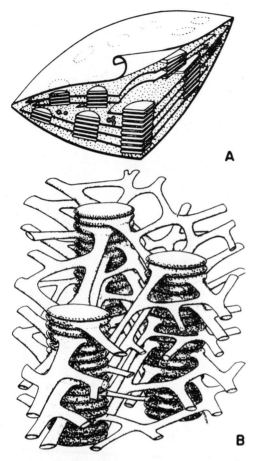

Figure 14–7 **A,** diagram of a chloroplast showing the inner structure with the grana disposed in stacks perpendicular to the surface. (From Erickson, G. A., Kahn, E., Wallis, B., and von Wettstein, D.) **B,** diagram of the ultrastructure of three grana showing the anastomosing tubules that join some of the membranous compartments of the grana. (From Weier, T. E., Stocking, C. R., Thompson, W. W., and Drever, H., *J. Ultrastruct. Res., 8*:122, 1963.)

structure of the chloroplast, with its three main components: the envelope, the stroma, and the thylakoids.

The *envelope* is made of a double limiting membrane across which the molecular interchange with the cytosol occurs. In contrast to the ultrastructure of mitochondria, the inner membrane of mature chloroplasts is not in continuity with the thylakoids (which could be compared with the mitochondrial crests). In spinach, the outer surface of the envelope is of the order of 500 cm² per gm of leaves. Isolated membranes have a yellow color due to the presence of small amounts of carotenoids. However, these areas lack chlorophyll and cytochromes and contain

only 1 to 2 per cent of the protein of the chloroplast.

The *stroma* fills most of the volume of the chloroplasts and is a kind of gel-fluid phase that surrounds the thylakoids. This component contains about 50 per cent of the chloroplast proteins and most of these are of the soluble type. It has ribosomes and also DNA, both of which are involved in the synthesis of some of the structural proteins of the chloroplast. As will be discussed later, the stroma is where CO_2 fixation occurs and where the synthesis of sugars, starch, fatty acids, and some proteins takes place.

The *thylakoids* consist of flattened vesicles arranged as a membranous network. The outer surface of the thylakoid is in contact with the stroma, and its inner surface encloses an *intrathylakoid space*. Thylakoids may be *stacked* like a pile of coins, forming the grana (Fig. 14–7,*B*), or they may be *unstacked (stroma thylakoids)*, forming a system of anastomosing tubules that are joined to the *grana thylakoids* (Fig. 14–8). The number of thylakoids per granum may vary from a few to 50 or more. The thylakoids contain about 50 per cent of the protein and all the components involved in the essential steps of photosynthesis, such as chlorophyll, the cytochromes, plastocyanine, and others (see later discussion).

Freeze-fracturing Best Reveals the Substructure of the Thylakoid Membrane

More detailed information about the macromolecular organization of the thylakoid membrane has been gained by the use of the technique of freeze-fracturing. The diagram of Figure 14–9 shows the nomenclature that has been adopted to interpret the images produced by freeze-fracturing.[18] It is based on the fact that all biological membranes consist of two leaflets, a protoplasmic (P) and an exoplasmic (E) (see also Fig. 14–10). Each leaflet has a true surface (S) and a fracture face (F). Thus, the designation PF refers to a fracture showing the protoplasmic leaflet, and ES corresponds to the true surface of the exoplasmic leaflet (in this case corresponding to the intrathylakoid space). The nomenclature also contains the subscripts *s* and *u*, referring to stacked or unstacked membranes. (Thus PFu refers to a fracture showing the protoplasmic face of an unstacked thylakoid membrane.)

With freeze-fracturing, the thylakoid membrane appears as a smooth continuum, representing one half of the lipid bilayer, with particles of various sizes believed to be protein macromole-

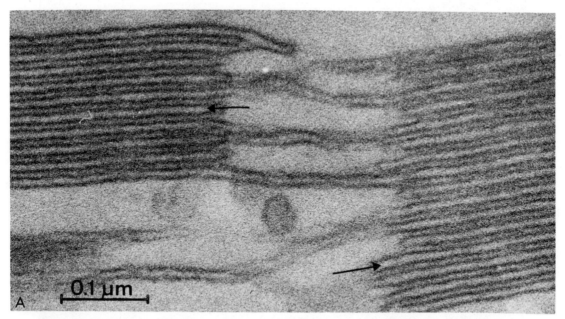

Figure 14–8 Electron micrograph of a section of two chloroplast grana and their connecting tubules (see Fig. 11–8). The arrows indicate the cavity of the membranous compartment of the grana (i.e., thylakoid). ×240,000. (Courtesy of I. Nir and D. C. Pease.)

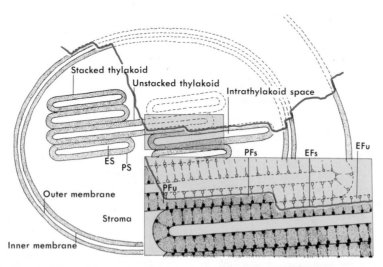

Figure 14–9 Nomenclature of freeze fracturing technique as applied to the chloroplast structure. The red line traces the course of the hypothetical fracture prior to etching. In the inset note that the cleavage plane tends to follow the hydrophobic region of the membrane. *P*, protoplasmic leaflet; *E*, exoplasmic leaflet; *S*, the membrane surface; *F*, fracture; *s*, stacked thylakoid; *u*, unstacked thylakoid. (Modified from Branton, D., et al., *Science, 190*: 54, 1975, Copyright 1975 by the American Association for the Advancement of Science.)

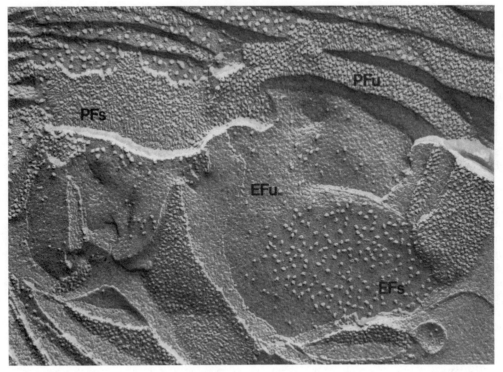

Figure 14–10 Electron micrograph of a freeze-fractured chloroplast of *Pisum sativum l.* showing the large particles observed in EFs (for a fracture showing the exoplasmic leaflet in the unstacked thylakoid, see Fig. 14–9) and the smaller particles in PFu and PSs. ×76,000. (Courtesy of Armond, P. A., Staehelin, L. A., and Arntzen, C. J., *J. Cell Biol., 73*:400, 1977.)

cules or aggregates which are asymmetrically localized within the lipid matrix. The freeze-cleaving of the membrane results in a selective segregation of the particles in either the PF or the EF leaflet. Observation of Figure 14–10 shows that PFu and PFs contain a large number of small particles, while in EFs the particles are larger and less numerous. Very few of these larger particles are found in the EFu face (corresponding to the unstacked intrathylakoid face). A possible interpretation of these observations will be given later on in this discussion of the chloroplast.

Chloroplast Development — The Formation of Granal and Stroma Thylakoids

Figure 14–11 traces the development of a chloroplast during the ontogenesis of a plant. Proplastids are limited by a double membrane. In the presence of light the inner membrane grows and gives off vesicles that arrange themselves to form larger discs. In the granal regions, piles of closely packed thylakoids are built. In the mature chloroplasts some compartments of the grana remain connected by intergranal tubules. When

plants are grown under low light intensity (*etiolation*), the vesicles formed in the proplastid aggregate, forming one or several *prolamellar bodies*. Sometimes the vesicles form a crystalline pattern consisting of regularly connected tubules (Fig. 14–11,*B*). When these plants are re-exposed to light, the vesicles may fuse into layers and again develop into grana.

14–4 MOLECULAR ORGANIZATION OF THYLAKOIDS

Modern ideas about the molecular organization of thylakoids are largely based on the fluid lipid-protein mosaic model of the membrane proposed by Singer and Nicolson in 1972 (Chapter 8–2). This basic concept includes several fundamental notions: (1) *fluidity*, which allows for the free lateral movement of components in the thylakoid membrane; (2) *asymmetry*, in the sense that the crossing of molecules from one half of the bilayer to the other is restricted, permitting the separation of different components; and (3) *economy*, since placing the components in a thin membrane eliminates random movements in the third dimension.[19, 20]

As will be mentioned later on in our study of photosynthesis, in thylakoids there are two *photosystems* (I and II), each consisting of an assembly of many chlorophyll and carotenoid molecules, and a reaction center where the primary conversion of light into chemical energy takes place. Each assembly of pigments is associated with structural proteins and a system of electron transport to constitute the photosystem (see Govindjee and Govindjee, 1975). The two photosystems act in series to transfer electrons from water to $NADP^+$. At the same time photophosphorylation is coupled to the electron transport. Here, we will only mention that *photosystem II* (PS II) is involved in the release of O_2 from H_2O, while photosystem I (PS I) plays the major role in the reduction of $NADP^+$(see Fig. 14–14).

Lipids represent about 50 per cent of the thylakoid membrane; this includes those directly involved in photosynthesis, such as *chlorophylls* (21 per cent), *carotenoids* (3 per cent), and *plastoquinones* (3 per cent). There are also *structural* lipids, which are mainly glycolipids (41 per cent), sul-

folipids (4 per cent), and a few phospholipids (10 per cent). Most of the structural lipids are highly unsaturated.

Chlorophyll is an asymmetrical molecule having a hydrophilic head made of four pyrrole rings bound to each other and forming a porphyrin (Fig. 14–12). This part of the molecule is similar to some animal pigments, such as hemoglobin and cytochromes. In chlorophyll, however, there is a Mg atom forming a complex with the four rings. In animal pigments the Mg is replaced by Fe. Chlorophyll has a long hydrophobic chain (phytol chain) attached to one of the rings.

In higher plants there are two types of chlorophyll — *a* and *b*. In chlorophyll *b* there is a —CHO group in place of the —CH₃ group, indicated by a circle in Figure 14–12. Three other types of chlorophylls are found in algae, diatoms, dinoflagellates, and in photosynthetic bacteria. Unlike the cytochromes, the chlorophylls are not bound to proteins.

A small amount of pigment absorbing at 700 nm has been found. This pigment is called P700

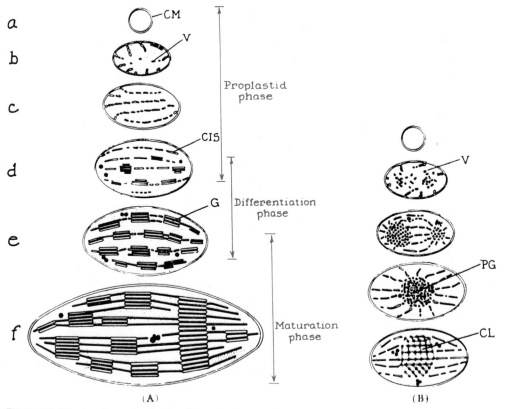

Figure 14–11 **A,** phases in the development of a proplastid into a chloroplast in the presence of light. **B,** same, but in the dark, showing the formation of the primary granum (*PG*), or prolamellar body. *CIS,* flattened cisternae; *CL,* crystal lattice; *CM,* double chloroplast membrane; *G,* granum; *V,* vesicles. (Modified from von Wettstein, D., *J. Ultrastruct. Res.,* 3:235, 1959.)

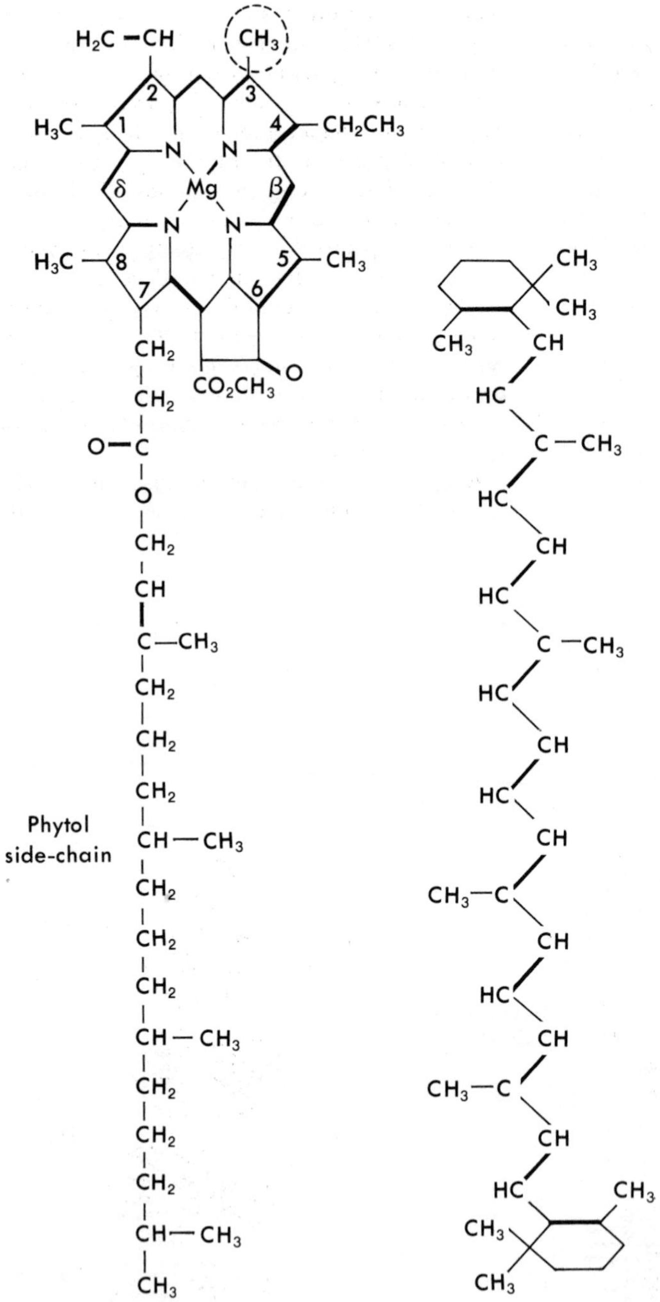

Figure 14–12 Structural formula of chlorophyll-*a* and β-carotene.

Chlorophyll-*a*

β-Carotene

Phytol side-chain

because it is bleached when it absorbs at that wavelength. (It will be shown later that P700 is related to one of the photosystems in photosynthesis.)

Pigments that belong to the group called *carotenoids* are masked by the green color of chlorophyll. In autumn, the amount of chlorophyll decreases and the other pigments become apparent. These belong to the *carotenes* and *xanthophylls*, which are both related to vitamin A. Carotenes are characterized chemically by the presence of a short chain of unsaturated hydrocarbon, which makes them completely hydrophobic (Fig. 14–12). Xanthophylls, on the contrary, have several hydroxyl groups.

Several Chlorophyll-protein Complexes in the Thylakoid Membrane

Most of the chlorophyll molecules are physically involved with intrinsic proteins, forming chlorophyll-protein complexes. The following complexes have been isolated: the *PS I complex*, containing the reaction center molecule P 700; the *PS II complex*, with the reaction center molecule P 680; and the *light-harvesting complex*. Small fragments of membranes rich in PS I and with a high chlorophyll a/chlorophyll b ratio have been isolated both from granal and stromal thylakoids; larger fragments rich in PS II and with a low chlorophyll a/chlorophyll b ratio are derived mainly from granal thylakoids[21, 22] (Table 14–1).

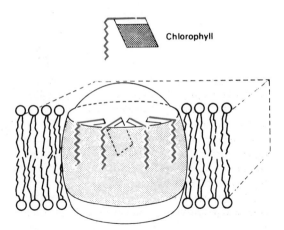

Figure 14–13 Diagram of a thylakoid in cross section showing an intrinsic protein extending across the membrane. It is proposed that the light-harvesting chlorophyll molecules are part of the boundary lipids. (Courtesy of J. M. Anderson.)

The *light-harvesting chlorophyll-protein complex (LHCP)* contains 40 to 60 per cent of the total chlorophyll of the mature thylakoid. This complex is photochemically inactive but transfers light energy to PS II and possibly to PS I, and it contains both types of chlorophyll molecules. It has been proposed that the light-harvesting chlorophylls may be located around their specific intrinsic proteins as shown in the model indicated in Figure 14–13.[19]

TABLE 14–1 SOME CHARACTERISTICS OF INTEGRAL PROTEIN COMPLEXES SEPARATE FROM CHLOROPLAST MEMBRANES*

Complex	Chlorophyll Content	Function
Photosystem II reaction center complex	a/b ratio >25, contains P680	Light-induced release of O_2 from H_2O
Light-harvesting complex	a/b ratio 1.3–1.5	No photochemical activity; binds reversibly to PS II.
Photosystem I reaction center complex	a/b >7.5, contains P700	Light-induced reduction of $NADP^+$
Cytochrome f-b_6 complex	Traces	Contains cytochromes f and b_6, non-heme iron, and bound plastocyanin. Electron carrier function.
Hydrophobic protein complex, CF_0 or HF_0	None	Proton translocation, binds CF_1 to form ATPase complex.

*From Arntzen, C. J., Dynamic structural features of chloroplast lamellae, in Current Topics in Bioenergetics, Vol. 8, p. 111. (Sanadi, D. R., and Vernon, L. P., eds.) Academic Press, Inc., New York, 1978.

The Photophosphorylation Coupling Factor and the Photosystems — Vectorially Disposed Across the Thylakoid Membrane

The *photophosphorylating system* comprises an ATPase similar to that of mitochondria, a *coupling factor* CF_1 located on the outer thylakoid surface, and a CF_0 segment across the membrane. Electron microscopic studies show that CF_1 is localized mainly in stromal thylakoids and is excluded from the stacked regions.[23] The CF_1 can diffuse in the lateral plane of the membrane.

Also in this system are the electron transport components of both photosystems. In PS I there is evidence of a vectorial disposition, with the acceptor-site ferredoxin and the ferredoxin-NADP$^+$-reductase situated toward the surface of the membrane.[24] The localization of the components of PS II is not well established; however, it is likely that this system also has a vectorial arrangement across the membrane. It is generally accepted that both photosystems I and II have their electron acceptors on the outer or stromal side of the thylakoid membrane and that the electron donors are inside (in the thylakoid space), so that the whole electron transport system is situated across the membrane.

SUMMARY:

Structure and Molecular Organization of Chloroplasts

Eukaryotic plant cells have specialized organelles — the *plastids* — which contain pigments and may synthesize and accumulate various substances. *Leukoplasts* are colorless plastids; amyloplasts produce starch; *proteinoplasts* accumulate protein; and *elaioplasts* produce fats and essential oils. *Chromoplasts* are colored plastids that contain less chlorophyll than the chloroplasts, but more carotenoid pigments, such as lycopene. Some plastids may store starch and protein at the same time.

Chloroplasts are the most important and most common plastids. Their shape, size, number, and distribution vary in different cells but are fairly constant for a given tissue. In higher plants they are discoid, 4 to 6 μm in diameter, and number about 20 to 40 per cell. The size and number are genetically controlled; in polyploids, therefore, they are larger. Chloroplasts multiply by division. The quantity of light available causes chloroplasts to undergo changes in shape and volume caused by contraction or swelling.

Many chloroplasts, under the light microscope, show small granules, called grana, of 0.3 to 1.7 μm embedded within the matrix. The electron microscope reveals their three main components: the envelope, stroma, and thylakoids. The envelope is made of two membranes, as in mitochondria; however, in the grana there is no continuity between the inner membrane and the thylakoids.

The stroma is a gel-fluid phase that contains 50 per cent of the chloroplast proteins (most are soluble). It has ribosomes and DNA. The thylakoids are flattened vesicles forming a membranous network. Within grana, thylakoids are stacked; those that are unstacked (stromal thylakoids) form a system of anastomosing tubules that connect the granal thylakoids. Thylakoids contain about 50 per cent of the chloroplast protein and all the components essential to photosynthesis.

The fine structure of the thylakoid membrane is best revealed by freeze-fracturing (Fig. 14–10), after which a smooth continuum corresponding to half of the lipid bilayer and containing particles of various sizes can be observed. In the PFu and PFs (protoplasmic leaflets unstacked and stacked), there are numerous small particles, and in the EFs (exoplasmic leaflet, stacked) the particles are larger and less numerous. In the exoplasmic unstacked leaflet (EFu), there are few of the large particles. Molecular organization of thylakoids is based on the fluid-mosaic model of the membrane, of which the main characteristics are fluidity, asymmetry, and economy (i.e., lack of movement in the third dimension).

In thylakoids, chlorophyll, carotenoid molecules, and a reaction center are assembled, forming two photosystems (I and II). Each photosystem is associated with an electron transport system and with structural proteins. Lipids represent 50 per cent of the thylakoid membrane. These include those involved in photosynthesis (i.e., chlorophyll, carotenoids, and plastoquinone) and structural lipids.

Chlorophyll, the main pigment, is an asymmetrical molecule with a porphyrin head composed of four pyrrole rings and forming a complex with a Mg atom. The molecule also has a long hydrophobic phytol chain. There are several types of chlorophylls (a, b, c, d, and e). Types a and b are in higher plants. This green pigment absorbs at 663 nm in the isolated condition; another pigment called P700 is bleached at 700 nm.

Chlorophyll molecules and structural proteins are assembled in the following complexes: the PS I complex, which has the reaction center molecule P700 and a high a/b ratio; the PS II complex, which has a low a/b ratio and is present mainly in grana thylakoids, and the light-harvesting chlorophyll-protein complex (LHCP), which contains 40 to 60 per cent of the chlorophyll in the thylakoid, but is photochemically inactive.

Chloroplasts have a photophosphorylation coupling factor (CF_1) similar to that of mitochondria. The CF_1 particles are localized mainly in the outer surface of stromal thylakoids. Both PS I and PS II have a vectorial arrangement across the membrane.

14–5 PHOTOSYNTHESIS

Photosynthesis is one of the most fundamental biological functions. By means of the chlorophyll contained in the chloroplasts green plants trap the energy of sunlight emitted as photons (i.e., excitons) and transform it into chemical energy. This energy is stored in the chemical bonds that are produced during the synthesis of various foodstuffs.

The previous chapter has emphasized how mitochondria can utilize and transform the energy contained in the foodstuffs by oxidative phosphorylation. Photosynthesis is somewhat the reverse process (Table 14–2). Chloroplasts and mitochondria have many structural and functional similarities, but there are also several differences.

The overall reaction of photosynthesis is:

$$nCO_2 + nH_2O \xrightarrow[\text{chlorophyll}]{\text{light}} (CH_2O)_n + nO_2 \quad \textbf{(1)}$$

This indicates that, essentially, photosynthesis is the combining of carbon dioxide and water to form various carbohydrates with loss of oxygen.

It has been calculated that each CO_2 molecule from the atmosphere is incorporated into a plant every 200 years, and that all the oxygen in the atmosphere is renewed by plants every 2000 years. Without plants there would be no oxygen in the atmosphere, and life would be almost impossible.

The carbohydrates first formed by photosynthesis are soluble sugars; these can be stored as granules of starch or other polysaccharides inside the chloroplasts or, more usually, inside the amyloplasts. After several steps involving different types of plastids and enzymatic systems, the photosynthesized material is either stored as a reserve product or used as a structural part of the plant (e.g., cellulose).

In early studies it was rightly suggested that in reaction (**1**) H_2O was the hydrogen donor in much the same way that, as shown in the following equation, H_2S is the donor in sulfur bacteria:

$$2H_2S + CO_2 \xrightarrow{\text{light}} (CH_2O) + 2S + H_2O \quad \textbf{(2)}$$

Thus, reaction (**1**) in higher plants can be written as follows:

$$2n\,H_2O + nCO_2 \xrightarrow{\text{light}} (CH_2O)_n + nH_2O + nO_2 \quad \textbf{(3)}$$

TABLE 14–2 DIFFERENCES BETWEEN PHOTOSYNTHESIS AND OXIDATIVE PHOSPHORYLATION

Photosynthesis	Oxidative Phosphorylation
Only in presence of light; thus periodic	Independent of light; thus continuous
Uses H_2O and CO_2	Uses molecular O_2
Liberates O_2	Liberates CO_2
Hydrolyzes water	Forms water
Endergonic reaction	Exergonic reaction
$CO_2 + H_2O +$ energy $\rightarrow$ foodstuff	Foodstuff $+ O_2 \rightarrow CO_2 + H_2O +$ energy
In chloroplasts	In mitochondria

Reaction (3) shows that water is the H_2 donor and all the O_2 liberated comes from water. Experiments using water labeled with heavy oxygen ($H_2^{18}O$) have confirmed this. In this process water participates primarily as a proton and electron donor. Reaction (3) involves a complex series of steps, of which some take place only in the presence of light and the others take place also in darkness — hence the names *light* and *dark reactions*. In the first, light is absorbed and used by chlorophyll; this is the *photochemical (Hill) reaction*. (In 1939, Robert Hill found that leaves ground in water, to which hydrogen acceptors were added [e.g., quinone], give off O_2 when exposed to light, without synthesizing carbohydrates.) In the second or dark reaction CO_2 is fixed and reduced by thermochemical mechanisms.

The Primary Reaction of Photosynthesis — The Photochemical Reaction

In the study of the photochemical reaction it is necessary to recall the process of oxidative phosphorylation in mitochondria (Chap. 12). In oxidative phosphorylation the flow of electrons is from $NADH_2$ to O_2, following the path of standard oxidation-reduction potentials (i.e., from -0.6 to $+0.81$ volts); in photosynthesis, the opposite process takes place. The electrons flow from H_2O to $NADPH_2$ (i.e., from $+0.81$ to -0.6 volts).

The *photochemical reaction*, the primary reaction of photosynthesis, takes place in the thylakoid membranes. When these are illuminated, there is a transfer of electrons from water ($E'_0 = 0.81$ volts) to the final acceptor ($E'_0 = -0.6$ volts). Transfer against such a negative electrochemical gradient uses the energy provided by the photons of light. Furthermore, the transfer of electrons is accomplished by a chain of *electron carriers* and is coupled with the *phosphorylation* of ADP to ATP. Within the thylakoid membrane, the energy of light is collected by the two *photosystems* PS I and PS II, which act as collector antennae.

Photosystems I and II. There is experimental evidence that the efficiency of photosynthesis is a function of the wavelength of the impinging light. For example, if chloroplasts are simultaneously excited at 680 and 650 nm, there is considerable stimulation of the photochemical reaction. The existence of two systems of light absorption has been postulated:

Photosystem I is stimulated at the longer wavelength and is not accompanied by the production of O_2, whereas photosystem II is activated at the shorter wavelength and yields O_2.

According to the diagram of the Figure 14–14, both photosystems operate in a sequential and interrelated fashion. When photosystem I is excited, electrons are boosted to a higher energy level, thereby reducing ferredoxin. This compound is then reoxidized by the transfer of electrons to $NADP^+$, forming $NADPH_2$. Two photons of light (i.e., two excitons) appear to be needed to produce one molecule of $NADPH_2$. This reducing component, having high energy electrons, can be used in a variety of biochemical reactions — particularly in the synthesis of carbohydrates.

From Figure 14–14 it is evident that to restore the electrons to photosystem I this process of light absorption must be coordinated with photosystem II. In photosystem II the quanta of light remove electrons from the hydrogen of water and boost them to a higher energy level, thus reducing plastoquinone. From here the electrons are brought back to photosystem I by way of an electron transfer system (i.e., cytochrome b559, cytochrome f, and plastocyanine). From this diagram it is evident that only the excited photosystem II results in the release of O_2 from water (the Hill reaction).

Electron Transport Systems. As shown in Figure 14–4, *ferredoxin* plays a role in the transport of electrons from photosystem I to $NADP^+$. Ferredoxin is a protein of about 11,600 daltons and contains iron and sulfur but lacks a porphyrin group. Another component is *ferredoxin-NADP oxireductase*, a flavoprotein that uses $NADP^+$ as an electron acceptor. Another electron acceptor, not yet identified, is indicated by Z in Figure 14–14 (see Lehninger, 1975). In the transfer of electrons from photosystem II to I there are other carriers. For example, there is *cytochrome f* (Latin, *frons*, leaf) a hydrophobic protein of 100,000 daltons isolated in non-polar solvents; other carriers include the *cytochromes* b_6 and b_3 (or *559*), tightly bound to the membrane structure. Chloroplasts also contain a copper-protein called *plastocyanine* and two quinones — *vitamin* K_1 and *plastoquinone* — the latter similar to ubiquinone in mitochondria. The sequence in the transfer of electrons from photosystem II to I begins with a non-identified carrier Q. This transfer is followed by the sequence: plastoquinone → cytochrome b559 → cytochrome f → plastocyanine.

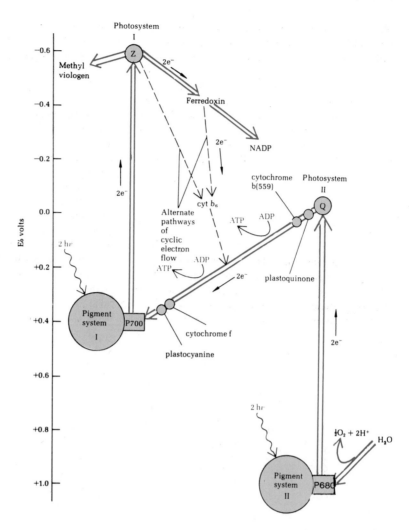

Figure 14–14 Diagram showing the arrangement of Photosystems I and II. These two systems of photosynthesis are connected by an electron transport chain situated between the acceptor Q and P700, a pigment in Photosystem I. Observe that starting from H_2O and with the release of O_2 electrons are boosted from Photosystem II toward Photosystem I, finally reaching NADPH + H$^+$. On the left the redox potential of the different components of the system is indicated. (See the description in the text.) (From Lehninger, A. L., *Biochemistry: The Molecular Basis of Cell Structure and Function*, New York, Worth Publishers, Inc., 1975.)

The mechanism of interaction between the two photosystems is still to be elucidated. Since there is a large pool of plastoquinone molecules, which are lipophilic and very mobile, the link may be not a direct, but a dynamic one.

Photophosphorylation. As in the case of mitochondria, in which the electron transport system is coupled with phosphorylation of ADP, in chloroplasts there is a light-induced phosphorylation that is coupled with the electron transfer chain of photosystem II (Fig. 14–14). This process can be uncoupled by 2,4-dinitrophenol, which also acts in mitochondria.

We mentioned before that, as in mitochondria, the coupling in chloroplasts involves an ATPase having the coupling factor CF_1 on the outer surface of the thylakoid membrane and a CF_0 hydrophobic segment across the membrane. The mechanisms possibly involved in this coupling are similar to those postulated for mitochondria (see Chapter 12–4). Here also, the *chemical-conformational* and the *chemiosmotic* coupling hypotheses have been proposed. The latter, however, is especially favored in view of the evidence provided by Jagendorf. In his experiments a suspension of spinach chloroplasts with an initial pH of 7 was transferred to a medium with an acid pH. When the chloroplasts had equilibrated to pH 4, they were put in a medium of pH 8, and ADP and Pi were added. Under these conditions there was ATP synthesis driven by the pH difference; this result conforms exactly to the postulate of Mitchell (see also Fig. 12–20).

In contrast to the oxidative phosphorylation of mitochondria, O_2 is not used in photophosphorylation of chloroplasts (Table 14–2). Green plants can produce 30 times as much ATP by photophosphorylation as by oxidative phosphorylation in their own mitochondria. In addition, these plants contain many more chloroplasts than mitochondria.

In the overall photosynthetic pathway, for every one molecule of O_2 liberated eight light quanta are needed, and two molecules of $NADPH_2$ are formed; the efficiency is close to 25 per cent. Along the electron transport there is the synthesis of three ATP molecules.

Photosynthetic Carbon Reduction — The Major Group of Chemical Reactions in Photosynthesis

The molecules of ATP and $NADPH_2$ produced in the thylakoids provide the chloroplast stroma with the energy needed to fix the CO_2 and to synthesize the carbohydrates.

The diagram in Figure 14–15 shows that

along with the energy provided by ATP, the reduced $NADPH_2$ can bring about the reduction of atmospheric CO_2 and combine it with the hydrogen to form the various carbohydrates. This process involves many steps, which have been elucidated principally by the use of radioactive CO_2. The reactions involved (Fig. 14–15) are so rapid that they appear one second or less after the addition of $^{14}CO_2$. These reactions occur in complete darkness if the plant has been previously exposed to light.[25]

In cells exposed to $^{14}CO_2$ for five seconds, the dominating compound is 3-phosphoglycerate; from this, all the compounds shown in the cycle in Figure 14–15 originate. Two triose phosphate molecules unite to form hexose (fructose) diphosphate, from which glucose phosphate is formed. Then, from glucose phosphate various disaccharides and polysaccharides are formed.

As shown in Figure 14–15, the initial en-

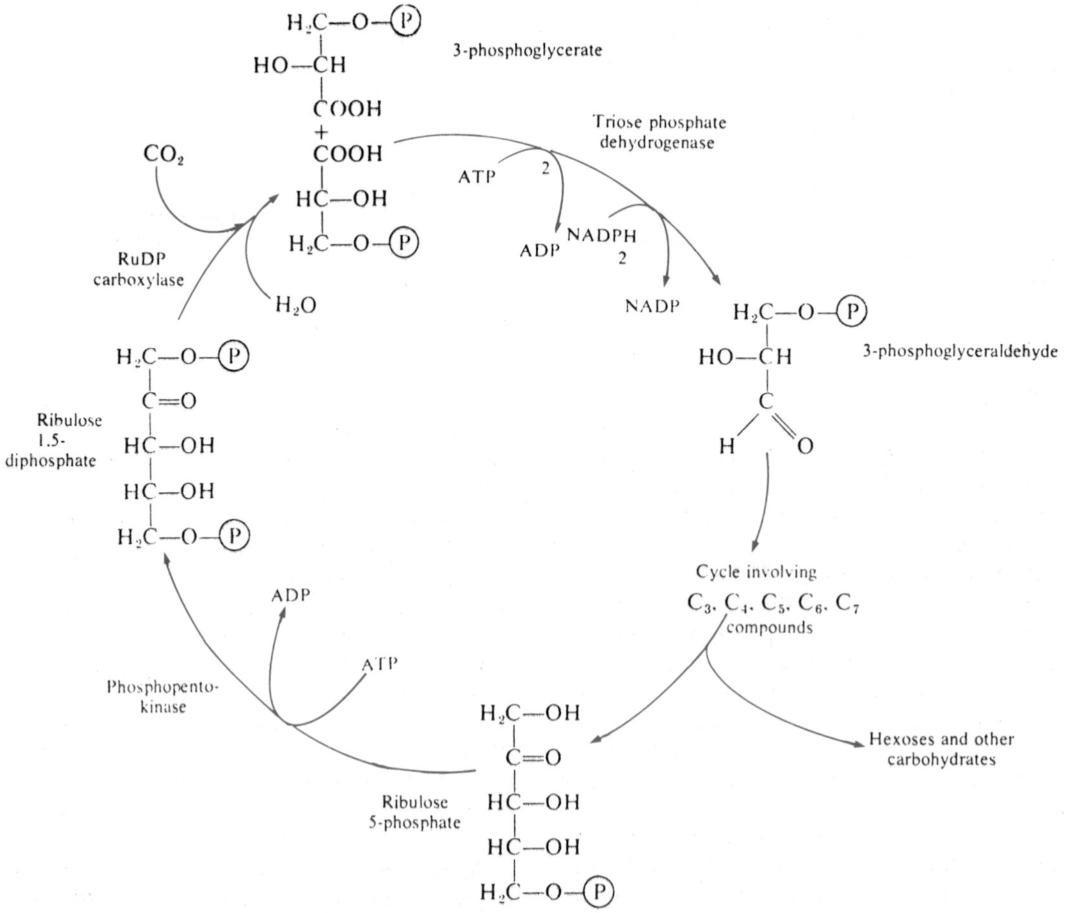

Figure 14–15 Diagram of the Calvin or C_3 cycle of photosynthesis in which CO_2 is reduced and carbohydrates are synthesized.

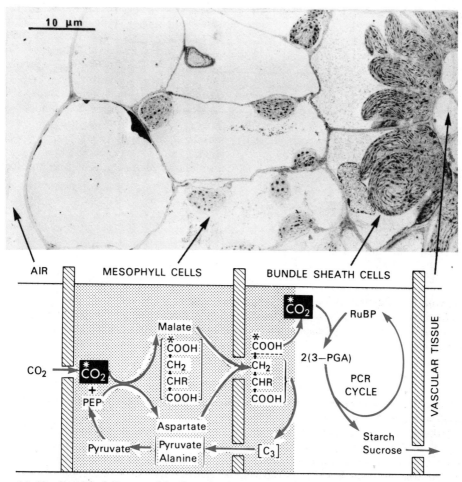

Figure 14–16 Simplified diagram of the C_4-pathway and the photosynthetic carbon reduction cycle in angiosperms. **Top,** electron micrograph of *Panicum miliaceum* leaf taken by S. Craig. (See the description in the text.) (Courtesy of M. D. Hatch, from *Trends Biochem. Sci.,* 2:199, 1977.)

zyme of the Calvin cycle is *ribulose 1,5-diphosphate carboxylase* (i.e., *carboxydismutase*), which has a large molecular weight (500,000 daltons), eight large subunits synthesized by the chloroplast, and eight small subunits produced in the cytoplasm. This enzyme, which constitutes half the stromal proteins, plays the central roles of integrating into the pentose (ribulose 1,5-diphosphate) a molecule of CO_2 from the air and of producing two molecules of *3-phosphoglycerate,* by the use of one molecule of H_2O.

The 3-phosphoglycerate (a triose sugar) is then phosphorylated by ATP and forms an activated molecule that is able to accept hydrogens and electrons from NADPH. This molecule is *3-phosphoglyceraldehyde,* which is then reduced to form hexose sugars and more complex carbohydrates.

Calvin and Benson proposed that in this cycle (also called the C_3 *cycle*) a molecule of ribulose 1,5-diphosphate is regenerated at each turn. Such regeneration is very complex and involves some twelve enzymatic reactions. To produce a hexose from the CO_2 fixation it is believed that six turns of the cycle are needed. The classical equation for the synthesis of one hexose is:

$$6CO_2 + 12\,H_2O \xrightarrow{\text{light}} C_6H_{12}O_6 +$$

$$6O_2 + 6\,H_2O \quad \textbf{(4)}$$

This represents the storage of 686 kcal/mole of hexose. This energy is provided by a total of 12 NADPH and 18 ATP molecules that themselves represent 750 kcal. Thus, the efficiency reached is as high as 90 per cent (686/750 × 100 = 90 per cent).

The C_4 Pathway is Found in Some Angiosperms

In addition to the cycle indicated in Figure 14–15, also known as the *photosynthetic carbon reduction* (PCR) or *Calvin cycle*, which is present in most higher plants, there is the so-called C_4 *pathway*, which is found in a small number of *angiosperm* species. From the viewpoint of cell biology it is interesting that this C_4 pathway is integrated with the PCR cycle in different cells of the same plant.

As shown in Figure 14–16, in the mesophyll cells the CO_2 is assimilated by carboxylation of phosphoenol pyruvate (PEP), which gives rise to C_4 acids, such as malate and aspartate. These are then transferred to the bundle sheath cells, probably by diffusion. Within these cells the CO_2 that is released by decarboxylation enters the PRC cycle, giving origin to 3-phosphoglyceric acid, and the C_3 products may go back to the mesophyll cell to enter the C_4 pathway again (see Hatch, 1977; for further details of the chemical reactions in photosynthesis refer to biochemistry textbooks).

SUMMARY:

Photosynthesis and the Chloroplast

Photosynthesis is the process by which chloroplasts trap light quanta (i.e., photons, excitons) and transform them into chemical energy. Photosynthesis is, in some ways, the reverse of oxidative phosphorylation in mitochondria. The overall reaction of photosynthesis is:

$$2nH_2O + nCO_2 \rightarrow nH_2O + nO_2 + (CH_2O)n$$

Thus, using H_2O as hydrogen donor and CO_2 from the atmosphere, carbohydrates are synthesized, and O_2 is released.

Photosynthesis consists of a *photochemical reaction* (Hill reaction), which occurs in the presence of light, and a *dark* or *thermochemical reaction*. In the first reaction, O_2 is released when the chloroplasts are exposed to light; in the second, CO_2 is fixed and carbohydrates are formed. In the photochemical reaction electrons flow from H_2O to $NADPH_2$ (i.e., from +0.81 to −0.6 volts of redox potential) because electrons are boosted to high energy levels by the absorbed light. In photosynthesis there are two photosystems (I and II) that are excited at different wavelengths. *Photosystem I* (excited at 680 to 700 nm) comprises 200 molecules of chlorophyll a, 50 of carotenoids, and one of P700. *Photosystem II* (excited at 650 nm) contains chlorophyll a and b; only this second system is associated with the release of O_2 from H_2O. Both photosystems operate in a sequential and interrelated fashion (Fig. 14–14). In photosystem I, two quanta of light, trapped by P700, boost electrons which reduce *ferredoxin*, a protein containing iron and sulfur. In turn, ferredoxin transfers the electrons to $NADP^+$, reducing it to $NADPH_2$. The electrons in photosystem I are restored by the second system. Here the light quanta remove the electrons from the hydrogen of water, releasing O_2. The electrons are transferred to P700 by a transport system which comprises several cytochromes and *plastocyanine* (a copper-containing protein). Two quinones — *vitamin K,* and *plastoquinone* — are also included in this electron transport system. As in mitochondria, the transfer of electrons is coupled with the phosphorylation of ATP at the level of photosystem II. The dark, or thermochemical, reaction involves many steps which start with the uptake of CO_2 and its reduction by $NADPH_2$ to form the various carbohydrates (Fig. 14–15). In several steps of this complex cycle of reactions, ATP is also used as an energy source. A dominating compound is 3-phosphoglyceric acid, which gives rise to glucose phosphate, from which come the various disaccharides and polysaccharides. The formation of phosphoglyceric acid depends on the initial enzyme, *carboxydismutase*. As in mitochondria, there is a strict structural-functional correlation in chloroplasts. All the enzymes of the dark reactions (like those of the Krebs cycle) are soluble and are contained in the stroma of the chloroplast. The photosystems, including electron transport and photophosphorylation, form part of the membranes of the chloroplasts.

14–6 A STRUCTURAL-FUNCTIONAL MODEL OF THE CHLOROPLAST MEMBRANE

We mentioned earlier that the use of biochemical and biophysical techniques, and especially the freeze-fracturing procedure with electron microscopy (Fig. 14–10), support the concept of a fluid mosaic membrane model for the thylakoid membrane. According to this concept, there is a lipid continuum in which protein molecules are partially or totally embedded.

We also mentioned that by solubilizing the membranes with non-ionic detergents, several integral protein complexes had been isolated and characterized. Three of them, the *photosystems I and II* and the *light-harvesting complex*, have already been described (Table 14–1). In addition, two more complexes have been separated. One corresponds to a *cytochrome f-b₆ complex* which, besides the cytochromes, contains non-heme iron, phospholipids, and carotenoids (but not chlorophyll), and which has a minimum molecular weight of 103,000 daltons.[26, 27] The other complex is the recently isolated hydrophobic protein component of the chloroplast ATPase (CF_0 or HF_0). This component has six major polypeptides with MW between 11,000 and 42,000 daltons. This HF_0 protein binds the CF_1 (coupling factor of ATPase) to reconstitute the enzyme, but is also capable of proton transfer by itself. It should be stressed here that the CF_1, which is analogous to the F_1 of mitochondria, is a peripheral enzyme with a MW near 350,000 and consisting of five polypeptides.[28] (A summary of the five intrinsic protein complexes is given in Table 14–1).

In trying to establish a structural-functional model of the chloroplast membrane, the data obtained by freeze-fracturing (Fig. 14–10) and the existence of the above-mentioned intrinsic protein complexes should be considered. Table 14–3 shows the distribution of particles in the PF and EF leaflets of the thylakoid membranes. The numbers indicate that the small PF particles are similarly distributed between the unstacked membranes (PFu) or stromal thylakoids

TABLE 14–3 DISTRIBUTION OF INTRAMEMBRANE PARTICLES AND PS II IN CHLOROPLASTS*

	Stromal Thylakoids	Granal Thylakoids
PF particles per μm^2	3409	3620
EF particles per μm^2	574	1495
Proportion of EF particles	20%	80%
Proportion of photosystem II	20%	80%

*Data from Arntzen, C. J., Dynamic structural features of chloroplast lamellae, in Current Topics in Bioenergetics, Vol. 8, p. 111. (Sanadi, D. R., and Vernon, L. P., eds.) Academic Press, Inc., New York, 1978.

and the stacked membranes (PFs), which correspond to the granal partitions (see Figs. 14–8 and 14–10). On the other hand, the large particles of the EF leaflet are more concentrated in the stacked region (EFs). Taking into consideration the surface area of the membranes in both regions, it can be calculated that 80 per cent of the EF particles are in the grana, and only 20 per cent in the stromal thylakoids. Several studies tend to identify the large EF particles with photosystem II and the associated light-harvesting complex (the complete PS II complex.[29] (see Arntzen, 1978). In fact, the proportion of PS II systems in both stacked and unstacked thylakoids is the same as that of the EF particles. (Table 14–3). Furthermore, by analyzing the size of EF particles during the development of chloroplasts it was found that while at the beginning both the PF and EF particles are small, with the greening process, the EF particles increase in size. These data were interpreted as indicating that the light-harvesting complexes were added to the basic "PS II core."

These concepts are summarized in the model shown in Figure 14–17, in which the various integral protein complexes are shown to be embedded in a lipid continuum. The diagram also emphasizes that the cleavage of the membrane during fracturing results in an EF leaflet

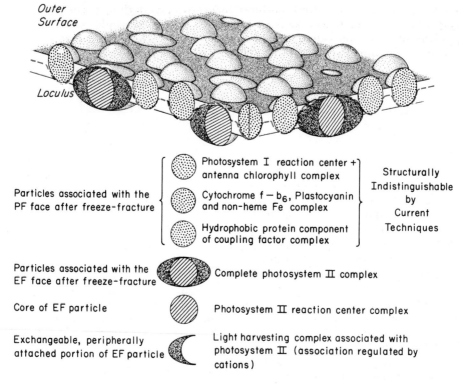

Outer Surface

Loculus

Particles associated with the PF face after freeze-fracture
- Photosystem I reaction center + antenna chlorophyll complex
- Cytochrome f − b_6, Plastocyanin and non-heme Fe complex
- Hydrophobic protein component of coupling factor complex

Structurally Indistinguishable by Current Techniques

Particles associated with the EF face after freeze-fracture — Complete photosystem II complex

Core of EF particle — Photosystem II reaction center complex

Exchangeable, peripherally attached portion of EF particle — Light harvesting complex associated with photosystem II (association regulated by cations)

---- Fracture line in freeze-fracture preparations

Figure 14–17 Model of the thylakoid membrane in which the various photosystems and protein complexes are embedded in a lipid bilayer. The diagram shows the cleavage plane of the freeze-fracturing in red and the relative number, position, and size of the particles in the P face (outer surface) and the E face (loculus). (Courtesy of C. J. Arntzen.)

with large particles and a PF leaflet with smaller and more numerous particles. From this model it is evident that while the EF particles can be identified with the complete PS II system, the small PF particles are more difficult to interpret, since they may represent any of the other protein complexes mentioned in Table 14–1.

Ion Fluxes and Conformational Changes—Caused by Light and Darkness

Structural changes in chloroplasts caused by the action of light or darkness have been observed both in vivo and in vitro. In many plant species chloroplasts flatten in the light and become more spherical in the dark.[30] Such changes appear to be mediated by ion fluxes across the membranes of individual thylakoids.[31] In fact, the size of the intrathylakoid space may decrease from 19.5 nm in the dark to 15.2 nm in the light, with a half-time of 1 min. Changes of this type are blocked by drugs that affect ion transport. Similar findings

have been observed in vitro in isolated chloroplasts.

Using freeze-fracturing it has been demonstrated that during illumination there are changes in the size and density of the particles within the membrane.[32] A more dramatic change was observed in chloroplasts suspended in a solution low in cations. There was unstacking of the membranes coupled with a dispersion of the EF particles, which are normally concentrated in the granal regions (Fig. 14–10).[33] This change is reversible, and upon re-addition of cations, stacking is re-established and the EF particles become concentrated in the grana.[34] These observations suggest that the protein complexes have mobility within the fluid membrane.

It was mentioned in Chapter 12 that mitochondria eject H^+ ions during the electron transport and that the interior becomes more alkaline. In the case of chloroplasts the opposite reactions occur; under the action of light, H^+ ions are absorbed, and the interior becomes acidified. At the same time K^+ and Mg^{2+} are ejected into the medium. In the dark, H^+ is ejected and K^+ and Mg^{2+} are reabsorbed by the chloroplasts. Thus, the chloroplast membrane has a vectorial organization that is the reverse of the mitochondrial membrane. In physical terms it is understood that dark reactions, being of a thermochemical nature, do not need an ordered structural arrangement (i.e., they are carried out by soluble enzymes). On the contrary, photophosphorylation requires the separation of positive and negative charges into specific pathways of electron flow.

As in mitochondria, according to the *chemiosmotic hypothesis* of Mitchell, the proton gradient may serve as a driving force for the formation of ATP at the CF_1 ATPase (see Barber, 1976).

SUMMARY:

A Structural-Functional Model and Conformational Changes

Several molecular complexes may be isolated by solubilizing thylakoid membranes with non-ionic detergents. In addition to photosystems I and II and the light-harvesting complex, are the cytochrome f-b_6 complex and the CF_0 or HF_0 (the hydrophobic component of ATPase). The CF_0 complex has six polypeptides and is capable of proton transfer. The CF_1 is similar to the F_1 of mitochondria and has five polypeptides. Based on the existence of these complexes and the data from freeze-fracturing, a functional model of the thylakoid membrane has been proposed (Fig. 14–17). The large EF particles are identified as PS II systems associated with light-harvesting complexes. The small PF particles are more difficult to interpret, since they may correspond to the other protein complexes mentioned in Table 14–1.

Under the influence of light, chloroplasts undergo conformational changes which are mediated by ion fluxes. Chloroplasts flatten in the light, and the intrathylakoid space is reduced. There are also changes in the size and density of the particles. In a solution low in cations there is unstacking of the membranes and dispersion of the EF particles. In the dark, the chloroplasts swell and the cavity of the thylakoid expands; H^+ is ejected, and K^+ and Mg^{2+} are reabsorbed. Under the action of light, the interior of the chloroplasts becomes more acid because of the uptake of H^+. At the same time, K^+ and Mg^{2+} are released and the cavity of the thylakoid becomes smaller.

14–7 CHLOROPLASTS AS SEMIAUTONOMOUS ORGANELLES

Chloroplasts, like mitochondria, exhibit a certain degree of functional autonomy within the intracellular environment. In fact, they undergo division and may contain some genetic information (cytoplasmic inheritance).

Chloroplasts contain their own DNA, different from nuclear DNA, special ribosomes, smaller than those present in the cytoplasm, and all the necessary molecular machinery to achieve protein synthesis.

Since the early work of Schimper and Meyer (1883) it has been accepted that plastids multiply by fission. This is easily observed in unicellular algae that contain only one chloroplast (Fig. 14–6). In the alga *Nitella* a division cycle of 18 hours has been recorded cinematographically. The division process is as orderly as chromosomal division. Plastid reproduction by fission implies a growth process of the daughter plastids. (The way in which plastids become differentiated during development and how they change in fine structure under the action of different external factors such as light have been mentioned earlier.)

There are several examples of inheritance that does not follow typical Mendelian segregation, thereby indicating an extrachromosomal type of heredity, which has been shown experimentally. When the unicellular green alga *Euglena* is grown in the dark it contains small, colorless proplastids. If under these conditions the nucleus is irradiated with an ultraviolet microbeam, the proplastids are still capable of developing into chloroplasts. However, if the cytoplasm is irradiated with the nucleus shielded, a considerable number of colorless colonies (lacking chloroplasts) develop. These results suggest that there is a cytoplasmic DNA in this alga. (For details on the genetic information of plastids see Stubbe, 1971; Bogorad, 1977.)

It is now generally accepted that a characteristic DNA occurs in chloroplasts of algae and higher plants. DNA regions resembling bacterial nucleoids were identified in thin sections under the electron microscope.[35] The quantity of DNA per chloroplast varies slightly with the species and is in the range of 2 to 5 × 10^{15} gm, or slightly less than in the bacterium *E. coli*. Segments of DNA as long as 150 μm have been separated from chloroplasts.[36] The replication of chloroplast DNA has been followed with ^{3}H-thymidine. The presence of ribosomes[5] within chloroplasts is easily demonstrated by electron microscopy. These ribosomes are smaller than cytoplasmic ribosomes. Polysomes have also been separated from chloroplasts.[37, 38]

The discovery of special ribosomes associated with chloroplasts provided evidence that these organelles contain a specific protein synthesizing system. In the presence of CO_2 as the sole source of carbon, chloroplasts actively incorporate amino acids into proteins. Protein synthesis is preferentially inhibited by chloramphenicol concentrations that do not affect protein synthesis in the cytoplasm.[39]

The involvement of the two types of protein synthesis in the assembly of chloroplasts may be studied by using cycloheximide to inhibit synthesis due to cytoplasmic ribosomes.[40]

Chloroplasts have been described as having many of the characteristics of a semiautonomous or symbiotic organism living within the plant cells. They divide, grow, and differentiate; they contain DNA, ribosomal RNA, and messenger RNA, and are able to conduct protein synthesis. It has been suggested that chloroplasts may have resulted from a symbiotic relationship between an autotrophic microorganism, one able to transform energy from light, and a heterotrophic host cell.

However, it is evident that chloroplasts have in their DNA the genetic information to code for only a limited number of proteins and the elements necessary to establish a DNA-RNA-directed protein synthesis (i.e., ribosomes, tRNA, and so forth).

In chloroplasts protein may be synthesized (1) by an exclusive chloroplastic mechanism, (2) by a mechanism involving nuclear genes and chloroplastic ribosomes, and (3) by nuclear genes and cytoplasmic ribosomes. Although the symbiont hypotheses is attractive, it is evident that synthesis and assembly of all the components integrating a chloroplast depend to a great extent on nuclear genes and on cytoplasmically directed protein synthesis (for further discussions, see Bogorad, 1977; Ohad, 1977).

SUMMARY:

Like mitochondria, chloroplasts exhibit a certain degree of functional autonomy. They undergo division and contain genetic information. They have DNA, ribosomes, and all the machinery for protein synthesis, which is inhibited by chloramphenicol. Protein in chloroplasts may be synthesized (1) by an exclusive chloroplastic mechanism, (2) by nuclear genes and chloroplastic ribosomes, and (3) by nuclear genes and cytoplasmic ribosomes. The symbiont hypothesis has also been applied to chloroplasts (see Chapter 12–6).

REFERENCES

1. Kiermayer, O., and Dobberstein, B. (1973) *Protoplasma, 77*:437.
2. Frey-Wyssling, A., López-Saez, J. F., and Mühlethaler, K. (1964) *J. Ultrastruct. Res., 10*:422.
3. Hepler, P. K., and Newcomb, E. H. (1967) *J. Ultrastruct. Res., 10*:497.
4. Juniper, B. E. (1977) *J. Theor. Biol., 66*:583.
5. Porter, K. R. (1966) In: *Principles of Biomolecular Organization*, p. 308. *Ciba Foundation Symposium*. (Wolstenholme, G. E. W., et al., eds.) J. & A. Churchill Ltd., London.

6. Millerd, A. (1975) *Annu. Rev. Plant Physiol., 26*:53.
7. Bailey, C. J., Cobb, A., and Boulter, D. (1970) *Planta* (Berl.), *95*:103.
8. Larkins, B. A., and Dalby, A. (1975) *Biochem. Biophys. Res. Commun., 66*:1048.
9. Burr, B., and Burr, F. A. (1976) *Proc. Natl. Acad. Sci. USA, 73*:515.
10. Chrispeels, M. J. (1976) *Annu. Rev. Plant Physiol., 27*:19.
11. Kagawa, T., Lord, J. M., and Beevers, H. (1973) *Plant Physiol., 51*:61.
12. Gonzalez, E., and Beevers, H. (1976) *Plant Physiol. 57*:406.

13. Northcote, D. H., and Pickett-Heaps, J. D. (1966) *Biochem. J.*, 98:159.
14. Dauwalder, M., Whaley, W. G., and Kephart, J. (1969) *J. Cell Sci.*, 4:455.
15. Amelunxen, F., and Groman, G. (1969) *Pflanzen Physiol.*, 60:156.
16. Packer, L. (1966) In: *Biochemistry of Chloroplasts*, Vol. 1, p. 233. (Goodwin, T. W. ed.) Academic Press, New York.
17. Ohnishi, T. (1964) *J. Biochem.* (Tokyo), 55:414.
18. Branton, D., et al. (1975) *Science*, 190:54.
19. Anderson, J. M. (1975) *Biochim. Biophys. Acta.*, 416:191.
20. Anderson J. M. (1977) *International Cell Biology*, p. 183. (Brinkley, B. R., and Porter, K. R., eds.) The Rockefeller University Press, New York.
21. Vernon, L. P., and Klein, S. M. (1975) *Ann. N. Y. Acad. Sci.*, 244:281.
22. Wessel, J. S. C., and Borchert, M. T. (1975) In: *Third International Congress on Photosynthesis* (Avron, M. ed.) Vol. 1, p. 473, Elsevier Publishing Company, Amsterdam.
23. Miller, K. R., and Staehelin, L. A. (1976) *J. Cell Biol.*, 68:30.
24. Trebst, A. (1974) *Annu. Rev. Plant Physiol.*, 25:423.
25. Calvin, M. (1962) *Science*, 135:879.
26. Nelson, N., and Neuman, J. (1972) *J. Biol. Chem.*, 247:1817.
27. Ke, B., Sugihara, K., and Shaw, E. R. (1975) *Biochim. Biophys. Acta*, 256:345.
28. Jagendorf, A. T. (1975) *Fed. Proc.*, 34:1718.
29. Armond, P. A., Staehelin, L. A., and Arntzen, C. P. (1977) *J. Cell Biol.*, 73:400.
30. Miller, M. M., and Nobel, P. S. (1972) *Plant Physiol.*, 49:535.
31. Murakami, S., Torres-Pereira, J., and Packer, L. (1975) In: *Bioenergetics of Photosynthesis*, p. 555. (Govinjee, ed.) Academic Press, New York.
32. Wang, A. V., and Packer, L. (1973) *Biochim. Biophys. Acta*, 347:134.
33. Ojakian, G. K., and Satir, P. (1974) *Proc. Natl. Acad. Sci. USA*, 71:2052.
34. Staehelin, L. A. (1976) *J. Cell Biol.*, 71:136.
35. Bisalputra, T., and Bisalputra, A. A. (1967) *J. Ultrastruct. Res.*, 17:14.
36. Woodcock, C. L. F., and Fernández Morán, H. (1968) *J. Molec. Biol.*, 31:627.
37. Stutz, E., and Noll, H. (1967) *Proc. Natl. Acad. Sci. USA*, 57:774.
38. Bager, R., and Hamilton, M. G. (1968) *J. Molec. Biol.*, 32:471.
39. Pogo, B. G. T., and Pogo, O. (1965) *J. Protozool.*, 12:96.
40. Hoober, J. K., Siekevitz, P., and Palade, G. E. (1969) *J. Biol. Chem.*, 244:2621.

ADDITIONAL READING

Albersheim, P. (1975) The walls of growing plant cells. *Sci. Am.*, 232:80.

Armond, P. A., and Arntzen, C. J. (1977). Localization and characterization of photosystem II in grana and stroma lamellae. *Plant Physiol.*, 59:398.

Arntzen, C. J. (1978) Dynamic structural features of chloroplast lamellae. In: *Current Topics in Bioenergetics*, Vol. 8, p. 111. (Sanadi, D. R., and Vernon, L. P., eds.) Academic Press, New York.

Avron, M. (1977) Energy transduction in chloroplasts. *Annu. Rev. Biochem.*, 46:143.

Avron, M. (1978) Energy transduction in photo phosphorylation. *FEBS Lett.*, 96:223.

Barber, J. (1976) *The Intact Chloroplast.* Elsevier Publishing Company, Amsterdam.

Barber, J. (1976) Cation control in photosynthesis. *Trends Biochem. Sci.*, 1:33.

Bogorad, L. (1977) Genes for chloroplast ribosomal RNAs and ribosomal proteins. In: *International Cell Biology*, p. 175. The Rockefeller University Press, New York.

Brown, R. M. Jr., and Romanovicz, D. K. (1976) Biogenesis and structure of Golgi-derived cellulosic scales in *Pleurochrysis. Appl. Polym. Symp.*, 28:537.

Brown, R. M., Jr., and Martin-Willison, J. H. (1977) Golgi apparatus and plasma membrane involvement in secretion and cell surface deposition, with special emphasis on cellulose biogenesis. *International Cell Biology* p. 267. The Rockefeller University Press, New York.

Buvat, R. (1961) Le réticulum endoplasmique des cellules végétales. *Ber. Dtsch. Bot. Ges.*, 74:261.

Chrispeels, M. J. (1977) The role of the endoplasmic reticulum in the biosynthesis and transport of macromolecules in plant cells. *International Cell Biology*, p. 284. The Rockefeller University Press, New York.

Douce, R., and Joyard, J. (1977) Le chloroplaste. *Recherche*, 8:527.

Gibbs, M. (1971) Structure and function of chloroplasts. Springer-Verlag, Berlin.

Govindjee, (1975) *Bioenergetics of Photosynthesis.* Academic Press, New York.

Govindjee, and Govindjee, R. (1975) Introduction to photosynthesis. In: *Bioenergetics of Photosynthesis*, Vol. 1, p. 50. (Govindjee, ed.) Academic Press, Inc., New York.

Granick, S. (1961) The chloroplasts: Inheritance, structure and function. In: *The Cell*, Vol. 2, p. 489. (Brachet, J., and Mirsky, A. E., eds.) Academic Press, New York.

Hatch, M. D. (1977) C_4 pathway photosynthesis: Mechanism and physiological function. *Trends Biochem. Sci.*, 2:199.

Hepler, P. K., and Palevitz, B. A. (1974) Microtubules and microfilaments. *Annu. Rev. Plant Physiol.*, 25:309.

Joliot, P. (1978) Photosynthesis. *Recherche*, 9:331.

Juniper, B. E. (1977) Some feature of secretory systems in plants. *Histochem. J.*, 9:659.

Ledbetter, M. C., and Porter, K. R. (1970) *Introduction to the Fine Structure of Plant Cells.* Springer-Verlag, New York.

Loewus, F., ed. (1973) *Biogenesis of Plant Cell Wall Polysaccharides.* Academic Press, New York.

Kirk, J. T. O., Tilney-Basset, R. A. E. (1978) The plastids: Their chemistry, structure, growth and inheritance. Elsevier North-Holland Publishing Co., Amsterdam.

Kreger, D. R. (1969) Cell walls. In: *Handbook of Molecular Cytology,* p. 1444. (Lima-de-Faria, A., ed.) North-Holland Publishing Company, Amsterdam.

Lehninger, A. L. (1975) Biochemistry: *The Molecular Basis of Cell Structure and Function,* 2nd Ed. Worth Publishers, Inc., New York.

Marchant, H. J. (1979) Microtubules, cell wall deposition and determination of plant cell shape. *Nature, 278:*167.

Morré, D. J., and Mollenhauer, H. H. (1976) Transport in plants, Ill. In: *Encyclopedia of Plant Physiology,* Vol. 3, p. 289. (Stocking, C. R., and Heker, U., eds.) Springer-Verlag, Berlin.

Newcomb, E. H. (1969) Plant microtubules. *Annu. Rev. Plant Physiol., 20:*253.

Northcote, D. H. (1974) Differentiation in higher plants. *Oxford Biology Readers,* No. 44.

Ohad, I. (1977) Ontogeny and assembly of chloroplast membrane polypeptides in *Chlamydomonas reinhardtii. International Cell Biology,* p. 193. The Rockefeller University Press, New York.

Preston, R. D. (1974) *The Physical Biology of Plant Cell Walls.* p. 491. Chapman & Hall, Ltd., London.

Robards, A. W., ed. (1974) *Dynamic Aspects of Plant Ultrastructure.* McGraw-Hill Book Co., New York.

Stubbe, W. (1971) Origin and continuity of plastids. In: *Origin and Continuity of Cell Organelles.* (Reinert, J., and Ursprung, H., eds.) Springer-Verlag, Berlin.

Willison, J. H. M., and Brown, R. M., Jr. (1978) Cell wall structure and deposition in *glaucocystis. J. Cell Biol., 77:*103.

Part 5

THE NUCLEUS AND CHROMOSOMES

In the following five chapters the nucleus and the chromosomes are presented as entities involved in genetic activity at the cellular level. This general topic is also called the *chromosomal basis of genetics*. The study, begun at the end of the last century, developed so rapidly that for many years it was the best known field of cytology. The development of cariology (Gr., *karion*, nucleus) was somewhat detrimental to the study of the cell as a whole and of its molecular and biochemical aspects, which are now included within the realm of cell biology.

Chapter 15 gives a general account of the progress made in recent years on knowledge of the ultrastructure of the interphase nucleus, the chromosomes, and the nucleolus. The special permeability properties of the nuclear envelope are also discussed here. Knowledge of the localization of the DNA, RNA, and nuclear proteins, as well as of the macromolecular organization of the chromosomes, is of paramount importance in the interpretation of their duplication. Eukaryotic DNA is not free but rather is assembled into chromatin. The repeating unit of chromatin is the *nucleosome*, which consists of an octamer of basic proteins called histones with 200 base pairs of DNA wrapped around them. It is astonishing to learn that within a single human chromosome several centimeters of DNA are tightly packed. The DNA is packed between 5000- and 10,000-fold in metaphase chromosomes. The way in which this is achieved is the main theme of this chapter.

Chapter 16 analyzes the life cycle of cells. The interphase period, which had been dismissed by classical cytologists because the chromosomes were not visible under the microscope, will be studied in its three periods: G_1, S, and G_2. During the S phase, the DNA complement of the cell is precisely duplicated, using several thousand starting points.

Chapter 17 is dedicated to *mitosis*, the basic process by which cellular material is equally divided between daughter cells. The process implies a series of complex changes involving the nucleus and the cytoplasm. It is directly related to the problem of continuity of chromosomes as entities capable of autoduplication and of maintaining their morphologic characteristics and function through successive cell divisions. The importance of isolating the mitotic apparatus, as a method of learning more about its composition and functioning, is emphasized. The main theories concerning the movement of chromosomes in mitosis are discussed.

Chapter 18 studies meiosis as the special type of cell division found in sexual cells that brings about the reduction in the number of chromosomes, and also the recombination and interchange of groups of genes by way of crossing over and the random assortment of the homologous chromosomes. To understand cytogenetics the student should have a clear comprehension of meiosis from the point of view of its morphology and biochemistry.

Chapter 19 presents in a very general way the chromosomal basis of Mendel's principles of heredity and the linkages between different genes, which depend on their position in the chromosome and the presence or absence of crossing over. The bases on which genetic maps of the chromosomes are built are mentioned in relation to these concepts. An important part of this chapter is devoted to the different chromosomal aberrations that can be produced either spontaneously or by radiation and chemical agents. The chromosomal aspects of evolution are considered briefly.

In the past decade the study of the normal and abnormal human karyotype has developed considerably and has acquired great importance in cytogenetics. This material has been incorporated in Chapter 20 along with a discussion of chromosomal sex determination. These studies have considerable theoretical and applied value because they include investigations of congenital and hereditary diseases and the varied sexual alterations that can be produced in man. This material is of great importance to students of medicine; with it they will be better able to interpret the pathogenic mechanisms of numerous hereditary diseases and congenital malformations. The latest knowledge achieved through the use of banding techniques and progress made on the human genetic map are included in this chapter.

THE INTERPHASE NUCLEUS, CHROMATIN, AND THE CHROMOSOMES

15–1   The Nuclear Envelope   334
    Nuclear Pores are not Wide-Open Channels; Pore Complexes
    Annulated Lamellae — Cytoplasmic Stores of Pore Complexes
    Nuclear Pores — Selective Diffusion Barriers Between Nucleus and Cytoplasm
    Nuclear Proteins Accumulate Selectively in the Nucleus
    Summary: *The Nuclear Envelope*

15–2   Chromatin   340
    Nuclei Contain a Constant Amount of DNA
    Thin-Section Electron Microscopy did not Reveal Details of Chromatin Structure
    Chromatin is a Complex of DNA and Histones
    Electron Microscopy of Chromatin Spreads Revealed a Beaded Structure
    The Nucleosome — An Octamer of Four Histones (H2A, H2B, H3 and H4) Complexed with DNA
    The 20 to 30 nm Fiber — A Further Folding of the Nucleosome Chain
    Summary: *Chromatin*

15–3   The Chromosomes   348
    Chromosomal Shape is Determined by The Position of the Centromere
    Chromosomal Nomenclature — Chromatid, Centromere, Kinetochore, Telomere, Satellite, Secondary Constriction, Nucleolar Organizer
    Karyotype — All the Characteristics of a Particular Chromosome Set
    Each Chromatid has a Single DNA Molecule
    Metaphase Chromosomes are Symmetrical
    Interphase Chromosomal Arrangement — Random or Non-Random?
    Chromosomes Undergo Condensation-Decondensation Cycles
    A Chromosomal Scaffold of Non-Histone Proteins

15–4   Heterochromatin   354
    Heterochromatin — Chromosomal Regions That do not Decondense During Interphase
    Heterochromatin can be Facultative or Constitutive
    Heterochromatin is Genetically Inactive
    Constitutive Heterochromatin Contains Repetitive DNA Sequences
    Summary: *The Chromosomes and Heterochromatin*

15–5   The Nucleolus   359
    The Nucleolus — A Fibrillar and a Granular Zone
    The Nucleolus is Disassembled During Mitosis
    Summary: *The Nucleolus*

About a century ago, in 1876, Balbiani described rod-like structures that were formed in the nucleus before cell division. In 1879 Flemming used the word "chromatin" (Gr., *chroma*, color) to describe the substance that stained intensely with basic dyes in interphase nuclei. He suggested that the affinity of chromatin for basic dyes was due to their content of "nuclein," a phosphorus-containing compound isolated from pus cells by Meischer in 1871 and later on from salmon sperm, a substance that is now called DNA. In 1888 Waldeyer used the word *chromosome* to emphasize the continuity between the chromatin of interphase nuclei and the rod-like objects observed during mitosis. In this chapter we will analyze the molecular organization of chromatin and of the chromosomes, as revealed by more recent studies. In addition, we will consider two other important components of the interphase nucleus: the nuclear membrane and the nucleolus.

15–1 THE NUCLEAR ENVELOPE

One of the main differences between prokaryotes and eukaryotes resides in the absence or presence of the nuclear envelope. This structure can be considered as a second permeability barrier that has arisen through evolution to regulate nucleocytoplasmic exchanges and to coordinate gene action with cytoplasmic activity.

As mentioned in Chapter 2–5, the light microscope gave little information about the *nuclear envelope*. One of the most interesting discoveries made with the electron microscope is that the nuclear envelope is a dependency of the cytoplasmic membrane system. This has been verified not only by observing the continuities of the two elements at many points but also by studying these structures in different stages of cellular activity.

Probably one of the clearest demonstrations that the nuclear envelope is a derivative of the endoplasmic reticulum is observed during mitosis. At telophase, cisternae of the endoplasmic reticulum collect around the chromosomes to reform the nuclear envelope.[1]

Nuclear Pores are not Wide-Open Channels; Pore Complexes

The nuclear envelope consists of two concentric membranes separated by a perinuclear cisterna 10 to 15 nm in width. These membranes have a basic unit structure similar to that of the plasma membrane. The nuclear envelope consists of flattened cisternae of the endoplasmic reticulum having ribosomes only on the outer surface.[2, 3] As shown in Figure 15–1, at certain points the nuclear envelope is interrupted by structures called *pores*. Around the margins of these nuclear pores both membranes are in continuity. At their nuclear side the pores are generally aligned with channels of nucleoplasm situated between more condensed lumps of chromatin which attach to the inner membrane (Fig. 15–1). The number of pores varies from 40 to 145 per square micrometer in nuclei of various plants and animals. At their highest density the pores are packed in hexagonal array with center-to-center pore distances of about 150 nm.

The nuclear pores are very large, 50 to 80 nm in diameter. In nuclei of *Mammalia* it has been calculated that nuclear pores account for 5 to 15 per cent of the surface area of the nuclear membrane. In amphibian oöcytes, certain plant cells, and protozoa the surface occupied by the pores may be as high as 20 to 36 per cent. Nuclear

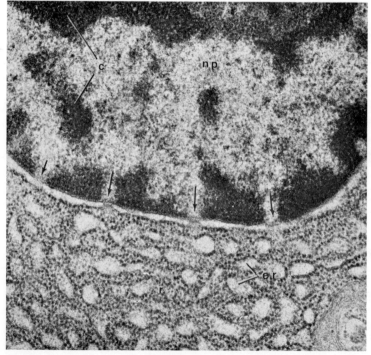

Figure 15–1 Electron micrograph of a portion of the nucleus and cytoplasm from a pancreatic cell of the mouse. The pores in the nuclear envelope are indicated by arrows within the interchromatin channels; *c,* chromatin; *er,* endoplasmic reticulum; *np,* nucleoplasm, ×48,000. Chromatin blocks are attached to the inner nuclear membrane but not to the nuclear pores. Ribosomes are attached to the outer nuclear membrane. (Courtesy of J. Andrè.)

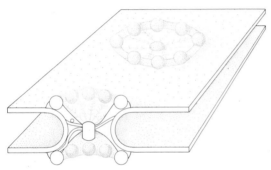

Figure 15-2 Diagram of the pore complex. According to Franke the octagonal structure of the pore is a result of the presence of eight granules of about 15 nm in diameter evenly spaced around the annulus on both the cytoplasmic and nuclear surfaces. Observe that the size of the pore (60 nm) is considerably reduced by the annulus. (From Franke, W. W., Z. Zellforsch. Mikrosk. Anat., 105:405–429, 1970.)

pores, however, are not wide-open channels, and the electron microscope has revealed that they are occluded by an electron-dense material. The pores are enclosed by circular structures called *annuli*. The pores and annuli are together designated the *pore complex*. Negative staining techniques have demonstrated that the annuli are octagonal in shape and are about 60 nm in diameter.[3]

Each annulus consists of eight *granules* of about 15 nm, which are present on both the nuclear and cytoplasmic surfaces (see Fig. 15–2), and a less defined, amorphous *annular material* in the opening itself. This material is digested by trypsin and remains unaffected when exposed to ribonuclease and deoxyribonuclease. These findings suggest that the annular material is protein in nature. The pore complex is apparently a rather rigid structure present in a fixed number according to cell type. In certain physiologic stages, however, they may change in number. For example, they are reduced in number in maturing erythroblasts and in spermatids, and this correlates with the low transcriptional activity of these cells. In the frog *Xenopus laevis*, oöcytes (which are very active in transcription) have 60 pores/μm^2 (and up to 30 million pore complexes per nucleus), whereas mature erythrocytes (inactive in transcription) have only about 3 pores/μm^2 (and a total of only 150 to 300 pores per nucleus).[4] The nuclear envelope has been isolated from a variety of cells and analyzed biochemically. The main conclusion to be derived from these studies is that the composition of the nuclear membrane is very similar to that of the rough endoplasmic reticulum (see Franke, 1977; Harris, 1978). This emphasizes the relationship between both membrane systems, which can also be visualized by their physical continuity (Fig. 2–6) and by the presence of ribosomes attached to the outer nuclear membrane.

Annulated Lamellae — Cytoplasmic Stores of Pore Complexes

In developing oöcytes and spermatocytes, as well as in certain embryonic and tumor cells, the so-called *annulated lamellae* may be found in the cytoplasm. As shown in Figure 15–3, these membranous structures appear as stacks of cisternae or flattened sacs with an intrasaccular space 10 to 20 nm wide. At regular intervals these sacs are traversed by pore complexes that are similar in morphology and structure to those found in the nuclear envelope.[5] These are closely spaced and may occupy up to 50 per cent of the surface of the membrane (Fig. 15–3). The lamellae have been interpreted as budding off the nuclear envelope. More precise work has demonstrated that first there is a budding of the outer membrane of the nuclear envelope into flattened sacs; later the pore complexes appear.[6] They are usually associated with rapidly proliferating cells.

Nuclear Pores — Selective Diffusion Barriers Between Nucleus and Cytoplasm

The materials exchanged between nucleus and cytoplasm must traverse the nuclear pore complexes. This exchange is very selective and allows passage of only certain molecules, of either low or very high molecular weight.

The permeability of the nuclear membrane with respect to small ions is reflected in some of the electrochemical properties of this structure, which can be investigated with fine microelectrodes.[7] As shown in Figure 15–4, when giant cells from the salivary gland of *Drosophila* are penetrated with a microelectrode, there is an abrupt change in potential at the plasma membrane (-12 mV); then, as the microelectrode enters the nucleus, there is another drop in negative potential at the nuclear membrane (-13 mV). These results suggest that the nuclear envelope is a diffusion barrier for ions as small as K$^+$, Na$^+$, or Cl$^-$.

On the other hand, very large structures such as ribosomal subunits, which are assembled in the nucleolus, are able to leave the nucleus. This can be visualized in Figure 23–16, where the ribonucleoprotein particles are seen traversing the nuclear pore complexes. Nevertheless, not all RNAs can exit freely; heterogeneous nuclear RNA (hnRNA, a large precursor of mRNA which contains intervening sequences— see Chapter 22) is never found in the cytoplasm, and the mRNA sequences can traverse the pores only after extensive processing has occurred within the nucleus.

Xenopus oöcytes are particularly useful for studying nucleocytoplasmic exchanges because these cells are very large (up to 1.2 mm diameter), and substances can be easily microinjected into the cytoplasm. One can then find out

whether or not these substances entered the nucleus by manually isolating the giant oöcyte nucleus (also called the germinal vesicle), which measures up to 0.4 mm in diameter. Studies involving the microinjection of dextrans,[8] colloidal gold particles, and labeled proteins[9, 10] suggest that the oöcyte nuclear membrane has functional pores with a radius of about 4.5 nm.[8] It was found that small proteins (such as lysozyme, MW 14,000; and ovalbumin, MW 44,000), were able to enter the nucleus, that bovine serum albumin (MW 62,500) entered extremely slowly, and that larger proteins did not enter at all. This upper limit for diffusion applies only to proteins

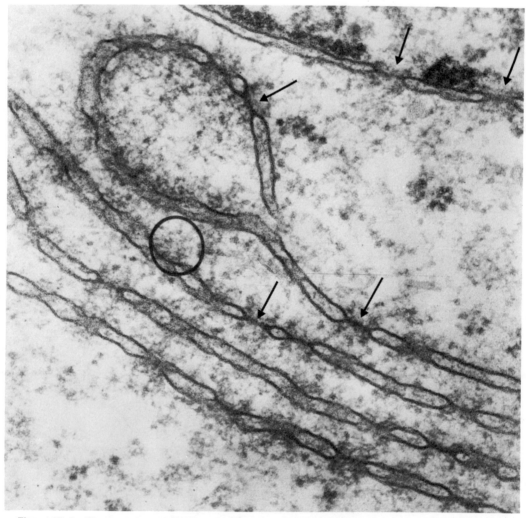

Figure 15–3 Electron micrograph of annulated lamellae observed in a human melanoma cell cultured in vitro. **Upper right,** the nucleus, with the nuclear envelope, showing two pore complexes (arrows). Similar complexes are observed in the lamellae present in the cytoplasm. ×80,000. (Courtesy of G. G. Maul.)

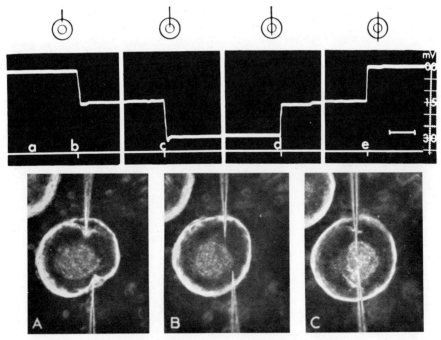

Figure 15–4 Experiment in microsurgery to study the potential of the nuclear envelope in giant nuclei of the salivary gland of *Drosophila*. **Above,** a diagram of the penetration of the microelectrode into the cell is shown along with the membrane or steady potentials registered at each position. **Below,** photomicrographs of cells penetrated by two microelectrodes. **A,** penetration of the membrane; **B,** into the cytoplasm; **C,** into the nucleus. (Courtesy of W. R. Loewenstein and Y. Kanno.)

that are not normally nuclear proteins for, as we will see in the following section, nuclear-specific proteins behave differently.

Nuclear Proteins Accumulate Selectively in the Nucleus

The cell nucleus contains a specific subset of proteins. For example, RNA polymerase, DNA polymerase, and histones are located specifically inside the nucleus. However, all these proteins are synthesized in the cytoplasm and must be subsequently transported into the nucleus. Here we will discuss experiments which show that nuclear proteins contain, within the structure of the mature protein, a signal that enables them to accumulate in the nucleus after microinjection into the cytoplasm.

This is known because when labeled proteins, such as histones[9] or a preparation of nucleoplasmic proteins, are microinjected into the cytoplasm of an oöcyte, they are not only able to enter the nucleus but also to *accumulate* in it (Fig. 15–5). The great degree of selectivity of this accumulation is illustrated in the experiment shown in Figure 15–6. When the nuclear pro-

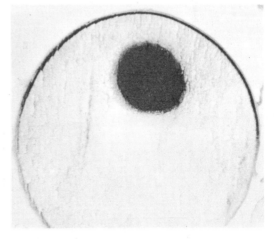

Figure 15–5 Migration of nuclear proteins. Autoradiograph of a *Xenopus* oöcyte that had been injected 24 hours earlier with a preparation of ^{35}S-methionine–labeled frog nucleoplasmic proteins. Injection was into the cytoplasm. The proteins accumulated in the nucleus. The black region observed in the periphery of the upper half of the oöcyte is composed of pigment granules, not autoradiographic grains. Oöcytes such as this one were separated manually into nucleus and cytoplasm and analyzed as shown in Figure 15–6, *B* and *C*. (From E. M. De Robertis et al., *Nature, 272*:254, 1978.)

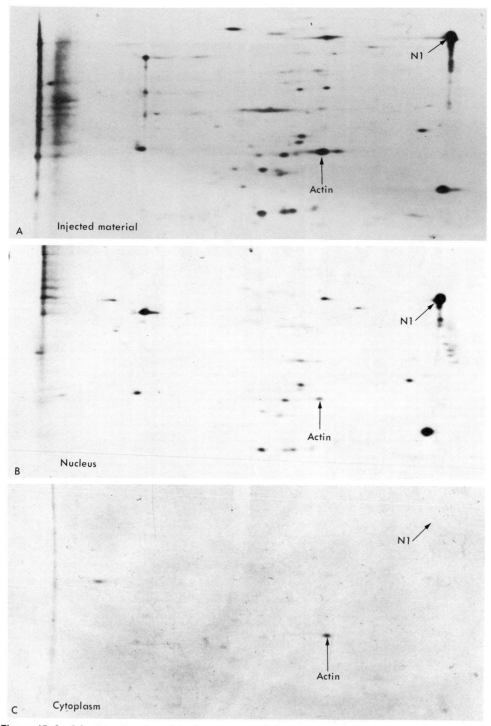

Figure 15–6 Selective migration of nuclear proteins. Two dimensional gel analysis of ^{35}S-methionine–labeled nucleoplasmic proteins microinjected into *Xenopus* oöcytes. An autoradiograph of this experiment is shown in Figure 15–5. **A,** A preparation of [35]S-*Xenopus* nucleoplasmic proteins that was injected into oöcytes. Note that it contains actin and many nuclear proteins. **B,** The nuclear fraction reisolated from unlabeled oöcytes injected 24 hours earlier into the cytoplasm with the ^{35}S proteins shown in **A.** Note that the nuclear proteins re-enter the nucleus, while actin decreases in relative amount. **C,** Cytoplasmic fraction from the same microinjected oöcytes. Note that actin is the most prominent protein, while the nuclear proteins are not detectable in the cytoplasm. (From E. M. De Robertis.)

teins of normal oöcytes are examined by two-dimensional gel electrophoresis, two types of proteins can be distinguished: (1) nucleus-specific proteins, and (2) proteins common to *both* nucleus and cytoplasm. Actin is the most conspicuous member of the "common-to-both" proteins, and it is one of the major constituents of the oöcyte nucleoplasm.[11, 12] A labeled preparation of *Xenopus* nucleoplasmic proteins (which contained actin and nuclear-specific proteins, such as the protein indicated as *N1* — see Figure 15–6,*A*) was injected into the cytoplasm of living oöcytes.[11] After 24 hours the oöcytes were manually separated into nucleus and cytoplasm, and the proteins were analyzed by two-dimensional gels (Fig. 15–6,*B* and *C*). It was found that nucleus-specific proteins were greatly concentrated in the nucleus (100-fold in the case of N1), while microinjected actin was found both in nucleus and cytoplasm (it is the major protein in Fig. 15–6,*C*). This is as expected, since actin is normally present in both cell compartments.

The nuclear-specific proteins have in their molecular structure some type of signal that enables them to accumulate selectively in the nucleus.[11] This signal must be absent in proteins which do not accumulate in the nucleus, such as actin. Some of the nuclear proteins are quite large (N1 has a MW of 120,000 daltons) but are still able to re-enter the nucleus rapidly; it therefore seems that when proteins are nucleus-specific, factors other than size determine nuclear pore permeability.

Cells have the capacity to place specific proteins into the corresponding cell compartments, as occurs with the mitochondria or chloroplast proteins that are synthesized in the cytoplasm, or with proteins that are exported from the cell. The mechanism by which proteins are secreted into the endoplasmic reticulum (Chap. 10) is very different from the one used for the accumulation of proteins inside the nucleus. The "signal peptide" enables the secreted proteins to traverse the endoplasmic reticulum membrane only as they are being synthesized, and it is rapidly removed thereafter (Fig. 10–12); the nuclear proteins, however, accumulate in the nucleus independently of protein synthesis, and this is not a transient property but is intrinsic to the structure of the mature protein.

SUMMARY:

The Nuclear Envelope

The nuclear envelope, a differentiation of the endoplasmic reticulum, is composed of two membranes and a perinuclear space. Nuclear pores represent openings in this envelope at sites where the two membranes are in contact. The pores are octagonal orifices about 60 nm in diameter. They are not, however, freely communicating openings, but are plugged by a cylinder of protein meterial — the so-called *annulus*. The pore and the annulus constitute a *pore complex*. In rapidly growing cells (oöcytes, embryonic cells, and cancer cells) preassembled pore complexes can be stored in the cytoplasm in membranous structures called *annulated* lamellae (Fig. 15–3).

The nuclear envelope regulates the passage of ions and small molecules. By way of the pore complex, the envelope may have an important role in the transfer of macromolecules between the nucleus and the cytoplasm — and vice versa. The differential permeability of the nuclear envelope to ions is reflected in a difference in electrical potential, the nucleus being more negative than the cytoplasm (Fig. 15–4). The passage of very large ribonucleoprotein particles (ribosomal subunits, mRNA, and others) may be seen by electron microscopy.

Nuclear proteins are able to accumulate inside the nucleus after microinjection. Nuclear proteins contain in their mature molecular structure a signal that enables them to accumulate inside the nucleus. This signal is absent in those proteins (such as actin) which are normally distributed in both nucleus and cytoplasm. The mechanism involved in this selective distribution of proteins is very different from that involved in the intracellular segregation of export proteins.

15–2 CHROMATIN

DNA is the main genetic constituent of cells, carrying information in a coded form from cell to cell and from organism to organism. Within cells, DNA is not free but is complexed with proteins in a structure called chromatin.

Nuclei Contain a Constant Amount of DNA

The demonstration that nuclei contain a constant amount of DNA[13, 14] was a landmark in cell biology because it suggested that DNA was the genetic material and also that genetic information is not lost during the differentiation of the various somatic tissues (for further details, see Chapters 21 and 26).

This was achieved using cytophotometric methods. The DNA in nuclei was stained using the Feulgen reaction, and the amount of stain was estimated by microcytophotometry (Fig. 4–15). Cytophotometry has produced accurate measurements in a large variety of cell types within the same species, showing that the DNA content is practically constant in the different tissues. This is certainly related to the number of chromosome sets *(n)*, which is 2n (diploid) in most somatic cells. However, certain tissues, like liver, have some cells that become polyploid (4n or 8n), and their nuclei have a correspondingly higher DNA content, as seen in Table 15–1.

Another interesting example is provided by the study of spermatogenesis (Table 15–2). Before meiosis occurs (see Fig. 18–1), there are two classes of cells (spermatogonia) having different DNA contents (2n and 4n). The early primary spermatocyte (meiotic prophase) has a tetraploid amount of DNA. After the first meiotic division, the secondary spermatocyte contains half the DNA content, corresponding to a diploid condition. Finally, the second meiotic division results in four spermatids; these have the DNA of only one chromosomal

TABLE 15–2 DNA CONTENT AT DIFFERENT STAGES IN SPERMATOGENESIS*

Cell Type		DNA-Feulgen
Premeiotic {	Class 2n	3.28 ± 0.07
	Class 4n	5.96 ± 0.07
Primary spermatocyte		6.28 ± 0.07
Secondary spermatocyte		3.35 ± 0.04
Spermatid		1.68 ± 0.02

*From Pollister, A. W., Swift, H., and Alfert, M., *J. Cell. Comp. Physiol., 38* (suppl. 1): 101, 1951.

set (haploid cell) (Table 15–2). Similar results have been observed in oögenesis.

Each species has a characteristic amount of DNA. Eukaryotes vary greatly in DNA content but always contain much more DNA than prokaryotes. Lower eukaryotes in general have less DNA, such as the nematode *C. elegans,* which has only 20 times more DNA than *Escherichia coli,* or the fruit fly *Drosophila melanogaster,* which has 40 times more. Vertebrates have greater DNA content, in general about 700 times more than *E. coli.* The highest DNA content is that of the salamander *Amphiuma,* which can contain 168 picograms of DNA in a single nucleus, i.e., nearly 40,000 times more than *E. coli.*

Considering that one picogram of DNA is equivalent to 31 cm of DNA, it is possible to calculate that there are about 174 cm of DNA in human diploid cells, 37 meters in *Trillium,* and 97 meters in polytenic chromosomes of *Drosophila.*[15] The DNA content in the 46 human chromosomes has been estimated by cytophotometry, and from these measurements it appears that the DNA content is proportional to the size of the chromosome. The largest chromosome (1), which is 10 μm long, should accommodate about 7.2 cm of DNA in a tightly packed form; i.e., the DNA is compacted about 7000-fold in a metaphase chromosome! In this chapter we will analyze how this folding is achieved.

Thin-section Electron Microscopy did not Reveal Details of Chromatin Structure

The optical microscope reveals that during interphase, chromatin can be present in a condensed state (heterochromatin) or in a more dispersed form (euchromatin). The heterochromatic regions tend to form aggregates, sometimes called chromocenters (Fig. 6–10).

Thin-section electron microscopy contributed little to our knowledge of the organization of

TABLE 15–1 DNA CONTENT AND CHROMOSOME COMPLEMENT*

Cells	Mean DNA-Feulgen Content	Presumed Chromosome Set
Spermatid	1.68	haploid (n)
Liver	3.16	diploid (2n)
Liver	6.30	tetraploid (4n)
Liver	12.80	octoploid (8n)

*From Pollister, A. W., Swift, H., and Alfert, M., *J. Cell. Comp. Physiol., 38* (suppl. 1): 101, 1951.

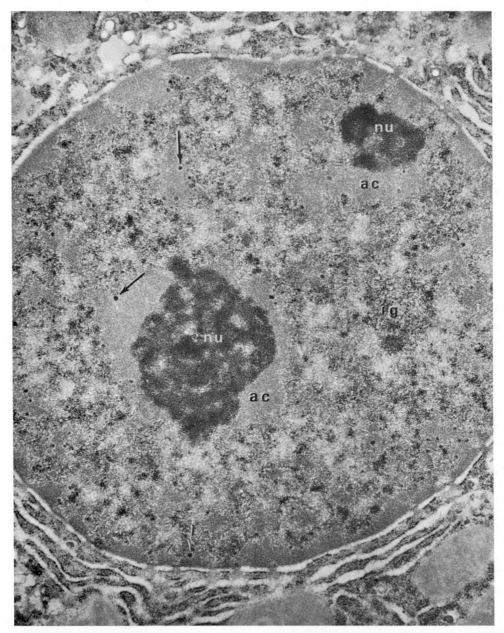

Figure 15–7 Electron micrograph of a normal rat liver nucleus stained with uranyl acetate and treated with EDTA to visualize the ribonucleoprotein components; *nu,* nucleolus; *ac,* associated chromatin; *ig,* interchromatin granules; perichromatin granules (arrows). Chromatin appears as pale areas because of the EDTA treatment (compare with Fig. 2–8). See the description in the text. ×25,000. (Courtesy of W. Bernhard.)

chromatin in the interphase nucleus, in contrast to the results obtained by electron microscopic analysis of *spread* preparations.

Figure 2–8 shows the distribution of chromatin in a nucleus stained with uranyl acetate. Both RNA and DNA are heavily stained, but DNA is most conspicuous. The placement of the chromatin in clumps near the nuclear envelope and the interchromatin channels leading to the nuclear pores are clearly visible. The mass of the nucleolus is of a lesser density and is surrounded by an associated chromatin of a greater density; the nucleoplasm is light and reveals fine granules.

In Figure 15–7 the nucleus has also been stained with uranyl acetate, but it was then treated with EDTA (ethylenediaminetetraacetic acid) which preferentially removes the uranyl ions from the DNA. EDTA is a chelating agent which binds to the phosphate groups of DNA, thus displacing the UO_2^{2+} ions. In this way, the RNA-

341

containing structures of the nucleus and the ribosomes in the cytoplasm become more conspicuous. By comparing Figure 15–7 with Figure 2–8, the student will recognize that the former is, in a certain manner, a reverse image of the latter. In Figure 2–8 the DNA structure shows a high electron density, whereas in Figure 15–7 the DNA structures are pale and the RNA-protein components appear more dense.

Chromatin is a Complex of DNA and Histones

Chromatin can be isolated biochemically by purifying nuclei and then lysing them in hypotonic solutions. When prepared in this way, chromatin appears as a viscous, gelatinous substance which contains DNA, RNA, basic proteins called histones, and nonhistone (more acidic) proteins (Table 15–3). The content of RNA and nonhistone protein is variable between different chromatin preparations, but histone and DNA are always present in a fixed ratio of about one to one (Table 15–3). The non-histone proteins are very heterogeneous; they vary in different tissues and include RNA and DNA polymerases, among other enzymes, and putative regulatory proteins. On the other hand, there are only five types of histones, each one present in large amounts (Fig. 15–8), and this suggested that they could have a structural role.

Table 15–4 shows some of the properties of histones. Histones are small proteins which are basic because they have a high content (10 to 20 per cent) of the basic amino acids arginine and lysine. Being basic, histones bind tightly to DNA, which is acid. The four main histones, H2A, H2B, H3, and H4, are very similar in different species.[16] For example, the sequence of histone H3 from the rat differs only in two amino acids from that of peas. These four histones are present in equimolar amounts (two of each every 200 base pairs of DNA). Histone H1 is not con-

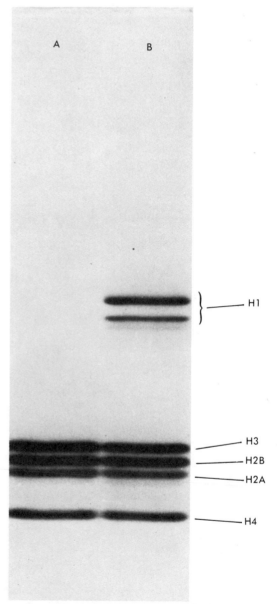

Figure 15–8 Histone proteins after electrophoresis in polyacrylamide gels containing sodium dodecyl sulphate. **A,** Histones from nucleosome "cores" which do not contain H1. **B,** Histones from total chromatin. Note that the four fundamental histones are present in about equal amounts. (Courtesy of Jean O. Thomas.)

TABLE 15–3 COMPOSITION OF CHROMATIN
(Data expressed as per cent dry weight)

Component	Liver*	Pea Embryo†
DNA	31	31
RNA	5	17.5
Histone	36	33
Non-histone proteins	28	18

*Data from Steele, W. J., and Busch, H., *Cancer Res.*, 23:1153, 1963.

†Data from Huang, R. C., and Bonner, J., *Proc. Natl. Acad. Sci. USA*, 48:1216, 1962.

served between species and has tissue-specific forms; is present only once per 200 base pairs of DNA; is rather loosely associated with chromatin (it can be eluted from DNA by adding low concentrations of salt); and is not a component of the nucleosome but is involved in the maintenance of a higher-order folding of chromatin.

Prokaryotes do not have histones, but a basic protein that might serve the same function has been isolated from *E. coli*.[17]

TABLE 15-4 SOME PROPERTIES OF HISTONES

Histone	Molecular Weight	Amino Acid Composition	Species Variation	Number of Molecules per 200 Base Pairs of DNA
H1	20,000	Lysine-rich	Wide	1
H2A	13,700 ⎫	Moderately lysine-rich	Fairly well-conserved	2
H2B	13,700 ⎭			2
H3	15,700 ⎫	Arginine-rich	Highly conserved	2
H4	11,200 ⎭			2

Electron Microscopy of Chromatin Spreads Revealed a Beaded Structure

When eukaryotic nuclei were *spread* on EM specimen holders (instead of prepared as thin sections), it was found that chromatin has a repeating structure of beads about 10 nm in diameter connected by a string of DNA.[18-20] Most, if not all, of the DNA is present in this form, as can be seen in Figure 15-9.

The beads-on-a-string appearance is not the true structure of chromatin but is an artifact arising from the way in which the sample is prepared, which results in a loss of histone H1. With more mild treatments (which do not remove H1), the beads are not stretched but can be observed as a 10 nm fiber, with the beads touching each other

(Fig. 15-10). The 10 nm fiber therefore represents the first level of organization of chromatin within cells.

The Nucleosome — An Octamer of Four Histones (H2A, H2B, H3, and H4) Complexed with DNA

The existence of a repeating unit of chromatin — called the *nucleosome* — was also predicted from biochemical studies independent of the EM results described above.[21]

Studies of digestion of chromatin by nucleases[22] showed that the DNA was cut into multiples of a unit size, which was later found to be about 200 base pairs of DNA in length;[23] i.e., the DNA was cut at intervals of 200, 600, 800

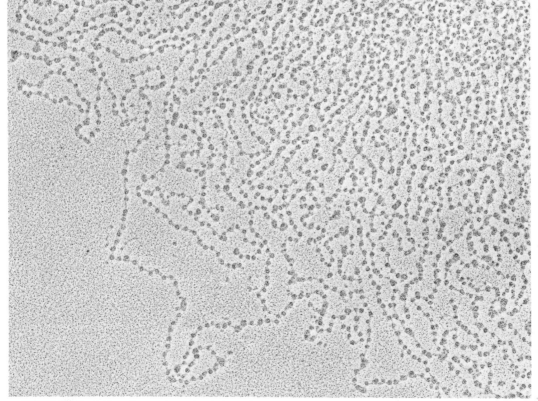

Figure 15-9 A *D. melanogaster* embryonic nucleus spread for electron microscopy. Most of the DNA, if not all, has a beaded structure. ×50,000. (Courtesy of O. L. Miller, Jr.)

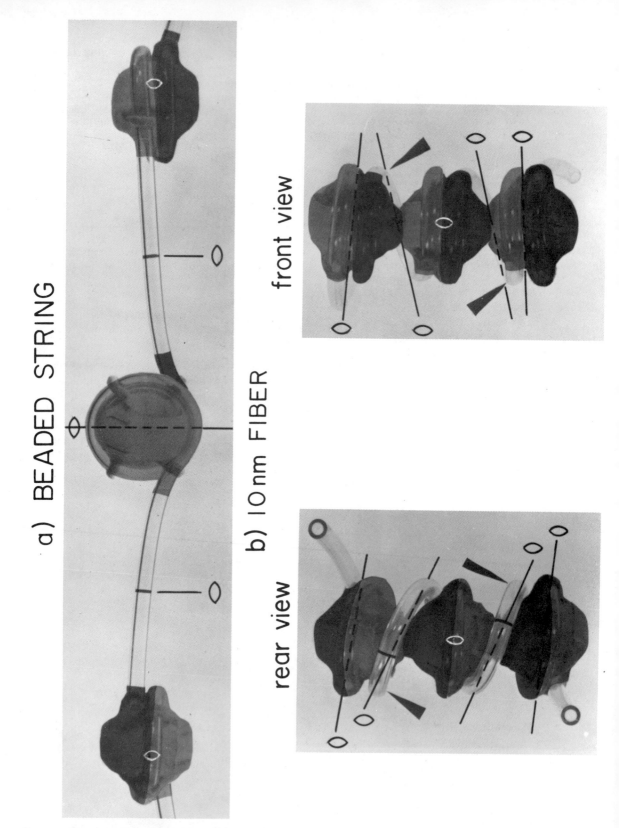

Figure 15-10 A model showing the two possible appearances of chromatin under the electron microscope. The nucleosome is represented by tennis balls, and the DNA, by plastic tubing. It is important to note that the DNA is wound on the *outside* of the nucleosome. **A,** The beads-on-a-string configuration that results from the unwinding of part of the DNA during sample preparation. **B,** The 10 nm fiber, in which nucleosomes are in close contact and is the natural configuration of chromatin. (Courtesy of A. Worcel, from *Cold Spring Harbor Symp. Quant. Biol.,* 42:313, 1977.)

base pairs and so forth. Other studies showed that histones H3 and H4 tend to associate in solution, forming tetramers consisting of two of each.[24] Since the four histones are in equimolar amounts (two of each per 200 base pairs of DNA — see Table 15–4), in 1974 R. Kornberg (son of Arthur Kornberg, the discoverer of DNA polymerase) proposed a model for the nucleosome, in which the four histones H2A, H2B, H3, and H4 are arranged in octamers containing two of each of them, every 200 base pairs of DNA.[21]

The histone octamers are in close contact (as in the 10 nm fiber), and the DNA is coiled on the *outside* of the nucleosome (see reviews by Chambon, 1977; Fenselfeld, 1978).

That the biochemical subunit is identical to the beads observed in the EM was demonstrated by the so-called *"one, two, three, four"* experiment shown in Figure 15–11. After partial digestion of chromatin with the enzyme micrococcal nuclease, monomer, dimer, trimer, and tetramer fractions were isolated on a sucrose gradient (Fig.

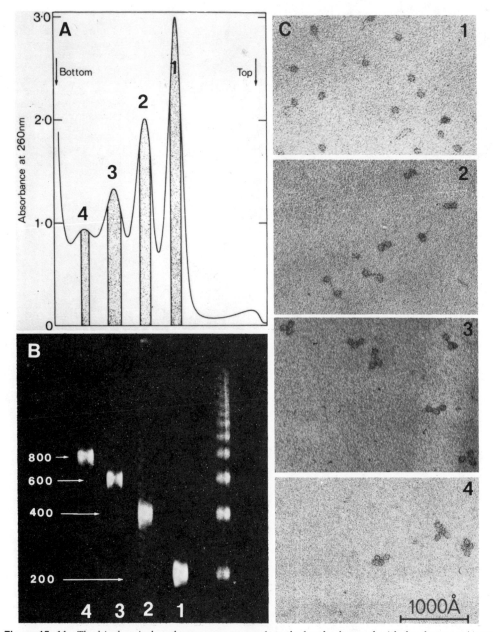

Figure 15–11 The biochemical nucleosome corresponds to the beads observed with the electron microscope. Chromatin was partially digested with micrococcal nuclease and centrifuged in a sucrose gradient. The fractions corresponding to the monomer, dimer, trimer, and tetramer peaks in **A** were analyzed by electrophoresis in agarose after extraction of the DNA in **B**; or examined directly under the electron microscope in **C**. (Courtesy of J. T. Finch, M. Noll, and R. D. Kornberg, from *Proc. Natl. Acad. Sci. USA*, 72:3320, 1975.)

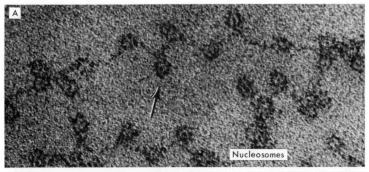

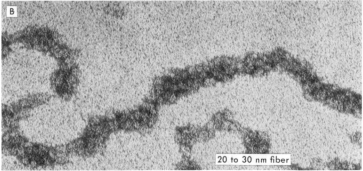

Figure 15–12 Chromatin from chicken erythrocyte nuclei, negative staining. **A,** the chromatin has stretched during spreading, and the nucleosomes can be seen as beads on a string of DNA. ×500,000. **B,** the structure of the 20 to 30 nm fiber has been preserved, and the nucleosomes can be visualized closely packed within it. ×250,000. (Courtesy of A. L. Olins.)

Figure 15–13 A possible model of the coiling of the 10 nm fiber into a 20 to 30 nm fiber. The model stresses the fact that the DNA (plastic tubing) is wound on the outside of the nucleosomes (tennis balls). Note that the histone H1 (indicated by arrowheads) is located in the inner part of the solenoid. (Courtesy of A. Worcel, from *Cold Spring Harbor Symp. Quant. Biol., 42:* 313, 1977.)

15–11,*A*). When analyzed by electrophoresis, these fractions contained DNA fragments 200, 400, 600, and 800 base pairs long respectively (Fig. 15–11,*B*) and one, two, three, and four beads long respectively under the EM (Fig. 15–11,*C*).

The DNA is cut into discrete fragments by micrococcal nuclease because the *linker* regions between two neighboring nucleosomes (Fig. 15–10,*A*) are more exposed to the enzyme. Very extensive digestion leads to the degradation of the linker DNA and generates a nucleosome *core particle*, which contains 140 base pairs of DNA and lacks the 60 base pairs of the linker DNA. Core particles lack histone H1 (Fig. 15–8,*A*), indicating that H1 is normally associated with the linker DNA.[26] Core nucleosomes have been crystallized, and x-ray diffraction and EM studies have shown that the nucleosome is a flat disk about 11 nm in diameter and 5.7 nm in height.[27] The DNA is on the outside and turns twice around each histone octamer.

The length of DNA per nucleosome can vary in different tissues.[28, 29] For example, chicken erythrocytes have a repeat length of 210 base pairs, and chicken liver has about 195 base pairs per repeat. The reason for this is not clear, but it seems to be related to the type of histone H1 in each tissue. The variability is restricted to the length of the linker segment, and all tissues have a constant amount of DNA (140 base pairs) in the nucleosome core.

When nucleosomes are in close apposition in the 10 nm filaments, the packing of DNA is about five- to seven-fold, i.e., five to seven times more compact than free DNA. However, this is still 1000-fold lower than the packing ratio of DNA in metaphase chromosomes, which is achieved by further folding the 10 nm fiber.

The 20 to 30 nm Fiber — A Further Folding of the Nucleosome Chain

Electron microscopic studies of chromatin fibers in interphase nuclei and mitotic chromosomes[30] have revealed a *"thick fiber"* of a diameter that varies from 20 to 30 nm and which probably represents the structure of inactive chromatin. Figure 15–12 shows that the 20 to 30 nm fiber consists of closely packed nucleosomes.

The 20 to 30 nm fiber probably arises from the folding of the nucleosome chain into a solenoidal structure having about six nucleosomes per turn.[31] Figure 15–13 shows a possible model explaining how this could occur. The histone H1 is located in the central "hole" of the solenoid, and the whole structure is stabilized by interactions between different H1 molecules. Solenoidal structures can be formed in vitro by adding Mg^{2+} to 10-nm fiber preparations, but they can not be formed when histone H1–depleted chromatin is treated in the same way.[31]

The DNA of a 20 to 30 nm solenoid has a packing that is about 40-fold.[31] However, the DNA of a metaphase chromosome is packed between 5000 and 10,000 times; i.e., the 20 to 30 nm fiber must be further folded more than 100-fold. At present, we know very little of how this feat is achieved.

SUMMARY:

Chromatin

The cells of eukaryotic organisms have a DNA content that is characteristic for each species. When packed in a metaphase chromosome, the DNA is compacted 5000- to 10,000-fold. Most, if not all, of the DNA is present in chromatin, which is a complex of DNA with an equal weight of basic proteins called histones.

Histones are small proteins which are basic because they have a high content of arginine or lysine. There are only five histones. The four fundamental histones, H2A, H2B, H3, and H4, are present twice every 200 base pairs of DNA (Table 15–4). Histones H3 and H4 were almost completely conserved during evolution; histone H3 from rat liver differs in only two amino acids from that of pea embryos. The fifth histone, H1, is present only once per 200 base pairs of DNA, and it varies considerably between species and even within tissues of the same species.

Chromatin is formed by a series of repeating units called *nucleosomes*. Under the EM, nucleosomes can be visualized as beads about 10 nm in diameter. During spreading of the preparation for electron microscopy, chromatin frequently becomes stretched, showing a beads-on-a-string configuration (Fig. 15–9). However, in the living cell the nucleosomes touch each other, forming a 10 nm fiber (Fig. 15–10).

Each nucleosome contains a histone octamer consisting of two of each of the four histones H2A, H2B, H3, and H4, with about 200 pairs of DNA coiled on the outside of the nucleosome. The whole structure has the shape of a flattened disk 11 nm in diameter and 5.7 nm in height. Histone H1 is bound to the DNA *linker* region, which extends between neighboring nucleosomes, and is involved in the production of higher orders of structure. The chains of nucleosomes can be folded into a 20 to 30 nm fiber, probably by forming a solenoidal structure having six nucleosomes per turn (Fig. 15–13). This structure is stabilized by interactions between different H1 molecules. The 20 to 30 nm fiber can be observed in metaphase chromosomes and in interphase nuclei, and it probably represents the natural conformation of transcriptionally inactive chromatin.

The packing of DNA in a chain of nucleosomes is about five to seven times more compact than free DNA, and in a 20 to 30 nm solenoid it is packed about 40-fold. This is still more than 100-fold less than the compaction of metaphase chromosomes (5000- to 10,000-fold).

15–3 THE CHROMOSOMES

At the time of cell division chromatin becomes condensed into the chromosomes. Since chromosomes can be clearly visualized with the optical microscope, they were the subject of much study soon after their discovery in 1876, and by 1910 it became clear that genetic phenomena could be explained in terms of chromosome behavior.[32] For this very same reason, the study of chromosomes remains of the utmost importance even today. In this section we will analyze the general morphology and composition of chromosomes, and their behavior during mitosis and meiosis will be treated in Chapters 17 and 18.

Chromosomal Shape is Determined by the Position of the Centromere

Chromosomes can be studied in tissue sections, but they are best visualized in *squash* preparations. Fragments of tissues are stained with basic dyes (e.g., orcein or Giemsa) and then squashed between slide and coverslip by gentle pressure. Sometimes hypotonic solutions are used prior to squashing to produce swelling of the nucleus and a better separation of the individual chromosomes. The morphology of the chromosomes is best studied during metaphase and anaphase, which are the periods of maximal contraction.

Chromosomes are classified into four types according to their shape, which, in turn, is determined by the position of the centromere (point of attachment to the mitotic spindle). As seen in Figure 15–14, *telocentric* chromosomes have the centromere located on one end; *acro-*

centric chromosomes have a very small or even imperceptible short arm; *submetacentric* chromosomes have arms of unequal length; and *metacentric* chromosomes have equal or almost equal arms. During anaphase movements the chromosomes bend at the centromere, so that metacentric chromosomes are v-shaped, and acrocentric chromosomes are rod-shaped. This angular deviation can also be observed at metaphase (provided that the cells are not treated with colchicine), as shown in Figure 15–17.

Chromosomal Nomenclature — Chromatid, Centromere, Kinetochore, Telomere, Satellite, Secondary Constriction, Nucleolar Organizer

Classical cytogeneticists produced many names to describe the various components that can be visualized in mitotic chromosomes. The student will probably find this terminology initially confusing, but it is important to become familiar with these names, for they not only define particular structures but also specific properties of the chromosomes.

Chromatid. At metaphase each chromosome consists of two symmetrical structures; the *chromatids* (Fig. 15–14), each one of which contains a single DNA molecule (see later discussion). The chromatids are attached to each other only by the centromere (Fig. 15–15) and become separated at the start of anaphase, when the sister chromatids migrate to opposite poles. Therefore, *anaphase chromosomes have only one chromatid, while metaphase chromosomes have two.*

Chromonema(ta). During prophase (and

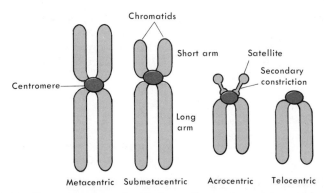

Figure 15–14 Chromosomes are classified according to the position of the centromere

sometimes during interphase) the chromosomal material becomes visible as very thin filaments, which are called chromonemata and which represent chromatids in early stages of condensation. "Chromatid" and "chromonema," therefore, are two names for the same structure: a single linear DNA molecule with its associated proteins.

Chromomeres. These components are bead-like accumulations of chromatin material that are sometimes visible along the chromonema. Chromomeres are especially obvious in polytene chromosomes, where they become aligned side by side, constituting the chromosome bands. These tightly folded regions of the chromonema are of considerable interest, because they are thought to correspond to the units of genetic function in the chromosome (Chapter 25). At metaphase the chromosome is tightly coiled, and the chromomeres are no longer visible.

Centromere. This is the region of the chromosome that becomes attached to the mitotic spindle (Fig. 15–14). The centromere lies within a thinner segment of the chromosome, the *primary constriction*. The regions flanking the centromere frequently contain highly repetitive DNA and may stain more intensely with basic dyes (heterochromatin).

Kinetochore. This component is a disc-shaped protein structure that is attached to the centromeric chromatin. Unlike the other chromosome components, the kinetochore can be studied best in thin sections with the electron microscope, which reveals a trilaminar structure.[33, 34] The kinetochore has an dense outer layer, a middle layer of low density, and a dense inner layer tightly attached to the centromere (Figs. 17–4 and 17–7). Between 4 and 40 microtubules become attached to the kinetochore and provide the force for chromosomal movement during mitosis. The function of the

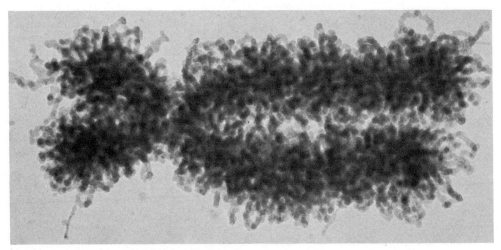

Figure 15–15 Electron micrograph of a whole mounted human chromosome 12 showing the two chromatids composed of fibrils 30 nm thick. The two chromatids are joined only at the centromere. (Courtesy of E. J. DuPraw.)

kinetochore is to provide a center of assembly for microtubules; when isolated metaphase chromosomes are incubated in a solution of tubulin with an appropriate buffer, microtubules are reassembled at the kinetochores of all the chromosomes.[35, 36]

The vast majority of chromosomes have only one centromere (*monocentric chromosomes*). Some species have diffuse centromeres, with microtubules attached along the length of the chromosome (*holocentric chromosomes*). In some chromosomal abnormalities (induced for example by x-rays), chromosomes may break and fuse with other ones, producing chromosomes without centromeres (*acentric*) or with two centromeres (*dicentric*). Both types of aberration are unstable; one because it cannot attach to the mitotic spindle and remains in the cytoplasm, the other because the two centromeres tend to migrate to opposite poles, thus leading to chromosomal fragmentation.

Telomere. This term applies to the tips of the chromosomes, which have distinct cytological properties. When chromosomes are broken by x-rays, the free ends without telomeres become "sticky" and fuse with other broken chromosomes; they will not, however, fuse with a normal telomere. Stable deletions in chromosomes are always interstitial and preserve the original telomere. In addition to promoting chromosomal stability, in some species telomeres mediate transient end-to-end association of chromosomes. A telomeric DNA sequence which is present at the tips of all the *Drosophila melanogaster* chromosomes has been isolated.[37] Telomeres contain the ends of the long linear DNA molecule contained in each chromatid.

Secondary Constrictions. Other morphologic characteristics are the *secondary constrictions*. Constant in their position and extent, these constrictions are useful in identifying particular chromosomes in a set. Secondary constrictions are distinguished from the primary constriction by the absence of marked angular deviations of the chromosomal segments.

Nucleolar Organizers. These areas are certain secondary constrictions which contain the genes coding for 18S and 28S ribosomal RNA and which induce the formation of nucleoli. The secondary constriction arises because the rRNA genes are transcribed very actively, interfering with chromosomal condensation. In man, the nucleolar organizers are located in the secondary constrictions of chromosomes 13, 14, 15, 21, and 22, all of which are acrocentrics and have satellites (Fig. 15–14).

Satellites. Another morphologic element present in certain chromosomes is the *satellite.** This is a rounded body separated from the rest of the chromosomes by a secondary constriction (Fig. 15–14). The satellite and the constriction are constant in shape and size for each particular chromosome.

Karyotype — All the Characteristics of a Particular Chromosomal Set

The name *karyotype* is given to the whole group of characteristics that allows the identification of a particular chromosomal set; i.e., the number of chromosomes, relative size, position of the centromere, length of the arms, secondary constrictions, satellites, and so forth. In Chapter 20 we will see that by the use of banding techniques, other morphologic criteria are available which have immensely increased the possibilities for identification of chromosomes and regions of chromosomes in the human and other species.

The karyotype is characteristic of an individual, species, genus, or larger grouping, and may be represented by a diagram in which the pairs of homologues are ordered in a series of decreasing size (Fig. 20–1). Some species may have special characteristics; for example, the mouse has acrocentric chromosomes, many amphibia have only metacentric chromosomes, and plants frequently have heterochromatic regions at the telomeres. Some species are particularly favorable for cytogenetic analysis: Salamanders and grasshoppers have high DNA content and very large chromosomes which produce very beautiful meiotic preparations; among plants, broad beans (*Vicia fava*) and onions (*Allium cepa*) have particularly large chromosomes which produce nice preparations of mitotic cells from root tips.

Each Chromatid has a Single DNA Molecule

Each chromatid represents a single linear DNA molecule with its associated proteins. This concept, sometimes called the *unineme theory*, is supported by several lines of evidence. As will be explained in Chapter 16, the replication of chromatids is consistent with the

---

*Chromosome satellites are a morphologic entity and should not be confused with satellite DNAs, which are highly repeated DNA sequences (Chapter 25).

semiconservative replication of a single DNA molecule (Fig. 16–9). When cells are labeled with ³H-thymidine or with bromodeoxyuridine (an analogue that becomes incorporated into DNA and prevents the staining of the chromatid with certain dyes) and then grown for two cycles in an unlabeled medium, only one of the granddaughter chromatids is labeled (Fig. 15–16), as expected for a single DNA molecule. Furthermore, exchanges between sister chromatids can be induced experimentally, and in this case every change in a chromatid is accompanied by a reciprocal change in the other (Fig. 15–16), a fact which is also consistent with the idea that there is only one DNA molecule per chromatid. Additional evidence comes with the finding that G_1 chromosomes (which have not yet replicated their DNA) have only one chromatid, and that G_2 chromosomes have two (see Fig. 16–4).

The unineme organization of the chromatid is supported directly by the isolation of DNA molecules that are long enough to contain all the DNA from a single chromatid. The length of these giant molecules can be estimated by viscoelastic measurements.[38]

Metaphase Chromosomes are Symmetrical

The two sister chromatids are mirror images one of the other. The morphologic characteristics of one chromatid, such as chromomeres, secondary constrictions, and satellites, always have their counterparts in the other (Fig. 15–14). The reason for this symmetry is that sister chromatids contain identical DNA molecules, and the morphologic features of chromosomes are ultimately determined by the DNA sequence they contain.

Interphase Chromosomal Arrangement — Random or Non-random?

The condensation and decondensation cycles of the chromatin fibers during mitosis and meiosis would be greatly facilitated if each chromosome were maintained in a definite space within the nuclear cavity during interphase, so as to avoid entanglements of the chromatin fibers. Some evidence exists in plant cells to indicate that the chromosomes retain during interphase the same spatial order they had at telophase. This was shown to be true in onion chromosomes, which have characteristic heterochromatic regions both at the centromeres and at the telomeres. Throughout interphase the centromeres remain on one side of the nucleus, whereas the telomeres are on the opposite side,[30] in the same orientation that occurs during telophase.

It is even possible that chromosomes may be arranged in a fixed pattern with respect to each other, as is suggested by the classical work of D. P. Costello, who analyzed the first embryonic division in a turbellarian that has chromosomes that can be recognized by particular morphologic features.[40] The chromosomes were arranged in a precise order during metaphase, as shown in Figure 15–17. The probability that such an arrangement would arise at random is less than 1 in 2^{48}.

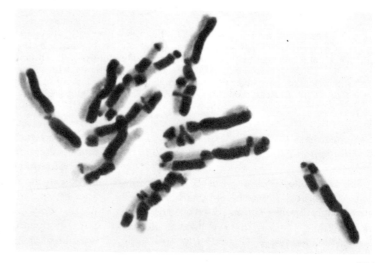

Figure 15–16 Chromosomes of onion root tips that had been treated two generations earlier with bromodeoxyuridine and that had experimentally induced sister chromatid exchanges. Note that each chromosome has two chromatids. The DNA molecule that incorporated bromodeoxyuridine does not stain with Giemsa. Observe that each change in one chromatid is accompanied by a reciprocal one in the other. These images are consistent with the idea that each chromatid contains a single DNA molecule. (Courtesy of J. B. Schvartzman.)

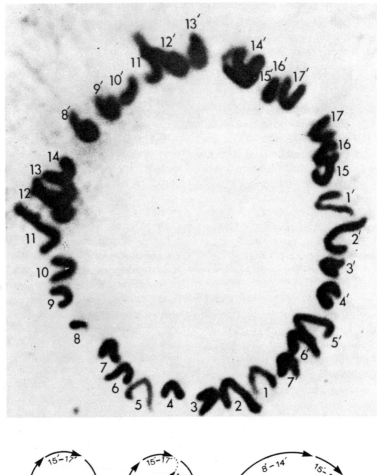

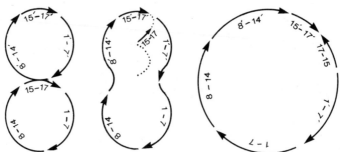

Figure 15–17 Section of a first cleavage metaphase in a *Turbellarian* embryo (polar view). The homologous chromosomes can be recognized by their particular morphology and are indicated by numbers and prime numbers. The chromosomes show an ordered pattern, which might have arisen in the way shown in the lower diagram, which assumes that the chromosomes of both gametes were in an identical linear order. (Courtesy of Mrs. D. P. Costello, from *Proc. Natl. Acad. Sci. USA, 67*:1951, 1970.)

Chromosomes Undergo Condensation-decondensation Cycles

The interphase chromatin becomes condensed at mitosis, and the chromatids become increasingly shorter from prophase to metaphase (Table 15–5), and they become decondensed at telophase (Fig. 15–18). In some favorable organisms, such as the plants *Trillium* and *Tradescantia,* the optical microscope shows that mitotic condensation involves a helical coiling of the chromatids (Table 15–5). In other species, including man, the coils become apparent only after certain pretreatments that partially open the tight helix.[41, 42]

Figures 15–15 and 15–19 show that mitotic

chromosomes consist of chromatin fibers 20 to 30 nm in diameter. In these whole mount chromosomal preparations the chromosomes are dispersed on an air-water interface and then dehydrated by the *critical-point drying* method of Anderson. In this process the water content is replaced by liquid CO_2 in a pressure chamber, and then the specimen is heated above the *critical temperature* (i.e., 31° C), at which liquid CO_2 is converted suddenly into gas. The disorderly folding of the 20 to 30 nm fibers that can be seen in these preparations is an artifact due to the

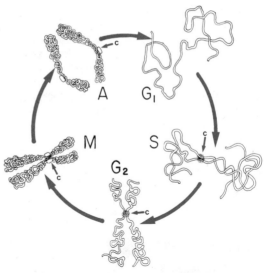

Figure 15–18 The condensation–decondensation cycle of chromosomes. G_1, chromosomes are completely dispersed; S, duplication occurs; and G_2, condensation starts. At metaphase, M, and anaphase, A, the condensation is maximal and the two centromeres are clearly visible.

TABLE 15–5 LENGTH IN MICROMETERS AND NUMBER OF COILS OF THE CHROMATIDS OF TRILLIUM GRANDIFLORUM DURING MITOSIS OF POLLEN GRAINS (MICROSPORES)*

Condition	Length of the Chromatid	Number of Turns
Prophase	346	554
Middle prophase	202	242
Middle prophase	205	276
Final prophase	173	151
Final prophase	154	170
Final prophase	142	187
Metaphase	77	130
Anaphase (15 cells, average)	95.0 ± 2.9	130 ± 3.3

*Data from Sparrow, A. H., Huskins, C. L., and Wilson, G., Studies on the chromosome spiralization cycle in *Trillium. Can. J. Res. C.,* 19:323, 1941.

spreading on water.[42] In the living cell, the chromatin fibers must be much more ordered, since an orderly structure can be packed into a much smaller space than a disorganized one.

As discussed earlier, a metaphase chromosome is 5000 to 10,000 times shorter than the equivalent

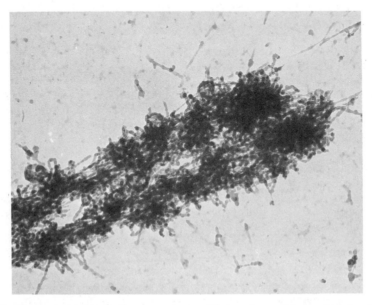

Figure 15–19 Electron micrograph of a chromosome with two chromatids in which bands are apparent as local accumulation of looping 20 to 30 nm chromatin fibers. (Courtesy of G. F. Bahr.)

amount of free DNA. By comparison, the DNA in the head of bacteriophage T_4 (Fig. 3–7) is packed at a ratio of 520/1. The DNA in mitotic chromosomes is contained in nucleosome chains,[43] which are in turn coiled into 20 to 30 nm chromatin fibers (Fig. 15–13); this represents a packing ratio of 40/1. The way in which the higher levels of packaging are achieved is not known, but some ideas about this are emerging from the study of histone-depleted chromosomes.

A Chromosomal Scaffold of Non-histone Proteins

Structural non-histone proteins could be involved in the higher-order coiling of the 20 to 30 nm chromatin fibers. It is possible to remove the histones from metaphase chromosomes (by adding the polyanion dextran sulfate, which competes with the DNA for the basic histones).[44, 45] As shown in Figure 15–20, histone-depleted chromosomes have a central core or *scaffold*, surrounded by a halo made of loops of DNA. The scaffold is made of non-histone proteins and retains the general shape of the metaphase chromosome. Each chromosome has two scaffolds, one for each chromatid, and they are connected together at the centromere region. When the histones are removed, the DNA, which was packed about 40-fold in the chromatin fibers, becomes extended and produces loops with an average length of 25 μm (75,000 base pairs). As can be seen in Figure 15–21,*A,* in each loop the DNA exits from the scaffold and returns to an adjacent point. On the basis of these observations a model of chromosome structure has been proposed (Fig. 15–21,*B*) in which the DNA is arranged in loops anchored to the non-histone scaffold. Since the lateral loops have 25 μm of DNA, after contracting 40-fold in the 20 to 30 nm fiber, they would be only about 0.6 μm long, a length consistent with the diameter of metaphase chromosomes (1 μm) and with micrographs such as the one shown in Figure 15–15.

Other work, using very different methodology, has suggested that it is possible that even in interphase nuclei the chromatin might be organized in "domains" or loops containing about 80,000 base pairs of DNA.[46, 47] This raises the possibility that the DNA domains or loops might persist throughout the cell cycle. More work will be required before we understand the molecular organization of chromosomes.

15–4 HETEROCHROMATIN

Heterochromatin – Chromosomal Regions That do not Decondense During Interphase

In 1928 Heitz defined *heterochromatin* as those regions of the chromosome that remain condensed during interphase and early prophase and form the so-called *chromocenters*. The rest of the chromosome, which is in a non-condensed state, was called *euchromatin* (Gr., *eu,* true). Heitz followed cells throughout the cell cycle and found chromosomal segments that do not decondense.

The heterochromatic segments tend to show preferential localizations in the centromeres of most plants and animals, at the telomeres (especially in plants), or intercalated in the chromosomal arms (frequently adjacent to the nucleolar organizers). In other cases, whole chromosomes become heterochromatic.

The heterochromatic regions can be visualized in condensed chromosomes as regions that stain more strongly or weakly than the euchromatic regions, showing what is called a *positive* or a *negative heteropyknosis* of the chromosomes (Gr., *hetero* + *pyknosis*, different staining).

It is thought that in heterochromatin the DNA remains tightly packed in the 20 to 30 nm fiber (Fig. 15–13), which probably represents the configuration of transcriptionally inactive chromatin.

Heterochromatin can be Facultative or Constitutive

Two types of heterochromatin are generally recognized: *constitutive* heterochromatin which is permanently condensed in all types of cells, and *facultative* heterochromatin, which is condensed only in certain cell types or at special stages of development. Frequently, in *facultative* heterochromatin one chromosome of the pair becomes either totally or partially heterochromatic. The best known case is that of the X chromosomes in the mammalian female, one of which is active and remains euchromatic, whereas the other is inactive and forms the sex chromatin, or Barr body, at interphase (Fig. 20–6).

Constitutive heterochromatin is the most common type of heterochromatin. Most chromosomes contain large blocks of heterochromatin at the centromeres, frequently comprising 5 to 10 per cent of the total chromosomal DNA and, in some extreme cases, such as in *Drosophila virilis,*

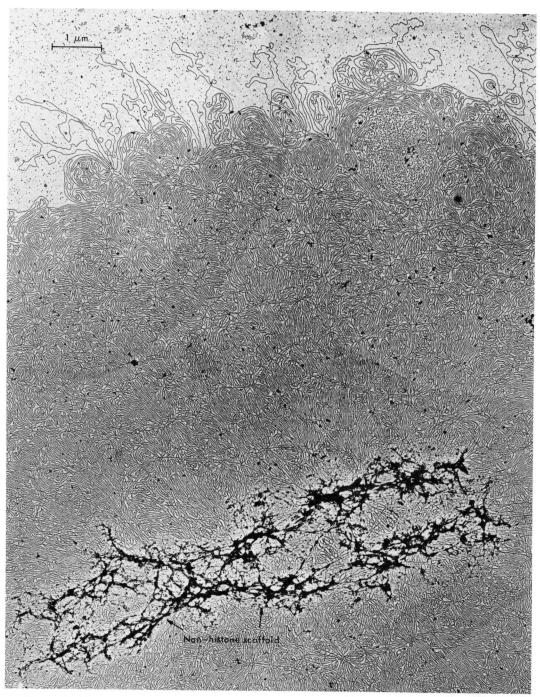

Figure 15–20 Electron micrograph of a histone-depleted human chromosome. The non-histone proteins form two scaffolds, one per chromatid, which are joined at the centromere. The scaffold retains the shape of an intact chromosome, while the naked DNA fibers form a halo around it. (Courtesy of U. K. Laemmli, from Paulson, J. R., and Laemmli, U. K., *Cell, 12*:817, 1977.)

A   B

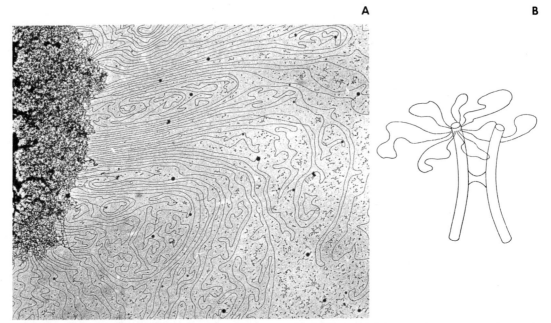

Figure 15–21   Loops of DNA attached to the scaffold of histone-depleted human chromosomes. Note in **A** that the two ends of each loop appear to be anchored at adjacent points of the scaffold. **B,** diagram of the non-histone scaffold with loops of DNA emerging from adjacent points. (Courtesy of U. K. Laemmli, from *Cold Spring Harbor Symp. Quant. Biol.,* 42:351, 1977.)

it can constitute up to 40 per cent of the chromosome. This type of heterochromatin contains highly repeated DNA sequences, which might have a structural role in chromosomes.

A property common to all types of heterochromatin is *late replication.*[48] When cells are given a brief pulse of ^{3}H-thymidine during the late S phase, the label is incorporated only into the heterochromatic segments, indicating that they replicate after the bulk of the DNA.

Heterochromatin is Genetically Inactive

It is generally agreed that condensed chromatin is inactive in RNA synthesis. For example, in the next chapter we will see that condensed mitotic chromosomes do not synthesize RNA (Fig. 16–6). There is good genetic evidence indicating that genes contained in heterochromatic segments are not expressed. Here, we will examine three of these lines of evidence.

Some cats have a striking spotted coloring of the coat. Because of the peculiar colored patches of the coat, these cats are known as *tortoiseshells.* Tortoiseshell cats are always female, never male. The reason for this became clear when Mary Lyon put forward the hypothesis that the patchy pigmentation is produced by a gene contained in the X chromosome, which becomes heterochromatic

and inactive in some groups of cells but not in others. One of the X chromosomes in mammalian females becomes inactivated early in embryogenesis on a random basis, so that adult animals are mosaics in which 50 per cent of the cells have the paternal X chromosome euchromatic and active, and the other 50 per cent have the maternal X chromosome in an active state (see Lyon, 1974, and Chapter 20–2).

Another striking example of facultative heterochromatin is found in the mealy bug (*Planococcus citri*), an insect in which the males have a haploid set of chromosomes that are entirely heterochromatic. The heterochromatic set is the paternal one. The genes in these paternal chromosomes are switched off, as demonstrated by the fact that it is not possible to induce lethal mutations in them, even by high doses of radiation, whereas dominant lethals can be readily obtained by irradiation of the euchromatic set. In female mealy bugs both sets of chromosomes are euchromatic and both are genetically active; i.e., lethal mutations can be induced in both the paternal and the maternal set (see Brown, 1966).

Genes can also be inactivated by translocation into heterochromatic regions of the chromosomes. This was first shown in *Drosophila melanogaster,*[49] in which a gene coding for eye color (*w*) can be inactivated by translocating it close to the centromeric heterochromatin.[50]

The biological significance of the switching off of genes in condensed chromatin might be more universal than indicated in the limited examples discussed here. Multi-cellular organisms have many specialized tissues, and in each one of them a particular set of genes is active and others are inactive. For example, globin genes are expressed in red blood cells but are not expressed in other tissues. We do not know how entire sets of genes are switched off, but many believe that variable degrees of chromatin condensation could be involved in the maintenance of the differentiated state (see Chapter 26).

Constitutive Heterochromatin Contains Repetitive DNA Sequences

Constitutive heterochromatin is related to repetitive DNA. In many cells, it is possible to demonstrate the presence of special DNA fractions using gradients of cesium chloride. For example, in Figure 15–22 most of the DNA of a salamander appears as a band with a buoyant density of 1.705, whereas a small peak, which

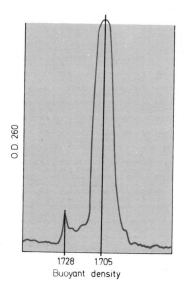

Figure 15–22 Density gradient in cesium chloride of the DNA extracted from cells of the salamander. Observe the main peak of DNA with a buoyant density of 1.705 and the small satellite peak of higher density (1.728). (From MacGregor, H. C., and Kezer, J., *Chromosoma, 33*:167–182, 1971.)

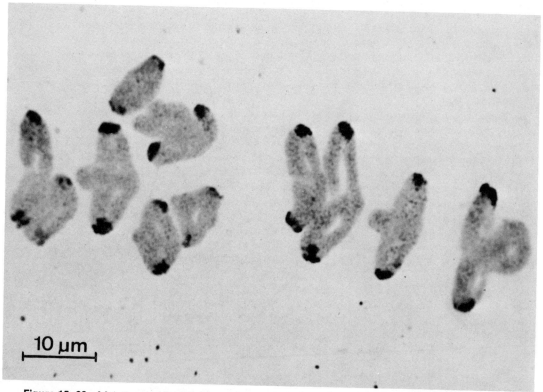

Figure 15–23 Meiotic metaphase I chromosomes of salamander processed for in situ hybridization. Incubation with tritiated RNA complementary to the heavy satellite DNA shown in Figure 15–22. Observe that the labeling has occurred in the centromeric region of the chromosomes. (Courtesy of H. C. MacGregor.)

constitutes only 2 per cent of the total DNA, has a higher density. This peak is the so-called *satellite DNA*. Its higher density is due to the higher content of guanine-cytosine base pairs (about 68 per cent). In Chapter 25 we shall see that the satellite DNA represents simple sequences of nucleotides that are repeated a large number of times in the genome (e.g., up to 10^7 times in mouse satellite DNA).

By using the satellite DNA as the template for in vitro synthesis, it is possible to obtain a ^{3}H-RNA complementary to the satellite DNA. This ^{3}H-RNA can be hybridized to the satellite DNA not only in the test tube but also directly on a slide with the cells that contain this type of DNA.[51, 52]

The hybridization, also called *in situ hybridization,* was detected by autoradiography and localized specifically at the centromeric regions of all the chromosomes (Fig. 15–23). This experiment demonstrates that the satellite DNA is present in the centromeric heterochromatin.

We shall see in Chapter 20 that constitutive heterochromatin (which can be either *centromeric, telomeric,* or *intercalary*) can be detected in chromosomes by the so-called banding technique (and in particular by the C-banding technique). These methods have greatly expanded our knowledge of chromosomal organization and have permitted the recognition of the many alterations that chromosomes may undergo, either spontaneously or by the effects of radiation or chemicals. These findings have had a tremendous impact on the study of the human karyotype in normal and pathologic conditions and on the mapping of genes in the chromosomes.

SUMMARY:

The Chromosomes and Heterochromatin

The study of the chromosomes is of the utmost importance in biology, because it allows one to follow the behavior of DNA molecules and genes in a visual way. The morphology of chromosomes can be best studied during metaphase and anaphase. Chromosomes are of four types: (1) telocentric, (2) acrocentric, (3) submetacentric, and (4) metacentric (Fig. 15–14), depending on the position of the centromere, which is at the primary constriction. Other morphologic characteristics include the secondary constrictions, the telomeres, the satellites, and the nucleolar organizers. The name karyotype is given to the whole group of characteristics that identify a particular chromosomal set.

Each metaphase chromosome has two chromatids, which are attached to each other only by the centromere (the region where the chromosome attaches to the mitotic spindle). Each chromatid consists of a single linear DNA molecule with its associated proteins (unineme theory). Sister chromatids are symmetrical in all respects, because they contain identical DNA molecules.

The chromosomes are probably arranged in a non-random manner in the interphase nucleus, each one maintaining a definite space in the nuclear cavity. It is possible that they might even be arranged in a certain order with respect to each other (Fig. 15–17).

Mitotic chromosomes are made of chromatin fibers of 20 to 30 nm (Fig. 15–15), which represent a condensation of 40 times, far less than the packing of DNA in metaphase chromosomes, which is between 5000- and 10,000-fold (compared with only 520-fold for the packaging of DNA in bacteriophage heads). Structural non-histone proteins could be involved in this higher-order coiling of chromatin fibers. This is suggested by studies in which histone-depleted chromosomes have been shown to have a central scaffold of non-histone proteins which anchors loops of DNA of about 25 μm (75,000 base pairs), as shown in Figures 15–20 and 15–21.

Some regions of the chromosome remain condensed during interphase and are stained differentially by basic dyes. These heterochromatic regions are late-replicating and genetically inert, and they probably consist of 20 to 30 nm chromatin fibers. Two types of heterochromatin can be recognized: *constitutive* and *facultative.*

Constitutive heterochromatin is permanently condensed in all types of cells, can be centromeric, telomeric, or intercalary, and is related to highly repetitive DNAs (satellite DNAs) (see Figure 15–23). *Facultative heterochromatin* is condensed only in certain cell types or at special stages of development. The genes contained in facultative heterochromatin are not expressed, as shown in the cases of X chromosome inactivation in mammals, of the inactive paternal chromosomal set of male mealy bugs, and in certain translocations in *D. melanogaster*. The switching-off of genes in condensed chromatin provides a mechanism that might explain the regulation of genes in cell differentiation.

15–5 THE NUCLEOLUS

The presence of nucleoli, which appear as dense granules within the nucleus, was first described by Fontana in 1781. By the end of the 19th century a relationship between the size of the nucleolus and the synthetic activity of the cell was postulated. It was found that nucleoli were small or absent in cells exhibiting little protein synthesis (sperm cells, blastomeres, muscle cells, and so forth), whereas they were large in oöcytes, neurons, and secretory cells — those in which protein synthesis is a prominent feature. In the living cell, nucleoli are highly refringent bodies (Fig. 2–4). This is the result of a large concentration of solid material, which may constitute 40 to 85 per cent of the dry mass. The light microscope generally reveals the nucleolus as structurally homogeneous, although small corpuscles or vacuoles are sometimes noted.

The nucleolus stains with pyronine and other stains (Fig. 23–7) and absorbs ultraviolet light at 260 nm (Fig. 5–12). Treatment with ribonuclease shows that the capacity to absorb this basophilic stain and ultraviolet radiation depends on the presence of RNA.

The nucleolus may be surrounded by a ring of Feulgen-positive chromatin (Fig. 6–9), which represents the heterochromatic regions of the chromosomes associated with the nucleolus. In large nucleoli some Feulgen-positive granules can be seen in portions of the chromosomes that penetrate the nucleolus.

In Chapter 23–3 we will see that the nucleolus is a factory for ribosomes. The nucleolus is formed at the nucleolar organizer, which is a chromosomal site that contains tandem repeats of the genes coding for 18S and 28S rRNA (Fig. 23–9). This DNA becomes uncoiled and penetrates the nucleolus, where it is actively transcribed. In addition, other ribosomal components, such as 5S RNA and the ribosomal proteins, which are synthesized in other parts of the cell, converge on the nucleolus, where the assembly of ribosomal subunits starts.

Electron Microscopy of the Nucleolus — A Fibrillar and a Granular Zone

Electron microscopy has confirmed the existence of a definite submicroscopic organization within the nucleolus. In some cells nucleoli have a compact structure (Fig. 23–16); in others the structure may be more or less open and include clear regions that communicate with the nucleoplasm (Fig. 15–24).

Two characteristic components may be recognized: the granular zone and the fibrillar zone.[53]

The *granular zone* consists of electron-dense granules of 15 to 20 nm, i.e., smaller in diameter than the ribosomes. Frequently the granular zone occupies the more peripheral region of the nucleolus, which is surrounded by the associated chromatin (Fig. 15–24). Both the granular and the fibrillar zones are most clearly distinguished in Figure 23–16. The *fibrillar zone* consists of fine fibers 5 to 10 nm in diameter. It generally occupies the central region of the nucleolus. Both the granular and fibrillar zones are digested by ribonuclease.

In some cells the nucleolus is surrounded by a ring of heterochromatin, called the *nucleolar-associated* chromatin (Figs. 2–8 and 15–24), which sometimes extends into the nucleolus.

Autoradiographic studies indicate that ^{3}H-uridine is incorporated first into the fibrillar area and is only later found in the granular part. This suggests the relationship:

nucleolar DNA→ fibrillar area→ granular area

The fibrillar area contains the ribosomal DNA and its initial transcripts (called 45S RNA), and the granular part contains ribosomal precursor particles at different stages of maturation.

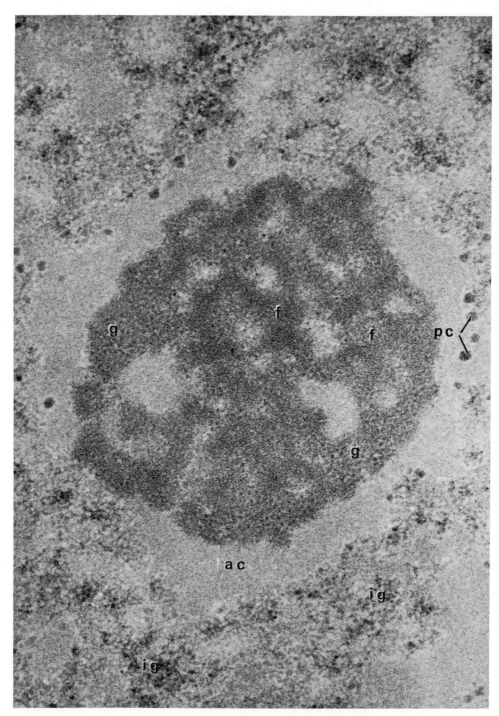

Figure 15–24 Nucleolus from a normal rat hepatocyte stained with uranyl acetate and treated with EDTA, which bleaches the DNA. In the nucleolus the fibrillar (*f*) and granular (*g*) portions and the associated chromatin (*ac*) are seen; *ig,* interchromatin granules; *pc,* perichromatin granules. ×60,000. (Courtesy of W. Bernhard.)

The Nucleolus is Disassembled During Mitosis

During mitosis nucleoli undergo cyclic changes. Nucleoli are formed around the DNA loop that extends from the nucleolar organizer. There may be several nucleoli per cell, but frequently they tend to fuse into one or few nucleoli. During late prophase the DNA loop containing the rRNA genes gradually retracts and coils into the nucleolar organizer of the corresponding chromosome. Since this DNA is very extended as a consequence of intense RNA synthesis, the nucleolar organizer region is one of the last to undergo condensation, thus producing a *secondary constriction* on the chromosome. The fibrillar and granular components are gradually dispersed into the nucleoplasm. After the cell divides, during telophase, the nucleolar organizer DNA uncoils and the nucleolus is reassembled.

SUMMARY:

The Nucleolus

The principal RNA-containing structure in the nucleus is the *nucleolus.* This is a highly refringent structure with a large concentration of solid material, in particular proteins. The RNA content ranges from 3 to 10 per cent. The nucleolus is frequently surrounded by a ring of heterochromatin which may penetrate into its structure. The fine structure of the nucleolus reveals the presence of two components: the *fibrillar* and *granular zones.* The fibrillar and granular zones are composed of ribonucleoproteins and are related to the biogenesis of the ribosomes. The nucleolus is formed around the nucleolar organizer region of the chromosomes, which contains the genes coding for 18S and 28S RNA. The fibrillar area consists of the rDNA and its initial transcripts, and the granular part contains ribosomal precursors at various stages of assembly. The nucleus undergoes cyclic changes during cell reproduction; it is disassembled during late prophase and reappears at telophase.

REFERENCES

1. Barer, R., Joseph, S., and Merck, G. A. (1959) *Exp. Cell Res.,* 18:179.
2. Gall, J. G. (1956) *Brookhaven Symp. Biol.,* 8:17.
3. Gall, J. G. (1967) *J. Cell Biol.,* 32:391.
4. Scheer, U. (1973) *Dev. Biol.,* 30:13.
5. Maul, G. G. (1970) *J. Cell Biol.* 46:604.
6. Kessel, R. G. (1968) *J. Ultrastruc. Res. Suppl.,* 10:1.
7. Loewenstein, W. R., Kanno, Y., and Ito, S. (1966) *Ann. N.Y. Acad. Sci.,* 137:708.
8. Paine, P. L., Moore, L. C., and Horowitz, S. B. (1975) *Nature,* 254:109.
9. Gurdon, J. B. (1970) *Proc. R. Soc. London [Biol.],* 176:303.
10. Bonner, W. M. (1975) *J. Cell Biol.,* 64:421, and 64:431.
11. De Robertis, E. M., Longthorne, R. F., and Gurdon, J. B. (1978) *Nature,* 272:254.
12. Clark, T. G., and Merriam, R. W. (1977) *Cell,* 12:883.
13. Boivin, A., Vendrely, R., and Vendrely, C. (1948) *C. R. Acad. Sci.* (Paris), 226:1061.
14. Mirsky, A. E., and Ris, H. (1948) *J. Gen. Physiol.,* 31:1.
15. Du Praw, E. J. (1968) *Cell and Molecular Biology.* Academic Press, Inc., N.Y.
16. De Lange, R. J., and Smith, E. L. (1971) *Annu. Rev. Biochem.,* 40:279.
17. Rouvière-Yaniv, J. (1977) *Cold Spring Harbor Symp. Quant. Biol.,* 42:439.
18. Olins, A. L., and Olins, D. E. (1974) *Science,* 183:330.
19. Woodcock, C. (1973) *J. Cell Biol.,* 59:368, abstract.
20. Oudet, P. M., Gross-Bellard, M., and Chambon, P. (1975) *Cell,* 4:218.
21. Kornberg, R. D. (1974) *Science,* 184:868.
22. Hewish, D. R., and Burgoyne, L. A. (1973) *Biochem. Biophys. Res. Commun.,* 52:504.
23. Noll, M. (1974) *Nature,* 251:249.
24. Kornberg, R. D., and Thomas, J. O. (1974) *Science,* 184:865.
25. Finch, J. T., Noll, M., and Kornberg, R. D. (1975) *Proc. Natl. Acad. Sci. USA,* 72:3320.
26. Noll, M., and Kornberg, R. D. (1977) *J. Molec. Biol.,* 109:393.
27. Finch, J. T., Lutter, L. C., Rhodes, D., Brown, R. S., Rushton, B., Levitt, M., and Klug, A. (1977) *Nature,* 269:29.
28. Morris, N. R. (1976) *Cell,* 9:627.
29. Compton, J. L., Bellard, M., and Chambon, P. (1976) *Proc. Natl. Acad. Sci. USA,* 73:4382.
30. Ris, H. (1975) *Ciba Found. Symp.,* 28:7.
31. Finch, J. T., and Klug, A. (1976) *Proc. Natl. Acad. Sci. USA,* 73:1897.
32. Morgan, T. H. (1910) *Science,* 32:120.
33. Stubblefield, E. (1973) *Int. Rev. Cytol.,* 35:1.
34. Kubai, D. F. (1975) *Int. Rev. Cytol.,* 43:167.

35. Telzer, B. R., Moses, M. J., and Rosenbaum, J. L. (1975) *Proc. Natl. Acad. Sci. USA*, 72:4023.
36. McGill, M., and Brinkley, B. R. (1975) *J. Cell Biol.* 67:189.
37. Rubin, G. M. (1977) *Cold Spring Harbor Symp. Quant. Biol.*, 42:1041.
38. Kavenoff, R., and Zimm, B. H. (1973) *Chromosoma*, 41:1.
39. Fussell, C. P. (1975) *Chromosoma*, 50:201.
40. Costello, D. P. (1970) *Proc. Natl. Acad. Sci. USA*, 67:1951.
41. Ohnuki, Y. (1968) *Chromosoma*, 25:401.
42. Ris, H. (1978) 9th International Congress of Electron Microscopy, Toronto. (Ed. by Microscopical Society of Canada, Toronto.)
43. Compton, J. L., Hancock, R., Oudet, P., and Chambon, P. (1976) *Eur. J. Biochem.*, 70:555.
44. Adolph, K. W., Cheng, S. M., and Laemmli, U. K. (1977) *Cell*, 12:805.
45. Paulson, J. R., and Laemmli, U. K. (1977) *Cell*, 12:817.
46. Cook, P. R., and Brazell, I. A. (1975) *J. Cell Sci.*, 19:261.
47. Benyajati, C., and Worcel, A. (1976) *Cell*, 9:393.
48. Lima de Faría, A., and Jaworska, H. (1968) *Nature*, 217:138.
49. Demerec, M. (1940) *Genetics*, 25:618.
50. Becker, H. J. (1978) *Results Probl. Cell Differ.* 9:29.
51. Pardue, M. L., and Gall, J. G. (1970) *Science*, 168:1356.
52. Jones, K. W. (1970) *Nature*, 255:912.
53. Bernhard, W., and Granboulan, N. (1968) In: *The Nucleus*, p. 81, Academic Press, Inc., New York.

ADDITIONAL READING

Back, F. (1976) The variable condition of euchromatin and heterochromatin. *Int. Rev. Cytol.*, 45:25.

Bahr, G. F. (1977) Chromosomes and chromatin structure. In: *Molecular Structure of Human Chromosomes*, p. 144. (Yunis, J. J., ed.) Academic Press, Inc., New York.

Birnstiel, M. (1967) The nucleolus in cell metabolism. *Annu. Rev. Plant Physiol.*, 18:25.

Bostock, C. J., and Summer, A. T. (1978) *The Eukaryotic Chromosome*. North-Holland Publishing Company, Amsterdam.

Brown, S. W. (1966) Heterochromatin. *Science*, 151:417. A classic paper. Highly recommended.

Callan, H. G., and Klug, A., eds. (1978) Structure of eukaryotic chromosomes and chromatin. *Philos. Trans. R. Soc. London [Biol. Soc.], 283*. An excellent meeting volume.

Chambon, P. (1977) The molecular biology of the eukaryotic genome is coming of age. *Cold Spring Harbor Symp. Quant. Biol.*, 42:1209. Highly recommended. Exceptionally clear summary of a most important meeting.

De Robertis, E. M., Longthorne, R. F., and Gurdon, J. B. (1978) Intracellular migration of nuclear proteins in *Xenopus* oöcytes. *Nature, 272*:254.

Feldherr, C. M. (1972) Structure and function of the nuclear envelope. In: *Advances in Cell and Molecular Biology*, 2:273.

Felsenfeld, G. (1978) Chromatin. *Nature, 271*:115.

Franke, W. W. (1974) Structure, biochemistry and function of the nuclear envelope. *Int. Rev. Cytol. Suppl.*, 4:71.

Franke, W. W. (1977) Structure and function of nuclear membranes. *Biochem. Soc. Symp.*, 42:125.

Ghosh, S. (1976) The nucleolar structure. *Int. Rev. Cytol.*, 44:1.

Goldstein, L. (1974) Movement of molecules between nucleus and cytoplasm. In: *The Cell Nucleus*, Vol. 1, p. 387. Academic Press, Inc., New York.

Harris, J. R. (1978) The biochemistry and ultrastructure of the nuclear envelope. *Biochim. Biophys. Acta, 515*:55.

Hsu, T. C. (1974) Longitudinal differentiation of chromosomes. *Ann. Rev. Genet.*, 7:153.

Kavenoff, R., Klotz, L. C., and Zimm, B. H. (1973) On the nature of chromosome-sized DNA molecules. *Cold Spring Harbor Symp. Quant. Biol.*, 38:1.

Kornberg, R. D. (1974) Chromatin structure: A repeating unit of histones and DNA. Chromatin structure is based on a repeating unit of eight histone molecules and about 200 base pairs of DNA. *Science, 184*:868.

Kornberg, R. D. (1977) Structure of chromatin. *Annu. Rev. Biochem.*, 40:931.

Laemmli, U. K., Cheng, S. M., Adolph, K. W., Paulson, J. R., Brown, J. A., and Baumbach, W. R. (1977) Metaphase chromosome structure: The role of nonhistone proteins. *Cold Spring Harbor Symp. Quant. Biol.*, 42:351.

Lewin, B. (1974) Eukaryotic chromosomes. In *Gene Expression*. Vol. 2. John Wiley & Sons, New York. A detailed account of pre-nucleosome days.

Lyon, M. (1974) Review lecture. Mechanisms and evolutionary origins of variable X chromosome activity in mammals. *Proc. R. Soc. London [Biol.], 187*:243.

Maul, G. G. (1977) Nuclear pore complexes. Elimination and reconstruction during mitosis. *J. Cell Biol.*, 74:492.

Maul, G. G. (1979) The nuclear and cytoplasmic pore complex. *Int. Rev. Cytol.*, Suppl. 6:76,

McKnight, S. L., Bustin, M., and Miller, O. L., Jr. (1977) Electron microscopic analysis of chromosome metabolism in the *Drosophila melanogaster* embryo. *Cold Spring Harbor Symp. Quant. Biol., 42*:741.

Monneron, A., and Bernhard, W. (1969) Fine structural organization of the interphase nucleus in some mammalian cells. *J. Ultrastruct. Res., 27*:266.

Schatten, G., and Thoman, M. (1978) Nuclear surface complex as observed with the high resolution scanning microscope. *J. Cell Biol., 77*:517.

Swift, H. (1973) The organization of genetic material in eukaryotes: Progress and prospects. *Cold Spring Harbor Symp. Quant. Biol., 38*:963.

Vincent, W. S., and Miller, O. L., eds. (1966) The nucleolus: Its structure and function. *Natl. Cancer Inst. Monogr., 23.*

White, M. J. D. (1972) *The Chromosomes*, 6th Ed. Chapman & Hall, London. Highly recommended for classical cytology.

Wischnitzer, S. (1973) The submicroscopic morphology of the interphase nucleus. *Int. Rev. Cytol., 34*:1.

Yunis, J. J., and Yasmineh, W. G. (1971) Heterochromatin, satellite DNA, and cell function. *Science, 174*:1200.

Yunis, J. J., and Yasmineh, W. G. (1972) Model of mammalian constitutive heterochromatin. In: *Advances in Cell and Molecular Biology*, Vol. 2. (Du Praw, E. J., ed.) Academic Press, Inc., New York.

16

THE CELL CYCLE AND
DNA REPLICATION

16–1 The Cell Cycle   364
  Interphase—The G_1, S, and G_2 Phases
  G_1—The Most Variable Period of the Cell
    Cycle
  Visualization of Chromosomes During G_1, S,
    and G_2 by Premature Condensation
  Condensed Chromosomes do not Synthesize
    RNA
  Several Molecular Events Occur at Defined
    Stages of the Cell Cycle
  Summary: The Cell Cycle

16–2 DNA replication   370
  DNA Replication is Semiconservative
  Replication of the E. coli Chromosome is
    Bidirectional
  DNA Synthesis is Discontinuous

Reverse Transcriptase can Copy RNA into
  DNA
Eukaryotic Chromosomes Have Multiple
  Origins of Replication
Eukaryotic DNA Synthesis is Bidirectional
The Number of Replication Units is De-
  velopmentally Regulated
Reinitiation within the Same Cycle is Pre-
  vented in Eukaryotic Cells
New Nucleosomes are Assembled Simul-
  taneously with DNA Replication
RNA Synthesis Continues During DNA Repli-
  cation
DNA Repair Enzymes Remove Thymine
  Dimers Induced By UV Light
Xeroderma Pigmentosum Patients Have De-
  fective DNA Repair
Summary: DNA Replication

The ability to reproduce is a fundamental property of cells. The magnitude of cell multiplication can be appreciated by realizing that an adult person is formed by 10^{14} cells, all derived from a single cell, the fertilized ovum. Even in fully grown adults, the amount of cell multiplication is impressive. A man contains 2.5×10^{13} red blood cells (five liters of blood, with 5,000,000 red blood cells/mm³), and the average life span of a red blood cell is 120 days (10^7 seconds). Therefore, to maintain a constant blood supply, 2.5×10^{13} cells must be produced every 10^7 seconds; i.e., 2.5 million new cells are required per second. Furthermore, cell reproduction is precisely regulated so that the production of new cells compensates exactly the loss of cells in adult tissues. In this chapter we will analyze the life cycle of eukaryotic cells.

16–1 THE CELL CYCLE

A growing cell undergoes a cell cycle that comprises essentially two periods; the *interphase*

(period of non-apparent division) and the period of *division*. Division may take place by mitosis, meiosis, or other mechanisms of cell replication. For many years the interest of cytologists was concerned mainly with the period of division in which dramatic changes visible under the light microscope could be observed, and interphase was considered a "resting" phase. However, cells spend most of their life span in interphase (Table 16–1), which is a period of intense biosynthetic activity in which the cell doubles in size and duplicates precisely its chromosome complement.

The cell cycle can be considered as the complex series of phenomena by which cellular material is divided equally between daughter cells. Cell division is only the final and microscopically visible phase of an underlying change that has occurred at the molecular level. Before the cell divides by mitosis, its main molecular components have already been duplicated. In this respect, cell division can be considered as the final separation of the already duplicated molecular units.

TABLE 16–1 MITOTIC AND INTERMITOTIC TIMES IN VARIOUS CELL TYPES*

Cell	Times in Minutes	
	Intermitotic	Mitotic
Vicia faba root meristem (19° C.)	1300	150
Pisum sativum (peas) root meristem (20 ° C.)	1350	177
Chick fibroblasts (38° C.)	660–720	23
Mouse spleen cultures	480–1080	43–90
Rat jejunum (in animal)	2000	28
Jensen's sarcoma (in animal)	720	27
Rat corneal epithelium (in animal)	14,000	70
Chrotophaga (grasshopper) neuroblast	27	181
Drosophila egg	2.9	6.2
Psammechinus (sea urchin) embryo, two to four cell stage (16° C.)	14	28

*From Mazia, D. *In* Brachet, J., and Mirsky, A. E. (eds.), *The Cell*, Vol. 3, New York, Academic Press, Inc., 1961.

Interphase — The G_1, S, and G_2 Phases

The introduction of cytochemical methods, such as the Feulgen stain, followed by a cytophotometric quantitative assay, first suggested that the doubling of DNA takes place during interphase. The studies done by autoradiography with labeled thymidine, however, were the most important in determining the exact period in which DNA replication takes place in a eukaryotic cell. These studies demonstrated that the synthesis occurs only in a restricted portion of the interphase — the so-called S period (*i.e., synthetic period*), which is preceded and followed by two "gap" periods of interphase (G_1 and G_2) in which there is no DNA synthesis. This led Howard and Pelc[1] to divide the cell cycle into four successive intervals: G_1, S phase, G_2, and mitosis; G_1 is the period between the end of mitosis and the start of DNA synthesis, S is the period of DNA synthesis, and G_2, the interval between the end of DNA synthesis and the start of mitosis (Fig. 16–1).

During G_2 a cell contains two times (4C) the amount of DNA present in the original diploid cell (2C). Following mitosis the daughter cells again enter the G_1 period and have a DNA content equivalent to 2C (Fig. 16–2).

G_1 — The Most Variable Period of the Cell Cycle

The duration of the cell cycle varies greatly from one type of cell to another. For a mammalian cell growing in culture with a generation time of 16 hours, the different periods would be as follows: G_1 = 5 hours, S = 7 hours, G_2 = 3 hours, and mitosis = 1 hour. Generally speaking, the S, G_2, and mitotic periods are relatively constant in the

cells of the same organism. The G_1 period is the most variable in length. Depending on the physiologic condition of the cells, it may last days, months, or years. Those tissues that normally do not divide (such as nerve cells or skeletal muscle), or that divide rarely (e.g., circulating lymphocytes), contain the amount of DNA present in the G_1 period. Cultured cells that stop multiplying because of density-dependent inhibition of growth (or contact inhibition — see Fig. 8–27) also stop at G_1.

The regulation of the duration of the cell cycle occurs primarily by arresting it at a specific point of G_1, and the cell in the arrested condition is said to be in the G_0 state.[2] In the G_0 state the cell may be considered to be withdrawn from the cell cycle (Fig. 16–1); when conditions change and growth is resumed, the cell re-enters the G_1 period (Fig. 16–1).

The different stages of the cell cycle were initially discovered by autoradiographic experiments.[1] In a typical experiment, cultured cells

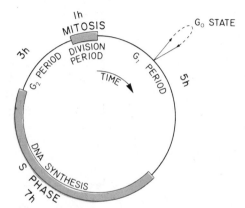

Figure 16–1 The cell cycle. The duration of each phase corresponds to a mammalian cell growing with a generation time of 16 hours. (After D. M. Prescott.)

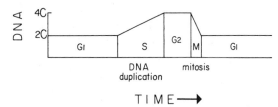

Figure 16–2 Life cycle of a cell showing the changes in DNA content during the various periods as a function of time. *2C* corresponds to the diploid content of DNA; *4C* corresponds to the tetraploid content.

are exposed for 10 minutes to ^{3}H-thymidine, a nucleoside that labels only the DNA molecule. The culture is then washed thoroughly and cultured in unlabeled medium; samples are fixed for autoradiographic studies at 1 to 2 hour intervals for 24 hours. The percentage of *mitotic* cells that are labeled is then used to measure the cell cycle stages (Fig. 16–3). The first labeled mitotic chromosomes will appear about three hours later, indicating that the G_2 period is about that long. The S phase spans the time between the increase in the labeled mitosis percentage and its decrease (Fig. 16–3). Eventually a new cell cycle starts, and the percentage of labeled mitosis rises again. The total duration of the cell cycle can be estimated from the interval between the two ascending slopes of the curve (or measured directly by counting the number of cells in the culture). The length of mitosis (usually about one hour) can be determined by microscopic observation (percentage of cells in mitosis) if the generation time of the culture is known. The length of G_1 can then be calculated by subtract-

ing the lengths of G_2, S, and mitosis from the total duration of the cell cycle.

Cell cycle studies can be greatly simplified by using *synchronized* cells, in which all the cells in the population are at the same stage of the cell cycle. There are many methods for synchronizing cells (reviewed by Mitchison, 1971, and Prescott, 1976). The simplest one is the *mitotic selection* of cultured cells, which takes advantage of the fact that cells in mitosis become rounded up and very loosely attached to the surface of the culture flask, while cells in interphase are flat and firmly attached to the culture vessel. By vigorously shaking the culture, it is possible to preferentially detach the mitotic cells, while the interphase ones remain attached, thus obtaining preparations in which up to 99 per cent of the cells are in mitosis.

Visualization of Chromosomes During G_1, S, and G_2, by Premature Condensation

During interphase the chromosomes are decondensed and cannot be distinguished under the microscope. However, using an experimental trick, it is possible to induce the condensation of chromosomes in all three stages of interphase.[3] Mitotic cells fused to interphase cells (using inactivated Sendai virus — see Fig. 26–4), are able to induce *premature chromosome condensation* in the interphase nuclei. As can be clearly seen in Figure 16–4, the prematurely condensed chromosomes from G_1 nuclei show only one chromatid (Fig. 16–4,*A*), while the ones from G_2 nuclei have two chromatids (Fig. 16–4,*C*).

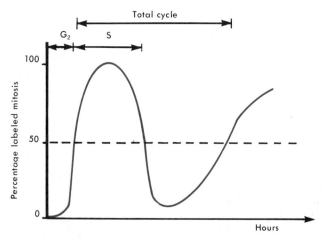

Figure 16–3 Determination of the duration of the cell cycle stages by the percentage of mitotic cells labeled after a ^{3}H-thymidine pulse. See description in the text. The length of each stage can be determined from the points at which the curve transects the 50 per cent line.

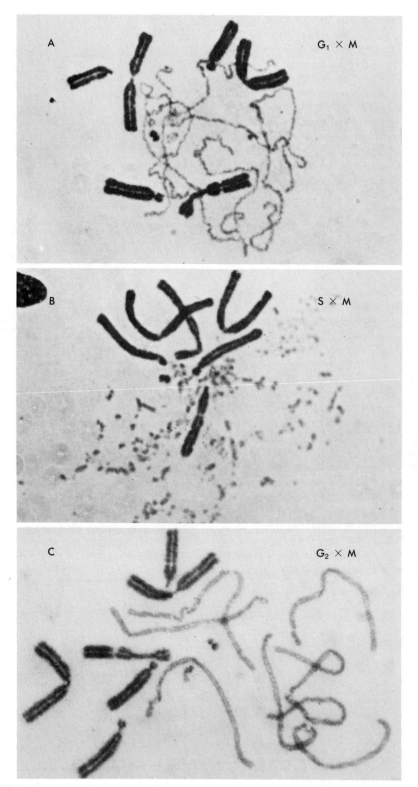

Figure 16–4 Premature chromosome condensation of G_1, S, and G_2 chromosomes, induced by fusion to mitotic cells. All cells are from the Indian muntjac; a deer that has a small number of chromosomes. The thick metaphase chromosomes are from the mitotic cells. **A,** $G_1 \times M$, note that the G_1 chromosomes have a single chromatid. **B,** $S \times M$, the S phase chromosomes have a fragmented appearance. **C,** $G_2 \times M$, the G_2 chromosomes have two chromatids. (Courtesy of D. Röhme.)

This shows visually that the G_1 phase corresponds to the period of interphase prior to DNA replication and that G_2 is the period of interphase that follows DNA replication.

When S phase nuclei are fused to mitotic cells, a more complex pattern is observed (Fig. 15–4,B) in which the chromosomes adopt a "pulverized" configuration. It is not yet known whether the replicating chromosomes actually break during premature condensation, or if the condensed segments are joined by threads of DNA that cannot be distinguished under the microscope.[4]

Condensed Chromosomes do not Synthesize RNA

Unlike DNA synthesis, RNA synthesis occurs throughout interphase. RNA synthesis takes place in the nucleus and then this molecule is transferred to cytoplasm. For example, if the protozoa *Tetrahymena* are incubated in ^{3}H-cytidine for 1.5 to 12 minutes, the labeled RNA appears in the nucleus (Fig. 16–5,A). After 35 minutes both the nucleus and cytoplasm contain about the same amount (Fig. 16–5,B). Finally, if the cell is subjected to ^{3}H-cytidine for a few minutes and then incubated with non-radioactive precursor (a chase) for a longer period, the cytoplasm is labeled, but the nucleus is not (Fig. 16–5,C).[5] The amount of mitochondrial RNA synthesis is too small to be detected in this type of experiment.

RNA synthesis stops during mitosis, as is shown dramatically in Figure 16–6. The rate of RNA synthesis declines rapidly in late prophase and stops in metaphase and anaphase. As occurs

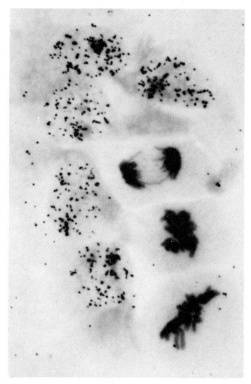

Figure 16–6 RNA synthesis stops during mitosis. Chinese hamster ovary (CHO) cells were labeled for 15 minutes with ^{3}H-uridine and then subjected to radioautography. (Courtesy of D. M. Prescott, from *Reproduction of Eukaryotic Cells.* Academic Press, Inc., New York, 1976.)

in other instances (Chapter 15–4), highly condensed chromatin cannot be transcribed, presumably because the DNA cannot be reached by RNA polymerase.

During the entire cell cycle the chromosome

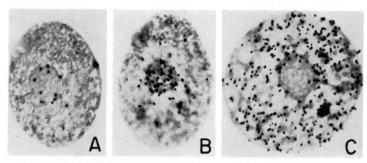

Figure 16–5 **A,** radioautograph of a *Tetrahymena* incubated in ^{3}H-cytidine for 1.5 to 12 minutes. Notice that all labeled RNA is restricted to the nucleus. **B,** the same as in **A,** after 35 minutes. RNA begins to enter the cytoplasm. **C,** the same, incubated for 12 minutes in ^{3}H-cytidine and then for 88 minutes in a nonradioactive medium. Notice that while the nucleus has lost all labeled RNA, the cytoplasm is heavily labeled. (Courtesy of D. M. Prescott, from *Reproduction of Eukaryotic Cells.* Academic Press, Inc., New York, 1976.)

participates in at least three different activities: (1) self-duplication, (2) transcription and transfer of genetic information to the rest of the cell, and (3) the coiling and uncoiling cycle associated with the separation of the duplicated chromosomes or daughter chromatids. Self-duplication and transcription occur at the moment when chromosomes are most dispersed (i.e., uncoiled).

Several Molecular Events Occur at Defined Stages of the Cell Cycle

Of the biochemical events that occur at defined stages of the cell cycle, the most noticeable one is DNA synthesis. The S phase cells contain a factor that induces DNA synthesis. This has been shown by cell fusion experiments, in which the onset of DNA replication in G_1 nuclei can be accelerated by fusion with S phase cells.[6] Similarly, DNA replication can be induced in nuclei that normally do not divide (such as those of neurons) by microinjection into *Xenopus* eggs.[7] Additional evidence of a factor inducing DNA synthesis comes from multi-nucleated cells, such as the plasmodium of the slime mold *Physarum polycephalum,* in which all the nuclei start DNA synthesis simultaneously.[8]

The S phase lasts for several hours, and during this period many units of replication are sequentially activated. In some cells, the CG-rich regions of the genome tend to replicate earlier than AT-rich ones.[9, 10] In all cells, the more condensed, heterochromatic regions of the genome replicate late during the S phase.[11] Autoradiographic studies have shown that the centromeric heterochromatin replicates later than the rest of the chromosome and, correspondingly, biochemical experiments have demonstrated that the highly repetitive satellite DNAs contained in the centromeres (Fig. 15–23) are also late replicating.[12] Sometimes whole chromosomes replicate asynchronously, as in the case of heterochromatic X chromosomes. In mammalian females one of the X chromosomes becomes heterochromatic and late replicating during the course of development, whereas the other one remains euchromatic and replicates earlier. During early development, before X inactivation occurs, both X chromosomes replicate early.[13]

Many other events occur at defined moments of the cell cycle, some of which are indicated in Figure 16–7. Histones are synthesized during the S phase, the period during which they become associated with the newly replicated DNA (Fig. 16–18). This temporal relationship can be determined biochemically, but it had been suggested previously by early cytochemical work

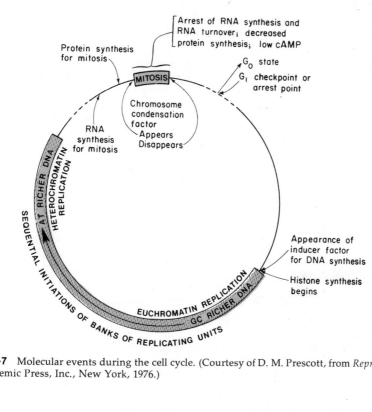

Figure 16–7 Molecular events during the cell cycle. (Courtesy of D. M. Prescott, from *Reproduction of Eukaryotic Cells.* Academic Press, Inc., New York, 1976.)

using the stain fast green, which stains basic proteins specifically.[14] During G_2 the factors necessary for chromosome condensation are synthesized. It is not known, in molecular terms, how chromosome condensation is achieved, but in many cells histone H1 becomes phosphorylated during this period.[15] During mitosis RNA synthesis stops (Fig. 16–6), and the rate of protein synthesis falls by about 75 per cent;[16, 17] both of these functions which recover rapidly in late telophase. The concentration of cyclic AMP is low during mitosis, and sometimes undergoes

a transient increase during G_1, but the significance of these changes is still obscure.

The most important point in the regulation of the cell cycle occurs in the G_1 phase, during which it must decided whether the cell will start a new cell cycle or become arrested in the G_0 state (Fig. 16–7). Once this G_1 checkpoint has been passed, the cell will go on to complete a new cycle. Unfortunately, we still know very little about the regulation of this fundamental step in cell proliferation.

SUMMARY:

The Cell Cycle

Mitosis represents only a small part of the life cycle of a cell (about one hour in most cells). The cell spends most of its lifetime in interphase, the period during which it doubles in size and replicates its DNA. The cell cycle can be divided into four periods: G_1, S, G_2, and mitosis. G_1 is the time "*gap*" between the end of mitosis and the start of DNA synthesis, S is the period of DNA synthesis, and G_2 is the time "*gap*" between the end of DNA synthesis and the beginning of mitosis (Fig. 21–1). For a mammalian cell with a generation time of 16 hours, the different periods, for example, could be: $G_1 = 5$ hours, S $= 7$ hours, $G_2 = 3$ hours, and mitosis $= 1$ hour. The most variable period is G_1; depending on the physiologic conditions of the cells, it may last days, months, or years. Cells that stop proliferating become arrested at a specific point of G_1 and remain withdrawn from the cell cycle in the G_0 state.

When interphase cells are fused to mitotic cells, the chromosomes become prematurely condensed, and it can be observed that G_1 chromosomes (unreplicated) have only one chromatid and that G_2 chromosomes (which have already replicated) have two (Fig. 16–4).

During mitosis RNA synthesis stops in the condensed chromosomes (Fig. 16–6), and the rate of protein synthesis decreases. During the S phase heterochromatin is late replicating. Sometimes whole chromosomes are late replicating (heterochromatic X chromosomes); and sometimes only part of a chromosome is late replicating, as occurs with centromeric heterochromatin, which contains satellite DNA. The most important point in the regulation of cell proliferation occurs during G_1, when the crucial decision of whether the cell undergoes a new division cycle or enters the G_0 state is taken, but we do not know how this is achieved.

16–2 DNA REPLICATION

The mechanism of DNA replication may be considered a direct consequence of the DNA structure presented in the molecular model proposed by Watson and Crick in 1953, with which the student should be familiar (Fig. 5–13). This mechanism involves the unwinding of the two polynucleotide strands, followed by the duplication of two new complementary

strands via a template mechanism (Fig. 16–8), in which each DNA strand acts as a template for the newly synthesized molecule. DNA replicates only once in the course of a cell cycle. The mechanism could be set into action simply by the separation of the two DNA strands, which permits the nucleotides to fall into place and to be linked by the action of enzyme DNA-polymerase. The two polynucleotide strands can be separated because they are joined by relatively weak hydrogen bonds.

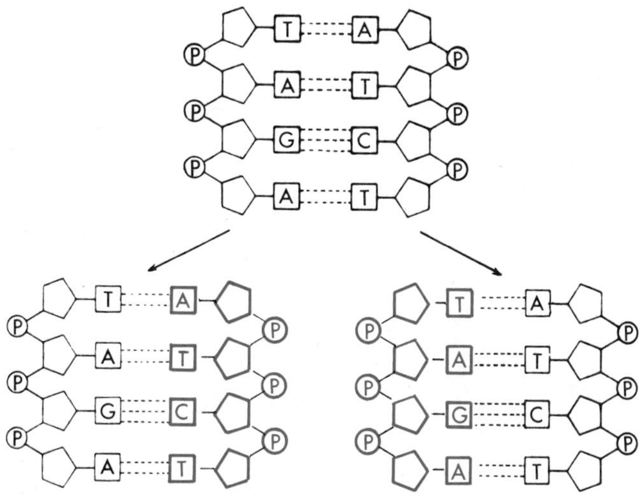

Figure 16–8 Diagram showing the mechanism of DNA duplication. **Above,** the two standard parent molecules, which separate by opening of hydrogen bonds. **Below,** the two new strands that have been synthesized and that have a complementary base composition with respect to the parent DNA strands are indicated by bold outlines. (From Kornberg, A., *Ciba Lecture in Microbial Biochemistry*, New York, John Wiley and Sons, 1962.)

DNA Replication is Semiconservative

The Watson-Crick model also suggests that replication is semiconservative, which means that half of the DNA is conserved (i.e., only one strand is synthesized; the other half of the original DNA is retained) (Fig. 16–9). This has been verified by several demonstrations. In their classic experiment Meselson and Stahl made use of the heavy isotope ^{15}N. The DNA containing ^{15}N (heavy-heavy, or HH DNA) is more dense than the DNA containing ^{14}N. E. coli was grown in a medium containing ^{15}N and was then passed to another medium containing normal ^{14}N. The DNA was isolated and its density was determined by ultracentrifugation on a cesium chloride gradient. It was found that after the first division cycle there is only one DNA peak corresponding to the heavy-light (HL) hybrid molecule (i.e., one strand is labeled with ^{14}N, the other with ^{15}N). At the second generation two peaks of DNA appear — one in which the two DNA strands contain ^{14}N (LL) and the other still corresponding to hybrid molecules (Fig. 16–9).[8]

Semiconservative DNA replication can also be demonstrated in higher organisms, since as explained in the previous chapter, each chromatid of a metaphase chromosome represents a single DNA molecule. In their classic experiment Taylor, Woods, and Hughes[19] labeled *Vicia fava* (broad bean) root tip cells with ^{3}H-thymidine and then allowed them to grow in unlabeled medium. When the metaphase chromosomes were analyzed by autoradiography one or two generations later (Fig. 16–9), it was found that both chromatids were labeled after one generation, but only one was radioactive after two cell cycles, as expected in semiconservative replication (Fig. 16–9).

More recently, techniques have been developed that allow visualization of semiconservative chromosome replication without autoradiography.[20] These methods are based on the use of 5-bromodeoxyuridine (BrdU), an analogue that is incorporated into chromosomes in place of thymidine. DNA that contains BrdU does not stain with a fluorescent dye (33258 Hoechst) or with certain modifications of the Giemsa staining technique. When cells labeled with BrdU are subsequently grown for two generations in a medium without the analogue, the metaphase chromosomes have the appearance shown in Figure 16–10.

Replication of the E. coli Chromosome is Bidirectional

Much of what we know about DNA replication comes from the study of E. coli. As mentioned in Chapter 2, this prokaryote has a single circular chromosome, which upon unfolding, has a length of about 1.1 mm (4.2×10^6 base pairs). The chromosome is made of a single molecular of DNA and is attached at a point of the cell membrane. Figure 16–11 shows an autoradiograph of an E. coli chromosome during

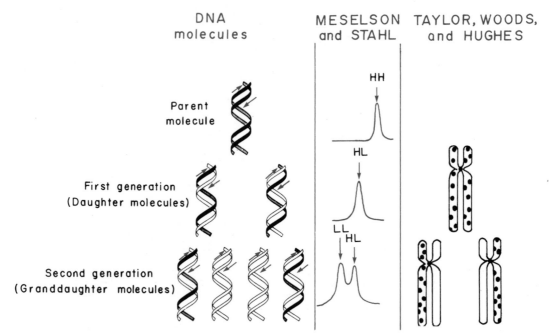

Figure 16–9 Semiconservative replication of DNA in *E. coli* and in eukaryotic chromosomes. Meselson and Stahl labeled the two chains of the parent DNA molecule with the heavy isotope ^{15}N and analyzed the results by CsCl density gradients, while Taylor et al. used ^{3}H-thymidine and radioautography. (See the description in the text.)

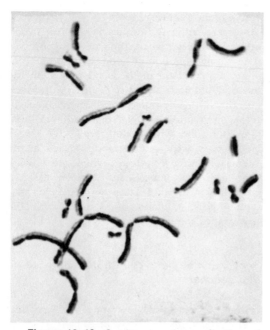

Figure 16–10 Semiconservative replication of DNA. Chinese hamster ovary cells (CHO) were labeled with bromodeoxyuridine for several generations and were then allowed to grow for two generations in unlabeled medium. Only one of the chromatids stains normally; the chromatid that contains BudR does not stain with Giemsa. Some sister chromatid exchanges, which arise from the experimental treatment, are apparent. (Courtesy of S. Latt.)

DNA replication. The cells were labeled with ^{3}H-thymidine and then lysed very gently upon a nitrocellulose filter, which retains the DNA molecule.[21] This technique, called DNA fiber autoradiography, has also been most useful in the analysis of eukaryotic DNA synthesis.

Replication may be visualized as an advancing fork forming a *Y* with the parent strands. In the chromosome of *E. coli* there are, in general, two replication forks, whereas in a eukaryotic cell there may be several thousand forks. From the autoradiographic work illustrated in Figure 16–11, it was initially concluded that the duplication of the DNA molecule proceeded unidirectionally. The two forks observed were interpreted by Cairns as follows: One represented the advancing replicating point, and the other represented the origin of replication, which remained at a fixed point.[21] Subsequent experiments, however, have shown that in *E. coli*, as in eukaryotes (see later discussion), DNA duplication proceeds bidirectionally[22, 23] starting from a fixed origin of replication. The entire replication of the *E. coli* chromosome takes 30 minutes. However, if the culture medium is enriched, the cells can divide in less time (every 20 minutes). This paradox is explained by the appearance of new replication forks even before the replication of the

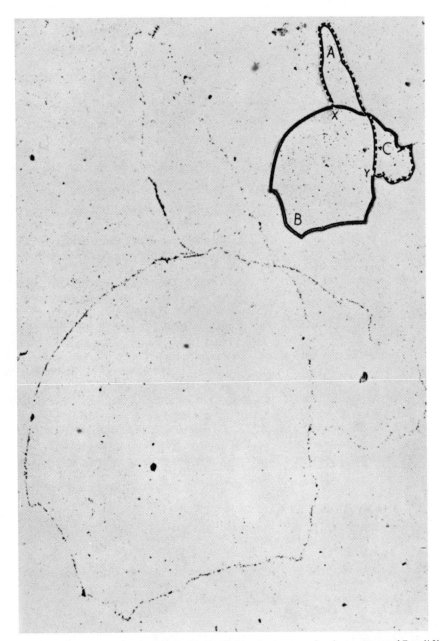

Figure 16–11 Radioautograph, observed with the light microscope, of a chromosome of *E. coli* K 12 labeled with tritiated thymidine for two generations. The bacterium has been gently lysed, and the entire duplicating chromosome is observed. Notice the circular disposition of the single DNA molecule that constitutes the chromosome and that is being duplicated. In segment *B* of the molecule there are about twice the number of grains per μm than in segments *A* and *C*. (See the inset.) (Courtesy of J. Cairns.)

molecule is completed.[24] Figure 16–12 shows an interpretation of these findings. From the initiation point of the circular chromosome, two replicating forks proceed bidirectionally; at a certain stage—also starting from the same point of origin—two new replication forks are initiated, thus producing four forks that advance bidirectionally.

DNA Synthesis is Discontinuous

DNA is synthesized by enzymes called DNA polymerases. *E. coli* has three different DNA polymerases, all of which are involved to some extent in DNA replication. Eukaryotic cells also have three DNA polymerases, but they are less well characterized.[25] These enzymes are re-

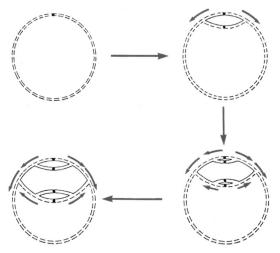

Figure 16–12 Diagram of the formation of a symmetric chromosome from a circular structure containing a single initiation site. The small arrows indicate the directions of movement of replication forks. (From Wake, R. G., *J. Molec. Biol.*, 68:501, 1972.)

quired not only for DNA replication but also for DNA repair (see the discussion that follows).

For its action DNA polymerase requires the presence of a *template DNA*, a *primer DNA*, and the four deoxynucleoside triphosphates (dATP, TTP, dGTP, and dCTP) that are used in the synthesis. It is supposed that the first of the binding sites of the enzyme attaches to the template, the second binds the triphosphate nucleotide, and the third binds to the 3′ —OH end of the primer (Fig. 16–13). The enzyme can add the new nucleotides only in the 5′ to 3′ direction of the polynucleotide clain. The action of the enzyme produces a linkage between the inner phosphate of the nucleotide and the oxygen attached to the 3′ carbon. At the same time the two terminal phosphates are hydrolyzed and released. Under this condition the enzyme is ready to proceed by linking the following nucleotides in a sequential manner. One of the problems, which is difficult to explain, concerns the fact that DNA polymerase acts only in the direction 5′ → 3′. (From Chapter 5 the reader should remember that both DNA strands are complementary and antiparallel.) When the two strands unwind, one of them will be facing the DNA polymerase in the correct direction (5′ → 3′), but in the other, the situation will be unfavorable. The solution to this paradox is that DNA synthesis is discontinuous, in at least one of the strands.

It is likely that DNA synthesis is discontinuous on both strands; in other words, after the unwinding of the DNA, the new strands are made in short segments, always in the 5′ → 3′ direction. These segments are then joined together by the action of another enzyme, *polynucleotide ligase*. (Fig. 16–14). The ligase can

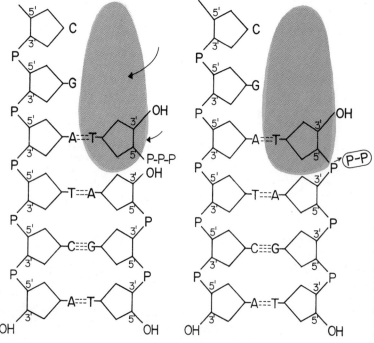

Figure 16–13 Diagram showing the action of DNA polymerase in DNA duplication. **A,** the template chain to the left and a piece of primer DNA to the right. A nucleotide (i.e., thymidine triphosphate) is being put in place along with the enzyme DNA polymerase. **B,** the enzyme has produced the linking of the nucleotide with release of P—P. Observe that the enzyme can add only nucleotides in the direction of 5′ → 3′.

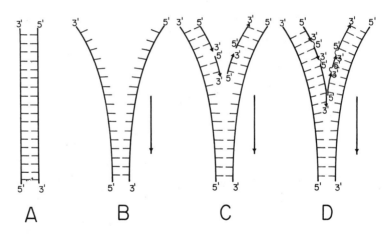

Figure 16–14 Diagram of DNA duplication according to the model of discontinuous synthesis. **A,** the two antiparallel strands of DNA; **B,** separation of the strands; **C,** beginning of replication by the synthesis of short chains in the 5′ → 3′ direction; and **D,** joining of the segments by a DNA ligase.

A   B   C   D

produce the linkage of the 3′ end of one strand with the 5′ end of another, provided the two are held in place in a double helix (see Kornberg, 1974).

The model of discontinuous synthesis of the DNA is supported by experimental evidence. If bacteria are exposed for a few seconds to ^{3}H-thymidine, only short pieces of DNA, having 1000 to 2000 nucleotides, are found. With longer periods of labeling the DNA is already in the form of high molecular weight strands, but some mutations in the DNA ligase enzyme induce the accumulation of short fragments even after longer periods of labeling. The short DNA pieces are called *"Okasaki fragments"* after their discoverer. It has been shown that Okasaki fragments sometimes start with a short segment of RNA, which acts as a primer for DNA synthesis and which is later removed.

Reverse Transcriptase can Copy RNA into DNA

The central dogma of molecular biology is:

$$DNA \xrightarrow[\text{transcription}]{} RNA \xrightarrow[\text{translation}]{} Protein$$

replication

However, there is an exception to this strict unidirectional flow of information, which is that RNA can sometimes be copied into DNA. In other words, *RNA may be transcribed into DNA.* This is done by the so-called *reverse transcriptase,* an *RNA-dependent DNA polymerase* that is able to synthesize DNA on an RNA template.

$$DNA \rightleftharpoons RNA \longrightarrow Protein$$

Tumor-producing (i.e., oncogenic) viruses containing an RNA genome have been found to act as templates for the synthesis of DNA. The tumor virion contains the reverse transcriptase by which an RNA/DNA hybrid molecule can be produced. In this way RNA viral genes can be integrated into the genome of the host cell. These findings have generated considerable interest in the entire field of RNA tumor viruses (see Temin, 1972).

Eukaryotic Chromosomes Have Multiple Origins of Replication

Eukaryotic chromosomes contain a large amount of DNA. The largest chromosomes, from the salamander *Amphiuma,* contain 2.5 meters of DNA, all of it in a single molecule! The broad bean chromosomes may have 800 cm of DNA, and the large human chromosomes have 7 cm. If these huge molecules were replicated from a single origin of replication, as in *E. coli* (in which the chromosome is only 1.1 mm), the S phase would be exceedingly long. Eukaryotic cells solve this problem by having multiple replication initiation sites in each chromosome (Fig. 16–15).

Eukaryotic DNA replication is best studied by the technique of DNA fiber autoradiography.[26] In this technique cells are labeled with ^{3}H-thymidine and placed in a dialysis chamber of which the walls are formed by nitrocellulose filters. The cells are gently lysed by adding a detergent and treated with pronase to digest the proteins, and the chamber is punctured and drained. Some of the DNA fibers adhere to the nitrocellulose filters. After autoradiography, the DNA molecules can be seen as parallel tracks of

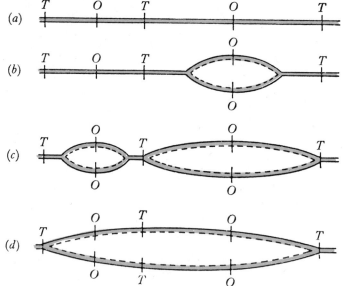

Figure 16–15 Bidirectional model of DNA replication in eukaryons. Each pair of horizontal lines represents a segment of a double helical DNA molecule with two strands. The newly formed chains are indicated by broken lines. *O* and *T* indicate sites of origin and termination of replication. (**a**), two adjacent replication units prior to replication; (**b**), replication started in the unit to the right; (**c**), replication started in the unit to the left; and (**d**), replication completed in both units. (From Huberman, J. A., and Riggs, A. D., *J. Molec. Biol., 32*:327, 1968.)

grains oriented in the direction in which the chamber was drained (Fig. 16–16).

In 1968 Huberman and Riggs[27] demonstrated that eukaryotic DNA has tandem units of replication (called replicons). The average replicons are 30 μm long, and there are about 30,000 of them per haploid mammalian genome. Each chromosome may have several thousand origins of replication. Each replication fork progresses at a speed of 0.5 μm/min, which is 50 times slower than in *E. coli.* As seen in Figure 16–15, two replication forks proceed in opposite directions from each origin (O) until they reach the terminal point (T), where they meet the neighboring replicon. The points of termination are probably not fixed, and they correspond simply to the point where two growing forks converge. (In fact, there is no direct evidence, as yet, for the use of specific DNA sequences as invariant origins of replication in eukaryotic chromosomes, although this has been clearly shown in viral DNAs).

Eukaryotic DNA Synthesis is Bidirectional

As in the case of *E. coli,* replication in eukaryotic chromosomes proceeds bidirectionally.[27–29] This was demonstrated by experiments in which cells were labeled with ³H-thymidine of two different specific activities. As shown in Figure 16–17, cells grown in highly radioactive thymidine and then transferred to less radioactive thymidine show two "tails" of diminishing

grain density after DNA fiber autoradiography. Conversely, when cells grown in moderately labeled thymidine are then placed in highly labeled thymidine, the two tails show increased grain density (Fig. 16–17).

The Number of Replication Units is Developmentally Regulated

The cell cycle is exceptionally rapid during the first cleavages of most egg cells, in which the blastomeres divide without any intervening growth of the cell. The S period is considerably shortened during cleavage. Cases in which the S phase is longer than usual also occur; the S phase immediately preceding meiosis is exceptionally prolonged in all organisms studied so far (Chap. 18). A good example is provided by the newt *Triturus cristatus,* which has an S phase that is 1 hour long in blastula cells, about 20 hours long in adult somatic cells, and 200 hours long in the premeiotic S phase of spermatogonia.[30] The differences in S phase duration are due to changes in the number of replication initiation sites. Cleavage cells have many more origins of replication, and the number of replicons is greatly reduced in the premeiotic chromosomes.[31]

A similar case occurs in *Drosophila,* in which the S period of embryonic nuclei lasts 3 to 4 minutes, while it lasts 600 minutes in adult somatic cells. The cleavage nuclei have origins of replication every 2 to 3 μm of DNA, and the

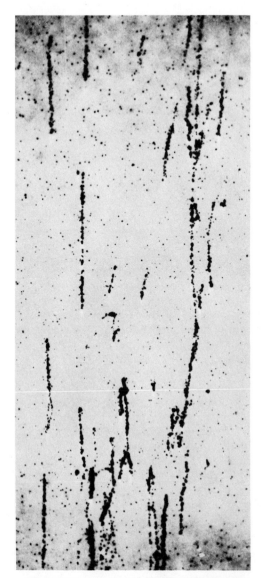

Reinitiation within the Same Cycle is Prevented in Eukaryotic Cells

Each chromosome set has 30,000 origins of replication which are successively activated in the course of several hours in S phase. However, in order to duplicate the DNA accurately, the cells must ensure that no initiation site is used more than once in each cell cycle. Prevention of reinitiation must therefore be a crucial feature of eukaryotic replication. We do not know how this is achieved, but some mechanism must exist to distinguish newly replicated chromatin from the unreplicated material; for example, different methylations in the bases of newly replicated DNA or specific modifications in the histones of the newly assembled nucleosomes could be utilized.

New Nucleosomes are Assembled Simultaneously with DNA Replication

Figure 16–18 shows a replication unit of *D. melanogaster* spread for electron microscopy under conditions that retain the chromosomal proteins. It can be observed that the newly replicated chromatin becomes rapidly associated with histones, giving a nucleosome configuration.[32] In eukaryotic cells we must therefore think in terms of replication of chromatin rather than of naked DNA.

The histones that are newly synthesized during the S phase (Fig. 16–7) become rapidly associated with DNA.[33] Some experiments seem to indicate that it is possible that the parental nucleosomes could go to one daughter molecule, indicate that it is possible that the parental nucleosomes could go to one daughter molecule,

Figure 16–16 Radioautograph of DNA replication in a cell grown in tissue culture. The various segments that are being duplicated are indicated by the parallel lines of silver grains of ³H-thymidine. The replicating regions are seen as single lines (instead of two replicating forks) because sister strand separation is not apparent in DNA fiber radioautography until at least 50 μm of DNA have replicated. (See the description in the text.) (Courtesy of H. G. Callan.)

distances are much longer in adult cells (see Blumenthal, Kriegstein, and Hogness, 1973).

Clearly, some of the replication origins active in embryonic cells are not utilized in adult cells. It has been suggested that this could be related to the formation of chromomeres, each containing a certain amount of condensed chromatin and in which the origins of replication become inactivated.

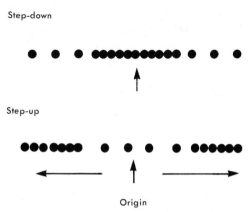

Figure 16–17 Bidirectionality of eukaryotic DNA synthesis, as revealed by DNA fiber radioautography. Cells were incubated first with highly labeled ³H-thymidine and then with moderately labeled thymidine (step-down), and vice versa (step-up), as described in the text.

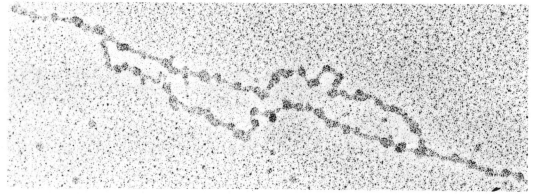

Figure 16–18 Replicating chromatin (Courtesy of S. L. McKnight and O. L. Miller, Jr.)

while new nucleosomes could go to the other daughter double helix.[34]

RNA Synthesis Continues During DNA Replication

Figure 16–19 shows a replication unit of *D. melanogaster* that is being transcribed. It is apparent that RNA synthesis can be initiated almost immediately after replication and that therefore transcription continues during the S phase.[35] The nascent RNA chains can be observed on both sister chromatids in a rather symmetrical arrangement (Fig. 16–19).

Electron microscopy of macromolecules spread by the technique devised by Oscar Miller is an enormously powerful tool for cell biologists, as can be readily realized from Figures 16–18 and 16–19, and from others throughout this book. The direct visualization of molecules can sometimes settle, in a single stroke, years of biochemical controversy.

DNA Repair Enzymes Remove Thymine Dimers Induced by UV Light

Although the gene is generally stable, it may change by a process called *mutation*. The capacity to mutate is a property of the genetic material that is as important as stability. Mutation takes place in all living organisms and is the origin of hereditary variations. A genetic mutation, which may occur spontaneously and without apparent cause, becomes incorporated in the population and is transmitted by sexual reproduction. Asexual unicellular organisms and somatic tissue may also undergo mutation; the rate of spontaneous mutation is generally low. For example, in each generation of *D. melanogaster*, mutations occur at the rate of 1:100,000 to 1:1,000,000. Unicellular organisms are more appropriate for the study of mutation frequency because large numbers of organisms can be handled in a short time.

A *mutation* may correspond to the change of a single nucleotide, of a long polynucleotide

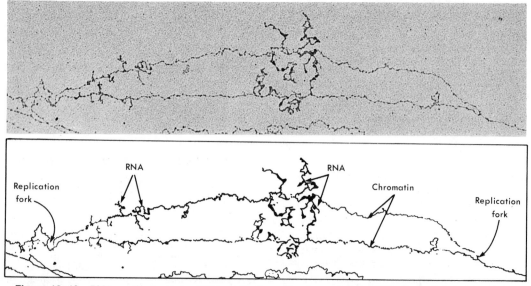

Figure 16–19 RNA synthesis during DNA replication. (Courtesy of O. L. Miller, Jr. and S. L. McKnight.)

chain, or even of an entire chromosome arm. When the locus is restricted, the result is called a *gene mutation;* when it is extended, involving the chromosome structure, it is considered a *chromosomal mutation* or *aberration.* In gene mutation the change in a single nucleotide may result in the *substitution* of one purine or pyrimidine for another, or even of a purine by a pyrimidine. *Insertions* or *deletions* of one or more nucleotides may also occur (Chap. 21).

In Chapter 19 the action of ionizing radiations on the chromosome will be presented; here the action of ultraviolet light on bacteria, and the molecular mechanism by which the alteration produced in the DNA molecule can be repaired will be considered. Some of these mechanisms of DNA repair also apply to ionizing radiations.

Mutations can be considerably increased by exposure of bacteria to ultraviolet light. This treatment tends to produce *dimerization* of adjacent pyrimidine bases; predominantly $T-T$ but also $C-C$ and $C-T$ *dimers* are produced within a single strand of DNA. Dimerization reduces the distance between nucleotides from 0.34 to 0.28 nm and produces other changes

that impair the mechanism of DNA duplication. In most cases this type of mutation is corrected by a mechanism of *DNA repair* which involves (Fig. 16–20): (1) the recognition and incision of the affected DNA strand by an endonuclease, (2) excision and broadening of the gap by an exonuclease (e.g., DNA polymerase), (3) filling of the gap by repair replication (this is also done by DNA polymerase), and (4) covalent joining of the polynucleotide by the ligase.

Xeroderma Pigmentosum Patients Have Defective DNA Repair

Xeroderma pigmentosum is a hereditary disease characterized clinically by severe intolerance to sunlight. The sun-exposed areas (face and hands) develop fibrosis, pigmentations, and more importantly, multiple skin cancers. The disease is due to a defect in the enzymes that repair damage due to ultraviolet light. The defective excision and repair of thymine dimers can be demonstrated cytologically by irradiating cultured fibroblasts with UV light and then incubating them with ^{3}H-thymidine for one to two

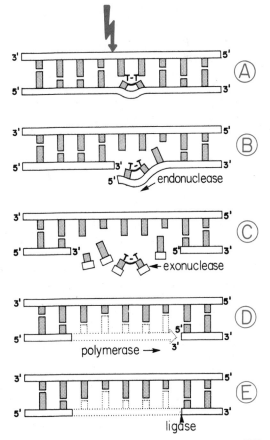

Figure 16–20 Diagram showing the effect of ultraviolet light on the DNA molecule and the mechanism of DNA repair. **A,** under the action of ultraviolet light a dimer of T—T is produced; **B,** the affected DNA strand is recognized and incised by a molecule of endonuclease; **C,** the strand segment is excised by a molecule of exonuclease; **D,** the gap is filled by DNA polymerase; and **E,** the synthesized segment is joined by a DNA ligase.

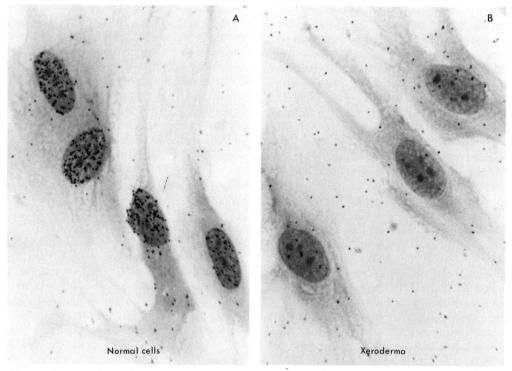

Figure 16–21 Repair of thymine dimers in normal and xeroderma pigmentosum human cultured cells. Cells were treated with ultraviolet light and then with ^{3}H-thymidine for one hour. The normal cells incorporate low amounts of thymidine in their nuclei, while xeroderma cells do not. None of the cells shown here was in S phase. (Courtesy of Professor Y. Okada.)

hours. As can be seen in Figure 16–21,*A,* the UV treatment induces normal cells to incorporate low levels of thymidine (when not replicating their DNA). This incorporation represents the repair of the excised thymine dimers. Fibroblasts from xeroderma pigmentosum patients are unable to repair the UV damage and therefore do not incorporate thymidine after the same treatment (Fig. 16–21*B*). This disease clearly links human cancer with somatic mutations and shows the benefit of having mechanisms to repair DNA defects.

SUMMARY:

DNA Replication

DNA replication is best understood in bacteria. *E. coli* has a circular chromosome containing 1.1 mm of DNA. DNA duplication is accomplished via a template mechanism by which the two DNA strands unwind and two new strands are formed by free nucleotides that fall into place and are linked by DNA polymerase. Replication may be visualized as an advancing fork forming a **Y** with the parent strands (Fig. 16–14).

DNA duplication is *semiconservative,* (i.e., only one strand is newly synthesized, the other half of the parental DNA is retained), as shown in Figure 16–9. Complete DNA duplication in *E. coli* takes place in 30 minutes using two replication forks that advance bidirectionally from a fixed origin. In rapidly growing cells there are times when four forks may be advancing concurrently.

DNA duplication is achieved on each of the parent polynucleotide strands by the intervention of DNA polymerases. The synthesis is in the $5' \rightarrow 3'$ direction and proceeds in short segments on both strands. DNA ligase is the enzyme involved in the subsequent covalent joining of the DNA segments.

In certain systems RNA may be transcribed into DNA. This is done by an RNA-dependent DNA polymerase (i.e., *reverse transcriptase*). Such an enzyme has been found in tumors produced by RNA viruses.

In *eukaryons* DNA duplication is also semiconservative, as can be demonstrated experimentally in plant cells (Fig. 16–9). While the DNA of *E. coli* duplicates at 20 to 30 micrometers per minute, in a cultured human cell duplication proceeds at only 0.5 μm per minute. It may be demonstrated that there are numerous replicating forks advancing bidirectionally on each chromatid. On the average, there are some 30,000 initiation points per mammalian genome.

The number of replication origins can be developmentally regulated. In the newt *Triturus*, the S phase is 1 hour long in cleaving embryos, about 20 hours long in adult somatic cells, and 200 hours long in the premeiotic S phase. These differences can be explained by variations in the number of replication units. The newly synthesized DNA is assembled into chromatin simultaneously with replication (Fig. 16–18), and RNA synthesis can be resumed immediately after replication (Fig. 16–19).

The mechanism of DNA replication is related to that of DNA repair. DNA repair occurs when an alteration has been produced in the DNA molecule, as, for example, under the action of radiation or certain chemicals. A well-known case is the DNA repair after UV irradiation of a bacterium. This treatment produces dimers between adjacent pyrimidine bases (e.g., T — T). In the dimer the distance between nucleotides is reduced from 0.34 nm to 0.28 nm. DNA repair involves: (1) an *endonuclease* that produces incision of the affected DNA strand, (2) an *exonuclease* that produces excision of the gap, (3) a *DNA polymerase* that fills the gap by repair replication, and (4) *a ligase* that causes covalent joining of the polynucleotide (Fig. 16–20). In the hereditary disease xeroderma pigmentosum there is a defect in the DNA repair enzymes which can be demonstrated cytologically (Fig. 16–21).

REFERENCES

1. Howard, A., and Pelc, S. R. (1953) *Heredity* (London) Suppl., 6:261.
2. Lajtha, L. G. (1963) *J. Cell Comp. Physiol.*, 62:143.
3. Johnson, R. T., and Rao, R. N. (1970) *Nature*, 266:717.
4. Röhme, D. (1975) *Hereditas, 80*:145.
5. Prescott, D. M. (1964) *Progr. Nucleic Acid Res. Mol. Biol.*, 3:33.
6. Rao, R. N., and Johnson, R. T. (1970) *Nature*, 225:159.
7. Gurdon, J. B. (1967) *Proc. Natl. Acad. Sci. USA*, 58:545.
8. Rusch, H. P. (1969) *Fed. Proc.*, 28:1761.
9. Tobia, A. M., Schildkraut, C. L., and Maio, J. J. (1970) *J. Molec. Biol.*, 54:499.
10. Flamm, W. G., Bernheim, N. J., and Brubaker, P. E. (1971) *Exp. Cell Res.*, 64:97.
11. Lima-de-Faría, A. (1969) In: *Handbook of Molecular Cytology*, p. 278. North-Holland Publishing Co., Amsterdam.
12. Tobia, A. M., Schildkraut, C. L., and Maio, J. J. (1971) *Biochim. Biophys. Acta*, 246:258.
13. Hill, R. N., and Yunis, J. J. (1967) *Science*, 155:1120.
14. Woodward, J., Rasch, E., and Swift, H. (1961) *J. Biophys. Biochem. Cytol.*, 9:445.
15. Bradbury, E. M., Inglis, R. J., and Matthews, H. R. (1974) *Nature*, 247:257.
16. Prescott, D. M., and Bender, M. A. (1962) *Exp. Cell Res.*, 16:279.
17. Martin, D. W., Tomkins, G. M., and Bresler, M. A. (1969) *Proc. Natl. Acad. Sci. USA*, 63:842.
18. Meselson, M., and Stahl, F. W. (1958) *Proc. Natl. Acad. Sci. USA*, 44:671.
19. Taylor, J. H., Woods, P. S., and Hughes, W. L. (1957) *Proc. Natl. Acad. Sci. USA*, 43:122.
20. Latt, S. A., Allen, J. W., and Stetten, G. (1977) In: *International Cell Biology*, (Brinkley, B. R., and Porter, K. R., eds.) p. 520. Rockefeller University Press, New York.
21. Cairns, J. (1963) *J. Molec. Biol.*, 6:208.
22. Masters, M., and Broda, P. (1971) *Nature* [New Biol.], 232:137.
23. Prescott, D. M., and Kuempel, P. L. (1972) *Proc. Natl. Acad. Sci.*, 69:2842.
24. Wake, R. G. (1972) *J. Molec. Biol.*, 68:501.
25. Weissbach, A. (1975) *Cell*, 5:101.
26. Cairns, J. (1966) *J. Molec. Biol.*, 15:372.
27. Huberman, J. A., and Riggs, A. D. (1968) *J. Molec. Biol.*, 32:327.
28. Huberman, J., and Tsai, A. (1973) *J. Molec. Biol.*, 75:5.
29. Kriegstein, J., and Hogness, D. (1974) *Proc. Natl. Acad. Sci. USA*, 71:135.
30. Callan, H. G., and Taylor, J. H. (1968) *J. Cell Sci.* 3:615.
31. Callan, H. G. (1974) *Cold Spring Harbor Symp. Quant. Biol.*, 38:195.
32. McKnight, S. L., and Miller, O. L., Jr. (1977) *Cell*, 12:795.
33. Crémisi, C., Chestier, A., and Yaniv, M. (1977) *Cold Spring Harbor Symp. Quant. Biol.*, 42:409.
34. Leffak, I. M., Grainger, R., and Weintraub, H. (1977) *Cell, 12*:837.
35. McKnight, S. L., and Miller, O. L., Jr. (1976) *Cell, 8*:305.

ADDITIONAL READING

Alberts, B., and Sternglanz, R. (1977) Recent excitement in the DNA replication problem. *Nature, 269*:655.

Blumenthal, A., Kriegstein, J., and Hogness, D. (1974) The units of DNA replication in *Drosophila melanogaster. Cold Spring Harbor Symp. Quant. Biol., 38*:205. A classic.

Callan, H. (1973) Replication of DNA in eukaryotic chromosomes. *Br. Med. Bull., 29*:192. Highly recommended.

Davidson, J. N. (1976) *The Biochemistry of Nucleic Acids,* 8th Ed. Chapman & Hall, London. A very nice advanced manual.

Edenberg, H., and Huberman, J. (1975) Eukaryotic chromosome replication. *Annu. Rev. Genet., 9*:245.

Hand, R. (1978) Eukaryotic DNA: Organization of the genome for replication. *Cell, 15:* 17.

Hartwell, L. H., Culotti, J., Pringle, J. R., and Reid, B. J. (1974) Genetic control of the cell division cycle in yeast. *Science, 183*:46–51.

Holley, R. W. (1975) Control of growth of mammalian cells in cell culture. *Nature, 258*:487.

Huberman, J., and Riggs, A. (1968) On the mechanism of DNA replication in mammalian chromosomes. J. Molec. Biol., 32:327.

Kornberg, A., (1974) *DNA Synthesis.* W. H. Freeman Co., San Francisco.

McKnight, S. L., Bustin, M., and Miller, O. L., Jr. (1977) Electron microscopic analysis of chromosome metabolism in the *Drosophila melanogaster* embryo. *Cold Spring Habor Symp. Quant. Biol., 42*:741.

Mitchison, J. M. (1971) *The Biology of the Cell Cycle.* Cambridge University Press.

Pardee, A. B., DuBrow, R., Hamlin, J. C., and Kletzien, R. F. (1978) Animal cell cycle. *Annu. Rev. Biochem., 47*:715.

Prescott, D. M. (1976) Reproduction of eukaryotic cells. Academic Press, Inc., New York. Highly recommended.

Rao, P. N., and Johnson, R. T. (1974) Induction of chromosome condensation in interphase cells. *Adv. Cell. Molec. Biol., 3*:135.

Ringertz, N. R., and Savage, R. E. (1976) *Cell Hybrids.* Academic Press, Inc., New York. Detailed account of cell fusion and the cell cycle.

Taylor, J. H. (1958) The duplication of chromosomes. *Sci. Am., 198*:36.

Taylor, J. H. (1974) Units of DNA replication in chromosomes of eukaryotes. *Int. Rev. Cytol., 37*:1.

Taylor, J. H., Wu, M., and Erickson, L. (1974) Functional sub-units of chromosomal DNA from higher eukaryotes. *Cold Spring Harbor Symp. Quant. Biol., 38*:225.

Temin, H. M. (1972) RNA-directed DNA synthesis. *Sci. Am., 226*:24.

MITOSIS AND CELL DIVISION

17–1   A General Description of Mitosis   383
      Prophase—Chromatid Coiling, Nucleolar Disintegration, and Spindle Formation
      Metaphase—Chromosomal Orientation at the Equatorial Plate
      Anaphase—Movement of Daughter Chromosomes Toward the Poles
      Telophase—Reconstruction of the Daughter Nuclei
      Summary: *Events in Mitosis*

17–2   Molecular Organization and Functional Role of the Mitotic Apparatus   387
      Centromere Association with the Kinetochore, Where Spindle Microtubules are Implanted

Spindle Microtubules—Kinetochore, Polar, and Free
Assembly of Spindle Microtubules—Control by Poles and Kinetochores
Metaphase Chromosomal Motion—Caused by Kinetochore and Polar Microtubular Interaction
Anaphase Chromosomal Movement—The Dynamic Equilibrium and the Sliding Mechanism Hypotheses
Cytokinesis in Animal Cells—A Contractile Ring of Actin and Myosin
Cytokinesis in Plant Cells—The Phragmoplast and Cell Plate
Summary: *The Mitotic Apparatus*

Cell division is the complex phenomenon by which cellular material is divided equally between daughter cells. This process is the final, and microscopically visible, phase of an underlying change that has occurred at molecular and biochemical levels. Before the cell divides by mitosis, its fundamental components have duplicated — particularly those involved in hereditary transmission. In this respect, cell division can be considered as the final separation of the already duplicated macromolecular units.

The essential features of cell division by mitosis were considered in Chapter 2. Here a detailed analysis of the process will be made.

Special consideration will be given to the *mitotic spindle,* the fibrous component made of microtubules (see Chapter 9–2) which is assembled every time the cell begins to divide and is disassembled at the end of mitosis. It will be seen that the mitotic spindle functions as a structural framework and force-generating system that sets the chromosomes in position and distributes them between the daughter cells. In our discussion of mitosis we will also deal with some of the structural differentiations of chromosomes, such as the *kinetochores* and *centromeres* to which the spindle microtubules are anchored, and the general changes in fine structure that chromosomes undergo during the mitotic cycle.

17–1 A GENERAL DESCRIPTION OF MITOSIS

In spite of the fact that the overall process of mitosis is similar in all eukaryotic cells, there are variations between different organisms — particularly between animal and plant cells — that make it difficult to give a general description. For example, it will be mentioned later that in higher plants the spindle has no *centrioles or asters* at the poles and, furthermore, that the separation of the daughter cells is a more complex phenomenon than in animal cells. Although plant *meristems* (e.g., young root tips) are particularly advantageous for the study of cell division, since 10 to 15 per cent of the cells are mitotic, our description will be based mainly on animal cells such as those in tissue culture or dividing eggs.

Figure 17–1 is a general diagram of the different stages of mitosis. These are considered as phases of a cycle that begins at the end of the intermitotic period *(interphase)* and ends at the beginning of a new interphase. The main divisions of this cycle are: *prophase, metaphase, anaphase,* and *telophase. Cytokinesis,* a process of separation of the two cytoplasmic territories, is simultaneous with anaphase or telophase, or it can occur at a later stage.

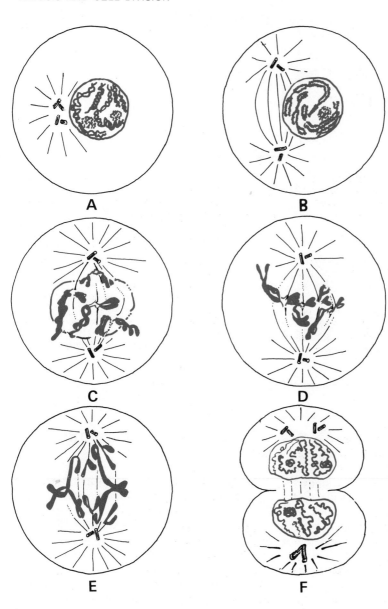

A

B

C

D

E

F

Figure 17–1 General diagram of mitosis. **A,** *prophase,* the nucleoli and chromosomes, shown as thin threads; in the cytoplasm the aster with the pairs of centrioles are shown. **B,** *prophase,* a more advanced stage of this phase in which the chromosomes have shortened. The primary constriction with the centromere is shown; in the cytoplasm the spindle is formed between the asters. **C,** *late prophase or prometaphase,* the nuclear envelope disintegrates and the chromosomes become attached to the spindle fibers. **D,** *metaphase,* the chromosomes are arranged along the equatorial plane. **E,** *anaphase,* the daughter chromosomes, preceded by the centromeres, are moving toward the poles. **F,** *telophase,* the daughter nuclei are in the process of reconstitution; cell cleavage has started.

The various phases of mitosis will be described in a way that should give an idea of the sequence of events that occur in the nucleus and the cytoplasm. Some of the underlying processes will be considered in terms of their fine structure, biochemistry, and physiologic significance in the second part of the chapter.

Prophase—Chromatid Coiling, Nucleolar Disintegration, and Spindle Formation

The beginning of *prophase* is indicated by the appearance of the chromosomes as thin threads inside the nucleus. In fact, the word "mitosis"

(Gr., *mitos,* thread) is an expression of this phenomenon, which becomes more evident as the chromosomes start to condense. The condensation occurs by a process of coiling or folding of the chromatin fibers (Fig. 17–1,*A*). At the same time the cell becomes spheroid, more refractile, and viscous.

Each prophase chromosome is composed of two coiled filaments, the *chromatids,* which are a result of the replication of the DNA during the S period. As the prophase progresses, the chromatids become shorter and thicker, and the primary constrictions, which contain the centromeres and kinetochores, become clearly visible (Fig. 17–1,*B*). During early prophase, the chromosomes are evenly distributed in the nuclear cavity; as prophase progresses, the chromosomes approach the nuclear envelope, causing the central space of the nucleus to become empty. The centrifugal movement of the chromosomes indicates that the disintegration of the nuclear envelope is approaching, and with it, the end of prophase. At this time each chromosome appears to be composed of two cylindrical, parallel elements that are in close proximity. Not only the primary constrictions, but also the secondary constrictions along some chromosomes may be observed (Fig. 17–1,B). Other changes taking place within the nucleus are the reduction in size of the nucleoli, and their disintegration within the nucleoplasm. At the end of prophase, with the rapid fragmentation and disappearance of the nuclear envelope, the nucleolar material is released into the cytoplasm.

. While these processes are taking place in the nucleus, in the cytoplasm the most conspicuous change is the formation of the spindle. In the diagram of Figure 17–1 the appearance of the spindle is related to the existence of centrioles. In early prophase there are two pairs of centrioles, each one surrounded by the so-called *aster*, composed of microtubules that radiate in all directions. (The name "aster" refers to the starlike aspect of this structure.) The two pairs of centrioles migrate along with the asters, describing a circular path toward the poles, while the spindle lengthens between. The migration of the asters continues until they become situated in antipodal positions (Fig. 17–1,C).

Centrioles replicate during interphase, generally during the S period. At the beginning of prophase there is a single aster surrounding the two pairs of centrioles. One of the pairs remains in position, with half the original aster, while the other pair, along with the other half aster, migrates about 180° around the periphery of the nucleus to reach the opposite pole.

Metaphase — Chromosomal Orientation at the Equatorial Plate

Sometimes the transition between prophase and metaphase is called *prometaphase* (Gr., *meta*, between). This is a very short period in which the nuclear envelope disintegrates and the chromosomes are in apparent disorder (Fig. 17–1,C). After that the spindle fibers invade the central area and their microtubules extend between the poles. The chromosomes become attached by the kinetochores to some of the spindle fibers, and they undergo oscillatory movements until they become radially oriented in the equatorial plane

and form the *equatorial plate* (Fig. 17–1,D). Those fibers of the spindle that connect to the chromosomes are called the *chromosomal fibers;* those that extend without interruption from one pole to the other are the *continuous fibers.*

Mitosis in which the spindle has centrioles and asters is called *astral* or *amphiastral* and is found in animal cells and some lower plants. Mitosis in which centrioles and asters are absent is called *anastral* and is found in higher plants, including all angiosperms and most gymnosperms (Fig. 17–2). Centrioles and asters are not indispensable to the formation of the spindle. In a certain way, in *astral mitosis* the formation of the spindle is a mechanism which leads to the distribution of the centrioles between the two daughter cells.

Anaphase — Movement of the Daughter Chromosomes Toward the Poles

At anaphase (Gr., *ana*, back) the equilibrium of forces that characterizes metaphase is broken by the separation of the centromeres — a process which is carried out simultaneously in all the chromosomes. The centromeres move apart, and the chromatids separate and begin their migration toward the poles (Fig. 17–1,E). The centromere always leads the rest of the chromatid or *daughter chromosome,* as if this is being pulled by the chromosomal fibers of the spindle. The chromosome may assume the shape of a V with equal arms, if it is metacentric, or with unequal arms, if it is submetacentric. During anaphase the microtubules of the chromosomal fibers of the spindle shorten one third to one fifth of the original length. Simultaneously, the microtubules of the continuous fibers increase in length. Some of these stretched spindle fibers now constitute the so-called *interzonal fibers.*

Telophase — Reconstruction of the Daughter Nuclei

The end of the polar migration of the daughter chromosomes marks the beginning of telophase (Gr., *telo*, end). The chromosomes start to uncoil and become less and less condensed by a process that in some ways, recapitulates prophase, but in the reverse direction. At the same time, the chromosomes gather into masses of chromatin which become surrounded by discontinuous segments of nuclear envelope. Such segments fuse to make the two complete nuclear envelopes of the daughter nuclei. During the

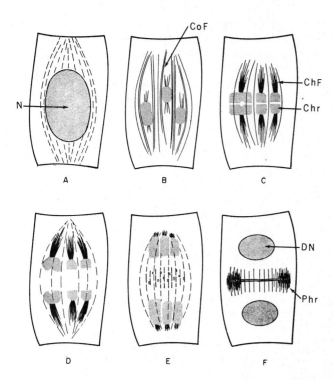

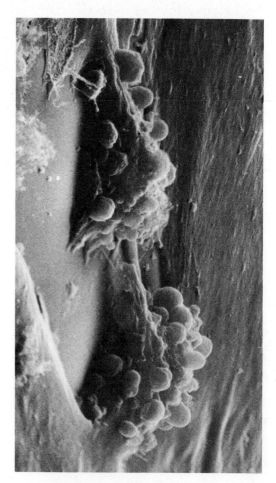

Figure 17–2 Diagram of mitosis in a plant cell showing the changes in birefringence of the spindle fibers. Note the absence of centrioles and asters. Abbreviations: *CoF,* continuous fibers; *ChF,* chromosome fibers; *Chr,* chromosome; *N,* nucleus; *Phr,* phragmoplast. (Courtesy of S. Inoué.)

Figure 17–3 Scanning electron micrograph of cultured normal hamster kidney cells. Two cells showing many blebs on the surface are in late cytokinesis, while the interphase cells remain flattened. ×1820 (Courtesy of R. D. Goldman.)

final stages the nucleoli reappear at the sites of the nucleolar organizers.

Simultaneous with the uncoiling of the chromosomes and the formation of the nuclear envelope, *cytokinesis* occurs. This is the process of segmentation and separation of the cytoplasm. In animal cells the cytoplasm constricts in the equatorial region, and this constriction is accentuated and deepened until the cell divides (Fig. 17–1,F).

In cells of higher animals the period of cytokinesis is marked by active movements at the cell surface that are best described as "bubbling." Some investigators suggest that bubbling may reflect the activity of a rapidly expanding membrane. In ameboid cells at telophase both daughter cells have active movements, which appear to pull them apart. This is best observed in films of dividing cells and in scanning electron micrographs (Fig. 17–3). The high viscosity of the cytoplasm, characteristic of metaphase and anaphase, decreases during telophase. In most cases asters become reduced and tend to disappear. During cytokinesis the cytoplasmic components are distributed, including the mitochondria and the Golgi complex. Cytokinesis in plant cells will be considered later on in this chapter.

SUMMARY:

Events in Mitosis

Mitosis is a mechanism by which the cell distributes, in equivalent amounts, the different components that have been duplicated during interphase. Prophase, metaphase, anaphase, and telophase are characterized by morphologic changes that take place in the nucleus and the cytoplasm (Fig. 17–1).

In *prophase* chromosomes appear as thin threads that condense by coiling and folding. Each chromosome contains two *chromatids* which will be the future *daughter chromosomes*. With condensation each chromatid shows the *centromere* and *kinetochore*. The nucleolus tends to disintegrate and disappears at the end of prophase. In the cytoplasm the spindle is formed between the asters (and centrioles), which move toward the poles. Centrioles replicate at interphase during the S period.

At the beginning of *metaphase* (prometaphase) the nuclear envelope disintegrates and there is mixing of the nucleoplasm with the cytoplasm. Chromosomes become attached to the microtubules of the spindle and are oriented at the *equatorial* plate. The spindle has both continuous and chromosomal microtubules. Animal cells have the type of spindle shown in Figure 17–1 *(astral mitosis)*. In plant cells centrioles and asters are absent *(anastral mitosis)* (Fig. 17–2).

In *anaphase* the daughter chromosomes, led by the centromere, move toward the poles. The spindle fibers shorten one third to one fifth the original length.

In *telophase* chromosomes again uncoil; the nuclear envelope is reformed from the endoplasmic reticulum; and the nucleolus reappears.

Cytokinesis is the process of separation of the cytoplasm. In animal cells there is a constriction at the equator that finally results in the separation of the daughter cells.

17–2 MOLECULAR ORGANIZATION AND FUNCTIONAL ROLE OF THE MITOTIC APPARATUS

The term mitotic apparatus has been applied to the *asters* that surround the centrioles and to the *mitotic spindle*.[1] In fixed preparations the aster appears as a group of radiating fibers that converge toward the centrioles. Around the centrioles there is often a clear zone called the *microcentrum* or *centrosome* into which the fibers of the aster do not penetrate.

The spindle has the so-called chromosomal fibers, joining the chromosomes to the poles; the continuous fibers, extending pole to pole; and the interzonal fibers, observed between the daughter chromosomes and nuclei in anaphase and telophase (Fig. 17–2). All these fibers, including those of the aster, are composed of microtubules.

The fine structure of the mitotic apparatus is studied principally with the electron microscope (Fig. 17–4). Interesting observations may also be made using polarization microscopy[2]

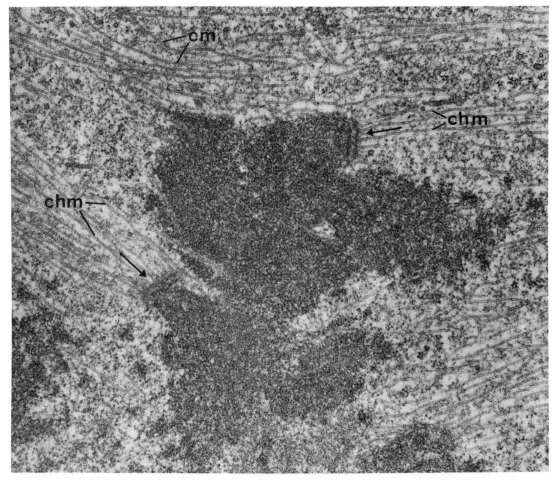

Figure 17–4 Electron micrograph of a Chinese hamster fibroblast in tissue culture at metaphase. The arrows point to kinetochores where the chromosomal microtubules (*chm*) are implanted. **Above,** interzonal or continuous microtubules (*cm*) are observed. The chromosomes show a tightly packed structure. ×40,000. (From Brinkley, B. R., and Cartwright, J., Jr., *J. Cell Biol., 50*:416–431, 1971.)

and the Nomarski interference–contrast microscope (Fig. 17–5).[3] Because of their positive birefringence, spindle and aster fibers are readily observed in living cells with the polarization microscope. In plant cells, which are devoid of centrioles and asters, the first spindle fibers appear at prophase in a clear zone surrounding the nucleus (Fig. 17–2,*A*). Birefringence is strongest near the kinetochores but becomes weaker toward the poles (Fig. 17–2,*B–D*). During anaphase, the chromosomes are led by intensely birefringent chromosomal spindle fibers (Fig. 17–2,*C* and *D*) The continuous fibers, in which birefringence is very low in early anaphase, become more conspicuous in late anaphase and telophase. In animal cells such fibers form a kind of bundle that main-

tains a connection between the two daughter cells for some time. Under the electron microscope this bundle of microtubules appears surrounded by a dense material and some vesicles that together constitute a structure called the *midbody*.[4]

The use of antibodies against tubulin (the main component of microtubules) has permitted interesting observations on the changes in the mitotic apparatus during mitosis in cultured cells. The degree of fluorescence is related to the number and condensation of the microtubules in the aster and in the various regions of the spindle. (It is a worthwhile exercise for the student to follow these changes throughout the phases of mitosis, shown in Fig. 17–6.)

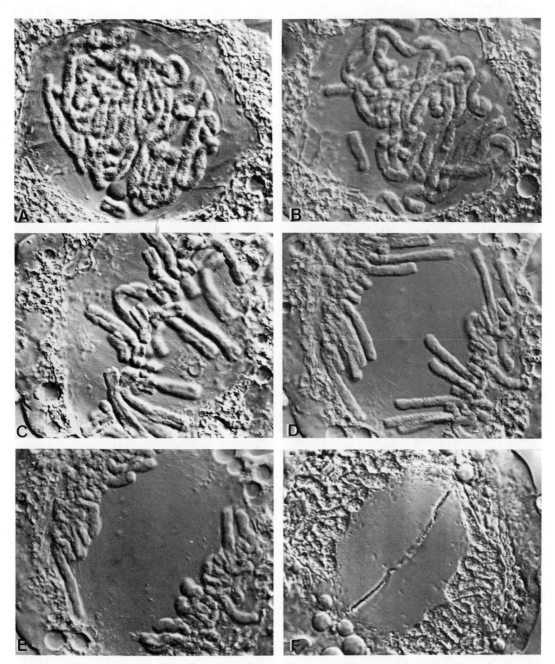

Figure 17–5 Series of micrographs taken with the Nomarski interference microscope of the same living cell of the endosperm of *Haemanthus*. The micrographs were taken at the following times: **A,** 5:19; **B,** 5:39; **C,** 7:03; **D,** 7:45; **E,** 7:53; **F,** 8:09.

A, *late prophase,* showing the condensed chromosomes; the nuclear envelope is breaking at certain points. **B,** *late prophase,* showing the disappearance of the nuclear envelope. **C,** *metaphase,* with the chromosomes in the equatorial plane; some spindle fibers are observed. Note that this is an anastral mitosis, lacking asters at the poles. **D,** *anaphase,* with the chromosomes moving toward the poles. **E,** *late anaphase,* showing some interzonal spindle fibers. **F,** *telophase,* showing the phragmoplast at the equatorial plane. ×2000. (Courtesy of A. S. Bajer.)

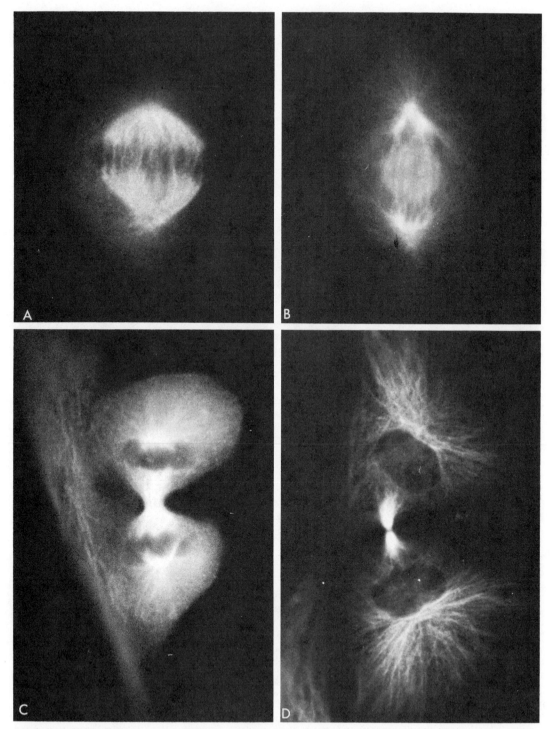

Figure 17–6 Cultured cells in mitosis stained by immunofluorescence with anti-tubulin antibodies. **A**, *Metaphase*. The stain is concentrated in the spindle fibers. **B**, *Anaphase*. Spindle fibers are apparent, but there is staining at the asters and the midzone of the spindle. **C**, *Telophase*. The asters and the midzone of the spindle are more heavily stained. **D**, *After telophase*. The tubulin fibers radiate from the asters, and the cytoplasmic bridge (*midbody*) between the two daughter cells is heavily stained. (Courtesy of B. R. Brinkley.)

Centromere Association with the Kinetochore, Where Spindle Microtubules are Implanted

In the primary constriction there is a special differentiation of the chromosome called the *centromere,* which is associated with a proteinaceous structure, the *kinetochore.* The kinetochore is the site of implantation of the microtubules that constitute the chromosomal spindle fibers; for this reason it has been considered as functionally related to the chromosomal movements during mitosis. Usually two kinetochores, one in each chromatid, are observed, but in some cases a large number may be found *(diffuse centromeres).* In these instances, the microtubules are implanted at many points along the chromosomal arms. Diffuse kinetochores have been found in groups of plants (e.g., *Luzula),* in insects (e.g., *Hemiptera),* and in some algae. The kinetochore may be detected during prophase, before there is a connection with the microtubules (Fig. 7–1,*B*). Under the electron microscope the kinetochore appears as a plate or cup-like disc, 0.20 to 0.25

μm in diameter, plastered upon the primary constriction (Fig. 17–7).

In a cross section the kinetochore consists of the following three layers (Fig. 17–4): (1) an *outer* layer of dense material 40 nm thick, (2) a *middle* layer of lower density, and (3) an *inner* dense layer in contact with the underlying chromatin fibers. Emanating from the convex surface of the outer layer, in addition to the microtubules, a "corona" of fine filaments has been observed[5, 6] (Fig. 17–7).

The exact chemical nature of the kinetochore is unknown, although it is thought that it represents a non-chromatin material added to the surface of the primary constriction. It has been suggested that the kinetochore is a gene product and that its activity is genetically controlled.[7] At present there is no evidence that the centromere itself contains genetic material.

The main function of kinetochores seems to be related to the attachment of the chromosomal microtubules. A related function may be to serve as nucleation centers for the polymerization of tubulin.[8]

Studies at the electron microscopic level

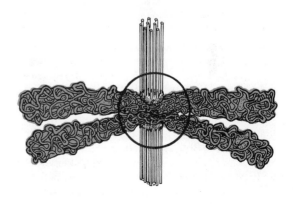

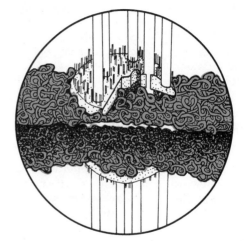

Figure 17–7 Diagram of a metaphase chromosome showing the folded-fiber structure and the centromere with implanted microtubules. **Below,** an inset at higher magnification, showing the convex electron-dense layer and the fibrillar material forming the "corona" of the kinetochore. Several microtubules of the spindle are shown penetrating the various layers reaching the chromosome fibers.

using anti-tubulin antibodies have revealed that tubulin is found not only in the spindle and centrioles (both of which contain microtubules) but also in kinetochores. This suggests that they contain the nucleating intermediates necessary for the assembly of tubulin monomers into microtubules.[9]

Spindle Microtubules—Kinetochore, Polar, and Free

In Chapter 9–2 we gave a general description of microtubules and tubulin which should be reviewed at this point. There may be as few as 16 microtubules in the spindle of yeast cells and as many as 5000 in the spindle of a higher plant cell. A precise study of the number of microtubules has been carried out in dividing animal cells in the various regions of the spindle.[4]

In a cultured cell line it has been found that about 34 ± 5 microtubules end at each kinetochore. These *kinetochore microtubules* correspond to the chromosomal fibers mentioned in the older literature. The microtubules point toward the poles, but not all of them are long enough to reach the pole. Quantitative measurements have also shown that the so-called continuous fibers are not actually so, since only a few microtubules may be so long as to span the two poles. As better names, in addition to kinetochore microtubules, the terms *polar* and *free* microtubules have been proposed.[10]

Several reports indicate that microtubules may have some kind of cross-bridges which may hold the spindle together. Fine microfilaments have been observed in between the microtubules[11] which may correspond to actin.[12] The presence of actin in parts of the spindle, especially near the poles, has been demonstrated by anti-actin antibodies. However, microfilaments do not appear to be connected to the kinetochores.[13] The spindle also contains ribonucleoprotein particles presumably derived from the nucleolus, small membrane vesicles, and a fine fibrous matrix.

Assembly of Spindle Microtubules— Control by Poles and Kinetochores

The cyclic changes in the birefringence of the spindle are interpreted as reflecting the systematic assembly and disassembly of the microtubules. Studies employing micromanipulation have shown that, within certain limits, the spindle fibers resist extension, maintain mechanical integrity, and are instrumental in the movement of the chromosomes.

Studies in living cells also reveal that the spindle fibers represent a very dynamic structure. Their birefringence is abolished in a matter of seconds by low temperature, but after return to normal temperature, the cell recovers in a few minutes, with continuation of the arrested mitosis. Intense hydrostatic pressure, microbeam ultraviolet irradiation, and certain drugs, such as colchicine, Colcemid, and others, also induce disappearance of the birefringence and the microtubules. One interesting change is produced with heavy water. When dividing sea urchin oöcytes are placed in 45 per cent D_2O, the birefringence increases twofold, and the volume of the spindle increases about tenfold in 1 to 2 minutes. After the eggs are returned to H_2O the birefringence reverts to normal within a few minutes.[2] These experiments imply that in cytoplasm there is a great excess of building material (see below) for the mitotic apparatus.

In Vitro Studies. By using gentle methods of cell fractionation on sea urchin eggs, Mazia and Dan (1952)[1] made the first isolation of the mitotic apparatus (Fig. 17–8). Slow progress was made throughout the years in separating this labile structure from various cells.

A new approach has recently been used based on the discovery of the conditions under which brain tubulin could be polymerized in vitro.[14] The technique consists of preserving the isolated spindle in a medium containing exogenous tubulin.[10, 14-16] With this method, some of the spindle functions are preserved. For example, in cells lysed at anaphase the chromosomes have been observed to move in normal fashion.

Control of the Assembly. Some of abovementioned observations suggest that the assembly of microtubules is *controlled* by the *poles* and also by the *kinetochores*. A third mechanism may involve the *lateral interaction* between the spindle microtubules. Although the exact mechanism is unknown, it is thought that it involves first the polymerization of tubulin subunits and then their assembly into microtubules by secondary bonds formation.

Normally in a cell there is a large *pool* of *tubulin monomers* which are in a kind of dynamic equilibrium with the polymerized microtubules. An equilibrium is also established with the *cytoplasmic microtubules* (see Chapter 9–2).

When cells enter prophase, these microtubules become depolymerized and are replaced by the mitotic spindle. At metaphase, only the spindle microtubules are present; at anaphase, with the movement of the chromosomes, the spindle becomes depolymerized, and at telophase the daughter cells are held by the *midbody* and the cytoplasmic microtubules reappear. (All these stages are clearly seen in the sequence shown in Fig. 17–6).[17]

Action of Ca²⁺. The finding that Ca^{2+} can inhibit the polymerization of tubulin[14] has led to the suggestion that the intracellular concentration of Ca^{2+} may play an important role in the assembly and disassembly of microtubules in vivo. The level of Ca^{2+} in the cytosol may be regulated through various mechanisms by which this cation is bound or sequestered in the cell (e.g., in mitochondria or the endoplasmic reticulum, see Fuller and Brinkley, 1976).

Metaphase Chromosomal Motion—
Caused by Kinetochore and Polar
Microtubular Interaction

Current hypotheses about the role of the mitotic apparatus are based on the idea that the microtubules can generate some sort of mechanical force either by "pushing" or by "pulling" the other cell components. These two mechanisms involve the elongation or the shortening of the microtubules. In prophase, the migration of centrioles toward the poles is probably caused by the "pushing" which is a result of the elongation of the continuous fibers. In fibroblasts, separation of centrioles occurs at a rate of 0.8 to 2.4 μm per minute. As soon as the nuclear envelope begins to disintegrate, the nuclear region is invaded by micro-

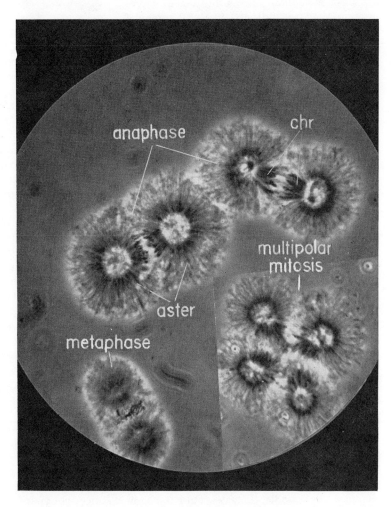

Figure 17–8 Mitotic apparatus isolated from the sea urchin egg in division. *chr,* chromosomes. (Courtesy of D. Mazia.)

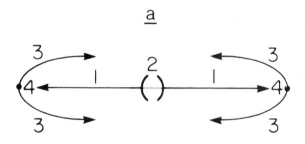

Figure 17–9 Diagram indicating the equilibrium of forces at metaphase: *1*, Pole-directed forces that pull sister kinetochores to opposite poles. *2*, forces that link the two kinetochores until anaphase; *3*, forces that pull the poles inward. *4*, resistance to the inward force. (From McIntosh, J. R., Cande, Z. W., and Synder, J., *in Molecules and Cell Movement*, p. 31, Inoué, S., and Stephens, R. E., eds. Raven Press, New York, 1975.)

tubules that establish pole-kinetochore attachments before the chromosomes move toward the metaphasic plate.

We have seen that the poles of the spindle and the kinetochore are the two main nucleation centers of the spindle microtubules. The motion of chromosomes toward the metaphase plate can be explained as the result of lateral interaction between the microtubules arising from the kinetochore and from the pole microtubules.

During metaphase an equilibrium of forces is established, as indicated in the diagram of Fig. 17–9. There are pole-directed forces that tend to pull the sister kinetochores toward the poles; however, since these are linked until the onset of anaphase, the opposing forces cancel each other.[10] It is possible that the pole microtubules could exert a counteracting force that would tend to pull the poles inward. (Such an equilibrium of forces could also explain the bowed shape of the spindle and the fact that the chromosomes remain stationary at metaphase.)

With the separation of the sister chromatids at anaphase, this equilibrium is broken. The chromosomes usually move toward the poles with a rate of about 1μm per min.

The characteristic shapes assumed by the chromosomes in metaphase and anaphase and the anaphase bridges that may result from the stretching of dicentric chromosomes suggest that the forces responsible for the pulling of the chromosomes toward the poles is transmitted to the kinetochores. These forces may be great enough to produce the rupture of these dicentric chromosomes. It has been calculated that to move a chromosome, a force of about 10^{-8} dynes is needed, and that the entire displacement — from the equator to the pole of a chromosome — may require the use of about 30 ATP molecules.

The chromosomes move at a constant rate until they are separated by a distance equivalent to the length of the metaphase spindle; then they slow down. It is important to remember that during anaphase the spindle elongates until it nearly

duplicates the length that it had at metaphase. The movement of the chromosomes is associated with a shortening of the kinetochore microtubules, which implies that there are pole-directed forces acting on the kinetochores.

Recently a better view of the spindle microtubules at different stages of mitosis has been obtained in thick sections of cultured cells[10] by high-voltage electron microscopy (1000 kvolts). The resultant images suggest that at anaphase there are two distinct events: (1) the motion of chromosomes toward the poles on each half spindle, and (2) the separation of the two half spindles by elongation of the free microtubules at the interzonal region. At the midregion of the interzonal fibers it is possible to identify a dense zone classically known as the *stem body* (Bélar, 1929). Electron microscopic observations have shown that in this region there is a high concentration of interdigitating microtubules.

Anaphase Chromosomal Movement— The Dynamic Equilibrium and the Sliding Mechanism Hypotheses

Recent theories trying to explain the role of the mitotic apparatus in mitosis have been influenced by the discovery of a cyclic mechanism in muscle and cilia in which there is a direct coupling between the hydrolysis of ATP and mechanical events. Such a coupling is based on an elaborate and rather stable molecular organization of these systems (see Chapters 9 and 27). On the other hand, the mitotic apparatus is very labile and is assembled and disassembled at every mitotic cycle.

Two main hypotheses have attempted to interpret the molecular mechanism responsible for the movement of chromosomes: (1) *the dynamic equilibrium hypothesis*, and (2) *the sliding mechanism hypothesis*.

The *dynamic equilibrium hypothesis* first proposed in 1949 by Ostergren has been established mainly by Inoué and Sato (1967)[2] as a consequence of their studies with the polarization microscope. This hypothesis postulates that the polymerization-depolymerization of microtubules is directly responsible for the movement of the chromosomes. While it is rather easy to see that the microtubules could have a "cytoskeletal" role, serving as a kind of guide for maintaining the shape of the cell and for "pushing" the centrioles, it is more difficult to understand how its depolymerization could provide a motive force. However, calculations of the cohesive forces that could be involved in assembling microtubules have led to the conclusion that they contain enough energy to move chromosomes through the cytoplasm.[18]

The *sliding hypothesis*, although recognizing that spindle microtubules assemble and disassemble, does not admit that they generate directly the driving force to move the chromosomes.[19] Rather, they are just the "railroad tracks" on which the chromosomes move. This hypothesis postulates the involvement of lateral interactions by way of molecules that could play a role analogous to *dynein* in flagella (see Chapter 9–2) or by interaction with other fibrous components, such as *actin*.[20] We mentioned earlier that actin may be demonstrated in certain parts of the spindle.

Furthermore, the use of antibodies against myosin has shown that *myosin* may be present in the spindle.[21] Thus, it is possible, *although the issue is not yet settled*, that the interaction of spindle actin with myosin could provide the motive force in chromosomal movement, and that the microtubules could play a mechanical role. Mitosis could be, in this way, a special case of *cell motility* (see Chapter 9). A molecular mechanism for the function of the mitotic apparatus still seems to be a long way ahead (see Nicklas, 1977).

Cytokinesis in Animal Cells—Contractile Ring of Actin and Myosin

The separation of the daughter nuclei and *cytokinesis* or *cell cleavage* may be two separate processes. For example, the eggs of most insects undergo division of the nucleus to form a multinucleate *plasmodium* without separation of the cytoplasmic territories. Cleavage of animal cells has been studied mainly in tissue culture and dividing eggs and is produced by the furrowing of the cell. The first visible changes consist of the appearance of a dense material around the microtubules at the equator of the spindle at either mid or late anaphase. Then, although spindle microtubules tend to disorganize and disappear during telophase, they usually persist, and

may even increase in number at the equator, frequently being intermingled with a row of vesicles and the dense material; the entire structure is called the *midbody* (Fig. 17–10). Simultaneously, there is a depression on the cell surface — a kind of constriction that deepens gradually until reaching the midbody. With the completion of the furrowing the separation of the cell is concluded.[22] The importance of the spindle was inferred by the study of multipolar cell divisions in which a furrow is formed between each pair of asters. Even the removal of the whole mitotic apparatus of the sea urchin egg, however, does not inhibit cell cleavage.

Current hypotheses suggest that cell cleavage by furrowing may be the result of a contractile-ring mechanism that involves the cell cortex.[22] A contractile protein exhibiting ATPase activity and properties similar to myosin was isolated from dividing sea urchin eggs.[23] These experiments should be correlated with other results showing that ATP may produce cytoplasmic contraction in different cell types. A system of fine microfilaments may be detected under the plasma membrane at the region of the furrow (Fig. 17–11). It can be demonstrated that such filaments are composed of an actin-like protein. Another indication that a contractile mechanism is acting in cell cleavage is provided by the action of the drug *cytochalasin*, which is able to inhibit such a process as well as other contractile mechanisms in cells.

Studies using antibodies against *actin* and *myosin* have shown that both these proteins are present in the contractile ring of cytokinesis (for a review, see Beams and Kessel, 1976).

Cytokinesis in Plant Cells—The Phragmoplast and Cell Plate

In plant cells at anaphase the interzonal region of the spindle becomes transformed into the *phragmoplast*, which in some way is similar to the midbody of animal cells (Fig. 17–2, E and F).

Under the electron microscope it is possible to observe that the interzonal microtubules at the equator plane have scattered patches of vesicles and of dense material applied to their surface. The vesicles are derived from the Golgi complexes which are found in the regions adjacent to the phragmoplast and which migrate into the equatorial region to be clustered around the microtubules. Although the phragmoplast initially is found as a ring in the periphery of the cell, with time it grows centripetally by addition of microtubules and vesicles until it extends across the

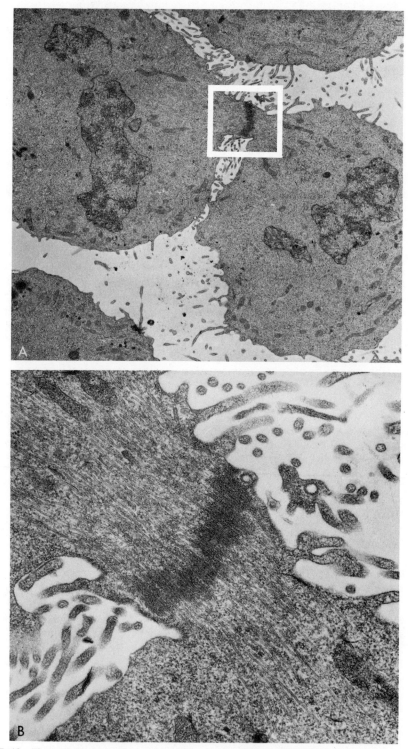

Figure 17–10 Electron micrograph of a HeLa cell at the completion of cytokinesis. **A**, the two daughter cells are still joined by a small bridge which contains the interzonal microtubules and the electron-dense midbody. **B**, an inset at higher magnification. **A**, ×10,000; **B**, ×30,000. (Courtesy of B. R. Brinkley.)

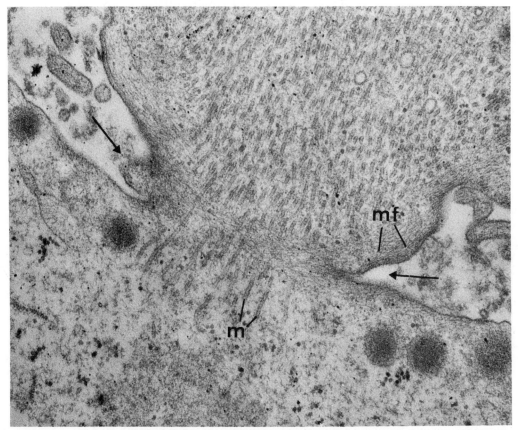

Figure 17–11 Electron micrograph of the advancing furrow (*arrows*) in a cleaving rat egg. Micotubules (*m*) of the interzonal fibers of the spindle are observed in the bridge between the daughter cells. Below the plasma membrane a network of fine microfilaments (*mf*) is observed. (See the description in the text.) ×34,000. (Courtesy of D. Szollosi.)

entire equatorial plane (Fig. 17–5, *E* and *F*). The vesicles increase in size and then fuse until the two daughter cells become separated by fairly continuous plasma membranes. At this time the phragmoplast has been transformed into the *cell plate*.

Thin cytoplasmic connections — the *plasmodesmata* — traverse the cell plate and remain in place for communication between the adjacent daughter cells.

The formation of the cell plate also leads to the synthesis of the *cell wall*. The *Golgi* vesicles in the phragmoplast are already filled with a secretory material consisting mainly of *pectin* (see Fig. 14–2). The fusion of the vesicles results in the combining of the pectin in the extracellular space between the daughter cells, thereby forming the main body of the *primary cell wall*. (As mentioned in Chapter 14 pectin is an amorphous polymer made of galacturonic acid.) Later on, microfibrils of cellulose are laid down in a semicrystalline lattice on the two surfaces of the facing daughter cells.

SUMMARY:

The Mitotic Apparatus

The *mitotic apparatus* comprises the spindle and the asters which surround the centrioles. The spindle is made of the *chromosomal* fibers, the *continuous* fibers, and the *interzonal* fibers; the latter are observed at anaphase and telophase between the daughter chromosomes.

The mitotic apparatus may be observed in the living cell using polarization microscopy or the Nomarski interference–contrast microscope. Anti-tubulin antibodies and fluorescence microscopy may provide information about the relative number and condensation of microtubules.

The *kinetochore* of the centromere is the site of implantation of the microtubules in the chromosome. There is usually one kinetochore per chromatid. It appears as a cup-like disk 0.2 to 0.25 μm in diameter. It has three layers: an outer dense layer, an inner dense layer, and a middle layer of lower density. Kinetochores also contain tubulin, and they act in the assembly of microtubules. About 34 microtubules attach to a single kinetochore.

Microtubules may be kinetochoric, polar, or free. Kinetochoric microtubules correspond to the chromosomal spindle fiber. Cross-bridges have been observed between microtubules and also microfilaments which may correspond to actin.

Microtubules are assembled and disassembled during the various phases of mitosis. The mitotic apparatus may be isolated, and best preservation is obtained in media containing exogenous tubulin.

The assembly of tubulin is controlled by the poles and by the kinetochore. There is a large pool of tubulin monomers in equilibrium with the microtubules. Ca^{2+} inhibits the polymerization of tubulin.

Elongation and shortening of microtubules seem to be the two major mechanisms by which chromosomes are moved toward the poles. At metaphase there is an equilibrium of forces (Fig. 17–9). In anaphase, chromosomes move at a rate of 1 μm per min. The motive force is transmitted by way of the kinetochores. Some 30 ATP molecules are required to move 1 chromosome to the pole. There is also an elongation of the spindle simultaneously with shortening of the kinetochore microtubules. In the midregion of the spindle, numerous microtubules appear.

Two main hypotheses have been devised to interpret the molecular mechanisms of chromosome movement, as follows:

(1) The *dynamic equilibrium hypothesis* postulates that the polymerization-depolymerization of microtubules is directly responsible for the movement. (2) The sliding hypothesis postulates that the driving force may originate from lateral interactions (as in the case of dynein in flagella) or by interaction with actin microfilaments. It is possible that actin and myosin in the spindle could provide the motive force.

Cytokinesis or cell cleavage differs considerably in animal and plant cells. In the former, separation of daughter cells is produced by an equatorial constriction which involves a contractile mechanism at the cell cortex. This is achieved by a system of actin-like microfilaments. A dense structure called the *midbody* may be formed.

In plant cells cytokinesis starts with the formation of the *phragmoplast*, which comprises the interzonal microtubules and Golgi vesicles. This structure is transformed into the *cell plate*, which separates the territories of the daughter cells. Within the cell plate the primary cell wall is produced by a secretory mechanism consisting mainly of the production of pectin, which is contained in Golgi vesicles (see Chapter 14).

REFERENCES

1. Mazia, D., and Dan, K. (1952) *Proc. Natl. Acad. Sci. USA, 38*:826.
2. Inoué, S., and Sato, H. (1967) *J. Gen. Physiol., 50*:259.
3. Bajer, A., and Allen, R. D. (1966) *Science, 151*:572.
4. Brinkley, B. R., and Cartwright, J. (1971) *J. Cell Biol., 50*:416.
5. Brinkley, B. R., and Stubblefield, E. (1966) *Chromosoma, 19*:28.
6. Jokelainen, P. T. (1967) *J. Ultrastruct. Res., 19*:19.
7. Luykx, P., ed. (1970) *Int. Rev. Cytol.,* Suppl. 2. Academic Press, Inc., New York.
8. Dietz, R. (1972) *Chromosoma, 38*:11.
9. Pepper, D. A., and Brinkley, B. R. (1977) *Chromosoma, 60*:223.
10. McIntosh, J. R., Cande, Z. W., and Snyder, J. (1975) In: *Molecules and Cell Movement,* p. 31. (Inoué, S., and Stephens, R. E., eds.) Raven Press, New York.
11. Forer, A., and Behnke, O. (1972) *Chromosoma, 39*:145.
12. Hinkley, R., and Telser, A. (1974) *Exp. Cell Res., 86*:161.

13. Ross, U. P. (1977) *J. Cell Biol.*, 70:283a.
14. Weisenberg, R. C. (1972) *Science*, 177:1104.
15. Inoué, S., Borisy, G. G., and Kierhart, D. P. (1974) *J. Cell Biol.*, 62:175.
16. Rebhun, L. I., Rosenbaum, J., Lefebvre, P., and Smith, G. (1974) *Nature*, 249:113.
17. Fuller, G. M., and Brinkley, B. R. (1976) *J. Supramol. Struct.*, 5:497.
18. Inoué, S., and Ritter, H., Jr. (1975) In: *Molecules and Cell Movement*. (Inoué, S., and Stephens, R. E., eds.) Raven Press, New York.

19. McIntosh, J. R., Hepler, P. K., and Van Wie, D. G. (1969) *Nature*, 224:659.
20. Forer, A. (1974) In: *Cell Cycle Controls*, p. 319. (Padilla, G. M., et al., eds.) Academic Press, New York.
21. Fujiwara, K., and Pollard, T. D. (1977) *J. Cell Biol.*, 70:181a.
22. Szollosi, D. (1970) *J. Cell Biol.*, 44:192.
23. Oknishi, T. (1962) *J. Biochem.*, 52:145.

ADDITIONAL READING

Bajer, A. S., and Molé-Bajer, J. (1972) Spindle dynamics and chromosome movements In: *International Review of Cytology*, Supplement 3. Academic Press, Inc., New York.

Beams, H. W., and Kessel, R. G. (1976) Cytokinesis. *Am. Sci.*, 63:279.

Berns, M. W., et al. (1977) The role of centriolar regions in animal cell mitosis. *J. Cell Biol.*, 72:351.

Cande, W. Z., Lazarides, E., and McIntosh, R. (1977) A comparison of the distribution of actin and tubulin in the mitotic spindle. *J. Cell Biol.*, 72:552.

Esponda, P. (1978) Cytochemistry of kinetochores under electron microscopy. *Exp. Cell Res.*, 114:247.

Forer, A., Chromosome movements during cell division. In: *Handbook of Molecular Cytology*, p. 553. (Lima-de-Faría, A., ed.) North-Holland Publishing Co., Amsterdam.

Forer, A. (1974) Possible roles of microtubules and actin-like filaments during cell division. In: *Cell Cycle Controls*, p. 319. (Padilla, G. M., et al., eds.) Academic Press, New York.

Fuller, G. M., and Brinkley, B. R., (1976) Structure and control of assembly of cytoplasmic microtubules in normal and transformed cells. *J. Supramol. Struct.*, 5:497.

Fuseler, J. W. (1975) Temperature dependence of anaphase chromosome velocity and microtubule depolymerization. *J. Cell Biol.*, 67:789.

Hartwell, L. H. (1978) Cell division from a genetic perspective. *J. Cell Biol.*, 77:627.

Heidemann, S. R., Sander, G., and Kirschner, M. W. (1977) Evidence for a functional role of RNA in centrioles. *Cell*, 10:337.

Inoué, S., and Ritter, H. (1975) Dynamics of mitotic spindles, Organization and Function. In: *Molecules and Cell Movement*. (Inoué, S., and Stephens, R. E., eds.) Raven Press, New York.

John, B., and Lewis, K. R. (1969) The chromosome cycle. *Protoplasmatologia, 6b*:1.

John, B., and Lewis, K. R. (1973) Somatic cell division. In: *Readings in Genetics and Evolution*, p.1. Oxford University Press, London.

Kubai, D. F. (1975) The evolution of the mitotic spindle. *Int. Rev. Cytol.*, 43:167.

Mazia, D. (1961) Mitosis and the physiology of cell division. In: *The Cell*, Vol. 3, p. 77. (Brachet, J., and Mirsky, A. E., eds.) Academic Press, Inc., New York.

Mazia, D. (1974) The cell cycle. *Sci. Am.*, 230(1):54.

McGill, M., and Brinkley, B. R. (1975) Human chromosomes and centrioles as nucleating sites for the in vitro assembly of microtubules from bovine brain tubulin. *J. Cell Biol.*, 67:189.

McIntosh, J. R., Cande, Z. W., and Snyder, J. A. (1975) Structure and physiology of the mammalian mitotic spindle. In: *Molecules and Cell Movement*, p. 31. (Inoué, S., and Stephens, R. E., eds.) Raven Press, New York.

Nicklas, R. B. (1974) Chromosome segregation mechanism. *Genetics*, 78:205.

Nicklas, R. B. (1977) Chromosome movement, facts and hypothesis. In: *Mitosis, Facts and Questions* (Little, M. et al., eds.) Springer-Verlag, Berlin.

Prescott, D. M. (1976) *Reproduction of Eukaryotic Cells*. Academic Press, New York.

Rappaport, R. (1975) Establishment and organization of the cleavage mechanism. In: *Molecules and Cell Movement*, p. 287. (Inoué, S., and Stephens, R. E, eds.) Raven Press, New York.

Schroeder, T. E. (1975) Dynamics of the contractile ring. In: *Molecules and Cell Movement*, p. 305. (Inoué, S., and Stephens, R. E. eds.) Raven Press, New York.

Sherline, P., and Schiavone, K. (1978) High molecular weight MAPs are part of the mitotic spindle. *J. Cell Biol.*, 77:R9.

Wolfe, S. L. (1972) *Biology of the Cell*. Wadsworth Publishing Co., Belmont, California.

Wolff, S. (1969) The strandedness of chromosomes. *Int. Rev. Cytol.*, 25:279.

18

MEIOSIS AND SEXUAL REPRODUCTION

18–1   A Comparison of Mitosis and Meiosis 400

18–2   A General Description of Meiosis   402
   Leptotene Chromosomes Appear to be Single and Have Chromomeres
   Zygonema — Pairing of Homologous Chromosomes and Synaptonemal Complex Formation
   Pachynema — Crossing Over and Recombination Between Homologous Chromatids
   The Synaptonemal Complex — Homologue Alignment and Recombination
   The Recombination Nodule is Probably Related to Crossing Over
   Diplonema — Separation of the Paired Chromosomes Except at the Chiasmata
   Diakinesis — Reduction in the Number of Chiasmata
   Meiotic Division I — Separation of the Homologous Centromeres

   Meiotic Division II — Separation of the Sister Centromeres
   Chromosome Distribution in Mitosis and Meiosis — Dependent on Kinetochore Orientation
   Summary: *Events in Meiosis*

18–3   Genetic Consequences of Meiosis and Types of Meiosis   413
   Meiosis is Intermediary or Sporic in Plants
   Meiosis may Last for 50 Years in the Human Female
   Meiosis Starts After Puberty in the Human Male
   Fertilization — A Species-Specific Interaction Between the Gametes
   Summary: *More About Meiosis*

18–4   Biochemistry of Meiosis   417
   Summary:

Meiosis is a special type of cell division that is found in organisms in which there is *sexual reproduction*. In many protozoa, algae, and fungi reproduction is asexual, i.e., by simple cell division or *mitosis*. In this form of reproduction all the descendant individuals have *uniparental inheritance*. On the other hand, in most multicellular organisms — animals or plants — the reproduction is sexual. This means that during their lifetime they produce *sexual cells* or *gametes* (*eggs* and *sperms*) which by their fusion in the *zygote* will originate the new organism. It is evident that in this process there is a *biparental inheritance* which implies a process of nuclear fusion called *fertilization*.

To understand the essentials of meiosis let us take the case of man, as indicated in Figure 19–1. The human karyotype, which will be studied in Chapter 20, has 46 chromosomes, 44 + XY in the male, and 44 + XX in the female. (The pairs XY and XX are the sex chromosomes.) If mitosis were the only type of cell division, then each

gamete would have 46 chromosomes and the zygote, 92. This would be repeated in the next generation leading to an increase according to the square of the number of generations!

Meiosis (Gr., *meioum*, to diminish) is the mechanism that prevents this from happening. By a series of two divisions, the number of chromosomes is reduced by half, which gives rise to four *haploid* cells (four spermatozoa in the male, and one ovum and three polar bodies in the female) (Fig. 18–1). The processes by which the gametes are produced are called *spermatogenesis* and *oögenesis* respectively, and they take place in the gonads (i.e., testicle and ovary). The type of meiosis found in man is characteristic of all animals and of a few lower plants, and it is called *terminal* or *gametic*, because the meiotic divisions occur just before the formation of the gametes. We shall see later that in most plants *meiosis* is *intermediary* or *sporic* and occurs at some time between fertilization and the formation of gametes. In other words, in this case

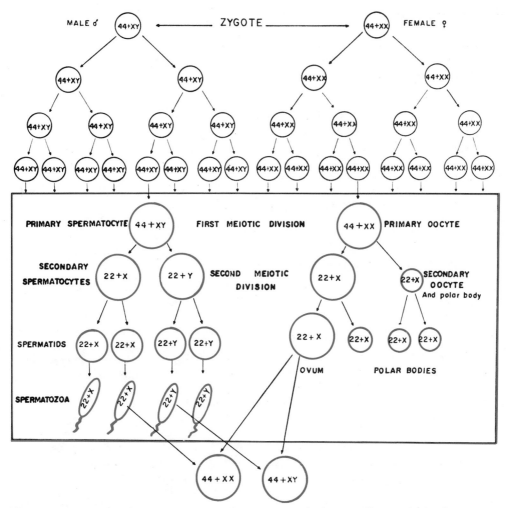

Figure 18–1 Diagram of spermatogenesis and oogenesis in the human. **Above,** mitosis of spermatogonia and oögonia. **Middle** (within the box), the meiotic divisions. **Below,** fertilization and zygote. Notice the 44 autosomes and the XY sex chromosomes.

meiosis is preceded and followed by mitosis. Cells undergoing meiosis are often called *meiocytes.*

18–1 A COMPARISON OF MITOSIS AND MEIOSIS

Although in Chapter 2 (Fig. 2–9) we compared the two types of cell division, it may be appropriate to reconsider this problem. Many of the events that we have studied in mitosis are also found in meiosis, for example, the sequence of changes in the nucleus and cytoplasm; the stages of prophase, metaphase, anaphase, and telophase; the formation of the mitotic apparatus with the asters and spindle; the condensation cycle of the chromosomes; and the structure and function of the centromeres. There are, however, essential differences between the two processes, some of which are the following:

(1) Mitosis occurs in all *somatic cells* of an individual, but meiosis is limited to only the *germinal cells.*

(2) In mitosis each replication cycle of DNA is followed by one cell division. The resulting daughter cells have a *diploid number* of chromosomes and the same amount of DNA as the parent cell. In meiosis one replication cycle of DNA is followed by two divisions, and the four cells are *haploid* and contain half the amount of DNA.

(3) In mitosis DNA synthesis occurs in the S period, which is followed by a G_2 phase before the onset of division (Fig. 2–7). In meiosis there

401

is *premeiotic DNA* synthesis which is much longer than that in mitosis[1, 2] and which is followed immediately by meiosis. In other words, the G_2 phase is short or non-existent.

(4) In mitosis every chromosome behaves independently; in meiosis the *homologous chromosomes* become mechanically related during the first meiotic division (i.e., *meiotic pairing*).

(5) While mitosis is rather brief (one or two hours), *meiosis is a long process*. For example, in the human male it may last 24 days, and in the female it may go on for several years (see later discussion).

(6) A fundamental difference between the two types of cell division is that in mitosis the genetic material remains constant (i.e., with only rare mutations or chromosomal aberrations), while *genetic variability* is one of main consequences of meiosis.

18–2   A GENERAL DESCRIPTION OF MEIOSIS

As shown in the diagram of Figure 18–1, after several mitotic divisions of the spermatogonia (or oögonia), meiotic division starts. A kind of switch shifts these cells from mitosis to meiosis (see Riley and Flavell, 1977). In the G_2 period of interphase there is apparently a decisive change that directs the cell toward meiosis instead of toward mitosis. Some experiments in cultured cells of lily anthers show that this change probably takes place at the beginning of G_2, but its exact nature is still unknown. In these cells it is possible to experimentally reverse the switch back to mitosis before the process becomes irreversible.

We shall see that the following are the essential processes of meiosis: (1) the *pairing* (i.e., *synapsis*) of the homologous chromosomes, (2) the formation of *chiasmata*, which represent the underlying *genetic recombination* (i.e., the interchange of genetic material), and (3) the *segregation of the homologous chromosomes*.

From the morphologic viewpoint the first meiotic division is characterized by a long prophase during which homologous chromosomes pair closely and interchange hereditary material. The classic stages of mitosis do not suffice to describe the complex movements of the chromosomes in meiosis. The successive meiotic stages are the following:

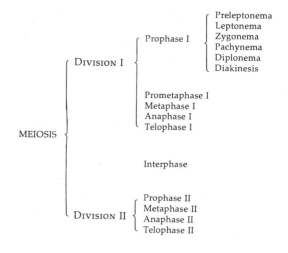

Leptotene Chromosomes Appear to be Single and Have Chromomeres

Preleptonema corresponds to the early prophase of meiosis. Chromosomes are extremely thin and difficult to observe. Only the sex chromosomes may stand out as compact heteropyknotic bodies.

Leptonema (Gr., *leptos*, thin + *nema*, thread) (Fig. 18–2), corresponds to a period in which the nucleus has increased in size and the chromosomes have become more apparent. The leptotene chromosomes differ in the following two ways from the prophase mitotic ones: (1) In spite of the fact that DNA duplication has already occurred and that they have two chromatids, leptotene chromosomes look single rather than double. (2) These chromosomes show bead-like thickenings, the so-called *chromomeres*, which appear at irregular intervals along their length (Fig. 18–3, *A*). Since these beads are characteristic in size, number, and position for a particular chromosome, they may be used as landmarks to identify a specific chromosome of an organism. With the further contraction of chromosomes during zygonema and pachynema, chromomeres become larger and fewer in number. Under the electron microscope the chromosomes show the unit fiber folded back and forth at the chromomeres (see Chapter 15).

Chromomeres were once thought to represent the genes. However, they are too few in number (1500 to 2500 in the lily) to include all the genes.

It is interesting to mention at this point that during the entire meiotic prophase (and also

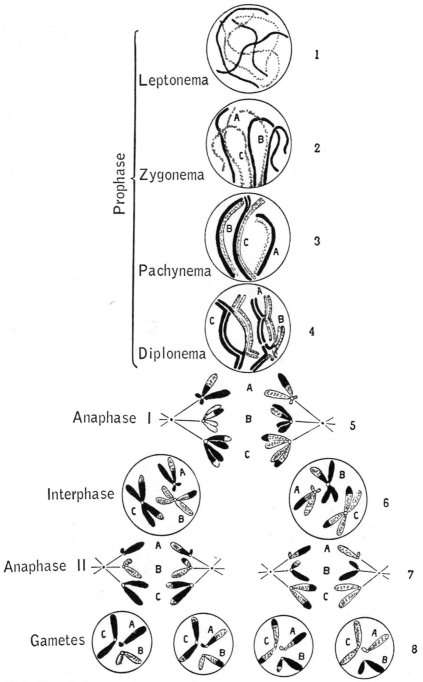

Figure 18–2 General diagram of meiosis, illustrating the union, separation, and distribution of the chromosomes.

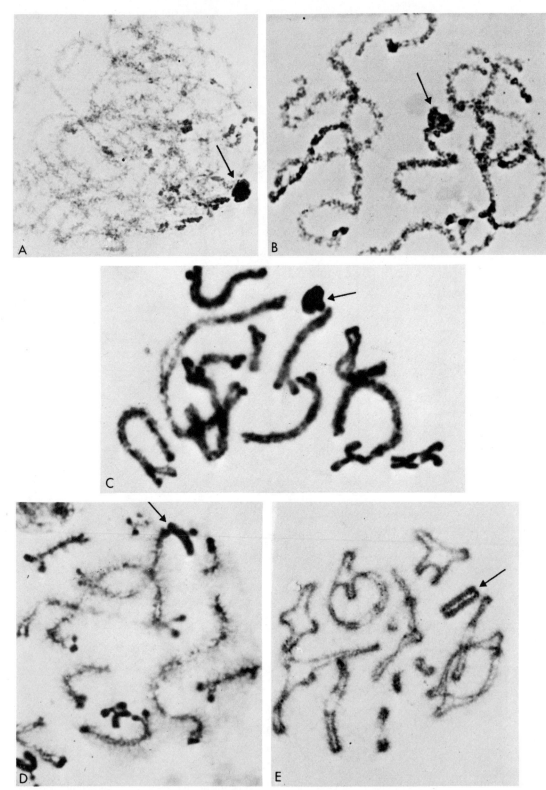

Figure 18–3 Stages of meiosis in the South American grasshopper. *Laplatacris dispar* (2n = 22 + X). **A,** *leptonema,* showing the long thin filaments with the chromomeres. (The X chromosome is indicated by an arrow in this and in subsequent micrographs.) **B,** *pachynema,* showing thick filaments in which the homologous chromosomes have paired. **C,** *early diplonema,* showing the way in which the homologous chromosomes have shortened considerably and have begun to separate. **D,** *mid-diplonema,* showing the configuration of the bivalents with the chiasmata. **E,** *late diplonema,* showing the chiasmata in distinct form. (Microphotographs taken by the late professor F. A. Saez.)

during the mitotic) the chromosomes become progressively shorter without becoming entangled, which suggests that during interphase the DNA strands are already in a three-dimensionally ordered pattern.

Frequently, leptonemic chromosomes have a definite polarization and form loops whose ends are attached to the nuclear envelope at points near the centrioles, contained within an aster. This peculiar arrangement is often called the "bouquet."

Zygonema — Pairing of Homologous Chromosomes and Synaptonemal Complex Formation

During *zygonema* (Gr., *zygon*, adjoining), the first essential phenomenon of meiosis occurs. The homologous chromosomes become aligned and undergo pairing in a process often called *synapsis of the chromosomes* (Fig. 18–2). Pairing is highly specific and involves the formation of a special structure which is observed under the electron microscope and is generally named the *synaptonemal complex* (SC).[3]

First described by Moses in 1956, this complex is composed of two *lateral components* or arms and a *central* or *medial* element which are both interposed in between the two pairing homologues (Fig. 18–4). The synaptonemal complex can be considered as the structural basis for pairing and synapsis of meiotic chromosomes. It is important to emphasize that at the end of leptonema the lateral elements of the SC have already appeared in the space between the two chromatids, while the medial component appears, with the pairing, at zygonema.[4]

Another important point is that when pairing starts, the packing of the chromosomes is already very high, on the order of at least 300/1 (i.e., the DNA is 300 times longer than the length of the chromosomes). In other words, only 0.3 per cent of the DNA of the homologous chromosomes is matched along the length of the lateral component of the SC. To explain this it is generally assumed that the unit fiber of the chromosome makes loops which at one point join the SC (Fig. 18–5). These points may function as matching sites for the synaptic alignment of the chromosomes. It is supposed that the prezygotene condensation is a highly ordered process

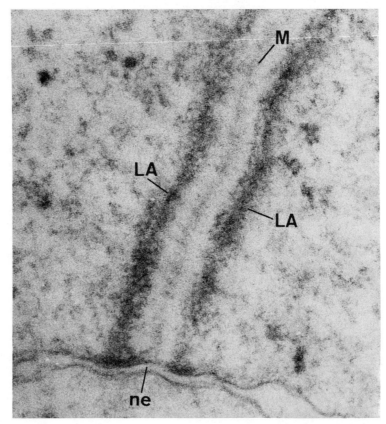

Figure 18–4 Electron micrograph of the synaptonemal complex in a dog spermatocyte. The two lateral arms (*LA*), corresponding to the homologous chromosomes, are shown parallel to each other and extending into the nuclear envelope (*ne*). The medial element (*M*) is simpler than in certain invertebrates. ×125,000. (Courtesy of J. R. Sotelo.)

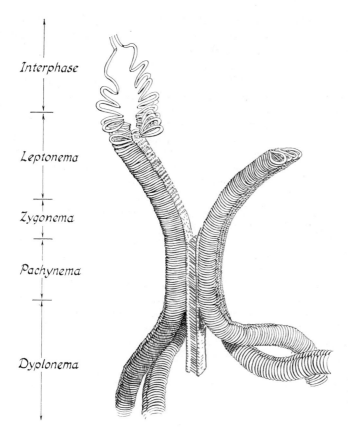

Interphase

Leptonema

Zygonema

Pachynema

Dyplonema

Figure 18–5 Three-dimensional diagram of the synaptonemal complex (*SC*) at different stages of meiosis. In pachynema the homologous chromosomes, now fully paired, are joined by the synaptonemal complex. This complex separates at diplonema; the chromatids show a relational coil. (From Roth, T. F., *Protoplasma, 61*:346, 1966.)

and that it follows a similar pattern in both homologues, so that the synapsis becomes specific for the corresponding set of genes.

During zygonema the homologous chromosomes become aligned. This is made easier by the fact that the telomeres of the chromosomes are frequently inserted in the nuclear envelope (Fig. 18–4).

The pairing does not have a special starting point. Sometimes the chromosomes unite at their polarized ends and continue pairing at the antipodal extremity; in other cases, fusion occurs simultaneously at various places along the length of the thread. Polarization seems to favor regularity in pairing. Pairing is remarkably exact and specific; it takes place point for point, and chromomere for chromomere, in each homologue. The two homologues do not fuse during pairing, but remain separated by a space of about 0.15 to 0.2 μm, which is occupied by the synaptonemal complex.

The morphology of the SC is very similar in plant and animal meiocytes. It has a tripartite structure, with the two lateral arms in each homologous chromosome and the central component. In cross section it can be observed that the SC is a flattened, ribbon-like structure. The lateral arms vary in width from 20 to 80 nm in various species. They are formed of electron-dense coarse granules or fibers. These arms are joined to the adjacent chromosomes by fine fibrils.

In most plants and animals — with the exception of the insects — the two side arms are separated by an axial space of lower density (Fig. 18–4). In insects the central component may be very complex. In this case, it has the aspect of a ladder, with three dense parallel lines and bridges crossing at intervals of 20 to 30 nm. These bridges are formed by fine fibrils that span the central and lateral components and are arranged perpendicularly to them.

By the use of serial sections it has been possible to reconstruct all the bivalent chromosomes of an oöcyte[5, 6] (Fig. 18–6).[7] Recently a special squashing technique has permitted the direct observation of the SC of all the bivalents in the karyotype (see Moses et al., 1975). For example, in Figure 18–7 all the SCs present in the karyotype of a chicken oöcyte are apparent. In this case, the cell is at the end of pachynema and pairing has been completed. (Observe that only the sex chromosomes — W an Z in this case — have an unpaired region.

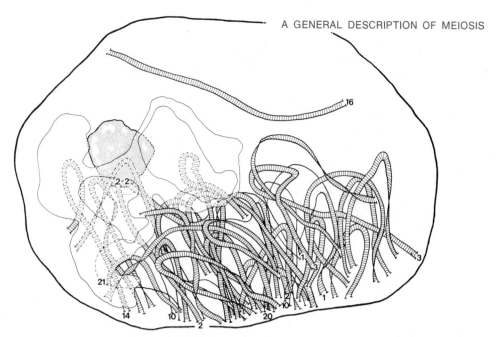

Figure 18–6 Reconstruction of the late zygonema stage of meiosis in the *Bombyx mori* female. It may be observed that all homologous chromosomes are paired, with the exception of pairs 1, 2, 5, 10, and 17. In those that are paired a full synaptonemal complex is present; in those with unpaired regions, the lateral arms of the *SC* are present. Note that most of the telomeres are attached to a region of the nuclear envelope, forming a *bouquet* figure. ×11,500. (Courtesy of S. W. Rasmussen.)

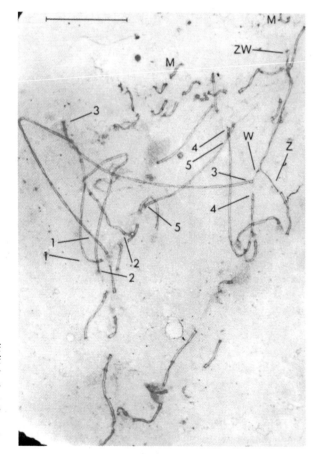

Figure 18–7 Synaptonemal complexes of all bivalents in a chicken oöcyte at the stage of pachynema. Only the SCs are preferentially stained, and the chromatin, which constitutes the bulk of the chromosomes, is not apparent. The largest bivalents are indicated by numbers (*1* through *5*). *M,* microchromosomes; *W* and *Z,* the sex chromosomes. ×2900. (Courtesy of A. J. Solari, *Chromosoma, 64*:155, 1977.)

In the human spermatocyte it has been found that the SC is present only in the region of pairing between the lateral componens of the X and Y chromosomes.[8, 9]

Researchers employing deoxyribonuclease to digest DNA, and the heavy metal indium to stain nucleic acids have suggested that the main component of the SC is protein in nature. This protein is basic and probably similar to the histones. The fine fibers that cross between the two lateral arms and connect with the chromosomes probably contain DNA (see Gillies, 1975). (For the possible functions of the SC see later discussion).

Pachynema—Crossing Over and Recombination Between Homologous Chromatids

During pachynema (Gr., *pachus,* thick), the pairing of the chromosomes reaches completion (Figs. 18–2 and 18–3,*B*). The chromosomes contract longitudinally, resulting in shorter and thicker threads. By middle pachynema, the nucleus contains half the number of chromosomes. Each unit is a *bivalent* or *tetrad* which is composed of two homologous chromosomes in close longitudinal union and which contains four chromatids.

Each homologous chromosome has an independent centromere; however, under the electron microscope it has been observed that each chromatid has its own centromere.[4] Thus, in a tetrad there are four centromeres, two homologous and two sister (Fig. 18–8, 2). (However, during the first meiotic division the centromeres of the two chromatids behave as a functional unit, Fig. 18–8, 3 through 5). In late pachytene a line of separation perpendicular to the plane of pairing appears, and the four chromatids become visible. The chromatids of each homologue are called *sister chromatids.* (In the diagram of Fig. 18–8, 1 and 2, the homologue and sister chromatids may be observed.) During pachynema, the space occupied by the SC is maintained, and all the homologous chromosomes have finished the pairing process.

Experimental evidence suggests that during pachynema two of the chromatids of the homologues exchange segments (i.e., recombine). It is thought that transverse breaks occur at the same level on each of the chromatids and that this event is followed by interchange and final fusion of the chromatid segments (Fig. 18–8, 2).

Pachynema is usually a long-lasting stage of prophase, whereas leptonema and zygonema may last only a few hours. Pachynema may last for days, weeks, or even years.

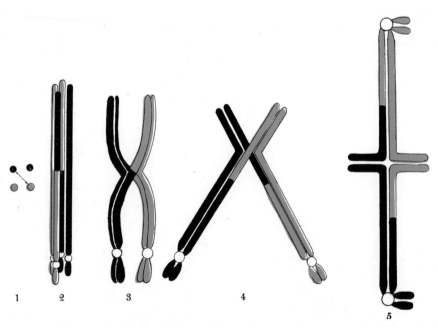

Figure 18–8 Above, 1 and 2, diagrams showing the process of crossing over; 3, formation of a chiasma; 4, terminalization; 5, rotation of the chromatids of one bivalent.

The Synaptonemal Complex— Homologue Alignment and Recombination

We have already mentioned that the lateral arms of the SC appear in leptonema, before pairing, and that completion of the SC occurs during zygonema and becomes more conspicuous at pachynema. After this stage, during diplonema, the SC disintegrates and usually disappears. Figure 18–5 is a three-dimensional diagram which interprets the appearance and disappearance of the SC in relation to the stages of meiotic prophase.[10]

One function of the SC appears to be that of stabilizing the pairing of the homologues; another, that of facilitating recombination. Because of the deposition of new protein molecules in the lateral arms of the SC the matching segments of DNA are placed in a way that allows interchange at the molecular level. The SC can be interpreted as a protein framework that permits not only the proper alignment of the homologues but also their recombination (see Stern and Hotta, 1976).

It must be assumed that the DNA fibers of the paired chromatids are within 1.0 nm of the central component of the SC in order for breakage and recombination to take place. This could be achieved by the fine DNA fibrils that are arranged like transverse bridges and that connect the homologous chromatids with the lateral and central elements of the SC (Fig. 18–6). It is thought that the homologous sequences of nucleotides search for each other and achieve a close pairing inside the central component of the SC. It is within these special regions that recombination, with the exchange of DNA segments, occurs.

The Recombination Nodule is Probably Related to Crossing Over

Electron microscopic studies of meiosis at the stage of pachytene have revealed the presence of dense nodules of about 100 nm, in intimate association with the synaptonemal complex. The number of these nodules and their location along the bivalents is related to the number and distributions of the genetic exchanges (crossovers). The name *recombination nodule* has been applied to these, with the suggestion that they may be involved in the recombination process taking place at pachytene of meiosis (see Carpenter, 1975; Zickler, 1977).

Diplonema—Separation of the Paired Chromosomes Except at the Chiasmata

At *diplonema* the intimately paired chromosomes repel each other and begin to separate (Fig. 18–3C, and E). However, this separation is not complete, since the homologous chromosomes remain united by their points of interchange, or *chiasmata* (Gr., *chiasma*, cross piece). Chiasmata are generally regarded as the expression of the phenomenon called *crossing over*, or *recombination*, by which chromosomal segments with blocks of genes are exchanged between homologous members of the pairs. With few exceptions, chiasmata are found in all plants and animals. At least one chiasma is formed for each bivalent. Their number is variable, since some chromosomes have one chiasma and others have several. During diplonema the four chromatids of the tetrad become visible; under the electron microscope the synaptonemal complex can no longer be observed (Fig. 18–9).

Diplonema is a long-lasting period. In the fifth month of prenatal life, for example, human oöcytes have reached the stage of diplonema and remain in it until many years later, when ovulation occurs. In most species there is uncoiling of the chromosomes during diplonema. In fish, amphibian, reptilian, and avian oöcytes, however, the uncoiling becomes so marked that the greatly enlarged nucleus assumes an interphase

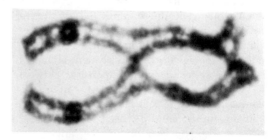

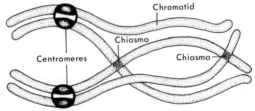

Figure 18–9 **Above,** photomicrograph of a single bivalent at diplonema in a salamander spermatocyte. **Below,** diagram interpreting the photomicrograph. The four chromatids are clearly apparent. Two chiasmata and the position of the centromeres are also seen. Observe that the two sister centromeres are side by side. (Courtesy of J. Kezer.)

appearance. In these cases the bivalent chromosomes may attain a special configuration known as the *lampbrush chromosome,* in which the chromonema uncoils into loops that converge upon a more coiled axis. It will be shown later that the presence of these lampbrush chromosomes is related to an intensive RNA synthesis and to the enormous growth of the oöcyte (see Chapter 25–3).

Diakinesis — Reduction in the Number of Chiasmata

In diakinesis (Gr., *dia,* across) the contraction of the chromosomes again becomes accentuated (Fig. 18–10,*B*). The tetrads are more evenly distributed in the nucleus, and the nucleolus disappears. During this period the number of chiasmata diminishes. By the end of diakinesis, in general the homologues are held together only at their ends (Fig. 18–8, *4,* and *5*), a process that has formerly received the name of *terminalization.*

Meiotic Division I — Separation of the Homologous Centromeres

In *prometaphase* I condensation of the chromosomes reaches its maximum. The nuclear envelope breaks down, and the spindle microtubules become attached to the centromeres. Each homologue is attached to one of the poles by the homologous centromere, and the two sister chromatids behave as a functional unit.

In *metaphase* I the chromosomes become arranged at the equator. If the bivalent is long, it presents a series of annular apertures between the chiasmata in perpendicularly alternating planes. If the chromosomes are short, they have a single annular aperture. Meiotic metaphase I can be distinguished under the microscope (from that occurring in mitosis or in metaphase II) by the fact that the homologues are still attached by chiasmata at their ends while the centromeres are being pulled toward the poles (Fig. 18–10,*C*).

In *anaphase* I the sister chromatids of each homologue, united by their centromeres, move toward their respective poles. The short chromosomes, generally connected by a terminal chiasma, separate rapidly. Separation of the long chromosomes, which have interstitial and unterminalized chiasmata, is delayed. In side view, anaphase chromosomes show different shapes, depending on the position of the centromere.

It should be recalled that by recombination, segments were transposed between two of the chromatids of each homologue. Thus, when the homologous paternal and maternal chromosomes separate in anaphase, their composition is different from that of the originals. Two of their chromatids are mixed; the other two maintain their initial nature with reference to a single locus (Figs. 18–2 and 18–8).

Telophase I begins when the anaphase groups arrive at their respective poles. Chromosomes may persist for some time in a condensed state, showing all their morphologic characteristics. Following telophase is a short *interphase* which has characteristics similar to mitotic interphase. Sometimes interphase may persist for a considerable length of time.

The result of the first meiotic division is the formation of the daughter nuclei, which in animals are called spermatocytes II (in the male) and oöcyte II plus the first polar body (in the female).

At the interphase between the two meiotic divisions *there is no replication of the chromosomes.* These are now haploid in number, although each one consists of two chromatids.

Meiotic Division II — Separation of the Sister Centromeres

The short prophase II is followed by the formation of the spindle, which marks the beginning of metaphase II. At metaphase II (Fig. 18–10,*D*) chromosomes become arranged on the equatorial plane, and the sister centromeres separate; the two sister chromatids go toward the opposite poles during anaphase II (Fig. 18–10,*E*). Since the longitudinal halves of each parental chromosome (chromatids) separate in this division each of the four nuclei of telophase II has one chromatid, and each nucleus has a haploid number of chromosomes (Figs. 18–2 and 18–10,*E*).

Chromosome Distribution in Mitosis and Meiosis — Dependent on Kinetochore Orientation

We have just described that in meiotic division I there is a separation of the homologous centromeres (and chromatids), while in meiotic division II, as in mitosis, separation of the sister centromeres (and chromatids) occurs. The differences observed are of great importance in ex-

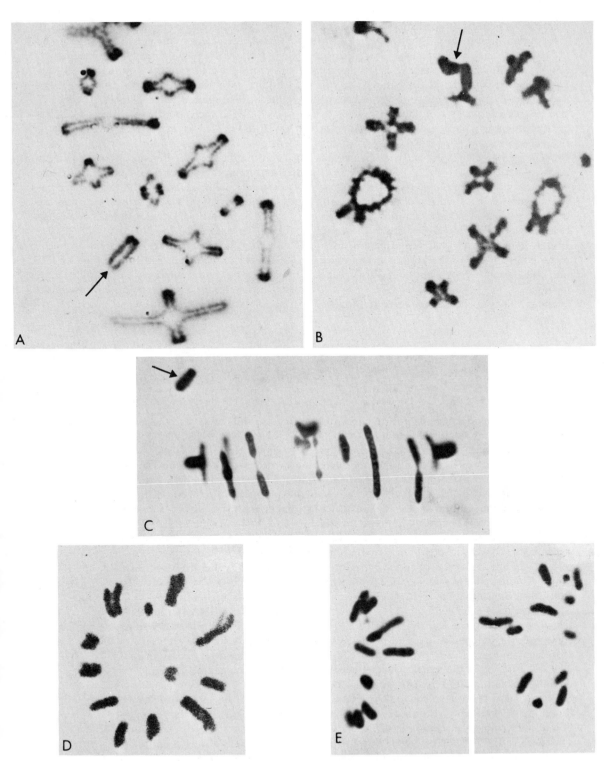

Figure 18–10 Continuation of Figure 18–3, showing other stages of meiosis in the grasshopper. **A,** *diplonema,* with Giemsa stain showing the constitutive heterochromatin localized at the centromere regions (C bands). **B,** *diakinesis,* showing the condensation of the bivalents. **C,** *metaphase I,* side view. The arrow points to the X chromosome that is advancing toward the pole; all the autosomes are still in the equatorial plane. **D,** *metaphase II,* polar view. **E,** *anaphase II,* side view. At this moment each chromosome consists of a single chromatid, and each daughter cell has the haploid number of chromosomes. (Micrographs taken by the late professor F. A. Saez.)

plaining the general mechanism of chromosome distribution in mitosis and meiosis.

Based on early observations by Schrader (1963), in 1951 Ostergren postulated that the differences in chromosome distribution in mitosis and meiosis resulted from the different orientation of the kinetochores. In fact, in mitosis the sister chromatids have their kinetochores arranged back to back (Fig. 17–7); in meiotic division I the kinetochores of the sister chromatids lie side by side (Fig. 18–9).

The factor that determines the distribution of the chromosomes is their initial interaction with the spindle just after disintegration of the nuclear envelope. This interaction occurs by way of spindle microtubules that run from the kinetochore toward one of the poles and that determine the particular movement of the given chromosome.

The problem of chromosome distribution has been studied in experiments using microsurgery in grasshopper spermatocytes. Meiocytes in the I and II divisions were fused by mechanical means to produce one cell containing the two spindles — one spindle with bivalents, the other with unpaired chromosomes. It was observed that in each spindle the chromosome behaved as in the original cells. Furthermore, it was possible to detach a bivalent from division II and to place it in the division I spindle. In this arrangement the attached bivalent also divided in the original fashion (i.e., the two sister chromatids moved together to each pole); in all the other chromosomes, single sister chromatids moved to each pole (see Nicklas, 1977). These and other experiments suggest that in the mechanism of chromosomal distribution the key role is played by the orientation of the kinetochore, which determines the nucleation and preferential polarization of microtubules.

SUMMARY:

Events in Meiosis

Meiosis is a special type of cell division present in germ cells of sexually reproducing organisms. Sexual reproduction occurs by way of sexual cells or gametes (eggs and sperms), which after fertilization comstitute the zygote. Meiosis consists of a single duplication of the chromosomes followed by two consecutive divisions. In animals and lower plants meiosis is *terminal or gametic* (i.e., it occurs before the formation of the gametes). In the male, four haploid sperms are produced; in the female, one ovum and three polar bodies. In most plants meiosis is *intermediary* or *sporic* (i.e., it occurs sometime between fertilization and the formation of gametes). Cells in meiosis are called *meiocytes*. The essential differences between mitosis and meiosis are: (1) Mitosis occurs in somatic cells, and meiosis is limited to germinal cells. (2) In mitosis, one replication cycle of DNA is followed by one division, resulting in diploid cells. In meiosis, one replication of DNA is followed by two divisions, resulting in haploid cells. (3) In mitosis DNA replication occurs in the S period, which is followed by G_2. In meiosis there is a premeiotic DNA synthesis which is very long and followed immediately by meiosis. (4) In mitosis each chromosome behaves independently, and in meiosis there is pairing of homologous chromosomes. (5) Mitosis lasts one to two hours, and meiosis lasts longer; in the male it may last 24 days, in the female, several years. (6) In mitosis the genetic material remains constant, and in meiosis there is genetic variability.

The essential processes of meiosis are (1) pairing of homologous chromosomes, (2) formation of chiasmata with underlying genetic recombination, and (3) the segregation of the homologous chromosomes.

Meiosis is divided into divisions I and II. In division I there is a long prophase of which the stages are preleptonema and leptonema, zygonema, pachynema, diplonema, and diakinesis. Leptotene chromosomes look single (in spite of their two chromatids), and they have chrommomeres that are characteristically placed for each chromosome. The unit fiber of chromatin is folded at the chromomeres. Sometimes leptotenes are polarized, forming the so-called "bouquet."

During *zygonema*, pairing and synapsis of the homologues occur. Pairing involves the formation of the *synaptonemal complex* (SC). This is composed of two lateral arms (which appear in each homologue at the end of leptonema) and a medial element. At the time of pairing the packing of DNA is 300/1 (i.e., there is only 0.3 per cent matching between homologous DNA). Pairing starts at random, but telomeres are generally inserted at the nuclear envelope. The 0.2 μm space between homologous chromosomes is occupied by the synaptonemal complex. This may have the appearance of a ladder, with bridges crossing the medial element. The main component of this complex is protein.

During *pachynema*, pairing is complete and chromosomes become shorter and thicker. The number of chromosomes has been halved (i.e., bivalents or tetrads). Each tetrad has four centromeres (two homologous and two sister). During pachynema, two homologous chromatids exchange segments at a molecular level (recombination). The SC appears to stabilize the pairing, thus enabling the interchange.

During *diplonema*, the paired chromosomes begin to separate, but they are held together at the *chiasmata* (points of recombination or *crossing over*). There is at least one chiasma per bivalent chromosome. At diplonema, the SC is shed from the bivalents. At each chiasma there is a piece of SC that ultimately disappears and is replaced by a chromatin bridge. Diplonema may last for months or years.

During *diakinesis* there is a reduction in the number of chiasmata and further contraction. This is followed by prometaphase I, metaphase I, anaphase I, and telophase I. In anaphase I the sister chromatids of each homologue move to the respective poles (segregation of the homologues).

During the interphase between the two mitotic divisions there is no replication of the chromosomes, and Division II is very similar to mitosis, by the end of which each nucleus contains one chromatid and has a haploid number of chromosomes.

The fact that in meiotic division I there is separation of the homologue centromeres (and chromatids) and that in division II the sister centromeres (and chromatids) are separated, can be explained by the orientation of the kinetochores, which determines nucleation and preferential polarization of microtubules.

18–3 GENETIC CONSEQUENCES OF MEIOSIS AND TYPES OF MEIOSIS

We said at the beginning of this chapter that the essential processes of meiosis were (1) *pairing* or *synapsis*, (2) *recombination* and *chiasmata*, and (3) *segregation of the homologous chromosomes*. After the description of the entire process of meiosis, the third point becomes apparent. Meiosis has not only resulted in the halving of the chromosome number but has also segregated each member (i.e., chromatid) of the homologous pair into four different nuclei. One point that is of special interest is that the homologous centromeres separate in anaphase I, and the sister centromeres, in anaphase II; but since crossing over mixes the homologous chromatids of the bivalents, both meiotic divisions are needed to segregate the genes contained in the chromatids. (The student can follow this process by studying closely the distribution of chromosome *a* in Fig. 18–11.)

We shall now see that segregation of the homologues and recombination have important genetic consequences. From the genetic viewpoint, meiosis can be considered as a mechanism for distributing the genes between the gametes, permitting their recombination and random segregation. Figure 18–11 is a diagram in which the genetic consequences of the meiosis of three pairs of chromosomes having one, two, and three chiasmata can be observed. The chiasmata are followed through diplonema, metaphase I, and anaphase I, and then into anaphase II. It is obvious that each one of the four gametes has a different genetic constitution. As a result of the crossing over, each chromosome does not consist solely of maternal or paternal material but of alternating segments of each.

The random assortment of the genes is due not only to crossing over but also to the fact that the distribution of the chromosomes in the first and second division is normally random. In Figure 18–11, to simplify the matter, the three paternal and maternal homologues are shown separating into different cells. Since this separation is a random process, the resulting cell will contain eight (i.e., 2^3) different

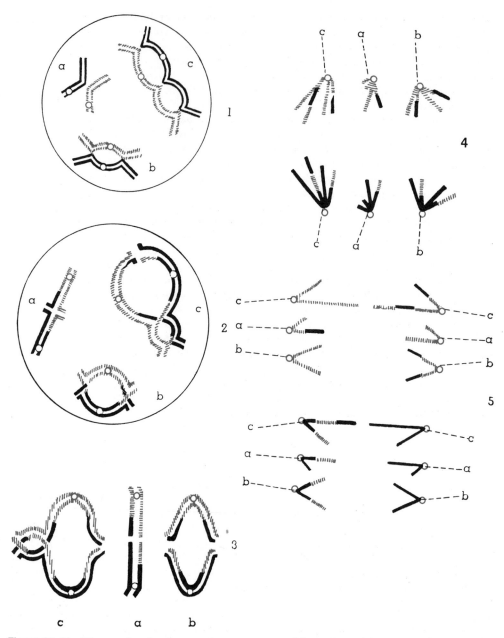

Figure 18–11 Diagram showing the genetic consequences of the meiosis of three pairs of chromosomes with (a) one chiasma, (b) two chiasmata, and (c) three chiasmata. **1**, diplonema; **2**, advanced diplonema showing the process of terminalization; **3**, metaphase I; **4**, anaphase I; **5**, anaphase II, showing the distribution of the chromosomes in the four nuclei formed. **Solid line**, the paternal chromosomes; **dashed line**, the maternal chromosomes. The centromere is represented by a circle.

chromosomal combinations, even in the absence of crossing over. In a human (with 23 pairs of chromosomes), the possible chromosomal combinations in gametes will be an immense number, 2^{23} or 8,388,608. Meiosis is the mechanism by which genetic variation is brought about, and knowledge of its mechanism is a prerequisite for the understanding of the chromosomal basis of genetics.

Meiosis is Intermediary or Sporic in Plants

We mentioned earlier that intermediary or sporic meiosis is characteristic of flowering plants. As shown in Figure 18–12, meiosis takes place at some time intermediate between fertilization and the formation of gametes.

In higher plants the reproductive organs — anthers in male and ovary or pistil in female —

produce microspores and megaspores, respectively. The cells that undergo meiosis to produce megaspores are called megasporocytes. Microspores are produced by microsporocytes (pollen mother cells). Each microsporocyte gives rise, by meiosis, to four functional microspores. Each megasporocyte produces four megaspores by meiosis, of which three degenerate. The remaining megaspore develops into the female gametophyte, which gives rise to the egg cell.

In plants, microspores and megaspores are not the final gametes. Before fertilization, they undergo two mitotic divisions in the anther or three in the ovary to produce the male and female gametophytes, respectively.

Fertilization in plants is a complex phenomenon. Each pollen grain or *microspore* (haploid) carries two sperm nuclei. One of them fertilizes the egg nucleus to give a *diploid zygote*, which will eventually form the embryo of a new plant (i.e., *sporophyte*). The other sperm nucleus fuses with two polar nuclei to form a *triploid endosperm nucleus*, which by mitotic divisions will give rise to the *endosperm*, which contains the nutritive material for the embryo.

Thus, the seed is a mosaic of tissues consisting of the diploid zygote, the triploid endosperm, and the diploid integuments, which are of maternal origin (Fig. 18–12).

Meiosis may Last for 50 Years in the Human Female

The primary female germ cells (i.e., *gonocytes*) appear in the human embryo in the wall of the yolk sac at about 20 days and during the fifth week migrate to the gonadal ridges. By mitosis, they form oögonia and become surrounded by the follicular cells, forming the *primary follicles*. At the end of the third month of prenatal development, oögonia enter meiosis, becoming oöcytes I and are then arrested at the stage of diplonema until sexual maturity at about 12 years. (This prolonged diplonema stage is sometimes called *dictiotene*.) During this long phase, the chromosomes remain in a rather uncondensed state and resemble the *lampbrush chromosomes* (see Chapter 25–3).

The number of oöcytes in a newborn female has been estimated at about 1,000,000; howev-

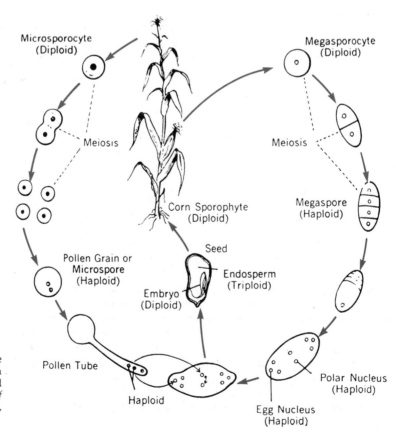

Figure 18–12 Life cycle of a plant. (Modified from Sinnott, E. W., Dunn, L. C., and Dobzhansky, T., *Principles of Genetics*, 5th ed., New York, McGraw-Hill Book Co., 1958.)

Microsporocyte (Diploid)

Megasporocyte (Diploid)

Meiosis

Meiosis

Megaspore (Haploid)

Corn Sporophyte (Diploid)

Pollen Grain or Microspore (Haploid)

Seed

Endosperm (Triploid)

Embryo (Diploid)

Pollen Tube

Polar Nucleus (Haploid)

Haploid

Egg Nucleus (Haploid)

er, most of these degenerate. By the age of 7 years, there are some 300,000 oöcytes, but only about 400 reach maturity between 12 and 50 years. Thus, meiosis may last as long as 50 years! This may explain the increase in the incidence of chromosomal aberrations with the increasing age of the mother (see Chapter 20).

When the ovum is released into the *oviduct*, the first meiotic division occurs, giving one polar body (Fig. 18–1). It is only when the ovum is fertilized by the spermatozoon that the second meiotic division takes place, giving the second polar body. As a result of oögenesis, only one viable egg is produced; the haploid nucleus is known as the *female pronucleus* (Fig. 18–1). The polar bodies degenerate and do not take part in embryogenesis.

Meiosis Starts after Puberty in the Human Male

In the human male the primitive gonad also starts with the migration of the primary gonocytes into the gonadal ridges, and later these are incorporated into the *seminiferous tubules*. These contain spermatogonia that enter meiosis at puberty, with the beginning of sexual life. In constrast to oögenesis in the female, spermatogenesis continues in the male until advanced age. Completion of this process takes about 24 days, and the meiotic prophase I occurs in 13 to 14 days. Four viable spermatozoa are formed from each meiotic cycle (Fig. 18–1). The transformation of spermatids into mature sperm cells occurs by a complex process called *spermiogenesis*, which can be studied in histology textbooks.

Fertilization — A Species-specific Interaction Between the Gametes

Fertilization is a specific process in the sense that, as a rule, the sperm of one species can not fertilize the egg of another. The egg produces a protein called *fertilizin* which reacts with an anti-fertilizin on the surface of the sperm. This is a species-specific interaction by which the cells are attached to each other. The sperm, together with one centriole, then penetrates the egg. After this has occurred, other sperm are prevented from entering by changes that take place at the *vitelline membrane* of the egg. The male nucleus swells and becomes the so-called *male pronucleus* and then fuses with the female pronucleus, thus beginning the first division of the egg. Thus, the sperm, in addition to acting as a stimulus of cell division, contributes (1) one set of haploid chromosomes carrying the paternal genes, and (2) a centriole, which will function in cell division during cleavage.

SUMMARY:

More About Meiosis

Meiosis not only results in the halving of the number of chromosomes but also in the segregation of each chromatid into a different nucleus. Meiosis is a mechanism for distributing the genes between gametes, enabling their recombination and random segregation. In this process, genetic variation is produced.

The study of meiosis is a prerequisite for the understanding of the chromosomal basis of genetics. Only after the process of meiosis is understood will its significance in hereditary phenomena become apparent (see Chapter 19).

In the human female, gonocytes appear in the embryo at 20 days and migrate into the gonadal ridges. At the third month, the oögonia enter meiosis, which is arrested at diplonema until sexual maturity (at about 12 years). The chromosomes remain uncondensed resembling lampbrush chromosomes. Most oöcytes degenerate, and only about 400 reach maturity between 12 and 50 years. Thus, meiosis may last for 50 years! When the ovum is released from the follicle, it produces the first polar body. The second meiotic division occurs only with fertilization.

In the human male, the primary gonocytes become incorporated into seminiferous tubules. Meiosis starts at puberty and ends in old age. Meiosis is completed in 24 days, and prophase I, in 13 to 14 days. Fertilization is an intraspecies process in which fertilizin and anti-fertilizin are involved. The sperm contributes with one set of haploid chromosomes and one centriole.

18–4 BIOCHEMISTRY OF MEIOSIS

Studies carried out mainly in *Lillium* (in which it is possible to isolate meiocytes in different stages of meiosis in large enough quantities for biochemical analysis) have permitted the identification of the distinctive metabolic characteristics of meiosis which are not found in mitosis.[12, 13] Recently, similar studies have been extended to meiocytes of male mammals using special fractionation techniques (see Hotta et al., 1977; Chandley et al, 1977). From the biochemical point of view the following four stages of prophasic meiosis have been specifically studied:

Prezygonema. We have already mentioned that the premiotic S phase is longer than in somatic cells. In fact, it may last 100 to 200 times longer; this is due primarily to the reduced frequency in the number of initiation points during DNA duplication.[14] Another important difference observed in microsporocytes of *Lillium* is that *0.3 to 0.4 per cent* of the DNA remains unreplicated.[12]

Zygonema. During zygonema the above-mentioned unreplicated DNA is synthesized. This is called *Z-DNA* because it occurs at this stage in coordination with chromosome pairing. This late-replicating Z-DNA appears to be essential to synapsis, since if it is inhibited, the cells are arrested at zygonema. This Z-DNA has a higher buoyant density than the rest of the DNA, and it is rich in GC bases. (Fig. 18–13). It is thought that Z-DNA contains highly repeated nucleotide sequences which are distributed over all *Lillium* chromosomes and may function in chromosome synapsis.[15]

Another biochemical characteristic of this period is the appearance of a *DNA binding protein* which is associated with a lipoprotein fraction and which catalyzes the reassociation of single-stranded DNA. This protein has been called *reassociation or r-protein*, and it has been suggested that it may play a role in the alignment of homologous stretches of DNA (see Stern and Hotta, 1977).

Pachynema. Protein synthesis seems essential to maintain chromosome pairing during pachynema. If protein synthesis is inhibited at late zygonema the chromosome pairs fall apart and are not reconstituted.[13] During pachynema, there is also a DNA synthesis which is of a smaller magnitude than that of zygonema. This P-DNA synthesis is apparently related to the process of recombination.

Figure 18–14 shows a molecular model based on a synthesis of repair-DNA that takes place when the two homologous chromatids recombine. It is thought that at the end of zygonema, an *endonuclease* is activated that produces nicks in the two DNA molecules that are aligned for recombination (Fig. 18–14,*A*). After recombination (Fig. 18–14,*B*), gaps and overlaps of short nucleotide sequences can occur (Fig. 18–14,*C*) The overlaps are excised by exonucleases, and the gaps are filled in by DNA polymerase and DNA ligase, as in the mechanism of DNA repair described in Chapter 16 (Fig. 18–14,*D*) This repair mechanism probably explains the small amount of DNA synthesis observed at pachynema.

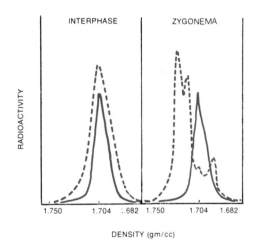

Figure 18–13 Ultracentrifugation patterns of DNA from meiotic cells of *Lillium* after equilibration in a CsCl gradient. Observe that the buoyant density of the DNA synthesized at zygonema is different from that at interphase (Courtesy of H. Stern and Y. Hotta.)

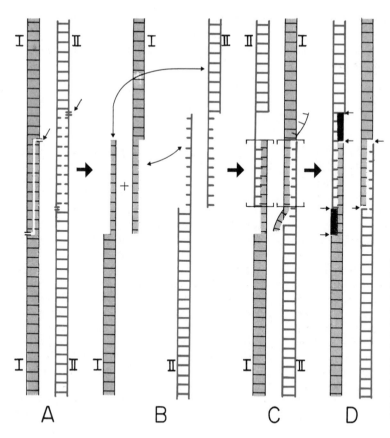

Figure 18–14 Molecular model of recombination during meiotic prophase. **A,** the DNA of the pairing chromatids (I and II) undergoing the effect of endonucleases, producing "nicks" on each of the strands; **B,** unraveling of the strands; **C,** rejoining of the opposite chromatids; and **D,** elimination of excess pieces of DNA and filling of gaps by a process similar to that of DNA repair (see Fig. 17–14). (Modified from a diagram of C. A. Thomas, Jr.)

SUMMARY:

Biochemical studies have been carried out in meiocytes of *Lillium* at different stages of meiosis.

At *prezygonema* there is a premeiotic DNA replication (S phase) which is 100 to 200 times longer than that in somatic cells. However, 0.3 to 0.4 per cent of the DNA remains unreplicated. During *zygonema*, replication of the previously unreplicated DNA occurs (Z-DNA), in coordination with pairing. Z-DNA has a higher buoyant density than other DNA and is made of highly repeated nucleotide sequences.

During *pachynema* there is protein synthesis that is essential for maintaining the chromosome pairing. There is also a DNA synthesis (P-DNA) of small magnitude that is probably related to the process of recombination. Enzymes related to nicking and repair of DNA also appear. In *postpachynema* all these metabolic changes disappear.

REFERENCES

1. Bennett, M. D., Chapman, V., and Riley, R. (1971) *Proc. R. Soc. London* [Biol.], *178*:259.
2. Callan, H. G. (1972) *Proc. R. Soc. London* [Biol.]., *181*:19.
3. Moses, M. J. (1968) *Annu. Rev. Genet.*, *2*:363.
4. Westergaad, M., and Von Wettstein, D. (1972) *Annu. Rev. Genet.*, *6*:71.
5. Von Wettstein, R., and Sotelo, J. R. (1967) *J. Microsc. (Paris)*, *6*:557.
6. Moens, P. B. (1968) *Chromosoma*, *28*:1.
7. Rasmussen, S. W. (1976) *Chromosoma*, *54*:245.
8. Solari, A. J., and Tres, L. L. (1970) *J. Cell Biol.*, *45*:43.
9. Solari, A. J., and Moses, M. J. (1973) *J. Cell Biol.*, *56*:145.
10. Roth, T. F. (1966) *Protoplasma*, *61*:346.

11. Stern, H., and Hotta, Y. (1977) *Philos. Trans. R. Soc. Lond.* [Biol.], *277*:277.
12. Hotta, Y., Ito, M., and Stern, H. (1966) *Proc. Natl. Acad. Sci. USA, 56*:1184.
13. Parchman, L. G., and Stern, H. (1969) *Chromosoma, 26*:298.
14. Callan, H. G. (1973) *Cold Spring Harbor Symp. Quant. Biol., 38*:195.
15. Hotta, Y., and Stern, H. (1975) In: *The Eukaryote Chromosome*, p. 283. (Peacock, W. J., and Brock, R. D., eds.) Australian University Press, Camberra.

ADDITIONAL READING

Beermann, W. (1972) Chromomeres and genes. In: *Developmental Studies on Giant Chromosomes* p. 41. Springer-Verlag, Berlin.

Carpenter, A. T. C. (1975) Electron microscopy of meiosis in *D. melanogaster* females *Proc. Natl. Acad. Sci. USA*, 72:3186.

Callan, G. H. (1969) Biochemical activities of chromosomes during the prophase of meiosis. In: *Handbook of Molecular Cytology*, p. 540. (Lima-de-Faría, A, ed.) North-Holland Publishing Co., Amsterdam.

Chandley, A. C., Hotta, Y., and Stern, H. (1977) Biochemical Analysis of meiosis in the male mouse. *Chromosoma, 62*:243 and 255.

Church, K. (1976) Arrangement of chromosome ends in early meiotic prophase. *Chromosoma, 58*:365.

Comings, D. E., and Okada, T. A. (1972) Architecture of meiotic cells and mechanisms of chromosome pairing. *Adv. Cell Mol. Biol.*, 2:310.

Gillies, C. B. (1975) Synaptonemal complex and chromosome structure. *Annu. Rev. Genet.*, 9:91.

Moens, P. B. (1973) Mechanism of chromosome synapsis at meiotic prophase. *Int. Rev. Cytol.*, 35:117.

Moens, P. B. (1978) Ultrastructural studies of chiasma distribution. *Annu. Rev. Genet., 12*:433.

Moses, M. J. (1968) Synaptonemal complex. *Annu. Rev. Genet.*, 2:363.

Moses, M. (1977) Synaptonemal complex karyotyping: I, II, and III. *Chromosoma, 60*:99, 127, and 345.

Moses, M., Counces, D., and Poulson, D. (1975) Synaptonemal complex complement of man in spreads of spermatocytes. *Science, 187*:363.

Nicklas, R. B. (1977) Chromosome distribution: Experiments on cell hybrids and in vitro. *Philos. Trans. R. Soc. Lond.* [Biol.], 277:267.

Pardue, M. L., Bonner, J. J., Lengyel, J. A., and Spradling, A. C. (1977) *Drosophila* salivary gland polytene chromosome studies by in situ hybridization. *International Cell Biology*, p. 509. Rockefeller University Press, New York.

Riley, R., and Flavell, R. B. (1977) A first view of the meiotic process. *Philos. Trans. R. Soc. Lond.* 277:191.

Rhoades, M. (1967) Meiosis. In: *The Cell*, Vol. 1 (Brachet, J., and Mirsky, A. E., eds.) Academic Press, Inc., New York.

Sotelo, J. R. (1969) Ultrastructure of chromosomes at meiosis. In: *Handbook of Molecular Cytology*, p. 412. (Lima-de-Faría, A., ed.) North-Holland Publishing Co., Amsterdam.

Stern, H., and Hotta, Y. (1974) Biochemical controls of meiosis. *Annu. Rev. Genet., 7*:37.

Stern, H. and Hotta, Y. (1977) Biochemistry of meiosis. *Trans. R. Soc. Lond.* [Biol.], 277:277.

Westergaard, M., and von Wettstein, D. (1972) The synaptonemal complex. *Annu. Rev. Genet.*, 6:71.

Zickler, D. (1977) The synaptonemal complex and recombination nodules. *Chromosoma, 61*:289.

19

CYTOGENETICS — CHROMOSOMES AND HEREDITY

19–1 Mendelian Laws of Inheritance 420
 The Law of Segregation — Genes are Distributed without Mixing
 Independent Assortment — Genes in Different Chromosomes are Distributed Independently During Meiosis
 Genes are Linked When They are on the Same Chromosome
 Linkage may be Broken by Recombination During Meiotic Prophase
 Neurospora — Ideal for Studying Recombination and Gene Expression
 Summary: *Fundamental Genetics*
19–2 Chromosomal Changes and Cytogenetics 426
 Euploidy — A Change in the Number of Chromosome Sets; Aneuploidy — Loss or Gain of Chromosomes
 Aneuploidy — The Result of Non-Disjunction at Meiosis or Mitosis

Chromosomal Aberrations — Structural Alterations; Gene Mutations — Molecular Changes
Radiation and Chemical Mutagens Act Mainly on the DNA Molecule
Germ Cell Mutations are Transmittable and Somatic are not
Specific Chromosomal Aberrations — Deletion, Duplication, Translocation, Inversion, and Sister Chromatid Exchange
Sister Chromatid Exchanges Increase in Diseases with Altered DNA Repair
19–3 Chromosomes Play a Fundamental Role in Evolution 433
 Summary: *Chromosomal Aberrations, the Action of Mutagens, and Cytogenetics and Evolution*

In Chapter 1 we mentioned the fundamental discoveries that have established a link between cell and molecular biology and genetics and have led to the emergence of *cytogenetics.* This discipline is concerned with the chromosomal and molecular basis of heredity and deals with important issues with applications in agriculture and medicine. In the chapter to follow we shall consider some of the recent advances in the application of cytogenetics to the human species, and later on in Chapter 21 we will study some of molecular aspects of genetics. However, since all these subjects and related problems (i.e., genetic variation, mutation, phylogeny, morphogenesis, and evolution of organisms) can be studied in textbooks of genetics and general biology, we shall consider here only that which concerns cell biology alone, particularly the relationship between chromosomes and heredity.

19–1 MENDELIAN LAWS OF INHERITANCE

The basis of the laws that rule the transmission of hereditary characters is to be found in a knowledge of the behavior of the chromosomes during meiosis and especially in what has already been said about the genetic consequences. However, in 1865 when Gregor Johann Mendel discovered the fundamental laws of inheritance, nothing was known about the chromosomes and meiosis. His discovery was based on the precise quantitation of experimental crosses and exceptional abstract thinking. He studied crosses between peas (*Pisum sativum*) which had pairs of differential or contrasting characteristics. For example, he used plants that have white and red flowers, smooth and rough seeds, yellow and green seeds, long and short stems, and so forth.

After crossing the parental generation (P$_1$), he observed the resulting *hybrids* of the first filial generation, F$_1$. Then he crossed the hybrids (F$_1$) among themselves and studied the result in the second filial generation, F$_2$.

The Law of Segregation — Genes are Distributed without Mixing

In a cross between parents with yellow and green seeds, in the first generation Mendel found that all the hybrids had yellow seeds and, thus, the characteristic of only one parent. In the second cross (F$_2$), the characteristics of both parents reappeared in the proportion of 75 per cent to 25 per cent, or 3:1.

Mendel postulated that the color of the seeds was controlled by a "factor" that was transmitted to the offspring by means of the gametes. This hereditary factor, which is now called the *gene*, could be transmitted without mixing with other genes. He postulated that the gene could be segregated in the hybrid into different gametes to be *distributed* in the offspring of the hybrid. For this reason this is called the *law* or *principle of segregation of the genes*. Mendel found that the plants with yellow seeds in F$_2$, in spite of showing the yellow color, had different genetic constitutions. One-third of this group always gave yellow seeds, but the other two-thirds of the F$_2$ generation produced plants with yellow and green seeds in the ratio of 3:1. When the 25 per cent of plants in F$_2$ with green seeds were crossed among themselves, they always produced green seeds. This shows that they were a pure strain for this character. If we represent the genes in the crossing by letters, designating by *A* the gene with yellow character and by *a* the gene with green character, we have the following:

In the first generation (F$_1$) both *A* and *a* genes are present, but only *A* is revealed because it is *dominant;* gene *a* remains hidden and is called *recessive.* In the hybrid F$_1$ both genes are segregated and enter different gametes. Half of them will have the gene *A,* and the other half, *a.* Since each individual produces two types of gametes in each sex, there are four possible combinations in F$_2$. This gives as a result the proportion 1:2:1, corresponding to 25 per cent of plants with pure yellow seeds (*AA*), 50 per cent with hybrid yellow seeds (*Aa*), and 25 per cent with pure green seeds (*aa*).

Mendel's results can now be explained in terms of the behavior of chromosomes and genes. The genes present in the chromosomes are found in pairs called *alleles.* In each homologous chromosome the gene for each trait occurs at a particular point called a *locus* (plural *loci*). In the case illustrated in Figure 19–1, the gray mouse will have two *GG* genes, one in each homologue. Since the two homologues pair and then separate at meiosis, the two *GG* genes must also separate to enter the gametes. The mechanism is the same in a dominant as in a recessive white mouse. In the hybrid F$_1$, one chromosome bears gene *G* and the homologous chromosome bears gene *g.* When hybrids are self-fertilized, the gametes unite in the combinations shown by the checkerboard method illustrated in Figure 19–1. The individuals having two similar alleles are called *homozygous* (i.e., *GG* or *gg*); those with different alleles, *heterozygous* (i.e., *Gg*).

Genotype and Phenotype. In 1911 Johanssen proposed the term *genotype* for the genetic constitution and *phenotype* for the visible characteristics shown by the individual. For example, in the case of the peas with green or yellow seeds there are two phenotypes in F$_2$: yellow seeds and green seeds in the ratio of 3:1, respectively. However, according to the genetic constitution,

Parental genes		$AA \times aa$	
F$_1$ generation		Aa	
F$_2$ generation	AA	$Aa \quad aA$	aa
Genotypic proportion	1	2	1
	Dominant homozygous	Heterozygous	Recessive homozygous
Phenotypic proportion	3		1
	Yellow		Green

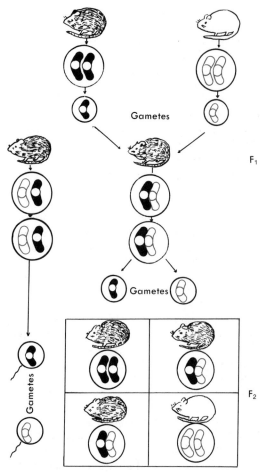

F₁

F₂

Gametes

Gametes

Gametes

Figure 19–1 A monohybrid cross between a gray mouse (dominant) and a white mouse (recessive). The parallelism between distribution of genes and chromosomes is indicated, as well as resulting phenotypes in the F₁ and F₂ generations.

there are three different genotypes: *AA, Aa,* and *aa* in the ratio of 1:2:1.

The phenotype includes all the characteristics of the individual that are an expression of gene activity. For example, in humans phenotypic characteristics include eye color, baldness, the different hemoglobins, the blood groups, and a difference in the ability to taste thiourea.

In crossings of certain plants that have white and red flowers, such as *Mirabilis jalapa,* it is possible to find in F₂ three phenotypes (red, pink, and white flowers), which correspond to the three genotypes. This is due to incomplete dominance. The rule of dominance and recessiveness is not always accomplished completely; dominance may be complete in most cases, but incomplete in others. In this case there is a mixture of characteristics, called *intermediary heredity.*

Independent Assortment — Genes in Different Chromosomes are Distributed Independently During Meiosis

Whereas the law of segregation applies to the behavior of a single pair of genes, the *law of independent assortment* describes the simultaneous behavior of two or more pairs of genes located in different pairs of chromosomes. Genes that lie in separate chromosomes are independently distributed during meiosis. The resulting offspring is a hybrid (also called a dihybrid) at two loci.

Figure 19–2 diagrams the cross between a black, short-haired guinea pig (*BBSS*) and a brown, long-haired guinea pig (*bbss*). The BBSS individual produces only *BS* gametes; the bbss guinea pig produces only *bs* gametes. At F₁ the offspring are heterozygous for hair color and hair length. Phenotypically they are all black and short-haired. However, when two of the F₁ dihybrids are mated, each produces four types of gametes (*BS, Bs, bS, bs*), which by fertilization result in 16 zygotic combinations. As shown in F₂ there are nine black, short-haired individuals; three black, long-haired; three brown, short-haired, and only one brown, long-haired individual. This phenotypic proportion (9:3:3:1) is characteristic of the second generation of a cross between two allelic pairs of genes.

Genes are Linked When They are on the Same Chromosome

All the above-mentioned examples of genetic crosses illustrated the fact that during meiosis there is a random distribution of the chromosomes which leads to the segregation of the genes in the gametes (see the previous chapter). However, when this type of study was carried out in the fruit fly *Drosophila melanogaster* by Morgan and collaborators (1910 to 1915), it became evident that the law of independent assortment was not universally applicable and that in certain crosses of two or more allelic pairs of genes, there was a certain limitation of free segregation. In each case there was a marked tendency for parental combinations to remain linked and for a lesser proportion of new combinations to be produced. In *Drosophila* there are only four pairs of chromosomes, and this increases the chances for genes to occupy loci in the same chromosome.

If two genes A and B, with the corresponding alleles a and b, are in the same chromosome, only two classes of gametes will be obtained,

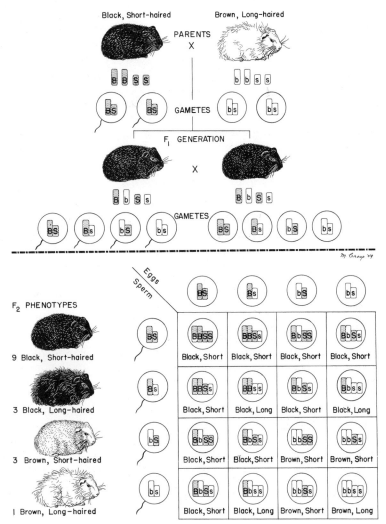

Black, Short-haired   Brown, Long-haired

PARENTS
X

GAMETES

F₁ GENERATION

X

GAMETES

m. Croup '49

F₂ PHENOTYPES

9 Black, Short-haired

3 Black, Long-haired

3 Brown, Short-haired

1 Brown, Long-haired

Figure 19-2 Diagram of a cross between black short-haired (dominant) and brown long-haired (recessive) guinea pigs. The independent assortment of genes is evident. (See the description in the text.) (From Villee, C. A., *Biology*, 6th ed., Philadelphia, W. B. Saunders Co., 1972.)

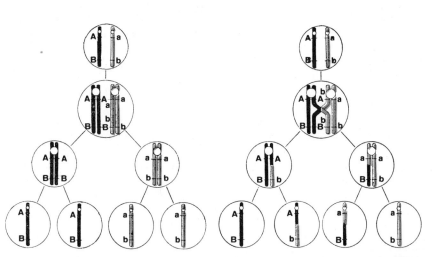

Figure 19-3 **Left,** diagram of the segregation of two pairs of allelic genes localized on the same pair of chromosomes without crossing over. The result is two types of gametes. AB and ab. A case of linkage. **Right,** diagram of the segregation of two pairs of allelic genes on the same chromosome betweeen which crossing over takes place during meiosis. Four types of gametes result: AB, ab, Ab, AB. A case of linkage with crossing over.

423

either AB or ab. Figure 19–3, *left* illustrates the mechanism of meiosis and the formation of the gametes in this hybrid. The coexistence of two or more genes in the same chromosome is called *linkage*.

After studying a considerable number of different crosses in *Drosophila*, Morgan reached the conclusion that all genes of this fly were clustered into four linked groups corresponding to the four pairs of chromosomes.

Linkage may be Broken by Recombination During Meiotic Prophase

Further studies along these lines revealed that the linkage is not absolute and that it may be broken with a certain frequency. This is explained in the diagram of Figure 19–3, *right*, in which there is a *recombination* between the genes A and B (i.e., *crossing over*) at meiosis. In this case four kinds of gametes are formed, in two of which the two genes are recombined. It is evident that in this example the linkage is broken in a certain proportion of the gametes by the interchange of segments of the homologous chromatids (see Fig. 18–8).

We have seen in the previous chapter that *genetic crossing over* or *recombination* takes place between the DNA molecules and is not visible microscopically. However, the *chiasmata* observed at diplonema morphologically represent in some way the sites of these molecular events. Thus, there should be a correlation between the two phenomena. It is also known that the frequency of recombination of two linked genes is a function of the distance which separates them along the chromosome. When two genes are close to one another, the probability of crossing over is less than when they are far apart. If the distance between genes is estimated by linkage analysis, it is possible to construct a map indicating the relative position of each gene along the chromosome.

An appraisal of the possible number of new recombinations may be obtained by counting the number of chiasmata during meiosis. The so-called *recombination index* is calculated by adding to the number of bivalents the number of chiasmata detected in the same cell at diplonema. In a species with a higher index, the possibility of new recombinations is higher, and this implies a greater possibility of variation.

There are several other pieces of evidence to support the relationship between chiasmata and crossing over. The number of chiasmata is related to the length of the chromosomes in the bivalent. The presence of one chiasma reduces the possibility of another occurring in the immediate vicinity. This phenomenon has been called *positive interference.* The chiasma frequency is roughly constant in a given species, but can be modified by genetic or environmental action. Chromosome pairing, chiasma formation, and crossing over are under genetic control. Lines of rye with high and low chiasma frequencies have been obtained by inbreeding. Among environmental factors that may affect the number of chiasmata are temperature, radiation, chemicals, and nutrition (see Henderson, 1969).

Neurospora — Ideal for Studying Recombination and Gene Expression

Among the different organisms studied in genetics, the mold *Neurospora* occupies a special place. The advantage of this material is twofold: (1) It is possible to identify and to follow the fate of each of the four chromatids present in the bivalent meiotic chromosome and thus to determine whether the crossing over involves two, three, or all four chromatids. (2) It is possible to make a close correlation between genetic constitution and biochemical expression of genes.

As shown in Figure 19–4, the four cells resulting from the two meiotic divisions undergo a mitotic division, which gives rise to eight haploid ascospores. Each of these ascospores can be isolated by dissection and cultured separately, giving rise to haploid individuals having the genetic constitution carried in each of the four original chromatids of the bivalent chromosome.

Figure 19–4 indicates a single crossing over between genes *a* and *b* and the resulting products. Analysis of the eight ascospores shows that only two of the chromatids interchange segments while the other two remain intact.

SUMMARY:

Fundamental Genetics

Mendel (1865) discovered the laws of heredity by studying crosses between peas having pairs of *contrasting* characteristics (i.e., allelic). In a cross between parents having yellow and green seeds he found that in F_1 all the *hybrids* had yellow seeds

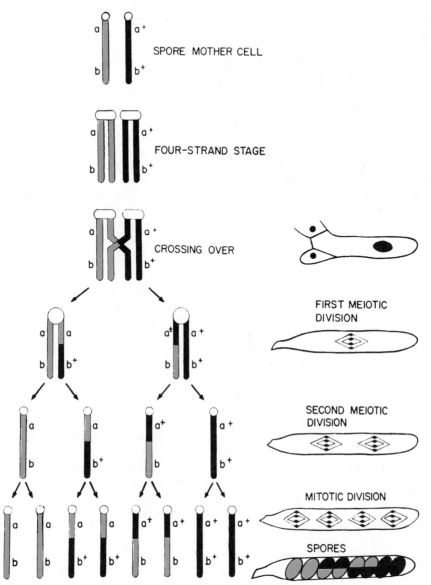

Figure 19-4 Diagram of the formation of ascospores in *Neurospora crassa*. A single crossing over between genes a and b, the behavior of one pair of chromosomes during the first and second meiotic divisions, and the division by mitosis of each of the four products are shown. The presence of a single chromatid in each spore is indicated diagrammatically.

(dominant gene). In the F_2 generation 75 per cent of the plants had yellow seeds and 25 per cent had green seeds (recessive gene). Mendel postulated that the genes are transmitted without mixing (i.e., via the *law of segregation*). He demonstrated that in F_2 there are 1 dominant homozygous, 2 heterozygous, and 1 recessive homozygous offspring. The *genotypic* segregation is 1:2:1 in spite of the fact that the *phenotypic* proportion is 3:1. The law of segregation can be explained in terms of the behavior of chromosomes during meiosis (see Chapter 18). At times there is incomplete dominance (i.e., *intermediary heredity*).

The behavior of two or more pairs of allelic genes follows the *law of independent assortment*. Genes that lie in different chromosomes are independently distributed during meiosis. For two pairs of alleles the phenotypic proportion is 9:3:3:1 (Fig. 19-2).

Studies by Morgan and collaborators (1910 to 1915) demonstrated that the law of independent assortment may be limited. In *Drosophila* it was found that all genes were clustered into four linked groups corresponding to the four chromosome pairs. *Linkage* is not absolute and may be broken by *recombination* (Fig. 19–3) during the meiotic prophase. There is a correspondence between the number of chiasmata at diplonema and that of recombinations (*crossing over*) taking place at the molecular level. The distance between genes in a chromosome may be measured in *units of recombination,* and *genetic maps* that represent the relative order of genes along the chromosomes may be constructed. The number of chiasmata (and recombinations) is related to the length of the chromosome. There is a *positive interference* that reduces the possibility of one chiasma occurring near another. In a species, chromosome pairing, chiasmata, and crossing over are rather constant and are under genetic control. The mold *Neurospora* is ideal for the study of recombination and of its relationship to the biochemical expression of genes. After the two meiotic divisions, there is a mitotic one, and eight ascospores are formed. Each one is haploid and contains a single chromatid in which the genetic recombination may be studied by separating and culturing each of the ascospores.

19–2 CHROMOSOMAL CHANGES AND CYTOGENETICS

The normal functioning of the genetic system is maintained by the constancy of the hereditary material carried in the chromosomes. We mentioned in Chapter 2 that each species has a diploid number of chromosomes which are in homologous pairs, and in Chapter 20 the normal human karyotype will be studied in detail. Under various conditions there may be changes in the karyotype, with different genetic consequences. Chromosomes may *change in number,* maintaining a normal structure, or they may have *structural alterations,* frequently called *chromosomal aberrations.* Such structural changes may appear spontaneously, or they may be caused by the action of ionizing radiations or by chemicals. Sometimes these changes may have a genetic component. For

example, in the hereditary disease *xeroderma pigmentosum,* chromosomes are extremely sensitive to ultraviolet radiation.

Euploidy — A Change in the Number of Chromosome Sets; Aneuploidy — A Loss or Gain of Chromosomes

Table 19–1 shows that there are two main kinds of change in the number of chromosomes. In *euploids,* the change is in the haploid number, but the entire set of chromosomes is kept balanced, whereas in *aneuploids* there is a loss or gain of one or more chromosomes, causing the set to become unbalanced. Chapter 20 will show that in humans aneuploidy may cause severe alterations of the phenotype.

Some exceptional plants and animals have a *monoploid* (or *haploid*) chromosome set. In these

TABLE 19–1 CHROMOSOME COMPLEMENTS IN EUPLOIDS AND ANEUPLOIDS

Type	Formula	Complement*
EUPLOIDS		
Monoploid	n	(ABCD)
Diploid	2n	(ABCD) (ABCD)
Triploid	3n	(ABCD) (ABCD) (ABCD)
Tetraploid	4n	(ABCD) (ABCD) (ABCD) (ABCD)
Autotetraploid	4n	(ABCD) (ABCD) (ABCD) (ABCD)
Allotetraploid	4n	(ABCD) (ABCD) (A'B'C'D') (A'B'C'D)
ANEUPLOIDS		
Monosomic	2n + 1	(ABCD) (ABC)
Trisomic	2n + 1	(ABCD) (ABCD) (B)
Tetrasomic	2n + 2	(ABCD) (ABCD) (B) (B)
Double trisomic	2n + 1 + 1	(ABCD) (ABCD) (AC)
Nullisomic	2n − 2	(ABC) (ABC)

*A, B, C, D are nonhomologous chromosomes.

organisms meiosis is irregular because of the absence of homologous chromosomes. As a result, gametes with varying numbers of chromosomes may be formed.

Polyploidy. A plant or animal that has more than two haploid sets of chromosomes is called a *polyploid* (Fig. 19–5). This change is common in nature, especially in the flowering plants. A diploid organism has two similar genomes; a triploid has three; an autotetraploid has four; and so on. Polyploids may originate either by reduplication of the chromosome number in somatic tissue with suppression of cytokinesis or by formation of gametes with an unreduced number of chromosomes.

Meiosis in a triploid is more irregular than in a tetraploid. In general, polyploids of uneven number are sterile because the gametes have a more unbalanced number of chromosomes.

The scarcity of polyploids among animals is due to the mechanism by which sex is determined. If polyploidy occurs, the genic balance between the sex chromosomes and the autosomes is disturbed, and the race or species may disappear because of sterility.

Several species of amphibians show spontaneous polyploidy. In such polyploids the DNA content is correspondingly increased.[1] Polyploids are useful in the study of the expression of genes in multiple dosage. Studies of this type have been made on amphibians for serum albumin, hemoglobin, and various enzymes.[2]

Polyploidy has been induced experimentally by temperature shock. It is also possible to induce polyploidy with substances such as colchicine, acenaphthene, heteroauxin, and veratrine.

These substances inhibit the formation of the spindle, and thus cell division is not completed. After a time the cells recover their normal activity, but have double the number of chromosomes. From the standpoint of pure and applied scientific work innumerable possibilities are offered by the experimental production of polypoids.

Allopolyploidy. This is a type of chromosomal variation that is produced in crosses between two species having different sets of chromosomes. The resulting hybrid has a different number of chromosomes than the parents. For example, the Argentine black *Sorghum* (*S. almum*) is an allotetraploid (2n = 4x = 40) originated in nature by an interspecific cross between *S. halepense* (2n = 4x = 40) and *S. sudanense* (2n = 2x = 20) (Fig. 19–6). In this case the fertilization occurred between one abnormal diploid gamete of *S. sudanense* and a normal gamete of the other species. In most cases crosses between distantly related species produce sterile diploid hybrids.

Aneuploidy — The Result of Non-disjunction at Meiosis or Mitosis

Aneuploidy is produced by a failure in the separation of chromosomes during meiosis, called *non-disjunction*. *Monosomic* individuals have lost one of the chromosomes of the karyotype, and in *trisomics* there is a gain of one chromosome. These changes are studied specifically in the following chapter on the human karyotype.

Aneuploidy may also occur in somatic cells in which the *non-disjunction* occurs at mitosis. This process may produce individual cells in various tissues or parts of the body that have different chromosome numbers, e.g., in mosaic individuals, variegations, and gynandromorphs.

In cultures of normal cells, chromosomal variations are common, particularly after several transplants. These changes may lead to malignancy, but this is not the case in all cultures. The cytogenetic analysis of mammalian tumors has led investigators to consider them as altered karyotypes, which are genetically and cytologically unstable.

Figure 19–5 Polyploid series in the plant *Crepis*. (After Nawashin.)

427

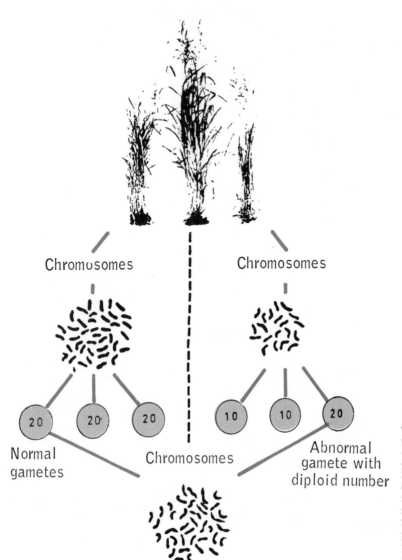

Chromosomes

Chromosomes

Normal gametes

Chromosomes

Abnormal gamete with diploid number

Figure 19–6 The origin of the Argentine *Sorghum almum* by crossing *Sorghum halepense* (2n = 4x = 40) with a diploid *Sorghum* (2n = 2n = 20) in which the fertilization occurred between one gamete not reduced (x = 20) and a normal gamete of *S. halepense*. Somatic chromosomes of the parents and the hybrid allo-polyploid are illustrated. (After Saez and Nuñez, 1949.)

Chromosomal Aberrations — Structural Alterations; Gene Mutations — Molecular Changes

To understand the mechanism by which chromosomal aberrations are produced the student should recall the folded fiber model of chromosome structure (see Chapter 15) and the concept of *mononemic* structure of chromatids. According to this concept, each chromatid is made of one molecule of DNA; thus, in the G_1 period there is a single DNA duplex per chromatid. It is only after DNA duplication that each chromosome contains two chromatids and two DNA molecules. In a chromosomal aberration there is an alteration in the structural or-ganization of the chromosome. For example, a break in the unit fiber may lead to a loss of material, called a *deletion*, and in cases in which there are two breaks, a *translocation*, an *inversion*, or an *exchange between sister chromatids* may be produced. In all these cases the chromosome change can be observed microscopically, especially by observing the chromosomes at the stages of condensation.

Chromosomal aberrations should be differentiated from *gene mutations* (also called *point mutations*), in which the changes are produced at the molecular level. In these, the alteration resides in the genetic code (i.e., in the sequence of DNA bases) and can be detected only by its genetic or molecular expression (see Chapter 21–2).

Radiation and Chemical Mutagens Act Mainly on the DNA Molecule

Both gene mutations and chromosomal aberrations may occur spontaneously, but their frequency increases by the action of so-called ionizing radiation (i.e., x-rays, β-rays, γ-rays and ultraviolet light, as well as accelerated particles, such as fast neutrons, protons, and so forth). Other *mutagenic agents* are chemical substances, viruses, mycoplasms, and temperature changes.

In various organisms it has been demonstrated that the number of mutations induced by radiation is proportional to the dose. In the case of *Drosophila*, the relation is a linear function in the range of 25 to 9000 roentgen units (R).*

The effects of radiation are cumulative over long periods of time. For example, 0.1 R per day for 10 years is enough to increase the mutation rate to about 150 per cent of the spontaneous level.

In contrast to gene mutations, chromosomal aberrations increase exponentially. More chromosomal aberrations are produced by continuous than by intermittent treatment.

A low dose of radiation may not be enough to cause fractures in one chromosome. As the dose increases, the number of breaks increases, and "aberrant" fusions become more and more likely. If the dose is intermittent or of low intensity, there is a greater chance that the broken ends of the chromosome will rejoin, or "heal," in the original chromosome structure before a second break could cause aberration.

---

*A roentgen (R) is the amount of radiation sufficient to produce 2×10^9 ion pairs per cm.[3]

The type of chromosomal aberration depends on the period of the cell cycle the cell was in at the time of irradiation. A *chromatid break* occurs in the G_2 period, and a *chromosomal break* (i.e., two chromatids) occurs if the cell was in the G_1 period, which is before DNA duplication. Irradiation generally produces localized lesions which may stabilize and form the so-called gaps (Fig. 19–7). After some time, such gaps may be repaired.[3] Such repair may result in complete restitution of the original structure (true repair). If there is no repair, the lesion becomes stabilized and cytologically visible. The process of restitution is inhibited by cold, cyanide, and dinitrophenol. Oxygen increases the number of fractures and chromosomal interchanges.

When a chromosome breaks, the two fragments may either reunite or remain separated permanently. If, instead of reconstituting at once, the broken ends reduplicate to form two chromatids, the fragments with the centromere migrate to different poles, and the fragments without a centromere (acentric) are eliminated in the cytoplasm (Fig. 19–8,*A*). Sometimes the sister fragments reduplicate and unite, forming a dicentric chromosome and another acentric chromosome. During mitosis, the centromeres of the dicentric chromosome move to opposite poles and form a "bridge" between the daughter nuclei that finally breaks at some point (Fig. 19–8,*B*).

In man, double or multiple chromosomal breaks are induced by acute exposures (e.g., heavy medical irradiations, nuclear accidents, or nuclear warfare), whereas single breaks are produced at low doses. Chromosomal aberrations have been observed in blood cultures of humans who have had radiation treatments or injections of radioactive substances. The genetic effect of radiation has been studied in space flights. After Geminis III and IV, no increase in chromosomal

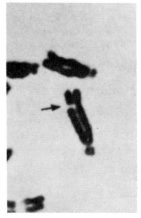

Figure 19–7   Chromatid gaps and breaks (arrows) induced by radiation. (From H. Evans.)

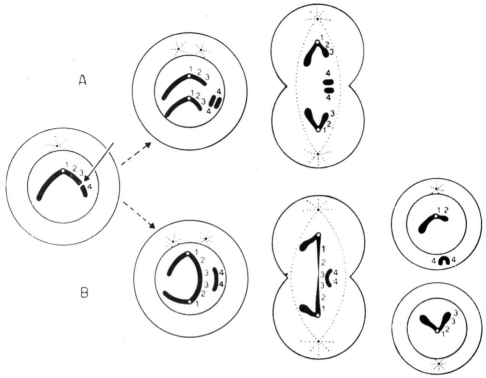

Figure 19–8 A single break in a chromosome between loci 3 and 4. **A,** the two parts of the broken chromosome reduplicate. The fragments with the centromere go to opposite poles. The fragments without the centromere remain in the equatorial region and are eliminated. **B,** the two parts of the broken chromosome reduplicate and the broken ends unite, forming a chromosome with two centromeres and another chromosome without a centromere. During mitosis the two centromeres move to opposite poles and the chromosome section between them breaks. The two daughter cells receive chromosomes with different constitutions. The fragment without a centromere is eliminated in the cytoplasm. (From Stern, C., *Principles of Human Genetics,* 3rd ed., San Francisco, W. H. Freeman & Co., 1973.)

aberrations was found in the blood cells of the astronauts who made the flights[4] (see Brinkley and Hittleman, 1975).

Germ Cell Mutations are Transmittable, Somatic are not

Mutations in *somatic cells* are not transmitted from generation to generation but may be cumulative and produce severe changes in the individual, depending on the type of cell affected and the time at which the mutation occurs. Radiation can affect tissues that undergo mitosis, as well as tissues in which cell division no longer takes place. If mutation occurs during early embryonic development, a large number of cells are affected. The majority of mutated genes are recessive and thus have no effect as long as the individual is heterozygous.

It is probable that some cases of cancer produced in irradiated individuals are caused by somatic mutation.

In contrast to somatic mutation, mutations in the germ cells may be transmitted to the offspring. However, in most cases of mutation caused by irradiation, both somatic and germ cells are frequently affected.

Even the lowest doses of radiation are genetically harmful, and the effects are dangerous to all organisms, from the simplest to man. In microorganisms and plants, however, a few useful mutations may be obtained by irradiation. For example, irradiation is one method of producing new antibiotics and plants with high economic value.

Specific Chromosomal Aberrations — Deletion, Duplication, Translocations, Inversion, and Sister Chromatid Exchange

Deficiency or Deletion. As shown in Figure 19–9, a deletion or loss of chromosomal material may be either *terminal* or *interstitial*, originating from one or two breaks, respectively, produced

430

in the G_1 period. Some interstitial deletions may originate both a ring chromosome and an acentric rod, which is eliminated. Figure 19–8 shows the above-mentioned example of a terminal deletion with formation of a dicentric chromosome.

In heterozygous deficiency, one chromosome is normal, but its homologue is deficient. Animals with a homozygous deficiency usually do not survive to an adult stage because a complete set of genes is lacking. This suggests that most genes are indispensable, at least in a single dose, to the development of a viable organism. Deficiencies are important in cytogenetic investigations of gene location for determination of the presence and position of unmated genes.

Duplication. Duplication occurs when a segment of the chromosome is represented two or more times. The duplicated fragment may be free, with a centromere of its own, or it may be incorporated into a chromosomal segment of the normal complement. If the fragment includes the centromere, it may be incorporated as a small extra chromosome. In general, duplications are less deleterious to the individual than deficiencies.

Translocation. In the so-called *reciprocal translocation*, segments are exchanged between non-homologous chromosomes. Such transloca-

tion may be homozygotic or heterozygotic (Fig. 19–10).

Cytologically, a translocated homozygote cannot be distinguished from a normal pair of chromosomes, but it can be detected by genetic experiments. Heterozygotic translocations give rise to special pairing configurations in meiosis. An interesting result occurs when during translocation, both chromosomes are broken very close to their centromeres. The fusion creates a metacentric chromosome with two arms in the form of a V, and a small fragment, which tends to be eliminated.

Centric fusion has occurred during the phylogeny of *Drosophila*, grasshoppers, reptiles, birds, mammals, and other groups. It is a process that establishes a new type of chromosome and reduces the somatic chromosome number of the species (Fig. 19–11). In the following chapter we will mention the importance of reciprocal translocations in human cytogenetics.

Inversion. An inversion is a chromosomal aberration in which a segment is inverted 180 degrees. Inversions are called pericentric when the segment includes the centromere, and paracentric if the centromere is located outside the segment. In these aberrations there is a typical configuration at pachynema consisting of a loop

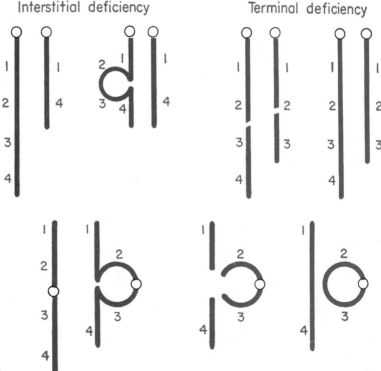

Interstitial deficiency   Terminal deficiency

Figure 19–9 Diagram illustrating the origin of various types of deficiencies.

Formation of an acentric rod and deletion ring

NORMAL | HOMOZYGOTIC TRANSLOCATION | HETEROZYGOTIC TRANSLOCATION

Figure 19-10 Schematic representation of homozygotic and heterozygotic reciprocal translocations compared with the normal arrangement.

that allows the pairing of the inverted segment. In the paracentric inversion acentric and dicentric chromatids are formed.

Isochromosomes. A new type of chromosome may arise from a break (i.e., a misdivision) at the centromere. As shown in Figure 19–12 the two resultant telocentric chromosomes may open up to produce chromosomes with two identical arms (i.e., isochromosomes). This type of chromosome has been produced in irradiated material. At meiosis they may pair with themselves or with a normal homologue.

Sister Chromatid Exchanges Increase in Diseases with Altered DNA Repair

A *sister chromatid exchange* is an interchange of DNA between sister chromatids in a chromosome, presumably involving DNA breakage followed by fusion. Sister chromatid exchanges are difficult to find using common cytologic methods because the chromatids are morphologically identical. Such chromatid exchanges were first described in studies in which ³H-thymidine was added during a replicating cycle which was followed by another cycle in a non-radioactive medium. As was mentioned in Chapter 16, this method demonstrated that the duplication of DNA is semiconservative. However, in a few mitoses observed by radioautography, an alter-

ation of the labeling along the chromatids was found, and this was interpreted as the result of chromatid exchange caused by the radioactivity of the tritiated label used.

Analysis of this phenomenon has been greatly facilitated by the use of *bromodeoxyuridine* (BrdU), a thymidine analogue that can be incorporated into the DNA of replicating cells instead of the original base. If BrdU is followed by a fluorescent dye (Hoechst 33528), the fluorescence of the segments that contain BrdU is greatly diminished in comparison with those with the original base.[5] Furthermore, there is also a similarly decreased staining with the Giemsa stain.

The use of this technique, however, has been unable to discover whether the chromatid exchange could occur spontaneously or whether it is induced by the BrdU. It has, though, been of great help in differentiating the various inherited diseases characterized by chromosome fragility, which have an increased frequency of sister chromatid exchanges and a tendency to have associated neoplasia. Some of these diseases (e.g., Bloom's syndrome, Fanconi's anemia, and ataxia-telangiectasia) are presumably related to defects in DNA repair.

Sister chromatid exchange has also been important in studying the effect of mutagens on the chromosomes. Various mutagenic drugs that are alkylating agents, such as *mitomycin C* and *nitrogen mustard*, produce a great number of

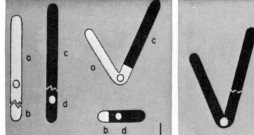
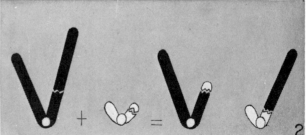

Figure 19–11 **1,** the origin of a new V-shaped (metacentric) chromosome by *centric fusion* of two non-homologous acrocentric chromosomes. Segment *bd* is lost. **2,** *dissociation.* A metacentric and a small, supernumerary chromosomal fragment undergo a translocation, which results in two chromosomes (acrocentrics or metacentrics).

breaks and chromatid exchanges (Fig. 19–13).[6] The intimate association of sister chromatid exchange with mutagenesis and carcinogenesis may have important medical implications (see Latt, 1978; Latt et al., 1977; Wolff, 1977).

19–3 CHROMOSOMES PLAY A FUNDAMENTAL ROLE IN EVOLUTION

The development of comparative cytology and cytogenetics has brought about great progress toward an understanding of evolution. McClung and S. Navashin were the first to emphasize the importance of cytogenetics to taxonomy and to the study of evolution by comparing genomes of related species. Systematics has been greatly advanced by cytogenetic investigation, which now provides many of the best methods for elucidating correlations between different taxonomic categories. In general, families, genera, and species are characterized by different genetic systems.

The study of the karyotypes of different species has revealed interesting facts about both the plant and animal kingdoms. It has been demonstrated that individuals in wild populations are, to some extent, heterogeneous cytologically and genetically. In some cases, even if the genes are identical they may be ordered in a different way, owing to alterations of the chromosomal segments. These changes have an important bearing on the evolution of species.

The majority of plant species originate from an abrupt and rapid change in nature, and aneuploidy and polyploidy are the prime sources of variation. In the animal kingdom polyploidy is not so important. Among vertebrates, different species of fishes have a different number of chromosomes. Amphibians are generally characterized by a specific number for each family. Reptiles and birds have large chromosomes (macrochromosomes) and small chromosomes (microchromosomes) that serve to differentiate them cytologically.

Matthey distinguishes between the basic chromosome number and the number of chromosomal arms, also called the fundamental number (FN). According to this concept, the metacentric chromosome has *two* arms and acrocentric and telocentric chromosomes have *one*. This is an important distinction in a group having both acrocentric and metacentric chromosomes, and the number of arms in each of the different species can be compared.

Another method used to study the cytogenetics of evolution is the application of measurements of total chromosomal area and DNA content.

With regard to the absolute size of their chromosomes, mammals and birds constitute two independent groups. These two orders have different DNA contents and different sex-determining mechanisms. Speciation depends more on chromosomal rearrangements and mutation of individual genes than on changes in the total amount of genetic content.

Two opposite changes in the number (and configuration) of chromosomes are of particular importance in evolution. In *centric fusion*, a process that leads to a decrease in chromosome number, two acrocentric chromosomes join together to produce a metacentric chromosome. In *dissociation*, or *fission*, a process that leads to an increase in chromosome number, a metacentric (commonly large) and a small supernumerary metacentric

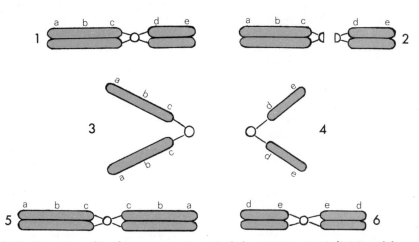

Figure 19–12 Formation of isochromosomes. **1**, original chromosome; **2**, misdivision of the centromere at the beginning of mitotic anaphase; **3** and **4**, the chromatids unfold into two isochromosomes; **5** and **6**, in the next division, two complete isochromosomes are present. Note that in each isochromosome, the arms exhibit genetic constitution.

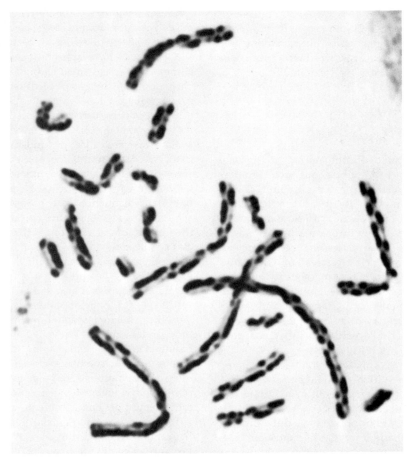

Figure 19–13 Cultured cell exposed to the mutagenic chemical agent 8-methoxypsoralen and to bromodeoxy-uridine. The regions stained only lightly with Giemsa and those in which the DNA was synthesized in the presence of BrdU. Observe the numerous sister chromatid exchanges in each chromosome. (Courtesy of S. Latt.)

fragment become translocated, so that two acrocentric or submetacentric chromosomes are produced.

Fusion and dissociation are the main mechanisms by which the chromosome number can be decreased and increased during evolution of the majority of animals and in some groups of plants (Fig. 19–11).

In recent years the karyotypes of the great apes, chimpanzee, gorilla, and orangutan, have been compared with that of man. It was found that these primates have 48 chromosomes, and attempts were made to correlate the chromosome pairs with those of the 46 human chromosomes. This analysis has been greatly facilitated by the use of banding techniques (see Chapter 20). Using the quinacrine technique it was demonstrated that human chromosomes 1, 3, 5, 6, 7, 8 10, 11, 13, 14, 15, 19, 20, 21, 22, X and Y had direct counterparts in the chimpanzee, but had less similarity with the pairs of the gorilla and orangutan (see Pearson, 1977). The difference between

the 48 pair karyotype and the 46 is due to the fusion of two acrocentric chromosomes, present in the hominoid apes, to make the human chromosome 2[7, 8] (see Dutrillaux, 1975 and 1977).

At present, pericentric inversion appears to be the main structural difference between individual chromosomes of the great apes and man.[9] It has been postulated that evolution of the human karyotype has occurred by a series of pericentric inversions which permitted genetic isolation of small breeding groups and selection of favorable gene combinations that gave rise to *Homo sapiens*. The homologies in morphology of the chromosomes are now being correlated with the mapping of genes in similar chromosomes of man and the great apes (see Pearson, 1977; Jones, 1977).

Studies of somatic and polytene chromosomes in several hundred species of *Drosophila* have elucidated the formation and evolution of this genus, which has been thoroughly analyzed from genetic, ecologic, and geographic standpoints.

Observation of chromosomal organization and of the different karyotypes in the individual, the species, genera, and the major systematic groups indicates that a chromosomal mechanism is involved in the process of evolution.

The problem of evolution, however, is very complex and should be considered from the different biochemical, cytologic, genetic, ecologic, and experimental aspects. All these methods and approaches should be used to analyze the intricate relationships between groups of organisms, particularly those that show marked variations. (These considerations are beyond the scope of this book.)

SUMMARY:

Chromosomal Aberrations, Action of Mutagens, and Cytogenetics and Evolution

Changes in the number and structure of chromosomes may occur spontaneously or experimentally by the action of radiation or chemicals. The number of chromosomes is generally constant for plant and animal species. *Chromosomal changes in number* are of two main kinds: in *euploids* the set is kept balanced; in *aneuploids* there is a loss or gain of one or more chromosomes. In exceptional cases there are *haploid* organisms. In plants, *polyploids* (triploid, tetraploid, etc.) are rather common. They originate by reduplication without cytokinesis. In animals, polyploids are scarce, because sex is frequently determined by a pair of different chromosomes (XY). Polyploidy can be induced by colchicine; this substance is used in agriculture to improve certain plant species. *Allopolyploidy* consists of the formation of a hybrid with different sets of chromosomes. Sometimes a diploid gamete fertilizes a normal haploid gamete, producing a new species with a triploid number.

Among the aneuploid organisms there are the *trisomic* (i.e., three similar chromosomes) and the *monosomic* (only one of the pair of chromosomes). These two conditions are important in the human (see Chapter 19) and may arise by non-disjunctional division.

Chromosomal aberrations, also called *structural alterations,* involve a change in the molecular organization of the chromosome. One or two breaks are produced at the level of the chromatin fiber, and the effect depends on whether it occurs in the G_1 period (one chromatid) or in the G_2 (two chromatids). Chromosomal aberrations should be differentiated from *gene* or *point mutations,* which occur at a molecular level.

Ionizing radiation — x-rays, γ-rays, β-rays, fast neutrons, slow neutrons, and ultraviolet light — can produce point mutations or chromosomal aberrations. The number of mutations increases proportionally with the dose of x-radiation. The effect of radiation is cumulative. An exposure of cultured cells to 20 R (roentgen units) is sufficient to produce one chromosome break per cell. In contrast to mutations, which increase proportionally with the dose, chromosomal aberrations increase exponentially with the dose. Breakage can be followed by "healing" of the broken end (see DNA repair in Chapter 16–2). (The breaks may be at the chromosome or the chromatid level.) Chromatid breaks are produced in cells irradiated after the S period. If two breaks are produced, translocations, inversions, and large deletions may be induced. Dicentric chromosomes may be produced, which form a bridge at anaphase. In the human, heavy medical irradiation, nuclear accidents, or radioactive substances may produce chromosomal aberrations. In the production of new antibiotics and plants, irradiation is being used to economic advantage. *Somatic mutations* are not transmitted from one generation to another, whereas germ cell mutations may be passed to the offspring.

Some of the main aberrations are: (1) *Deficiency* or *deletion,* in which a part (either *interstitial* or *terminal*) of the chromosome is missing. The parts of the chromosome lacking the centromere generally are lost. Deficiency may be *heterozygous* (one chromosome is normal) or *homozygous* (both chromosomes are deficient). The latter generally do not survive. (2) *Duplication,* in which a chromosome segment is represented two or more times (tandem duplication). (3) *Translocation,* in which there is an exchange of segments between nonhomologous chromosomes (*reciprocal* type) or between different parts of the same chromosome (simple type). Sometimes *centric* fusion occurs when the two chromosomes are broken near the centromere and form a metacentric (**V**-shaped) chromosome. (4) *Inversion,* in which there is breakage of a segment, followed by its fusion in a reverse position. It is *pericentric* if it includes the centromere, and *paracentric* if the centromere is outside. *Isochromosomes* may arise from a break at the centromere, resulting in two chromosomes with identical arms.

Sister chromatid exchange consists of the interchange of DNA segments between sister chromatids in a chromosome. First described in studies using ^{3}H-thymidine, it was interpreted as the result of radiation from ^{3}H. This aberration can be studied with bromodeoxyuridine (BrdU) which is incorporated instead of thymidine and which produces changes in fluorescence (by Hoechst 33528) or in the Giemsa stain. This technique permits the detection of diseases characterized by chromosome fragility and enables the study of the action of mutagenic drugs which increase the number of breaks.

Cytogenetic studies have provided excellent methods for establishing taxonomic interrelationships and have, thereby, contributed to studies of *evolution* and *systematics.* One of the most frequent causes of evolution is a change in the order of genes as a result of chromosomal aberrations. In plants, aneuploidy and polyploidy are frequent sources of variation, whereas in mammals and birds, speciation depends more on chromosomal rearrangement and point mutations.

In the study of evolution, the number of chromosomes, the characteristics of the karyotype, the total chromosomal area, and the content of DNA are investigated. The presence of metacentric chromosomes may, in some cases, result from fusion of two acrocentric chromosomes. The contrasting phenomenon (i.e., *dissociation*) may lead to an increase in chromosome number. The problem of *evolution* is very complex and beyond the scope of this book; however, knowledge of cytogenetics is most fundamental to its understanding. One example of the importance of cytogenetics in evolution is given in a comparison of the chromosome pairs of the primates and man.

REFERENCES

1. Beçak, W. (1969) *Genetics, 61*:183.
2. Beçak, W., Beçak, M. L., and Rebello, M. N. (1967) *Chromosoma, 22*:192.
3. Evans, H. J. (1967) *Radiation Research.* North-Holland Publishing Company, Amsterdam.
4. Bender, M. A., Gooch, P. C., and Kondo, S. (1968) *Radiat. Res., 34*:228.
5. Latt, S. A. (1973) *Proc. Natl. Acad. Sci. USA, 70*:3395.
6. Lin, M. S., and Alfi, O. S. (1976) *Chromosoma, 57*:219.
7. de Grouchy, J., Turleau, C., Roubin, M., and Chavin-Colin, F. (1973) In: *Chromosome Identification,* p. 127. (Caspersson, T., and Zech, L., eds.) Academic Press, Inc., New York.
8. Lejeune, J., Dutrillaux, B., Rethoré, M., and Prieur, M. (1973) *Chromosoma, 43*:423.
9. Turleau, C., and de Grouchy, J. (1973) *Humangenetik, 20*:151.

ADDITIONAL READING

Ayala, F. J. (1977) Variation genetique et evolution. *La Recherche, 8*:736.

Brinkley, B. R., and Hittelman, W. N. (1975) Ultrastructure of mammalian chromosome aberrations. *Int. Rev. Cytol., 42*:49.

Darlington, G. D. (1978) A diagram of evolution. *Nature, 276*:447.

de Grouchy, J., Turleau, C., and Finaz, C. (1978) Chromosomal phylogeny in the primates. *Annu. Rev. Genet., 12*:289.

Dobzhansky, T. (1958) *Genetics and the Origin of Species,* 3rd Ed. Columbia University Press, New York.

Du Praw, E. J. (1968) *Cell and Molecular Biology.* Academic Press, New York.

Dutrillaux, B. (1975) Sur la nature et l'origine des chromosomes humains. *Monogr. Ann. Genet.,* p. 102. Expansion scientifique française, Paris.

Dutrillaux, B. (1977) New chromosome techniques. In: *Molecular Structure of Human Chromosomes*, p. 233. (Yunis, J. J., ed.) Academic Press, Inc., New York.

Dutrillaux, B. (1978) Which primate is closest to man? *La Recherche, 9*:596.

Dutrillaux, B. (1979) Chromosomal evolution in primates. *Human Genet., 48*:251.

Evans, H. J. (1967) *Radiation Research*. North-Holland Publishing Co., Amsterdam.

Hamkalo, B. A., and Papconstantinou, J., eds. (1973) *Molecular Cytogenetics*. Plenum Press, New York.

Henderson, S. A. (1969) Chromosome pairing, chiasmata, and crossing over. In: *Handbook of Molecular Cytology*, p. 327. (Lima-de-Faría, A., ed.) North-Holland Publishing Co., Amsterdam.

Hollaender, A. (1954) *Radiation Biology*, 3 volumes. McGraw-Hill Book Co., New York

Jacobs, P. A., Price, W. H., and Lou, P. (1970) *Population Cytogenetics*. Edinburgh University Press, Edinburgh.

John, B., and Lewis, K. R. (1973) The meiotic mechanism. In: *Readings in Genetics and Evolution*, p. 2. Oxford University Press, London.

Jones, K. W. (1977) Repetitive DNA in primate evolution. In: *Molecular Structure of Human Chromosomes*, p. 295. (Yunis, J. J., ed.) Academic Press, Inc., New York.

Latt, S. A. (1976) Optical studies of metaphase chromosome organization. *Annu. Rev. Biophys. and Bioeng., 5*:1.

Latt, S. A., Allen, J. W., and Stetten, G. (1977) In vitro and in vivo analysis of chromosome structure, replication, and repair using BrdU–33258 Hoechst techniques. In: *International Cell Biology*, p. 520. Rockefeller University Press, New York.

Lewis, K. R., and John, B. (1963) *Chromosome Marker*. J & A. Churchill, Ltd., London.

Lewis, K. R., and John, B. (1964) *The Matter of Mendelian Heredity*. J. & A. Churchill, Ltd., London.

Pearson, P. L. (1977) Banding patterns, chromosome polymorphism, and primate evolution. In: *Molecular Structure of Human Chromosomes*, p. 267. (Yunis, J. J., ed.) Academic Press, Inc., New York.

Stern, C. (1973) *Principles of Human Genetics*. W. H. Freeman & Co., San Francisco.

Taylor, J. H. (1963 and 1967) *Molecular Genetics*, Parts I and II. Academic Press, Inc., New York.

Taylor, J. H. (1967) Meiosis. In: *Encyclopedia of Plant Physiology*, Vol. 18, p. 344. Springer-Verlag, Berlin.

White, M. J. D. (1973) *Animal Cytology and Evolution*, 3rd Ed. Cambridge University Press.

Wolff, S. (1977) Sister chromatid exchange. *Annu. Rev. Genet., 11*:183.

Yunis, J. J. (1977) *Molecular Structure of Human Chromosomes*. Academic Press, Inc., New York.

20

HUMAN CYTOGENETICS

20–1 The Normal Human Karyotype 439
    Karyotype Preparation – Chromosome
      Ordering by Size and Position of the
      Centromere
    Numerous Banding Techniques Reveal
      Structural Details of Chromosomes
    Banding Provides New Features for
      Identification of Human Chromosomes
    Summary: *The Human Karyotype*
20–2 Sex Chromosomes and Sex
    Determination 444
    X and Y Sex Chromatin in Interphase
      Nuclei
    X Chromatin – Facultative Heterochro-
      matin Bodies Equaling nX – 1 in Number
    Y Chromatin – The Heterochromatic
      Region of the Y Chromosome
    Sex is Determined by Sex Chromosomes,
      but Hormones Influence Differentiation
    Summary: *Sex Chromosomes and Sex*
      Determination
20–3 Human Chromosomal
    Abnormalities 448
    Aneuploidy – Caused by Non-disjunction
      of Chromosomes

    Reciprocal Translocations – Identified
      by Banding
    Sex Chromosome Abnormalities in
      Human Syndromes
    Mongolism – The Best Known
      Autosomal Abnormality
    Structural Aberrations May Cause Other
      Syndromes Besides Aneuploidy
    Banding Techniques Have Detected
      More Than 30 New Syndromes
    Certain Human Tumors Show Specific
      Chromosomal Aberrations
    Summary: *Chromosomal Abnormalities*
20–4 Human Chromosomes and the
    Genetic Map 455
    Daltonism and Hemophilia – The Best
      Known Sex-Linked Diseases
    Chromosome Mapping – Linkage Analysis,
      Somatic Cell Hybridization, and In Situ
      Hybridization
    Localization of Genes in Chromosomes
      and Regions of Chromosomes
    Summary: *The Human Genetic Map*

In the last two decades great advances have been made in the study of human chromosomes in the areas of their identification in the karyotype and the localization of genes. Such studies have had important biological and medical implications in light of the discovery that many congenital diseases and syndromes are related to chromosomal aberrations. Today, *human cytogenetics* has become a specialized science in itself, the wide-ranging interests of which far exceed the limits of a text on cell biology. Here, we shall present to the student an elementary account which could be supplemented by reading some of the general references.

Historical Aspects. The first step in this field was made by Tjio and Levan in 1956, with the final demonstration that the correct *diploid number* of human chromosomes is 46 (i.e., 44 autosomes + XY in the male, and 44 + XX in the female) (Fig. 20–1).[1] The field of human cytogenetics, however, started to receive great

attention three years later with the discovery by Lejeune et al. of a *trisomy* (that is an extra chromosome) in patients affected by *mongolism* or *Down's syndrome*.[2] This finding led to rapid advances, with the identification in the same year of a series of aberrations of the sex chromosomes, such as *Klinefelter's syndrome* with XXY, and *Turner's syndrome* with XO (i.e., without Y),[3, 4] and later on a series of autosomal aberrations was discovered.

After 1968 a new era was initiated with the demonstration by Casperson et al. of chromosome banding using a fluorescent dye (i.e., quinacrine mustard).[5] This was followed by the development of a number of banding techniques which, by demonstrating a substructure in chromosomes, have permitted a more precise identification not only of individual chromosomes but also of their parts. These methods have increased the precision of cytogenetic diagnosis by allowing the study of finer chromosomal aberra-

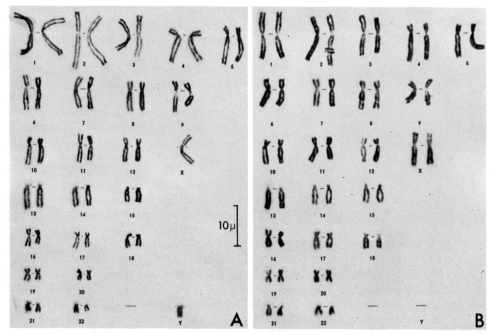

Figure 20–1 Human male (**A**) and female (**B**) karyotypes from a blood culture. (Courtesy of M. Drets.)

tions, such as deletions, translocations, inversions, and so forth, of individual chromosomes (see Chapter 19–2). The impact of such technical advances has been so great that from the point of view of clinical application, the field can be divided historically into two major periods, i.e., one prior to the discovery of the banding techniques and the other after.[6] These advances have permitted the observation of new chromosomal defects involving almost every chromosome of the human karyotype and have increased the possibility of a genetic mapping of the human chromosomes (see McKusick and Ruddle, 1977).

20–1 THE NORMAL HUMAN KARYOTYPE

Tissue cultures of fibroblasts, bone marrow, skin, and peripheral blood combined with the action of colchicine and hypotonic solutions to block mitosis at metaphase and to separate the chromosomes were used to study the human karyotype. An important technical advance has been the introduction of *phytohemagglutinin*, which induces lymphocytes to transform into lymphoblast-like cells that start to divide 48 to 72 hours after exposure. The strong mitogenic properties of this substance allowed the development of microtechniques which employ small amounts of blood. Spreading the cells on a slide

causes them to burst and to display all the chromosomes which are usually studied in metaphase. Recently, a higher resolution of the chromosome structure has been achieved by using banding techniques to study mitosis in prophase and prometaphase.[7]

Karyotype Preparation—Chromosomal Ordering by Size and Centromere Position

A *karotype* of human metaphase chromosomes is usually obtained from microphotographs. The individual chromosomes are cut out and then lined up by size with their respective partners. The technique can be improved by determining the so-called *centromeric index*, which is the ratio of the lengths of the long and short arms of the chromosome. More recently, a system has been introduced that involves a computer-controlled microscope and several accessories that permit (1) scanning of slides, (2) location of cells in metaphase, (3) counting chromosomes, and (4) transmission of digitally expressed images for computation and storage. All these steps, which can be carried out automatically, may help in making karyotypes more rapidly and in determining chromosomal aberrations.

Table 20–1 shows the classification of human chromosomes now generally used. The 23 pairs are disposed in 7 groups (A through G) decreasing in size and having the characteristics

TABLE 20–1 CHARACTERISTICS OF THE CHROMOSOMES IN THE HUMAN KARYOTYPE

Group	Pairs	Description
A	1–3	Large almost metacentric chromosomes
B	4–5	Large submetacentric chromosomes
C	6–12 + X	Medium-sized submetacentric chromosomes
D	13–15	Large acrocentric chromosomes with satellites
E	16–18	No. 16, metacentric; Nos. 17–18, small submetacentric chromosomes
F	19–20	Small metacentric chromosomes
G	21–22 + Y	Short acrocentric chromosomes with satellites. (The Y chromosome belongs to this group but has no satellites.)

described. For example, group A includes pairs 1 to 3, large almost metacentric chromosomes; group B comprises pairs 4 and 5, large submetacentric chromosomes, and so forth. The X chromosome is in group C (pairs 6 to 12), medium-sized submetacentric chromosomes; on the other hand, chromosome Y is in group G, together with pairs 21 and 22, small acrocentric chromosomes. It is important to remember that pairs 21 and 22 have satellites that correspond to nucleolar organizers, while the Y chromosome lacks satellites. Furthermore, group D (pairs 13 to 15) also contains acrocentric chromosomes with satellites that correspond to nucleolar organizers (see Fig. 20–3).

Banding Techniques Reveal Structural Details of Chromosomes

The main banding techniques are identified by letters, i.e., Q, G, C, R, T, and so forth, which are related to the method used or the results obtained.

Q Banding. We mentioned earlier that the introduction of certain fluorescent dyes (i.e., proflavine and quinacrine) and especially an alkylating derivative, *quinacrine mustard,* led to the discovery of a substructure along the length of the chromosomes (Fig. 20–2). The bands stained with quinacrine were named *Q bands.*

Later on, other techniques, which involve a prestain treatment that tends to denature the DNA (such as exposure to alkali, increased temperature, or formamide), followed by various staining techniques, were used.

G Banding. One of the best known techniques is that using the Giemsa stain (originally used to stain blood smears). The bands stained

with Giemsa were designated *G bands.*[8] Both Q and G bands (Fig. 20–3) correlate with the *chromomeres* observed in leptotene and pachytene chromosomes during meiosis. This means that the bands represent a fundamental subdivision of the chromosome.[9] On the basis of the banding pattern, the following three major types of chromatin may be identified in chromosomes: (1) *centromeric constitutive heterochromatin,* (2) *intercalary heterochromatin,* and (3) *euchromatin.* The Q and G banding patterns are generally similar and correspond to the second type, intercalary heterochromatin.

C Banding. Pretreatment of chromosomes with procedures that denature DNA, followed by Giemsa staining,[10] results in an intense staining of the centromeric region corresponding to the localization of *constitutive heterochromatin.* This differential staining is the result of the rapid reannealing of satellite DNA, which is located mainly in the centromeric region. Satellite DNA renatures easily because it is composed of short highly repetitive nucleotide sequences. In addition to the pericentromeric regions, the secondary constrictions of chromosomes 1, 9, and 16 and the distal segment of the long arm of the Y chromosome are stained. This method has provided the interesting information that there may be a *polymorphism* in human chromosomes (see Sanchez and Yunis, 1977). In fact, the size of the C bands may vary from person to person, and this may correspond to differing amounts of *satellite DNA.*

R Banding. With minor variations in pretreatment of the chromosomes, a banding pattern that is the reverse of that found with the Q and G band techniques is obtained.[11]

T Banding. Another modification has allowed the staining of certain telomeric regions of the chromosomes.

Other banding techniques use the Feulgen stain (F bands), and one selectively stains the *nucleolar organizers* (N bands). These are localized in the satellites of chromosomes 13, 14, 15, 21, and 22 (Fig. 20–3).

Banding Provides New Features for Identification of Human Chromosomes

The recognition of chromosome bands has provided new morphologic features for use in the identification of human chromosomes and chromosomal parts. In addition to the telomere, centromere, and arms, special landmarks (i.e., well-defined bands) were selected to subdivide the arms into regions. The regions are designat-

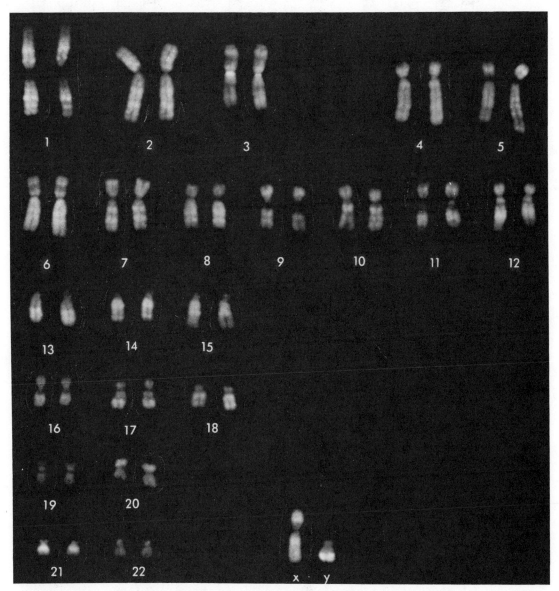

Figure 20–2 Human karyotype showing the fluorescent bands produced by staining with quinacrine mustard. Q bands are shown here. (Courtesy of T. Caspersson.)

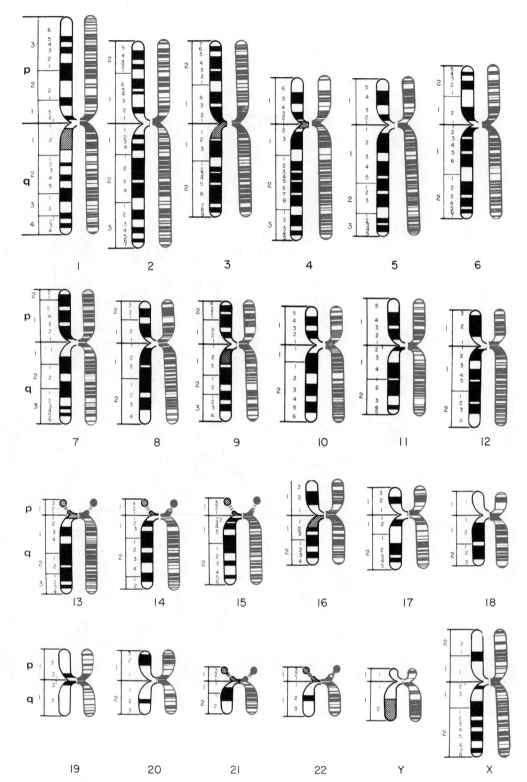

Figure 20–3 Human karyotype studied with the banding techniques. Chromosomes are disposed following the Paris conference (see Table 20–1). Left chromatid, in black, represents banding pattern at midmetaphase; right chromatid, in red, represents G-banding pattern at late prophase. (Courtesy of J. J. Yunis.)

ed 1, 2, 3, and 4, moving from the centromere toward the telomere. Using the nomenclature established at the Paris conference in 1971, any particular band or segment of a chromosome can be identified according to the following parameters: (1) chromosome number, (2) an arm symbol (p = short arm, q = long), and (3) the region and band numbers (Fig. 20–4). (For example, 6p23 indicates band number 3 in region 2 of the short arm of chromosome 6.) (For further details, see Sanchez and Yunis, 1977.)

With the use of the banding techniques, particularly the G and Q banding, some 320 bands per haploid set can be identified at metaphase. However, if mitosis is followed through late prophase and prometaphase, up to 1256 bands may be recognized.[7] This is indicated in the idiogram of Figure 20–3, in which the left chromatid shows the bands observed at midmetaphase, and the right chromatid represents the G banding in late prophase.

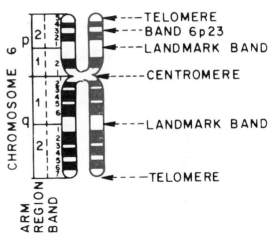

Figure 20–4 Nomenclature set at the Paris conference (1971) to identify any particular region and band in a chromosome. *p,* short arm; *q,* long arm. Landmark bands divide each arm into regions which are composed of bands. (Courtesy of O. Sanchez and J. J. Yunis. From Sanchez, O., and Yunis, J. J., New chromosome techniques and their medical applications, *in New Chromosome Syndromes,* p. 1. New York, Academic Press, Inc., 1977.)

SUMMARY:

The Human Karyotype

The correct diploid number of human chromosomes (i.e., 44 autosomes + XY in the male, and 44 + XX in the female) was established in 1956. Rapid advances were made in human cytogenetics and several cases of aneuploidy were discovered (mongolism, Turner's and Klinefelter's syndromes and so forth). After 1968 a new era was begun with the use of the *banding* techniques which permitted a more precise identification of individual chromosomes and their parts. Thus, chromosomal aberrations could be demonstrated.

In karyotyping cells in tissue culture, blastlike transformation by phytohemagglutinin and hypotonic treatment with the use of colchicine to arrest mitosis at metaphase have been important technical advances. Recently chromosomes at prophase and prometaphase have been studied with banding techniques, thus achieving higher resolution. The karyotype is generally made by cutting the microphotographs of individual chromosomes and lining them up by size with their respective partners. The *centromeric index* refers to the ratio of the lengths of the long and short arms of the chromosome. Computer-controlled microscopes are now being used in karyotyping.

The 22 pairs of autosomes are in 7 groups (A to G) with differing morphology (i.e., metacentric, submetacentric, acrocentric with satellites, and others) (see Table 20–1). The X chromosome is submetacentric and belongs to group C. The Y chromosome falls into group G, as a small acrocentric, but it lacks satellites.

The first banding technique (Q banding) used fluorescent dyes (quinacrine) and resulted in the so-called Q bands. The other techniques generally involve a prestain treatment followed by various stains. The G banding uses Giemsa stain. With G banding, three major types of chromatin may be recognized (i.e., euchromatin and centromeric and intercalary heterochromatin). C banding stains specifically centromeric heterochromatin. R banding gives a pattern that is the reverse of that of Q and G banding. T banding stains telomeres. With banding, the chromosomes show special regions, subregions, and bands. Some 320 bands per haploid set have been identified, and at prometaphase, as many as 1256 bands.

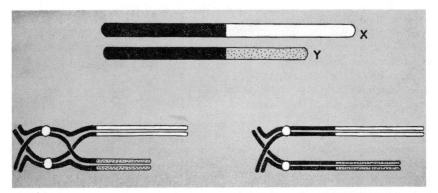

Figure 20–5 Diagram of a pair of XY sex chromosomes of a mammal. In black, the pairing or homologous segments; in white, the differential segments of the X chromosome; stippled, the differential segments of the Y chromosome. The configuration of the bivalent depends on the position of the chiasmata, which is produced only in the homologous segment.

20–2 SEX CHROMOSOMES AND SEX DETERMINATION

The well-known fact that male and female individuals are statistically found in about equal members suggests that sex is determined by a hereditary mechanism. There is physiologic and cytologic proof that sex is determined as soon as the egg is fertilized and that it depends on the gametes. Among the physiologic evidence is the finding that identical twins — which originate from a single zygote — are always of the same sex. Cytologic evidence was first obtained by McClung, who demonstrated that the karyotype of a cell is composed not only of common chromosomes (*autosomes*) but also of one or more special chromosomes that are distinguished from the autosomes by their morphologic characteristics and behavior. These were called *accessory chromosomes, allosomes, heterochromosomes,* or *sex chromosomes.*

The majority of organisms have a pair of sex chromosomes, which, in the course of evolution, have been specialized for sex determination. One of the sexes has a pair of identical sex chromosomes (XX); the other may have a single sex chromosome, which may be unpaired (XO) or paired with a Y chromosome (XY). The XY pair is also called *heteromorphic* because of the different morphology of the chromosomes.

In the human the gametes are not identical with respect to the sex chromosomes. The male is heterozygous (XY) and produces two types of spermatozoa in similar proportions, i.e., one carries the X chromosome, and the other, the Y. The female, being homozygous (XX), produces only one kind of gamete (the ovum).

444

Therefore, only two combinations of gametes are possible at fertilization, and the result is 50 per cent males and 50 per cent females (Fig. 18–1). This type of sex determination is found in all mammals and in certain insects, such as *Drosophila.* In other vertebrates and certain invertebrates, the female is heterogametic, and the male, homogametic. In some animals the sex chromosomes cannot be distinguished from the others. In such cases the sex-determining genes are confined to a region in a pair of chromosomes.

In the human the Y chromosome contains an essential testis-determining factor and determines the male sex. The genes involved are situated in a region near the centromere. An individual with XO (lacking the Y chromosome) will resemble a female but will have atrophic ovaries (Turner's syndrome).

We have seen earlier that the X and Y chromosomes differ in size and morphology. In the XY pair there is a *homologous region* in which recombination may take place during meiosis and a *differential region* that is unpaired (Fig. 20–5).

X and Y Sex Chromatin in Interphase Nuclei

In 1949 Barr and Bertram made the important discovery that in the interphase nucleus of females there is a small chromatin body (i.e., a chromocenter) which was lacking in males.[12] This was called *sex chromatin or the Barr body,* and more recently, after the 1971 Paris conference, *X chromatin.*

The X chromatin can be found as a small

body in different positions within the nucleus. For example, in nerve cells it may be near the nucleolus (Fig. 20–6,A), in the nucleoplasm, or near the nuclear envelope (Fig. 20–6, B and C). In cells of the oral mucosa it is generally attached to the nuclear envelope (Fig. 20–6, F), and in neutrophil leukocytes it may appear as a small rod called the *drumstick* (Fig. 20–6, D).

The study of·sex chromatin has a wide field of medical applications and offers the possibility of relating the origin of certain congenital diseases to chromosome anomalies. Among these applications is the diagnosis of sex in intersexual states in postnatal and even in fetal life.

The frequency with which sex chromatin can be detected in the female varies from tissue to tissue. In nervous tissue the frequency may be 85 per cent, whereas in whole mounts of amniotic or chorionic epithelium it may be as high as 96 per cent. In oral smears the frequency may vary between 20 and 50 per cent in normal females (see Hamerton, 1969).

X Chromatin — Facultative Heterochromatin Bodies Equaling nX −1 in Number

The relationship between sex chromatin and sex chromosomes has been elucidated. Sex chromatin is derived from only one of the two X chromosomes; the other X is not heteropyknotic at interphase. The number of corpuscles of sex chromatin at interphase is equal to nX −1. This means that there is one Barr body fewer than the number of X chromosomes. This relationship between sex chromatin and sex chromosomes is particularly evident in some humans who have an abnormal number of sex chromosomes (see Table 20–2).

The differential behavior of the two X chromosomes in the female led Lyon (see Lyon, 1972) to the so-called *inactive X hypothesis* by which: (1) only one of the X chromosomes is genetically active; (2) the X undergoing heteropyknosis may be either of maternal or paternal origin, and the decision by which X becomes

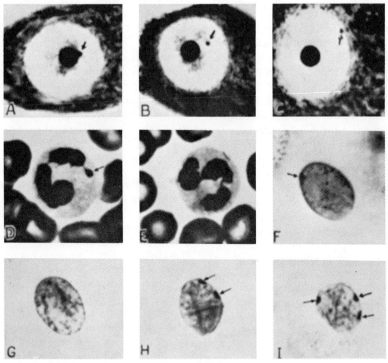

Figure 20–6 Sex chromatin in a nerve cell of a female cat. **A,** near the nucleolus; **B,** in the nucleoplasm; **C,** under the nuclear membrane (from M. L. Barr); **D,** normal leukocyte with a drumstick nuclear appendage from a human female. ×1800; **E,** same as D, in a male. (Remember that 90 per cent of females also lack the drumstick, as in the male.) ×1800. **F,** one sex chromatin corpuscle (*arrow*) in a nucleus from an oral smear. ×2000; **G,** same as F, from a male. Notice the lack of sex chromatin. ×1800. **H,** nucleus from the XXX female with two sex chromatin bodies. Vaginal smear. ×2000. **I,** similar, from an XXXX female. The three Barr bodies are indicated by arrows. ×2000. (From Barr, M. L., and Carr, D. H., *in* Hamerton, J. L. (Ed.), *Chromosomes in Medicine,* Medical Advisory Committee of the National Spastics Society in association with Wm. Heinemann, Little Club Clinics in Developmental Medicine, No. 5, 1963.)

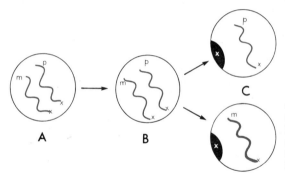

Figure 20–7 Diagram of the evolution of XX chromosomes. **A**, in the zygote both the paternal (*p*) and maternal (*m*) X are euchromatic. **B**, in the early blastocyst the same is true as in **A**. **C**, in late blastocyst 50 per cent of the cells have a maternal heterochromatic X chromosome, and the other 50 per cent have the paternal heterochromatic X chromosome.

inactive is taken at random; and (3) the inactivation occurs early in embryonic life and remains fixed. It is now admitted that only a part of the X chromosome condenses into a Barr corpuscle; this condensed chromosome is said to contain *facultative heterochromatin,* as opposed to the *constitutive heterochromatin* found in other chromosomes. Inactivation of the X chromosome takes place even in 3X and 4X individuals, in which only one X remains uncondensed (Fig. 20–6, *H* and *I*).

The inactivation starts in the human in the late blastocyst stage, on about the sixteenth day of embryonic life; the inactivated X chromosome then remains heterochromatic in the somatic cells (Fig. 20–7).[13]

The fact that X inactivation occurs at random has been demonstrated in human diseases linked to the X chromosome. The *Lesch-Nyhan syndrome,* in which a deficiency of one enzyme of the purine metabolism (i.e., hypoxanthine-guanine phosphoribosyl transferase) produces mental retardation and increased uric acid levels, results from a recessive mutation in the X chromosome. When fibroblasts of heterozygous women are cultured in vitro, two types of cell clones are obtained. Half the clones contain the enzyme, whereas the other half lack the enzyme.[14]

During the cell cycle the heterochromatic X chromosome is characterized by its late replication. This can be clearly observed by the use of 5-bromodeoxyuridine, followed by the fluorescent dye Hoechst 33258 or the Giemsa stain (Fig. 20–8).[15]

Y Chromatin — The Heterochromatic Region of the Y Chromosome

The above-mentioned fluorescence method for the study of Q banding led to the demonstration that a large portion of the Y chromosome is heterochromatic and appears, at interphase, as a strongly fluorescent body, the so-called *Y chromatin* (Fig. 20–9).[16] While the X chromatin is 1.0 to 1.2 μm in diameter, the Y chromatin is only about 0.7 μm in diameter and generally is not attached to the nuclear envelope. The number of Y chromatin bodies is identical to the number of Y chromosomes. For example, individuals with XYY have two Y chromatin bodies.

The human Y chromosome consists of two segments, one that does not fluoresce and is genetically active, and a second that fluoresces and is genetically inactive (Fig. 20–2). It is this latter segment that may be polymorphic, showing wide variations in size between individuals (see Sanchez and Yunis, 1977). Among mammals, only man and the gorilla share this fluorescent region of the Y chromosome, suggesting that it is of recent evolutionary origin.

This region corresponds to a highly repetitive (satellite) DNA, which is specific for the Y chromosome and probably has no genetic function.

Sex is Determined by Sex Chromosomes, but Hormones Influence Differentiation

Although the primary determination of sex is made at fertilization (Fig. 18–1) the embryo acquires its definite sex characteristics by a more complex mechanism. An epigenetic factor (i.e., hormonal) may assume control of the genetic determination during development, thereby changing the phenotypic direction of sex. Among vertebrates a condition of bisexuality may exist (e.g., the coexistence of structures of the functional sex, together with primordia of the heterologous sex). For example, male amphibians have a rudimentary ovary (Bidder's organ) and vestigial oviducts.

In the human embryo until the sixth week the gonads and the primordia of the urogenital tract are identical in males and females. At this time the gonad has already been invaded by the primary XX or XY germ cells. At this point,

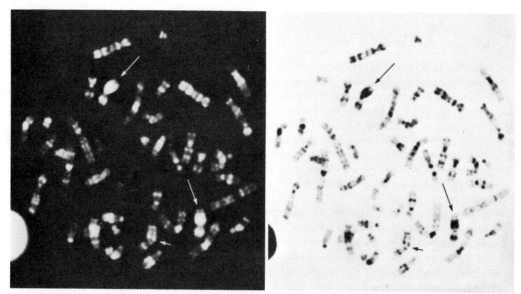

Figure 20–8 Late replication in a lymphocyte with a karyotype 44 + XXX (i.e., with an extra X chromosome). This cell initiated replication (S period) in 5-bromodeoxyuridine and then completed it in deoxythymidine. The late-replicating X chromosomes have strong fluorescence (**left**) and an intense stain (**right**). The early-replicating X chromosome is indicated by a short arrow, and the two late-replicating X chromosomes, by long arrows. (Courtesy of S. A. Latt, H. F. Willard, and P. S. Gerald, 1978.)

a gene (or set of genes) present in the Y chromosome causes the undifferentiated gonad to differentiate into a testis, and the absence of this gene allows the gonad to become an ovary. The development into a testis starts as soon as the *gonocytes* (i.e., primordial germ cells) from the yolk sac have finished their migration into the *gonadal ridge*. Gonocytes of the male (XY) migrate deeper into the gonadal blastema, and female gonocytes (XX) remain at the periphery, forming a thick cortical layer.

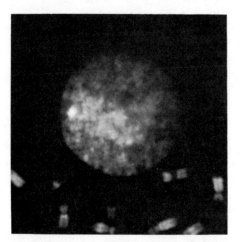

Figure 20–9 Staining, by quinacrine mustard, of the Y chromosome in a human interphase cell. The strongly fluorescent body is clearly distinguished within the nucleus. (Courtesy of T. Caspersson.)

H-Y Antigen. It is known that the Y chromosome directs the organogenesis of the testis by inducing the production of a protein (or proteins) that becomes localized at the plasma membrane and determines, by cell recognition, the formation of the seminiferous tubule. It has been proposed that a Y-linked determining gene of mammals produces this *H-Y antigen*, which is found in various cell types in males but is absent in females (see Ohno, 1976). This antigen is also a histocompatibility factor (H) that is found, for example, in the skin of male mice and that determines the rejection of female skin transplants.

Action of Testosterone. The gonad differentiates into a definite testicle at the seventh week, whereas the female gonad differentiates between the eighth and ninth weeks of development. At the time of differentiation an important epigenetic factor is the production of androgens by somatic cells in the embryonic male gonad.

In mammals administration of *testosterone* (the male hormone) to the mother produces in the fetus a shift in the differentiation of XX genitalia to a male type, producing what is called *feminine masculine pseudohermaphroditism*. This hormone acting locally accelerates the development of the testis, whereas in the female the absence of the hormone permits the slower development characteristic of the ovary (see Jost, 1970; Ohno, 1976).

447

SUMMARY:

Sex Chromosomes and Sex Determination

Sex determination is transmitted from one generation to the next by a hereditary mechanism. In many species, as in the human, the gametes are not identical with respect to the sex chromosomes. One of the sexes is heterozygous; the other is homozygous. Upon fertilization the result is 50 per cent males and 50 per cent females. In most cases a pair of sex chromosomes is involved in sex determination (XX in females, XY or XO in males). Frequently the male is heterogametic, but in certain cases (birds, fishes) the female is heterogametic. In the human and most mammals the male character is determined by the Y chromosome. Sex chromosomes have a homologous region in which recombination may occur and a differential region that is related to sex determination.

X chromatin (i.e., sex chromatin or the Barr body) is a small chromatin body observed in the interphase nucleus of females and has a medical application in the diagnosis of intersexual states. Sex can be determined in the fetus by the study of smears of amniotic epithelium. Sex chromatin is derived from only one of the two X chromosomes, which becomes heteropyknotic. The number of chromatin bodies is nX − 1. It is thought that the X chromosome forming the sex chromatin is genetically inactive — an inactivation which occurs early in embryonic life (i.e., *facultative heterochromatin*). This inactivation occurs at random, as can be demonstrated in cell clones from an individual with hereditary disease involving an enzyme deficiency in purine metabolism.

Y chromatin is a strongly fluorescent body observed in the interphase nucleus of males by using Q banding techniques. Individuals with XYY have two Y chromatins. This body is generally not attached to the nuclear envelope and represents a large heterochromatic portion of the Y chromosome. This part is genetically inactive and is found only in man and gorilla, and it contains repetitive (satellite) DNA.

Although sex is determined primarily by the sex chromosomes, during embryonic development an epigenetic factor (i.e., hormonal) may be important in sex differentiation.

Until the sixth week of embryonic life, both male and female gonads are identical. At this time they are invaded by XX or XY germ cells. The testicle differentiates in the seventh week, whereas the ovary begins to develop one or two weeks later. The early differentiation in the male depends on a local production of androgen. In the female the lack of androgen results in the slower development of the ovary.

The organogenesis of the testis is directed by the Y chromosome, which carries the gene for the H-Y antigen and induces its production. The H-Y antigen is a protein of the cell membrane that is involved in cell recognition and the formation of the seminiferous tubules. The H-Y is absent in females. The same antigen (histocompatibility factor) is found in the skin of male mice, where it determines the rejection of female skin transplants. Testosterone given to a mother may shift differentation in the fetus from an XX to a male type (*masculine pseudohermaphroditism*).

20–3 HUMAN CHROMOSOMAL ABNORMALITIES

The study of chromosomal abnormalities in man, like that of normal cytogenetics, can be divided into periods before and after the development of the banding techniques. In the earlier period it was found that 0.5 per cent of all newborns have gross chromosomal abnormalities, about half of which are in the sex chromosomes and the rest of which are in the autosomes.

These abnormalities generally consist of aneuploidy, such as monosomy or trisomy. Structural aberrations, such as translocation, deficiency, duplication, and other more complex alterations, have also been observed (see Chapter 19–2).

Diagnosis of chromosomal aberrations can be made even before a child is born. This is done by removing amniotic fluid by puncturing the mother's abdomen, a procedure called *amniocentesis*, which does not entail risks to the fetus.

The amniotic fluid contains cells shed from the fetus and which can be cultured. This technique is useful for karyotyping the babies of mothers with a history of chromosomal abnormality, or children of those who are carriers of a balanced translocation, in whom the risk of conceiving an affected child is high. In addition, this technique permits the *antenatal* determination of the fetal sex, which may be of importance in certain sex-linked diseases.

The improved resolution provided by the banding techniques has not only confirmed the previously observed abnormalities but has also led to the discovery of over 30 new chromosomal syndromes and several malignancies (i.e., cancers) involving chromosomal defects (see Lewandoski and Yunis, 1977).

Aneuploidy—Caused by Non-disjunction of Chromosomes

In aneuploidy the genetic message contained in each chromosome is maintained intact. The alteration is quantitative; it resides in a disequilibrium established by the excess (trisomy) or deficit (monosomy) in the amount of genetic material. Such a dosage effect may be dangerous to the organism and may produce severe anatomical and functional anomalies (i.e., malformations). It will be shown later that trisomy produces changes that are characteristic for each chromosome present in excess; however, in all instances of trisomy the tendency is toward

involution of the nervous system, resulting in a more or less severe mental defect.

Another important consequence of aneuploidy is spontaneous abortion. A large proportion of aborted fetuses show trisomy in one of the larger chromosomes in the karyotype. Aneuploidy of one of the larger chromosomes is generally lethal, being more severe than an equivalent alteration in a smaller chromosome. Malformation, mental retardation, sterility, and spontaneous abortion operate as strong selective mechanisms tending to eliminate from the general population those individuals carrying deleterious genetic imbalances.

Aneuploidy originates by the mechanism called *non-disjunction*. The immediate cause of non-disjunction is the lagging, during anaphase, of one sister chromatid, which during telophase remains in one of the cells together with the other sister chromatid (Fig. 20–10). This change gives rise to a cell line that lacks one chromosome or has one chromosome in excess for that pair (monosomy and trisomy).

Non-disjunction may occur during *meiosis*, giving rise to an aneuploid *ovum*. This, when fertilized by a normal spermatozoon, results in a zygote and later an organism in which all cells are aneuploid. In some cases the alteration may be in the male gamete or in both, thus producing more complex types of abnormalities. If non-disjunction occurs in a *mitotic division*, which precedes the formation of germ cells, the effects are similar to those of *meiotic non-disjunction*. If mitotic non-disjunction occurs during embryon-

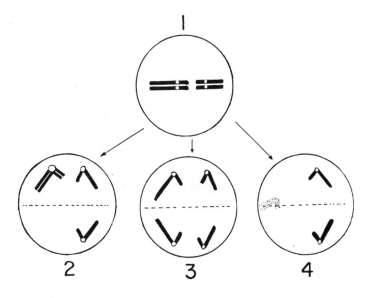

Figure 20–10 Mitotic non-disjunction and chromosome loss. **1,** normal metaphase. **2,** non-disjunction anaphase giving rise to monosomic and trisomic nuclei. **3,** normal anaphase. **4,** a chromosome loss results in two monosomic nuclei.

Reciprocal Translocations – Identified
by Banding

We mentioned earlier the various structural aberrations that are found in human chromosomes (i.e., translocation, deficiency, and duplication) (see Chapter 19–2). The most common of these is the so-called *reciprocal translocation*, in which chromosome segments are exchanged between non-homologous chromosomes. Some of these translocations may result in a chromosome having a length and configuration that may be easily identified. However, in other cases the apparent morphology changes very little. Banding techniques are of fundamental importance in detecting such cases. This is clearly illustrated in the diagram of Figure 20–11, in which there is a *balanced translocation* between a segment of the long arm of chromosome 2 and a segment of the short arm of chromosome 6 (a 2/6 translocation).

Sex Chromosome Abnormalities in Human
Syndromes

Table 20–2 shows some of the most common abnormalities found in sex chromosomes together with the clinical syndromes and an indication of the presence or absence of sex chromatin. The most important aberrations of this type follow.

Klinefelter's Syndrome. Most affected individuals are practically normal except for minor phenotypic anomalies. They have small testes, frequent gynecomastia (enlarged breasts), tendency to tallness, obesity, and underdevelop-

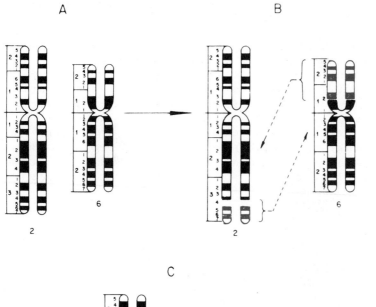

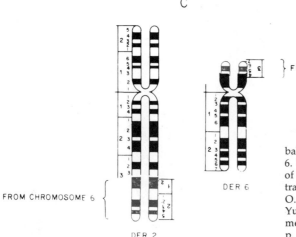

Figure 20–11 Schematic representation of a balanced translocation between chromosomes 2 and 6. The banding technique facilitates the recognition of the two segments that were translocated. The translocated segments are in red. (Courtesy of O. Sanchez and J. J. Yunis. From Sanchez, O., and Yunis, J. J., New chromosome techniques and their medical application, *in New Chromosome Syndromes,* p. 1. New York, Academic Press, Inc., 1977.)

TABLE 20–2 SEX ANEUPLOIDS IN MAN

X \ Y	O	Y	YY	Sex Chromatin
X	Monosomic XO Turner's syndrome 2X − 1 2n = 45	Disomic XY Normal 2X 2n = 46	XYY	0
XX	Disomic XX Normal 2X 2n = 46	Trisomic XXY Klinefelter's syndrome 2X + 2 2n = 47	Tetrasomic XXYY Klinefelter's syndrome 2X + 2 2n = 48	1
XXX	Trisomic XXX Metafemale 2X + 1 2n = 47	Tetrasomic XXXY Klinefelter's syndrome 2X + 2 2n = 48		2
XXXX	Tetrasomic XXXX Metafemale 2X + 2 2n = 48	Pentasomic XXXXY Klinefelter's syndrome 2X + 3 2n = 49		3
Phenotype	♀	♂	♂	

ment of secondary sex characteristics. Spermatogenesis does not occur, thereby resulting in complete infertility. These individuals have a positive sex chromatin and 47 chromosomes (44 autosomes + XXY).

Males with 48 chromosomes (44 autosomes + XXXY) and two Barr corpuscles have also been described. These individuals have features of Klinefelter's syndrome and are mentally retarded. Persons with 49 chromosomes (44 autosomes + XXXXY) have also been reported. They display extensive skeletal anomalies, extreme hypogenitalism, and severe mental deficiency. These persons have three X chromatin bodies (Table 20–2).

XYY Syndrome. Males having two Y chromosomes have been identified in the past in maximum security institutions.[17, 18] It was proposed that such individuals had a strong tendency toward anti-social behavior and aggression (see Hook, 1973). More recently, XYY individuals have been found in the normal population in a proportion of 1 in 650 male infants, suggesting that the correlation with violence may not be as strong as was previously thought.

Turner's Syndrome (Gonadal Dysgenesis). Patients with Turner's syndrome usually have a female appearance with short stature, webbed neck (folds of skin extended from the mastoid to

the shoulders), and generally infantile internal sexual organs. The ovary does not develop and shows complete absence of germ cells. As a result of this ovarian dysgenesis, menstruation does not occur, and secondary sexual characteristics do not develop. The karyotype shows 45 chromosomes (44 autosomes + X), and there is no sex chromatin. In this syndrome there is little difference in the gonads up to the third month of gestation; the ovaries contain approximately the normal number of germ cells (about 2 or 3 × 10^6 cells). Later on there is a rapid atresia of the germ cells, leading to their virtual disappearance after puberty. It is probable that the lack of one X chromosome determines the progressive ovarian atresia by failure of primordial follicles.

Females with X Polysomy. Triplo-X constitution (47 chromosomes: 44 autosomes + XXX) was detected in phenotypically near-normal females. A number of these females are mentally subnormal or psychotic, and some of them menstruate. Two sex chromatin bodies are found in cells from these women. A few severely retarded patients with three corpuscles of sex chromatin and 48 chromosomes (44 autosomes + XXXX) have been found. These persons are also called metafemales (Table 20–2).

Mixed Chromosomal Aberrations. Klinefelter's syndrome may be found combined with

451

mongolism. Such a person has 48 chromosomes: 45 autosomes (including a trisomic pair 21) +XXY.

Mosaics. Sometimes chromosomal aberrations are produced during development of the embryo. One interesting example is induced by the loss of one Y chromosome in the first division of the zygote. This may result in twins, of which one has *normal* male characters and the other has Turner's syndrome.

Sometimes chromosomal aberrations due to non-disjunction are produced during development of the embryo. These individuals possess different chromosomal complements from cell to cell within the same tissue or between tissues, depending on the embryonic stage at which non-disjunction occurred. Table 20–3 illustrates the most frequently found mosaics of the sex chromosomes in females and males.

Sex Chromosomes and True Hermaphroditism. A true hermaphrodite is an individual who has both ovarian and testicular tissue. The two types of gonadal tissue may be separated or in close proximity (ovotestis). Frequently, true hermaphrodites have a 46, XX chromosome complement and are chromatin-positive. Others may have a 46, XY chromosome set. However, 46, XX/46, XY mosaicism has also been detected in this condition.

Mongolism — The Best Known Autosomal Abnormality

Before the development of the banding techniques about 12 classic syndromes had been described, the majority of which were caused by an extra chromosome or a loss of a chromosome

TABLE 20–3 SEX CHROMOSOME MOSAICS

	Clinical Syndrome	Sex Chromatin
Females		
XO/XY	Turner	−
XO/XX	Turner	−
XO/XYY	Turner	−
XO/XXX	Variable	−
Males		
XX/XXY	Klinefelter	+
XY/XXY	Klinefelter	+
XXXY/XXXXY	Small gonads and immature sexual characteristics, mental disorder	3+
XO/XY	Hermaphrodite	

segment. Except for the X monosomy (Turner's syndrome), any other monosomy is believed to be inviable (i.e., leads to death of the embryo). This also holds true for all trisomies other than those of chromosomes 13, 18, or 21.

21-Trisomy (Mongolism). Among the most important autosomal aberrations is *mongolism*, which is characterized by multiple malformations, mental retardation, and markedly defective development of the central nervous system. It was discovered that the mongoloid has an extra chromosome. Pair 21 is trisomic instead of normal. This aberration probably originates from non-disjunction of pair 21 during meiosis.

The extra chromosome of pair 21 in some cases may become attached to another autosome (translocation), usually to pair 22.

The phenotype of a mongoloid is recognizable at birth in most cases. The face of such a patient has a special moon-like aspect, with oblique palpebral fissures, increased separation between the eyes, and a skin fold (epicanthus) at the inner part of the eyes. The nose is flattened, the ears are malformed, the mouth is constantly open, and the tongue protrudes.

Mongolism is the most common congenital disease and is present in more than 0.1 per cent of births. Its frequency increases as the mother's age exceeds 35. The occurrence of this trisomy is sporadic, and in general there is no recurrence in the family. However, in the rarer cases of mongolism by translocation, the disease may affect siblings and may appear in successive generations. Fortunately this "translocation trisomy" represents only 3 to 4 per cent of all cases of mongolism. In this type there is no change in frequency with the age of the mother, and when the aberration is properly determined by karyotype analysis, the parents should be warned of a repetition of this defect.

21-Monosomy. Complete deletion of one of the chromosomes in pair 21 is apparently lethal, but there is a syndrome in which a large part of one is lacking. Children with this condition have a morphologic aspect which is, to some extent, the opposite of mongolism. The nose is prominent, the distance between eyes shorter than normal, and the ears are large and the muscles contracted. It seems that in trisomy and monosomy of the 21st, the phenotype shifts to one or the other side of normal.

18-Trisomy. In this case the child is small and weak, the head is laterally flattened, and the helix of the ear scarcely developed. The hands are short and the digital imprints are rather simple. These children are very retarded mentally and usually die before one year of age.

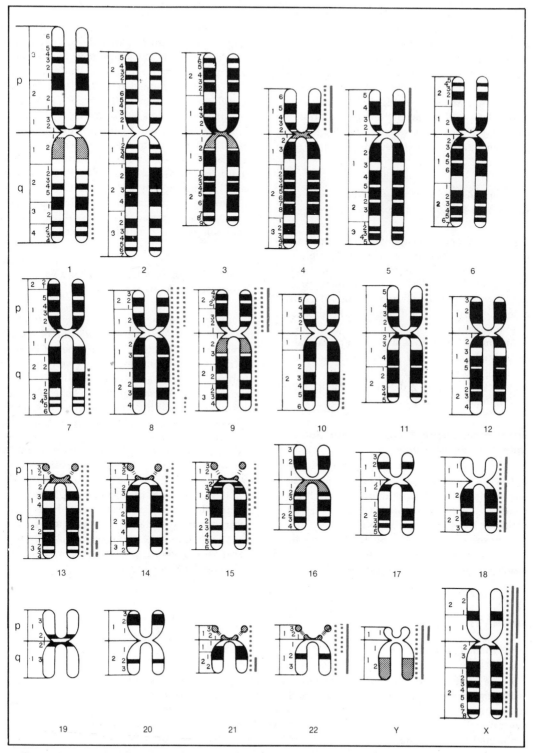

Figure 20–12 Human karyotype, as in Figure 20–3, showing the syndromes of total or partial trisomy of various chromosomes (indicated by dotted lines) and total or partial monosomy (solid lines). (Courtesy of J. J. Yunis.)

18-Monosomy. This is the opposite syndrome in which a partial deletion of one chromosome of the pair occurs.[19] The ears are voluminous, the fingers long, and the digital imprints are complex and convoluted.

13-Trisomy. Multiple and severe body malformations, as well as profound mental deficiency, are characteristic features. The head is small and the eyes are often small, or absent. Hare-lip, cleft palate, and malformations of the brain are frequent. The internal organs are also severely malformed, and in most cases death occurs soon after birth.

Structural Aberrations May Cause Syndromes Besides Aneuploidy

One of the most remarkable phenotypes associated with a chromosome structural change is the condition known as *cri du chat* syndrome.[20] The affected baby has a strange mewing cry, multiple malformations, and mental retardation associated with partial deletion of the short arm of chromosome 5.

Another type of patient showing facial alteration and skeletal and ophthalmologic abnormalities, along with profound mental retardation, is the carrier of a deletion of the long arm of chromosome 18.

Banding Techniques Have Detected More Than 30 New Syndromes

The use of the banding techniques has permitted the detection of more than 30 new chromosomal syndromes which result from a partial trisomy or deletion from each of almost all the chromosomes (Fig. 20–12). Such alterations have been found in children with mental retardation or minor congenital defects and also in some with normal intelligence and minor anomalies which were earlier considered as deviations of normality.[6]

The technical advances have permitted the identification of chromosome defects even when there are few clinical symptoms. Furthermore, these methods have allowed the study of cases of balanced chromosomal translocations (Fig. 20–11), which, as carriers of this type of syndrome, are important to detect. In such cases, transmission can be prevented by proper counseling and prenatal diagnosis. In Figure 20–12, syndromes of total or partial trisomy of various chromosomes are indicated by dotted lines, and total or partial monosomy, by solid lines. (Consideration of these new syndromes, although of considerable medical importance, is beyond the scope of this book; see the general references.)

Certain Human Tumors Show Specific Chromosomal Aberrations

The development of the banding techniques has also resulted in the demonstration of specific chromosomal defects in certain types of tumors. Here only a few examples will be given. The most characteristic of these is the association of *chronic myelogenous leukemia* with the so-called *Philadelphia chromosome* (Ph 1). This involves a balanced translocation between chromosomes 9 and 22. The Ph 1 chromosome is acquired with the disease and is not found in cultured fibroblasts or in the identical twins of affected patients. In a *retinoblastoma* (i.e., a cancer of the retina) a deletion in the long arm of chromosome 13 has been identified. Other abnormalities have been observed in lymphomas, meningiomas, and many other tumors. Cancer cells are usually associated with severe chromosomal abnormalities, such as polyploidy and aneuploidy.

SUMMARY:

Chromosomal Abnormalities

In the early period of human cytogenetics about 0.5 per cent of newborns were found to have gross chromosomal abnormalities. With the development of the banding techniques, that finding was duplicated and some 30 new syndromes and several malignancies have been shown to be associated with chromosomal defects. Abnormalities may consist of aneuploidy (monosomy or trisomy), in which case the genetic message in the chromosome is intact, but there is a *dosage effect* that may produce severe anatomical and functional anomalies (mental defect, in trisomy).

Aneuploidy is produced by non-disjunction, which may occur at meiosis or mitosis. In non-disjunction one sister chromatid does not reach the pole and remains with the other sister chromatid (one cell is monosomic, the other trisomic). Several structural aberrations (i.e., translocation, deficiency, or duplication) are found in human chromosomes. In the *reciprocal* translocation segments are exchanged between non-homologous chromosomes. The banding techniques are very helpful in detecting cases of balanced translocation (Fig. 20–11).

Aberrations of the sex chromosomes lead to the production of several syndromes. In *Klinefelter's* syndrome there are 47 chromosomes (44 + XXY), and the sex chromatin is positive. The individual has male characteristics, but has a small testis and underdeveloped secondary sex characteristics. Cases with 48 and 49 chromosomes (44 + XXXY and 44 + XXXXY) and having 2 and 3 corpuscles of sex chromatin, respectively, have been described (Table 20–2).

In Turner's syndrome the individual has a female appearance with dwarfism. The ovaries do not develop; there are 45 chromosomes (44 + X), and sex chromatin is lacking. Examples of numerical autosomal aberrations are: *mongolism,* generally caused by a trisomy of the 21st pair (sometimes the extra chromosome is translocated upon another autosome); *monosomy of the 21st pair,* which produces individuals in whom the appearance of the face is approximately opposite to that of a mongoloid; *trisomy* and *monosomy* of the 18th pair; and *trisomy* of the 13th pair. All these cases have profound alterations of the CNS and lethal malformations.

The 30 new chromosomal syndromes discovered by using the banding techniques are the result of either partial trisomy or a deletion in each of almost all the chromosomes. These alterations are found in children with mental retardation or minor congenital defects and also in others with minor deviations from normality. Special consideration is given to balanced translocations because their transmission can be prevented by prenatal diagnosis or proper counseling.

Several types of cancer show chromosomal defects. The best known is the balanced translocation 9/22 (*Philadelphia chromosome*) found in *chronic myelogenous leukemia.* This chromosome is acquired with the disease. In *retinoblastoma* there is a deletion of the long arm of chromosome 13.

20–4 HUMAN CHROMOSOMES AND THE GENETIC MAP

In recent years important advances have been made in the localization of specific genes to a particular chromosome or even to a chromosomal region (i.e., arm or band). In man the number of *structural genes* (i.e., those that determine the polypeptide sequence of the different proteins) could be on the order of 50,000. Before the era of molecular genetics, the presence of a gene was inferred, as in Mendel's experiments, by the existence of two alternatives of a given trait (see Chapter 19–1). In man, more than 1200 genes have been identified by this method and by the occurrence of diseases resulting from mutations.

Daltonism and Hemophilia – The Best Known Sex-Linked Diseases

The easiest of the genes to identify are the sex-linked genes, which are present either in the X chromosome or in the non-homologous region of the Y chromosome (Fig. 20–5). In 1911 Wilson assigned the specific gene for *daltonism* (i.e., red-green color blindness) to the X chromosome. This mutation is expressed in males (8 per cent of which have daltonism) by a recessive mechanism, but it is very rare in females. This is because males have only one X chromosome and therefore only one allele for this gene. As a result, any mutation on this chromosome will produce a phenotypic change. On the other hand, females have two copies of the same gene, and in the heterozygous condition color blindness (a recessive gene) is not expressed. However, women are genetic *carriers* and can transmit color blindness to their progeny.

Hemophilia (a disease with a defect in blood clotting) is also inherited by an X-linked recessive gene. This disease is transmitted by females but is expressed in males (Fig. 20–13). The few cases of daltonism or hemophilia in females are due to a homozygous condition in which both X chromosomes are altered at the same locus. At present about 100 genes have been assigned to the X chromosome.

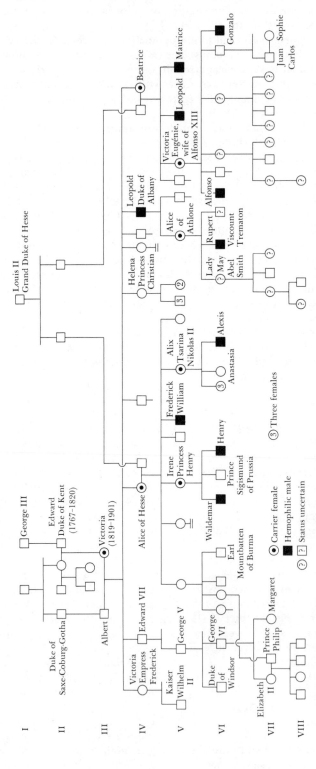

Figure 20–13 Pedigree of the descendants of Queen Victoria showing carriers and affected males who possessed the X-linked gene conferring the disease hemophilia. (From Mc Kusick, V. A., *Human Genetics*, 2nd ed., Englewood Cliffs, N. J., 1969.)

We mentioned earlier that the Y chromosome carries a *testis-determining factor* that produces the *H-Y antigen*. A few diseases also follow the male line and pass directly from father to son. However, there are other mutations that are *partially sex-linked* because the genes are localized in the homologous segments of both the X and Y chromosomes (Fig. 20–5).

Chromosome Mapping — Linkage Analysis, Somatic Cell Hybridization, and In Situ Hybridization

Linkage Analysis. As described in the previous chapter, genes on the same chromosome behave as though genetically linked. In man numerous genes have been mapped to the sex chromosomes using standard methods of linkage analysis. The most common approach consists of following a certain variation in a family. In Figure 20–13 is shown the pedigree of the descendants of Queen Victoria and the occurrence of hemophilia among them.

Somatic Cell Hybridization. At present the most widely used technique for gene mapping is *somatic cell hybridization*. In this technique the first step consists of the fusion of a cultured human cell with a mouse or hamster cell. Fusion is facilitated by the use of inactivated Sendai virus (parainfluenza virus) or by chemical agents such as lysolecithin or polyethyleneglycol.

Genetic analysis is possible because the resulting hybrid cells, with subsequent cell divisions, tend to preferentially lose the human chromosomes. Sometimes the loss of a certain genetic trait can be correlated with the elimination of a particular chromosome, thus allowing the assignment of a gene to a specific chromosome.

Another experimental strategy consists of selecting those hybrid cells that retain a specific human chromosome. This is done by using mouse cells which are unable to grow in a particular selective medium unless the deficiency is compensated by a human gene.

Let us consider the case of a mouse cell deficient in *thymidine kinase (TK⁻)* which is fused to a normal human cell. The enzyme TK is required for the utilization of thymidine added to the culture medium. TK⁻ cells can survive because they are able to synthesize thymidine by a pathway which involves *de novo* synthesis (Fig. 20–14).

A selective medium called HAT (containing hypoxanthine, aminopterin, and thymidine) is a very useful tool in somatic cell hybridization. TK⁻ cells are unable to grow in this medium because aminopterin blocks the *de novo* synthesis of nucleotides (Fig. 20–14). When TK⁻ cells are hybridized with human cells, the human TK gene can compensate for the defect by producing human thymidine kinase.

As the hybrid cells divide, human chromosomes are lost at random. When the chromosome containing the TK gene is lost, the cells in the HAT medium die. In this way, cell lines are eventually obtained which have the mouse chromosome and only one human chromosome (chromosome 17), the one which contains the TK gene. Once this is achieved, it is possible by gel electrophoresis to identify isoenzymes and other proteins encoded by this human chromosome.

To select other chromosomes similar procedures can be used with other *auxotrophic mutations*, i.e., mutants that require a particular nutrient. It is hoped that eventually geneticists

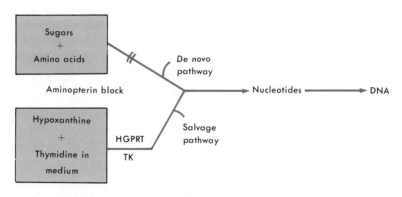

Figure 20–14 The nucleotides required for DNA replication may come from two biochemical pathways. The *de novo* pathway is blocked by the drug aminopterin present in the HAT medium. Hypoxanthine (a purine precursor) and thymidine (a pyrimidine precursor) are utilized through the salvage pathway provided that the enzymes thymidine kinase (TK) and hypoxanthine-guanine phosphoribosyl transferase (HGPRT) are present.

will have a bank of cell lines containing each one of the individual human chromosomes, thus enabling the assignment of genes to every one.

Somatic cell hybridization has so far allowed the mapping of more than 70 genes on autosomes. Furthermore, by using translocated chromosomes it has been possible to localize genes to a particular region of the chromosome.

In Situ Hybridization. The method of *in situ hybridization* makes use of a radioactive complementary RNA to localize a gene in specific chromosomal region. With this method genes which are present in multiple copies are more easily localized. For example, the genes for the 18S and 28S ribosomal RNA (i.e., nucleolar organizers) are localized in the satellites of chromosomes 13, 14, 15, 21, and 22 (Fig. 20–3), and the 5S ribosomal RNA is at the distal end of the long arm of chromosome 1.

Localization of Genes in Chromosomes and Regions of Chromosomes

At present, structural genes have been assigned to each one of the 22 autosomes and to the X and Y chromosomes. In the large chromosome 1, more than 20 genes have been localized. About 110 genes have been assigned to specific autosomes. Among recent localizations are the genes for the ABO blood group, at the distal end of the long arm of chromosome 9; the Rh blood factor, in the short arm of chromosome 1; and the histone genes, in the short arm of chromosome 7. Of the 100 genes known to be present in the X chromosome, at least 16 are localized in a definite arm or region of the chromosome. For example, the loci of the hemophilia and color blindness genes are in the distal portion of the long arm. (For further details on human gene maps, see McKusick and Ruddle, 1977.)

SUMMARY:

The Human Genetic Map

In the human there may be some 50,000 structural genes. Before the times of molecular genetics, the presence of a gene was inferred by the existence of two alternatives of a given trait (as in Mendel's experiments). By this method some 1200 genes were identified in man. The easiest to identify are the sex-linked genes (X or Y chromosome). The first of these to be assigned to the X chromosome was *daltonism*. *Hemophilia* is also X-linked (transmitted by females but expressed in males). Some 100 genes have been assigned to the X chromosome. The Y chromosome carries the testis-determining factor with the H-Y antigen.

The most common methods for gene mapping are linkage analysis, somatic cell hybridization, and *in situ* hybridization. In *somatic cell hybridization* a hybrid between human and rodent cultured cells is made by cell fusion (Sendai virus or chemicals). The clones of the hybrid cells produced tend to eliminate the human chromosomes. The remaining chromosomes may express their genes by the production of human proteins (enzymes). Some 70 genes have been localized by this method.

In situ hybridization uses labeled RNA to localize a specific chromosome region. Genes with multiple copies are easily localized; for example, 18S and 28S ribosomal RNA are in the satellites of chromosomes 13, 14, 15, 21, and 22 (nucleolar organizers). The 5S ribosomal RNA is in chromosome 1.

At present, structural genes have been assigned to the 22 autosomes and sex chromosomes. Twenty genes have been assigned to the large chromosome 1. The ABO blood group is in chromosome 9, the Rh blood factor in 1, and the histone genes in 7. The exact localization in a chromosomal arm or band has been achieved for certain genes.

REFERENCES

1. Tjio, J. H., and Levan, A. (1956) *Hereditas, 42*:1.
2. Lejeune, J., Turpin, R., and Gautier, M. M. (1959) *Ann. Génét., 1*:41.
3. Jacobs, P. A., and Strong, J. A. (1959) *Nature, 183*:302.
4. Ford, C. E., Jones, K., Polani, P., De Almeida, J. C., and Briggs, J. (1959) *Lancet, 1*:711.
5. Caspersson, T., Zech, L., Johansson, C., and Modest, E. J. (1970) *Chromosoma, 30*:215.
6. Lewandowski, R., and Yunis, J. J. (1975) *Amer. J. Dis. Child., 129*:515.
7. Yunis, J. J. (1976) *Science, 191*:1268.

8. Drets, M. E., and Shaw, M. W. (1971) *Proc. Natl. Acad. Sci. USA, 68*:2073.
9. Comings, D. E. (1975) *J. Histochem. Cytochem., 23*:461.
10. Arrighi, F. E., and Hsu, T. C. (1971) *Cytogenetics* (Basel), *10*:81.
11. Dutrillaux, B., and Lejeune, J. (1971) *C. R. Hebd. Seances Acad. Sci., 272*:2638.
12. Barr, M. L., and Bertram, E. G. (1949) *Nature, 163*:676.
13. Ohno, S. (1967) *Sex Chromosomes and Sex-linked Genes.* Springer-Verlag, Berlin.
14. Migeon, B., Der Kaloustian, V., Nyham, W.,

Young, W., and Childs, B. (1968) *Science, 160*:425.
15. Latt, S. A., Huntington, F. W., and Park, S. G. (1976) *Chromosoma, 57*:135.
16. Pearson, P. L., Bobrow, M., and Vosa, L. G. (1970) *Nature, 226*:78.
17. Jacobs, P. A., et al. (1965) *Nature, 208*:1351.
18. Price, W. W., and Whatmore, P. B. (1967) *Nature, 213*:815.
19. Grouchy, J., Bonnette, J., and Salmon, C. (1966) *Ann. Génét., 9*:19.
20. Lejeune, J., Lafourcade, J., Berger, R., and Rethoré, M. O. (1965) *Ann. Génét., 8*:11.

ADDITIONAL READING

Caspersson, T., and Zech, L. (1973) *Chromosome Identification,* p. 355. Nobel Foundation, Academic Press, Inc., New York.

Comings, D. E. (1978) Mechanism of chromosome banding and implications for chromosome structure. *Annu. Rev. Genet., 12*:25.

Dutrillaux, B. (1977) New chromosome technique. In: *Molecular Structure of Human Chromosomes,* p. 233. (Yunis, J. J., ed.) Academic Press, Inc., New York.

Epstein, C. J., Smith, S., Travis, B., and Tucker, B. (1978) Both X chromosomes function before visible X chromosome inactivation in female mouse embryos. *Nature, 274*:500.

Ford, C. E. (1977) Twenty years of human cytogenetics. *Hereditas, 86*:5.

Ford, E. H. R. V. (1973) *Human Chromosomes,* p. 381. Academic Press, Inc., London.

Greagan, R. P., and Ruddle, F. H. (1977) New approaches to human gene mapping by somatic cell genetics. In: *Molecular Structure of Human Chromosomes,* p. 90. (Yunis, J. J., ed.) Academic Press, New York.

Hamerton, J. L. (1971) *Human Cytogenetics,* Vol. 1, p. 422; Vol. 2, p. 474. Academic Press Inc., New York.

Harris, H. (1975) *The Principles of Human Biochemical Genetics.* 2nd Ed. Elsevier Publishing Co., Amsterdam.

Hook, E. B. (1973) Behavioral implication of the human XYY genotype. *Science, 179*:139.

International workshop in human genetic mapping. IV. (1978) *Ogtogenet. Cell Genet., 22*:1.

Jost, A. D. (1970) Development of sexual characteristics. *Sci. J.* (London), *6*:67.

Levan, A., Levan, G., and Mitelman, F. (1977) Chromosomes and cancer. *Hereditas, 86*:15.

Lewandowski, R. C., and Yunis, J. J. (1977) Phenotypic mapping in man. In: *New Chromosomal Syndrome,* p. 369. (Yunis, G. J., ed.) Academic Press, Inc., New York.

Lyon, M. F. (1972) X chromosome inactivation and developmental patterns in mammals. *Biol. Rev., 47*:1.

McKusick, V. A. (1971) The mapping of human chromosomes. *Sci. Am., 224*:104.

McKusick, V. A., and Ruddle, F. H. (1977) The status of the gene map of the human chromosomes. *Science, 390*:405.

Mittwoch, U. (1967) *Sex Chromosomes.* Academic Press, Inc., New York.

Ohno, S. (1976) Major regulatory genes for mammalian sexual development. *Cell, 315*:321.

Ohno, S. (1976) La differenciation sexuelle. *La Recherche 7*(63):5.

Paris Conference on standardization in Human Cytogenetics (1973). Birth Defects Atlas and Compendium, Vol. 8, pp. 1–43. National Foundation on Birth Defects Original Article Series. The Williams & Wilkins Co., Baltimore.

Rothwell, N. V. (1977) *Human Genetics.* Prentice-Hall, Inc., Englewood Cliffs, N.J.

Sanchez, O., and Yunis, J. J. (1977) New chromosome techniques and their medical application. In: *New Chromosome Syndromes,* p. 1. (Yunis, G. J., ed.) Academic Press, Inc., New York.

Srivastave, P. K., and Lucas, F. V. (1976) Evolution of human cytogenetics, an encyclopedic essay. *J. Genet. Hum., 24*:235 and 337.

Stern, C. (1973) *Principles of Human Genetics.* 3rd Ed. W. H. Freeman, San Francisco.

Yunis, J. J., Tsai, M. Y., and Willey, A. M. (1977) *Molecular Organization and Function of the Human Genome.* Academic Press, Inc., New York.

Yunis, J. J. (1977) *Molecular Structure of Human Chromosomes.* Academic Press, Inc., New York.

Yunis, J. J., Sawyer, J. R., and Ball, D. W. (1978) The characterization of high resolution G-banded chromosomes of man. *Chromosoma, 67*:293.

Part 6

GENE EXPRESSION

The following six chapters comprise the main topics currently included within the general heading of molecular biology. These are the fields in which progress has been rapid and remarkable in recent years and where the future of many aspects of cell biology lies. These areas of study are intimately related to molecular genetics, a discipline that attempts to explain hereditary phenomena as the result of specific chemical components localized or formed in the chromosomes. Experimental studies have provided conclusive evidence that genetic information is initially dictated by the arrangement of bases in deoxyribonucleic acid (DNA), and that this information is then transcribed in the various molecules of ribonucleic acid (RNA) (i.e., messenger RNA, transfer RNA, and ribosomal RNA). The genetic information is finally translated into various specific proteins and enzymes. Knowledge of the Watson-Crick model of DNA presented in Chapter 5 and of the disposition of the base pairs in this molecule is prerequisite to understanding this part of the book.

Chapter 21 analyzes the way in which genetic information is stored in DNA. In the *genetic code,* each amino acid is specified by a group of three nucleotides (a *codon*). The experiments which led to deciphering this code will be reviewed. The genetic code is universal and is shared by all living organisms.

Chapter 22 analyzes *transcription,* i.e., the process by which genes are copied into RNA to produce messenger RNAs which in turn code for the proteins. RNA synthesis is carried out by special enzymes called RNA polymerases which copy the nucleotide sequence of one of the strands of DNA, which is used as a template. Eukaryotes have three different RNA polymerases, each one transcribing a special set of genes. RNA is frequently produced as a long precursor molecule, which is then cleaved to produce the mature mRNA, a phenomenon that is examined under the heading of RNA processing.

Messenger RNAs are very short-lived in prokaryotes, but are stable in higher cells. Other special characteristics of eukaryotic mRNA will also be examined, such as the blockage of the starting end by a methylated guanosine nucleotide (known as a "cap"), and the presence of a polyadenylic acid (poly A) tail at the other end of the molecule.

Chapter 23 is focused on the study of ribosomes as macromolecular organelles presiding over the many steps of protein synthesis. Knowledge of the chemical organization of ribosomes has progressed considerably now that these organelles have been separated into their RNA and protein components and the functional units have been reconstituted. Another fundamental point elucidated is the biogenesis of ribosomes from the nucleolar organizer. The function of the nucleolus should become clear when it is recognized that this nuclear element is the site of the origin and processing of ribosomes.

In Chapter 24 the structure and function of transfer RNA (tRNA) will be examined. Transfer RNAs are small molecules (80 nucleotides) which function as adaptors during protein synthesis, carrying the amino acids to the ribosomes and recognizing the specific codons in the mRNA. The fidelity of protein synthesis relies to a great extent on the correct attachment of the amino acids to the corresponding tRNA.

Translation, the last step in the central dogma of molecular biology (DNA → RNA → protein), is explained in Chapter 24, when the relationship between genes and protein synthesis is considered. The role of ribosomes and polysomes in protein synthesis will be analyzed. The presence of special initiation, elongation, and termination factors, which play specific roles in the synthesis of proteins, is emphasized.

Chapter 25 discusses the way in which gene expression is regulated. In prokaryotes, genes are turned on and off at the level of transcription. The role of *repressors* will be examined in detail. Repressors are proteins which bind at the beginning of genes, thus preventing transcription by RNA polymerase. In eukaryotes, gene regulation is poorly understood, partly because they have a much higher DNA content than prokaryotes. A mammalian cell has 700 times more DNA than *Escherichia coli,* much of which is repetitive DNA. Some DNA sequences, called satellite DNAs, are present in millions of copies per nucleus. Only a small fraction of the eukaryotic genome is transcribed into RNA in a given cell type. Giant chromosomes (polytene and lampbrush) provide the opportunity of visualizing transcription under the microscope. The studies with polytene chromosomes indicate that in eukaryotes the turning on and off of genes is also regulated at the level of RNA synthesis.

Chapter 26 deals with the fundamental problem of cell differentiation. This chapter is of special importance in cell biology because it brings together the recent knowledge of how regulation of genic action takes place in higher cells. Experiments on nuclear control of the cytoplasm and cytoplasmic action on the nucleus illustrate the continuous interrelationship that exists between the two main territories of the cell. In higher organisms the different tissues have specialized cells which express different genes. These changes in gene activity are very stable and persist throughout many cell generations. The way in which the first differences between cells arise in early embryos will be analyzed.

THE GENETIC CODE AND GENETIC ENGINEERING

21–1   The Genetic Code   464
    Genes Code for Proteins—Inborn Errors
      of Metabolism
    Mutations Produce Amino Acid Changes
    Genes are Made of DNA—Genetic Trans-
      formation
    Three Nucleotides Code for One Amino
      Acid
    Artificial mRNAs Were Used to Decipher
      the Genetic Code
    Synonyms in the Code—61 Codons for
      20 Amino Acids
    AUG—The Initiation Codon; UAG, UAA,
      and UGA—Termination Codons
    The Genetic Code is Universal
    Summary: *The Genetic Code*
21–2   Mutations and the Genetic Code   472
    Suppressor Mutations—Reversal of the
      Effect of Point Mutations

    Chemical Mutagens can be Very Specific
    Summary: *Mutations and the Genetic*
      Code
21–3   DNA Sequencing and the Genetic
    Code   474
    The 5375-Nucleotide Sequence of ϕX174
      Revealed Overlapping Genes
    Nucleic Acid Sequencing Confirmed the
      Genetic Code
    Summary: *DNA Sequencing and the Con-*
      firmation of the Genetic Code
21–4   Genetic Engineering—Restriction
    Endonuclease   478
    Restriction Enzyme Recognize Specific
      DNA Sequences
    Eukaryotic Genes can be Introduced
      into Plasmids and Cloned in E. coli
    Summary: *Genetic Engineering*

We now know that deoxyribonucleic acid (DNA) acts as the genetic material of cells, carrying information in a coded form from cell to cell and from parent to offspring. When a gene is expressed, its information is copied into another nucleic acid, *ribonucleic acid* (RNA), which in turn directs the synthesis of the ultimate gene products, the specific *proteins*. RNA is also the genetic material of some viruses which contain RNA instead of DNA. These concepts, which constitute *the central dogma of molecular biology,* were summarized by F. Crick in the following diagram:

template mechanism; (2) *transcription* of this information into RNA molecules; and (3) *translation* of this information into the various protein components of a cell (including the enzymes).

The student should remember that in molecular biology the word *transcription* is used as a synonym for RNA synthesis, and *translation,* as a synonym for protein synthesis.

An exception has been found to the flow diagram of the central dogma. Transcription may occasionally be reversed in some RNA-containing viruses. In other words, *RNA can sometimes be transcribed into DNA,* by the action

REPLICATION   DNA   TRANSCRIPTION→   RNA   TRANSLATION→   PROTEIN

The diagram shows that there are three main steps in the flow of genetic information from the genome (DNA): (1) *replication* of the DNA molecule, and thus of its genetic information, by a

of the enzyme *reverse transcriptase* (Chap. 16). Translation of RNA into protein is unidirectional and cannot be reversed.

The basic mechanisms by which genetic information is expressed were initially worked out using simple organisms, such as bacteria and viruses. The best understood cell is *Escherichia coli*, a bacterium that lives in the human intestine and is well suited for genetic and biochemical studies. The same genetic principles that operate in *E. coli* are also true for higher plant and animal cells. For this reason, in the next few chapters, reference will be made frequently to studies using *E. coli*, although the subject of this book is the eukaryotic cell.

In this chapter we will analyze the way in which DNA (which is made of four types of nucleotides) codes the information for proteins (which are made of 20 amino acids). We shall see that in the *genetic code,* each amino acid is specified by a group of three nucleotides. The genetic code is universal; i.e., it is the same for all living organisms. Deciphering the genetic code was one of the most important scientific achievements of this century; we will briefly review here the way in which this was done.

The last part of this chapter is devoted to two important topics: (1) DNA sequencing and (2) the isolation of eukaryotic genes using "genetic engineering" techniques by which eukaryotic genes can be grown in *E. coli*. These subjects are included because studies using these techniques have produced a revolution in our knowledge of how genes are organized.

21–1 THE GENETIC CODE

Genes Code for Proteins—Inborn Errors of Metabolism

A fundamental step forward in our understanding of molecular genetics was the *one gene– one enzyme* hypothesis put forward by Beadle and Tatum in 1948. They studied metabolic pathways in *Neurospora crassa* and found that any step of a metabolic chain could be blocked by a mutation in a specific enzyme, each enzyme representing a different gene product. We now know that one enzyme may be formed by several different polypeptides, so the modern version of Beadle and Tatum's hypothesis is *one gene–one polypeptide chain.* Not all genes code for proteins, however; some code for RNAs, such as ribosomal and transfer RNAs.

In man, the relationship between genes and specific proteins is clearly shown by the hereditary diseases known as *inborn errors of metabolism.*

By 1930 it was discovered that certain patients who had a severe mental disorder excreted an abnormal compound in the urine *(phenylpyruvic acid)*. The disease was called *phenylketonuria* and was found to be associated with a recessive gene. In fact, this disease manifests itself only in homozygous individuals; the probability of this

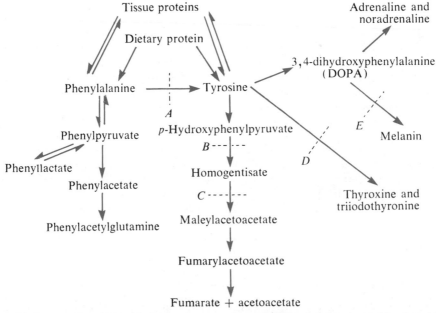

Figure 21–1 Possible genetic blocks in the normal metabolism of the amino acids phenylalanine and tyrosine in humans. These blocks lead to the production of the following genetic syndromes: *A,* phenylketonuria; *B,* tyrosinosis; *C,* alkaptonuria; *D,* goitrous cretinism; *E,* albinism. (From Harris)

condition appearing is increased by consanguineous unions (e.g., cousin-cousin or uncle-niece). In this disease, phenylalanine, a normal amino acid of the diet, cannot be oxidized to tyrosine and is transformed into phenylpyruvic acid. (Fig. 21–1, *A*) The primary action of the mutation has been to produce an *absence of the enzyme* needed for normal metabolism. The mental disorder, which can lead to idiocy or imbecility, is the result of the accumulation of phenylalanine and phenylpyruvic acid, both of which reach a toxic level in the central nervous system. If the disorder is discovered early enough, the mental disease can be prevented by a special diet that is low in phenylalanine. There are several other "inborn errors of metabolism," such as *tyrosinosis, alkaptonuria, goitrous cretinism,* and *albinism,* in which blocks in the metabolism of phenylalanine and tyrosine are involved.

Table 21–1 shows some of the human diseases caused by mutations in which the altered protein is known. Cultured fibroblasts from these patients can be very useful in cell biology studies; for example, *Lesch-Nyhan* cells allow the

study of X chromosome inactivation (Chap. 20–2) and *xeroderma pigmentosum* cell (Fig. 16–21) facilitate the analysis of DNA repair.

Mutations Produce Amino Acid Changes

Direct proof that genes determine the structure of proteins was provided by Ingram in 1957 when he showed that the mutation in sickle cell anemia causes the replacement of a specific amino acid in hemoglobin. An inherited disorder found principally in Blacks, sickle cell anemia is characterized by a change in the shape of the red blood cells as they travel through the veins. Owing to the decrease in oxygen tension, the erythrocytes become sickle-shaped (Fig. 21–2), a condition which may cause rupture of the cell membrane, followed by severe hemolytic anemia.

The study of the distribution of this disease within a family shows that it is caused by a recessive gene. A homozygous recessive individual has sickle-shaped cells and suffers from anemia, whereas a heterozygous individual has

TABLE 21–1   SOME HEREDITARY DISORDERS IN MAN IN WHICH THE SPECIFIC LACKING OR MODIFIED ENZYME OR PROTEIN HAS BEEN IDENTIFIED*

Disorder	*Affected Enzyme or Protein*
Acatalasemia	Catalase
Afibrinogenemia	Fibrinogen
Agammaglobulinemia	γ-Globulin
Albinism	Tyrosinase
Alkaptonuria	Homogentisic acid oxidase
Analbuminemia	Serum albumin
Galactosemia	Galactose-1-phosphate uridyl transferase
Glycogen storage diseases:	
Type I (von Gierke's)	Glucose-6-phosphatase
Type III	Amylo-1, 6-glucosidase
Type IV	Amylo-(1, 4 → 1, 6)-transglycosylase
Type V (McArdle's)	Muscle phosphorylase
Type VI (Hers')	Liver phosphorylase
Goiter (familial)	Iodotyrosine dehalogenase
Hemoglobinopathies	Hemoglobins
Hemophilia A	Antihemophilic factor A
Hemophilia B	Antihemophilic factor B
Histidinemia	Histidase
Hyperbilirubinemia (Gilbert's disease)	Uridine diphosphate glucuronate transferase
Hypophosphatasia	Alkaline phosphatase
Lesch-Nyhan syndrome	Hypoxanthine-guanine phosphoribosyl transferase
Maple syrup urine disease	Amino acid decarboxylase
Methemoglobinemia	Methemoglobin reductase
Phenylketonuria	Phenylalanine hydroxylase
Wilson's disease	Ceruloplasmin
Xanthinuria	Xanthine oxidase
Xeroderma pigmentosum	DNA repair enzymes

*Modified from White, A., Handler, P., and Smith E., *Principles of Biochemistry,* New York, McGraw-Hill Book Co., 1964.

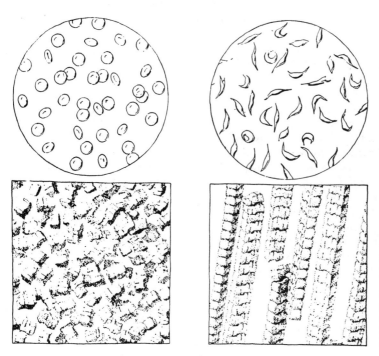

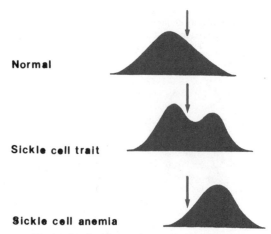

Figure 21–2 Above, left, normal erythrocytes; right, erythrocytes from venous blood of a patient with sickle cell anemia. Below, left, the hemoglobin molecules are randomly distributed in a normal individual; right, disposition of molecules of the venous blood in sickle cell anemia. In this case the hemoglobin molecules are in a crystalline array, which produces birefringence and deformation of the erythrocytes. (From L. Pauling.)

sickling, but no other symptoms of the disease.[1]

The molecular basis of this genetic disease was discovered through studies of the hemoglobin (Hb) molecule.[2] Abnormal hemoglobin was found to have a different electrophoretic behavior. Normal hemoglobin (HbA) and sickle cell hemoglobin (HbS) differ in their net surface charge and thus move differently in an electric field. Note in Figure 21–3 that a heterozygous individual has both HbA and HbS in about equal quantity.

Hemoglobin is a tetrameric protein with a molecular weight of 64,500 and contains 600 amino acids arranged in four polypeptide chains: two identical α-chains and two identical β-chains. The amino acid sequence of the hemoglobin is controlled by two structural genes, α and β. It has been demonstrated that the chemical abnormality of HbS resides in a change of a single amino acid. As shown in Table 21–2, the sixth amino acid from the N-terminus of the β-peptide chain *glutamic acid* is replaced by *valine*.[2] In another abnormal hemoglobin (HbC), glutamic acid is replaced by *lysine* in the same position. In this case the abnormality leads only to mild anemia. From this analysis it is evident that, at the genetic level, the mutation has occurred only in the β structural gene for hemoglobin. As shown in the same table, the DNA codons coding for the amino acid have changed only in a

single letter (see below). In the heterozygous individual, one β structural gene is abnormal, and the other is normal. This type of mutation — and others that can be found in hemoglobin in which only one amino acid is replaced — correspond to the so-called *point mutations* of bacterial genetics.

Normal

Sickle cell trait

Sickle cell anemia

Figure 21–3 Electrophoretic behavior of various human hemoglobins. **Upper diagram,** normal homozygous dominant. **Center diagram,** sickle cell trait heterozygous. **Lower diagram,** sickle cell anemia. The arrows indicate the reference point of origin of the electrophoretic pattern. (After L. Pauling.)

TABLE 21–2 CHEMICAL DIFFERENCES IN THE SEQUENCE OF AMINO ACIDS IN HUMAN HEMOGLOBINS*

	Amino Acid Sequence†	*Codon*
HbA	$\overset{+}{}\ \overset{+}{}\ \overset{-}{}\ \overset{-}{}\ \overset{+}{}$ NH₃-Val-His-Leu-Thr-Pro-*Glu*-Glu-Lys. . . .	GAA, GAG
HbS	$\overset{+}{}\ \overset{+}{}\ \overset{-}{}\ \overset{+}{}$ NH₃-Val-His-Leu-Thr-Pro-*Val*-Glu-Lys. . . .	GUA, GUG
HbC	$\overset{+}{}\ \overset{+}{}\ \overset{+}{}\ \overset{+}{}$ NH₃-Val-His-Leu-Thr-Pro-*Lys*-Glu-Lys. . . .	AAA, AAG

*From Ingram, V. M., *The Biosynthesis of Macromolecules*. Menlo Park, Calif., W. A. Benjamin, Inc., 1965, p. 160. Copyright © 1965, W. A. Benjamin, Inc., New York and Amsterdam.

†Glu, glutamic acid; His, histidine; Leu, leucine; Lys, lysine; Pro, proline; Thr, threonine; Val, valine.

Genes are Made of DNA—Genetic Transformation

Long after Mendel discovered that genes were inherited as discrete factors, cytogenetic observations suggested that the hereditary material was located in the chromosomes. In 1924 Feulgen developed his well-known histochemical reaction and showed that chromosomes contain DNA (Chap. 15). However, for a long time it was thought that the chromosomal proteins would turn out to be the genetic material, mainly because the composition of DNA (four nucleotides) was considered far too simple to code for the great diversity of proteins. The cytophotometric studies on Feulgen-stained cells by Pollister, Mirsky, and others (Chap. 15) showed that eukaryotic nuclei have a *constant* amount of DNA (Tables 15–1 and 15–2) and drew attention to the role of DNA in heredity. However, by then the demonstration that DNA is the genetic material had already been made in experiments with microorganisms.

In 1928 Griffith found that non-pathogenic strains of *Pneumococcus* could be *transformed* into virulent bacteria if injected into mice together with a dead (heated) culture of a pathogenic *Pneumococcus*. Avery and collaborators purified the transforming factor and showed in 1944 that a preparation of pure *Pneumococcus* DNA had this activity.[3]

The molecular events involved in *Pneumococcus* transformation are now known in detail. Figure 21–4,*A* shows that in *transformation* DNA liberated from one bacterium can be taken up by another, where it may recombine with the host chromosome. In Avery's experiment the DNA transferred the gene coding for an enzyme (UDP-glucose-dehydrogenase) required for synthesis of the extracellular polysaccharide capsule that is present in virulent *Pneumococci* (and absent in the non-pathogenic strains).

The ability for DNA to be transferred from one bacterium to another is essential in genetics, for it allows one to study the frequency of recombination between different mutations and to construct genetic maps. (The closer two genes are, the less frequently they will recombine; see Chapter 19.) Transformation has also made possible modern genetic engineering technology, whereby foreign fragments of DNA are joined in vitro into small bacterial chromosomes (called *plasmids*) and then incorporated into *E. coli* by

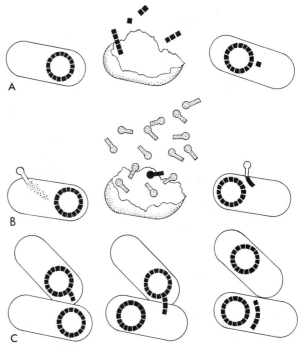

Figure 21–4 Diagram showing the three methods by which DNA can be transferred from one bacterium to another. **A**, in *transformation* the bacteria are destroyed and the DNA that is liberated penetrates another bacterium. **B**, in *transduction* the bacteriophage carries DNA from one bacterium to another. **C**, in *conjugation* the DNA is transferred directly by pairing and sexual recombination.

transformation, as explained later on in this chapter. Figure 21–4 shows two other ways of gene transfer. In *transduction,* fragments of bacterial DNA become incorporated into bacteriophage (bacterial virus) particles, which can then inject that DNA into other bacteria (Fig. 21–4,*B*). In *conjugation* (Fig. 21–4, *C*) bacteria of different sexes pair and recombine sexually. These and other mechanisms of recombination analysis are widely used in molecular genetics and have permitted the establishment of very detailed genetic maps of bacterial chromosomes.

It has recently been found that even cultured mammalian cells can take up DNA added to the culture medium, which is subsequently integrated into the chromosomes (Wigler et al., 1979). However, this can be achieved only in a small proportion of the cells in the culture, and it is necessary to select for the few successful transformants. This can be done by cotransformation with DNA containing the genetic coding for the enzyme thymidine kinase, using the HAT selection method discussed in Figure 20–14. It is expected that transformation of cultured mammalian cells will permit genetic manipulations in vertebrates that were previously possible only in bacteria.

Although DNA is the genetic material in most organisms, some viruses code their information in RNA. In both cases the same four-letter code is used to store the information for protein synthesis.

Three Nucleotides Code for One Amino Acid

The *codons;* i.e., the hereditary units that contain the information coding for one amino acid, consist of three nucleotides (a triplet). This information is first *transcribed* into the messenger RNA, which has a sequence of bases complementary to the DNA from which it is copied. DNA and mRNA have only 4 different bases, whereas proteins contain 20 different amino acids. Deciphering how cells translate a 4 letter code into one of 20 components was a major scientific achievement.

The code is read in groups of three bases, each triplet representing one amino acid. Three is the minimum number of bases needed to code for 20 amino acids; the possible permutations of the four bases are $4^3 = 64$. If the genetic code consisted of doublets, the number of codons would be insufficient ($4^2 = 16$), and if groups of four bases were utilized, the possibilities would be many more than necessary ($4^4 = 256$).

The length of the coding portion of a gene depends on the length of the message to be translated, i.e., the number of amino acids in the protein. For example, a sequence of 1500 nucleotides may contain 500 codons that code for a protein having 500 amino acids. The message is read in groups of three from a *fixed starting point.* The initial amino acid of a protein is determined by special *initiation codons* (see later discussion).

The sequence of triplets determines the sequence of amino acids in a protein. Amino acids, however, cannot by themselves recognize a given triplet of mRNA; in order to do this, each amino acid is first attached to an *adaptor* molecule, which is called *transfer* RNA (tRNA). The structure of tRNA is described in detail in Chapter 23; here it is enough to note that every tRNA molecule has an *amino acid attachment site* and a site for the recognition of the triplets in mRNA (*anticodon*). Each tRNA has an *anticodon* of three nucleotides which can base-pair with the complementary *codon* of mRNA. The translation of the message into protein occurs in the *ribosomes,* which ensure the ordered interaction of all the components involved in protein synthesis (Chap. 23).

Artificial mRNAs Were Used to Decipher the Genetic Code

The breakthrough that led to the solution of the genetic code came in 1961 when Nirenberg and Mattaei discovered that synthetic polyribonucleotides used as artificial mRNAs stimulated the incorporation of amino acids into polypeptides in a cell-free protein-synthesizing system.[4] The first RNA used was polyuridylic acid (poly U), and the result was synthesis of polyphenylalanine (a polypeptide made only of phenylalanine). Thus, it was deduced that the codon for phenylalanine was made exclusively of U (UUU).

The cell-free protein-synthesizing system was an extract of *E. coli* which had been broken open and from which the cell walls had been removed by centrifugation. This extract contained the ribosomes, tRNA, aminoacyl-tRNA synthetases, and other factors required for protein synthesis; and radioactive amino acids were incorporated after the addition of ATP, GTP and exogenous mRNA. (The endogenous mRNAs from *E. coli* are short-lived and are degraded after preincubation of the extract at 37°C for a few minutes.) In addition to poly U, other homopolymers were tested, poly A stimulated the uptake of lysine (code word AAA), and poly C, that of proline (codon CCC). (Poly G forms complex triple-stranded structures in solution and is therefore unable to function as a synthetic mRNA.)

The use of synthetic RNAs of known composition was made possible by the earlier discovery by Ochoa that the enzyme polynucleotide phosphorylase could polymerize nucleoside diphosphates, producing artificial RNAs of composition similar to the ratio of the different nucleotides in the incubation mixture. *Polynucleotide phosphorylase* found in vivo is not used for synthesizing RNA but rather for degrading it. RNA is synthesized by a different enzyme, *RNA polymerase,* which uses nucleoside triphosphates as substrates and which requires DNA as a template (polynucleotide phosphorylase uses diphosphates and does not copy a template).

The base composition of many codons was found using polynucleotides formed by two bases. Thus, a random copolymer of U and G contains eight different triplets (UUU, UUG, UGU, GUU, UGG, GUG, GGU, and GGG) and directs the synthesis of several amino acids, as indicated in Table 21–3.

The use of copolymers allowed the determination of the *composition* of codons but not the order

in which the nucleotides were placed in the codon;[5] for example, UUG could not be distinguished from UGU or GUU by this method. The recognition of the *sequence* of codons was later made possible by the use of trinucleotide templates of known base composition. In 1964 Leder and Nirenberg[6] found that trinucleotides induced the binding of the specific AA-tRNAs (aminoacyl-tRNAs) to ribosomes. The complexes formed when ribosomes are incubated with the various triplets and ^{14}C-AA-tRNA could be easily detected, since they were retained by nitrocellulose filters, while unbound AA-tRNA passes through the filter. The short triplets could be made using standard organic chemistry methods of synthesis, and about 50 codons were soon deciphered using the binding assay. For example, UUG induced the binding of leucine-tRNA, UGU bound cysteine-tRNA, and GUU, valine-tRNA.

Another method which allowed the recognition of the *sequence* of codons was developed in the laboratory of Khorana.[7] Using chemical synthetic and enzymatic methods, these workers were able to produce long polyribonucleotides with alternating doublets or triplets of known sequence. These synthetic mRNAs were then used to direct protein synthesis in a cell-free system. When an alternating doublet was used, a

TABLE 21–3 THE GENETIC CODE

1st Base									3rd Base
	2nd Base								
	U		**C**		**A**		**G**		
U	UUU	Phe	UCU	Ser	UAU	Tyr	UGU	Cys	U
	UUC	Phe	UCC	Ser	UAC	Tyr	UGC	Cys	C
	UUA	Leu	UCA	Ser	UAA	Termination	UGA	Termination	A
	UUG	Leu	UCG	Ser	UAG	Termination	UGG	Trp	G
C	CUU	Leu	CCU	Pro	CAU	His	CGU	Arg	U
	CUC	Leu	CCC	Pro	CAC	His	CGC	Arg	C
	CUA	Leu	CCA	Pro	CAA	Gln	CGA	Arg	A
	CUG	Leu	CCG	Pro	CAG	Gln	CGG	Arg	G
A	AUU	Ile	ACU	Thr	AAU	Asn	AGU	Ser	U
	AUC	Ile	ACC	Thr	AAC	Asn	AGC	Ser	C
	AUA	Ile	ACA	Thr	AAA	Lys	AGA	Arg	A
	AUG	Met	ACG	Thr	AAG	Lys	AGG	Arg	G
	AUG	fMet							
G	GUU	Val	GCU	Ala	GAU	Asp	GGU	Gly	U
	GUC	Val	GCC	Ala	GAC	Asp	GGC	Gly	C
	GUA	Val	GCA	Ala	GAA	Glu	GGA	Gly	A
	GUG	Val	GCG	Ala	GAG	Glu	GGG	Gly	G

polypeptide chain formed by two alternating amino acids was formed. For example, with the alternation of U and G the result was:

GUG UGU GUG UGU GUG
Val — Cys — Val — Cys — Val

This showed that GUG and UGU code for valine (Val) and cysteine (Cys) and confirmed that each codon is a triplet. When the template consisted of three alternating bases, various polypeptide chains containing only one type of amino acid each were obtained. For example, the following alternation of UUGs was used:

UUG UUG UUG UUG UUG UUG

Since this message can be read in three "frames," each one out of phase by one base, three homo-polypeptide chains result as follows (see Table 21–3):

UUG   UUG   UUG   UUG   UUG
Leu — Leu — Leu — Leu — Leu

U   UGU   UGU   UGU   UGU   UG
    Cys — Cys — Cys — Cys

UU   GUU   GUU   GUU   GUU   G
     Val — Val — Val — Val

Synonyms in the Code — 61 Codons for 20 Amino Acids

By 1966 all 64 possible codons had been deciphered. As shown in Table 21–3, 61 codons correspond to amino acids, and three represent signals for the termination of polypeptide chains. Since there are only 20 amino acids, it is clear that several triplets can code for the same amino acid; i.e., some triplets are synonyms. (This fact is sometimes called *degeneracy of the genetic code.*) For example, proline is encoded by CCU, CCA, CCG, and CCC. Note that in most cases the synonymous codons differ only in the base occupying the third position of the triplet and that the first two bases of the codon are more important in coding. As a result, mutations that change the third base frequently go unnoticed, since they may not change the amino acid composition of the protein.

The DNA sequencing studies have confirmed that all the possible codons are utilized in vivo. The fact that all the possible sequences are used probably has selective advantages, since it minimizes the effect of harmful mutations. If codons had been left without a meaning, muta-

tions giving rise to them would interfere severely with the normal sequence of protein synthesis.

The use of 61 different codons to code for 20 amino acids posed an additional question: Is each triplet recognized by a special tRNA molecule? It was found that there are less than 61 tRNA species and that one tRNA could recognize more than 1 codon (although always for the same amino acid). Crick[8] proposed that the third base of the anticodon (the one that is less important in coding) could have some degree of "wobble," i.e., a certain amount of movement that allows this base to establish hydrogen bonds with bases other than the normal complementary ones. This prediction was found to be correct; for example, G in the third position of an anticodon can base-pair with U or C in the mRNA, while U in the anticodon can interact with A or C in the third position of the codon.

AUG — The Initiation Codon
UAG, UAA, and UGA — Termination Codons

Figure 21–5 shows that mRNA is translated in the 5'→3' direction (which is also the direction of protein synthesis). The polypeptide chain is always assembled sequentially, starting from the end bearing the NH_2-terminus.

The *initiation signal* for the synthesis of a protein is the AUG codon. This codon has a dual function. When the AUG is at the beginning of a bacterial message *(starting codon)*, it will code for the incorporation of N-formylmethionine, and AUG in any other position of the message will code for normal methionine. There is a special fMet-tRNA which is different from the tRNA for normal methionine. In order to bind to ribosomes, fMet-tRNA requires a special protein (initiation factor 2), while Met-tRNA uses the same elongation factor used by all other tRNAs. (The initiation of protein synthesis is dealt with in detail in Chapter 24.) In most cases AUG is the initiation codon, but less frequently protein synthesis can start with GUG. Initiation with GUG uses the same fMet-tRNA as an initiator AUG. In eukaryotes there is also a special initiation tRNA, but it directs the incorporation of methionine instead of formylmethionine.

The *termination signal* is provided by the codons UAG, UAA, and UGA (sometimes called *nonsense* codons, see Table 21–3, Fig. 21–5,C). When the ribosome reaches a termination codon, the completed polypeptide chain is released. Unlike all other codons, UAG, UAA, and UGA are

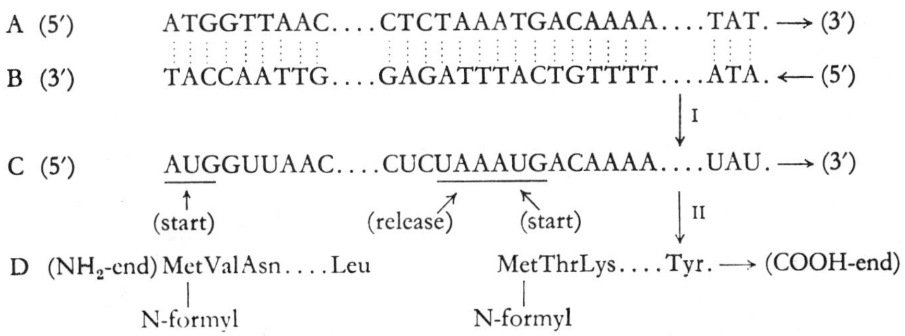

A (5')   ATGGTTAAC....CTCTAAATGACAAAA....TAT. → (3')

B (3')   TACCAATTG....GAGATTTACTGTTTT....ATA. ← (5')

                                                                $\downarrow$ I

C (5')   AUGGUUAAC....CUCUAAAUGACAAAA....UAU. → (3')
              ↑   ↗   ↖
            (start)   (release) (start) $\downarrow$ II

D (NH₂-end) MetValAsn....Leu   MetThrLys....Tyr. → (COOH-end)
              | |
            N-formyl   N-formyl

Figure 21–5 Diagram illustrating the transcription and translation steps in the expression of genetic information. **I**, transcription; **II**, translation. **A**, DNA strand 5'→3'; **B**, DNA strand 3'→5'; **C**, polycistronic messenger RNA 5'→3' copied from **B**; **D**, polypeptide chains. The starting and termination codons are underlined. Note that mRNA is translated in the 5' to 3' direction and that proteins are synthesized starting with the amino terminus and ending with the carboxyl end.

not recognized by special tRNAs, but by special proteins, the *releasing factors* of protein synthesis (Chap. 24–2).

The Genetic Code is Universal

Although most of our knowledge about the genetic code comes from experiments with *E. coli*, essentially similar results have been obtained with other systems, such as amphibian, mammalian, and plant tissue. It may be said that the *genetic code* is universal, i.e., there is a single code for all living organisms.

The genetic code developed at the same time as the first bacteria, some three billion years ago, and since then it has changed relatively little throughout evolution of living organisms. Once the initial code evolved, there were strong selective pressures to maintain it invariant, for the change in a single codon assignment would disrupt too many pre-existing proteins and would therefore be lethal.

SUMMARY:

The Genetic Code

The flow of genetic information is as shown in the following diagram:

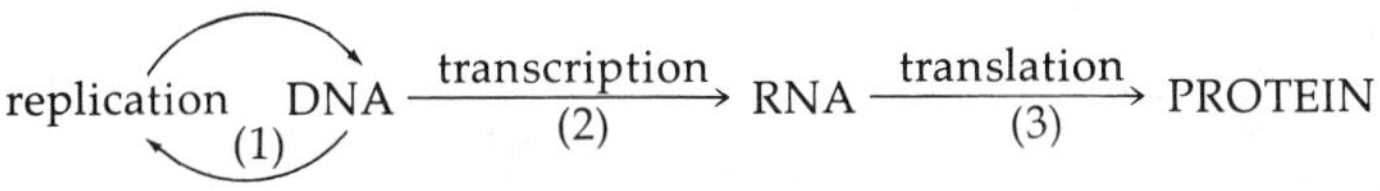

replication DNA $\xrightarrow[\text{(2)}]{\text{transcription}}$ RNA $\xrightarrow[\text{(3)}]{\text{translation}}$ PROTEIN
        (1)

Step (3) is undirectional, but step (2) can be reversed (RNA→DNA) by means of the enzyme *reverse transcriptase.* The information contained in DNA and in RNA is based on four nucleotides, while proteins are made of 20 amino acids. Cells can translate this four-letter code into one of 20 components.

The demonstration that DNA is the informational molecule came mainly from the experiments of Avery, in which DNA from virulent *Pneumococci* penetrated the nonvirulent type and produced its *transformation.* Pieces of DNA can also be incorporated into bacteria by way of plasmids, by *transduction,* and by *conjugation.*

The *one gene-one enzyme* hypothesis is now better represented by *one gene–one polypeptide chain* and is illustrated by some of the human hereditary diseases, also called *inborn errors of metabolism.* For example, phenylketonuria, albinism, galactosemia, and cretinism are caused by the absence of single proteins (enzymes) (Table 21–1). Furthermore, mutation of a single amino acid in hemoglobin can cause sickle cell anemia.

The genetic code is read in groups of three nucleotides. The units of genetic information are the *codons,* i.e., the triplets of bases that specify single amino acids. The genetic code consists of $4^3 = 64$ codons, and the length of a gene is related to the number of amino acids in the protein (e.g., 1500 nucleotides constitute 500 codons coding for 500 amino acids). The DNA code is transcribed into a RNA code which is complementary to one DNA strand, and on the ribosomes the message is read by an adaptor molecule, the transfer RNA (tRNA), which carries the amino acid.

The genetic code was deciphered by the use of synthetic mRNAs made of a single type of nucleotide, of polynucleotides containing two or three types of bases, and of trinucleotide templates of known base composition. Chemical synthesis and enzymatic methods produced alternating doublets or triplets forming polyribonucleotides that yielded various polypeptides and permitted further studies of the code. Of the 64 codons, 3 represent termination signals, and 61 code for the 20 amino acids (Table 21–3). Thus, some triplets are synonyms (i.e., the code is degenerate). The number of tRNAs that carry the anticodons is less than 61. This discrepancy is explained by the wobble of the third base in the anticodon of tRNA, which can base-pair with more than one base.

The initial or starting codon is AUG, which codes for N-formylmethionine in prokaryotes and for methionine in eukaryotes. The termination or nonsense codons are UAG, UAA, and UGA. The genetic code is universal, i.e., there is a single code for all living organisms.

21–2 MUTATIONS AND THE GENETIC CODE

There is much genetic evidence that base changes in the DNA sequence produce mutations such as those described for the human hemoglobins (Table 21–2). Mutations may involve several base pairs, as in large deletions of segments of DNA, or they may alter the genetic code by a *point* or *single base* mutation. *Point* mutations may be classified in the following four groups: (1) *transitions,* in which a pyrimidine base is replaced by another pyrimidine base (i.e., C $\rightleftharpoons$ T) or a purine is replaced by another purine (i.e., A $\rightleftharpoons$ G); (2) *transversions,* in which a purine is substituted by a pyrimidine or vice versa (i.e., A $\rightleftharpoons$ C or G $\rightleftharpoons$ T); (3) *single base deletions;* and (4) *single base insertions.*

Single base changes may have a very profound effect on protein synthesis. *Missense mutations* change the meaning of a codon from one amino acid to another; for example, in sickle cell anemia a codon changes from GAA to GUA, and a valine is incorporated instead of a glutamic acid residue (Table 21–2).

Frameshift mutations arise from the insertion or deletion of single bases and cause the rest of the message downstream of the mutation to be read out of phase, producing an incorrect protein. For example, in a hypothetical mRNA the codons would normally be translated as follows:

UAU   CCA   UAU   CCA   UAU

Tyr — Pro — Tyr — Po — Tyr

The insertion of a single G residue between the third and fourth bases produces a completely different protein from there on:

inserted base
↓

UAU   GCC   AUA   UCC   AUA   U

Tyr — Ala — Ileu — Ser — Ileu

Another type of mutation that was very important in understanding the genetic code is the case of chain termination or *nonsense* mutations. As shown in Figure 21–6, these mutations arise when a codon coding for an amino acid is changed into a chain termination codon (UAG, UAA, or UGA), resulting in the production of a much shorter protein. Nonsense mutations very rarely go unnoticed, since the incomplete protein is in general inactive, while in *missense* mutations the change in one amino acid is frequently compatible with some biological activity.

Suppressor Mutations — Reversal of the Effect of Point Mutations

The phenotypic effects of point mutations can sometimes be reversed by a second mutation in a different gene, which is called a *suppressor* mutation. There are several types of suppressor mutations, but the most interesting are tRNA mutations that act as *nonsense* (chain termination) *suppressors.* These mutations substitute a base in the anticodon of a tRNA, so that it will now incorporate an amino acid at

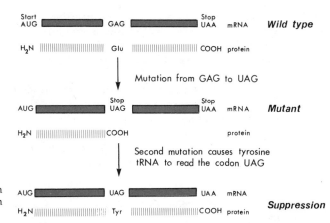

Figure 21–6 Suppression of a chain termination mutation by a second mutation on a tRNA molecule.

the codon that has mutated to a chain termination codon (see Garen, 1968). In the example shown in Figure 21–6, a mutation in the tyrosine tRNA causes it to read a UAG (chain termination) codon and insert tyrosine at its site, thus allowing completion of the polypeptide chain, although with one amino acid change. The tRNA mutations that suppress UAG codons are sometimes called *amber* suppressors for historical reasons (due to the aspect of the plaques of the T_4 bacteriophage mutants in which the phenomenon was initially found,[9] and UAA suppressors are sometimes called *ochre* suppressors.[10]

A very interesting case is that of mutant tRNAs that can suppress frameshift mutations caused by insertions of single bases.[11] *Frameshift suppressor* tRNAs have an extra base in the anticodon (for example CCCC instead of CCC) and read a four-letter word instead of a triplet, thereby restoring the correct reading frame.[12]

Chemical Mutagens Can be Very Specific

Some chemical mutagens can induce base changes in DNA in a rather specific way. Some mutagens can change one base into another and may act even on isolated DNA. For example, *nitrous acid* can deaminate adenine in DNA to hypoxanthine, which is then complementary to cytosine (instead of thymine). Similarly, *hydroxylamine* reacts with cytosine, giving a derivative that pairs with adenine rather than guanine. Other mutagens are base analogues (e.g., *5-bromouracil, 2-aminopurine, 5-bromodeoxyuridine,* and *5-fluorodeoxyuridine*) that are incorporated only during the period of DNA replication. Some alkylating agents, such as *ethylmethane sulfonate,* can produce ethylation in the purine ring. Others may *intercalate* between the DNA bases (e.g., fluorescent dyes such as acridine and proflavine) and interfere with DNA replication, producing the *insertion* or *deletion* of single bases.

Mutagens are very important from the medical point of view, because most of them are also powerful carcinogens. Bruce Ames has developed a very sensitive test for the detection of mutagens using *Salmonella* mutants which are defective in histidine biosynthesis. The bacteria cannot grow on culture plates without histidine unless a second mutation that restores histidine biosynthesis is produced. Many substances are not carcinogenic or mutagenic as such but are converted into mutagens by hydroxylating enzymes in the liver. Ames therefore added liver homogenate to his cultures and greatly increased the mutagen sensitivity of the test.[13, 14] These methods are very useful for detecting the mutagens that man has introduced into his diet and daily life. Some products, such as commonly used hair dyes and flame retardants used in children's sleepwear, were unexpectedly found to be highly mutagenic. In other cases, the mutagenic effect was more predictable, as with tobacco smoke condensate.

SUMMARY:

Mutations and the Genetic Code

Changes in the base sequence of DNA produce gene mutations. These may involve a *deletion* of a DNA segment or a change in a single base *(point mutation)*. In *missense mutations* one amino acid is replaced by another (e.g., sickle cell anemia, Table

21–2). *Frameshift* mutations arise through insertion or deletion of single bases, which causes the message to be read out of phase. *Chain termination* or *nonsense mutations* occur when mutation of a codon results in a termination single, and, therefore in the synthesis of shorter protein. A point mutation may sometimes be reversed by a *suppressor mutation* (Fig. 21–6).

Mutations can be produced by the action of chemicals (i.e., *mutagens*) which may change one base into another (e.g., nitrous acid and hydroxylamine). Others are base analogues (e.g., 5-bromouracil, 2-aminopurine, 5-bromo- and 5-fluorodeoxyuridine) that act on DNA replication. Some mutagens are used in cancer chemotherapy. The test developed by Ames is important in detecting mutagenic agents that are frequently also carcinogenic (i.e., they induce cancer).

21–3 DNA SEQUENCING AND THE GENETIC CODE

The genetic code can also be deciphered by comparing the amino acid sequence of proteins with the nucleotide sequences of the genes that code for them. However, the methods for sequencing nucleic acids had not yet been developed in the 1960s, and, as we have seen, the code was solved using cell-free systems with artifical mRNAs. Methods that allow very rapid sequencing of DNA have been developed recently and have produced a revolution in molecular biology. The complete nucleotide sequences of living organisms such as bacteriophage ϕX174[15] (a virus that infects *E. coli*) and SV40[16, 17] (a small DNA virus that infects monkey cells) have been established.

The nucleotide sequences obtained by the newer methods confirmed the genetic code and established that all of the possible codons are used in vivo. They also provided completely unexpected insights into the way genes are organized. Because of its great importance, we will briefly analyze here how DNA is sequenced and the main conclusions derived from the ϕX174 work.

There are several DNA sequencing methods, all of them based on the generation of DNA fragments of different lengths which start at a fixed point and terminate at specific nucleotides. The DNA fragments are separated by size on polyacrylamide gels, and the nucleotide sequence is read directly from the gel.[18] The DNA fragments can be produced by chemical cleavage[19] or by enzymatic copying of single-stranded DNA.

Figure 21–7 outlines the different steps involved in Sanger's chain-termination method. Single-stranded DNA is copied using *DNA polymerase I* from *E. coli*. This enzyme will not work on single-stranded DNA unless a short primer is annealed to it, producing a stretch of double-stranded DNA (Fig. 21–7). The primer has on its deoxyribose a free 3'OH group to which the next nucleotide can be attached; this provides a fixed starting point. The short primers are generated using restriction endo-nucleases, which are enzymes that cut DNA at specific sites. The primer is then extended using DNA polymerase and radioactive nucleotides. The chains are terminated at specific bases by adding a 2', 3'-dideoxynucleotide. Once a 2',3'-dideoxynucleotide is incorporated, DNA elongation stops because it lacks a 3' OH group in the sugar moiety, thus preventing the attachment of the next nucleotide. Four parallel reactions are performed, each containing a small amount of one chain-terminating agent (either ddATP, ddGTP, ddCTP, or ddTTP). These produce DNA chains of various lengths that start at a unique site and end at specific bases (Fig. 21–7).

To read the sequence, the four reactions are electrophoresed on a polyacrylamide gel which separates the radioactive fragments according to size. The radio-autograph of one such gel is shown in Figure 21–8. The fragments are visualized as a series of bands, each differing in length by one nucleotide. Each of the four tracks in the gel indicates chains terminating at one of the four nucleotides. The DNA sequence is read directly from the gel. For example, the band in position 30 in Figure 21–8 is under the track labeled C and therefore represents a cytosine in the DNA sequence.[20]

Figure 21–8 is a good example of a sequencing gel, and the student is advised to read the sequence unassisted and compare the results with the sequence indicated in the figure legend. The sequence between positions 30 and 70 can be read rather easily.

The number of nucleotides that can be sequenced with these rapid methods is limited only by the resolution of the polyacrylamide gels; in a good experiment 200 nucleotides can be sequenced in only 1 day's work. These methods, together with new genetic engineering methods that allow the preparation of large amounts of purified genes (later discussion), have led to important advances in our understanding of how genes are organized, as exemplified by the work with φX174.

The 5375-Nucleotide Sequence of φX174 Revealed Overlapping Genes

φX174 is a small virus having a circular chromosome of single-stranded DNA. After it infects *E. coli,* the complementary strand is synthesized, and the now double-stranded chromosome replicates to produce more phage progeny. The sequence of the 5375 nucleotides of φX174, established by Fred Sanger and colleagues[20], unexpectedly revealed that sometimes genes can overlap, producing two proteins from the same stretch of DNA.

φX174 codes for 10 proteins (Fig. 21–9), and it was estimated from the molecular weight of the proteins that more than 6000 nucleotides would be required to code for them. The reason for the discrepancy between the size of the genome and the amount of DNA required to code for the ten proteins was explained when the complete nucleotide sequence became known.

Three of the genes of φX174 (genes E, B, and K in Fig. 21–9) overlap with other genes. In these regions, two proteins with different amino acid sequences are coded by the same stretch of DNA. This can be achieved because the proteins are encoded in different reading frames.[21, 22]

Figure 21–10 shows the nucleotide sequence at the start of gene E, which is completely contained within gene D. There is a ribosome recognition site and an ATG initiation codon present within the D gene, which are read by the ribosomes one nucleotide out of frame, thus producing two proteins from a single segment of mRNA.[21] The fact that genes E and D overlap was confirmed by sequencing mutations that generate chain termination codons when the E frame is read, producing a shorter E protein while not significantly affecting the D gene (see Fig. 21–10, bottom line).

The most impressive case of genetic coding

(1) Primer (restriction fragment) is annealed to single-stranded DNA template.

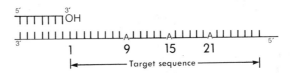

(2) Extension with DNA polymerase and four ^{32}P nucleoside triphosphates. Termination occurs at specific nucleotides because a small amount of one of the four dideoxynucleotides is added to each reaction mixture. For example, ddTTP stops the copy at T.

(3) Gel

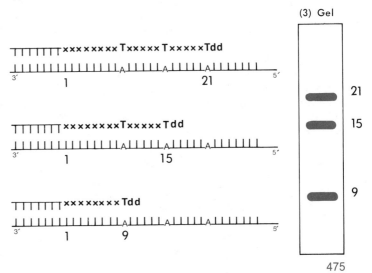

Figure 21–7 Steps in Sanger's chain-termination sequencing method. (1) Single stranded DNA is copied by DNA polymerase I from *E. coli.* (2) Extension of DNA copy is terminated, by adding a 2', 3' dideoxynucleotide which lacks the 3'-OH group and prevents the attachment of further nucleotides. (3) To read the sequence the labelled DNA is denatured and electrophoresed in a polyacrylamide gel (see Fig. 21–8).

economy is found in gene K (see Fig. 21–9), which overlaps with genes A and C. In this region, four nucleotides are read in the three possible translational phases,[22]

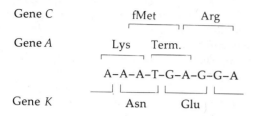

A sequence of 140 base pairs in which 2 protein coding genes overlap 1 nucleotide out of phase has also been found in simian virus 40 (SV40).[16, 17] It is possible that overlapping genes are peculiar to small viruses which can store only a limited amount of DNA within the viral capsid and which are under a strong selective pressure to use their DNA as efficiently as possible. On the other hand, cells from vertebrates contain large amounts of DNA and are probably not under the same constraints as viruses. Whether overlapping genes exist in animal cells or not will be known only as more gene sequences become available.

Nucleic Acid Sequencing Confirmed the Genetic Code

The knowledge of the nucleotide sequence in ϕX174 and SV40 and also of the amino acid sequence of some of their proteins has provided a beautiful confirmation of the genetic code.

The different codons utilized in ϕX174 and in SV40 are shown in Tables 21–4 and 21–5. The main conclusion to be derived from this data is that all 61 codons that code for amino acids are utilized in vivo. Since most amino acids are coded by several triplets (six in the case of leucine, see Table 21–3), it was conceivable that some were not utilized at all. Although all codons are utilized, some are used more than others, depending on the organism. ΦX174 uses more codons having U in the third position, probably reflecting the fact that its DNA has a high T content. In SV40, as in many eukaryotic mRNAs, there is a bias against the presence of a C followed by a G residue (CpG). (The reason for this is unknown, but it may be related to DNA methylation, for eukaryotes have enzymes that methylate cytosines preferentially when they are followed by a G residue.)

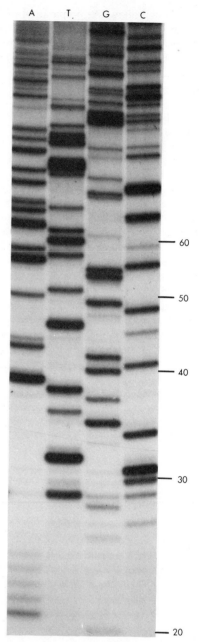

Figure 21–8 Polyacrylamide gel showing the sequence of a segment of bacteriophage ϕ × 174. Each channel represents DNA chains terminated at a specific nucleotide. The sequence can be read clearly (except for weaker G bands at positions 33, 47, 52 and 99). The sequence is:

GAAAAAGCGT	CCTGCGTGTA	GCGAACTGCG	ATGGGCATAC
20	30	40	50
TGTAACCATA	AGGCCACGTA	TTTTGCAAGC	TATTTAACTG
60	70	80	90

(Courtesy of W. Barnes)

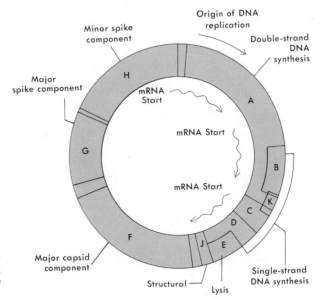

Figure 21–9 Genetic map of bacteriophage ϕX174. The position of the ten protein coding genes (A through K) is indicated. Note that genes E, B, and K are overlapping.

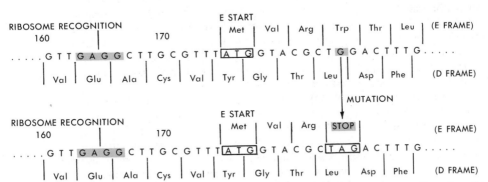

Figure 21–10 Overlap of D and E genes of ϕX. In the midst of the D-gene sequence there is a ribosome recognition signal and then an ATG initiation codon in a new reading frame, which begins the E gene. The E-gene frame was identified by the discovery of the mutation of a G to an A in a mutant virus (bottom line), changing the codon for the amino acid tryptophan (Trp) to a premature termination codon, TAG, that stops E-gene translation. The reading frame for D gene is not affected by the mutation since both CTG and CTA specify leucine (Leu). Since the DNA sequence is indicated, T is used (in an RNA sequence U would be used instead). (Redrawn from Fiddes, J. C., Sci. Am., 237(6):54, 1977.)

TABLE 21–4 CODON UTILIZATION IN φX174*

First Position	Second Position†				Third Position
	U	C	A	G	
U	39	35	36	12	U
	26	9	15	10	C
	19	16	3	5	A
	29	14	0	16	G
C	36	34	16	40	U
	15	6	7	29	C
	3	6	27	4	A
	24	21	34	8	G
A	45	40	37	9	U
	12	18	25	5	C
	2	13	47	6	A
	42	19	31	1	G
G	53	64	44	38	U
	14	17	35	28	C
	10	12	27	13	A
	11	12	34	3	G

*Data from Sanger, F., et al., *Nature*, 265:687 1977.
†The numbers indicate how many times each codon is present in the φX174 genome.

TABLE 21–5 CODON UTILIZATION IN SV40 mRNA*

First Position	Second Position†				Third Position
	U	C	A	G	
U	60	29	32	14	U
	4	9	23	18	C
	40	15	3	1	A
	39	0	0	26	G
C	23	38	21	2	U
	3	14	10	1	C
	16	25	42	0	A
	20	0	31	2	G
A	45	40	44	32	U
	0	24	25	12	C
	21	26	70	36	A
	40	0	35	26	G
G	38	72	58	20	U
	2	12	31	15	C
	21	21	60	35	A
	31	0	41	24	G

*Data from Reddy, V., et al., *Science, 200*:494, 1978.
†The numbers indicate how many times each codon is present in the SV40 genome.

SUMMARY:

DNA Sequencing and the Confirmation of the Genetic Code

Methods have been developed that allow the rapid sequencing of DNA and that have produced a revolution in molecular biology. One of the various sequencing methods is illustrated in Figure 21–7. All the rapid sequencing methods are based on the production of DNA fragments of different lengths which start at a fixed point and terminate at specific nucleotides. The reading of the sequence is done on polyacrylamide gels that separate the DNAs by size (Fig. 21–8). These methods have allowed the determination of the complete DNA sequence of two small viruses, φX174 and SV40. (SV40 can transform normal cells into cancer cells.)

The determination that the φX174 genome contains 5375 nucleotides has unexpectedly revealed that sometimes genes can overlap. In fact, about 6000 nucleotides would be needed to code for the 10 proteins produced by this virus if the overlaps did not exist. In certain regions the same segment of DNA codes for two protein sequences by use of a different reading frame (Fig. 21–10). In gene K, which overlaps with genes A and C, four nucleotides are read in the three possible translational phases.

Knowledge of the nucleotide sequences of φX174 and SV40 and of the amino acid sequences of some of the viral proteins has provided a beautiful confirmation of the genetic code (Tables 21–4 and 21–5). It has been found that all 61 codons are used, although some are preferred over others in each of the two organisms.

21–4 GENETIC ENGINEERING — RESTRICTION ENDONUCLEASES

The studies on φX174 and SV40 were possible because their chromosomes are small and can be obtained in pure form. A eukaryotic chromosome, however, contains a DNA molecule that is very large and consequently is much more difficult to study. This problem has now been overcome by the development of techniques (known as *genetic engineering*) which allow short segments of eukaryotic DNA to be excised and ligated into small bacterial chromosomes (plasmids), which can then be grown in large amounts in E. coli. Although genetic engineering is mainly a technological achievement, it will be considered here because it is now very widely used and will have practical applications in industry and medicine in the large-scale production of valuable proteins.

Restriction Enzymes Recognize Specific DNA Sequences

Genetic engineering is based on the use of *restriction endonucleases*, which are enzymes that recognize specific nucleotide sequences and cut DNA. There are two main types of restriction endonucleases (see Nathans and Smith, 1974; Roberts, 1976). Class I enzymes recognize a specific sequence but cut the DNA at nonspecific sites, which can be many nucleotides away, and they are therefore not useful for genetic engineering. On the other hand, class II enzymes provide a molecular scalpel for cutting DNA at the specific recognition site, and they are most useful in DNA research. So far, more than 80 class II enzymes have been isolated.

The different restriction endonucleases are named according to the microorganisms from which they are isolated (see Table 21–6). For example, Eco RI is the enzyme isolated from E.

coli containing the drug-resistance plasmid RI. Restriction enzymes usually recognize nucleotide sequences four or six nucleotides long (Table 21–6). Those that recognize four nucleotides produce shorter fragments (on average, 250 bases long) than those that recognize six (on average, 4000 bases long). An important feature of these recognition sequences is that they are symmetrical (Table 21–6). For example, the following recognition sequence has an axis of symmetry from which the nucleotide sequence reads the same on both strands in the 5' to 3' direction:

$$5' \quad G \quad A \quad A \; | \; T \quad T \quad C \quad 3'$$
$$3' \quad C \quad T \quad T \; | \; A \quad A \quad G \quad 5'$$

Some enzymes cut in the middle of the recognition sequence (producing blunt-ended fragments, see case of Hae III in Table 21–6), but in many other enzymes the breaks in the two strands are separated by several nucleotides. This results in single-stranded segments at the ends of the cleaved DNA that can base-pair between themselves (Table 21–6). These *"sticky" ends* can anneal with any DNA fragment cut by the same enzyme, thus providing a very good tool for genetic engineering.

Most bacteria contain restriction endonucleases, and their function is to provide protection against foreign DNA. If foreign DNA (for example from a bacteriophage), invades the cells, it is cleaved by the restriction enzymes. Bacteria protect their own DNA from being

TABLE 21–6 RECOGNITION SEQUENCES OF SOME RESTRICTION ENDONUCLEASES

Name	Recognition Sequence	Ends After Cleavage*		Source
Eco RI	↓ – G A A T T C – – C T T A A G – ↑	– G – C T T A A	|A A T T C – | G –	E. coli containing drug-resistance plasmid RI
Hind III	↓ – A A G C T T – – T T C G A A – ↑	– A – T T C G A	|A G C T T – | A –	Hemophilus influenzae, serotype D
Bam I	↓ – G G A T C C – – C C T A G G – ↑	– G – C C T A G	|G A T C C – | G –	Bacillus amyloliquefaciens
Hae III	↓ – G G C C – – C C G G – ↑	– G G – C C	|C C – |G G –	Hemophilus aegyptius

*Note that some sequences produce "sticky" ends.

cleaved by having *modification enzymes* which methylate DNA in the same DNA sequences recognized by their particular restriction enzyme, thus rendering it resistant to cleavage. The *restriction-modification phenomenon* was first discovered[23, 24] in studies of the behavior of two strains of *E. coli* (called K12 and B) which have different restriction enzymes. It was found that bacteriophage lambda grown in *E. coli* K12 could infect *E. coli* B only very poorly (while being very active in *E. coli* K12), and vice versa. We now know that the DNA grown in each bacterial type is methylated at different sites. If the DNA is not methylated at the proper sites, it is cleaved by the restriction enzymes as it enters the heterologous cells. The study of this apparently esoteric phenomenon of *restriction* of bacteriophage growth[23, 24] led to the discovery of restriction enzymes and to a revolution in modern biology. This is a good example of how the beneficial results of scientific research cannot be predicted beforehand.

Eukaryotic Genes can be Introduced into Plasmids and Cloned in E. coli

Bacteria sometimes contain small plasmids (i.e., small circles of DNA which replicate autonomously) in addition to their chromosome. Some of the best known plasmids are the resistance factors (R factors), which carry the genes for resistance to various antibiotics. In 1973 it was shown that fragments of foreign DNA could be introduced in vitro into plasmids with the help of restriction endonucleases and that these recombinant molecules could replicate when introduced into *E. coli*.[25-27]

Figure 21–11 shows the different steps involved in the production of a recombinant DNA. The circular plasmid is cut open at a single site with a restriction enzyme. Eukaryotic DNA is also cleaved with the same enzyme (which generates sticky ends). The two DNAs are allowed to anneal to each other at their sticky ends and are then joined together with the enzyme *DNA ligase*. The recombinant plasmids are next introduced into *E. coli* made permeable to DNA by incubation in calcium chloride. Afterwards, the cells that have taken up a plasmid molecule are selected by the use of antibiotics. For example, if the plasmid carries the gene for tetracycline resistance, those cells that do not have such a plasmid can be killed by adding tetracycline to the culture medium. Once an *E. coli* colony is obtained, millions of copies of the eukaryotic DNA segment can be grown. Since all these copies are derived from an *individual* hybrid molecule, this procedure is also called *gene cloning* (see Cohen, 1975; Murray, 1976). In addition to plasmids, bacteriophage lambda can also be used to propagate eukaryotic genes in *E. coli*.

Genetic engineering will make possible the use of bacteria to produce proteins of medical importance, such as insulin and growth hormone of human origin. Another possibility is to introduce nitrogen fixation genes into bacteria that live in the roots of non-leguminous plants, to eliminate the need for nitrogen fertilizers.

Many eukaryotic genes have been cloned in *E. coli*, but so far their proteins do not seem to be expressed in *E. coli* in a correct way. This is probably related to the need for prokaryotic promoters (i.e., start signals for RNA polymerase) and ribosome binding sites in order for the proteins to be expressed in bacteria. This problem can be circumvented by cloning eukaryotic genes within well-characterized bacterial genes. For example, the sequence coding for the hormone somatostatin was inserted within the gene for β-galactosidase (the lactose operon is described in detail in Chapter 25), and the synthesis of biologically active somatostatin protein was obtained using the β-galactosidase promoter and ribosome binding site (see Itakura et al., 1977). Another way to produce eukaryotic proteins in bacteria is to clone the eukaryotic gene within the gene coding for penicillinase, an enzyme that is secreted outside of the cell and into the surrounding medium, where it can degrade the antibiotic penicillin. When an insulin gene was inserted within the penicillinase gene, a hybrid protein was obtained which had amino acid sequences for insulin and penicillinase within the same protein chain (see Villa-Komaroff et al., 1978). The recombinant protein is exported to the outside of the cell (as penicillinase normally is) and can be purified directly from the culture medium.

Genetic engineering will be invaluable in understanding how genes are controlled in eukaryotic organisms. Genes can not only be grown in large quantities and sequenced, but their *expression* in eukaryotic cells can also be studied by injection into the nucleus of living *Xenopus* oöcytes (see Chapter 26). Figure 21–12 shows the result of one such experiment, in which genes coding for frog 5S ribosomal RNA were cloned in an *E. coli* plasmid and then injected into the nucleus of living oöcytes. It may be observed that 5S RNA was indeed correctly

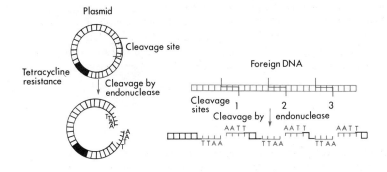

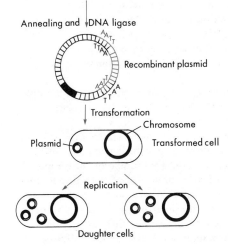

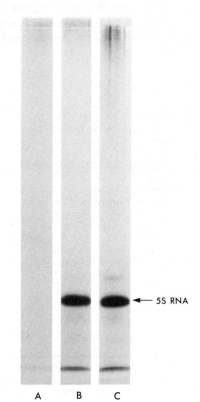

Figure 21–11 Production of a recombinant DNA molecule by genetic engineering (Courtesy of S. Cohen.)

Figure 21–12 Xenopus oöcytes injected into the nucleus with an *E. coli* plasmid containing eukaryotic 5S ribosomal RNA genes. The oöcytes were labeled with radioactive RNA precursors, and the RNA was electrophoresed in a polyacrylamide gel. **A,** no DNA. **B,** injected with recombinant DNA. **C,** injected with recombinant DNA and 1 μg/ml of α-amanitine. Results indicate that the genes are transcribed by the correct eukaryotic RNA polymerase (see Chapter 22). (From E. M. DeRobertis.)

transcribed in large quantities within the oöcyte. The availability of a functional test for cloned genes is very valuable because it allows a new type of genetics in which mutations (e.g., deletions) are introduced at known sites of cloned DNA in the test tube, and then the effect of the mutation on the expression of the gene is analyzed. It is expected that this approach will help to identify DNA sequences which are important for the control of gene activity.

SUMMARY:

Genetic Engineering

The techniques of genetic engineering are based on the use of *restriction endonucleases* from bacteria. These enzymes recognize specific DNA nucleotide sequences and provide a molecular scalpel for cutting DNA at these sites (Table 21–6). Some restriction enzymes produce "sticky" ends, which can anneal with any DNA fragment cut by the same enzyme. The normal function of the bacterial endonucleases is that of protection against invasion by foreign DNA (e.g., a bacteriophage). The DNA of the bacterium is rendered resistant to these endonucleases by a *restriction-modification* process in which methylation of the specific DNA sequences is produced.

Genetic engineering techniques permit the cloning (i.e., the replication of a single molecule) of genes by inserting them into plasmids (Fig. 21–11) or into bacteriophage lambda.

Plasmids are small circles of DNA that replicate autonomously, and many of them carry genes for resistance to antibiotics. The recombinant plasmids carrying eukaryotic genes are introduced into *E. coli* and are then selected by the use of the proper antibiotic. Colonies may be obtained in which millions of copies of the same molecule are produced (gene clones). These techniques make possible the use of bacteria to produce proteins of medical importance (e.g., human insulin or the growth hormone). However, the expression of eukaryotic genes in *E. coli* is difficult to obtain, probably because of the need for prokaryotic promoters and ribosome binding sites. A method already proven successful has been to include the eukaryotic DNA fragment within a bacterial gene. Expression of cloned genes in eukaryotes can be studied by injecting them into the nucleus of *Xenopus* oöcytes.

REFERENCES

1. Pauling, L. (1952) *Proc. Am. Philos. Soc., 96*:556.
2. Ingram, V. M. (1966) *The Biosynthesis of Macromolecules.* W. A. Benjamin, Inc., New York.
3. Avery, O. T., Macleod, C. M., and McCarty, M. (1944) *J. Exp. Med., 79*:137.
4. Nirenberg, M., and Mattaei, H. (1961) *Proc. Natl. Acad. Sci., USA, 47*:1588.
5. Nirenberg, M., and Ochoa, S. (1963) *Cold Spring Harbor Symp. Quant. Biol., 28*:549.
6. Leder, P., and Nirenberg, M. (1964) *Proc. Natl. Acad. Sci., USA, 52*:420.
7. Khorana, H. G. (1965) *Fed. Proc., 24*:473.
8. Crick, F. H. C. (1966) *J. Molec. Biol., 19*:548.
9. Benzer, S., and Champe, S. P. (1962) *Proc. Natl. Acad. Sci. USA, 48*:1114.
10. Brenner, S., and Beckwith, J. R. (1965) *J. Molec. Biol., 13*:629.
11. Riddle, D. L., and Roth, J. R. (1970) *J. Molec. Biol., 54*:131.
12. Riddle, D. L., and Carbon, J. (1973) *Nature (New Biol.), 242*:230.
13. Ames, B. N., Kammen, H. O., and Yamasaki, E. (1975) *Proc. Natl. Acad. Sci. USA, 72*:2423.
14. McCann, J., and Ames, B. N. (1976) *Proc. Natl. Acad. Sci. USA, 73*:950.
15. Sanger, F., Air, G. M., Barrell, B. G., Brown, N. L., Coulson, A. R., Fiddes, J. C., Hutchinson, C. A., Slocombe, P. M., and Smith, M. (1977) *Nature, 265*:687.
16. Fiers, W., Contreras, R., Haegeman, G., Rogiers, R., Van de Voorde, A., Van Heuverswyn, H., Van Herregweghe, J., Volckaert, G., and Ysebaert, M. (1978) *Nature, 273*:113.
17. Reddy, V. B., Thimmappaya, B., Dahr, R., Subramanian, B., Zain, S., Pan, J., Gosh, P. K., Celma, M. L., and Weissman, S. M. (1975) *Science, 200*:494.
18. Sanger, F., and Coulson, A. R. (1975) *J. Molec. Biol., 94*:441.
19. Maxam, A. and Gilbert, W. (1977) *Proc. Natl. Acad. Sci. USA, 74*:560.
20. Sanger, F., Nicklen, S., and Coulson, A. R. (1977) *Proc. Natl. Acad. Sci. USA, 74*:5463.
21. Barrell, B. G., Air, G. M. and Hutchison, C. A. (1976) *Nature, 264*:34.
22. Shaw, D. C., Walker, J. E., Northrop, F. D., Barrell, B. G., Godson, G. N. and Fiddes, J. C. (1978) *Nature, 272*:510.
23. Luria, S. E. (1953) *Cold Spring Harbor Symp. Quant. Biol., 18*:237.
24. Arber, W. (1974) *Prog. Nucleic Acid Res. Mol. Biol., 14*:1.

25. Cohen, S. N., Chang, A. C., Boyer, H. W., and Helling, R. B. (1973) *Proc. Natl. Acad. Sci. USA, 70*:3240.
26. Morrow, J., Cohen, S. N., Chang, A. C., Boyer, H. W., Goodman, H. M., and Helling, R. (1974) *Proc. Natl. Acad. Sci. USA, 71*:1743.
27. Kedes, L. H., Cohen, S. N., Houseman, D. and Chang, A. C. (1975) *Nature, 255:533.*

ADDITIONAL READINGS

Ames, B. N. (1979) Identifying environmental chemicals causing mutations and cancer. *Science, 204*:537.

Arber, W. (1975) DNA modification and restriction. *Prog. Nucleic Acid Res. Mol. Biol., 14*:1.

Cohen, S. N. (1975) The manipulation of genes. *Sci. Am., 233*:24.

Cold Spring Harbor Laboratory (1966) *The Genetic Code.* Cold Spring Harbor Symposium of Quantitative Biology Volume 31.

Crick, F. H. C. (1966) The genetic code, III. *Sci. Am., 215*(4):55.

Crick, F. H. C. (1968) The origin of the genetic code. *J. Molec. Biol., 38*:367.

Crick, F. H. C. (1970) Central dogma of molecular biology. *Nature, 227*:561.

Davis, B. D. (1977) The recombinant DNA scenario. *Am. Sci., 65*:547.

Dawid, I. B., and Wahli, W. (1979) Application of recombinant DNA technology to questions of developmental biology: A review. *Dev. Biol.,* in press.

Fiddes, J. C., (1977) The nucleotide sequence of a viral DNA. *Sci. Am., 237*(6):54.

Garen, A. (1968) Sense and nonsense in the genetic code. *Science 160*:149.

Godson, G. N., Barrell, B. G., Staden, R. and Fiddes, J. C. (1978) Nucleotide sequence of bacteriophage G4 DNA. *Nature, 276*:236.

Gurdon, J. B., Melton, D. M., and De Robertis, E. M. (1979) Genetics in an oöcyte. In: Ciba Foundation Symposium No. 66. Elsevier Publishing Company, New York.

Itakura, K., Hirose, T., Crea, R., Riggs, A. D., Heyneker, H. L., Bolivar, F., and Boyer, H. W. (1977) Expression in *E. coli* of a chemically synthesized gene for the hormone somatostatin. *Science, 198*:1056.

Maniatis, T., Kee, S. G., Efstratiadis, A., and Kafatos, F. C. (1976) Amplification and characterization of a β-globin gene synthesized in vitro. *Cell, 8*:163.

Maniatis, T., Hardison, R. C., Lacy, E., Lauer, J., O'Connell, C., Quon, D., Sim, G. K., and Efstratiadis, A. (1978) The isolation of structural genes from libraries of eukaryotic DNA. *Cell 15*:687.

Maxam, A. M., and Gilbert, W. (1977) A new method for sequencing DNA. *Proc. Natl. Acad. Sci. USA, 74*:560.

Murray, K. (1976) Biochemical manipulation of genes. *Endeavour, 35*:129.

Nathans, D., Smith, H. O. (1975) Restriction endonucleases in the analysis and restructuring of DNA molecules. *Annu. Rev. Biochem., 44*:273.

Roberts, R. J. (1976) Restriction endonucleases. *CRC Crit. Rev. Biochem., 4*:123.

Roth, J. R. (1974) Frameshift mutations. *Annu. Rev. Genet., 8*:317.

Sanger, F., Air, G. M., Barrell, B. G., Brown, N. L., Coulson, A. R., Fiddes, J. C., Hutchinson, C. A., III, Slocombe, P. M., and Smith, M. (1977) Nucleotide sequence of bacteriophage ϕX174 DNA. *Nature, 265*:687.

Sanger, R., Nicklen, S., and Coulson, A. R. (1977) DNA sequencing with chain-terminating inhibitors. *Proc. Natl. Acad. Sci. USA, 74*:5463.

Shaw, D. C., Walker, J. E., Northrop, F. D., Barrell, B. G., Godson, G. N., and Fiddes, J. C., (1978) Gene K, a new overlapping gene in bacteriophage G4. *Nature, 272*:510.

Sinsheimer, R. L. (1977) Recombinant DNA. *Annu. Rev. Biochem., 46*:415.

Stryer, L. (1975) *Biochemistry,* W. H. Freeman & Co., San Francisco.

Villa-Komaroff, L., Efstratiadis, A., Broome, S., Lomedico, P., Tizard, R., Naber, S. P., Chick, W. L., and Gilbert, W. (1978) A bacterial clone synthesizing proinsulin. *Proc. Natl. Acad. Sci. USA, 75*:3727.

Watson, J. D. (1975) *Molecular Biology of the Gene.* W. A. Benjamin, Inc., Menlo Park, Calif.

Wigler, M., Sweet, R., Sim, G. K., Lacy, E., Maniatis, T., and Axel, R. (1979) Transformation of mammalian cells with genes of prokaryotes and eukaryotes. *Cell, 16*:777.

Wu, R. (1978) DNA sequence analysis. *Annu. Rev. Biochem., 47*:607.

Ycas, M. (1969) The biological code. John Wiley & Sons, Inc., New York.

22

TRANSCRIPTION AND PROCESSING OF RNA

22–1   Messenger RNA in Prokaryotes   484
            Prokaryotic RNAs are Transcribed by
            a Single RNA Polymerase
            Transcription has Three Stages —
                Initiation at Promoters, Elongation,
                and Termination
            In Prokaryotes Transcription is Coupled
                to Protein Synthesis
            Summary: *RNA Transcription in*
            Prokaryotes
22–2   Transcription in Eukaryotes   490
            Three Different RNA Polymerases
                Transcribe the Various Eukaryotic RNAs
            Transcription may be Visualized by
                Electron Microscopy
            Summary: *Transcription in Eurkaryotes*

22–3   Eukaryotic Messenger RNA   492
            The 5' End of Eukaryotic mRNA is
                Blocked by 7-Methyl G
            The 3' End of Eukaryotic mRNA — A
                Poly A Segment
            Eukaryotic Genes Frequently Contain
                Insertions of Non-Coding DNA
            Heterogeneous Nuclear RNA — mRNA
                Precursors Containing Interventing
                Sequences
            Eukaryotic mRNAs are Associated with
                Proteins
            Summary: *Eukaryotic Messenger RNA*

The flow of genetic information follows the pathway shown below:

$$DNA \xrightarrow[\text{transcription}]{} RNA \xrightarrow[\text{translation}]{} Protein$$

The genetic information contained in DNA cannot act directly as a template for protein synthesis but must be first transcribed into messenger RNA. In addition to this informational RNA, other RNAs are required for the very complex process of translating a 4 base code into a sequence of 20 amino acids. These other nucleic acids are ribosomal RNA (rRNA) and transfer RNA (tRNA), which account for most of the RNA contained in cells (Table 22–1).

In this chapter we will analyze how the information contained in DNA is *transcribed* into RNA. RNA is synthesized by the enzyme RNA polymerase. This step is very important in gene regulation since the rate of expression of a particular gene is in general controlled by the frequency with which RNA polymerase starts the transcription of that gene.

The transcripts synthesized by RNA polymerase are frequently not the final products utilized by the cell. To become functionally competent the primary transcripts must be modified by a series of chemical alterations known as *processing*. There are three main types of modifications in the processing of RNA: (1) *cleavage of large precursor RNAs* into smaller RNAs, (2) *terminal addition of nucleotides*, as occurs in eukaryotic mRNAs with the addition of poly A residues to the 3' end and the "cap" nucleotides to the 5' end, and (3) *nucleoside modifications*, such as the methylations that are very common in transfer and ribosomal RNA. The processing of each major RNA species has its own peculiarities. The processing of mRNA will be analyzed here and that of rRNA and tRNA in the following chapters. RNA synthesis and processing differ in many respects in prokaryotes and eukaryotes and will therefore be treated separately but compared and contrasted whenever possible.

22–1 MESSENGER RNA IN PROKARYOTES

The demonstration that a template RNA carries genetic information from DNA emerged

TABLE 22–1   RNA MOLECULES IN *E. COLI*

Type	Relative Amount (%)	Sedimentation Coefficient	Number of Nucleotides
Ribosomal RNA (rRNA)	80	23S	3700
		16S	1700
		5S	120
Transfer RNA (tRNA)	15	4S	~75
Messenger RNA (mRNA)	5	Heterogeneous	

through work with bacteria. The term "messenger" RNA (mRNA), proposed by Jacob and Monod in 1961, refers to the fact that this molecule is a template copied from DNA and having a rapid turnover.[1, 2] As shown in the experiment of Figure 22–1,I, the rapidly synthesized mRNA could be detected after five seconds of incubation with ¹⁴C-uridine.[3] After longer time periods (15 minutes), the mRNA disappeared and the ribosomal RNAs became labeled (Fig. 22–1,II).

The average life span of most mRNAs in *E. coli* is about two minutes, after which the molecules are broken down by ribonucleases. In fact, in bacteria mRNA is translated into protein on one end while the other end is still being transcribed.

Another characteristic of mRNA is that it is heterogeneous. The size of the molecule varies considerably, since it is adapted to the length of the polypeptide chain for which it will code. In *E. coli* the average size of a mRNA cistron (i.e., the length of DNA coding for one polypeptide chain) is between 900 and 1500 nucleotides, a length corresponding to peptide chains containing between 300 and 500 amino acids. However, when several adjacent cistrons are copied at the same time the mRNAs are much longer. These are the so-called polycistronic messengers. For example, the ten specific enzymes involved in the metabolism of histidine are encoded in a single mRNA molecule. In higher cells most mRNAs are monocistronic.

Messenger RNA is complementary to chromosomal DNA; it forms RNA-DNA hybrids after separation of the two DNA strands.[4] Synthesis

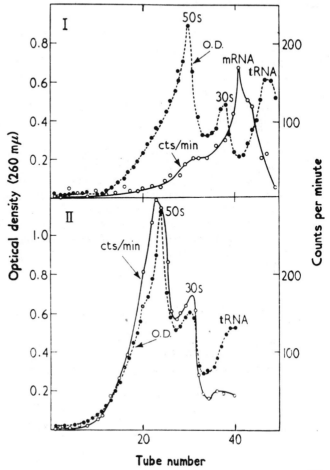

Figure 22–1 Diagram showing the rapid turnover of messenger RNA (mRNA). *E. coli* were incubated for 5 seconds with ¹⁴C-uridine and then washed in non-radioactive uracil. In **I** the cells were rapidly frozen and in **II** they were incubated for 15 minutes at 37° C prior to freezing. Ribonucleic acids were extracted in both experiments and ultracentrifuged. The optical density at 260 nm indicated the concentration of different RNA molecules, and the counts per minute indicated the incorporation of ¹⁴C-uridine. Note in **I** that the only labeled RNA is mRNA, whereas in **II** radioactivity is found in two peaks of ribosomal RNA (rRNA) (50S and 30S subunits) and in transfer RNA (tRNA). (From Gros., F., Hiat, H., Gilbert, W., Kurland, C. G., Risebrough, R. W., and Watson, J. D., *Nature* (London), *190*:581, 1961.)

of mRNA is accomplished with only one of the two strands of DNA, which is used as a template; for this reason transcription is said to be *asymmetrical*. The enzyme RNA polymerase joins the ribonucleotides, thus catalyzing the formation of the 3'-5'-phosphodiester bonds that form the RNA backbone. In this synthesis the AU/GC ratio of RNA is similar to the AT/GC ratio of DNA. Another criterion by which mRNA can be recognized is that is becomes rapidly attached to ribosomes and forms part of the polyribosomes.

The following properties may, therefore, be used to identify mRNA in prokaryons: rapid turnover, heterogeneity, complementary base composition with DNA, and attachment to ribosomes. The most important of these properties is that mRNA functions as a template in protein synthesis.

Messenger RNAs of eukaryotes differ in many respects from those of bacteria. For example, eukaryotic mRNAs are *stable*, with half-lives that can be measured in hours or days (instead of minutes); in general they are monocistronic, and they frequently have special types of post-transcriptional modifications, such as the addition of a methylated "cap" structure at the 5' end and a tail of poly A added to the 3' end. Eukaryotic mRNAs will be analyzed in detail later on in this chapter.

Prokaryotic RNAs are Transcribed by a Single RNA Polymerase

In prokaryotes all types of RNAs are transcribed by the same enzyme. Using DNA as the template, the enzyme catalyzes the formation of RNA from the four ribonucleotide triphosphates. The enzyme copies the base sequence of one of the strands of DNA, according to the Watson-Crick base pairing rules. The polymerization reaction is as follows:

In addition to copying the nucleotide sequence of the DNA template precisely, RNA polymerase is able to recognize a variety of genetic signals on the chromosome, such as the signals for starting and stopping RNA synthesis at precise sites. There are an estimated 3000 genes in *E. coli*, and they are all transcribed by the same enzyme. Considering these multiple functions, it is not surprising that RNA polymerase is a large and complex enzyme. A complete molecule (holoenzyme) is formed by several polypeptides: 2α of 40,000 daltons, 1β of 155,000, $1\beta'$ of 165,000 and 1σ of 90,000. The total molecular weight of the enzyme is therefore 490,000 daltons.[5] The sigma (σ) factor is only loosely bound to the rest of the enzyme and can be separated by chromatography on phosphocellulose. The $\beta\beta'\alpha_2$ enzyme (without sigma) is called *core polymerase*. The sigma factor is required for the enzyme to recognize the correct start signals on DNA (the binding sites or *promoters*).[6] Figure 22–2 shows that as soon as the RNA chain is started, the sigma factor is released from the core enzyme and can be used in the transcription of other specific RNA molecules.[7][8]

Transcription has Three Stages—Initiation at Promoters, Elongation, and Termination

The following three stages are commonly distinguished in the transcription by RNA polymerase: (1) binding to promoters and RNA chain initiation, (2) elongation, and (3) termination (Fig. 22–2).

The start signals on DNA are called *promoters*. The nucleotide sequence of several of these RNA polymerase binding sites is now known and several features common to all promoters are emerging.

Figure 22–3 shows that all promoters have an

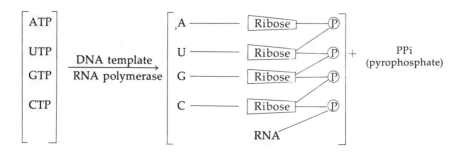

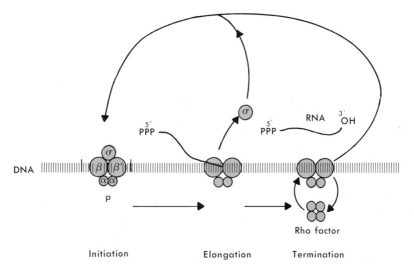

Figure 22–2 Diagram showing the three stages of transcription by RNA polymerase. At *initiation* the enzyme composed of subunits β, β', 2α and σ is at the site of the promoter (p). During *elongation* the enzyme transcribes the gene and subunit σ is eliminated. *Termination* occurs at the termination site and the termination factor *rho* causes the release of the transcribed RNA molecule.

AT-rich region some ten bases before the start of the mRNA, with the common sequence TATAATG or slight variations of it.[9] Mutations in this sequence severely affect RNA production in vivo. This AT-rich region probably favors local separation of the DNA strands, a step required for RNA polymerase to gain access to the DNA bases. In addition to the TATAATG sequence, the enzyme also recognizes a DNA region located about 35 bases before the start of the mRNA (Fig. 25–3). Single base mutations in this region can inactivate promoters, but different promoters have no obvious common sequences in this region (see Gilbert, 1976). Individual variations between promoters are not surprising, since different genes are expressed with varying efficiencies; i.e., some promoters are "stronger" than others.

The binding of RNA polymerase to its promoter is a crucial step in the regulation of gene expression. *E. coli* has proteins called *repressors* which can bind to specific sequences of DNA (called *operators*), turning off the expression of a given set of genes. Repressors work by binding to a DNA site that overlaps with the promoter, thereby preventing the binding of RNA polymerase. Other proteins affect transcription as *positive* regulators, e.g., some promoters of *E. coli* cannot be recognized by RNA polymerase unless a protein called *CAP factor* (a cAMP binding protein) is bound to a nearby sequence of DNA. (Transcriptional regulation is covered in detail in Chapter 25.)

During *elongation*, RNA polymerase copies the DNA sequence accurately, progressing at a speed of about 30 nucleotides per second. Only

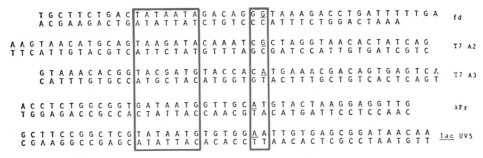

Figure 22–3 Promoter sequences which are protected by RNA polymerase from digestion by DNAase. Observe that all promoters (large box) have a common sequence, TATAATG, or a close derivative of it. This sequence is situated about ten bases before the start of the messenger RNA. The base in which transcription starts is outlined by the small box.

....-A-G-G-U-A-A-U-A-G-U-U-A-G-A-G-C-C-U-G-C-A-U-A-A-C-G-G-U-U-U-<u>C-G-G-G</u>-A-U-U-U-U-U-A-OH   λ 6S

....-U-U-U-G-U-U-G-C-C-G-C-C-U-U-A-G-A-A-C-G-C-U-C-G-G-U-U-<u>G-C-C-G-G-G-C</u>-U-U-U-U-U-A-OH   λ <u>oop</u>

....-C-A-A-U-C-A-G-A-U-A-C-C-C-A-G-C-C-U-A-A-U-G-A-<u>G-C-G-C-C-C-G-C-G-G-G-C</u>-U-U-U-U-U-U-U-(U)-OH   trp
leader

....-G-U-G-C-A-U-A-A-G-G-G-C-C-U-U-U-U-G-A-A-G-U-U-A-C-C-G-C-U-U-C-A-A-<u>G-G-G-C</u>-U-U-U-U-U-U-(A)-(A)-OH   ϕ80 M₃

Figure 22–4 Stop signals for RNA polymerase in four prokaryotic RNAs. Observe the presence of a GC-rich region (underlined) followed by between five and eight consecutive U residues.

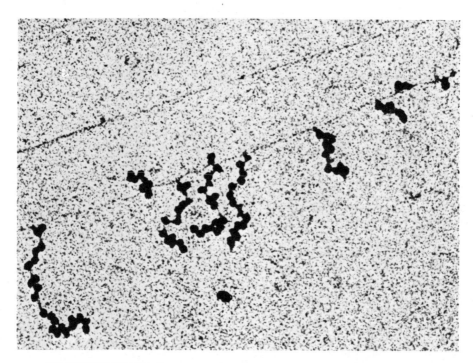

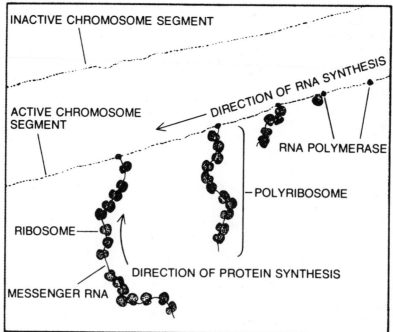

Figure 22–5 Electron micrograph of the transcription and translation processes in bacteria. Two stretches of DNA, one naked, the other with nascent messenger RNA arranged at right angles, are observed. The bottom diagram facilitates the interpretation of the electron micrograph. (See the description in the text.) (Courtesy of O. L. Miller, Jr.)

one strand of the DNA is transcribed. Elongation of RNA proceeds only in the 5'→3' direction (which is the same direction used in DNA replication and protein synthesis). As a result, the first nucleotide incorporated into a RNA chain retains all three phosphates (Fig. 22–2). This is because nucleotides have the triphosphate joined to the 5' carbon of the ribose molecule. The last nucleotide incorporated will have a free OH group on the 3' carbon of the ribose. RNA synthesis is therefore said to progress in the 5' → 3' direction. The first nucleotide incorporated is always A or G.

Termination of the transcription of the RNA molecule occurs when the enzyme arrives at a stop signal on the DNA. A *termination factor* or *rho factor* causes the release of the transcribed RNA molecules[10, 11] (Fig. 22–2). Rho factor is a protein that has monomeric molecular weight of 50,000 daltons and that normally works as a tetramer. Purified rho factor has an ATPase activity which seems to be required for proper termination[12]; termination is affected in rho mutants with poor ATP hydrolyzing activity.[13]

Gene expression in *E. coli* can also be regulated at the level of transcription termination, as will be seen in Chapter 25 in the discussion of the tryptophan operon.

The termination signal recognized by RNA polymerase on DNA is a GC-rich region followed by a series of A residues. Figure 22–4 shows the end sequences of four prokaryotic RNAs, which stop after a GC-rich region followed by five to eight consecutive U residues. In eukaryotes transcription termination also seems to occur in AT-rich stretches of DNA, at least in the case of those tRNA and 5S RNA genes that have been sequenced.[14 15]

In Prokaryotes Transcription is Coupled to Protein Synthesis

Since bacteria do not have a nuclear membrane, the ribosomes are free to attach to the mRNA molecules as they are being synthesized (this is possible because both RNA and protein synthesis start at the 5' end of the mRNA).

Experiments performed on bacteria treated with lysozyme to render them osmotically sensitive have permitted the visualization of the bacterial chromosomes during the process of transcription. The DNA fibers have stretches in which ribosomes appear attached to nascent mRNA chains positioned at right angles. As shown in Figure 22–5, in such regions there is a gradient of short to long polyribosomes, indicative that the transcription of the mRNA is coupled with translation. In other words, as soon as the mRNA transcription by RNA polymerase begins, the ribosomes become attached to the mRNA to initiate protein synthesis. Some electron micrographs show the presence of the RNA polymerase molecule as a granule smaller than the ribosome attached to the DNA template (Fig. 22–5).[16, 17] Possible degradation of the mRNA is suggested by the finding of shorter polyribosomes distal to the longest ones in the gradient. In such cases mRNA degradation must have started even before its synthesis was complete. There are long stretches of DNA that have no polyribosomes but that do show the granules corresponding to RNA polymerase; therefore, it has been suggested that these regions may be associated with inactive parts of the genome awaiting a proper initiation signal. Furthermore, these findings suggest that only a small portion of the bacterial genome is active in transcription at any one moment.[18]

SUMMARY:

RNA Transcription in Prokaryotes

The genetic information contained in DNA (Chap. 21) is transcribed into messenger RNA (mRNA) by RNA polymerase. To become functionally competent the transcripts are processed by (1) cleavage of large precursors, (2) terminal addition of nucleotides (poly A at the 3' end and the cap structure at the 5' end), and (3) nucleoside modifications (e.g., methylation). The DNA segment corresponding to a polypeptide chain is called a cistron. In bacteria many mRNAs are polycistronic and have a rapid turnover (the average life span is two minutes).

Transcription copies only one of the DNA strands and RNA is synthesized in the 5' → 3' direction. A single RNA polymerase with five subunits (1β, $1\beta'$, 2α, and the sigma [σ] factor) binds to the *promoter*, and RNA synthesis starts. In elongation the DNA template is copied and the nucleotide triphosphates (ATP, UTP, GTP, and CTP) are added sequentially. The promoter has special nucleotide sequences recognized by the enzyme. Overlapping with the promotor is sometimes an *operator,* which is the binding site for proteins acting as *repressors* in gene regulation (Chap. 25).

As soon as the RNA chain starts, the sigma factor is released and the rest of the polymerase (core enzyme) continues the *elongation. Termination* occurs at the stop signal and involves a *rho factor,* which has ATPase activity. The stop signal is a region rich in GC that is followed by series of AT residues.

In prokaryotes transcription is coupled with protein synthesis, and under the electron microscope it is possible to visualize ribosomes attached to nascent mRNA chains being transcribed from DNA (Fig. 22–5). The degradation of mRNA may start even before the translation is completed.

22–2 TRANSCRIPTION IN EUKARYOTES

Three Different RNA Polymerases Transcribe the Various Eukaryotic RNAs

In *E. coli* all the genes of the chromosome are transcribed by the same enzyme. In higher organisms the transcription process is more complex. Three different nuclear RNA polymerases have been identified, each one of which transcribes different classes of genes. Figure 22–6 shows how the three enzymes can be separated from each other by ion-exchange chromatography. Nuclear extracts of all eukaryotic cells show these three peaks of activity. The enzymes have been classified as I, II, or III according to the order of elution from the column. Another useful criterion for distinguishing the various polymerases is their sensitivity to inhibition by *α-amanitine,* the toxin of the poisonous mushroom *Amanita phalloides.* This toxin is of medical interest because accidental ingestion of *Amanita* is the most frequent cause of fungal intoxication in man.

Table 22–2 summarizes the properties and functions of nuclear RNA polymerases. RNA polymerase I is not affected by α-amanitine and is localized in the nucleolus.[19] This enzyme transcribes the large ribosomal RNAs (18S and 28S).[20] RNA polymerase II is completely inhibited by low concentrations (1 $\mu g/ml$) of α-amanitin and is responsible for the synthesis of mRNA. RNA polymerase III is inhibited by high concentrations of α-amanitin (100 $\mu g/ml$) and synthesizes tRNA and 5S RNA.[21]

Eukaryotic RNA polymerases are very complex enzymes, each one having between six and ten protein subunits (see Chambon, 1975; and Roeder, 1976). The enzyme structure consists of two large subunits and several small ones, resembling in this respect the structure of bacterial RNA polymerase. The protein subunits of each enzyme type are different, except for two small polypeptides of 29,000 and 19,000 daltons that are common to all three types of RNA polymerase.[22, 23]

Eukaryotic cells also contain DNA in mitochondria and chloroplasts, which is transcribed by a different enzyme. In contrast to nuclear enzymes, mitochondrial RNA polymerase is a very simple enzyme formed by a single peptide. It is not inhibited by α-amanitine but is sensitive to *rifampicin,* an antibiotic that inhibits prokaryotic RNA polymerases; this repre-

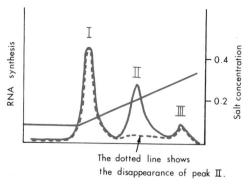

The dotted line shows the disappearance of peak II.

Figure 22–6 Separation of RNA polymerases I, II, and III by chromatography in DEAE-Sephadex. The peaks are eluted by increasing the salt concentration. *Solid line,* without α-amanitine; *dotted line,* assays performed with 0.1 $\mu g/ml$ α-amanitine showing selective inhibition of polymerase II.

TABLE 22–2 PROPERTIES AND FUNCTIONS OF EUKARYOTIC RNA POLYMERASES

Enzyme	Localization	Gene Transcripts	Inhibition by α-Amanitine
I	Nucleolus	18S and 28S rRNAs	Insensitive
II	Nucleoplasm	mRNA	Sensitive to low concentration
III	Nucleoplasm	tRNA, 5S RNA	Sensitive to high concentration

sents another property of mitochondria that suggests that their origin was prokaryotic (see Chapter 12–6).

Although eukaryotic transcription is still incompletely understood, it is clear that the different RNA species are transcribed by distinct forms of RNA polymerase which differ in their molecular structure and possibly in their mechanisms of regulation.

A serious obstacle in the study of eukaryotic transcription is that once purified, nuclear RNA polymerases are unable to recognize promoters in vitro. This is possibly due to the loss of essential subunits during the purification process or to the fact that the normal template in eukaryotic nucleus is a DNA-protein complex and not naked DNA.[24] This obstacle can be circumvented by injecting purified DNA into the nucleus of amphibian oöcytes.[25, 26] DNAs injected into the nucleus of these cells are transcribed accurately, thereby providing a very useful assay system for eukaryotic RNA synthesis.

Transcription may be Visualized by Electron Microscopy

Oscar Miller has devised a remarkable technique by which transcription complexes can be visualized directly under the electron microscope. In this method, isolated nuclei are lysed gently, and the chromosomal material is centrifuged directly on top of an electron microscope specimen holder. Transcribing genes are

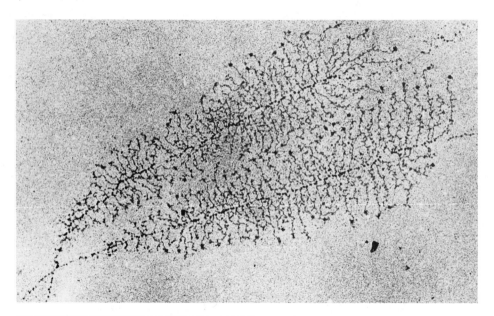

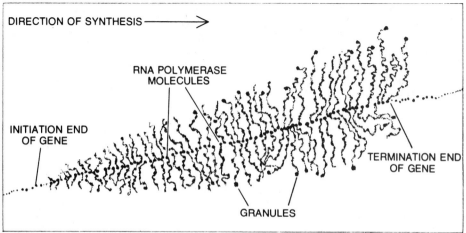

Figure 22–7   **Above,** electron micrograph showing two nucleolar genes in the process of transcription of the ribosomal RNA; **below,** labels are self-explanatory. ×35,000. (Courtesy of O. L. Miller and B. R. Beatty.)

491

spread together with their attached RNA polymerases and nascent RNA chains. Figure 22–7 shows ribosomal RNA genes during transcriptions, obtained by spreading nucleoli of frog cells.

Several conclusions about eukaryotic transcription can be drawn from this type of study.[27, 28]

1. The template for transcription is a complex of DNA and protein that has the characteristic beaded appearance of chromatin nucleosomes (Fig. 16–19).

2. Eukaryotic RNA synthesis starts and ends at precise sites on DNA (*promoter* and *terminator* sequences). Between these two sites, the RNA chains gradually increase in length. As a result, genes that are very actively transcribed show the so-called *"fern leaf"* or *"Christmas tree"* configuration seen in Figure 22–7 in rRNA genes and in Figure 25–14 in lampbrush chromosomes.

3. At any one time, only a very small fraction of the total chromatin is transcribed.

4. In fully transcribing genes the RNA polymerase molecules are packed very close together, with one polymerase more often than every 200 base pairs of DNA. This information allows one to make some intersting calculations. Since the elongation rate of RNA synthesis is 20 to 30 nucleotides per second, it follows that a maximally expressed gene can complete one RNA chain about every 10 seconds.

5. The nascent RNA becomes associated with protein as it is being transcribed, producing ribonculeoprotein particles (RNP) rather than free RNA. Processing of the initial transcripts into smaller RNAs (Fig. 25–18) can also start before the RNA chains are completed.

6. In eukaryotes the nuclear envelope introduces a barrier between transcription and protein synthesis. The mRNA must be transported into the cytoplasm before it is utilized. This is in contrast with the type of organization in bacteria, which is shown in Figure 22–5.

SUMMARY:

Transcription in Eukaryotes

Transcription in eukaryotes in more complex and is carried out by three RNA polymerases (I, II, and III). RNA polymerase I is located in the nucleolus and transcribes the 18S and 28S ribosomal RNAs, polymerase II transcribes mRNA and is very sensitive to α-amanitine, and polymerase III synthesizes tRNA and 5S rRNA. All these enzymes contain between six and ten subunits. In mitochondria and chloroplasts DNA transcription is carried out by a simpler RNA polymerase that is sensitive to *rifampicin* (as is prokaryotic RNA polymerase). Eukaryotic transcription can also be visualized directly under the electron microscope. It is possible to observe the following: (1) During transcription the template has the characteristic beaded structure of nucleosomes. (2) The transcripts gradually increase in size, producing the typical "Christmas tree" configuration, as seen in rRNA (Fig. 22–7) and in the mRNA of lampbrush chromosomes (Fig. 25–14). (3) Only a small fraction of the total chromatin is transcribed at any one time. (4) In fully transcribing genes, RNA polymerases may be very close together. (5) The nascent RNA becomes rapidly associated with proteins. (6) mRNA must be transported into the cytoplasm through the nuclear envelope before it is translated.

22–3 EUKARYOTIC MESSENGER RNA

The origin and fate of messenger RNA in eukaryotic cells is more complex than in bacteria. The formation of a functionally active mRNA is the consequence of a complex series of steps that comprise: (1) the actual transcription of DNA into mRNA precursors, (2) the intranuclear processing of these precursors, and (3) the transport of the mRNAs into the cytoplasm and their association with ribosomes to initiate the process of translation of protein synthesis.

One of the main features of bacterial mRNA is its very short half-life (on the order of minutes). However, *metabolically stable* mRNAs are found in eukaryotes. The best example of a long-lived mRNA is that in the mammalian reticulocyte, an immature red blood cell that has lost its nucleus but still retains the ribosomes and other components of the protein synthesizing machinery. Reticulocytes contain mRNA which continues to synthesize hemoglobin for many hours and even days after the nucleus (and therefore the globin genes) has been lost.

Several eukaryotic mRNAs have been purified. This has been done by taking advantage of differentiated tissues that synthesize large amounts of a given protein and therefore accumulate considerable amounts of stable specific mRNAs. Examples of such tissues are mammalian reticulocytes, in which 90 per cent of the protein synthesized is α and β globin, the protein subunits of hemoglobin; chick oviduct, which produces ovalbumin; and the silk gland of the moth *Bombyx mori*, which secretes silk fibroin. The availability of pure mRNAs has allowed the study of their structure. In addition, *radioactive probes* for specific genes can be prepared from the mRNA by a variety of methods, such as using the enzyme *reverse transcriptase* to copy a labeled *complementary DNA* (see Chapter 16); this probe in turn allows one to study the genes at the DNA level.

Here we will analyze the structure of globin mRNA, which is the best known mRNA. The complete nucleotide sequence of globin mRNA is known,[29] and certain features which are common to all mRNAs have emerged.

When the RNAs extracted from rabbit reticulocyte polysomes (Fig. 24–4) are centrifuged in a sucrose gradient, globin mRNA sediments as a small 9S peak, that can be separated from ribosomal and transfer RNAs (i.e., 28S, 18S, 5S, and 4S) by repeated cyles of centrifugation.[30] The estimated length of the 9S mRNA is between 650 and 700 nucleotides. Of these, only 589 are coded by the DNA; the rest are accounted for by the poly A tail that is attached to the mRNA after transcription is completed (see later discussion). The mRNA molecule is much longer than is required to code for the globin protein, which has 146 amino acids and is therefore coded by only 438 nu-

cleotides. The fate of the extra nucleotides is shown in Figure 22–8.

In addition to the sequence coding for protein, all eukaryotic mRNAs have two regions which are not translated into protein. The 5' non-coding region, which is located before the start of the region coding for protein, has about 50 nucleotides in most mRNAs[31] and contains the ribosome binding site (Fig. 22–8). The function of the 3' non-coding region in protein synthesis is unknown, even though it is about twice the size of the 5' non-coding region.[32] Some mRNAs have a very large 3' non-coding region; in ovalbumin mRNA it is 600 nucleotides long.[33]

The non-coding regions are thought to be functionally important, because their sequence has tended to be conserved through evolution (i.e., globin mRNAs from different species have similar non-coding sequences[31, 32]). In addition, all mammalian mRNAs have the sequence AAUAAA close to the start of the poly A sequence (indicated in Fig. 22–8). The significance of this hexanucleotide is not known, but it could be a signal for mRNA processing or for the addition of poly A.

Two other structural features are present in eukaryotic mRNAs: the starting end (5') is protected by an unusual methylated "cap," and the 3' end has a sequence of poly A 100 to 200 nucleotides long.

The 5' End of Eukaryotic mRNA is Blocked by 7-Methyl-G

Most eukaryotic mRNAs have their starting ends blocked by the post-transcriptional addition of a "cap" of 7-methyl-guanosine. Normally the phosphodiester bonds of a RNA molecule go from the 3' position of the ribose of one nucleotide to the 5' position of the ribose of the next one (Fig. 22–9). The linkage of the cap to mRNA is very different, because the riboses of 7-methyl-G and the terminal nucleo-

Figure 22–8 Rabbit globin mRNA. Note that eukaryotic mRNAs have two non-coding regions, a 5' cap, and a 3' poly A tail. The number of nucleotides in each section is indicated.

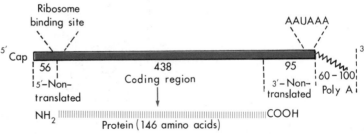

tide are linked by a 5' to 5' triphosphate bridge, as shown in Figure 22–9.

The cap structure m[7]-G5'ppp5'Xp has no free phosphates, and this protects mRNA against attack by phosphatases and other nucleases.[34] Bacterial mRNA is not capped and therefore starts with pppXp. The cap is important in mRNA translation; when it is removed mRNA is unable to form an initiation complex with 30S ribosomal subunits. Similarly, cap analogues, such as m[7]-G-5' monophosphate, are potent inhibitors of protein synthesis initiation.[35] Bacterial mRNAs are translated poorly by eukaryotic ribosomes, but if they are artificially capped in vitro, they become very active templates for mammalian ribosomes. However, the dependence of protein synthesis on caps is not absolute; some mRNAs (like poliovirus mRNA) do not have caps but are still translated normally.

The 3' End of Eukaryotic mRNA — A Poly A Segment

Most mRNAs contain a sequence of polyadenylic acid attached to their 3' end. This poly-A segment is 100 to 200 nucleotides long and is added sequentially to the mRNA after transcription is completed, by the action of a poly A synthetase that uses ATP as substrate. Bacterial mRNAs do not have poly A.

The finding of poly A tails in mRNA represented a major advance in molecular biology, because it has made possible several methods for the rapid separation of mRNA from other contaminant RNAs. Poly A + RNA will bind to poly U or polydeoxythymidine (oligo dT) columns, and, under certain conditions, to millipore filters.

Not all mRNAs are poly adenylated; up to 30 per cent of mRNAs in polysomes lack poly A. Among these, the most prominent ones are the histone mRNAs. The function of poly A is still unknown. It has been suggested that it could be involved in nuclear processing of mRNA precursors and in transport between the nucleus and the cytoplasm, but these hypotheses lack experimental evidence (see Perry, 1976).

Poly A does have a role in promoting mRNA stability. When globin mRNA is microinjected into living *Xenopus* oöcytes, it is translated into globin. Fully adenylated globin mRNA is very stable in oöcytes, as shown by the fact that it is translated continuously for up to two weeks after injection, a period during which each globin mRNA molecule can produce 100,000 protein molecules.[39] It has been found that if the poly A segment is removed from globin mRNA (using the enzyme polynucleotide phosphorylase, which degrades RNA from the 3' end), the half-life of the mRNA in oöcytes is greatly decreased. This effect is reversible, since when poly A is added back using an RNA-adenyltransferase from *E. coli*), the stability of the mRNA is restored.[40] Similarly, nonadenylated histone mRNAs have a short halflife after injection into oöcytes, but become stable if poly A tails are added to them before the injection.[41] It seems that a critical length of poly A is required for mRNA stability: molecules with less than 10 to 20 adenylate residues

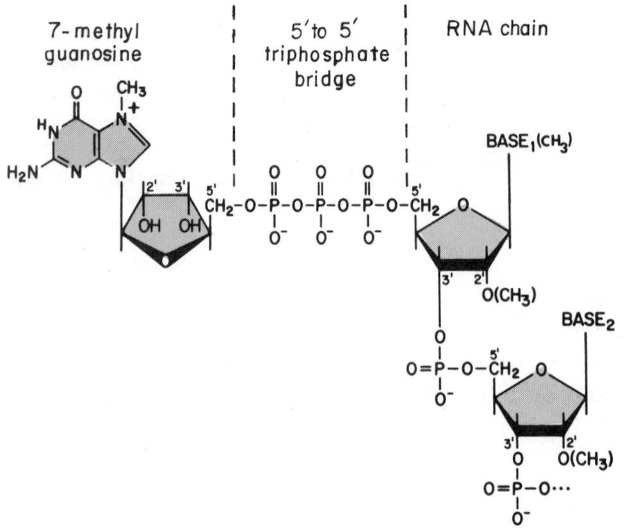

Figure 22–9 Structure of the 5' cap of messenger RNA. Observe that the linkage of the 7-methyl guanosine to the terminal nucleotide is by a 5' to 5' triphosphate bridge.

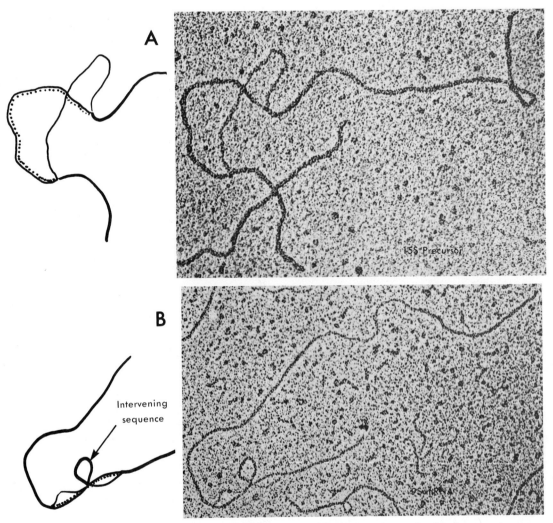

Figure 22-10 Visualization of the intervening sequence of the globin gene under the electron microscope. A cloned segment of DNA containing the mouse β-globin gene was hybridized with **A**, 15S globin mRNA precursor; **B**, mature 9S globin mRNA. The hybridized RNA is represented by a dotted line in the diagram. The 15S precursor hybridizes in a continuous way, showing that the intervening sequence is transcribed into RNA. Note in **B** that the mature globin mRNA hybridizes to two discontinuous regions of the DNA, and the intervening sequence remains as a loop of double-stranded DNA. It was from electron micrographs such as these that intervening sequences were first discovered. (Courtesy of Phillip Leder.)

are degraded, while those with more than 10 to 20 are not.[41]

Eukaryotic Genes Frequently Contain Insertions of Non-coding DNA

Segments of eukaryotic DNA can be grown in *E. coli* plasmids using the genetic engineering technology described in Chapter 21-4. Those plasmids containing a given eukaryotic gene can be identified by using radioactive hybridization probes prepared from a purified mRNA. For example, a labeled *complementary*

DNA copy of globin mRNA can be prepared with the help of the enzyme *reverse transcriptase,* and used to identify plasmids containing the globin genes. When a variety of such cloned eukaryotic genes became available during 1977, molecular biologists were in for a big surprise.

Unexpectedly, it was found that in eukaryons the information for covalently continuous mRNA is frequently found in non-contiguous DNA segments. In other words, genes are interrupted by insertions of non-coding DNA. These inserted DNA sequences, which are absent in the mature mRNA, are called *intervening sequences.* Insertions have been found in globin,[42, 43] ovalbu-

495

min,[44] immunoglobulin,[45] tRNA,[14, 46] and many other genes. Not all eukaryotic genes are interrupted; some genes, such as those coding for histones and some tRNAa, are continuous.

The β-globin gene is interrupted by a DNA fragment 600 base pairs long inserted within the sequence coding for protein. A second, much shorter intervening sequence is found closer to the 5' end of the coding sequence. The insertion is therefore of about the same length as the total mature globin mRNA (589 nucleotides, see Fig. 22–8).

The globin intervening sequence can be visualized in Figure 22–10,B. The electron micrograph shows a cloned segment of mouse DNA which was hybridized to globin mRNA. When RNA hybridizes to the DNA, a "bubble" is seen in which one strand is a RNA-DNA hybrid, and the other is single-stranded DNA. Globin mRNA hybridizes to two discontinuous regions of the genomic DNA, while the intervening sequence, which is not present in the 9S mRNA, remains as a loop of double-stranded DNA (see an interpretation of the structure in the line drawing of Fig. 22–10,B).

Intervening sequences are transcribed into precursor RNAs which are larger than the mature mRNAs. To obtain a functional mRNA these internal sequences must be precisely excised and the molecule re-ligated. It is possible that the non-coding inserts produce "hairpin loops," which can then be cleaved and covalently joined by processing enzymes. This mechanism, shown schematically in Figure 22–

11, has been called RNA *splicing*. An enzyme which is able to remove intervening sequences from tRNA precursors has been isolated from yeast.[47]

The β-globin gene is not transcribed initially as a 9S molecule, but rather as a precursor that sediments at 15S in sucrose gradients.[48, 49] This sedimentation value represents a molecule two to three times longer than globin mRNA, and this extra length is due to the presence of the intervening sequence, as can be visualized in Figure 22–10,A. Mature 9S mRNA is the result of removing the intervening sequence, a process that the cell does rapidly, since the 15S precursor has a half-life of less than two minutes.[48]

While it is clear that several genes are interrupted at the DNA level, we do not know why this process would be beneficial to eukaryotic cells. It has been suggested[50] that some evolutionary advantage could be gained through this process because small errors or mutations in the RNA splicing mechanism could result in the deletion or addition of whole sequences of amino acids (for example from a neighboring gene). In this way novel proteins could evolve more rapidly than by single nucleotide changes, but this explanation is speculative (see Darnell, 1978; and Grick, 1979). In the case of the immunoglobulin heavy chain it has been shown that the intervening sequences are inserted in regions separating protein functional units. The heavy chain protein has three regions of globular shape called *domains,* and in the genome each domain is separated by an intervening sequence

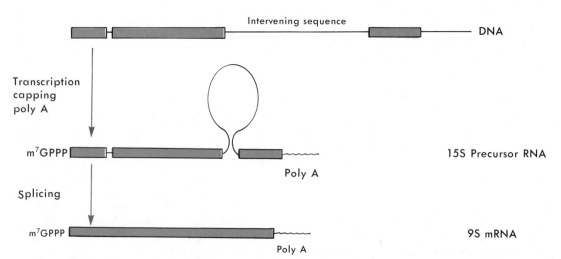

Figure 22–11 Diagram showing the processing of globin mRNA transcripts. The globin gene has two intervening sequences (one is very short) which are transcribed into the 15S precursor RNA. The intervening sequences are excised and the molecule religated (a process called *splicing*), producing the mature 9S globin mRNA.

(see Sakano *et al.,* 1979). These findings are consistent with the view that in evolution new proteins might be constructed from parts of old ones brought together by the splicing mechanism. The splicing mechanism has not yet been detected in prokaryotes.

Heterogeneous Nuclear RNAs may be mRNA Precursors Containing Intervening Sequences

When cells are treated with radioactive RNA precursor for short periods, most of the labeled molecules are incorporated into a population of large nuclear RNAs of variable length, the so-called *heterogeneous nuclear* RNAs (hnRNAs). These rapidly labeled nuclear RNAs are not related to rRNA or to tRNA, they contain 3' poly A tails and 5' caps, and they are larger than mature mRNAs. The average length of eukaryotic mRNA is 1800 nucleotides, and that of hnRNA is 4000, but some molecules can be 20,000 nucleotides long. Most of the hnRNA is never released into the cytoplasm and is degraded within the nucleus (see Perry, 1976).

Nucleic acid hybridization studies have shown that hnRNA contains more sequences than mRNA. It is estimated that 6 to 20 per cent of the information available in the genome is represented in hnRNA, while only 1 per cent is represented in cytoplasmic mRNA.[51]

It has been suggested that hnRNA could be a mRNA precursor. At least some of these hnRNAs must be primary transcripts from which the intervening sequences have not yet been removed, as in the case of the globin 15S precursor (Fig. 22–11). It is not clear, however, if all the very large hnRNAs (20,000 nucleotides) are true mRNA precursors or not.

The reason that cells transcribe excess DNA that is not required for mRNA activity is at this moment obscure. Some authors believe that the RNA that is transcribed but never leaves the nucleus may code for regulatory functions. It has been suggested that fragments cleaved from hnRNA could interact with other regions of the genome, regulating their activity.[52, 53]

In prokaryotes mRNAs are polycistronic (i.e., a single RNA molecule codes for several proteins) and undergo a minimal amount of processing. Eukaryotic mRNAs are monocistronic, but might be transcribed as larger units which are then processed. However, in some cases mRNA processing also occurs in bacteria. During infection of *E. coli* with bacteriophage T_7, a large "early RNA"

is produced which contains information for five proteins. This transcript is then cut into the five individual mRNAs by the action of ribonuclease III, an enzyme with preference for double-stranded segments of RNA.[54, 55] Mutant bacteria which are deficient in RNAse III accumulate the large precursor mRNA and also ribosomal RNA precursors.

Eukaryotic mRNAs are Associated with Proteins

Eukaryotic mRNAs in living cells are not present as naked RNA strands but in association with proteins, forming ribonucleoprotein particles (RNP). Although cells have large numbers of different mRNAs, only between two and five proteins are associated with mRNP, and therefore the same proteins are bound to many different kinds of RNA. Globin mRNA is associated with two proteins of 78,000 and 52,000 daltons. The 78,000 protein is associated specifically with the poly A segment of globin and of other mRNAs.[56]

Some RNP complexes may remain free in the cytoplasm without being associated with ribosomes — a characteristic specifically associated with mRNA. Spirin has coined the term *informo-*

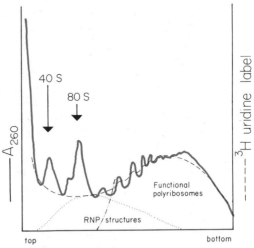

Figure 22–12 Distribution of ribosomes, functional ribosomes, and ribonucleoprotein (RNP) structures in a sucrose gradient. The distribution is demonstrated by absorption at 260 nm (A_{260}) and by radioactivity with ^{3}H-uridine. (See the description in the text.) (From Perry, R. P., Greenberg, J. R., Kelley, D. E., La Torre, J., and Schochetman, G., Messenger RNA: Its origin and fate in mammalian cells, in Kenney, F. T., et al., eds., *Gene Expression and Its Regulation,* Vol. 1, New York, Plenum Publishing Corporation, 1973, p. 149.)

somes to refer to these free nucleoprotein complexes to convey the idea that they may represent mRNAs kept in the cytoplasm for long periods of time before being genetically expressed. Some of the studies were carried out on eggs of fish or sea urchins at a stage of rapid cleavage when no ribosomal RNA was being synthesized. This RNA is uridine-labeled, and the complex is preserved by stabilization with formaldehyde. In a sucrose gradient these ribonucleoprotein structures settle in the region of free ribosomes, and some of them settle in that of polyribosomes (Fig. 22–12). The difference between ribosomes and the RNP structures is in the higher protein:RNA ratio (i.e., 4:1) and the buoyant density (1.55 for ribosomes, and about 1.40 to 1.47 for the RNP structures). In the RNP structures the content of poly A is considerably lower than in the functional polyribosomes. Studies suggest that some of the free RNP structures normally found in the cell cytoplasm probably do not constitute true mRNA (see Perry et al., 1973).

SUMMARY:

Eukaryotic Messenger RNA

In eukaryotes transcription of mRNA is more complex than in prokaryotes and involves (1) precursor transcripts of large size, (2) intranuclear processing of the transcripts, and (3) transport into the cytoplasm. Eukaryotic mRNAs are stable (e.g., globin mRNA in reticulocytes). They can be purified from differentiated cells (e.g., globin, ovalbumin, and silk fibroin mRNAs), and labeled complementary DNA can be produced by using *reverse transcriptase.*

The entire sequence of the globin mRNA is known. All eukaryotic mRNAs have a sequence coding for protein, and two non-coding regions at the 5′ and 3′ ends (Fig. 22–8). The region near the 5′ end contains the ribosome binding site. The region at the 3′ end may be very long and includes a constant hexanucleotide sequence (AAUAAA) and a long poly A tail. Most eukaryotic mRNAs have a special cap of 7-methyl-guanosine (Fig. 22–9) at the 5′ end. In prokaryons mRNA is not capped and starts with a triphosphate. The cap is important in mRNA translation.

The polyadenylic acid sequence at the 3′ end may be 100 to 200 nucleotides long and is added after transcription by a poly-A synthetase. Histone mRNAs are non-adenylated. Poly A makes the globin mRNA more stable; this has been demonstrated by injection of polyadenylated and non-adenylated mRNA into oöcytes.

By using cloned eukaryotic genes in plasmids (Chapter 21), it was unexpectedly found that the mRNA chain could be derived from non-contiguous DNA segments. In globin, ovalbumin, immunoglobulin, tRNA, and in other genes there are *intervening sequences* which are transcribed but are absent in the mature RNAs. Intervening sequences can be visualized with the electron microscope (Fig. 22–10). The presence of such intervening (non-coding) sequences explains why the mRNA precursors are larger than the mature mRNA. The non-coding inserts are removed by a process called RNA splicing (Fig. 22–11), which has not yet been detected in prokaryotes.

Within the nucleus there are large molecules of RNA, the so-called *heterogeneous nuclear RNAs* (hnRNAs) that contain 3′ poly A tails and 5′ caps. These molecules are rapidly labeled and range between 4000 and 20,000 nucleotides. It is thought that hnRNA contains mRNA precursors. While 6 to 20 per cent of the genome is represented in hnRNA, only about 1 per cent is present in cytoplasmic mRNA. Most of the hnRNA is degraded and metabolized within the nucleus. Some hnRNAs represent the large transcripts with intervening non-coding sequences mentioned earlier. The meaning of the excess RNA is at present obscure, although a regulatory function has been suggested.

In prokaryotes mRNAs are polycistronic and undergo little processing, and in eukaryotes mRNAs are monocistronic and processing is complex.

Eukaryotic mRNAs are associated with proteins, forming RNP particles. There are between two and five proteins associated with the many different mRNAs. Some of these RNP complexes may remain in the cytoplasm and might be kept for long periods of time *(informosomes)* before being expressed.

REFERENCES

1. Jacob, R., and Monod, J. (1961) *J. Molec. Biol.,* 3:318.
2. Brenner, S., Jacob, F., and Meselson, M. (1961) *Nature,* 190:576.
3. Gros, F., Hiat, H., Gilbert, W., Kurland, C., Risebrough, R., and Watson, J. D. (1961) *Nature,* 190:581.
4. Hall, B. D., and Spiegelman, S. (1961) *Proc. Natl. Acad. Sci.,* 47:137.
5. Chamberlin, M. J. (1976) In: *RNA polymerase.* (Losick, R., and Chamberlin, M., eds.) Cold Spring Harbor Laboratory, Cold Spring Harbor, New York.
6. Burgess, R. R., Travers, A., Dunn, J., and Bautz, E. K. F. (1969) *Nature,* 221:5175.
7. Travers, A., and Burgess, R. R. (1969) *Nature,* 222:537.
8. Burgess, R. R. (1976) In: *RNA Polymerase,* p. 69. (Losick, R., and Chamberlin, M., eds.) Cold Spring Harbor Laboratory, Cold Spring Harbor, New York.
9. Pribnow (1975) *J. Molec. Biol.,* 99:419.
10. Roberts, J. W. (1969) *Nature,* 224:1168.
11. Roberts, J. W. (1976) In: *RNA polymerase,* p. 247. (Losick, R., and Chamberlin, M., eds.) Cold Spring Harbor Laboratory, Cold Spring Harbor, New York.
12. Howard, B., and de Crombrugghe, B. (1976) *J. Biol. Chem.,* 251:2520.
13. Richardson, J. P., Grimlay, C., and Lowery, C. (1975) *Proc. Natl. Acad. Sci.,* 72:1725.
14. Goodman, H. M., Olson, M. V., and Hall, B. D. (1977) *Proc. Natl. Acad. Sci. USA,* 74:5453.
15. Valenzuela, P., Bell, G. I., Masiarz, F. R., De Gennaro, L. J., and Rutter, W. J. (1977) *Nature,* 267:641.
16. Miller, O. L., Jr., and Beatty, B. R. (1969) *Science,* 164:995.
17. Miller, O. L., Jr., Hamkalo, B. A. and Thomas, C. A. (1970) *Cold Spring Harbor Symp. Quant. Biol.,* 35:505.
18. Hamkalo, B. A., and Miller, O. L., Jr. (1973) In: *Gene Expression and Its Regulation.* (Kenney et al., eds.) Plenum Press, New York.
19. Roeder, R. G., and Rutter, W. J. (1970) *Proc. Natl. Acad. Sci. USA,* 65:675.
20. Reeder, R. H., and Roeder, R. G. (1972) *J. Molec. Biol.,* 67:433.
21. Weinmann, R., and Roeder, R. G. (1974) *Proc. Natl. Acad. Sci. USA,* 71:1790.
22. Sklar, V. F., Schwartz, L. B., and Roeder, R. G. (1975) *Proc. Natl. Acad. Sci.,* 72:348.
23. Buhler, J. M., Iborra, F., Sentenac, A., and Fromageot, P. (1976) *J. Biol. Chem.,* 251:1712.
24. Parker, C. S., and Roeder, R. G. (1977) *Proc. Natl. Acad. Sci. USA,* 74:44.
25. Mertz, J. E., and Gurdon, J. B. (1977) *Proc. Natl. Acad. Sci. USA,* 74:1502.
26. De Robertis, E. M., and Mertz, J. E. (1977) *Cell,* 12:175.
27. McNight, S. L., and Miller, O. L., Jr. (1976) *Cell,* 8:305.
28. Chooi, W. Y., and Laird, C. D. (1976) *Chromosoma,* 58:193.
29. Efstratiadis, A., Kafatos, F. C., and Maniatis, T. (1977) *Cell,* 10:571.
30. Chantrenne, H., Burny, A. and Marbaix, G. (1967) *Progr. Nucleic Acid Res. Mol. Biol.,* 7:173.
31. Baralle, F. E., and Brownlee, G. G. (1978) *Nature,* 274:84.
32. Proudfoot, N. J., and Brownlee, G. G. (1976) *Nature,* 263:211.
33. McReynolds, L., O'Malley, B. W., Nisbet, A. D., Fothergill, J. E., Fields, S., Robertson, M., and Brownlee, G. G. (1978) *Nature,* 273:724.
34. Shatkin, A. J. (1976) *Cell,* 9:645.
35. Hickey, E. D., Weber, L. A., and Baglioni, C. (1976) *Proc. Natl. Acad. Sci. USA,* 73:19.
36. Darnell, J. E., Wall, R., and Tushinski, R. J. (1971) *Proc. Natl. Acad. Sci. USA,* 68:1321.
37. Edmonds, M., Vaughan, M. H., and Nakazoto, H. (1971) *Proc. Natl. Acad. Sci. USA,* 68:1336.
38. Lee, S. Y., Mendecki, J., and Brawerman, G. (1971) *Proc. Natl. Acad. Sci. USA,* 68:1331.
39. Gurdon, J. B., Lingrel, J. B., and Marbaix, G. (1973) *J. Molec. Biol.,* 80:539.
40. Marbaix, G., Huez, G., and Soreq, H. (1977) *Trends Biochem. Sci.,* 2:N106.
41. Huez, G., Marbaix, G., Gallwitz, D., Weinberg, E., Devos, R., Hubert, E., and Cleuter, Y. (1978) *Nature,* 271:572.
42. Tilghman, S. M., Tiemeier, D. C., Seidman, J. G., Peterlin, B. M., Sullivan, M., Maizel, J. V. and Leder, P. (1978) *Proc. Natl. Acad. Sci. USA,* 75:725.
43. Jeffreys, A. J., and Flavell, R. A. (1977) *Cell,* 12:1097.
44. Breathnack, R., Mandel, J. L., and Chambon, P. (1977) *Nature,* 270:314.
45. Brack, C., and Tonegawa, S. (1977) *Proc. Natl. Acad. Sci. USA,* 74:5652.
46. Valenzuela, P., Venegas, A., Weinberg, F., Bishop, R., and Rutter, W. (1978) *Proc. Natl. Acad. Sci. USA,* 75:190.
47. Knapp, G., Beckmann, J. S., Johnson, P. F., Fuhrman, S. A., and Abelson, J. (1978) *Cell,* 14:221.
48. Curtis, P. J., Mantei, N., Van den Berg, J., and Weissmann, C. (1977) *Proc. Natl. Acad. Sci. USA,* 74:3184.

49. Ross, J., and Knecht, D. A. (1978) *J. Molec. Biol.*, *119*:1.
50. Gilbert, W. (1978) *Nature, 271*:501.
51. Lewin, B. (1975) *Cell, 4*:77.
52. Davidson, E. H., and Britten, R. J. (1973) *Quart. Rev. Biol., 48*:565.
53. Robertson, H. D. and Dickson, E. (1974) *Brookhaven Symp. Biol. 26*:240.

54. Dunn, J. J., and Studier, F. W. (1973) *Proc. Natl. Acad. Sci. USA, 70*:1559.
55. Dunn, J. J., and Studier, F. W. (1973) *Proc. Natl. Acad. Sci. USA, 70*:3296.
56. Bloebel, G. (1973) *Proc. Natl. Acad. Sci. USA, 70*:924.

ADDITIONAL READING

Adha, S., and Gottesman, M. (1978) Control of transcription termination. *Annu. Rev. Biochem., 47*:967.

Chamberlin, M. J. (1974) The selectivity of transcription. *Annu. Rev. Biochem., 43*:721.

Chambon, P. (1975) Eukaryotic nuclear RNA polymerases. *Annu. Rev. Biochem., 44*:613.

Crick, F. (1979) Split genes and RNA splicing. *Science, 204*:264.

Darnell, J. E. (1978) Implication of RNA-RNA splicing in evolution of eukaryotic cells. *Science, 202*:1257.

Dawid, I. B., and Wahli, W. (1979) Application of recombinant DNA technology to questions of developmental biology. *Dev. Biol., 69*:305. Good review of intervening sequences.

Dugaiczyk, A., Woo, S. L. C., Lai, E. C., Mace, M. L., McReynolds, L., and O'Malley, B. (1978) The natural ovalbumin gene contains seven introns. *Nature, 274*:328.

Gilbert, W. (1976) Starting and stopping sequences for the RNA polymerase. In *RNA Polymerase*, pp. 193–205. (Losick, R., and Chamberlin, M., eds.) Cold Spring Harbor Labortory, Cold Spring Harbor, New York.

Küpper, H., Sekiya, T., Rosenberg, M., Egan, J., and Landy, A. (1978) A rho-dependent termination site in the gene coding for tyrosine tRNA suIII of *E. coli. Nature, 272*:423.

Lewin, B. (1975) Units of transcription and translation: The relationship between nRNA and mRNA. *Cell, 4*:11.

Lewin, B. (1975) Units of transcription and translation: Sequence components of hnRNA and mRNA. *Cell, 4*:77.

Mandel, J. L., Breathnach, R., Gerlinger, P., Le Meur, M., Gannon, F., and Chambon, P. (1978) Organization of coding and intervening sequences in the chicken ovalbumin split gene. *Cell, 14*:641.

McKnight, S. L., Bustin, M., and Miller, O. L., Jr. (1977) Electron microscopic analysis of chromosome metabolism in the *Drosophila melanogaster* embryo. *Cold Spring Harbor Symp: Quant. Biol., 42*:741.

Miller, O. L., Jr. (1973) The visualization of genes in action. *Sci. Am., 228*(3):34.

Perry, R. P. (1976) Processing of RNA. *Annu. Rev. Biochem., 45*:605.

Preobrazhensky, A. A., and Spirin, A. S. (1978) Informosomes and their protein components. *Progr. Nucleic Acid, Res. Mol. Biol., 21*:1.

Processing of RNA (1974) *Brookhaven Symp. Biol., 26.*

Reanney, D. (1979) RNA splicing and polynucleatide evolution. *Nature, 277*:598.

Roberts, J. W. (1976) Transcription termination and its control in *E. coli.* In: *RNA Polymerase*, pp. 247–271. (Losick, R., and Chamberlin, M., eds.) Cold Spring Harbor Laboratory, Cold Spring Harbor, New York.

Roeder, R. G. (1976) Eukaryotic nuclear RNA polymerases. In: *RNA Polymerase*, pp. 285–329. (Losick, R., and Chamberlin, M., eds.) Cold Spring Harbor Laboratory, Cold Spring Harbor, New York.

Robertson, H. D., and Dickson, E. (1974) RNA processing and the control of gene expression. *Brookhaven Symp. Biol., 26*:240.

Sakano, H., Rogers, J., Hüppi, K., Breck, C., and Tonegawa, S. (1979) Domains of an immunoglobulin heavy chain are encoded in separate DNA segments. *Nature, 277*:627.

Schibler, U., Marcu, K. B., and Perry, R. P. (1978) Synthesis and processing of the mRNAs specifying heavy and light chain immunoglobulins. *Cell, 15*:1495.

Tilghman, S. M., Curtis, P. J., Tiemeier, D. C., Leder, P., and Weissman, C. (1978) The intervening sequence of mouse β-globin gene is transcribed within the 15S β-globin mRNA precursor. *Proc. Natl. Acad. Sci. USA, 75*:1309.

Transcription of Genetic Material (1970) *Cold Spring Harbor Symp. Quant. Biol., 35.*

23

RIBOSOMES AND
NUCLEOLAR FUNCTION

23–1   Ribosomes 501
            *Ribosomes may be Free or Membrane-
                Bound*
            *Ribosome Composition — A Large and a
                Small Subunit*
            *Ribosomal RNAs — 18S, 28S, 5.8S and 5S
                in Eukaryotes*
            Summary: *Structures of Ribosomes and
                rRNAs*
23–2   Ribosomal Proteins 506
            *Dissociation and Reconstitution of
                Ribosomal Proteins and RNA*
            *Prokaryotic and Eukaryotic Ribosomes —
                Little Homology, but the Same
                Basic Function*
            Summary: *Organization of Ribosomes*

23–3   Biogenesis of Ribosomes and
            Nucleolar Function 508
            *The Nucleolar Organizer Contains
                Ribosomal DNA*
            *Ribosomal Genes are Tandemly Repeated
                and Separated by Spacer DNA*
            *5S Genes — Tandem Repeats with
                Spacers, but Outside the Nucleolar
                Organizer*
            *Ribosomal DNA is Highly Amplified in
                Oöcytes*
            *Ribosomal RNAs Undergo a Complex
                Processing in the Nucleolus*
            *Ribosomal Biogenesis can be Followed
                Under the Electron Microscope*
            Summary: *Biogenesis of Ribosomes*

In this chapter we will start to analyze protein synthesis, the last step in the flow of genetic information. The translation of the 4 base code into a 20 amino acid sequence in proteins is a formidable task involving many cellular components. In addition to mRNA, protein synthesis requires the ribosomes, which are complex cellular particles formed by 54 proteins and 3 RNA molecules; about 60 tRNA molecules which act as molecular adaptors; and many soluble enzymes.

23–1 RIBOSOMES

Ribosomes are used by the cell for protein synthesis. First observed by Palade under the electron microscope as *dense particles* or *granules*,[1] ribosomes were then isolated and shown to contain about equal amounts of RNA and protein. Ribosomes are universal components of cells, in which they provide a scaffold for the ordered interaction of all the molecules involved in protein synthesis.

Cells devote great effort to the production of these essential organelles. An *Escherichia coli* cell contains 15,000 ribosomes, each one with a molecular weight of about 3 million daltons. Ribosomes therefore represent 25 per cent of the total mass of bacterial cells.

We will first analyze the morphology, structure, and biogenesis of ribosomes. Even though this book is devoted to the eukaryotic cell, many aspects of the structure and function of ribosomes will be discussed in relation to the bacterial cell, since the molecular organization of ribosomes has been studied principally in prokaryotes.

Ribosomes may be Free or Membrane-bound

Eukaryotic ribosomes are either free in the cytoplasmic matrix or attached to the membranes of the endoplasmic reticulum (see Chapter 10). It will be shown later that these two types are interchangeable; subunits of both types of ribosome are stored in a common pool after the ribosomes complete a cycle of protein synthesis (see Fig. 24–7).

In prokaryotes the ribosomes are free in the

protoplasm. In eukaryotes, the proportion of membrane-bound ribosomes varies widely. In cells engaged in producing proteins for export, such as those of the exocrine pancreas, which synthesize various digestive enzymes (Fig. 11–10), or plasma cells, which produce immunoglobulins, most ribosomes (more than 90 per cent) are associated with membranes. The opposite situation exists in reticulocytes, meristematic plant tissues (Fig. 14–4), and embryonic nerve cells (Fig. 9–1), which have mainly free ribosomes. In HeLa cells, a human tumor cell line, only 15 per cent of the ribosomes are membrane-bound. The liver is an intermediate type of cell in which about 75 per cent of the ribosomes are bound, with the other 25 per cent remaining free in the cytoplasmic matrix. As described in Chapter 10–1, bound ribosomes are attached to the membrane via the large ribosomal subunit (Fig. 10–6).

In chick embryo cells subject to hypothermia, free ribosomes become organized into sheets with a crystal-like arrangement with a square unit cell. These crystals are interpreted as groups of inactive ribosomes that are not associated with messenger RNA and which lack nascent protein.[2] Similar crystalline sheets of ribosomes occur naturally in the oöcytes of lizards (Fig. 23–1), which store large number of inactive ribosomes for use during early development. In this case, however, the ribosomes are not free but attached to membranes via the large ribosomal subunit.[3]

Ribosome Composition — A Large and a Small Subunit

The ribosome is a spheroid particle of 23 nm which is composed of a large and a small subunit (Fig. 23–2). At low concentrations of Mg^{2+} (0.001M) eukaryotic ribosomes sediment in sucrose gradients with a sedimentation coefficient of 80S. Prokaryotic ribosomes are smaller and sediment at 70S (Table 23–1). When the Mg^{2+}

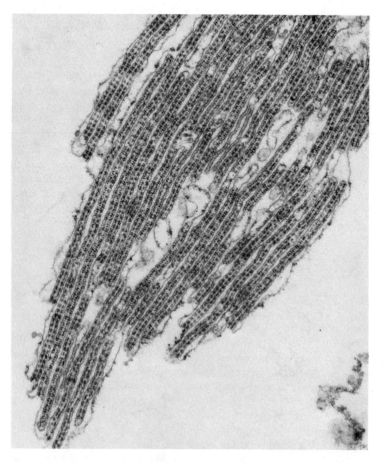

Figure 23–1 Electron micrograph of a lizard oöcyte showing ribosome crystals. These inactive ribosomes are attached to membranes and are stored for use during early development. (Courtesy of P. N. Unwin.)

concentration is lowered, ribosomes dissociate reversibly into *subunits* (Fig. 23–2). The 80S (eukaryotic) ribosomes have subunits of 40S and 60S; 70S (prokaryotic) ribosomes have subunits of 30S and 50S (Table 23–1).

During protein synthesis many ribosomes become attached to one mRNA molecule (Figs. 23–2), forming a *polyribosome* or *polysome*. Eukaryotic cells also contain ribosomes inside mitochondria and chloroplasts. These ribosomes are always smaller than the 80S cytoplasmic ribosomes (Table 23–1). Although the sedimentation values for mitochondrial ribosomes seen to vary in different phyla, they tend to be comparable to prokaryotic ribosomes in both size and sensitivity to antibiotics.[4]

Negative staining of ribosomes and their subunits allows one to visualize structural features that are at the limit of resolution of the electron microscope. Figure 23–3 shows a three-dimensional model of an *E. coli* ribosome. The 30S subunit is elongated and has an indentation positioned one third of the way down its length (Fig. 23–3,*A*); a dense region called the *twist* is also observed in the small subunit (Fig. 23–3,*C*). The large subunit has a concave surface that accommodates the small subunit and presents three protuberances called the *crown region* (Fig. 23–3,*C*) (see Boublik et al., 1977).

The mRNA is positioned in the gap between both ribosomal subunits; as a result the ribosome protects a stretch of some 25 nucleotides of mRNA from degradation by ribonuclease (Fig. 23–4),[5, 6] It has been suggested that the large ribosomal subunit may have a groove or tunnel through which the nascent protein chain grows (Fig. 23–4). This is based on experiments in which ribosomes still associated with nascent polypeptide chains were digested with proteolytic enzymes. It was found that the ribosome is able to protect a segment of 30 to 40 amino acids from degradation.[7]

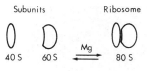

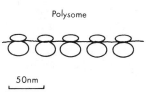

Figure 23–2 Diagram of the subunit structure of the ribosome and the influence of Mg. A polyribosome formed by five ribosomes is indicated. The filament uniting the ribosomes represents messenger RNA. The sedimentation constants (S) of the different particles are indicated.

Ribosomal RNAs—18S, 28S, 5.8S, and 5S in Eukaryotes

The major constituents of ribosomes are RNA and proteins present in approximately equal amounts; there is little or no lipid material. The positive charges of proteins are not sufficient to compensate for the many negative charges in the phosphates of RNA, and for this reason ribosomes are strongly negative and bind cations and basic dyes. Ribosomal RNA represents more than 80 per cent of the total RNA present in cells (Table 22–1). Due to the high RNA content, the cytoplasm of cells that contain many ribosomes is intensely stained with basic dyes, such as toluidine blue.

Prokaryotic ribosomes have three RNA molecules, 16S RNA in the small subunit, and 23S and 5S in the large subunit (Table 23–1). In eukaryotes these RNAs are larger (Table 23–2) and

TABLE 23–1 SIZES OF VARIOUS RIBOSOMES

Ribosomes	Size	Subunits		RNAs		Number of Proteins	
Eukaryotes	80S	60S	40S	28S + 5.8S 5S	18S	40	30
Bacteria	70S	50S	30S	23S 5S	16S	34	21
Mitochondria (in mammals)	55S	35S	25S	21S 3S	12S	—	—

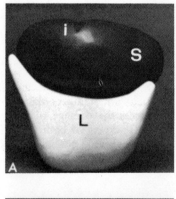

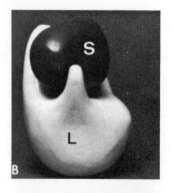

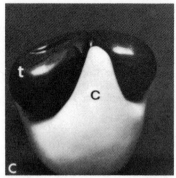

Figure 23–3 Three-dimensional model of the *E. coli* ribosome observed in four different positions. The small subunit is in black, and the large one in white. *S*, small subunit; *L*, large subunit; *c*, crown; *t*, twist. (See further description in the text.) (From Boublik, M., Hellman, W., and Kleinschmidt, A. K., *Cytobiologie, 14*:293, 1977.)

are represented by the 18S in the small subunit, and 28S and 5S in the large subunit. In addition, eukaryotic ribosomes have an extra RNA species which went unnoticed for some time. When 28S RNA is heated briefly or otherwise denatured, a small non-covalently attached component is released. This small RNA is called 5.8S RNA (Table 23–1) and is transcribed as a single unit in the nucleolus along with 18S and 28S RNAs, while

5S RNA is synthesized outside the nucleolus (see later discussion).

Ribosomal RNA has a high degree of secondary structure; about 70 per cent of it is double-stranded and helical, due to base-pairing. These double-stranded regions are formed by "hairpin loops" between complementary regions of the same linear RNA molecule. Ribosomal subunits contain a highly folded RNA molecule, to which

Figure 23–4 Left, diagram of a ribosome showing the two subunits and the probable position of the messenger RNA and the transfer RNA. The nascent polypeptide chain passes through a kind of tunnel within the large subunit. **Right,** diagram of the relationship between the ribosomes and the membrane of the endoplasmic reticulum and the entrance of the polypeptide chain into the cavity. *m*, membrane of endoplasmic reticulum. (Courtesy of D. D. Sabatini and G. Blobel.)

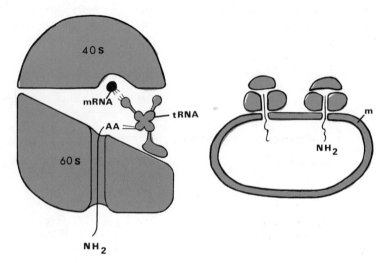

TABLE 23–2 MOLECULAR WEIGHTS OF RIBOSOMAL COMPONENTS*

	Prokaryotes			Eukaryotes		
Ribosomal Particle	70S	50S	30S	80S	60S	40S
Particle MW (RNA + Protein)	2,700,000	1,800,000	900,000	4,500,000	3,000,000	1,500,000
RNA MW		23S 1,100,000 5S 36,000	16S 600,000		28S 1,700,000 5.8S 50,000 5S 36,000	18S 700,000
Number of Proteins	54	34	21	70	40	30

*Compiled from data of Van Holde, K. E., and Hill, W. E., and of Wool, I. G., and Stoffler, G., *in* Nomura, M., Tissieres, A., and Lengyel, P., (eds.), *Ribosomes,* Cold Spring Harbor Laboratory, Cold Spring Harbor, New York, 1974.

the various proteins adhere. Ribosomal RNA genes are rich in guanine and cytosine, and this is the basis for their isolation in cesium chloride gradients.

In addition to maintaining ribosome structure, ribosomal RNA also participates in protein synthesis by virtue of its base-pairing properties. The 3′ end of 16S RNA has a sequence which is complementary to the ribosome bind-ing site of most prokaryotic mRNAs (Shine and Dalgarno, 1976). The interaction of the 16S RNA and the mRNA helps the 30S subunit to recognize the starting end of the mRNA when binding to it. Similarly, 5S RNA has a sequence complementary to the tetranucleotide TψCG (ψ stands for pseudouridine), which is present in all tRNAs and which is essential for the binding of tRNA to ribosomes.

SUMMARY:

Structure of Ribosomes and rRNAs

Ribosomes are submicroscopic particles composed of ribonucleic acid and protein that have a fundamental role in protein synthesis. In prokaryotes ribosomes are free in the protoplasm, and in eukaryotes a certain proportion of the ribosomes are membrane-bound, depending on the cell type. In both cases there is a pool of ribosomal subunits which are used in each cycle of protein synthesis. The ribosomes bound to the endoplasmic reticulum have been studied in Chapter 10 in relation to their main function of synthesizing exportable and membrane proteins.

In eukaryotes each ribosome is a spheroid of about 23 nm and 80S, and each is composed of a large subunit (60S) and a small one (40S). Prokaryotic ribosomes are smaller (70S) and dissociate into subunits of 50S and 30S. During the cycle of protein synthesis, several ribosomes become attached to a molecule of mRNA, forming a *polyribosome* or *polysome.* Within mitochondria and chloroplasts there are ribosomes of a small size (see Chapters 12 and 14).

A model of a ribosome showing the finer details of the two subunits is given in Figure 23–4. It is thought that the mRNA is in a gap between the two subunits and that there is a groove or tunnel for the nascent protein chain.

Ribosomal RNA represents more than 80 per cent of the total RNA in the cell (Table 22–1). The large subunit of eukaryotes contain 28S, 5.8S, and 5S RNAs. The small subunit contains 18S RNA. Only the 5S RNA is synthesized outside the nucleolus.

About 70 per cent of rRNA is double-stranded and helical, due to base-paired hairpin loops. The proteins adhere in an organized manner to the various RNAs. Ribosomal rRNA genes have a high GC content. The rRNA participates in protein synthesis; the 3′ end of the 16S RNA of bacteria is complementary to the ribosome binding site in mRNA, and a sequence in 5S RNA is complementary to the conserved sequence TψGC in tRNA.

23–2 RIBOSOMAL PROTEINS

The *E. coli* ribosome contains 21 proteins in the small subunit (designated S1 through S21) and 34 in the large (L1 through L34) (Table 23–2). All the proteins are different with the exception of one that is present in both subunits (S20 and L26). Therefore, the total number of different proteins in one ribosome is 54. With the exception of S1 (molecular weight 65,000), they range between 7000 and 32,000 daltons. Most of them are rich in basic amino acids.

All the ribosomal proteins have been isolated, and specific antibodies against them have been produced. Such antibodies, in conjunction with electron microscopy and use of chemical reagents that can cause cross-linking between two neighboring proteins or between a protein and RNA, have permitted work to begin on determining the topography of the proteins within the ribosomal subunits. Maps such as that shown in Figure 23–5 have been produced in this way.

These methods have also shown that some proteins, the so-called *primary binding proteins,* are attached to specific regions of the RNA strands. Six proteins bind to 16S RNA, 3 to 5S RNA, and 11 to 23S RNA; these primary bind-ing proteins in turn enable the attachment of additional proteins. This binding occurs in a precise order (see Wittmann, 1976; Kurland, 1977).

Dissociation and Reconstitution of Ribosomal Proteins

Of particular interest, from the standpoint of cell and molecular biology, is the work in which ribosomal proteins have been partially or totally dissociated. This dissociation has been followed by the reconstitution of their structure and function. These studies have led to knowledge of the role played by some of the proteins in ribosomal function. For example, if ribosomes or ribosomal subunits are centrifuged in a gradient of 5 M cesium chloride they lose 30 to 40 per cent of their proteins. In this way, both the 50S and the 30S subunits may be dissociated into two inactive core particles which contain the RNA and some proteins (*core proteins*), at the same time, several other proteins — the so-called *split proteins* (SP) — are released from each particle (Fig. 23–6). There are SP50 and SP30 proteins which may reconstitute the functional subunit when added to the corresponding core. This partial reconstitution is rapid and may be achieved in a few minutes at 37° C.[8]

In reconstitution experiments in which one protein of each subunit at a time is omitted, some of these proteins have been recognized to be essential for ribosomal function. As will be shown later the testing of ribosomal function can be carried out on a cell-free system containing ribosomes, messenger RNA, amino acids, and a supernatant containing transfer RNA and amino acid-activating enzymes.

Functional 30S subunits have been reconstituted by the addition of 16S rRNA to a mixture of proteins from the 30S subunit.[9] Similar results were also obtained with 50S subunits (see Nomura and Held, 1974). Interestingly, the total *reconstitution* of both ribosomal subunits and the complete ribosome takes place spontaneously. The subunits are re-formed by the principle of "self-assembly," as are the subunits of some proteins (see Chapter 7–1). The complex organization of this organelle may be dissociated and regenerated by simple physicochemical interactions of the component macromolecules; this may be performed in vitro and without pre-existing cell structures.

Dissociation of the 30S subunit may be achieved by treatment with 4 M urea and 2 M

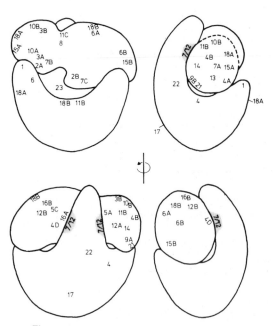

Figure 23–5 Four views of a model of ribosome from *E. coli*. The numbers indicate the topography of individual ribosomal proteins as determined by immuno–electron microscopy. (From Wittman, H. G., *Eur. J. Biochem., 61*:1, 1976.)

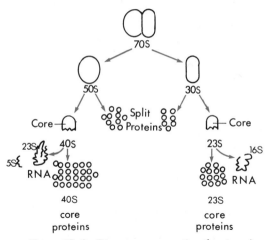

Figure 23-6 Diagram representing the stepwise dismantling of the two subunits of a 70S ribosome. Note that the proteins may be separated into split and core proteins. The 50S subunit contains 23S and 5S RNAs, and the 30S subunit has 16S RNA.

lithium chloride, which separate all the proteins. If the 16S, previously extracted with phenol, is placed in the presence of the two proteins obtained from the 30S, the reconstitution takes place in two steps.

The following reconstitution scheme has been proposed:

$$16S \text{ RNA} + \xrightarrow[\text{proteins}]{R_1} \text{RI particles} \xrightarrow{\text{heat}}$$

$$\text{RI*} \text{ particles} + \xrightarrow{\text{S proteins}} 30S \text{ ribosomal} \atop \text{subunit}$$

In the first step, performed at a low temperature, the 16S RNA binds some of the 30S ribosomal proteins, forming an RI particle (i.e., a reconstitution intermediate) that is inactive. In the second step, the RI particles are heated at 40° C in the presence of the other proteins that have remained in the supernatant (i.e., S proteins) thereby forming an excited intermediate, RI*. Within 20 minutes fully active 30S ribosomal subunits are formed. Certain proteins require the prior attachment of other proteins in order to be incorporated in a stepwise and

cooperative manner. (In other words, the previous attachment of certain proteins facilitates the binding of others.)

This type of analysis has demonstrated that whereas some of the proteins probably play a structural role in the assembly of the ribosome, others are definitely engaged in the specific ribosomal functions.

Prokaryotic and Eukaryotic Ribosomes — Little Homology but the Same Basic Function

Eukaryotic ribosomes do not differ functionally from those in prokaryotes in a fundamental way; both perform the same functions, by the same set of chemical reactions. The genetic code is the same for all living organisms, and eukaryotic ribosomes are able to correctly translate bacterial mRNAs.[10] However, the eukaryotic ribosomes are much larger than the bacterial (Table 23-2), and most of their proteins are different. Several antibiotics, such as *chloramphenicol,* inhibit bacterial but not eukaryotic ribosomes (this is the basis of the use of many antibiotics in medical treatment). Protein synthesis by eukaryotic ribosomes is inhibited by *cycloheximide.*

Ribosomes from mitochondria and chloroplasts resemble those in bacteria. They are inhibited by chloramphenicol, and hybrid ribosomes containing one bacterial subunit and one subunit from chloroplast ribosomes are fully active in protein synthesis. In eukaryotes hybrid ribosomes with subunits from both plants and mammals are active in protein synthesis, but hybrids are inactive if one subunit is dervied from bacteria. Some similarities must also exist, for ribosome reconstitution experiments have shown that proteins L7 and L12 from *E. coli* can replace the homologous proteins in mammalian ribosomes. These studies suggest that there is little homology between ribosomes of prokaryotes and eukaryotes, although the basic function in all organisms remains the same (see Wittmann 1976, 1977).

SUMMARY:

Organization of Ribosomes

In *E. coli* there are 21 proteins in the small subunit (S1 to S21) and 34 in the large one (L1 to L34) all but one of which are different and range between 7000 and 32,000 daltons. The positions of many of these have been mapped.

About 40 per cent of the proteins can easily be separated from both subunits (split proteins), leaving a core with the RNAs and the more tightly bound proteins. Reconstitution of functionally active subunits can readily be obtained by addition of the split proteins to the core. Total reconstitution of each subunit may be achieved by addition of the corresponding RNA and a pool of the proteins. The reconstitution of the 30S subunit is as follows:

$$16S \text{ RNA} + \xrightarrow[\text{proteins}]{R_1} \text{RI particles} \xrightarrow{\text{heat}}$$

$$\text{RI* particles} + \xrightarrow[\text{proteins}]{S} 30S$$

A reconstitution intermediate (RI) is formed in the cold; the final reconstitution needs activation by heat. By performing reconstitution experiments in which one protein is omitted at a time, it is possible to gain information about the role of each protein in the ribosome. The reconstitution of ribosomes is an excellent example of the principle of self-assembly, which was discussed in Chapter 7.

Although eukaryotic ribosomes do not differ functionally from those in prokaryotes, they are much bigger (Table 23–2), and most of the proteins are different. The action of certain inhibitors is also different; e.g., chloramphenicol inhibits bacterial ribosomes, and cycloheximide acts on eukaryotic protein synthesis. Ribosomes from mitochondria and chloroplasts resemble those in bacteria. Hybrid ribosomes made of bacterial and chloroplast subunits are active in protein synthesis, and a similar activity is observed with plant and mammalian ribosome hybrids. However, hybrids of prokaryotic and eukaryotic subunits are inactive.

23–3 BIOGENESIS OF RIBOSOMES AND NUCLEOLAR FUNCTION

The synthesis of a eukaryotic ribosome is a complex phenomenon, in which several regions of the cell are involved. The 18S, 5.8S, and 28S RNAs are synthesized as part of a much longer precursor molecules in the nucleolus, 5S RNA is synthesized on the chromosomes outside the nucleolus, and the 70 ribosomal proteins are synthesized in the cytoplasm. All these components migrate to the nucleolus, where they are assembled into ribosomes and transported to the cytoplasm. Ribosome biogenesis is a striking example of coordination at the cellular and molecular levels.

Historically, the concept that the nucleolus is a ribosome factory is related to the early cytochemical work of Caspersson with ultraviolet absorption and to that of Brachet with specific staining for ribonucleic acids. The work of both researchers demonstrated the presence of large nucleoli, rich in RNA, in protein-synthesizing cells. The discovery of ribosomes and their role in protein synthesis promptly suggested a relationship between the nucleolus and these organelles.

The Nucleolar Organizer Contains Ribosomal DNA

Nucleoli disappear during mitosis, but at telophase new nucleoli are formed at specific chromosomal sites called *nucleolar organizers* (located in *secondary constrictions* on the chromosomes). These sites are now known to contain the genes coding for 18S, 28S, and 5.8S rRNAs.

Direct evidence that the nucleolus is responsible for the synthesis of rRNA was obtained in 1964, when it was discovered that an anucleolate mutant of the South African frog *Xenopus laevis* was unable to synthesize rRNA.[11] Figure 23–7 shows that diploid cells of the wild-type *Xenopus* have two nucleoli (2-*nu*); the heterozygous mutant has only one nucleolus (1-*nu*) and the homozygous mutant (0-*nu*) is anucleolate. When two 1-*nu* heterozygotes are crossed, 25 per cent of the progeny are 0-*nu*. This condition is lethal and the tadpoles die after one week. Up to this stage the embryos rely on maternal ribosomes inherited from the egg cytoplasm (one *Xenopus* egg contains 10^{12} ribosomes). The 0-*nu* embryos do not synthesize 18S, 28S, or 5.8S rRNA,[11] and DNA-RNA hybridization studies showed that

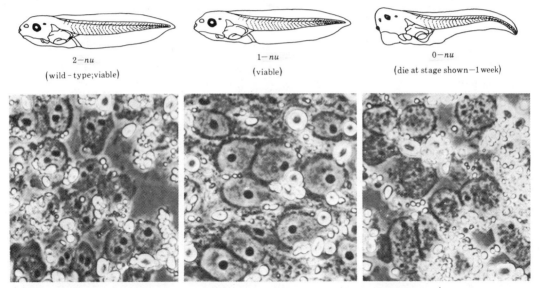

2-*nu*
(wild-type;viable)

1-*nu*
(viable)

0-*nu*
(die at stage shown—1 week)

Figure 23-7 Number of nucleoli in mutants of *Xenopus laevis*. The wild-type tadpole has two nucleoli (*2-nu*); the heterozygous mutant has one nucleolus (*1-nu*), and the homozygous anucleolate mutant has none (*0-nu*). The 0-nu tadpoles do not synthesize ribosomal RNA. (Courtesy of J. B. Gurdon.)

they do not contain the genes for rRNA[12] (i.e., the mutation is a deletion of the rRNA genes). The fact that 0-*nu* embryos can synthesize 5S RNA[13] shows that these genes are not located in the nucleolar organizer.

A similar type of experiment was performed with *Drosophila melanogaster*, in which the nucleolar organizer is contained in a heterochromatic region of the X or Y chromosome.[14] Cells with organizers that were either deficient or duplicated were analyzed with the DNA-RNA hybridization technique. As shown in Figure 23-8 the saturation level for the wild type is about 0.27 per cent. Half the saturation level was found in mutants having only one organizer, and this was proportionally higher in those *Drosophila* mutants with three or four organizers. Some 130 rDNA cistrons per organizer were calculated in the haploid nucleus of the wild type. In the so-called "bobbed" mutants of *Drosophila*, the number of rDNA cistrons may be reduced, and it can be calculated that death will occur when fewer than 40 copies of rDNA cistrons are present in the nucleolar organizer.

Ribosomal Genes are Tandemly Repeated and Separated by Spacer DNA

Table 23-3 shows that all organisms have multiple copies of rRNA genes. Few copies are

present in bacteria, but in eukaryotes rDNA is more repetitive. In the case of *Xenopus*, each nucleolar organizer contains 450 rRNA genes. These genes are *tandemly repeated* along the DNA molecule (head to tail) and are separated

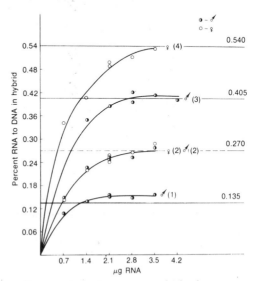

Figure 23-8 Experiment which demonstrates that the DNA complementary with ribosomal RNA is associated with the nucleolar organizer. The saturation level for the DNA-rRNA hybrids increases with the number of nucleolar organizers. Strains of *Drosophila* having one, two, three, and four nucleolar organizers were used. (From Ritossa, F. M., and Spiegelman, S., *Proc. Natl. Acad. Sci. USA, 53*:737, 1965.)

TABLE 23–3   REPETITIVE rDNA CISTRONS IN VARIOUS ORGANISMS

Organism	% rDNA	rDNA Cistrons Per Haploid Genome
E. coli	0.42–0.65	8–22
B. subtilis	0.38	9–10
HeLa cells	0.005–0.02	160–640
Drosophila (wild type)	0.27	130
Xenopus (wild type)	0.2	450

from each other by stretches of *spacer* DNA which is not transcribed. These linear repeats of genes can be visualized very clearly in Figure 23–9, which shows a nucleolar organizer spread for electron microscopy.[15, 16] These rRNA genes are being actively transcribed, and the nascent RNA chains are spread perpendicularly to the DNA axis. Each gene is transcribed into a long RNA molecule (which varies in size from 40S to 45S, according to the

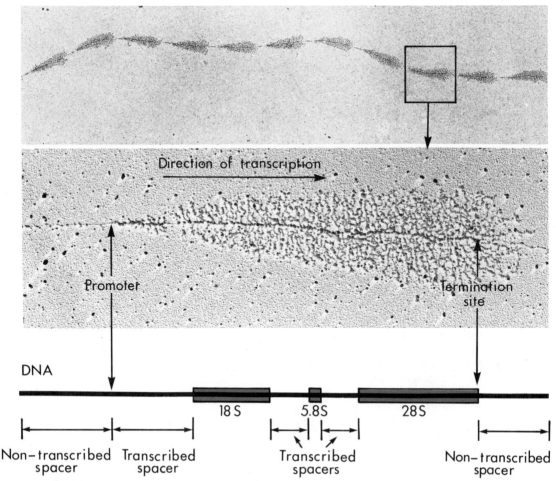

Direction of transcription

Promoter

Termination site

DNA

18 S   5.8S   28 S

Non–transcribed spacer   Transcribed spacer   Transcribed spacers   Non–transcribed spacer

Figure 23–9   Ribosomal RNA genes are tandemly repeated (head-to-tail) and separated from each other by non-transcribed spacers. A nucleolar organizer was spread for electron microscopy and 11 consecutive rRNA genes can be seen. Note the "Christmas tree" or "fern leaf" configuration of the nascent RNA precursor molecules. The long precursor molecule contains 18S, 5.8S and 28S rRNAs, and is associated with proteins forming a nucleoprotein complex. Bottom, The genetic map of an rDNA unit. (Courtesy of U. Scheer, M. F. Trendelenburg and W. W. Franke).

species) which will eventually be processed giving rise to 18S, 28S, and 5.8S RNA (see later discussion). Since each gene has a fixed initiation site (promoter) and a fixed termination site, the transcripts adopt the characteristic "Christmas tree" or "fern leaf" configuration (see Chapter 22–2). Nucleolar rRNA genes are transcribed by RNA polymerase I (the α-amanitine-resistant enzyme). The polymerase I molecules (about 100 per gene) can be visualized at the origin of each nascent RNA chain (Figs. 23–9 and 22–7).

The tandemly repeated rDNA genes may be isolated by the use of ultracentrifugation techniques on density gradients of cesium chloride, followed by hybridization with ribosomal RNA labeled with radioisotopes (Fig. 23–10). In the case of X. laevis it is possible to demonstrate the presence of a small fraction of DNA having a higher buoyant density than the major band (i.e., 1.723, as compared to 1.698). This minor fraction, which represents only 0.20 per cent of the total DNA, is the so-called satellite ribosomal DNA. This fraction is completely absent in the case of the anucleolate mutant (see earlier discussion) of X. laevis. The higher buoyant density is related to the higher guanine-cytosine content (70 per cent) in the satellite DNA. By using the hybridization technique with ribosomal RNA it is possible to demonstrate that this hybridization coincides with the satellite DNA (Fig. 23–10).[17, 18] In Chapter 15 another type of satellite DNA, corresponding to the centromeric heterochromatin, was mentioned.

The availability of purified rDNA has allowed more refined studies of its structure, mainly through the use of restriction endonucleases. It was found that although the length of the gene (transcribed segment) is constant, the length of the non-transcribed spacer (Fig. 23–9) can vary considerably between adjacent repeats[19] and between different Xenopus individuals.[20] The non-transcribed spacer is itself formed by short repeated sequences about 50 base pairs in length, and differences in the spacer length are due to variation in the number of these repeating units. It is thought that the spacer heterogeneity might arise from unequal crossing-over events (see Fig. 25–7).

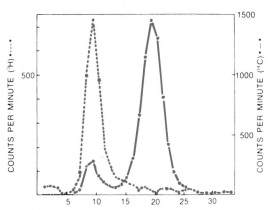

Figure 23–10 Experiment to demonstrate ribosomal DNA in the ovary of Xenopus. Solid lines represent the ultracentrifugation pattern of the DNA labeled with ^{14}C in a CsCl gradient. Note the main peak and the small satellite peak. The dotted lines represent hybridization experiments performed with ribosomal RNA labeled with tritium. Hybridization coincides with the minor DNA peak. (From Gall, J. G., Proc. Natl. Acad. Sci. USA, 60:553, 1968.)

in the Xenopus anucleolate mutant. The 5S genes are present in multiple copies; Xenopus has 24,000 5S genes which are located in the tips (telomeres) of most chromosomes, as shown by in situ hybridization of [3]H-5S-RNA to chromosome preparations.[21] In Drosophila all the 5S genes are located in a single cluster[22] in one of the autosomes (while the nucleolar organizer is in the X and Y chromosomes).

As with rDNA, the 5S genes are arranged in tandem repeats of genes separated by segments of spacer DNA. Most of the spacer is formed by repeating segments of a short sequence of 15 base pairs.[23] This sequence is very rich in AT, which explains why 5S DNA migrates as a satellite band lighter than bulk DNA in cesium chloride gradients.[24] Adjacent 5S repeats can be heterogeneous in length owing to variation in the numbers of 15 base pair subrepeats in each spacer,[25] which probably arise by unequal crossing-over events. Tandem repeats of genes separated by spacers seem to be a general feature of the organization of the eukaryotic genome.

The 5S RNA is transcribed by RNA polymerase III (the enzyme sensitive to high levels of α-amanitine), and is then transported to the nucleolus, where it is incorporated into the immature large ribosomal subunits.

5S Genes—Tandem Repeats with Spacers, but Outside the Nucleolar Organizer

The genes coding for 5S RNA are not located in the nucleolus, and therefore are not deleted

Ribosomal DNA Is Highly Amplified in Oöcytes

Gene amplification is the process by which one set of genes is replicated selectively while the rest

TABLE 23–4   RIBOSOMAL GENES OF *XENOPUS LAEVIS* IN SOMATIC CELLS AND AFTER AMPLIFICATION IN OÖCYTES*

Cell Type	Nuclear DNA (pg)	Number of Chromosome Sets per Nucleus	Chromosomal DNA (pg per Nucleus)	Ribosomal DNA (pg per Nucleus)	Number of rDNA Genes	Number of Nucleoli
Somatic	6	2	6	0.012	900	2
Oöcyte (after amplification)	37	4	12	25	2,000,000	1000–1500

*Oöcytes are in meiotic prophase and therefore contain a tetraploid amount of DNA; ribosomal DNA is amplified to 1000 times more than expected for that amount of chromosomal DNA. (Modified from Gurdon, J. B.), the Control of Gene Expression in Animal Development. Oxford University Press, Oxford, 1974).

of the genome remains constant. The clearest example of gene amplification is seen in the rDNA of amphibian oöcytes.

Xenopus eggs are very large cells (1.3 mm in diameter) that accumulate large numbers of ribosomes (10^{12}) for use during early development. Anucleolate embryos can survive throughout the first week of development using these maternal ribosomes (Fig. 23–7). Oöcytes must synthesize all these ribosomes using the genes contained in a single nucleus, and they achieve this by amplifying the number of rDNA genes 1000-fold (Table 23–4).

Amplification takes place in very small oöcytes which are in the early stages of meiotic prophase. During pachytene, excess DNA begins to accumulate on one side of the nucleus, forming a Feulgen-positive cap, as shown in Figure 23–11, *A* and *B*. This cap incorporates ^{3}H-thymidine intensely (while the chromosomes do not), and by the end of the amplification process it contains 25 pg of DNA, equivalent to 2,000,000 rRNA genes[17, 18] (Table 23–4). As oöcytes grow, the extra

DNA is accommodated in between 1000 and 1500 extrachromosomal nucleoli.[26] The amplified DNA is not inherited by the embryo, but is lost in the course of development.

The newly synthesized DNA is indeed rDNA. This can be shown by hybridizing ^{3}H-ribosomal RNA to cytological preparations of *Xenopus* oöcytes that have been previously treated with alkali to denature the DNA,[27] a technique called "*in situ*" hybridization. The hybridized material can then be located by radioautography. Figure 23–11, *C* shows that rRNA hybridizes to the cap of amplified DNA.

The initial demonstration that the amplified material is rDNA was dependent on the purification of these genes as a satellite band in cesium chloride gradients[17, 18] (Fig. 23–10). Ribosomal DNA amplification also occurs in oöcytes of some insects,[28] as demonstrated in Figure 23–12, which shows that the large oöcytes of the cricket *Acheta domesticus* contain more rDNA than other tissues of the same animal.

Unlike the genes coding for the large ribosomal

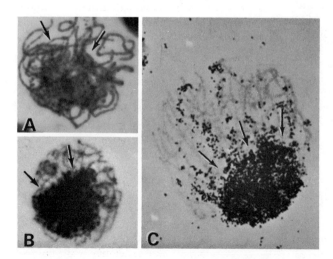

Figure 23–11 *Xenopus* oöcytes during the period of nucleolar DNA synthesis. **A,** at late pachynema, the excess DNA begins to accumulate as granules around the nucleolus (arrows). **B,** later on, the excess DNA appears as a dense mass (arrows). **A** and **B,** Feulgen reaction, × 1700. **C,** large pachynema oöcyte prepared with the ^{3}H-rRNA hybridization technique. Observe that the silver grains are deposited mainly on the mass of excess DNA. (See the description in the text.) × 1200. (Courtesy of J. G. Gall.)

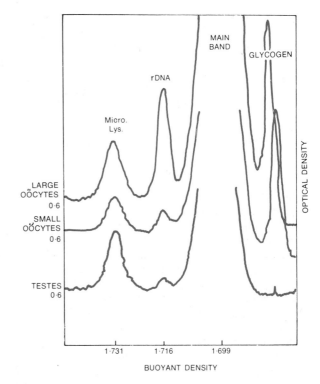

Figure 23–12 DNA components in different tissues of *Acheta* demonstrated by analytical ultracentrifugation. The ribosomal DNA (rDNA), with a buoyant density of 1.716, is most prominent in large oöcytes. The density of rDNA is compared with that of the main DNA band (1.699), DNA from a bacterium, and the glycogen band. *Micro. Lys. Micrococcus lysoseikticus.* (Courtesy of A. Lima-De-Faría.)

RNAs, the 5S genes are not amplified during oögenesis. Oöcytes meet the increased demands for ribosomes by activating all 24,000 5S genes. (Somatic cells express only the fraction of the 5S genes called "somatic genes;" while in oöcytes, "oöcyte active" genes having a slightly different nucleotide sequence are also activated.[29]

The amplification of specific genes is not a common event. The only instance in which it is known to occur other than rDNA in oöcytes, is in the DNA puffs of *Sciarid* flies (Chap. 25). In both cases, the amplified material is not passed on to future cell generations.

Ribosomal RNAs Undergo a Complex Processing in the Nucleolus

The biogenesis of rRNA provides the clearest example of RNA processing. As shown in Figure 23–9, the rRNA genes are transcribed into a long precursor RNA which must be cleaved into 18S, 28S, and 5.8S RNA. In the cleavage process about 50 per cent of the precursor RNA is degraded within the nucleus.

The steps involved in the processing of the ribosomal RNAs can be studied best after isolation of the nucleolus from cultured cells.[30] Figure 23–13 shows the variety of ribosomal RNA fractions that are present in such isolated nucleoli. The predominating fractions are 45S, 32S, and 28S, whereas there is very little 18S RNA. This last finding is interpreted as an indication that the 18S RNA is rapidly released into the cytoplasm and is not retained in the nucleolus. To study the time-course of the RNA processing, the cultured cells are examined at various intervals following the application of a pulse with a radio-labeled RNA precursor (e.g., ³H-uridine). From this study the following series of events may be recognized:

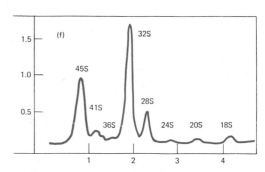

Figure 23–13 Main peaks of ribosomal RNA observed in the nucleolus of *HeLa* cells. (See the description in the text.) Nucleolar RNA was submitted to electrophoresis in polyacrylamide gel. (From Weinberg, A., Loening, V., Willems, M., and Penman, S., *Proc. Natl. Acad. Sci. USA, 58*:1088, 1967.)

TABLE 23-5 APPROXIMATE MOLECULAR WEIGHTS AND NUMBER OF NUCLEOTIDES IN DIFFERENT rRNAs FROM HeLa CELLS

rRNA	MW in Daltons	Nucleotides
45S	4.5×10^6	14,000
32S	2.2×10^6	7000
28S	1.6×10^6	5000
18S	0.6×10^6	2000
5.8S	5.0×10^4	150

(1) The first ribosomal RNA in HeLa cells is a large 45S molecule which has 14,000 nucleotides (Table 23-5). Other organisms have shorter precursors (40S for *Xenopus*). Within this molecule the rRNAs are separated by stretches of spacer RNA, and the order of transcription is: 5' end — 18S — 5.8S — 28S — 3' end (Fig. 23-14). On a fully active gene about 100 RNA polymerases are transcribing simultaneously on the ribosomal DNA cistron.

(2) In the nucleolus, 45S RNA is rapidly methylated, even before transcription is completed. Methylations occur mostly on the ribose moiety (producing 2'-O-methylribose), and occur only in the 18S (46 methylations) and 28S (71 methyl groups) sequences that will be conserved, while those segments that will be degraded remain non-methylated.[31]

(3) The 45S RNA has a lifetime of about 15 minutes and is then cleaved into smaller components, according to the following general pattern, and also as indicated in Figure 23-14:

(4) The 20S RNA is rapidly processed into 18S rRNA, and presumably for this reason the small ribosomal subunits appear in the cytoplasm before the large ribosomal subunits, which have a slower RNA processing.

(5) The 32S RNA remains in the nucleolus for about 40 minutes and is then cleaved into 28S RNA and 5.8S RNA. These remain in the nucleolus for another 30 minutes before entering the cytoplasm as part of the large ribosomal subunit (Fig. 23-15).

From the data shown in Table 23-5 it is evident that about half the 45S molecule is lost by the successive degradations. This occurs in the regions that are non-methylated and have a higher content of GC. Thus, the processing of the ribosomal RNA leads to an increase in methyl groups and to a decrease in GC content.

In *E. coli* the 16S and 23S rRNAs are also transcribed as a single unit but are processed so rapidly that the precursor rRNAs cannot be detected. However, a 30S precursor accumulates in mutants defective in ribonuclease III, which is the processing enzyme.[32]

In eukaryotes ribosomes are assembled in the nucleolus. The 45S RNA becomes rapidly associated with protein, forming a particle of about 80S. All the transitions between 45S and the final 28S and 18S occur within nucleolar ribonucleoprotein particles[33] and are indicated in Figure 23-15.

Ribosomal Biogenesis can be Followed Under the Electron Microscope

In Chapter 15 we analyzed the ultrastructure of the nucleolus. Most nucleoli have a *granular region* and a *fibrillar region* that forms a central core, as can be clearly observed in Figure 23-16.

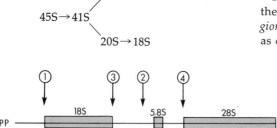

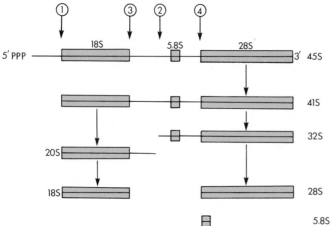

Figure 23-14 Diagram showing the processing of ribosomal RNA in HeLa cell nucleoli. The orientation of the various rRNAs in the 45S precursor and the four main sites of processing are indicated.

SYNTHESIS AND MATURATION OF RIBOSOMES IN THE NUCLEOLUS

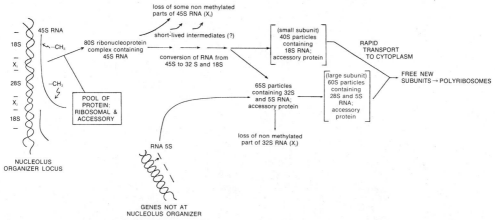

Figure 23–15   Stages in the biogenesis of ribosomes. (From Perry, R. P., *in* Lima-De-Faría, A. (Ed.), *Handbook of Molecular Cytology,* Amsterdam, North-Holland Publishing Co., 1969.)

Cytochemical studies have suggested that the following dynamic relationship exists between the different portions of the nucleolus:

nucleolar DNA → fibrillar area → granular area.

When cells are labeled with ³H-uridine for five minutes and followed by a chase (in cold uridine), the fibrillar part of the nucleolus first incorporates the precursor, which is only later found in the granular part.[34] The fibrillar part contains the rDNA and the nascent 45S RNA, and when spread under the EM (rather than sectioned), it has the appearance shown in Figure 23–9. The granular part represents ribosome precursor particles at various stages of assembly and processing. With the introduction of thin sectioning to electron microscopy, observation of dense particles on both sides of the nuclear envelope of different cells suggested the passage of ribosomes from the nucleus to the cytoplasm.[35] The transport of ribonucleoprotein particles into the cytoplasm is easily observed in amphibian oöcytes[36] in which material of nucleolar origin may be seen passing through the pores of the nuclear envelope into the cytoplasm (Fig. 23–16).

SUMMARY:

Biogenesis of Ribosomes

Biogenesis of ribosomes is the result of the coordinated assembly of several molecular products that converge upon the nucleolus. The genes coding for 18S, 28S, and 5.8S RNA are present in the nucleolar organizer; the 5S genes are located in other regions of the genome, and the ribosomal proteins are manufactured in the cytoplasm. The 5S gene products and the proteins migrate toward the nucleolus where they are processed and assembled into active ribosomes.

Direct evidence of the function of the nucleolus as a ribosome factory was obtained with the discovery of an anucleolate mutant of *Xenopus laevis* (Fig. 23–8). Different numbers of ribosomal genes were also found in *Drosophila* mutants with varying numbers of nucleolar organizers.

While in bacteria there are few copies of rRNA genes, in eukaryotes they are repetitive, and the genes are arranged in tandem and are separated by non-transcribed spacers. Under the electron microscope the process of transcription can be visualized (Fig. 23–9). The rDNA genes can be separated in a density gradient of cesium chlo-

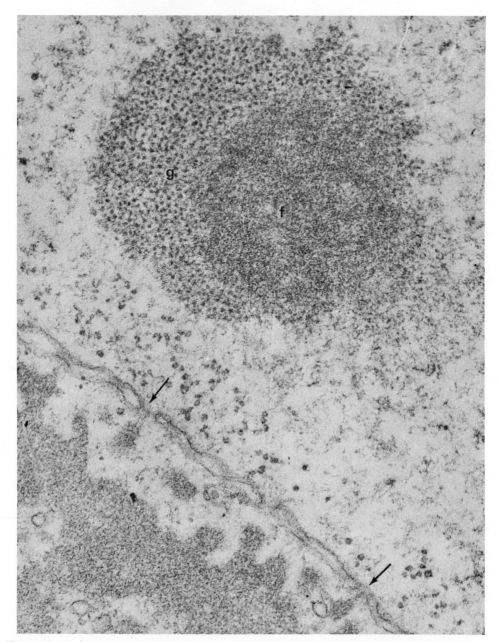

Figure 23–16 Electron micrograph of an oöcyte of *Rana clamitans* showing one of the peripheral nucleoli (containing amplified DNA) with a fibrillar (*f*) central portion and a granular (*g*) peripheral portion. Arrows indicate material entering the cytoplasm through the nuclear pores. × 70,000. (Courtesy of O. L. Miller.)

ride because of their higher GC content. The spacer is made of short repeated sequences and may vary in length between individuals.

In *Xenopus* there are 450 rRNA genes in each nucleolar organizer. In the same species the 5S RNA genes are located at the telomeres of most chromosomes. There are some 24,000 genes arranged in tandem spacers rich in AT. The 5S RNA is transcribed by RNA polymerase III and is then transported to the nucleolus.

Gene amplification is the process by which a set of genes is selectively replicated. The rDNA in the amphibian oöcyte undergoes this process to accumulate in the egg the huge numbers of ribosomes that are used in the first stages of development. During pachytene there is an active replication of the nucleolar organizers, and the rDNA is amplified 1000-fold. In a *Xenopus* egg 25 pg of extra DNA with 2,000,000 rDNA genes is accommodated by between 1000 and 1500 nucleoli. This amplified DNA is then lost during development.

The 5S rDNA genes are not amplified during oögenesis, and thus all the 24,000 original genes become active to meet the need for ribosomes.

Processing of the rRNA occurs in the nucleolus. As in the case of other RNAs, longer transcripts are made which are then partially degraded. There is a single precursor for 18S, 28S, and 5.8S rRNAs, which has 14,000 nucleotides and a sedimentation coefficient of 45S (Table 23–5).

Within the rDNA the order of transcription is 5' end — 18S — 5.8S — 28S — 3' end, and each rRNA is separated by a spacer. Some 100 RNA polymerases are active on a single rDNA cistron. The 45S precursor is methylated in the regions to be conserved, and its lifetime is about 15 minutes. It then is cleaved as follows:

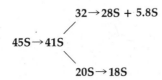

Processing of the rRNA from 20S to 18S is rapid, and the small ribosomal subunit appears in the cytoplasm before the large one. The 32S RNA remains in the nucleolus for about 40 minutes before it is processed into 28S + 5.8S RNAs, and these in turn take about 30 minutes before entering the cytoplasm as the large subunit.

The different steps in the processing of the ribosomes can be followed by cytochemical studies in the fibrillar and granular regions of the nucleolus (Chap. 15). The electron microscope permits the observation not only of the transcription of the ribosomal subunits but also of their transport into the cytoplasm.

REFERENCES

1. Palade, G. E. (1955) *J. Biophys. Biochem. Cytol.,* 1: 59.
2. Byers, B. (1967) *J. Molec. Biol., 26*:155.
3. Unwin, P. N. (1977) *Nature, 269*:118.
4. O'Brien, T. W. (1977) In: *International Cell Biology,* p. 245. (Brinkley, B. R., and Porter, K. R., eds.) The Rockefeller University Press, New York.
5. Nonomura, Y., Bloebel, G., and Sabatini, D. D. (1971) *J. Molec. Biol., 60*:303.
6. Lake, J., Sabatini, D. D., and Nonomura, Y. (1974) In: *Ribosomes,* p. 543. (Nomura, M., Tissieres, A., and Lengyel, P., eds.) Cold Spring Harbor Laboratory, Cold Spring Harbor, New York.
7. Blobel, G., and Sabatini, D. D. (1970) *J. Cell Biol., 45*:130.
8. Nomura, M., and Traub, P. (1968) *J. Molec. Biol., 34*:609.
9. Traub, P., and Nomura, M. (1969) *J. Molec. Biol., 40*:391.
10. Anderson, C. W., Atkins, J. F., and Dunn, J. J. (1976) *Proc. Natl. Acad. Sci. USA, 73*:2752.
11. Brown, D. D., and Gurdon, J. B. (1964) *Proc. Natl. Acad. Sci. USA, 51*:139.
12. Wallace, H., and Birnstiel, M. L. (1966) *Biochim. Biophys. Acta, 114*:296.
13. Miller, L. (1973) *J. Cell Biol., 59*:624.
14. Ritossa, F. M., and Spiegelman, S. (1965) *Proc. Natl. Acad. Sci. USA, 53*:737.
15. Miller, O. L., Jr., and Beatty, B. R. (1969) *Science, 164*:955.
16. Scheer, U., Trendelenburg, M. F., and Franke, W. W. (1976) *J. Cell Biol., 69*:465.
17. Brown, D. D., and Dawid, I. B. (1968) *Science, 160*:272.
18. Gall, J. G. (1968) *Proc. Natl. Acad. Sci. USA, 60*:553.
19. Wellauer, P. K., Dawid, I. B., Brown, D. D., and Reeder, R. H. (1976) *J. Molec. Biol., 105*:461.
20. Reeder, R. H., Brown, D. D., Wellauer, P. K., and Dawid, I. B. (1976) *J. Molec. Biol., 105*:507.
21. Pardue, M. L., Brown, D. D., and Birnstiel, M. L. (1973) *Chromosoma, 42*:191.
22. Artavanis, S., Schedl, P., Tschudi, C., Pirrota, V., Steward, R., and Gehring, W. (1977) *Cell, 12*:1057.
23. Miller, J. R., Cartwright, E. M., Brownlee, G. G., Fedoroff, N. V., and Brown, D. D. (1978) *Cell, 13*:717.
24. Brown, D. D., Wensink, P. C., and Jordan, E. (1971) *Proc. Natl. Acad. Sci. USA 68*:3175.
25. Carrol, D., and Brown, D. D. (1976) *Cell, 7*:477.
26. MacGregor, H. C. (1972) *Biol. Rev., 47*:177.
27. Gall, J. G., and Pardue, M. L. (1969) *Proc. Natl. Acad. Sci. USA, 63*:378.
28. Lima-de-Faría, A., Birnstiel, M., and Jaworska, H. (1969) *Genetics, 61*:145.
29. Ford, P. J., and Southern, E. M. (1973) *Nature (New Biol.), 241*:7.
30. Weinberg, R. A., and Penman, S. (1970) *J. Molec. Biol., 47*:169.

31. Maden, B. E. H. (1977) *Trends Biochem. Sci., 1*:196.
32. Schlessinger, D. (1974) In: *Ribosomes,* p. 393. Cold Spring Harbor Laboratory, Cold Spring Harbor, New York.
33. Warner, J. R. (1974) In: *Ribosomes,* p. 461. Cold Spring Harbor Laboratory, Cold Spring Harbor, New York.
34. Bernhard, W., and Granboulan, N. (1968) In: *The Nucleus,* p. 81. Academic Press, Inc., New York.
35. De Robertis, E. D. P. (1954) *J. Histochem. Cytochem., 2*:341.
36. Miller, O. L. (1969) In: *Handbook of Molecular Cytology* (Lima-de-Faría, A., ed.) North-Holland Publishing Co., Amsterdam.

ADDITIONAL READING

Bielka, H. (1978) The eukaryotic ribosome. *Trends Biochem. Sci., 3*:153.

Boublik, M., Hellman, W., and Kleinschmidt, A. K. (1977) Size and structure of *E. coli* ribosomes by electron microscopy. *Cytobiologie, 14*:293.

Brimacombe, R., Stoffler, G., and Wittman, H. G. (1978) Ribosome structure. *Annu. Rev. Biochem., 47*:217.

Boseley, P., Moss, T., Mächler, M., Portmann, R., and Birnstiel, M. (1949) Sequence organization of the spacer DNA in a ribosomal gene unit of *X. laevis. Cell, 17*:19.

Brown, D. D., and Dawid, I. B. (1968) Specific gene amplification in oöcytes. *Science, 160*:272.

Brown, D. D. (1973) The isolation of genes. *Sci. Am., 229*(2):21.

Carroll, D., and Brown, D. D. (1976) Repeating units of *Xenopus laevis* 5S DNA are heterogeneous in length. *Cell, 7*:467 and 477.

De Wilde, M., Cabezon, T., Herzog, A., and Bollen, A. (1977) Apports de la génétique à la connaissance du ribosome bactérien. *Biochimie (Paris), 59*:125.

Emuanilov, I., Sabatini, D. D., Lake, J. A., and Freienstein, C. (1978) Localization of eukaryotic initiation factor three on native small ribosomal subunits. *Proc. Natl. Acad. Sci. USA, 75*:1389.

Fedoroff, N. V. (1979) On spacers. *Cell, 16*:997.

Gall, J. G. (1968) Differential synthesis of the genes for ribosomal RNA during amphibian oögenesis. *Proc. Natl. Acad. Sci. USA, 60*:553.

Kurland, C. G. (1977) Structure and function of the bacterial ribosome. *Annu. Rev. Biochem., 46*:173.

Lake, J. A. (1976) Ribosome structure determined by electron microscopy of *E. coli* small subunits, large subunits and monomeric ribosomes. *J. Molec. Biol., 105*:131.

Maden, B. E. H. (1977) Ribosomal precursor RNA and ribosome formation in eukaryotes. *Trends Biochem. Sci., 1*(3):196.

Miller, O. L. (1973) Visualization of genes in action. *Sci. Am., 228*:34.

Nomura, M. (1969) Ribosomes. *Sci. Am., 221*:(4):28.

Nomura, M. (1973) Assembly of bacterial ribosomes. *Science, 179*:864.

Nomura, M., and Held, W. A. (1974) Reconstitution of ribosomes: Studies of ribosome structure, function and assembly. In: *Ribosomes.* pp. 193–223. (Nomura, M., Tissieres, A. and Lengyel, P., eds.) Cold Spring Harbor Laboratory, Cold Spring Harbor, New York.

Pardue, M. L., Brown, D. D., and Birnstiel, M. L. (1973) Location of the genes for 5S ribosomal RNA in *Xenopus laevis. Chromosoma, 42*:191.

Perry, R. P. (1972) Regulation of ribosome synthesis. *Biochem. J., 129*:35.

Reeder, R. H. (1974) Ribosomes from eukaryotes: Genetics. In: *Ribosomes,* pp. 489–518. (Nomura, M., Tissieres, A., and Lengyel, P., eds.) Cold Spring Harbor Laboratory, Cold Spring Harbor, New York.

Scheer, U., Trendelenburg, M. F., and Franke, W. W. (1976) Regulation of transcription of genes of rRNA during amphibian oögenesis. *J. Cell Biol., 69*:465.

Shine, J., and Dalgarno, L. (1974) The 3′ terminal sequence of *E. coli* 16S ribosomal RNA: complementarity to ribosome binding sites. *Proc. Natl. Acad. Sci. USA, 71*:1342.

Speirs, J., and Birnstiel, M. (1974) Arrangement of the 5.8S RNA cistrons in the genome of *Xenopus laevis. J. Molec. Biol., 87*:237.

Warner, J. R. (1974) The assembly of ribosomes in eukaryotes. In: *Ribosomes,* pp. 417–488. (Nomura, M., Tissieres, A., and Lengyel, P., eds.) Cold Spring Harbor Laboratory, Cold Spring Harbor, New York.

Wittmann, H. G. (1976) Structure, function and evolution of ribosomes. *Eur. J. Biochem., 61*:1.

Wittmann, A. G. (1977) Structure and function of *E. coli* ribosomes. *Fed. Proc., 36*:2075.

PROTEIN SYNTHESIS

24-1 Transfer RNA 519
  tRNA Transcription—Long Precursors with Complex Post-transcriptional Modifications
  The Fidelity of Protein Synthesis Depends on the Correct Charging of tRNA by the AA-tRNA Synthetases
  Summary: *Transfer RNA*

24-2 Ribosomes and Protein Synthesis 524
  Polysomes Consist of Several Ribosomes Attached to the Same mRNA
  During Protein Synthesis There is a Ribosome-Polysome Cycle
  The Small Subunit Binds to the mRNA Ribosome Binding Site
  Prokaryotic Proteins Start with Formyl-methionine

  IF_1, IF_2, and IF_3—Protein Factors That Initiate Protein Synthesis
  Eukaryotic Initiation Factors are More Complex than Those in Prokaryotes
  Eukaryotic IF_2 is Phosphorylated by a Cyclic AMP-Dependent Protein Kinase
  EFTu + EFTs, and EFG are Protein Elongation Factors
  The Large Ribosomal Subunit is Involved in the Process of Elongation
  Chain Termination Involves UAA, UGA, and UAG codons, and Releasing Factors R_1 and R_2
  Antibiotics and Toxins are Useful in Molecular Biology and Medicine
  Secretory Proteins have a Hydrophobic Signal Peptide
  Summary: *Protein Synthesis*

We have already examined the genetic code (Chap. 21), the synthesis of mRNA (Chap. 22), and the structure of ribosomes (Chap. 23). In the first part of this chapter we will discuss the structure and biogenesis of tRNA. In the second part we will analyze how all these components come together to accomplish the final step in the flow of genetic information, the synthesis of specific proteins.

24-1 TRANSFER RNA

Transfer RNAs are a group of small ribonucleic acids that act as interpreters of the genetic code. In fact, tRNAs are able to read the message expressed as codons in mRNA, and at the same time are able to recognize the amino acids that the codons specify. The need for this transfer RNA molecule (or adaptor) is easily understood, since there is no specific affinity between the side groups of many amino acids and the bases in the mRNA. This fundamental step in translation is performed by specific aminoacyl-activating enzymes (AA-tRNA synthetases) that attach the specific amino acids to one end of the respective tRNA molecule, and

by specific triplets of bases (*anticodons*) that are able to bind to the corresponding codons of mRNA with hydrogen bonds. For each of the 20 naturally occurring amino acids there must be at least 1 or more specific tRNAs.

In *E. coli* there are about 50 tRNAs which utilize almost all the possible codons (triplets) of the genetic code. The only exceptions are the codons for termination of protein synthesis, which are recognized by protein factors (releasing factors) instead of tRNA. There are 20 isoacceptor tRNA families, one for each amino acid. In addition, for each amino acid there is one synthetase enzyme, which is able to charge the various tRNAs of the corresponding isoacceptor family.

Transfer RNA accounts for 10 to 15 per cent of the total RNA in *E. coli*. Each tRNA has a sedimentation constant of 4S and contains between 75 to 85 nucleotides. The sequence of alanine tRNA was elucidated by Holley and collaborators in 1965.[1]

The primary structure of over 100 tRNAs from *E. coli* to mammalian and plant cells is known (see Rich, 1978), and all of them have common features and fit the *cloverleaf* model of secondary structure shown in Figure 24–1.

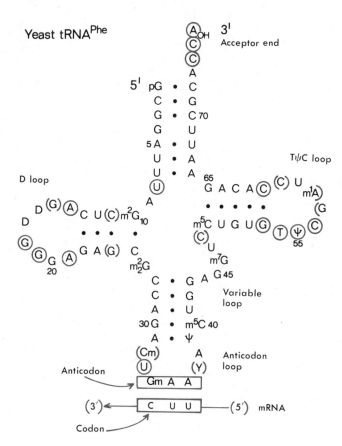

Figure 24–1 Diagram of the cloverleaf model of transfer RNA showing the amino acid acceptor end (CCA); the anticodon loop with the three bases that read the genetic message; the D loop, which contains several dihydrouridines (D); the TψC loop, which is thought to interact with the 5S ribosomal RNA; and a variable loop. Bases that are invariant in all tRNAs are indicated by circles; bases that tend to be moderately conserved (e.g., always a purine or pyrimidine) are indicated by parentheses. These constant bases interact with each other via hydrogen bonds, producing the three dimensional folding of the molecule (Fig. 24–2).

One characteristic in common is the presence of several unusual bases, most of which are derived from methylation; pseudouridine (abbreviated ψ), inosinic acid, methylguanine, methyl aminopurine, methylcytosine, ribothymine and others. Some of the common features of tRNAs are the 3' end of the chain which always terminates in CCA and attaches to the specific amino acid; and the 5' end which frequently terminates with guanine (Fig. 24–1). In all tRNAs there is a sequence TψCG in a special region, and other constant residues, the dihydrouridines, are in a different loop (Fig. 24–1).

The main feature of the cloverleaf diagram is a series of loops separated by short stems of helical double-stranded regions between four and seven complementary bases long. Figure 24–1 shows the principal elements of this structure: (1) the acceptor end (CCA), where the specific amino acid becomes attached; (2) the *anticodon loop*, directly opposite from the acceptor end and having the three bases that recognize and form hydrogen bonds with the mRNA, thus reading the genetic message; (3) the *D loop* (named for dihydrouridine), which is important for the recognition by the specific aminoacyl-tRNA synthetase; and (4) the *T loop* (named for TψCG), which has the conserved sequence TψCG, which is thought to interact with a complementary region of 5S ribosomal RNA during protein synthesis (see Kurland, 1977). There is also a highly variable region (variable loop), which differs greatly in length in different tRNAs.

The tertiary structure of tRNA has been determined by x-ray diffraction studies (see Rich, 1978). As shown in Figure 24–2 tRNA has a compact configuration in which the side-arms are folded together, rather than the open configuration shown in Figure 24–1. The molecule has the shape of an L. The anticodon is on one end of the L, and the acceptor end is on the other. The D and T loops are hydrogen-bonded together at the corner of the structure. All tRNAs share these elements of tRNA folding.

The gene coding for bacterial *tyrosine*-tRNA (including its promoter and terminator sequences) has been synthesized by Khorana and associates using organic chemistry methods.[2] This technological feat was made possible by previous knowledge of the complete nucleotide sequence of these tRNAs.

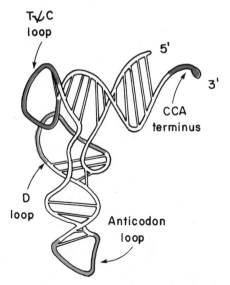

Figure 24–2 Tertiary structure of yeast phenylalanine transfer RNA showing a compact configuration with an L form. The position of the various parts shown in Figure 24–1 are indicated. Numbers refer to nucleotide residues in the sequence. Note that the D and TψC loops are hydrogen bonded together. (Redrawn from A. Rich, 1978.)

tRNA Transcription—Long Precursors with Complex Post-transcriptional Modifications

As in the case of the ribosomal RNAs, tRNA is not transcribed directly into its final form but rather into a longer precursor. For example, *E. coli* tyrosine tRNA is synthesized as a precursor having 39 extra nucleotides at the 5′ end and 3 at the 3′ end. These nucleotides must be removed before an active tRNA molecule is obtained.[3] Other *E. coli* tRNAs are synthesized as very long precursors that contain several tRNAs (as many as 5 to 7) in a single RNA chain.[4] These precursor tRNAs are cleaved by a specific enzyme, *ribonuclease P* (P for processing). *E. coli* mutants defective in ribonuclease P accumulate the large precursor molecules instead of the mature tRNAs (see Smith, 1976).

Other post-transcriptional modifications important in tRNA function are the base modifications (such as methylations) which lead to the unusual bases found in tRNA, and the terminal addition of CCA. The 3′ end of tRNA always terminates with CCA, and its integrity is essential because it is at this end that the amino acid is attached. The 3′ end is occasionally degraded in vivo and bacteria have an enzyme capable of terminal addition of CCA, which regenerates the active end. In fact, some tRNA genes do not contain in their DNA the information for the CAA end which is added post-transcriptionally.

In eukaryotes, an even more bewildering form of tRNA processing has been found. Some (but not all) yeast tRNA genes contain in their DNA sequence insertions of between 10 and 40 base pairs close to the anticodon site (Fig. 24–3).[5, 6] As seen in Chapter 22, many other eukaryotic genes (such as globin genes) are also interrupted by intervening sequences. The extra nucleotides are transcribed into precursor RNAs and are excised, and the rest of the molecule is then rejoined (by "splicing"), as shown in Figure 24–3. An enzyme capable of splicing tRNA precursors has been isolated from yeast.[7] Yeast mutants defective in RNA processing accumulate large amounts of precursor tRNAs containing these inserts. The function of these intervening sequences is still unknown.

The splicing enzyme probably recognizes the general shape of the folded tRNA molecule, notices that there is an extra sequence, excises it, and then rejoins the ends. This is thought to be the case because *Xenopus* oöcytes can splice yeast tRNA[Tyr] precursors, in spite of the fact that the intervening sequence of the corresponding gene from frogs has a totally different nucleotide sequence (DeRobertis and Olson, 1979).

Frog oöcytes microinjected with cloned genes are useful for the analysis of RNA processing because up to 80 per cent of the RNA synthesized may come from the injected tRNA genes, and the different tRNA precursors can be isolated and studied before extensive processing has occurred. A possible function for the 5′ leader sequence became apparent when the order in which the various base modifications appear was examined. The RNA precursors having a leader sequence have particular base modifications added, while other modifications are not present at all and can only be added after the leader sequence is removed. In other words, the precursor RNA with a 5′ leader sequence is a substrate for some base modification enzymes, and the RNA without it is a substrate for others. It therefore seems that the leader sequence introduces a precise order in the intricate processing events involved in tRNA production (Melton, Cortese, and De Robertis, 1979).

The Fidelity of Protein Synthesis Depends on the Correct Charging of tRNA by the AA-tRNA-Synthetases

The tRNA molecule is essential to the two steps that determine the fidelity of protein synthesis. The first step is the selection of the correct amino acid for attachment to tRNA to form an amino-

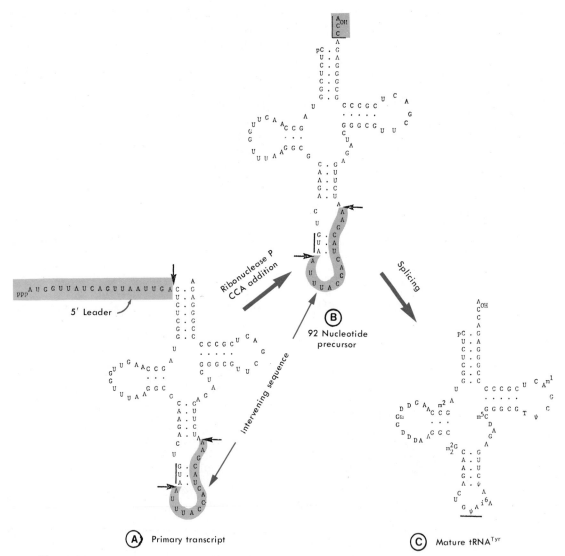

Figure 24–3 Processing of yeast tyrosine-tRNA. *A*, primary transcript with a 5' leader sequence of 19 nucleotides and an intervening sequence of 14 nucleotides. *B*, the 5' leader sequence is cleaved by Ribonuclease P (in a two step process, not shown), and the CCA end is added post-transcriptionally. *C*, the intervening sequence is spliced and mature tRNA^{Tyr} is formed. The position of the anticodon is underlined. The primary transcript was detected by microinjecting cloned yeast tRNA genes into the nucleus of *Xenopus* oöcytes. (From DeRobertis, E. M. and Olson, M. V., *Nature*, 278:134–143, 1979.

acyl-tRNA. The second one occurs in the ribosome and involves the selection of the correct tRNA in response to a codon in mRNA.

Amino acids are attached to tRNAs by specific enzymes called *aminoacyl-tRNA synthetases* (or *activating* enzymes). The accuracy of protein synthesis depends on the correct charging of tRNA by the synthetase, for the ribosome and mRNA only recognize the tRNA (not the amino acid attached to it), and if a tRNA is mischarged it will be incorporated incorrectly into the protein molecule. A crucial feature of protein synthesis is that the translation from the 4 base code

to 20 amino acids is performed by the tRNA synthetases.

For each amino acid there is a specific aminoacyl-tRNA synthetase which activates the carboxyl group of that amino acid for covalent bonding with the adenylic acid residue of the 3' end of tRNA. The first step of the reaction involves the use of ATP:

$$AA + ATP \xrightleftharpoons[\text{synthetase}]{\text{aminoacyl}}$$

$$AA \sim AMP + P \sim P \qquad \textbf{(1)}$$

The AA~AMP intermediate formed remains bound to the enzyme until the proper tRNA arrives, and at that moment the formation of the AA~tRNA complex occurs:

$$AA{\sim}AMP + tRNA \xrightarrow[\text{synthetase}]{\text{aminoacyl}} \quad \text{(2)}$$

$$AA{\sim}tRNA + AMP$$

In order to perform these two functions, the activating enzyme has two sites: one to recognize the amino acid and another to recognize the specific tRNA.

The bond between the amino acid and the CCA end of tRNA is a high-energy one which stores the energy from the ATP used in reaction (1). The energy thus stored is used in the ribosome to drive the formation of the peptide bond.

The great accuracy of tRNA synthetases depends on having a two-step reaction. The AA ~ AMP must remain attached to the enzyme until the tRNA arrives; if an incorrect amino acid were selected initially, during this period it would have a higher chance of falling off since it would have a decreased affinity for the enzyme. This results in a double check of the amino acid, the first when it is initially bound to the enzyme, and the second during the period when it must remain attached. This process is called "proofreading" and is essential to avoid mistakes in the translation of the genetic code.

SUMMARY:

Transfer RNA

The study of the molecular steps of protein synthesis involves previous knowledge of the main actors: the genetic code (Chap. 21), the messenger RNA (Chap. 22), the ribosome (Chap. 23), and the transfer RNAs (tRNAs).

Transfer RNAs are a group of small RNAs (4S) that act as interpreters of the genetic code. They are able to read the codons by way of the *anticodon* (a triplet of bases) and to carry specific amino acids. In *E. coli* there are about 50 tRNAs that recognize codons and transfer the 20 amino acids. There are no tRNAs for the termination codons, which are recognized by termination factors of a protein nature. Ten to 15 per cent of the RNA of a cell is tRNA, and the sequence of between 75 and 85 nucleotides has been determined in more than 100 tRNAs from bacteria, plants, and animals.

The secondary structure of tRNA is represented by the *cloverleaf* model (Fig. 21–1,A), which consists of (1) the *acceptor end* at the 3' end which receives the amino acid and which terminates in CCA: (2) the *anticodon loop* with the triplet; (3) the *D loop* (named for dihydrouridine), which is recognized by the specific aminoacyl-tRNA synthetases; and (4) the *T loop* (named for TψCG), which interacts with the 5S ribosomal RNA. By x-ray diffraction studies the cloverleaf has been shown to be a folded and compact L-shaped structure (Fig. 21–1,B).

The tRNA gene is transcribed as a long precursor with extra nucleotides at the 5' end that are later removed by ribonuclease P. Post-transcriptional modifications include methylations of some bases and addition of the CCA at the 3' end. In some yeast tRNAs there are intervening sequences 10 to 40 nucleotides long, which are excised by the splicing enzyme.

Amino acids are attached to the 3' end of tRNA by specific aminoacyl-tRNA synthetases which recognize the tRNAs and charge them correctly. There is a two-step reaction which involves the use of ATP and two sites in the enzyme, one for the amino acid and the other for the specific tRNA. Thus, a crucial step in the translation of the 4 base code into 20 amino acids is carried out by the tRNA synthetase. The high-energy bond between the 3' end and the amino acid provides the energy for the formation of the peptide bond. The accuracy of the enzyme depends on a two-step "proofreading" reaction.

24–2 RIBOSOMES AND PROTEIN SYNTHESIS

Polysomes Consist of Several Ribosomes Attached to the Same mRNA

The early electron microscopic observations of cell sections revealed that ribosomes were frequently associated in groups, occasionally forming recurrent patterns (Fig. 10–1). It was not until 1962 that the function of these *polyribosomes*, or *polysomes*, in protein synthesis was discovered.

After treating reticulocytes with ^{14}C-labeled amino acids and using gentle methods for disruption (e.g., lysis in hypotonic solutions) researchers found that in addition to the typical sedimentation band of single ribosomes (see Fig. 24–4) larger units were present. These particles ranged from 108S to 170S, or more. At the same time, the maximum radioactivity, indicating the synthesis of hemoglobin, was detected in the 170S fraction; this corresponded to a polyribosome of five units (a pentamer) (Fig. 24–4). It was confirmed by electron microscopy that about 75 per cent of the ribosomes in the 170S peak were present as pentamers. In a fully active mRNA there is a ribosome about every 80 nucleotides, and longer mRNAs therefore have longer polysomes.

Figure 22–12 shows the pattern of ribosomes obtained from cultured eukaryotic cells upon ultracentrifugation on a continuous gradient of sucrose. The absorbance at 260 nm shows the high peak of the single ribosomes (i.e., 80S) preceded by the 60S and 40S subunits, and followed by a series of peaks of larger size, corresponding to the functional polyribosomes.

The best way of visualizing polysomes is to spread them for electron microscopy using the methods described in Chapter 22. Figure 24–5 shows a molecule of silk moth fibroin mRNA with its associated ribosomes. Each ribosome has a nascent protein chain which gets longer in the direction of protein synthesis (5′ → 3′).[9] Fibroin mRNA is very long, and each polysome may contain up to 80 ribosomes; this means that there can be up to 80 polypeptides translated simultaneously from the same mRNA molecule.

In prokaryotes protein synthesis starts even before mRNA synthesis is completed (chap. 22, Fig. 22–5), because there is no nuclear envelope interposed between the genome and the ribosomes.[10] Bacterial polysomes are therefore visualized with the nascent mRNA still attached to the DNA via the RNA polymerase (Fig. 24–6).

During Protein Synthesis There is a Ribosome-Polysome Cycle

When not engaged in protein synthesis, most ribosomal subunits exist in a cytoplasmic pool of *free* ribosomal subunits. The subunits associate during protein synthesis and dissociate at the end of this process (Fig. 24–7). This ribosomal cycle has been demonstrated by ex-

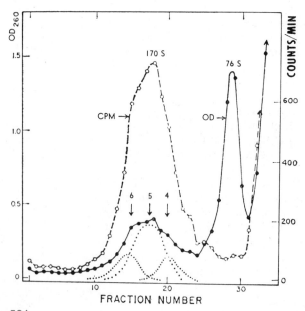

Figure 24–4 Experiment done to demonstrate that the synthesis of hemoglobin occurs in polyribosomes. Reticulocytes were incubated for 45 seconds in the presence of a pool of amino acids labeled with [14]C. The cells were lysed osmotically and the soluble part (hemoglobin) was centrifuged for two hours on a continuous density gradient of sucrose. After this the tube was punctured and thirty fractions were collected and observed under the electron microscope. Similar fractions were analyzed for optical density at 260 nm for RNA and for the number of counts. While the optical density (OD) shows a peak at 76S (sedimentation constant corresponding to single ribosomes), the peak of radioactivity corresponds to the ribosomal tetramers, pentamers, and hexamers indicated with arrows. (From Warner, J. R., Rich, A., and Hall, C. E., *Science*, 138:1399–1403, 1962.)

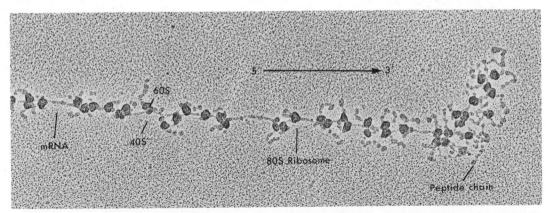

Figure 24–5 Silk fibroin mRNA visualized during protein synthesis. A nascent peptide can be observed emerging from each ribosome and increasing in length in the 5' to 3' direction. The nascent peptide may be visualized in this case because silk fibroin is a fibrous protein and remains extended, while most other proteins adopt a globular conformation as they are being synthesized. (Courtesy of O. L. Miller, Jr.)

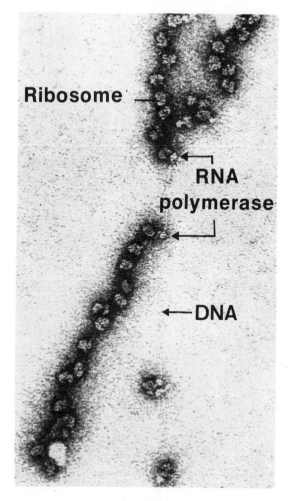

Figure 24–6 Electron micrograph of bacterial DNA with messenger RNA and polyribosomes. (Courtesy of O. L. Miller, Jr.)

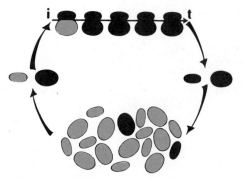

Figure 24–7 Ribosomal cycle in protein synthesis. (See the description in the text.) (From Kaempfer, R., *Proc. Natl. Acad. Sci. USA, 61*:106, 1968.)

periments which show that ribosomal subunits can be exchanged. Figure 24–7 is a diagram of an experiment in which two types of ribosomes were allowed to translate mRNA. The black subunits came from heavy ribosomes tagged with ^{13}C, ^{15}N, and ^{2}H, whereas the white subunits contained the normal ^{12}C, ^{14}N, and ^{1}H atoms. Hybrid ribosomes were detected, indicating that ribosomal subunits come apart after protein synthesis.[11] Subunit exchange depends on protein synthesis, as demonstrated by the fact that it does not take place in the presence of translation inhibitors.

The ribosomes are separated into subunits at the end of the ribosome cycle by a *dissociation factor* which binds to the 30S subunit.[12, 13] This

dissociation protein was later identified as initiation factor IF$_3$, which is also required for the binding of mRNA to the 30S subunit. Thus, the same event that dissociates ribosomes enables the 30S subunit to start a new cycle of protein synthesis (Fig. 24–8). The IF$_3$ protein is released after the 30S subunit binds to mRNA.

The Small Subunit Binds to the mRNA Ribosome Binding Site

The first step in protein synthesis is the binding of the 30S subunit (containing IF$_3$) to a precise location of the mRNA, the *ribosome binding site*, which contains the AUG codon at which protein synthesis starts. The start signal for protein synthesis is complex and involves other sequences in addition to the AUG codon. These sequences differentiate the starting AUG codon from the other AUG codons located within the message, which are not utilized as initiation signals. To bind to the correct starting site the ribosome must therefore recognize other structural features in mRNA that precede the AUG codon and that together represent the ribosome binding site.

The nucleotide sequence of many ribosome binding sites is known. (This information can be gained by binding ribosomes to mRNA and then adding ribonuclease to digest all nucleotides except those protected by ribosome at-

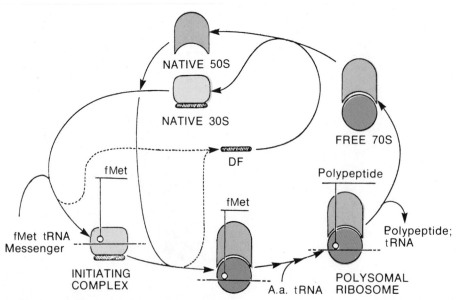

Figure 24–8 A model of the ribosome-polysome cycle in which the action of a dissociation factor (*DF*) is indicated. DF is identical to initiation factor 3. (See the description in the text.) (From Subramanian, A. R., Ron, E. Z., and Davis, B. D., *Natl. Acad. Sci. USA, 61*:761, 1968.)

tachment.) It was found that ribosome binding sites in prokaryotic mRNAs have between three and eight bases that are complementary to the sequence of the 3' end (tail) of the 16S RNA of the small ribosomal subunit.[14] This complementary region is placed about ten nucleotides before the start of the protein, and is found in all prokaryotic mRNAs so far studied (see Table 24–1). It is thought that the end of the 16S RNA hybridizes to the mRNA sequences,[15] thus aligning the initiation codon into the correct position to start protein synthesis.

The secondary structure of mRNA may also be important for the correct initiation of protein synthesis. Some regions of the mRNA chain have *hairpin loops* of double-stranded RNA which are held together by hydrogen bonds (Fig. 5–20). If this secondary structure is destroyed (e.g., by the action of formaldehyde) the specificity of initiation is lost and abnormal proteins are synthesized in vitro.

TABLE 24–1 RIBOSOME BINDING SITES IN BACTERIOPHAGE G₄*

Gene	Nucleotide Sequence
A	A-T-C-A-A-A-C-[G-G-A-G-G]-C-T-T-T-T-C-[A-T-G]
B	A-G-G-C-T-[A-G-G]-G-A-A-T-A-A-A-G-A-A-[A-T-G]
K	C-A-A-G-T-A-C-[G-G-A]-T-A-T-T-T-C-T-G-[A-T-G]
C	G-T-G-G-A-C-T-G-C-T-G-G-T-[G-G-A]-A-A-[A-T-G]
D	A-C-C-A-C-A-[A-A-G-G-A]-A-A-C-T-G-A-A-[A-T-G]
E	C-T-A-T-C-[G-A-G-G]-C-T-T-G-C-G-T-A-T-[A-T-G]
J	C-T-T-T-[T-A-A-G-G-A-G]-T-T-A-T-G-T-A-[A-T-G]
F	C-T-A-T-T-[T-A-A-G-G-A]-T-A-C-A-A-A-A-[A-T-G]
G	A-G-C-C-A-A-[A-A-G-G-A]-C-T-A-A-C-A-T-[A-T-G]
H	T-G-A-A-A-[T-A-A-G-G-A]-T-T-A-T-C-C-T-[A-T-G]
16S rRNA 3' end	ₕₒA-U-U-C-C-U-C-C-A-C-U-A-G

*The boxes indicate the initiation codon (ATG) and the bases that are complementary to the 3' end of 16S ribosomal RNA. The DNA sequence is depicted, and therefore T is used instead of U. G₄ is a bacteriophage closely related to φX174 (Chap. 21). (Data from Godson, G. N., Barrell, B. G., Staden, R., and Fiddes, J. C., *Nature*, 276:236, 1978.)

Prokaryotic Proteins Start with Formylmethionine

Bacterial proteins start with the codon AUG, and the first AA-tRNA to initiate protein synthesis is always a formyl (—CHO) derivative of methionine (fMet-tRNA).

Since in protein synthesis the peptide chain always grows sequentially from the free terminal —NH₂ group toward the —COOH end, the function of formylmethionine tRNA (fMet-tRNA) is to ensure that proteins are synthesized in this direction. In fMet-tRNA the —NH₂ group is blocked by the formyl group, leaving only the —COOH available to react with the —NH₂ of the second amino acid. In this way the synthesis follows in the correct sequence. Later, the first amino acid is separated from the protein by a hydrolytic enzyme. fMet-tRNA has been found in mitochondria, but it is not involved in the general cytoplasmic protein synthesis of eukaryotes.[16]

The position of an AUG codon within mRNA can be discriminated by the ribosome. When this codon is at the start of mRNA, fMet is incorporated, but when the same AUG codon is in the middle of the message, normal methionine is incorporated instead. There are two different tRNAs that respond to the codon AUG, an initiator fMet-tRNA and one for normal methionine. Methionine is formylated only after it has been bound to this special initiator tRNA. The fMet-tRNA is carried to the ribosome as a complex together with *initiation factor 2* and GTP, while normal methionine tRNA (as well as all other AA-tRNAs) is carried to the ribosome as a complex together with *elongation factor* Tu and GTP (see later discussion).

In eukaryotic cells the initiation of the synthesis is by a special met-tRNA which is not formylated. There is another met-tRNA which is specific for the placement of methionine inside the polypeptide. In both cases, the coding triplet is AUG. In the case of eukaryotes the initial methionine is generally split off from the finished polypeptide (see Ochoa, 1977).

IF₁, IF₂, and IF₃ — Protein Factors That Initiate Protein Synthesis

Three initiation factors from *E. coli* (IF₁, IF₂, and IF₃) have been purified in Ochoa's laboratory. These initiation factors are proteins loosely associated with 30S subunits and can be isolated by washing ribosomes with NH_4Cl (Table 24–2). All three factors are essential for

TABLE 24–2 PROTEIN FACTORS IN PROTEIN SYNTHESIS

Phase	Factor	Source	Function
1° Initiation	IF$_3$	High salt 30S	Dissociation of ribosomal subunits.
	IF$_1$ IF$_2$ GTP IF$_3$	ribosomal wash	Binding of mRNA and initiator tRNA to 30S subunit.
2° Elongation	Tu Ts T + GTP	Supernatant fraction	Binding of aminoacyl-tRNA Tu-GTP complex to ribosome.
	Peptidyl transferase	50S ribosomal subunit	Peptidyl transfer from peptidyl-tRNA to aminoacyl-tRNA.
	G + GTP	Supernatant fraction	Translocation of peptidyl-tRNA; release of free tRNA.
3° Termination	R$_1$ R$_2$	Supernatant fraction	Release of protein at UAA, UAG, or UGA codons.

*From Lipmann, F., "What Do We Know About Protein Synthesis?" *in* Kenney, F. T., Hamkalo, B., Favelukes, G., and August, J. T. (eds.), *Gene Expression and Its Regulation*, New York, Plenum Publishing Corp., 1973. (See Fig. 24–8.)

initiation of protein synthesis when natural mRNAs are used, but they are not required for artificial mRNAs, such as poly U.

IF$_3$ has the dual function of acting as a dissociation factor and as a stimulatory factor for the binding of mRNA to the 30S subunit. The IF$_2$ forms a ternary complex with fMet-tRNA and

GTP, which then binds to the 30S mRNA, forming an *initiation complex*. The IF$_1$ helps in the interaction between IF$_2$ and the initiator tRNA. Once the 50S subunit binds to the 30S initiation complex the initiation factors are released and can be reutilized.

The upper part of Figure 24–9 shows the

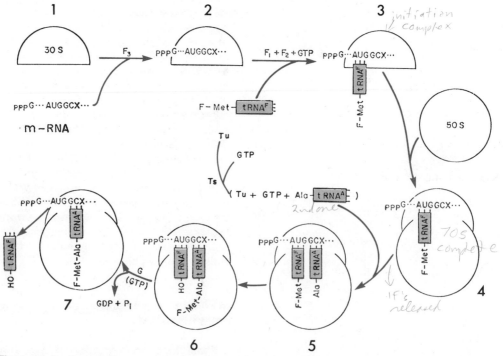

Figure 24–9 General diagram of the initiation steps in protein synthesis involving 30S and 50S ribosomal subunits, messenger RNA, the initiation factors F$_3$ and F$_1$ + F$_2$, GTP, the elongation factors T and G, formylmethionine-tRNA (fMet-tRNA) and alanine-tRNA. **1,** isolated 30S; **2,** binding of m-RNA and F$_3$ to the 30S; **3,** binding of the F$_1$ + F$_2$ + GTP and fMet-tRNA to make the initiation complex; **4,** binding of the 50S subunit to make the complete 70S ribosome; **5,** binding of the second aminoacyl-tRNA; **6,** synthesis of the first peptide bond; **7,** liberation of the free tRNA after translocation.

steps involved in the formation of the initiation complex in *E coli*.

Eukaryotic Initiation Factors are More Complex Than Those in Prokaryotes

In eukaryotes protein synthesis initiation factors have also been isolated, but more than the three factors of prokaryotes are required to perform the same functions (reviewed by Weissbach and Ochoa, 1976).

Eukaryotic initiation factors are more complex than those in prokaryotes. For example, eIF$_3$ is a large complex formed by at least 10 different polypeptide chains and having a molecular weight greater than 500,000 daltons. The large size of eIF$_3$ makes it possible to visualize it under the electron microscope when it is bound to the small ribosomal subunit. In Figure 24–10, the upper row shows negatively stained 40S subunits, while the lower row shows the eIF$_3$ bound to a defined region of the 40S subunit (Emanuilov et al., 1978).

Eukaryotic IF$_2$ is Phosphorylated by a Cyclic AMP-dependent Protein Kinase

It has recently been found that the eukaryotic initiation factor 2 (eIF$_2$, similar in function to the *E. coli* IF$_2$) can become phosphorylated under certain circumstances.[17] Phosphorylated eIF$_2$ is inactive in protein synthesis, and the degree of phosphorylation can be regulated by cyclic AMP in a cascade mechanism very similar to the one indicated in Figure 6–9. The eIF$_2$ is phosphorylated by a specific kinase which is normally inactive but which becomes active when phosphorylated by another enzyme, a cyclic AMP-dependent protein kinase[18] (see Fig. 6–9). The eIF$_2$ phosphorylation provides a

mechanism by which the *total* rate of protein synthesis could be regulated in eukaryotes.

EFTu + EFTs, and EFG are Protein Elongation Factors

In Lipmann's laboratory soluble protein fractions present in the post-ribosomal supernatant were found to be active in the elongation of the polypeptide chain (Table 24–1). At first, only the EFT and EFG factors were isolated, but later the *EFT factor* was found to comprise two proteins: Tu, which is temperature-unstable, and Ts, which is temperature-stable. The function of the EFTu + EFTs factors is in the binding of the aminoacyl-tRNA to the ribosome (Fig. 24–9,5). Initially, EFTu + GTP form a complex with the aminoacyl-tRNA prior to its binding to the aminoacyl or acceptor site; the formation of this complex is catalyzed by factor Ts. EFTu is released from the ribosome after GTP is hydrolyzed.

The *EFG factor*, also called *translocase*, is involved in the translocation of the mRNA (Fig. 24–9,6 and 7). This factor is required to split GTP in the ribosome, thereby yielding GDP and inorganic phosphate. The energy released is needed for the removal of the deacylated tRNA from the ribosome and for the translocation process (Fig. 24–9,7).

Chain elongation can, therefore, be considered as a kind of cyclic reaction in which T + GTP stimulates the binding of aminoacyl-tRNA and G + GTP promotes the translocation of the newly elongated peptidyl-tRNA.

The energy for peptide bond formation is stored in the AA ~ tRNA and comes from an ATP molecule used during the activation of tRNA. Thus, protein synthesis is expensive from the point of view of energy: For each amino acid incorporated, one ATP and two GTP molecules are required.

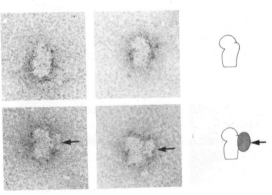

Figure 24–10 Electron micrographs of small ribosomal subunits of eukaryotes. **Upper row,** 40S subunit alone. **Lower row,** the eIF$_3$ factor has attached to the 40S subunit (arrows), and the complex is ready for initiation. (Courtesy of I. Emanuilov, D. D. Sabatini, J. A. Lake, and C. Freinstein.)

The Large Ribosomal Subunit is Involved in the Process of Elongation

The large subunit contains the enzyme *peptidyltransferase,* or peptide synthetase, which is involved in the formation of the peptide bond. Another function of the 50S subunit is to provide two binding sites for the two tRNA molecules, i.e., the *aminoacyl* or *acceptor site* and the *peptidyl* or *donor site* (Fig. 24–11). These two sites should be next to each other to permit the formation of the peptide bond.

The stepwise growth of the polypeptide chain involves (1) the entrance of an aminoacyl-tRNA into the aminoacyl site; (2) the formation of a peptide bond and consequent ejection of the tRNA that was in the peptidyl site; and (3) the movement of the tRNA (now carrying the peptide chain) from the aminoacyl to the peptidyl site. This process should be coupled with the simultaneous movement of the mRNA to place the following codon in position (Fig. 24–11). This translocation, in which the ribosome moves along the mRNA in the $5' \rightarrow 3'$ direction, requires the G factor and GTP.

The velocity with which these coordinated processes are produced may be illustrated by the fact that it takes only about 1 minute to construct a single hemoglobin chain carrying 150 amino acids.

Chain Termination Involves UAA, UGA, and UAG Codons and Releasing Factors R_1 and R_2

The termination of the polypeptide chain occurs when the 70S ribosome carrying the peptidyl-tRNA reaches the termination codon located at the end of each cistron. Chain termination leads to the release of the free polypeptide and tRNA, and to the dissociatión of the 70S ribosome into 30S and 50S subunits (Fig. 24–8).

The chain termination codons are UAA, UGA, and UAG (Table 21–3). Unlike all other triplets, these codons are not recognized by tRNAs but rather by two specific proteins, the releasing factors R_1 and R_2.

R_1 is specific for UAG and UAA, and R_2, for codons UAA and UGA (Table 24–1). The complex formed by the termination factor with the codon in some manner induces the peptidyl transferase to catalyze the transfer of the peptide chain to water, rather than to an amino group,

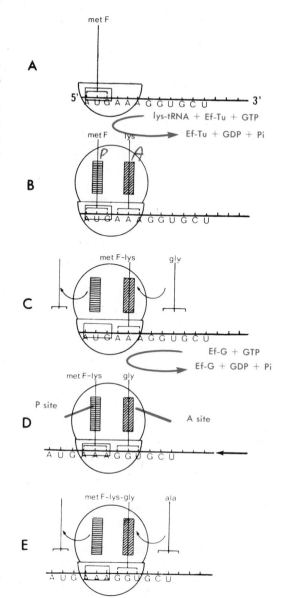

Figure 24–11 Diagram representing the early stages of translation of messenger RNA (5'–3'). The *initiation site* in the 30S subunit is indicated by a white rectangle. The *aminoacyl* site is indicated by oblique stripes and the *peptidyl* site by horizontal stripes. **A,** initiation complex in which formylmethionine-tRNA (fMET-tRNA) binds to the first codon in mRNA (AUG). **B,** the 70S ribosome has been formed, and the second amino-acyl-tRNA (lys-tRNA) binds the second codon (AAA). The function of Ef-Tu and GTP is indicated. **C,** the tRNA is eliminated from the peptidyl site, and the first peptide bond is formed, the function of Ef-G and GTP is indicated. **D,** translocation of the mRNA and of the peptidyl-tRNA has occurred, and a new aminoacyl-tRNA (gly-tRNA) binds to the third codon (GGU). **E,** the molecular events of **C** are now repeated. (Adapted from Ochoa, S.)

thereby releasing the polypeptide chain from the terminally added tRNA (see Lipmann, 1973).

Antibiotics and Toxins are Useful in Molecular Biology and Medicine

Antibiotics are very useful tools in the study of protein synthesis, because many of them work at very precise steps of the process. For example, *puromycin* is an analogue of AA-tRNA that can bind to the A site. The peptidyl transferase uses it as a substrate, forming a polypeptide-puromycin chain, which is released. *Fusidic acid* blocks the translocation induced by G factor. *Streptomycin* causes misreading of the genetic message, and the errors accumulated in the proteins kill the bacteria. *Tetracycline* inhibits AA-tRNA binding. *Sparsomycin* inhibits peptidyl transferase. *Chloramphenicol* inhibits protein synthesis of bacteria, chloroplasts, and mitochondria, but not the general cytoplasmic ribosomal system. *Cyclohexi-mide,* on the other hand, affects eukaryotic cells.

Many antibiotics (such as chloramphenicol, streptomycin, and tetracycline) inhibit bacterial ribosomes but not eukaryotic ribosomes. This is the molecular basis for the use of these antibiotics in medical treatment (Table 24–3).

Other drugs are useful for the study of RNA synthesis. *Actinomycin D* inhibits RNA synthesis by intercalating between the DNA bases, especially in GC-rich regions. *Rifampicin* works by binding to bacterial RNA polymerase and inhibiting initiation of transcription. Rifampicin does not inhibit eukaryotic RNA polymerases and is therefore also used in medical treatment. The toxic peptide *α-amanitine* is isolated from the fungus *Amanita phalloides* (the main cause of human intoxication by poisonous mushrooms), and it inhibits eukaryotic RNA

polymerases II and III but not RNA polymerase I (see Chapter 22).

Another toxin of which the mode of action is known in molecular terms is *diphtheria toxin.* Caused by *Corynebacterium diphtheriae,* diphtheria was a major cause of death in children before the advent of vaccination. The bacteria grow in colonies on the throats of patients and release a potent toxin into the bloodstream. The toxin, a protein of 65,000 daltons, affects a specific step in protein synthesis in eukaryotes. It inhibits translocation by inactivating the translocase. In this process, the toxin catalyzes a covalent modification in which an ADP-ribose residue is added to the human translocase.

Secretory Proteins have a Hydrophobic Signal Peptide

As explained in Chapter 10, in eukaryotes the proteins that are destined for *export* outside the cell are synthesized on the membranes of the endoplasmic reticulum. The nascent protein chains traverse the membrane and are deposited inside the ER cavity. The cytoplasmic endomembrane system then transports and secretes the proteins outside the cell. In addition to export proteins, membrane proteins are also synthesized in the ER (Fig. 10–14) (See Lodish and Rothman, 1979).

The information for determining translation by free or membrane-bound polysomes is somehow coded in the mRNA itself. This is shown clearly by experiments in which various mRNAs were microinjected into *Xenopus* oöcytes.[19] Oöcytes injected with mRNAs coding for proteins that are normally secreted (such as immunoglobulin or albumin) accumulate the protein products within membrane vesicles, while the injection of globin mRNA (which is normally not secreted) results in the accumulation of soluble proteins unassociated with membranes.

The best explanation for why some mRNAs are translated on membranes, while others are not, is provided by the *signal hypothesis,*[20, 21] seen in detail in Chapter 10. Briefly, secreted proteins contain in their starting (amino) end a *signal peptide* composed of between 15 and 30 amino acids, most of which are hydrophobic (such as leucine, isoleucine, or valine). Protein synthesis starts in free ribosomes (Fig. 24–12,1 and 2), and when the signal peptide emerges from the large ribosomal subunit, it is recognized by a receptor protein in the ER membrane.

TABLE 24–3   MODE OF ACTION OF SOME ANTIBIOTICS USED IN MEDICAL PRACTICE

Antibiotic	Action on Bacterial Cells
Streptomycin	Misreading of mRNA
Tetracycline	Binds to small ribosomal subunit and inhibits binding of AA-tRNA.
Chloramphenicol	Inhibits peptidyl transferase in large ribosomal subunit.
Rifampicin	Inhibits bacterial RNA-polymerase
Penicillin and ampicillin	Inhibit cell wall formation

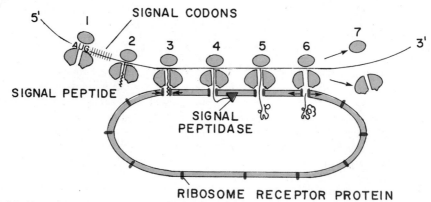

Figure 24-12 The signal hypothesis for the synthesis of exported proteins. See the description in the text. (Courtesy of G. Blobel.)

The ER receptor protein attaches to the large ribosomal subunit and forms a tunnel through which the nascent peptide traverses the membrane (Fig. 24-12,3 and 4). The initial transport through the membrane is facilitated by the hydrophobicity of the signal peptide. Once inside the lumen, the signal peptide is rapidly clipped off by a specific peptidase (Fig. 24-12,4). The completed protein is finally released into the ER cavity, and the ribosome detaches from the membrane and joins the cytoplasmic pool of ribosomal subunits (Fig. 24-2,6 and 7) (see Blobel, 1977).

The signal hypothesis is supported by a considerable amount of experimental evidence. Signal peptides have been found in the starting ends of many secreted proteins, such as immunoglobulin, pancreatic enzymes, insulin, albumin, parathyroid hormone, and others.[22-24] These signal peptides have been sequenced, and all of them were found to contain mostly hydrophobic amino acids. Furthermore, a peptidase that is able to excise signal peptides has been isolated from microsomal membranes.[25] However, it has been shown that chicken ovalbumin, which is normally secreted, does not have a signal peptide;[26] it is therefore possible that alternatives to the signal hypothesis might exist.

Bacteria do not have endoplasmic reticulum but nevertheless are able to secrete proteins to the outside of the cell. For example, bacteria resistant to penicillin secrete the enzyme penicillinase into the surrounding medium, and *C. diphtheriae* exports diphtheria toxin. Bacterial extracellular proteins also start with a hydrophobic signal peptide, which facilitates the movement of the nascent peptide through the plasma membrane.

Direct proof that the signal peptide is indeed required for traversing the membrane came from genetic experiments in *E. coli*, which showed that single amino acid changes in the signal peptide can abolish transport through the membrane (Bassford and Beckwith, 1979). As expected, transport was most affected when hydrophobic amino acids in the signal peptide were substituted by charged (hydrophilic) amino acids.

In addition to secretory and membrane proteins, other cellular proteins are also segregated into special compartments. A large proportion of the proteins of mitochondria and chloroplasts are synthesized in cytoplasmic ribosomes and must gain access to the interior of the organelles (Chap. 12). These proteins are synthesized on free polysomes, the completed polypeptides are released into the cytoplasm and only then traverse the organelle membrane. Mitochondrial and chloroplast proteins are initially synthesized as longer precursors with extra amino acids, which are removed when they enter the organelles (Maccecchini et al., 1979; Highfield and Ellis, 1978). We do not yet know whether these extra amino acids are hydrophobic or not, but it is already clear that this mechanism differs from that of secreted proteins in that protein segregation can be achieved after protein synthesis is completed (secretory proteins cross the ER membrane only as they are being synthesized; see Fig. 24-12). Nuclear proteins (Chap. 15) can accumulate very selectively in the correct compartment independently of protein synthesis. In this case, the segregation property resides in the mature protein, and longer precursor proteins are not involved (Fig. 15-16). The segregation of cellular components to their final destination seems to be achieved by several different mechanisms; this is not entirely surprising, in view of the complexity of living cells.

SUMMARY:

Protein Synthesis

Ribosomes are macromolecular structures that guide the multiple interactions involved in the synthesis of proteins. To direct these interactions ribosomes function in groups called *polyribosomes* or *polysomes.* The number of ribosomes in a polysome is related to the length of the protein to be synthesized. For example, for the polypeptides of hemoglobin, a pentamer (i.e., five ribosomes) is used. Each ribosome carries a nascent polypeptide chain. In the case of hemoglobin, therefore, five chains are formed at the same time along the mRNA.

The polysomes and their mRNA may be spread and observed under the electron microscope (Fig. 24–5). In bacteria the process of transcription of mRNA is simultaneous with the translation; i.e., ribosomes become attached to mRNA while this is being transcribed. Under the electron microscope the points of attachment of RNA polymerase at the DNA template may be observed (Fig. 24–6).

In polysomes of bacteria and eukaryotes a *ribosome-polysome cycle* may be demonstrated. The two ribosomal subunits associate at initiation and dissociate at termination. Thus, subunits are exchanged and there is a pool of free subunits. A dissociation factor, protein in nature, is involved in the separation of the two subunits at the end of the translation. This factor is identical to initiation factor IF_3.

The two ribosomal subunits perform different functions. The 30S subunit with the attached IF_3 factor binds to the mRNA at the *ribosome binding* site, which contains the AUG codon. In prokaryotes, the correct alignment for starting the synthesis depends on a sequence of between three and eight bases in the ribosomal binding site which are complementary to the 3' end of the 16S rRNA. Hairpin loops in mRNA may also be important in the formation of the *initiation complex.*

The first codon to be read is always AUG, which in bacteria and mitochondria codes for formylmethionine, and in eukaryotes, for normal methionine. These initial amino acids are later split off by enzymes.

Three protein factors, IF_1, IF_2, and IF_3, loosely associated with the small ribosomal subunit, are involved in the *initiation* of protein synthesis. IF_1 + IF_2 and GTP are required to bind fMet-tRNA to the small subunit in response to codon AUG. IF_3 factor stimulates the binding of mRNA to the initiation site of the small subunit. The initiation complex of the 40S subunit with eIF_3 may be observed with the electron microscope. In eukaryotes several initiation factors have also been separated. Recently phosphorylation of eIF_2, which is regulated by cAMP has been found. This mechanism may regulate the overall rate of protein synthesis.

Protein factors used in the *elongation* of the polypeptide chain are found in the post-ribosomal supernatant. Factor EFT comprises proteins Tu and Ts. EFTu + GTP forms a complex with the aminoacyl-tRNA, and this complex is catalyzed by factor EFTs. The G factor, *translocase,* which is in contact with the ribosome, splits GTP into GDP and P_1. The energy released is used for the removal of the tRNA and the translocation process. Protein synthesis is expensive from a point of view of energetics: For each amino acid incorporated, two GTPs are used in elongation, and one ATP is used during tRNA aminoacylation.

The large ribosomal subunit contains the enzyme *peptidyltransferase,* involved in the formation of the peptide bond. This subunit also carries two binding sites, i.e., the *aminoacyl* or *acceptor site,* and the *peptidyl* or *donor site.* After the binding of fMet-tRNA to the AUG codon on the small subunit, the large subunit of the first ribosome becomes attached. The growth of the polypeptide chain involves: (1) the entrance of an aminoacyl-tRNA into the aminoacyl site, (2) the formation of the peptide bond and the release of the free tRNA that was present at the peptidyl site, (3) the movement of the peptidyl tRNA from the aminoacyl to the peptidyl site, and (4) the entrance of a new aminoacyl tRNA into the aminoacyl site (Fig. 24–11). Simultaneously, the mRNA moves to the following codon in the 5' → 3' direction. All these coordinated steps are accomplished at a fast rate (1 globin chain of 150 amino acids per minute). Termination of the polypeptide chain occurs when a termination codon (UAG, UAA, or UGA) is reached. Two termination protein factors, R_1 and R_2, are involved in this process.

Antibiotics interfere with precise steps of protein synthesis, and this is the molecular basis for their use in medical practice. Other drugs act on RNA synthesis by intercalating between DNA bases (e.g., *actinomycin D*) or by interfering with RNA polymerases (e.g., *α-amanitine*).

In eukaryotes secretory and membrane proteins are synthesized on membrane-bound polysomes. Export proteins have a hydrophobic *signal peptide* (encoded by the starting codons in mRNA), which binds to protein receptors in the endoplasmic reticulum membrane, thus traversing the membrane. Once inside the endoplasmic reticulum cavity, the signal peptide is cleaved by a specific peptidase (see Chapter 10).

REFERENCES

1. Holley, R. W., Apgar, J., Everett, G., Madison, J., Marquisee, M., Merrill, S., Penswick, J., and Zamir, A. (1965) *Science*, 147:1462.
2. Khorana, H. G. (1975) A bacterial gene for tyrosine tRNA. In: *International Symposium on Macromolecules*, p. 371, IUPAC. (Mano, E. B., ed.) North-Holland Publishing Co., Amsterdam.
3. Smith, J. D. (1973) In: *Gene Expression and Its Regulation.* Plenum Publishing Co., New York.
4. Schedl, P., Primakoff, P., Roberts, J. (1975) *Brookhaven Symp. Biol.*, 26:106.
5. Goodman, H. M., Olson, M. V., and Hall, B. D. (1977) *Proc. Natl. Acad. Sci. USA*, 74:5453.
6. Valenzuela, P., Venegas, A., Weinberg, F., Bishop, R., and Rutter, W. J. (1978) *Proc. Natl. Acad. Sci. USA*, 75:190.
7. Knapp, G., Beckmann, J. S., Johnson, P. F., Fuhrman, S. A., and Abelson, J. (1978) *Cell*, 14:221.
8. Warner, J. R., Rich, A., and Hall, C. E. (1962) *Science*, 138:1399.
9. McKnight, S. L., Sullivan, N. L., and Miller, O. L., Jr. (1976) *Progr. Nucleic Acid Res. Mol. Biol.* 19:313.
10. Miller, O. L., Jr., Hamkalo, B. A., and Thomas, C. A. (1970) *Cold Spring Harbor Symp. Quant. Biol.*, 35:505.
11. Kaempfer, R. (1975) In: *Ribosomes*, p. 679. Cold Spring Harbor Laboratory, Cold Spring Harbor, New York.
12. Algranati, I. D., Gonzalez, N. S., and Bade, E. G. (1969) *Proc. Natl. Acad. Sci. USA*, 62:574.
13. Subramanian, A. R., Ron, E. Z., and Davis, B. D. (1969) *Proc. Natl. Acad. Sci. USA*, 61:761.
14. Shine, J., and Dalgarno, L. (1974) *Proc. Natl. Acad. Sci. USA*, 71:1342.
15. Steitz, J. A., and Jakes, K. (1975) *Proc. Natl. Acad. Sci. USA*, 72:4734.
16. Clark, B. F. C., and Marcker, K. (1968) *Sci. Am.*, 218(1):36.
17. Farrell, P., Balkow, K., Hunt, T., and Jackson, R. (1977) *Cell*, 11:187.
18. Datta, A., De Haro, C., Sierra, J. M., and Ochoa, S. (1977) *Proc. Natl. Acad. Sci. USA*, 74:1463.
19. Lane, C. D., and Zehavi-Willner, T. (1977) *Cell*, 11:683.
20. Blobel, G., and Sabatini, D. D. (1971) *Biomembranes*, 2:193.
21. Blobel, G., and Dobberstein, B. (1975) *J. Cell Biol.*, 67:852.
22. Milstein, C., Brownlee, G. G., Harrison, T., and Matthews, M. B. (1972) *Nature (New Biol.)*, 239:117.
23. Devillers-Thiery, A., Kindt, T., Scheele, G., and Blobel, G. (1975) *Proc. Natl. Acad. Sci. USA*, 72:5016. ·
24. Chan, S. J., and Steiner, D. F. (1977) *Trends Biochem. Sci.*, 2:254.
25. Jackson, R. C., and Blobel, G. (1977) *Proc. Natl. Acad. Sci.*, 74:5598.
26. Palmiter, R. D., Gagnon, J., and Walsh, K. A. (1978) *Proc. Natl. Acad. Sci.*, 75:94.

ADDITIONAL READING

Anderson, W. F., Bosch, L., Gros, F., Grunberg-Manago, M., Ochoa, S., Rich, A., and Staehelin, O. (1974) Initiation of protein synthesis in prokaryotic and eukaryotic systems. *FEBS Lett.*, 48:1.

Bassford, P., and Beckwith, J. (1979) *E. coli* mutants accumulating the precursor of a secreted protein in the cytoplasm. *Nature*, 277:538.

Blobel, G. (1977) Synthesis and segregation of secretory proteins: The signal hypothesis. In: *International Cell Biology.* (Brinkley, B. R., and Porter, K. R., eds.) The Rockefeller University Press, New York.

Clark, B. F. C. (1977) Correlation of biological activities with structural features of tRNA. *Progr. Nucleic Acid Res. Mol. Biol.*, 20:1.

De Robertis, E. M., and Olson, M. V. (1979) Transcription and processing of yeast tyrosine tRNA genes microinjected into *Xenopus* oöcytes. *Nature*, 278:137.

Haselkorn, R., and Rothman-Denes, L. B. (1973) Protein synthesis. *Annu. Rev. Biochem.*, 42:397.

Highfield, P. E., and Ellis, R. J. (1978) Synthesis and transport of the small subunit of chloroplast ribulose biphosphate carboxylase. *Nature, 271*:420.

Khorana, H. G. (1979) Total synthesis of a gene. *Science, 203*:614.

Kozak, M. (1978) How do eukaryotic ribosomes select initiation regions in mRNA? *Cell, 15*:1109.

Lewin, B. (1974) *Gene Expression,* Vol. 1. John Wiley & Sons, Inc., New York.

Lodish, H. F. (1976) Translational control of protein synthesis. *Annu. Rev. Biochem., 45*:39.

Lodish, H., and Rothman, J. E. (1979) The assembly of cell membranes. *Sci. Am., 240*(1):38.

Lipmann, F. (1969) Polypeptide chain elongation in protein synthesis. *Science, 164*:1024.

Lipmann, F. (1973) What do we know about protein synthesis? In: *Gene Expression and Its Regulation*, pp. 1–12. (Kenney, F. T., Hamkalo, B., Favelukes, G., and August, J. T., eds.) Plenum Publishing Co., New York.

Lucas-Lenard, J., and Lipmann, F. (1971) Protein biosynthesis. *Annu. Rev. Biochem., 40*:409.

Maccecchini, M. L., Rudin, Y., Blobel, G., and Schatz, G. (1979) Import of proteins into mitochondria: Precursor forms of F_1 ATPase subunits. *Proc. Natl. Acad. Sci. USA, 76*:343.

Ochoa, S. (1977) Initiation of protein synthesis in prokaryotes and eukaryotes. *J. Biochem., 81*:1.

Rich, A. (1978) Transfer RNA: Three-dimensional structure and biological function. *Trends Biochem. Sci., 3*(1):34.

Rich, A., and Kim, S. H. (1978) The three-dimensional structure of transfer RNA. *Sci. Am., 238*(1):52.

Shine, J., and Dalgarno, L. (1974) The 3' terminal sequence of *E. coli* 16S ribosomal RNA: Complementarity to ribosome binding sites. *Proc. Natl. Acad. Sci., 71*:1342.

Smith, J. D. (1976) Transcription and processing of transfer RNA precursors. *Progr. Nucleic Acid Res., 16*:25.

Söll, D. (1971) Enzymatic modification of transfer RNA. *Science, 173*:293.

Spirin, A. S. (1978) Energetics of the ribosome. *Progr. Nucleic Acid Res. Mol. Biol., 21*:39.

The Mechanism of Protein Synthesis. (1969) *Cold Spring Harbor Symp. Quant. Biol., 34.*

Watson, J. D. (1975) *Molecular Biology of the Gene,* W. A. Benjamin Inc., Menlo Park, Calif.

Weissbach, H., and Ochoa, S. (1976) Soluble factors required for eukaryotic protein synthesis. *Annu. Rev. Biochem., 45*:191.

25

GENE REGULATION

25–1 Gene Regulation in Prokaryotes 537
   Control of Enzyme Induction and Repression — Repressors Bind to Operators
   The Lac Operon Codes for β-Galactosidase, Lac Permease, and Transacetylase
   The i Gene Codes for the Repressor Protein, Which Binds to the Inducer
   The Operator is a DNA Sequence that has Two-Fold Symmetry
   Repressors Bind within Promoters and Prevent Attachment of RNA Polymerase
   The Jacob-Monod Model is Based on a Mechanism of Negative Control
   Positive Control of Lac Transcription is Regulated by the CAP-cAMP Complex
   The Tryptophan Operon — Regulation of Transcription at Initiation and Termination
   A Finer Control of Metabolism Occurs at the Level of Enzyme Activity
   Summary: *Gene Regulation in Prokaryotes*
25–2 Gene Regulation in Eukaryotes 544
   Heterochromatin is not Transcribed

   Repetitive DNA Sequences are Characteristic of the Eukaryotes
   Satellite DNAs Contain the Most Highly Repetitive Sequences
   Moderately Repetitive Genes — rDNA, 5S DNA, and Histone Genes
   Summary: *Gene Regulation in Eukaryotes and Repetitive DNA*
25–3 Regulation at the Chromosome Level — Giant Chromosomes 548
   Polytene Chromosomes Have a Thousand DNA Molecules Aligned Side by Side
   The Bands Represent Chromomeres Aligned Side by Side
   Puffs in Polytenes are the Sites of Gene Transcription
   Puffs can be Induced by Ecdysone and by Heat Shock
   Lampbrush Chromosomes — In Oöcytes at the Diplotene Stage of Meiosis
   The Lateral Loops are Sites of Intense RNA Synthesis
   Summary: *Gene Regulation in Eukaryotes at the Chromosome Level*

In the previous chapters we have seen that the information encoded in DNA is transcribed into RNA and then translated into protein. Here we will analyze the ways in which a cell regulates the expression of genes. The ability to switch genes on and off is of fundamental importance to cells, since it enables them to respond to a changing environment and is the basis for the processes of cell growth and differentiation.

The regulation of gene expression in bacteria will be analyzed in the first part of this chapter. Much of our knowledge of how genes work comes from study of the genes required for the utilization of the sugar lactose (i.e., the *lac operon* of *Escherichia coli),* which will therefore be described in some detail.

The rate of expression of a bacterial gene is controlled at the level of mRNA synthesis. Regulation can occur at the *initiation* of mRNA synthesis (mediated by proteins called *repressors*

which prevent RNA polymerase from binding to DNA) or at the *termination* of the RNA chains (as described below in the case of the genes required for the synthesis of tryptophan).

Since bacteria are unicellular organisms that obtain their food from the medium that immediately surrounds them, bacterial gene regulation is designed to adapt quickly to changes in the environment. In higher organisms cells are surrounded by a constant internal milieu, and other types of adaptations, such as responding to hormones and to impulses from the nervous system, become more important.

The genome of eukaryotic cells is much more complex than in bacteria. A human cell contains about 700 times more DNA than *E. coli*. Since a human genome (haploid) contains 3×10^9 base pairs, its coding potential is very large. If its DNA sequence were written in a book with a page of standard size, the book would have approximately one million pages! Much of the human genome, however, is in the form of repetitive DNA.

The complexity of the genome explains why our knowledge of how genes are controlled in eukaryotes is still very incomplete. However, valuable information has been obtained from studies of polytene and lampbrush chromosomes, in which the transcription process can be visualized directly under the microscope. These studies will be analyzed in the last part of the chapter.

25-1 GENE REGULATION IN PROKARYOTES

Control of Enzyme Induction and Repression — Repressors Bind to Operators

The circular chromosome of E. coli contains the information for a total of about 3000 proteins. Those genes that code for a protein molecule are called structural genes. The DNA is transcribed in discrete segments that cover 1 to 20 structural genes at the same time and in which the initiation and termination points are fixed. The initiation point, corresponding to the site of attachment of RNA polymerase, is called the promoter.

Normally, only some genes of the total genome are active. In fact, if all of them were operative at the same time, the cell would become filled with unnecessary proteins. By changing the nutrients in the culture medium it is possible to activate or repress certain genes and in this way, the cell adapts to the change in environment. For example, E. coli growing on glycerol increases its production of enzymes breaking down this substrate, whereas other enzymes are kept at a minimum. On the other hand, if the bacteria are exposed to lactose, the enzyme β-galactosidase, which hydrolyzes this disaccharide into galactose and glucose, may increase 1000-fold. These changes, which involve a true enzyme synthesis, are referred to as enzyme induction.

Regulation by enzyme induction is found in many catabolic systems, which degrade sugars, amino acids, and lipids. The best known induction system is that of β-galactosidase, which will be discussed in detail later on.

Working in opposition to enzyme induction is the process of enzyme repression, in which

the synthesis of a certain enzyme is selectively inhibited by the end product of its metabolic chain. An example of enzyme repression is provided by the five enzymes required for the synthesis of the amino acid tryptophan. When tryptophan is available from the surrounding culture medium, it represses the expression of the five tryptophan genes. In this way, E. coli synthesizes the enzymes only when they are required, thus saving precious energy.

Many anabolic systems involved in the synthesis of amino acids or nucleic acid precursors are regulated by enzyme repression.

Both enzyme repression and induction are controlled by the action of specific proteins called repressors, which can prevent mRNA synthesis by binding to DNA sequences called operators, located near the beginning (i.e., the promoter) of the genes they control. Unlike the mRNA chains in eukaryotes, most of those in prokaryotes contain the information for many proteins (i.e., they are polycistronic mRNAs, see Fig. 25–1). Since repressors control the synthesis of these mRNAs, each repressor controls the synthesis of several enzymes in a metabolic chain.

The Lac Operon Codes for β-Galactosidase, Lac Permease, and Transacetylase

In 1961 Jacob and Monod[1] postulated the existence of a new genetic unit, the operon. This is a group of genes that are clustered in the chromosome and that can be controlled (i.e., activated or inactivated) in a coordinated way. The best known operon is the lac operon, involved in the utilization of lactose. As shown in Figure 25–1 the lac operon comprises three structural genes (z, y, and a) and produces a polycistronic mRNA (lac mRNA).

This messenger RNA codes for β-galactosidase, lac permease, and transacetylase. β-galactosidase is a tetrameric enzyme (with subunits of 130,000 daltons each) which cleaves the disaccharide lactose into galactose and glucose. The lac permease is a single-unit protein that acts as a carrier for the entrance of lactose into the cell. Transacetylase is an enzyme of two subunits that catalyzes the transfer of one acetyl group from acetyl-CoA to galactose.

The group of structural genes is controlled by three DNA segments that act as regulatory elements. In Figure 25–1 these are shown as the i gene or regulator gene which codes for the lac repressor, the promoter, and operator.

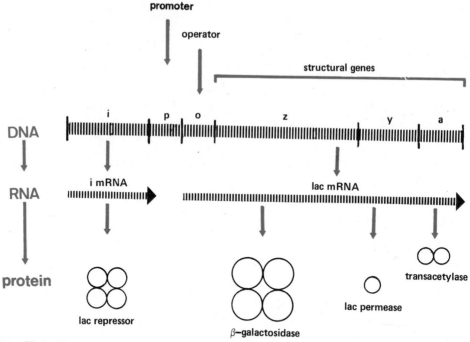

Figure 25–1 Diagram representing the *lac operon*. *i*, the regulatory gene that produces the *i mRNA* that codes for the *lac repressor*, a protein with four subunits. The *promoter* (*p*) is the region of attachment of RNA polymerase; the *operator* (*o*) is the region where the repressor binds (see Fig. 25–24). The *i gene*, the promoter, and the operator are regulatory elements. The structural genes, *z, y,* and *a,* produce a polycistronic lac mRNA, and the three proteins indicated below.

The i Gene Codes for the Repressor Protein, Which Binds to the Inducer

A key element of the operon model is the existence of *regulator* genes which control the rate of information transfer from structural genes to proteins. The rate of expression of the *lac* operon is regulated by the product of the *i* gene, which codes for the repressor. The *lac repressor* is a protein which has been purified,[2, 3] and found to be composed of 4 identical subunits of 40,000 daltons each (Fig. 25–1). There are only about ten lac repressor molecules in each *E. coli* cell.

The repressor binds strongly and specifically to a short DNA segment called the *operator* (*o*), which is located very close to the start of the structural genes. As can be seen in Figure 25–2, the repressor bound at the operator site prevents the synthesis of lac mRNA.

The affinity of the binding of repressor to the operator is regulated by the presence of the *inducer*. The natural inducer of the *lac* operon is lactose (a disaccharide formed by galactose and glucose), which upon entering the cell is modified to *allolactose*, which in turn binds to the repressor.[4]

Experimentally, it is much better to use an artificial inducer such as isopropylthiogalactoside (IPTG), which without further change binds to the repressor and acts as an inducer. Each of the subunits of the repressor has one binding site for the inducer, and upon binding of IPTG, the repressor undergoes a conformational change by which it becomes unable to bind to the operator (Fig. 25–2), thus inducing the transcription of the lac operon.

The effect of this conformational change is dramatic, while in the absence of lactose *E. coli* cells have an average of only 3 molecules of β-galactosidase enzyme per cell, after induction of the lac operon 3000 molecules of β-galactosidase are present in each cell, representing 3 per cent of the total protein.

The Operator is a DNA Sequence that has Two-fold Symmetry

The operator (*o*) is the segment of DNA upon which the repressor binds. The operator requires a close linkage with the structural genes under its control, whereas the regulator gene may be situated at a distance from them.

NO INDUCER

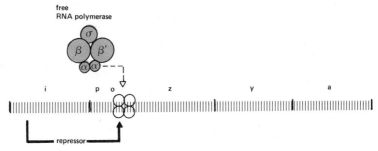

Figure 25-2 Diagram representing the regulation of the lactose operon in the absence and in the presence of the inducer. In the absence of the inducer, the *lac repressor* binds tightly to the operator, interfering with the transcription of the structural genes. The binding sites for the repressor and for RNA polymerase overlap, and RNA polymerase is therefore unable to bind when the repressor occupies the operator. The repressor is a tetramer with subunits of 40,000 daltons. When the inducer (IPTG) is present, it binds to the repressor, eliciting a conformational change that prevents its binding to the operator; as a consequence, RNA polymerase is free to transcribe the structural genes. Observe that *E. coli* RNA polymerase has five subunits of different size: σ (95,000 daltons); β/β' (about 160,000); and two α (40,000). Note that after transcription has started the sigma (σ) subunit is released and only the core enzyme remains bound to the DNA.

INDUCTION

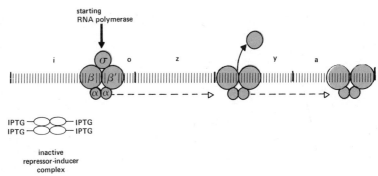

In the lac operon the operator is situated to the left of the structural genes *z, y,* and *a.* (Fig. 25–1).

The lac repressor binds to the following 21 base pair sequence of DNA.[5]

AATTGT	G	A	G	C	G	G	A	T	A	ACAATT
TTAACA	C	T	C	G	C	C	T	A	T	TGTTAA

The sequence contains regions of two-fold symmetry. Some sequences (boxed areas) on the left side of the operator are also present on the right side but on the opposite strand. Similar symmetrical regions have been found in other operators[6] and in DNA binding proteins (such as restriction enzymes), and probably reflects the fact that most proteins that recognize specific sequences in DNA are multimeric. The two symmetrical sites probably represent recognition sites for different subunits of the repressor. This 21 base pair sequence is sufficient to bind the *lac* repressor, as demonstrated by the fact that a chemically synthesized fragment with the same sequence has operator activity both in vivo and in vitro.[7]

Repressors Bind within Promoters and Prevent Attachment of RNA Polymerase

The *promoter (p)* is the DNA segment to which the RNA polymerase first becomes attached when initiating the transcription of the structural genes.[8, 9] The nucleotide sequence of the *lac* promoter-operator region is shown in Figure 25–3. RNA polymerase will bind to the promoter only if a *cyclic-AMP binding protein (CAP protein)* is bound to its left side. The function of CAP protein and cyclic AMP will be discussed shortly.

Two regions of the promoter seem to be important for RNA polymerase binding:[10] (1) an AT-rich region situated 25 to 35 bases before the start of the z gene and (2) a TATGTTG sequence located 6 to 12 bases before the z gene (Fig. 25–3). Mutations in the first region severely inhibit *lac* expression. The AT-rich region probably facilitates local separation of DNA strands by RNA polymerase, a process required for RNA synthesis. The TATGTTG sequence is conserved in most prokaryotic promoters (Fig. 22–3).

Figure 25–3 shows that the *lac* repressor bound to DNA covers the initial region that

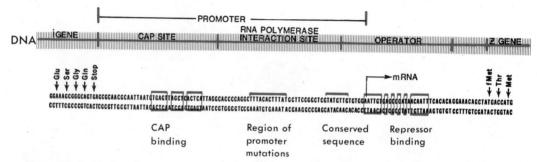

Figure 25–3 Nucleotide sequence of the *lac* promoter, showing its relation to the overlapping operator sequences. Note that the repressor and CAP protein bind at regions of symmetry. The sequence TATGTTG, which is conserved in prokaryotic promoters, and the region where several promoter mutations occur are indicated. (Data from Dickson, R. C., Abelson, J., Barnes, W. M., and Reznikoff, W. S., *Science, 187*:27, 1975. Copyright 1975 by the American Association for the Advancement of Science.)

must be transcribed into messenger RNA. In fact, the RNA polymerase binding site (promoter) overlaps with the repressor binding site (operator). In vitro experiments have shown that repressors block the binding of RNA polymerase to promoters.[6, 11] *The way in which repressors work is therefore very simple: They bind within promoter sites and prevent the attachment of RNA polymerase.*

The Jacob-Monod Model is a Mechanism of Negative Control

The Jacob and Monod[1] model of the operon is based on a mechanism of *negative control* that operates at the level of transcription. In its active form, the repressor has an inhibitory effect upon RNA synthesis. For example, if a culture of *E. coli* is growing in a medium without lactose, the repressor molecules, being in an active conformation, will be able to bind to the operator and will prevent the RNA polymerase from transcribing the structural genes (Fig. 25–2). The result is that *β-galactosidase* and the other lac proteins will not be synthesized. On the other hand, if a culture of *E. coli* is exposed to lactose or IPTG, the *inducer* binds to the *lac* repressor, which by a conformational change loses its affinity for binding the operator. In this condition the operator is free, and RNA polymerase can progress to transcribe all the *lac* genes, thereby producing a polycistronic *lac* mRNA that will code for the three enzymes of the system (Fig. 25–2). Under these conditions large amounts of β-galactosidase, permease, and transacetylase are produced, and the bacteria are able to metabolize lactose.

These adaptations can occur very rapidly; most bacterial mRNAs have a half-life of only two minutes (see Chapter 22–2), and therefore

any change in the rate of mRNA synthesis is rapidly reflected in the rate of protein synthesis.

Positive Control of Lac Transcription is Regulated by the CAP-cAMP Complex

Cyclic AMP (cAMP) has a regulatory role in bacteria, activating inducible operons at the level of transcription. *E. coli* has a cyclic AMP receptor protein that is able to bind cAMP with high affinity. This protein is also known as "Catabolite gene *A*ctivator *P*rotein." CAP is a dimeric protein (with subunits of 45,000 daltons each) which when complexed with cAMP, is able to bind to a specific site within the *lac* promoter region (Fig. 25–3). As in the case of other DNA binding proteins, CAP protein binds to a symmetrical DNA sequence (indicated in Fig. 25–3). In this case, also, each subunit probably binds to half the binding sequence.

RNA polymerase will recognize the *lac* promoter only if the CAP-cAMP complex is already bound to it (Fig. 25–4). Therefore, in the *lac* operon, in addition to *negative control* by the repressor, there is *positive control* by the CAP-cAMP complex.

The latter mechanism is also required for the expression of many other inducible operons, such as those involved in the utilization of maltose, galactose, and arabinose. However, cAMP is not necessary for the synthesis of those enzymes required for the utilization of glucose as an energy source. Thus, cAMP is not essential for survival, and mutants that cannot synthesize cAMP (i.e., that lack the enzyme adenylate cyclase) can grow on a glucose substrate but not on one having lactose, maltose, or galactose as the sole energy source.

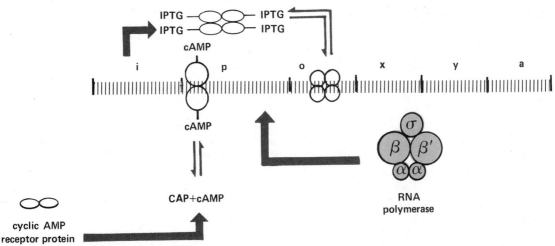

Figure 25–4 Diagram representing the mechanism by which cyclic AMP regulates the lac operon. The cyclic receptor protein (CAP), with its two subunits of 45,000 daltons, is indicated. This protein binds cAMP, forming the CAP-cAMP complex, which favors the action of RNA polymerase by binding to the promoter. In addition to the *negative control* by the repressor (see Fig. 25–2) there is this *positive control* by the CAP-cAMP complex.

What is the purpose of a second regulatory mechanism for the lac operon in *E. coli*? The answer to this question lies in a long-known phenomenon: When bacteria are grown in glucose, the rate of synthesis of inducible enzymes (such as those of the lactose, maltose, or arabinose operons) is very low. This phenomenon is known as *catabolite repression* and can now be explained in molecular terms. Bacteria grown in glucose have a lower cAMP content than those cultured in a poorer energy source, such as lactose.[13] When the intracellular cAMP level is low, the cAMP receptor protein will not bind to the promoter of the *lac* operon, which therefore will not be turned on, even in the presence of lactose (see Fig. 25–4). This situation can be reversed if exogenous cAMP is added to the culture medium.[14] If *E. coli* is grown in the presence of glucose and lactose, the cAMP levels will be low, and the bacteria will utilize only glucose, which is the richest and most efficient energy source. It is understandable that by this mechanism of regulation *E. coli* can adapt more efficiently to a changing environment, such as that of its natural habitat, the human intestine.

The Tryptophan Operon — Regulation of Transcription at Initiation and Termination

The *tryptophan operon* codes for the five enzymes that are required for the synthesis of tryptophan (Fig. 25–5). It has been known for many years that the expression of the operon is regulated by the availability of tryptophan (*trp*) in the culture medium, and that the presence of tryptophan *represses* the synthesis of *trp* enzymes. *Enzyme repression* can also be explained on the basis of a model similar to the one depicted for enzyme *induction* in the *lac* operon.

In enzyme repression the regulatory gene produces a repressor protein which is normally in the *inactive* form. The repressor, upon binding with a metabolite called a *co-repressor* (which in this case is the amino acid tryptophan), undergoes a conformational change by which it can bind to the operator and thereby inhibit the binding of RNA polymerase to the *trp* promoter[15] (Fig. 25–5). The affinity of the repressor for binding to the operator is normally low, but it increases by the action of the co-repressor. (This is in contrast to induction, in which the repressor is active on its own and loses affinity for the operator when bound to the inducer.)

A second mechanism by which the expression of the *trp* genes is regulated has recently been found by Yanofsky.[16] This additional control mechanism operates at the level of termination of RNA synthesis. The essential finding was that small deletions to the right of the *trp* operator (see Fig. 25–5) resulted in the increased production of *trp* enzymes. The deletions remove a termination sequence, called the *"attenuator,"* which is within the "leader" sequence located between the promoter-operator region and the first *trp* gene (*trp* E, see Fig. 25–5).

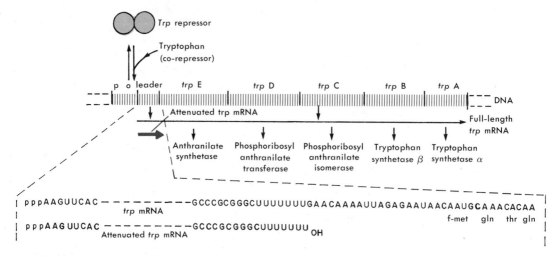

Figure 25–5 The tryptophan operon of *E. coli* is regulated at transcription initiation (by the *trp* repressor) and at transcription termination. The leader RNA has a termination sequence called the attenuator, and two types of mRNA can be produced: a short leader RNA or a full-length mRNA that codes for the five enzymes of the operon. The proportion of the two RNAs is regulated by the tryptophan concentration in the cell. Note that as with many other terminators, the attenuated mRNA ends in a row of Us. (Data from Bertrand et al., *Science, 189*:22, 1975.)

Normally 90 per cent of the RNA polymerase molecules that start transcription at the promoter terminate at the attenuator site, yielding a short RNA (140 nucleotides) that does not code for *trp* enzymes. Only ten per cent of the molecules continue on to produce a full-length *trp* mRNA. Termination at the attenuator site requires termination factor rho[17] and is regulated by the level of tryptophan in the cells. When the tryptophan level is low, the proportion of RNA molecules giving full-length *trp* mRNA increases. When it is high, most of the mRNAs terminate prematurely. The way in which this is achieved is not known, but it is thought to be mediated by an additional *anti-terminator* control element which becomes activated when the levels of tryptophan are low. Other repressible operons, such as the operon for histidine and phenylalanine biosynthesis also show similar regulation at the level of RNA termination.[18]

The mechanism by which the metabolic end products (tryptophan, or histidine, or phenylalanine) regulate termination of transcription of their operons is now starting to emerge from nucleotide sequencing data. All three leader (for *trp, his,* and *phe*) RNAs contain a ribosome binding site and an AUG starting codon and can produce a short peptide between 14 and 16 amino acids long. The striking feature of this sequence is that most of its codons are for the amino acid synthesized by the operon. For example, the *his* operon leader has 7 histidine

codons in a row (out of 16 codons for the complete peptide) (see Barnes, 1978); and the *phe* operon leader has 7 phenylalanine codons (out of 15 codons) (see Zurawski et al., 1978). This suggests a regulatory mechanism in which protein synthesis regulates transcription termination. (The student should remember that in prokaryotes transcription and translation are simultaneous, see Fig. 22–5.) Two situations can be envisaged: (1) In the presence of the amino acid, ribosomes could be expected to translate the complete peptide and then be released; in this situation the leader RNA terminates at the attenuator. (2) When the cells are starved for the amino acid, the ribosomes would remain attached to the leader RNA, because there would be an insufficient concentration of the corresponding AA-tRNA to rapidly complete the synthesis of the peptide. The attached ribosome presumably then has an anti-terminator effect on RNA polymerase, and the long mRNA is produced (Fig. 25–5). (The exact way in which this effect is achieved is not known, but it could, for example, occur by prevention of the formation of hairpin loops on the leader RNA.) Thus the anti-terminator control element is apparently a ribosome arrested on the leader RNA (see Barnes, 1978).

In summary, *E. coli* uses two distinct mechanisms to ensure that the *trp* enzymes are not synthesized unless tryptophan is required. The use of two mechanisms which operate at different tryptophan levels probably permits a finer control of *trp* enzyme synthesis, over a wider range of tryptophan concentration than would be possible by the repression mechanism alone.

A Finer Control of Metabolism Occurs at the Level of Enzyme Activity

Enzyme induction and repression provide a *coarse control* of metabolism by switching off the *synthesis* of enzymes when they are not required. However, cells also have *fine controls* which operate directly at the level of enzyme activity and which were analyzed in Chapter 6. The most common mechanism is *feedback inhibition*, in which the end product of a metabolic pathway acts as an *allosteric inhibitor* of the first enzyme of its metabolic chain. This process is shown in Figure 6–7, which the student is advised to study in detail.

While genetic control (induction and repression) saves precious energy preventing synthesis of unnecessary enzymes, control of enzyme activity by allosteric regulation allows an almost instantaneous fine tuning of catalytic activity.

SUMMARY:

Gene Regulation in Prokaryotes

Gene regulation is a fundamental property of living matter by which the organism adapts to a changing environment. In *E. coli* the circular DNA codes for some 3000 proteins, and the mRNAs are frequently polycistronic. Each group of genes has an initiation point or *promoter,* i.e., the point of attachment of RNA polymerase. Normally, not all genes are fully active, and some of them become activated when there is a change in the medium. For example, if *E. coli* is grown on lactose there is an induction of β-galactosidase. *Enzymatic induction* implies an actual synthesis of the corresponding enzyme. In *enzymatic repression* there is an inhibition of the synthesis of the enzyme by the end product of its metabolic chain (e.g., as in the case of the tryptophan operon). Enzyme repression and induction are controlled by specific proteins, the *repressors,* which are produced by *regulator* genes. The repressor binds to DNA sequences called *operators,* thus preventing mRNA synthesis.

A group of structural genes controlled by the same promoter and operator constitutes an *operon.* In the *lac* operon there are three structural genes (i.e., for β-galactosidase, permease, and transacetylase) that are regulated by the same operator, and all of the enzymes they code for are involved in the utilization of lactose (Fig. 25–1). In the operon the repressor binding site (operator) overlaps with the promoter. The way in which repressors work is therefore very simple: They bind within promoter sites and prevent the attachment of RNA polymerase.

Regulation of the lac operon occurs by way of the repressor which binds to the operator, thereby interfering with the transcription by RNA polymerase. In this condition the *lac* operon is inactive and no enzymes are produced. In the presence of an inducer (suitable metabolite), the repressor is inactivated and the operator becomes free. In this condition the entire operon may be transcribed by the RNA polymerase (Fig. 25–2). The *lac* repressor is a protein of 160,000 daltons with 4 subunits of 40,000 daltons, each of which has a binding site for the inducer.

The operator is a 21 base pair sequence with a two-fold symmetry which may correspond to the binding site of each repressor subunit. In the promoter there is an AT-rich region and a TATGTTG sequence that are both important for the binding of RNA polymerase.

The regulation thus far described for the *lac* operon is based on negative control. If lactose is low in the medium, the repressor molecules are in an active conformation and prevent mRNA synthesis. In a lactose-rich medium the inducer (allolactose) binds to the lac repressor and the operator becomes free, and RNA polymerase can bind.

The *lac* operon also has a positive control mechanism provided by cAMP, which interacts with a *C*atabolite gene *A*ctivator *P*rotein (CAP). The CAP-cAMP complex binds to the promoter and facilitates the attachment of RNA polymerase, thus activating the transcription of the operon (*positive control*). The cAMP molecule is required for the utilization of several sugars but not glucose. If glucose is present in the medium, cAMP levels are low and consequently the synthesis of inducible enzymes (i.e., β-galactosidase and others) is reduced; this phenomenon is called *catabolite repression*. By using both negative and positive regulation *E. coli* rapidly adapts to changes of nutrients in the medium.

Another mechanism of gene regulation is exemplified by the *tryptophan operon*, which codes for the five enzymes required to synthesize tryptophan (Fig. 25–5). In this case there is dual regulation, at both the *initiation* and *termination* of transcription. If tryptophan is available in the medium the *trp* genes are repressed by a *repressor* protein, as in the *lac* operon. In this case, however, tryptophan acts as a *co-repressor* and induces a conformational change by which the repressor binds to the operator (the case of the *lac* repressor is the other way around).

RNA chain termination is also very important in the *trp* operon because transcription from its promoter leads (in 90 per cent of the chains) to premature termination at the so-called attenuator site, producing a short leader RNA without mRNA activity. The proportion of RNA chains that terminate prematurely depends on the tryptophan concentration in the cell. If it is high, most RNA chains terminate prematurely. At low tryptophan levels premature termination is inhibited and full-length mRNA is produced. This dual mechanism, which is also present in other operons (such as those of histidine and phenylalanine), permits a more finely tuned gene regulation.

25–2 GENE REGULATION IN EUKARYOTES

One of the greatest challenges of modern biology is interpreting higher cells and organisms in the light of molecular mechanisms known to act in viruses and bacteria. The simpler structural organization of these lower organisms is conducive to study by genetic and biochemical methods under precise and reproducible conditions. As seen in the previous chapters, these investigations have disclosed the fundamental processes by which genes are duplicated and the genetic expression is regulated. Such mechanisms also function, in a general sense, in higher cells; and

the fact that a single genetic code exists for all living organisms has already been mentioned.

The much greater complexity exhibited by a plant or animal cell, relative to a bacterium suggests that they could contain a much larger store of genetic information.

The DNA of *E. coli* is about 1 mm long and contains approximately 4.5×10^6 base pairs. Table 25–1 shows that the genomes of eukaryotic cells contain much more DNA. For example, a bull spermatozoon (haploid) contains 3.2×10^9 base pairs, which corresponds to a 700-fold increase in information relative to the size of the *E. coli* genome. This quantitative difference could be reflected, at least in principle, in an increased number of mRNAs and proteins. However, as will be shown later, the increase

TABLE 25–1 GENOME SIZE AND REPETITIVE DNA IN VARIOUS ORGANISMS

	Species	Base Pairs in Haploid Genome	Single Copy (%)	Moderately Repetitive (%)	Highly Repetitive (%)
Bacterium	*E. coli*	4.5×10^6	100	—	—
Fruit fly	*Drosophila melanogaster*	1.4×10^8	74	13	13
Frog	*Xenopus laevis*	3.1×10^9	54	41	6
Mouse	*M. musculus*	2.7×10^9	70	20	10
Cow	*B. domesticus*	3.2×10^9	60	38	2

in genome complexity is not as great as indicated by the cell DNA content. For example, in the salamander *Amphiuma*, the creature having the largest known genome (168 pg of DNA and 8×10^{10} base pairs), the DNA content is 80 per cent *repetitive*. Even *Drosophila melanogaster*, which is one of those eukaryotes having a relatively small DNA content (only 40 times that of *E. coli*), contains significant amounts of repetitive DNA (Table 25–1).

Heterochromatin Is Not Transcribed

One of the main conclusions drawn from modern studies of cell differentiation (Chap. 26) is that in any given eukaryotic cell most of the DNA does not function in transcription; i.e., it is not active in RNA synthesis. Some genes are transcribed in only certain specialized tissues, and not in others; for example, hemoglobin mRNA is produced only in blood cells, while it is not expressed in any other tissues. This type of persistent inactivation could be related to the structure of chromatin.

Regulation of gene function must necessarily be much more complex in higher cells than in bacteria. Only the proper balance of genes can produce the normal development of a complex organism. At the chromosome level, gene regulation may be related to the existence of proteins tightly bound to certain regions of DNA (both histone and non-histone proteins) which could prevent transcription of extensive segments of the genome, as in the case of heterochromatin.

As seen in Chapter 15, at interphase a portion of the chromatin is present in a dense coiled heterochromatic state; the rest exists as more loosely extended euchromatin. Using radioautography with ³H-uridine it was demonstrated with the electron microscope that most of the RNA is synthesized in euchromatin, whereas the condensed heterochromatin is virtually inactive in RNA synthesis.[19] During spermiogenesis RNA synthesis ceases almost completely. In fact, the spermatozoon can be thought of as a device that injects tightly packed and inactive DNA into the egg, very much like a phage that injects DNA into a bacterium. Other examples of DNA inactivation and chromatin condensation are found in metaphase chromosomes, in which RNA synthesis stops (Fig. 16–6); and in *X chromosome inactivation*, in which one of the two X

chromosomes in females becomes heterochromatic and genetically inactive.

Repetitive DNA Sequences Are Characteristic of the Eukaryotes

The concept of *repetitive DNA* was introduced in Chapter 23, when we mentioned that ribosomal genes are present in many copies per cell (Table 23–3). Repetitive DNA is a constant characteristic of all eukaryotic cells, from protozoa to the higher plants and animals; this represents another important difference between eukaryons and prokaryons, in which there are few repeated DNA sequences in the genome (Table 25–1).

The discovery of repeated sequences of DNA in higher cells was a consequence of the remarkable ability of separated complementary strands of DNA to recognize each other and to reassociate. This process can be measured by the techniques of denaturation and renaturation (Chap. 5).[20] Essentially, these methods are based on the fact that the greater the number of repeated sequences in the molecule, the faster the association of DNA strands (see Fig. 5–17). In the DNA extracted from mouse cells, 10 per cent of the total was found as short nucleotide sequences that were repeated in about 1 million copies. Another 20 per cent of the genome consisted of repeated sequences of 1000 to 100,000 copies; the remaining 70 per cent was non-repetitious DNA; i.e., it was represented in only one copy (Fig. 25–6).

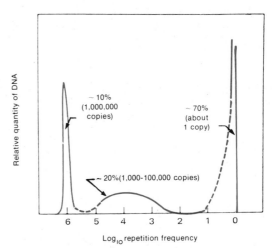

Figure 25–6 Frequency of repeated DNA sequences present in DNA from a mouse. (See the description in the text.) (From R. J. Britten and D. E. Kohne.)

Satellite DNAs Contain the Most Highly Repetitive Sequences

The most highly repetitive DNAs in eukaryons are also called *satellite* DNAs, because they can frequently be separated from the bulk of the DNA by equilibrium centrifugation in cesium chloride (Fig. 4–7). (Cesium chloride gradients separate DNAs according to their buoyant densities, which in turn depend on their base composition; DNA rich in AT is less dense than GC-rich DNA.) Some repetitive genes can also be purified as satellite bands in cesium chloride gradients, as shown in Figures 23–10 and 23–12 for rRNA genes. All eukaryotes contain highly repetitive DNAs which are present in many thousands or millions of copies per cell.[21]

The heterochromatin regions located near the centromeres have been shown to contain satellite DNA in a variety of organisms. This has been established by hybridizing labeled nucleic acids in situ to chromosome preparations[22, 23] (Fig. 15–23). The centromeric heterochromatin is particularly obvious in *Drosophila virilis,* a species which has large heterochromatic segments flanking the centromeres of mitotic chromosomes and which has approximately half its DNA devoted to three main satellite DNAs. *D. melanogaster* has smaller heterochromatic regions and consequently less satellite DNA (Table 25–1).

In general, satellite DNAs are found in all the chromosomes of a cell, but sometimes preferential localizations can be observed.[24] The best example is provided by the human chromosome Y, which is mostly heterochromatic and has a special satellite DNA found only in male cells.[25]

Satellite DNAs are formed by short nucleotide sequences which are tandemly repeated (head to tail) in DNA.[26] The three satellites of *D. virilis* have a very simple base composition, each one resulting from repeating units of seven nucleotides:

Satellite	Nucleotides
I	— ACAAACT —
II	— ATAAACT —
III	— ACAAATT —

All three basic sequences are related, differing by only one nucleotide. Since the repeating sequence is so short, it is repeated many times. For example, *D. virilis* has 12 million copies per haploid genome of the 7-nucleotide repeat unit of satellite I (25 per cent of the total DNA content).[27]

The function of the highly repetitive DNAs is unknown, and most of them are not transcribed into RNA. From an evolutionary standpoint, satellite DNAs tend to change rapidly; there is little selective pressure to keep the sequences constant, and as a result satellites vary widely between species.

A major problem in the study of the evolution of repeated DNAs is to explain how they evolve together. In other words, how it is that millions of repeats in the genome change their sequence in a coordinated manner and not independent of each other. It has been suggested that this could occur by unequal crossing-over events, which are particularly frequent in repeating sequences, and as shown in Figure 25–7, could lead to duplication or deletion of repeating units (see Smith, 1976, 1978).

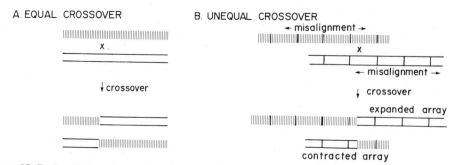

Figure 25–7 Possible mechanism for the evolution of repetitive DNAs by unequal crossing over. **A,** in equal crossing over the DNA molecules are correctly aligned. **B,** in unequal crossing over tandem repeats (shown by the vertical lines) become misaligned. This results in recombinant molecules having duplications or deletions of repeating units. By this mechanism mutations in repeating sequences can spread throughout the genome, and satellite DNAs can change between closely related species. (Modified from Smith, G. P., *Trends, Biochem. Sci.,* 3:N34.

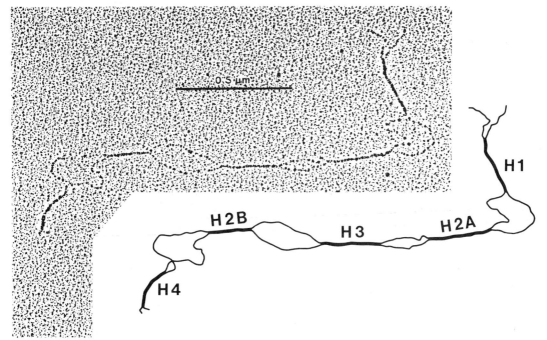

Figure 25–8 Electron micrograph of a partially denatured DNA molecule containing the five histone genes. The genes are separated by spacers that are AT-rich and therefore melted at a lower temperature than the gene regions. This molecule was cloned in *E. coli* and cut with a restriction enzyme. In the cell, many units of these five genes are tandemly repeated. (Courtesy of R. Portmann and M. L. Birnstiel.)

Moderately Repetitive Genes — rDNA, 5S DNA, and Histone Genes

In Chapter 23 we saw that some genes are present in multiple copies. For example, the frog *Xenopus laevis* has 450 ribosomal RNA genes and 23,000 5S RNA genes. These genes are clustered together in the chromosomes, separated from each other by segments of *spacer* DNA (Fig. 23–9). The spacers themselves consist of tandem repeats of shorter DNA sequences.

Most protein-coding genes (such as hemoglobin and ovalbumin) are present only once per haploid genome, but some are reiterated. The genes coding for the 5 histone proteins are repeated about 400 times in the sea urchin and between 10 and 20 times in man.

Cells have five different histone proteins, and the genes for all of them are arranged in repeating clusters, separated by segments of spacer DNA of a higher AT content. The histone genes have been cloned in *E. coli* plasmids. In Figure 25–8 one such set of genes has been partially denatured to reveal the AT-rich spacers and the five genes.[29, 30]

Histone genes tend to be clustered in a few chromosome loci, where many of the five-gene units are tandemly repeated (head to tail). In *Drosophila* all the histone genes are located in a single chromosome band. The fact that all the histone genes are clustered together suggests that their expression is coordinately regulated. Tandem repeats of genes, alternating with stretches of spacer DNA, seem to be a central feature in the organization of the eukaryotic genome.

SUMMARY

Gene Regulation in Eukaryotes and Repetitive DNA

Although the genetic code is the same for all living organisms, the eukaryotic genome exhibits a much greater complexity, which is reflected in the DNA content. For example, while the *E. coli* DNA has 4.5×10^6 base pairs, the haploid DNA of a bull contains 3.2×10^9 base pairs, which corresponds to a 700-fold increase in DNA. However, this high DNA content is not all genetically active, and as will be explained in Chapter 26, in a differentiated cell some genes are preferentially transcribed (e.g., hemoglobin gene in blood cells), while others are inactive. At the chromosome level, the heterochromatic regions are virtually inactive in RNA synthesis.

A characteristic of all eukaryotic cells is the presence of repetitive DNA, which may be highly or moderately repetitive (Table 25–1). For example, in the mouse 70 per cent of the genome is made of single copy, 10 per cent of highly repetitive, and 20 per cent of moderately repetitive DNA.

Highly repetitive DNA is also called *satellite DNA* because it can frequently be separated by buoyant density centrifugation. Centromeric heterochromatin (Chap. 15) contains satellite DNA that can be demonstrated by in situ hybridization. In the human Y chromosome there is a large heterochromatic region containing a special type of satellite DNA.

In *D. virilis* three satellites with repeating units of seven nucleotides are found which differ by only a single nucleotide. One of these units is repeated 12 million times in the genome. The function of this type of DNA is unknown, but in most cases it is not transcribed into RNA. Satellite DNA varies considerably in related species, and an unequal crossing-over mechanism (Fig. 25–7) has been suggested as an explanation of this.

Among the moderately repetitive genes are those for ribosomal RNAs and the 5S rRNA (Chap. 23). These genes are clustered together in tandem repeats that are separated by spacers of short repeating DNA sequences. The genes coding for the five histones are also clustered and tandemly repeated. Tandemly repeated genes, separated by spacers, are a central feature of eukaryotic gene organization.

25–3 REGULATION AT THE CHROMOSOME LEVEL — GIANT CHROMOSOMES

In certain cells, particularly at certain stages of their life cycle, special types of giant chromosomes may be observed. These are characterized by their enormous size and by a corresponding increase in volume of the nucleus and the cell. These special chromosomes include the so-called *polytenes* found in dipteran larvae — particularly in the salivary glands — and the *lampbrush chromosomes* observed in oöcytes of most vertebrates and invertebrates.

Discussion of these two types of chromosomes in a chapter on gene regulation is justified, because the study of polytene and lampbrush chromosomes has provided the best evidence thus far that gene activity in eukaryotes is regulated at the level of transcription.

Polytene Chromosomes have a Thousand DNA Molecules Aligned Side by Side

We will first analyze the morphology of polytene chromosomes and, subsequently, how this relates to their function. In tissues of dipteran larvae, such as the salivary glands, gut, trachea, fat body cells, and malpighian tubules, large cells can be found in which the chromosomes are strikingly different from the somatic chromosomes of the same organisms. First observed by Balbiani in 1881, polytene chromosomes received little attention until after 1930, when their cytogenetic importance was demonstrated.

In *D. melanogaster* the volume of polytene chromosomes is about 1000 times greater than that of the somatic chromosomes. The total length of the 4-paired chromosome set is 2000 μm, compared to 7.5 μm in somatic cells. Figure 25–9,7 indicates (at the same magnification) the entire somatic set as compared with the smallest pair (IV) of giant chromosomes.

When a chromosome becomes polytenic, the DNA replicates by *endomitosis,* and the resulting daughter chromatids do not separate and remain aligned side by side, as shown in Figure 25–9. A giant chromosome of *Drosophila* salivary glands contains about 1000 DNA fibers and arises from 10 rounds of DNA replication (i.e., $2^{10} = 1024$).[31]

Although in general most of the DNA sequences undergo the same number of rounds of DNA replication, there are two exceptions. The satellite DNA of *Drosophila* (which is located in the centromeric heterochromatin) undergoes

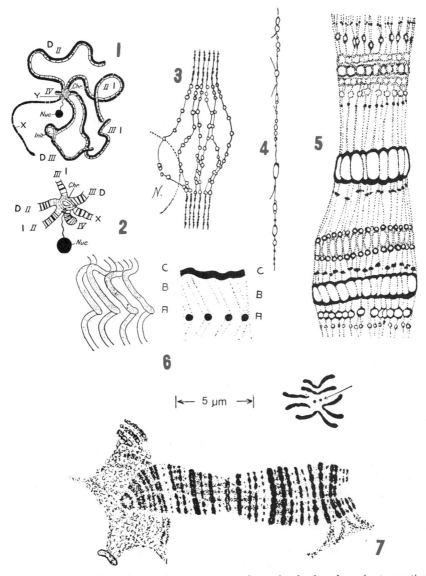

Figure 25–9 Structure of the polytene chromosomes, as drawn by the founders of cytogenetics.

1, general schematic aspect of the chromosomes of the salivary gland of a male of *Drosophila melanogaster* after they have been spread out by crushing the nucleus. The paternal chromosome (in white) and the maternal one (in black) are paired. Note that in *Drosophila* the centromeres of all the chromosomes coalesce in a heterochromatic structure called the chromocenter. *Chr.,* chromocenter; *D II* and *II I,* right and left arms of chromosome II; *D III* and *III I,* right and left arms of the third chromosome; *IV,* the fourth chromosome; *Inv.,* an inversion in the right arm of the third chromosome—inversions (and deletions) in polytene chromosomes can be visualized as loops because both homologues are paired; *Nuc.,* nucleus; *X* and *Y* indicate the sex chromosomes respectively.

2, the chromocenter (*chr.*) formed by the union of the heterochromatic parts of all the chromosomes in a female of *D. melanogaster.* (The other symbols are the same as for **1.**)

3, a heterochromatic region of the X chromosome of *D. pseudoobscura,* showing its relations with the nucleolus (*N.*) and the filamentous (chromonemic) constitution of the chromosome.

4, detail of a component chromonema of the polytene chromosome in which the different chromomeres are seen.

5, schematic structure of the chromosome of *Simulium virgatum,* showing the organization of the chromonemata, chromomeres, and vesicles, which together give the appearance of the bands. The segment drawn corresponds to a euchromatic zone.

6, diagram to illustrate the interpretation of the helicoidal chromonema and the false chromomeres produced by the turns of the spiral. A zone (*B*) with four chromonemata is shown between two consecutive bands (at the left). To the right is the aspect of the same region when observed in a different focusing plane. *A* has a granular aspect, which simulates chromomeres. *C* appears as a continuous solid line.

7, the fourth polytene chromosome of *D. melanogaster,* adhering to the chromocenter, which is at the left. Above, at the right, the somatic chromosomes of the same fly as they appear in mitosis. The difference in size between the giant chromosome IV and the somatic chromosome IV is indicated by the arrow and drawn to the same scale. (**1** and **2,** after White, 1942; **3,** after Bauer, 1936; **4,** after Painter and Griffen, 1937; **5,** after Painter, 1946; **6,** after Ris and Crouse, 1945; **7,** after Bridges, 1935.)

very little replication, if any, in polytene chromosomes (see Gall, 1974). The other exception occurs in the dipteran *Rhynchosciara angelae,* in which the DNA of some special puffs (DNA puffs) is replicated more than the rest of the chromosome, providing one of the few examples of *gene amplification.*

Polytene cells are unable to undergo mitosis and are destined to cell death. Not all the cells in a dipteran larva have polytene chromosomes; those destined to produce the adult structures after metamorphosis (*imaginal disks*) remain diploid. Not all cells in a larva have the same degree of polyteny; Figure 25–10 shows that the saliva-producing cells have a higher DNA content than the excretory duct cells, which are less polytenic.

Another characteristic of polytene chromosomes is that the homologous pairs are closely associated, as in meiotic prophase. This phenomenon is called *somatic pairing,* and the chromosomes are considered to be in a permanent interphase (see Fig. 25–9,1).

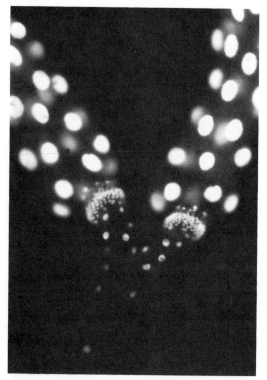

Figure 25–10 Salivary glands of a *Drosophila melanogaster* larva showing variable degrees of polyteny. The cells were stained with ethidium bromide, a fluorescent compound that binds to DNA. Note that the secretory cells (top) are larger and contain more DNA than those of the excretory duct (bottom). The rings of small nuclei located between the excretory ducts and the secretory cells are the *imaginal discs,* which are diploid and destined to produce the salivary glands of the adult fly. (Courtesy of M. Jamrich.)

550

The Bands Represent Chromomeres Aligned Side by Side

Along the length of the chromosome a series of dark *bands* alternates with clear zones called *interbands.* The bands represent regions where the DNA is more tightly coiled (chromomeres), and in the interbands (interchromeric regions) the DNA fibers are less folded.[3]

The dark bands stain intensely, are Feulgen-positive, and absorb ultraviolet light at 260 nm. The interbands do not stain with basic stains, are Feulgen-negative and absorb little ultraviolet light because their DNA content is lower.

The constancy in number, localization, and distribution of the bands in the two homologous (paired) chromosomes is notable. It is easy to construct, from a giant chromosome, topographic maps of the bands and interbands and to verify any disarrangement or alteration in the order of their linear structure. There are 5000 bands in the 4 chromosomes of *Drosophila,* and it is possible that each one represents a single genetic unit. Genetic studies indicate that there are about 5000 genes in *Drosophila;* and it has therefore been proposed that each band may correspond to a gene.[32, 33]

Before the recognition of these giant chromosomes, genetic maps of the chromosomes of *Drosophila* were made by crosses. After their discovery, it was possible to compare the rather abstract genetic map with the topographic map of each chromosome. Combined genetic and cytologic studies have permitted the identification of many of the bands with specific genetic loci.

More recently, the cytogenetic technique of *in situ hybridization* of nucleic acids to chromosome preparations has allowed individual RNAs,[34] and recombinant plasmids containing *Drosophila* DNA fragments to be mapped[35] to specific chromosome bands.

Polytene chromosomes are very suitable for in situ hybridization because in them 1000 DNA molecules are aligned side by side, thereby greatly increasing the possibility of detecting single-copy genes. (At present this cannot be achieved in non-polytene chromosomes.)

Puffs in Polytenes are the Sites of Gene Transcription

One of the most remarkable characteristics of polytene chromosomes is that it is possible to visualize in them the genetic activity of specific chromosomal sites. The morphologic expression of such sites is represented by local enlargements of certain regions called *puffs* (Fig. 25–11). A puff

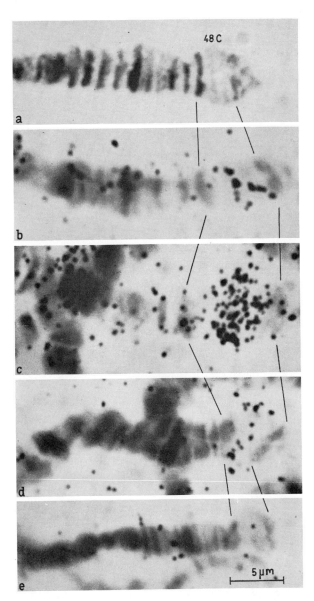

Figure 25-11 Experiment to demonstrate how ³H-uridine is incorporated into puff 48 C of *Drosophila hydei*. **a,** control; **b,** after 3 minutes of temperature shock (a few silver grains have been deposited); **c,** 15 minutes after temperature shock. Observe the greatly increased incorporation; **d,** 15 minutes after injection of actinomycin D in larvae having puffs as in **c; e,** 30 minutes after injection of the antibiotic in animals with puffs as shown in **c.** These results indicate that in 15 to 30 minutes most of the synthesized RNA has been eliminated from the puff. (Courtesy of H. D. Berendes.)

can be considered as a band in which the DNA unfolds into open loops (as in the lampbrush chromosomes) that actively synthesize RNA.[36, 37] The very largest puffs are called *Balbiani rings* and have maximally extended DNA fibers which form lateral loops. Puffs are the morphologic expression of gene transcription and provide a unique opportunity to study the regulation of eukaryotic genes.

Polytene chromosomes have provided the strongest evidence we have showing that in eukaryotes gene activity is regulated at the level of RNA synthesis. They constitute very valuable material for studying gene regulation, because in these chromosomes gene transcription can be visualized directly under the microscope by the appearance of puffs.

In 1952 W. Beerman compared the polytene chromosomes of different tissues of *Chironomus* larvae and showed that although the pattern of bands and interbands was similar in all tissues, the distribution of puffs differed from one tissue to another.[38] He correctly interpreted the puffing of polytene chromosomes as an expression of intense gene transcription.

In salivary glands the appearance of some puffs has been correlated with the production of specific proteins which are secreted in large amounts in the larval saliva[31, 39] (see Grossbach, 1977). For example, *Chironomus* has at the base of the salivary glands four specialized cells which contain cytoplasmic granules of a special secretory protein. The gene for this protein is located in a distinct puff that appears only in the four specialized cells.[31] These results clearly show that cell specialization results from variable gene transcription.

A very special type of puff has been described in *Rhynchosciara angelae* by Pavan and Breuer.[40] The so-called DNA puffs are able to incorporate ³H-thymidine, and they represent regions of the chromosomes which have extra replication of the DNA. This phenomenon, by no means common, is a case of gene amplification (Chap. 26). The DNA puffs have also been correlated with the production of specific salivary proteins.[41]

Puffs Can be Induced by Ecdysone and by Heat Shock

Puffing is a cyclic and reversible phenomenon; at definite times, and in different tissues of the larvae, puffs may appear, grow, and disappear. Puff formation can be studied experimentally using factors that will induce their formation. The steroid hormone *ecdysone*, which induces molting in insects, will induce the formation of specific puffs when injected into larvae[42] or when added to salivary glands in culture.[37]

Puffing can also be induced by temperature shock. When *Drosophila* larvae, normally grown at 25° C, are exposed to a temperature of 37° C, a series of specific genes is activated, while most other genes are repressed.[43] Five minutes after heat shock nine new puffs are seen on the giant chromosomes of the salivary glands (Fig. 25–11).

These puffs are very active in RNA synthesis, as can be demonstrated by ³H-uridine labeling and radioautography (Fig. 25–11,*B* and *C*). The newly made RNA is released from the puffs (Fig. 25–11,*D* and *E*) and accumulates in the cytoplasm.[44] New kinds of mRNA can be isolated from heat-treated cells,[34] and these mRNAs give rise to eight specific heat-shock proteins (see Moran et al., 1978). It is clear, once again, that in this experimental situation the induction of specific proteins is due to an increased rate of transcription of individual genes.

One of the earliest events in the induction of heat-shock puffs is the accumulation of RNA polymerase II in the puff loci. RNA polymerase can be located on polytene chromosome preparations by immunofluorescence using specific antibodies.[45] By this technique RNA polymerase II is found to be normally located in the interbands and the puffs (uncondensed regions of the chromosome) (Fig. 25–12). When cells are placed at 37° C (Fig. 25–13), RNA polymerase II rapidly accumulates in the heat-shock puffs, while it disappears from other regions of the chromosome.

Lampbrush Chromosomes in Oöcytes at the Diplotene Stage of Meiosis

Lampbrush chromosomes were first observed by Flemming in 1882 and were described in detail in shark oöcytes by Ruckert in 1892. He coined this

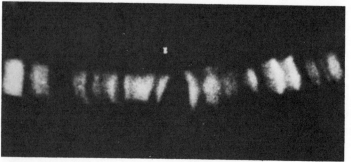

Figure 25–12 Visualization of RNA polymerase II distribution in polytene chromosomes of *Drosophila melanogaster*; *A*, orcein stain; *B*, immunofluorescence using an anti-RNA-polymerase-II antibody. Note that the fluorescence is located at the interbands (Courtesy of M. Jamrich and E. K. F. Bautz.)

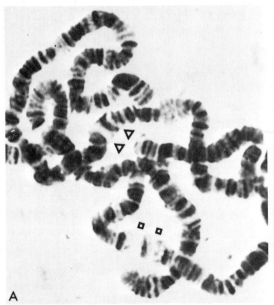
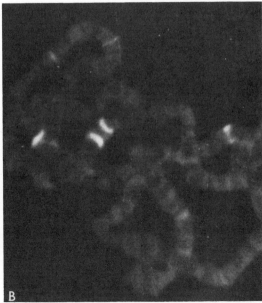

Figure 25–13 Localization of RNA polymerase II in heat-shock puffs. Larvae were placed at 37° C for 45 minutes before preparation of chromosomes. **A,** orcein stain; **B,** immunofluorescence. Triangles indicate the largest heat-shock puffs, 87 A and 87 C. The squares indicate pre-existing puffs. RNA polymerase II accumulates in the heat-shock puffs, while it disappears from other regions of the chromosome (even from the pre-existing puffs). (Courtesy of E. K. F. Bautz.)

particular name because they looked like the brushes used in those times to clean the chimneys of oil lamps. However, it is only recently the structure of these chromosomes has been interpreted in functional terms, largely through the work of H. G. Callan.

Lampbrush chromosomes occur at the diplotene stage of meiotic prophase in oöcytes of all animal species, in spermatocytes of several species, and even in the giant nucleus of the unicellular alga *Acetabularia*.[46] Lampbrush chromosomes have many fine lateral projections (Fig. 25–14), giving them the characteristic "hairy" appearance. They are best visualized in salamander oöcytes because they have a high DNA content and consequently very large chromosomes. In some newts the total length of a lampbrush chromosome set may be 5900 μm, which is 3 times longer than the polytene chromosomes of *Drosophila*.[47] Species with lower DNA content have small lampbrushes; therefore, strictly speaking, lampbrush chromosomes are not always giant.

Since the lampbrush chromosomes are found in meiotic prophase, they are present in the form of *bivalents* in which the maternal and paternal chromosomes are held together by chiasmata at those sites where crossing over has previously occurred (Fig. 25–14,*A*). Each bivalent has four

chromatids, two in each homologue. The axis of each homologue consists of a row of granules or chromomeres, each one of which has two looplike lateral extensions, one for each chromatid (Fig. 25–14,*B*).

The Lateral Loops are Sites of Intense RNA Synthesis

The chromomeres represent regions in which the chromatids are tightly folded and inactive in transcription; in the loops the DNA is extended as a result of intense RNA synthesis.[48]

The diplotene stage lasts for about 200 days in salamander oöcytes, and this is a period of great synthetic activity during which a single nucleus produces all the RNA required to build an egg cell more than 1 mm in diameter. There is evidence that some of the mRNAs synthesized on these chromosomes are stored in the cytoplasm during oögenesis and are used later during embryonic development. Lampbrush loops may therefore be considered as an adaptation which allows intense transcription of prophase chromosomes by producing localized uncoiling of the chromatin fibers.

It has been estimated that 5 per cent of the DNA in lampbrush chromosomes is present in loops,

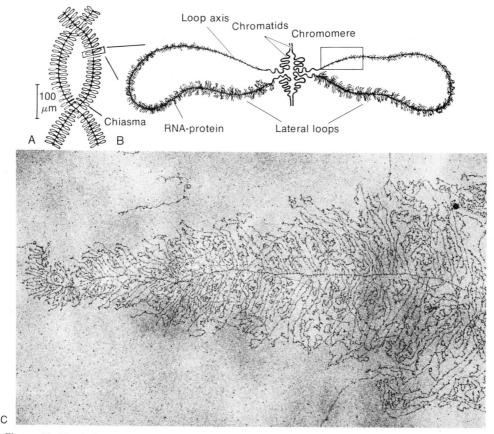

Figure 25–14 A diagram of the lampbrush chromosomes of *Triturus* oöcytes. **A,** low magnification showing two chiasmata. **B,** higher magnification showing two lateral loops and the folding of the chromatids in the chromomere. **C,** electron micrograph of a region corresponding to the inset in **B.** Note the DNA axis upon which fine fibrils of ribonucleoprotein are inserted perpendicularly. Observe that they increase in length toward the right. ×15,000 (Courtesy of O. L. Miller, Jr..)

the rest being coiled in chromomeres. In *Triturus* the length of all the loops has been calculated to be about 50 cm, and the total DNA length, 10 meters per oöcyte.

The various loops can be distinguished by size, thickness, and other morphologic characteristics. Each loop appears at a constant position in the chromosome, and detailed chromosome maps can be drawn.

There are about 10,000 loops per chromosome set, and most of them correspond to a single unit of transcription. Although it was previously thought that each loop had only one transcriptional unit, more recent work has shown that some loops may contain more than one (see Scheer et al., 1976).

Each loop has an axis formed by a single DNA molecule, which is coated by a matrix (Fig. 25–14). Cytochemical studies using ribonuclease and protease showed that this matrix is made of RNA and protein.[49] Furthermore, when deoxyribonuclease is used, the loops break and disappear,

confirming that they are held together by an axis of DNA.

In general the ribonucleoprotein matrix is asymmetrical, being thicker at one end of the loop than at the other. RNA synthesis starts at the thinner end and progresses toward the thicker end, as can be seen in preparations spread for electron microscopy.[50] Figure 25–14,C shows the nascent ribonucleoprotein chains inserted perpendicular to the DNA axis. The RNA molecules become longer as they progress farther away from the transcription initiation site (promoter). The appearance is very similar to the "Christmas tree" images observed during ribosomal RNA synthesis (Fig. 23–9). These studies show that transcription starts and stops at precise points in the DNA and also that the initial RNA transcripts are exceedingly long.[51]

Study of these lampbrush chromosomes has led to some interesting conclusions about the nature of eukaryotic RNA synthesis. It has been found that the units of transcription in eu-

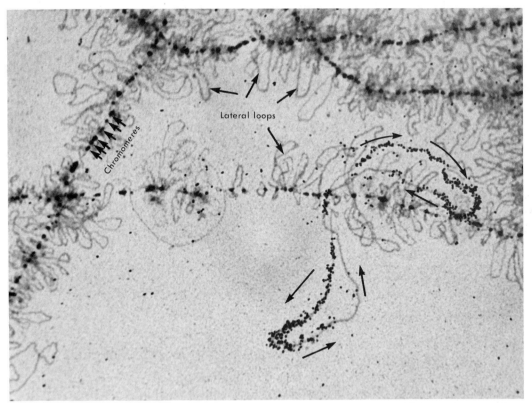

Figure 25–15 Lampbrush chromosomes of *Triturus cristatus* oöcytes hybridized with labeled cloned genes under conditions in which hybridization to nascent RNA chains could be detected. One pair of loops is labeled. Lateral loops are always in pairs, one for each sister chromatid. The arrows indicate the direction of transcription. The hybridization ceases abruptly before the end of the loop. This mechanism is interpreted in Figure 25–16. ×1000. (Courtesy of H. G. Callan.)

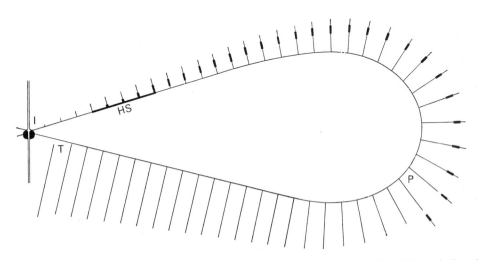

Figure 25–16 Diagram interpreting the experiment described in Figure 25–15. The DNA in the lampbrush loop is drawn as a single line that is thicker in the region corresponding to the hybridizing sequences (*HS*). The RNA transcripts project perpendicular to the loop and hybridize with the labeled DNA, indicated by a heavier tracing. *I*, initiation site; *T*, termination site; *P*, site where processing of the mRNA occurs. At this site the mRNA is cleaved off, before the transcription of the loop is completed. (Courtesy of H. G. Callan, from Old, R. W., Callan, H. G., and Gross, K. W., *J. Cell Sci.*, 27:57, 1977.)

karyotes are very large, involving DNA segments between 10 and 30 μm long. The very large RNA molecule produced in this way probably represents the so-called *heterogeneous nuclear RNA* (hnRNA, see Chapter 22), which is thought to be a precursor of messenger RNA.

Another important conclusion is that RNA processing starts even before transcription is terminated. This can be inferred from the experiments shown in Figures 25–15 and 25–16. Lampbrush chromosomes were hybridized in situ to a labeled cloned DNA segment under conditions in which the segment could hybridize to only the nascent RNA chains.[52] Figure 25–15 shows a pair of labeled loops, one for each sister chromatid. It is important to note that the label-

ing does not continue throughout the loops but rather ceases suddenly before the end of the loop. Figure 25–16 shows a diagram which explains these results on the basis of RNA processing. According to the diagram, the hybridizing sequence is removed from the nascent RNA chain at a specific site before the transcription of the loop is completed.

It is interesting to note that the loops of the lampbrush chromosomes and the so-called *puffs* of the polytene chromosomes have the following features in common: (1) There is a dispersion of the chromosome structure at the site where genes are activated. (2) Transcription (i.e., synthesis of RNA) is preceded by uncoiling. (3) They represent units of gene activity.

SUMMARY:

Regulation in Eukaryotes at the Chromosome Level

The fact that gene activity is regulated at the transcriptional level is best illustrated by the polytene and lampbrush chromosomes.

Polytenes are giant chromosomes that are found in tissues of certain insects and in which there have been about ten reduplications by endomitosis, resulting in the production of about 1000 threads (*chromatids*). These chromosomes are in a permanent interphase, and the homologous chromosome pairs are closely associated (somatic pairing).

It seems that most of the DNA sequences undergo the same number of rounds of replication, but there are two exceptions: (1) the satellite DNA of the centromeric region is under-replicated and (2) in *Rhyncosciara* are found *DNA puffs*, which are a special case of DNA amplification in which some genes are replicated more than the rest.

Polytene cells do not divide and finally die. Not all have the same degree of polyteny (Fig. 25–10). In polytene chromosomes dark bands (chromomeres) alternate with interbands, where the chromatin fibers are less folded. There are some 5000 bands in *Drosophila* that may correspond to individual genes. Genetic maps constructed by crosses have been matched with cytologic maps showing the location of the bands. Hybridization with mRNAs and recombinant plasmids has permitted precise localization of certain genes. This process has been facilitated by the fact that each gene is represented in 1000 copies aligned side by side.

Genetic activity can be visualized by the so-called *puffs* and *Balbiani rings,* in which the bands enlarge because they are actively synthesizing RNA, and the DNA has unfolded into loops. Some puffs have been correlated with the production of specific proteins in the larval saliva.

Specific puffs can be induced experimentally by the steroid hormone *ecdysone* and by temperature shock (Fig. 25–11). Five minutes after heat shock, nine new puffs appear; the RNA made is then released to the cytoplasm, and eight specific proteins appear.

The studies on chromosome puffs have shown that the induction of specific proteins is due to an increased rate of transcription of individual genes. Thus, eukaryotic genes are regulated at the level of RNA synthesis.

Lampbrush chromosomes are observed during the long period of diplonema in the meiosis of oöcytes of all animals. Each bivalent is made of two homologues held together by chiasmata. The four chromatids have characteristic loops that extend laterally. Each loop contains a DNA molecule and a matrix of ribonucleoprotein fibrils inserted perpendicular to the DNA (Fig. 25–14). Within the loop the matrix is thin at one end and thick at the other; this is caused by the synthesis of RNA molecules. At the chromosome axis, chromatids are tightly folded, forming chromomeres that are inactive in transcription.

In *Triturus* the total length of the DNA contained in the lateral loops is about 50 cm, and this represents 5 per cent of the genome (10 meters). There are about 10,000 loops, most of which probably correspond to a single transcriptional unit.

The RNA transcripts increase in size from the initiation site (promoter) and become exceedingly long (10 to 30 μm). These RNA molecules probably represent hnRNA that is then processed to make mRNA (Chap. 22). Processing starts even before transcription is terminated, as can be inferred from the experiments shown in Figures 25–15 and 25–16.

Both lampbrush loops and polytene puffs represent units of genetic activity in which DNA is uncoiled at sites of intense RNA synthesis.

REFERENCES

1. Jacob, F., and Monod, J. (1961) *J. Molec. Biol., 3*:318.
2. Gilbert, W., and Muller-Hill, B. (1966) *Proc. Natl. Acad. Sci. USA, 56*:189.
3. Gilbert, W., and Muller-Hill, B. (1967) *Proc. Natl. Acad. Sci. USA, 58*:2415.
4. Jobe, A., and Baurgeois, S. (1972) *J. Molec. Biol., 69*:397.
5. Gilbert, W., Maizels, N., and Maxam, A. (1974) *Cold Spring Harbor Symp. Quant. Biol., 38*:845.
6. Ptashne, M., Backman, K., Humayun, M. Z., Jeffrey, A., Maurer, R., Meyer, B., and Sauer, R. T. (1976) *Science, 194*:156.
7. Heyneker, H. L., Shine, J., Goodman, H. M., Boyer, H. W., Rosemberg, J., Dickerson, R. E., Narang, S., Itakura, K., Lin, S., and Riggs, A. D. (1976) *Nature, 264*:748.
8. Ippen, K., Miller, J., Scaife, J., and Beckwith, J. (1968) *Nature, 217*:825.
9. Beckwith, J. (1967) *Science, 156*:597.
10. Dickson, R. C., Abelson, J., Barnes, W. M., Reznikoff, W. S. (1975) *Science, 187*:27.
11. Majors, J. (1975) *Proc. Natl. Acad. Sci., 72*:4394.
12. De Crombrugghe, B., Chen, B., Gottesman, M., Varmus, H., Emmer, M., and Perlman, R. (1971) *Nature (New Biol.), 230*:37.
13. Makman, R., and Sutherland, E. W. (1965) *J. Biol. Chem., 240*:1309.
14. Pastan, I., and Perlman, R. (1970) *Science, 169*:339.
15. Squires, C. L., Lee, F., Yanofsky, C. (1975) *J. Molec. Biol., 92*:93.
16. Bertrand, K., Korn, L., Lee, F., Platt, T., Squires, C. L., Squires, C., and Yanofsky, C. (1975) *Science, 189*:22.
17. Korn, L., and Yanofsky, C. (1976) *J. Molec. Biol., 106*:231.
18. Kasai, T. (1974) *Nature, 249*:523.
19. Frenster, J. H., Allfrey, V. G., and Mirsky, A. E. (1963) *Proc. Natl. Acad. Sci. USA, 50*:1026.
20. Britten, R. J., and Kohne, D. E. (1968) *Science, 161*:529.
21. Skinner, D. M. (1978) *Bioscience, 27*:790.
22. Pardue, M. L., and Gall, J. G. (1970) *Science, 168*:1356.
23. Jones, K. W. (1970) *Nature, 255*:912.
24. Peacock, W. J., Appels, R., Dunsmuir, P., Lohe, A. R., and Gerlach, W. L. (1977) In: *International Cell Biology*, p. 494. (Brinkley, B. R., and Porter, K. R., eds.) The Rockefeller University Press, New York.
25. Cooke, H. J., and McKay, R. D. G. (1978) *Cell, 13*:453.
26. Southern, E. M. (1975) *J. Molec. Biol., 94*:51.
27. Gall, J. G., and Atherton, D. D. (1974) *J. Molec. Biol., 85*:633.
28. Kedes, L. H. (1976) *Cell, 8*:321.
29. Portmann, R., Schaffner, W., and Birnstiel, M. (1976) *Nature, 264*:31.
30. Birnstiel, M. L., Kressmann, A., Schaffner, W., Portmann, R., and Busslinger, M. (1978) *Philos. Trans. R. Soc. London [Biol. Sci.], 283*:319.
31. Beerman, W. (1972) *Developmental Studies on Giant Chromosomes: Results and Problems in Cell Differentiation*. Springer-Verlag, Berlin.
32. Judd, B. H., Shen, M. W., and Kaufman, T. C. (1972) *Genetics, 71*:139.
33. Garcia-Bellido, A., and Ripoll, P. (1978) *Nature, 273*:399.
34. Pardue, M. L., Bonner, J. J., Lengyel, J. A., and Spradling, A. O. (1977) In: *International Cell Biology*, p. 509. Rockefeller University Press, New York.
35. Wensink, P. C., Finnegan, D. J., Donelson, J. E., and Hogness, D. S. (1974) *Cell, 3*:315.
36. Berendes, H. D. (1973) *Int. Rev. Cytol., 35*:61.
37. Ashburner, M. C., Chihara, C., Metzer, P., and Richards, G. (1973) *Cold Spring Harbor Symp. Quant. Biol., 38*:655.
38. Beerman, W. (1952) *Chromosoma, 5*:139.
39. Daneholt, B., Case, S. T., Lamb, M. M., Nelson, L., and Weislander, L. (1978) *Philos. Trans. R. Soc. London [Biol. Sci.], 283*:383.
40. Breuer, M., and Pavan, C. (1955) *Chromosoma, 7*:371.
41. Winter, C. E., de Bianchi, A. G., Terra, W. R., and Lara, F. J. (1977) *Chromosoma, 61*:193.
42. Clever, U., and Karlson, P. (1960) *Exp. Cell Res., 20*:263.
43. Ritossa, F. M. (1962) *Experientia, 18*:571.
44. Danehott, B. (1975) *Cell, 4*:1.
45. Jamrich, M., Greenleaf, A. L., and Bautz, E. K. F. (1977) *Proc. Natl. Acad. Sci. USA, 74*:2079.

46. Spring, H., Scheer, U., Franke, W. W., and Trendelenburg, M. F. (1975) *Chromosoma*, 50:25.
47. Gall, J. G., and Callan, H. G. (1962) *Proc. Natl. Acad. Sci. USA*, 48:562.
48. Callan, H. G. (1963) *Int. Rev. Cytol.*, 15:1.
49. Gall, J. G. (1956) *Brookhaven Symp. Biol.*, 8:17.
50. Hamkalo, B. A., and Miller, O. L. (1973) Electronmicroscopy of genetic activity. *Annu. Rev. Biochem.*, 42:379.
51. Scheer, U., Franke, W. W., Trendelenburg, M. F., and Spring, H. (1976) *J. Cell Sci.*, 22:503.
52. Old, R. W., Callan, H. G., and Gross, K. W. (1977) *J. Cell Sci.*, 27:57.

ADDITIONAL READINGS

Appels, R., and Peacock, W. J. (1978) The arrangement and evolution of highly repeated (satellite) DNA sequences, with special reference to *Drosphila*. *Int. Rev. Cytol. Suppl.*, 8:70.

Ashburner, M., and Bonner, J. J. (1979) The induction of gene activity in *Drosophila* by heat shock. *Cell*, 17:241.

Barnes, W. M. (1978) DNA sequence from the histidine operon control region: Seven histidine codons in a row. *Proc. Natl. Acad. Sci. USA*, 75:4281.

Beckwith, J. (1967) Regulation of the lac operon. *Science*, 156:597.

Beckwith, J., and Zipser, D., eds. (1968) *The Lactose Operon*. Cold Spring Harbor Laboratory, Cold Spring Harbor, New York.

Berendes, H. D. (1973) Synthetic activity of polytene chromosomes. *Int. Rev. Cytol.*, 35:61.

Bertrand, K., Korn, L., Lee, F., Platt, T., Squires, C. L., Squires, C., and Yanofksy, C. (1975) New features of the regulation of the tryptophan operon. *Science*, 189:22.

Bridges, C. B. (1935) Salivary chromosome maps. *J. Hered.*, 26:60.

Britten, R. J., and Kohne, D. E. (1968) Repeated sequences in DNA. *Science*, 161:529.

Daneholt, B. (1975) Transcription in polytene chromosomes. *Cell*, 4:1.

Davidson, E. H., and Britten, R. J. (1973) Organization, transcription, and regulation in the animal genome. *Quart. Rev. Biol.*, 48:565.

Dickson, R. C., Abelson, J., Barnes, W. M., and Reznikoff, W. S. (1975) Genetic regulation: The lac control region. *Science*, 187:27.

Eron, L., Arditti, R., Zubay, G., Connaway, S., and Beckwith, J. R. (1971) An adenosine 3':5'-cyclic monophosphate-binding protein that acts on the transcription process. *Proc. Natl. Acad. Sci. USA*, 68:215.

Fedoroff, N. V. (1979) On spacers. *Cell*, 16:997.

Gall, J. G., and Callan, H. G. (1962) ^{3}H-uridine incorporation in lampbrush chromosomes. *Proc. Natl. Acad. Sci. USA*, 48:562.

Gall, J. G., Cohen, E. H., and Atherton, D. D. (1973) The satellite DNAs of *Drosophila virilis*. *Cold Spring Harbor Symp. Quant. Biol.*, 38:417.

Gall, J. G. (1974 Repetitive DNA in Drosophila. In: *Molecular Cytogenetics*, p. 59. (Hamkalo, B. A., and Papaconstantinow, J., eds.) Plenum Publishing Co., New York.

Grossbach, U. (1977) The salivary gland of *Chironomus*: A model system for the study of cell differentiation. In: *Results and Problems in Cell Differentiation*. Vol. 8, p. 147. (Beerman, W., ed.) Springer-Verlag, Berlin.

Jacob, F., and Monod, J. (1961) Genetic regulatory mechanisms in the synthesis of proteins. *J. Molec. Biol.*, 3:318. Highly recommended.

Jamrich, M., Greenleaf, A. L., Bautz, F. A., and Bautz, E. K. F. (1977) Functional organization of polytene chromosomes. *Cold Spring Harbor Symp. Quant. Biol.*, 42:89.

Maniatis, T., and Ptashne, M. (1976) A DNA operator-repressor system. *Sci. Am.*, 234(1):64.

Moran, L., Mirault, M. E., Arrigo, A. P., Goldschmidt, M., and Tissieres, A. (1978) Heat shock of *Drosophila melanogaster* induces the synthesis of new messenger RNAs and proteins. *Philos. Trans. R. Soc. London [Biol. Sci.]*, 283:391.

Old, R. W., Callan, H. G., and Gross, K. W. (1977) Localization of histone gene transcripts in lampbrush chromosomes. *J. Cell Sci.*, 27:57.

Pardue, M. L. (1973) Localization of repeated DNA sequences in *Xenopus* chromosomes. *Cold Spring Harbor Symp. Quant. Biol.*, 38:475.

Pardue, M. L., Kedes, L. H., Weinberg, E. S., and Birnstiel, M. L. (1977) Localization of sequences coding for histone messenger RNA in the chromosomes of *Drosophila melanogaster*. *Chromosoma*, 63:135.

Pastan, I., and Perlman, R. (1970) Cyclic adenosine monophosphate in bacteria. *Science*, 169:339.

Pavan, C., and Da Cunha, A. B. (1969) Chromosomal activities in *Rhynchosciara* and other *Sciaridae*. *Annu. Rev. Genet.*, 3:425.

Peacock, W. J., Appels, R., Dunsmuir, P., Lohe, A. R., and Gerlach, W. L. (1977) Highly repeated DNA sequences: Chromosomal localization and evolutionary conservatism. In: *International Cell Biology*, pp. 494–506. (Brinkley, B., and Porter, K., eds.) Rockefeller University Press, New York.

Platt, T. (1978) Regulation of gene expression in the tryptophan operon of *E. coli*. In: *The Operon*, p. 263. (Miller, J. H., and Reznikoff, W. S., eds.) Cold Spring Harbor Laboratory, Cold Spring Harbor, New York.

Ptashne, M., and Gilbert, W. (1970) Genetic repressors. *Sci. Am.*, *222*(6):36.

Schaffner, W., Kung, G., Daetwyler, H., Telford, J., Smith, H. O., and Birnstiel, M. L. (1978) Genes and spacers of cloned sea urchin histone DNA analysed by sequencing. *Cell, 14*:655.

Schedl, P., Artavanis-Tsakonas, S., Steward, R., Gehring, W. J., Mirault, M. E., Goldschmidt-Clermont, M., Moran, L., and Tissieres. A. (1978) Two hybrid plasmids with *Drosophila melanogaster* DNA sequences complementary to mRNA coding for the major heat shock protein. *Cell, 14*:921.

Scheer, U., Franke, W. W., Trendelenburg, M. F., and Spring, H. (1976) Classification of loops of lampbrush chromosomes according to the arrangement of transcriptional complexes. *J. Cell Sci., 22*:503.

Smith, G. P. (1976) Evolution of repeated DNA sequences by unequal crossing over. *Science, 191*:528.

Smith, G. P. (1978) What is the origin of repetitive DNAs? *Trends Biochem. Sci., 3*:N34.

Southern, E. M. (1975) Long-range periodicities in mouse satellite DNA. *J. Molec. Biol., 94*:51.

Spradling, A., Pardue, M. L., and Penman, S. (1977). Messenger RNA in heat-shocked *Drosophila* cells. *J. Molec. Biol., 109*:559.

Sommerville, J., Malcom, D. B., and Callan, H. G. (1978) The organization of transcription on lampbrush chromosomes. *Philos. Trans. R. Soc. London [Biol. Sci.], 283*:359.

Swift, H. (1973) The organization of genetic material in eukaryotes: Progress and prospects. *Cold Spring Harbor Symp. Quant. Biol., 38*:963.

Wensink, P. C., Finnegan, D. J., Donelson, J. E., and Hogness, D. S. (1974) A system for mapping DNA sequences in the chromosomes of *Drosophila melanogaster. Cell, 3*:315.

Wischnitzer, S. (1976) The lampbrush chromosomes. *Endeavour, 35*:27.

Zurawski, G., Brown, K., Killingly, D., and Yanofsky, C. (1978) Nucleotide sequence of the leader region of the phenylalanine operon of *E. coli. Proc. Natl. Acad. Sci. USA, 75*:4271.

26

CELL DIFFERENTIATION

26–1   General Characteristics of Cell
       Differentiation   562
           The Differentiated State is Stable
           Determination can Precede Morphological
           Differentiation
26–2   Nucleocytoplasmic Interactions   562
           Acetabularia is Well Suited for the study
           of Nucleocytoplasmic Interactions
           Red Blood Cell Nuclei can be
           Reactivated by Cell Fusion
           Cytoplasts and Karyoplasts can be Pre-
           pared by Cytochalasin B Enucleation
           Cell Fusion Yields Pure Antibodies of
           Medical Importance
           Gene Expression by Somatic Nuclei is
           Reprogrammed in Xenopus laevis
           Oöcytes
       Summary: *Nucleocytoplasmic*
           Interactions

26–3   Molecular Mechanisms of Cell
       Differentiation   571
           The Genome Remains Constant During
           Cell Differentiation — Nuclear Trans-
           plantation
           Gene Amplification Cannot Explain Cell
           Differentiation
           Cell Differentiation Cannot be Explained
           by Translational Control
           Cell Differentiation is Probably Con-
           trolled at the Level of Transcription
           Cytoplasmic Determinants Localized in
           Egg Cytoplasm are Important in Early
           Development
           Cell Differentiation is Intimately Related
           to Cellular Interactions
       Summary: *Mechanisms of Cell*
           Differentiation

Cell differentiation is the process by which stable differences arise between cells. All higher organisms develop from a single cell, the fertilized ovum, which gives rise to the various tissues and organs. The question of how an apparently structureless egg converts itself into a complex and highly organized embryo has interested scientists since the time of Aristotle 2000 years ago, and this still remains one of the major unanswered questions of biology.

In most animal species females produce large unfertilized *eggs* which contain most of the materials and nutrients required to form an embryo. Development is triggered by *fertilization.* The sperm contributes a small, condensed nucleus (*male pronucleus*), which rapidly enlarges in the egg cytoplasm, fuses with the *female pronucleus,* and finally divides. The fertilized egg then undergoes a series of very rapid cycles consisting of DNA synthesis followed by cell divisions. These divisions are called *cleavage,* because unlike normal cell division, the cytoplasm is partitioned without growth. Then the cells form a hollow sphere (*blastula*) in which tissues are not yet evident (Fig. 26–1). Some of the cells then invaginate in a series of cell movements known as *gastrulation,* and the first signs of mor-

phologic differentiation appear (Figs. 26–1 and 26–2). These very complex changes take place in a comparatively short time. For example, in the South African frog *Xenopus laevis,* a swimming tadpole containing most differentiated tissues (such as blood, nerve, eye, muscle, and so forth) hatches only 72 hours after fertilization.

In molecular terms, cell differentiation means *variable gene activity* in different cells of the same organism. Cell specialization involves the preferential synthesis of some specific proteins (e.g., hemoglobin in erythrocytes, antibodies in plasma cells, and ovalbumin in oviduct). Each eukaryotic cell expresses only a small percentage of the genes it contains, and cells of different tissues express different sets of genes. However, it should be remembered that some genes are expected to be active in all types of cells, such as those genes required for building membranes, ribosomes, mitochondria, and the cytoskeleton, components which are essential to all cells. It is therefore clear that cell differentiation will be explained in molecular terms only when the complex mechanisms of gene regulation in higher cells are understood.

In this chapter we will analyze the possible mechanisms by which the initial differences be-

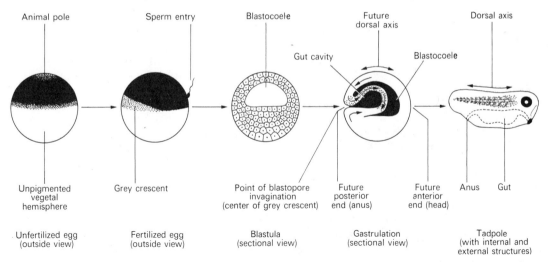

Figure 26-1 Development of *Xenopus laevis*. During gastrulation cells invaginate at the blastopore, and extensive morphogenetic movements of cells occur (arrows). The blastopore always forms at the center of a pigmented area called the *grey crescent*, which appears in eggs opposite to the point of sperm entry shortly after fertilization. Thus, the orientation of the dorsal axis of the embryo is determined by the point of sperm entry. (Courtesy of J. B. Gurdon, from *Gene Expression During Cell Differentiation*, Carolina Biology Readers, Vol. 25, Carolina Biological, 1978.)

tween cells arise in early embryos. The first part of the chapter will be devoted to the general characteristics of cell differentiation and to *nucleocytoplasmic interactions*. This is because a large part of our knowledge of the mechanisms by which differentiation is established and maintained comes from experiments in which nuclei are experimentally placed in a foreign cytoplasm. Nucleus and cytoplasm are interdependent and cell differentiation results from their interaction. We will analyze the three experimental systems which have contributed most to our knowledge in this field: (1) the unicellular alga *Acetabularia*, (2) cells fused with Sendai virus, and (3) frog (*Xenopus laevis*) oöcytes and eggs injected with somatic nuclei. In the final part of the chapter we will analyze the level at which differential gene expression is controlled in early development and we will examine the possible role that is played in this process by

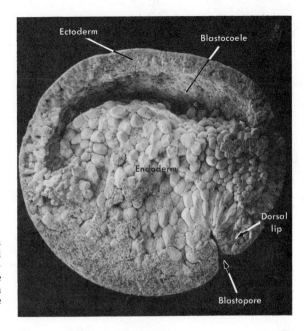

Figure 26-2 Scanning electron micrograph of a frog embryo during gastrulation. The dorsal lip of the blastopore is also called the *organizer*, because the invaginated cells produce a diffusible substance (of unknown chemical nature) which induces the ectoderm cells to differentiate into the neural cord. ×65. (Courtesy of J. Herkovits.)

"determinants," i.e., molecules localized in specific regions of the egg cytoplasm. Whenever possible, we will refer to experimental work using *Xenopus laevis,* for it is in this animal that the biochemistry of early development is best understood.

26–1 GENERAL CHARACTERISTICS OF CELL DIFFERENTIATION

The Differentiated State is Stable

One of the principal characteristics of cell differentiation in higher cells is that once established, the differentiated state is very stable and can persist throughout many cell generations. For example, a neuron will persist as such throughout the lifetime of an individual, and a cell committed to become a skin cell will gradually keratinize and eventually die. These persistent changes are very different from the type of regulation involved in enzyme induction and repression in bacteria, which is specially designed to respond rapidly to changes in the environment. As seen in the previous chapter, *E. coli* responds to the addition of lactose to the culture medium by synthesizing β-galactosidase a few minutes later, and when lactose is exhausted, enzyme synthesis is stopped with equal speed.

Cell differentiation is induced in the organism by various stimuli, but once it has been established, it can persist even in the absence of the initial stimulus. This is demonstrated, for example, by the fact that differentiated cloned cell lines, such as steroid-secreting cell lines and neuroblastoma cell lines (which differentiate into neurons), are able to grow indefinitely in vitro.[1, 2] Without this type of data it could have been argued that differentiation persists in the organism because the stimulus that has given rise to it also persists. In vitro, although such a stimulus cannot be present, the differentiated state is maintained over many cell generations.

Determination can Precede Morphological Differentiation

In many cases, before it is possible to recognize morphologically that a cell has differentiated, there is a period during which the cell is already committed to a particular change. In this interval, called *determination,* the cell has already focused its capacity for differentiation on a specific pathway.

One of the classic examples of determination is provided by the *imaginal discs* of *Drosophila.* The discs are groups of cells which are present in the larva in an undifferentiated form, but which upon metamorphosis will give rise to legs, wings, antennae, and so forth. Each disc is predetermined to become a particular type of adult structure. This commitment is very stable and heritable, as shown by the disc transplantation experiments carried out by Hadorn. Discs transplanted into the abdomen of adult flies remained undifferentiated but could be stimulated to differentiate by transplantation into a larva where the hormonal conditions would induce metamorphosis. Discs transplanted serially in the abdomens of adult flies for 9 years (1800 cell generations) were still able, after metamorphosis, to differentiate into structures belonging to the original disc.

26–2 NUCLEOCYTOPLASMIC INTERACTIONS

Nucleus and cytoplasm are interdependent; one cannot survive without the other. The cytoplasm provides most of the energy for the cell through oxidative phosphorylation (in mitochondria) and anaerobic glycolysis, and the cytoplasmic ribosomes contain most of the "machinery" for protein synthesis. On the other hand, the nucleus provides templates for specific synthesis (mRNA) and also supplies the other important RNA molecules (rRNA and tRNA). Any discussion of nucleocytoplasmic interrelations must consider (1) the mechanisms by which the genes contained in the chromosomes exert their control on the metabolic processes of the cytoplasm and (2) the mechanisms by which the cytoplasm influences gene activity.

Nuclear control of the cytoplasm was first considered during the last century by Balbiani in the so-called *merotomy* experiments (Gr., *meros,* part), in which protozoa were enucleated and studied. Such enucleated fragments are able to sustain most cellular activities; e.g., they can form a cellulose membrane and carry on photosynthesis (plant cells), react to stimuli and ingest food (amebae), activate cilia (ciliated cells), undergo cytoplasmic streaming, and so forth. However, these cells generally survive only a short time and are incapable of growth and reproduction. Within 5 to 10 minutes after being enucleated by micromanipulation[3, 4] an ameba loses its surface tension, becomes spheroid, and develops numerous blunt pseudopodia. Its movements are slowed down and it can no longer digest foods. In this state it may survive for about 20 days. If a nucleus is successfully implanted within three days, the ameba becomes extended, starts to move normally, and digests foods. Finally, it may even divide and eventually produce a culture mass.[1]

Acetabularia is Well Suited for the Study of Nucleocytoplasmic Interactions

The unicellular marine alga *Acetabularia* has been invaluable for the analysis of nucleocytoplasmic interactions. This giant cell has a stem between 3 and 5 cm long and a cap 1 cm in diameter (Fig. 26–3). The unique nucleus is located in the basal or *rhizoid* end of the cell. The great experimental advantage of *Acetabularia* is that the nucleus can be removed simply by cutting off the rhizoid. The resulting enucleated cells are active in photosynthesis and can survive for many weeks.[5]

Each *Acetabularia* species has a particular cap morphology (Fig. 26–3). If a nucleus from *A. crenulata* is implanted into an enucleated *A. mediterranea*, the resulting cell will eventually develop a cap of the *A. crenulata* type. If two nucleate cells of different species are grafted together, a hybrid having a cap with intermediate morphology is formed (Fig. 26–3). This clearly shows that the shape of the cap is determined by the nucleus.

Although dependent on the nucleus, the cytoplasm has some degree of functional autonomy. The information required to form the cap passes to the cytoplasm before cap formation, where it can be stored in an inactive form. *A. crenulata* does not form a cap in the darkness, only an elongated stem. If such a stem is enucleated and then exposed to light, a cap will develop. Therefore, the cap formation triggered by light is not dependent on new nuclear activity, and all the necessary information is present in the cytoplasm but is not converted to the final active form until a light stimulus is present. The information for cap production is very stable; enucleated stalks maintained in the dark for weeks are still able to produce a cap upon illumination. This experimental result is probably due to the presence of stable mRNA molecules stored in the cytoplasm. Stable mRNAs are known to occur in other eukaryotic cells, for example, in the mammalian reticulocyte, which continues to produce hemoglobin for several days after the nucleus is eliminated in the course of normal erythropoiesis.[6]

The nucleus itself is dependent on the cytoplasm. *Acetabularia* has a giant nucleus in its rhizoid, but its volume decreases markedly when cytoplasmic energy production is impaired by darkness or by amputation of a large part of the cytoplasm.

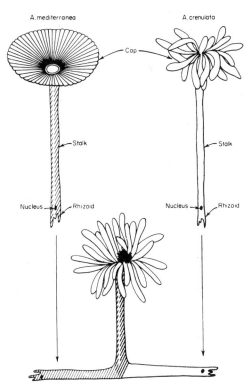

Figure 26–3 Experiment in nuclear grafting between two species of the unicellular alga *Acetabularia*. (See the description in the text.) The type of cap formed appears to depend on substances synthesized by the implanted nucleus. (From Hämmerling, J., *Ann. Rev. Plant Physiol.*, 1963.)

Red Blood Cell Nuclei can be Reactivated by Cell Fusion

The fusion of cells by using inactivated *Sendai virus* (a member of the parainfluenza viruses) provides the opportunity of placing a nucleus in a different cytoplasmic environment. Cell fusion can also be induced with other agents that affect membrane structure, such as polyethyleneglycol and lysolecithin. When different cells are fused, initially a *heterokaryon* is formed (i.e., a single cell containing nuclei of two types), as seen in Figure 26–4. Eventually, both nuclei might enter mitosis synchronously, form a single metaphase plate, divide, and produce a *hybrid cell line* (also known as a *synkaryon*), as shown in Figure 26–4. The cells of a hybrid cell line have a single nucleus containing chromosomes of both parental nuclei.

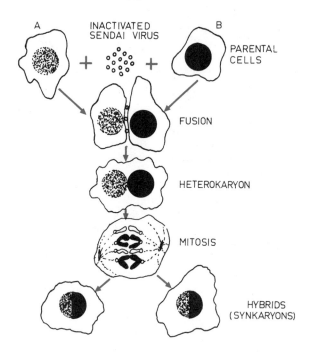

A   INACTIVATED   B
     SENDAI VIRUS

PARENTAL CELLS

FUSION

HETEROKARYON

MITOSIS

HYBRIDS (SYNKARYONS)

Figure 26–4 Cell fusion with Sendai virus results first in the production of a *heterokaryon* with two nuclei, one from each parent. A hybrid cell line (synkaryon) arises when the two nuclei undergo mitosis synchronously. The hybrid cell line has one nucleus which contains chromosomes from both parents, and is capable of multiplying in cell culture. (Modified from Ringertz, N. R., and Savage, R. E., *Cell Hybrids,* Academic Press, Inc., New York, 1976.)

In 1965 Henry Harris[7] found that chick erythrocyte nuclei are reactivated when fused to HeLa cells (an undifferentiated human cell line so named because it was derived from the uterine carcinoma cells of a patient named *He*nrietta *La*cks). These heterokaryons are of interest because the erythrocyte nucleus does not synthesize RNA or DNA. Chick erythrocytes are terminally differentiated cells which have a very condensed nucleus and which are destined to die (Fig. 26–5,*inset*). (Mammalian erythrocytes normally eliminate their nuclei during red blood cell maturation, while the red cells of birds, am-

phibians, and reptiles retain the nucleus, which becomes inactivated.)

When fused to HeLa cells, the chick erythrocyte nucleus (Fig. 26–5) increases 20 times in volume, disperses its chromatin, reassumes RNA synthesis, develops a nucleolus, and eventually replicates its DNA. This reactivation process is accompanied by the uptake of large amounts of human nuclear proteins which are thought to reactivate the erythrocyte nucleus. This was shown by experiments in which heterokaryons were stained by immunofluorescence with antisera obtained from patients suf-

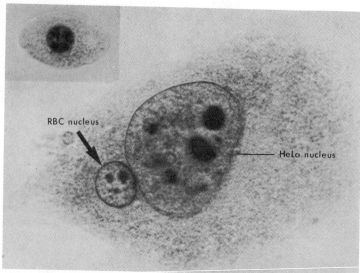

RBC nucleus

HeLa nucleus

Figure 26–5 Heterokaryon formed by fusing a chick erythrocyte with a human HeLa cell. The inset shows a normal chick erythrocyte with condensed inactive chromatin. Three days after fusion the red blood cell nucleus has enlarged, and its chromatin has dispersed. (Courtesy of N. R. Ringertz.)

fering from *lupus erythematosus*.[8] (Lupus is a human autoimmune disease characterized by the production of antibodies against one's own nuclear proteins.) After fusion with HeLa cells, the chick nuclei stained intensely with the human nuclear antibodies.

From these experiments it is clear that the synthesis of macromolecules in a nucleus is controlled by the cytoplasmic environment. Even though erythrocytes are terminally differentiated cells, they can reassume RNA and DNA synthesis. A similar result is obtained with macrophage–HeLa cell heterokaryons. Rabbit macrophages do not synthesize DNA but will do so if fused to HeLa cells.[7]

Cytoplasts and Karyoplasts can be Prepared by Cytochalasin B Enucleation

Only the cell cytoplasm is required to reactivate the chick erythrocyte nucleus. This was established by fusing erythrocytes to enucleated HeLa cells.[9, 10] Populations of enucleated cultured cells can be obtained with the use of *cytochalasin B*, a drug that affects predominantly actin microfilaments (see Chapter 9–4). Cells exposed to this drug cease moving and extrude their nuclei, which remain attached to the rest of the cell by only a cytoplasmic stalk.[9] Figure 26–6 shows the method used to obtain large numbers of enucleated cells.[11] Cultured cells that have been grown attached to a plastic surface are centrifuged in the presence of cytochalasin B, and the nuclei form a pellet, while the cytoplasm of the cells remains attached. The detached nucleus is surrounded by the cell membrane and a small amount of cytoplasm and is called a *karyoplast*. The enucleated cytoplasm is called a *cytoplast*.

Enucleated cells are viable for at least two days after enucleation and perform many cell functions normally. For example, cell movements, pinocytosis, and contact inhibition are unaffected, and if cells are detached from the plastic surface by trypsinization, they can reattach and spread on a new Petri dish.[12] Clearly, many cytoplasmic functions are independent of the cell nucleus.

Since karyoplasts are still surrounded by the cell membrane it is possible to fuse them to a different cytoplasm by using Sendai virus.[13] *Reconstituted cells*, which arise from the fusion of a karyoplast and a cytoplast, can survive longer than enucleated cytoplasm, can synthesize RNA, and in some cases can undergo cell division.[13, 14]

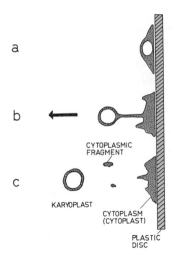

Figure 26–6 Diagram showing the enucleation of cells adhering to plastic disks. **a**, at the beginning of the experiment the cells are attached to the plastic surface. **b**, the nuclei are pulled out into protrusions by the combined action of cytochalasin and centrifugal force. **c**, the stalks connecting the nuclei are broken by centrifugation, and karyoplasts (nuclei surrounded by a rim of cytoplasm) form a pellet, while the cytoplasts (enucleated cells) remain attached to the plastic disk. (Courtesy of N. R. Ringertz, from Ringertz, N. and Savage, R., *Cell Hybrids*, Academic Press, Inc., New York, 1976.)

Figure 26–7 shows the experimental design used in the first cell reconstruction experiment.[13] Karyoplasts and cytoplasts from L cells (a mouse cell line) were prepared by centrifugation in the presence of cytochalasin B and were fused with Sendai virus. To distinguish true reconstituted cells from cells that failed to enucleate (karyoplast preparations are always contaminated by small numbers of intact cells), the labeling procedure shown in Figure 26–7 was adopted. The nucleus was labeled with ^{3}H-thymidine, and the cytoplasm, with latex beads of various diameters. (Cultured cells incubated in the presence of latex spheres incorporate the beads by phagocytosis, thus providing a very useful cytoplasmic marker.)

Cytochalasin enucleation and cell reconstitution are very powerful techniques because they provide the opportunity of placing a nucleus (containing very little of its own cytoplasm) in an entirely new cytoplasmic environment.

In a few cases it has been possible to activate latent genes coding for specific proteins by fusion to differentiated cells. For example, albumin-secreting rat hepatoma cells fused with mouse lymphocytes (which do not produce albumin) will occasionally give rise to clones of hybrid cells (synkaryons) which are able to pro-

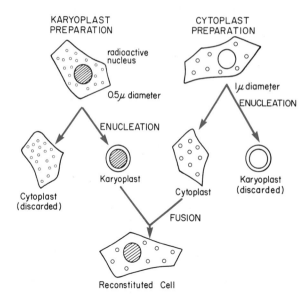

Figure 26-7 Diagram representing the reconstitution of a cell from karyoplasts and cytoplasts (see Fig. 26-4.) Karyoplasts derived from cells in which the nucleus was labeled with ^{3}H-thymidine and the cytoplasm was labeled with small latex spheres (0.5 μm in diameter). The cytoplasts came from cells in which the cytoplasm had incorporated large latex spheres (of 1 μm). Fusion resulted in cells with a labeled nucleus and with large spheres in the cytoplasm. (Courtesy of D. M. Prescott, from Veomett, G., Prescott, D., Shay, J., and Porter, K., *Proc. Natl. Acad. Sci. USA, 71*:1999, 1974.)

duce both rat and mouse albumin.[15] Somatic cell hybridization experiments show that the patterns of nucleic acid synthesis and gene expression by a nucleus can be modified by substances present in the cell cytoplasm. Similar conclusions can be drawn from work with nuclei transplanted into *Xenopus laevis* oöcytes.

Cell Fusion Yields Pure Antibodies of Medical Importance

When animals are injected with a macromolecule of a shape that is recognized as foreign to that individual (e.g., a protein or complex polysaccharide), *antibodies* appear in the serum several days later. Antibodies are protein molecules that are able to bind specifically to the foreign *antigen*. Many different antibodies appear in the serum of an immunized animal, each one recognizing a different part of the antigen's shape (or different antigens if more than one was injected). Furthermore, individuals of the same species have different immunological responses, so that two antisera directed against the same antigen can in fact be very different. This is a major problem in medicine because, for example, success or failure of an organ transplantation depends on whether the donor and the recipient patient have the correct *histocompatibility antigens* on the cell surface. Clearly, antibodies that can be used throughout the world as standardized diagnostic reagents are highly desirable. This has now been made possible by using cell fusion techniques to create cell lines that produce only a single type of antibody, which can then be obtained in unlimited amounts.

Antibodies are produced by the lymphocytes. (When producing large amounts of antibody, lymphocytes adopt a special morphology and are also called *plasma cells*.) *Each antibody-producing cell can synthesize only one type of antibody.* Lymphocytes do not multiply in culture, but C. Milstein has developed a technique whereby a *single* antibody-producing cell can be propagated indefinitely in culture by hybridization with a tumor cell.

Figure 26-8 shows the way in which these pure "monoclonal" antibodies are made. Initially a mouse is immunized against the desired antigen. By using Sendai virus, the lymphocytes from the spleen of this mouse are then fused to a mouse cell line derived from plasma cell tumors (called *myelomas* because they invade the bone marrow). The myeloma cell carries a mutation in the enzyme *hypoxanthine-guanine phosphoribosyl transferase* and is therefore unable to grow in HAT medium, as explained in Figure 20-14. However, the resulting hybrid cells are able to grow in HAT medium (Fig. 26-8), and the clones that secrete the desired antibody have been identified and grown. The fusion with the tumor cell "immortalizes" the spleen lymphocyte, which can then be grown indefinitely in culture or injected into mice, where the cells produce secreting tumors that can be maintained by serial transplantation. Since all of the cells of one clone are derived from a single lymphocyte, a *monoclonal antibody* of high purity is produced.

Monoclonal antibodies are currently used as anti-viral antibodies (e.g., for diagnosis of influenza virus type and for the treatment of rabies), as diagnostic reagents in clinical bio-

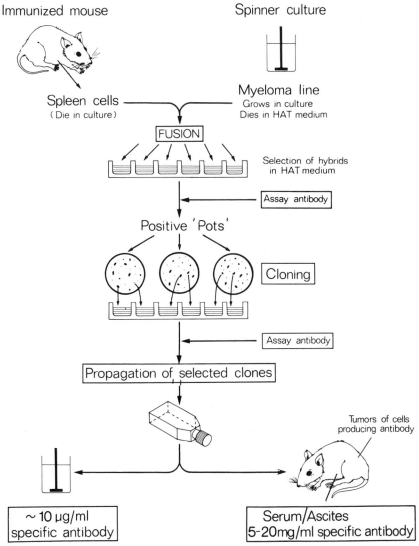

Figure 26–8 Production of a monoclonal antibody. The hybrid cell line produces only one type of antibody because a single colony (a clone) of cells is propagated. The resulting hybrid cell line can be grown in suspension (spinner) culture, or as a tumor in mice. (Courtesy of C. Milstein.)

chemistry, and for the classification of human histocompatibility antigens. Monoclonal antibodies provide a good example of the fact that the practical applications of science cannot be predicted beforehand. Cell fusion was initially developed to study the seemingly specialized cell biology problem of nucleocytoplasmic interactions but eventually produced immunological tools of practical importance to all mankind.

Gene Expression by Somatic Nuclei is Reprogrammed in Xenopus Oöcytes

Oöcytes used for experiments in differentiation are growing egg cells originally located in the ovary of the female frogs. Each oöcyte re-

quires three months or more to grow into a very large cell 1.2 mm in diameter. During this period the oöcyte accumulates proteins for use during early embryogenesis, and its nucleus (also called the *germinal vesicle*) is very active in RNA synthesis. This high transcriptional activity can be visualized morphologically in the form of lampbrush chromosomes that have lateral loops packed with nascent RNA molecules (Fig. 25–14). During this growth period, the oöcyte nucleus is arrested at the diplotene stage of meiotic prophase and therefore does not synthesize DNA. The stimulus to continue meiosis is provided by gonadotrophic hormones released from the frog's pituitary gland. In the laboratory this stimulus can be mimicked by the injection of mammalian hormones (or urine from a pregnant woman, thus providing a test that was much used for the detection of pregnancy). Hormone treatment rapidly induces maturation; the oöcytes are released from the ovary, progress through the first meiotic division while traversing the oviduct, and 15 hours later are laid as *eggs,* which are at meiotic metaphase II and ready to start development upon fertilization.

If nuclei are transplanted into oöcytes, the pattern of nucleic acid synthesis closely resembles that of the host cell. In the experiment shown in Figure 26–9 brain nuclei (which synthesize RNA but not DNA) were injected into oöcytes at different stages of their life cycle.[16] If injected into growing oöcytes (which synthesize RNA but not DNA) (Fig. 26–9), the brain nuclei incorporate ^{3}H-uridine but not ^{3}H-thymidine in radioautographic studies. When nuclei are injected into maturing oöcytes isolated from the oviduct after hormone stimulation (such oöcytes are undergoing meiotic division I and therefore have condensed chromosomes), the chromosomes become condensed and mitotic spindles are formed. If brain nuclei are injected into eggs (which do not synthesize RNA but actively synthesize DNA after penetration of sperm or the microinjection needle), they do not synthesize RNA, but will start replicating their DNA (Fig. 26–9).

The oöcyte cytoplasm not only affects the pattern of macromolecular synthesis but is also able to *reprogram* the expression of individual genes in transplanted nuclei. Figure 26–10 shows *Xenopus* oöcytes that were injected with a suspension of about 200 HeLa nuclei. The nuclei survive for several weeks inside the oöcytes and in that period synthesize substantial amounts of RNA.[17] In the first few days after injection the somatic nuclei enlarge to about ten times their original volume (Fig. 26–10, *A* and *D*). The injected material can also be introduced into the oöcyte nucleus (Fig. 26–10, *B*); in this case the somatic nuclei swell to about 100 times their initial volume. These dramatic changes are accompanied by chromatin dispersion and a mas-

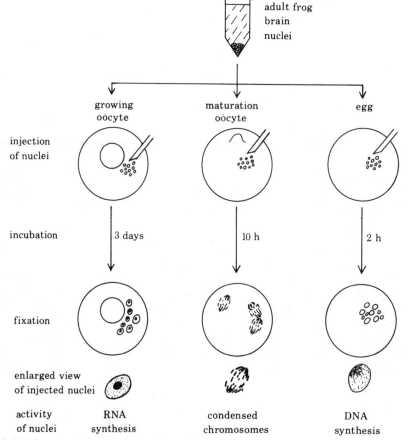

Figure 26–9 The cytoplasm controls nuclear activity. A suspension of adult brain nuclei was injected into oöcytes or eggs at different stages of maturation. The pattern of nucleic acid synthesis of the injected nuclei resembles that of the host cell. (Courtesy of J. B. Gurdon.)

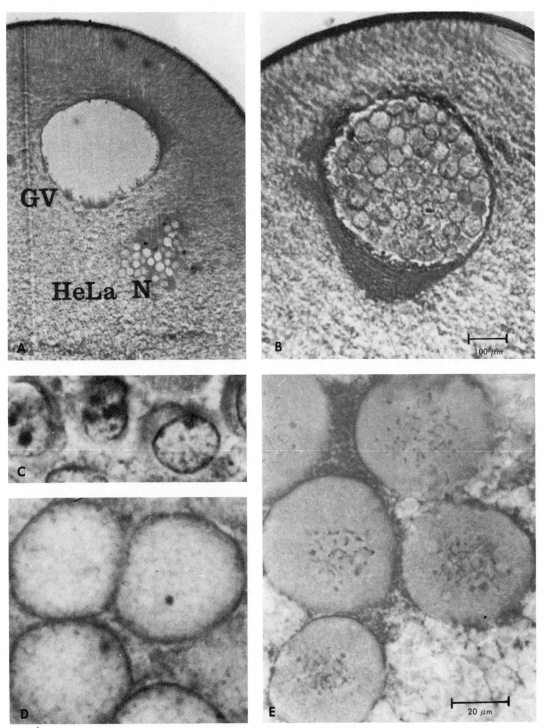

Figure 26–10 *Xenopus* oöcytes injected with human HeLa nuclei. *GV,* germinal vesicle or nucleus of the oöcyte. **A,** HeLa nuclei injected into the cytoplasm three days previously. **B,** HeLa nuclei injected into the oöcyte nucleus enlarge further. **C,** HeLa nuclei immediately after injection into the cytoplasm. **D,** three days later, the nuclei have swollen more than 10-fold. **E,** one week after injection the HeLa chromosomes become visible, and tend to resemble the nucleus of the oöcyte, which is in meiotic prophase (late lampbrush stage). (From E. M. De Robertis.)

sive uptake of proteins from the surrounding cytoplasm. The injected nuclei resemble the oöcyte's own nucleus morphologically,[18] and occasionally the human chromosomes form structures reminiscent of lampbrush chromosomes (Fig. 26–10, *E*).

The resemblance is not only morphological. The oöcyte cytoplasm reprograms the gene expression of the injected nuclei in such a way that only those genes which are normally active in oöcytes are expressed. This can be determined by taking advantage of the fact that the RNAs synthesized by the injected nuclei in the course of several days accumulate in the cytoplasm, where they code for new proteins.[19] De Robertis and Gurdon injected *Xenopus* cultured kidney cell nuclei into oöcytes of a different amphibian species (the salamander *Pleurodeles*, whose proteins can be readily distinguished from those of *Xenopus* by two-dimensional elec-

trophoresis). Those genes that are normally expressed in kidney cells but not in oöcytes became inactive after injection into *Pleurodeles* oöcytes. More importantly, some oöcyte-active genes (which were not expressed by the kidney cell nuclei) were activated by the oöcyte cytoplasm. It is therefore clear that the oöcyte contains components which can determine that a particular spectrum of protein-coding genes will be active and that others will be inactive.

The work with cell fusion and *Xenopus* oöcytes indicates that the cytoplasm of all cells contains components which determine the state of activity of nuclear genes. In the rest of the chapter we will see that cytoplasmic gene-controlling substances (or "determinants"), which sometimes become unequally distributed in the cytoplasm, play a crucial role in establishing the initial steps of cell differentiation in early embryos and adult tissues.

SUMMARY:

Nucleocytoplasmic Interactions

The problem presented by cell differentiation is to discover how an apparently structureless egg gives rise to a complex embryo having many specialized tissues.

Cell differentiation is a persistent change involving the preferential synthesis of some specific proteins (e.g., hemoglobin and gamma globulin) by the specialized cells (e.g., the red blood cell and the plasma cell). These persistent changes imply a complex mechanism of gene regulation which is different from that which operates in bacteria (i.e., induction and repression by metabolites). Cell differentiation can be induced by various stimuli, but it persists even in the absence of the initial stimulus. Prior to the overt differentiation there may be a period of *determination* during which the capacity to differentiate in a particular direction is already established. For example, the imaginal discs of *Drosophila* are pre-determined but differentiate only during metamorphosis in the larva; transplanted into the adult they remain in a permanently determined but undifferentiated state.

In eukaryotic cells there is a continuous nucleocytoplasmic interrelationship. This can be demonstrated in *protozoa* that are enucleated in the so-called *merotomy* experiments. The alga *Acetabularia* is invaluable for these studies. It is a giant cell with a stem, a cap, and a rhizoid containing the nucleus. Appropriate experimentation (i.e., cutting off the rhizoid, grafting of two species, exposure to light or darkness, and so forth) demonstrates the existence of nuclear substances diffusing into the cytoplasm and controlling morphogenesis of the cap (i.e., stable mRNAs).

Cell fusion by Sendai virus allows a nucleus to be placed in a different cytoplasm (i.e., a single cell with two nuclei, called a *heterokaryon*). Cell nuclei that no longer synthesize RNA or DNA (e.g., chicken erythrocytes) increase in volume, resume RNA synthesis, and may replicate their DNA when fused with HeLa cells. Reactivation is probably due to the entrance of proteins from the cytoplasm. *Karyoplasts* and *cytoplasts* may be obtained by the action of cytochalasin B, and can be reconstituted into cells by Sendai virus fusion, in which the nucleus and the cytoplasm come from different cells.

Frog oöcytes and eggs are most useful for the study of differentiation. Growing oöcytes have very active RNA synthesis, and eggs actively synthesize DNA. Nuclei from different tissues (e.g., brain) may be transplanted into oöcytes or eggs. When injected into growing oöcytes, the foreign nucleus undergoes stimulation of RNA metabolism, while in eggs brain nuclei will start replicating DNA. Thus the pattern of nucleic acid synthesis depends on the host cytoplasm.

The oöcyte cytoplasm is able to reprogram the expression of genes in the transplanted nuclei. This was shown by injecting kidney nuclei of one species (*Xenopus*) into oöcytes of the salamander *Pleurodeles;* it was found that there was selective activation and inactivation of certain genes.

26-3 MOLECULAR MECHANISMS OF CELL DIFFERENTIATION

We will now analyze at which level cell differentiation is controlled and we will examine the experimental evidence indicating that *the genome remains constant throughout cell differentiation.* Specialized cells from different tissues contain the same kind and number of genes, and therefore the differences between specialized cells must be explained in terms of variable rates of gene expression. We will arrive at the conclusion that cell differentiation is most probably controlled at the transcriptional level.

The Genome Remains Constant During Cell Differentiation — Nuclear Transplantation

Much attention was devoted by nineteenth century embryologists to the possibility that cell differentiation could be due to the loss of those genes which are not expressed.[20] However, massive loss of chromosomal material cannot be the cause of cell differentiation, since we know that the DNA content of diploid cells is constant in all tissues of an organism[21] (Chap. 15).

Rigorous proof that during embryonic differentiation there is no loss of genetic information came from the classical nuclear transplantation experiments performed by J. B. Gurdon. As shown in Figure 26–11, *Xenopus laevis* unfertilized eggs can be irradiated with ultraviolet light to destroy the endogenous nucleus and can then be injected with a single *Xenopus* diploid nucleus. Nuclei obtained from *Xenopus* tadpole intestinal cells (which are clearly differentiated cells, since they have a "brush border" of microvilli) are able to sustain development of normal adult frogs[22] which are fertile.[23] This demonstrates that the intestine cells retained all the genes required for the complete life cycle of a frog. In addition, since many frogs can be obtained from the intestine of the same tadpole, nuclear transplantation allows a

clone of genetically identical twins to be created. Development up to the stage at which the tadpole swims has been obtained using a variety of adult tissues, such as keratinizing skin cells[24] and lymphocytes,[25] thus demonstrating that the genes required to make nerve, blood, muscle, cartilage, and other tadpole tissues were not irreversibly inactivated in the donor nuclei.

Similar conclusions have been obtained using plant cells. It is common knowledge that whole plants can be grown from cuttings. In some cases, such as in carrots, a complete plant can be grown from a single cultured cell.[26]

Gene Amplification Cannot Explain Cell Differentiation

One way of obtaining differential gene activity would be to increase the number of copies of a given gene. There are two circumstances in which *gene amplification* is known to occur. One is ribosomal DNA amplification observed in the oöcytes of amphibians and insects. As seen in Chapter 23, *Xenopus* oöcytes selectively replicate their rDNA genes; a mature oöcyte has 2 million copies of them (compared to 900 for a diploid somatic nucleus) and 1000 nucleoli in order to produce the vast amount of ribosomes (10^{12}) contained in a single egg (see Table 23–4). The other case of amplification occurs in the DNA replication observed in certain puffs of *Rhyncosciara* salivary gland polytene chromosomes[27] (Chap. 25). However, *in both cases the specially amplified DNA is not passed on to future cell generations.* The salivary gland cells die at metamorphosis, and the amplified oöcyte rDNA is not inherited by the embryo.

Since some differentiated cells produce large amounts of certain gene products (90 per cent of the protein synthesized in a reticulocyte is globin), it seemed possible that similar phenomena could occur in other genes. However, nucleic acid hybridization experiments have

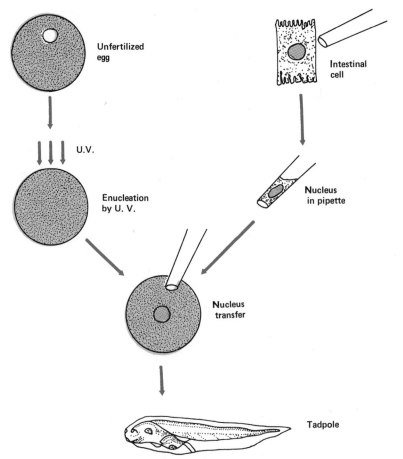

Figure 26–11 Schematic diagram of the experiment on nuclear transplantation in an unfertilized egg of *Xenopus laevis*. The egg is treated with ultraviolet irradiation to destroy the nucleus. The nucleus of an intestinal cell is drawn into a micropipette and then implanted in the egg. The result is a normal tadpole. (From J. B. Gurdon.)

shown that there is only one copy per haploid genome of globin or chicken ovalbumin genes in all tissues, regardless of whether the gene is preferentially expressed or not. In other words, a single globin gene, when fully activated, can give rise to all the globin required by a red blood cell. Gene amplification does not seem to be a widespread phenomenon.

We have seen that differential gene expression does not result from changes in the *kind* or *number* of genes. Another possibility is that it could arise from changes in the *location* of genes in the chromosomes. Gene rearrangements and translocations occur in the case of antibody-producing cells.[28] However, the DNA sequences adjacent to globin and ovalbumin genes have been analyzed by cleaving cell DNA with restriction enzymes and were found to be the same regardless of the tissue from which the DNA was extracted.[29, 30] Similarly, no differences were found in the restriction en-

zyme fragments of *Drosophila* DNA extracted from embryonic and adult tissues (Potter and Thomas, 1977).

Cell Differentiation Cannot be Explained by Translational Control

Since *the genome remains constant throughout development*, the differences between cells must be due to differential gene activity; i.e., some genes must be expressed at a higher rate than others. This could be explained by differential *transcription* (RNA synthesis or processing) or by differential *translation* (protein synthesis) of genes. One possibility is that cells may have mechanisms by which some mRNAs are translated in a given cell type but not in others. Experiments in which mRNA is injected into living cells argue strongly against this possibility.

Xenopus oöcytes, due to their large size (1.2

mm) and great hardiness (they can be injected with one twentieth of their volume without adverse effects) are very suitable for microinjection. Figure 26–12 shows that when purified mRNA coding for rabbit globin is injected into *Xenopus* oöcytes, it is translated into globin protein.[31] A great number of mRNAs of animal and plant origin are efficiently translated in oöcytes (such as those coding for immunoglobulins, thyroglobulin, interferon, collagen, tobacco mosaic virus coat proteins, and many others). This shows that oöcytes do not have a mechanism that excludes the translation of certain mRNAs.

Globin mRNA in oöcytes is very stable and can be continually translated for weeks; translation is extremely efficient, and each mRNA molecule can give rise to up to 100,000 molecules of globin protein.[32] In fact, living oöcytes translate mRNA much more efficiently than

any cell-free system and have therefore been widely used as a system for assaying mRNA. It is interesting to note both that purified mRNA requires great care when handled in vitro to avoid degradation by ribonucleases, which are very active and ubiquitous enzymes, and that *homogenates* of *Xenopus* eggs and oöcytes have a great deal of RNAse activity. However, injected *living* oöcytes do not degrade mRNA. It is therefore clear that living cells can sometimes provide very favorable conditions for biochemical analysis (reviewed by De Robertis et al., 1977).

Although oöcytes are able to translate many different mRNAs, it could still be argued that this is because an oöcyte is a non-specialized cell and that a specialized cell might not translate heterologous mRNAs. This proposition was tested in the elegant experiment shown in Figure 26–13.[33] Rabbit globin mRNA was in-

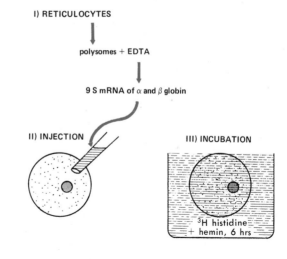

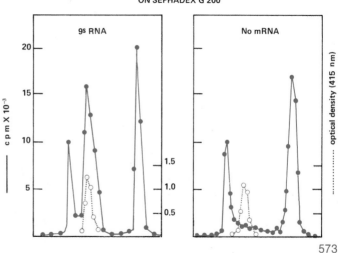

Figure 26–12 Diagram of the experiment of the injection of hemoglobin messenger RNA into an amphibian egg and demonstration of its translation into hemoglobin molecules. (See the description in the text.) (From J. B. Gurdon.)

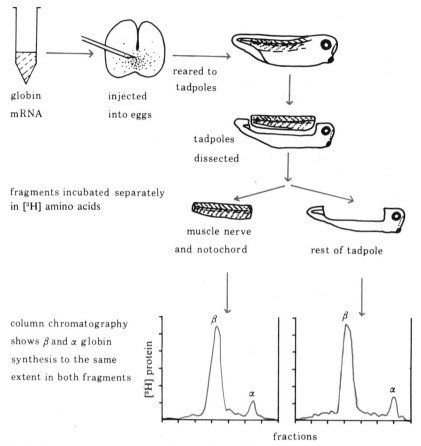

Figure 26–13 Experiment demonstrating that globin mRNA can be translated in specialized cells (muscle and nerve cells). (See the description in the text.) (From Gurdon, J. B., *Proc. R. Soc. London* [Biol.], *198*:211, 1978.)

jected into fertilized eggs, which were then allowed to develop to the tadpole stage. The tadpoles were dissected into an axial fragment (which contained almost exclusively muscle, nerve, and notochord) and the rest of the tadpole (which contained many tissues, including blood-forming cells). After labeling with amino acids, both fragments were analyzed for synthesis of rabbit globin. Since the axial fragment (which does not contain blood-forming tissues) was found to synthesize globin to the same extent as the rest of the tadpole, it can be concluded that differentiated cells (muscle, nerve, and notochord) can also translate heterologous mRNAs.

The experiments thus far described suggest that cells cannot choose which *kinds* of specific mRNAs they will translate. However, cells can regulate the *amount* of proteins they synthesize. One example of this *quantitative* regulation is seen in sea urchin eggs, which undergo a ten-fold increase in protein synthesis after fertilization. This increase is due to the transla-
574

tion of mRNAs which were already present in the unfertilized egg, since the increase also occurs if new RNA synthesis is eliminated (using actinomycin D or enucleation). Although the *amount* of protein synthesized changes, the *kinds* of proteins synthesized before and after fertilization are the same.[34] Thus, protein synthesis is subject to *quantitative* control, but does not seem to be regulated in the *qualitative* way required to explain the expression of hemoglobin in red blood cells but not in other tissues.

Cell Differentiation is Probably Controlled at the Level of Transcription

Several lines of evidence suggest that eukaryotic gene expression is regulated at the level of RNA synthesis. The clearest example is provided by polytene chromosomes in which transcription of genes can be visualized in the form of puffs. As seen in Chapter 24, specialized cells have distinct patterns of puffing (and therefore of transcription) which vary between different tissues and respond to the hormone *ecdysone*.

Molecular hybridization studies have shown that only a fraction of the total genes are expressed in any one cell (only five to ten per cent of the single-copy genes). Furthermore, different sets of genes are transcribed in each specialized cell type (reviewed by Davidson, 1976). Although many RNAs are different, many others are shared by different cell types.[35] The shared sequences represent the so-called *housekeeping genes* required for the survival of all cells, such as, for example, genes coding for membrane proteins, glycolytic enzymes, ribosomal and mitochondrial proteins, and so forth. The unshared sequences represent the so-called *luxury functions* which are characteristic of specialized cells, for example, hemoglobin in red blood cells, ovalbumin in oviduct, keratin in skin.

Some of these proteins are expressed in very large amounts, and it is possible to prepare the corresponding mRNAs in pure form. These pure mRNAs can then be used to prepare *radioactive probes* for hybridization. (Highly labeled copies of the mRNA can be prepared with the enzyme *reverse transcriptase*. Since the enzyme requires a primer, oligo dT is first hybridized to the poly A tail of the mRNA, and then a *complementary*-H³ DNA copy or cDNA is obtained.) The ³H-cDNA probe can then be used to measure the concentration of mRNA for that gene in various types of cells. For example, ovalbumin sequences are present in hen oviduct RNA but are not detectable in brain RNA. When hen oviduct is stimulated to secrete ovalbumin by the injection of estrogen, the increase in protein synthesis is paralleled by an increase in the specific mRNA.[36, 37] For all differentiated genes which have been examined, the experimental evidence is consistent with the view that *production of specialized proteins is due to differential gene transcription*.

Understanding the way in which eukaryotic transcription is regulated is of great importance in cell biology. Unfortunately, in eukaryotes we lack the sophisticated genetics which led to discovering how the *lac* operon works in *E. coli* (Chap. 25). One way in which progress might be made is provided by *Xenopus* oöcytes microinjected with cloned eukaryotic DNAs (see De Robertis et al., 1977; Gurdon et al., 1979). DNA injected into *Xenopus* oöcytes is transcribed only when introduced into the oöcyte nucleus[38] (see Fig. 26–10,*B*). DNA injected into the cytoplasm is not transcribed and is rapidly degraded. DNA injected into the nucleus seems to be transcribed accurately (see Fig. 21–12) and to associate with histones to form chromatin, and in some cases the mRNAs synthesized are translated into the correct protein products.[39] For example, Figure 26–14 shows the synthesis of an SV40 protein in oöcytes injected with DNA from SV40 virus. It is possible that living oöcytes injected with various purified eukaryotic DNAs (or with somatic nuclei) might provide a good assay system for factors (such as proteins or RNAs) presumed to be of importance in controlling chromatin transcription.

Cytoplasmic Determinants Localized in Egg Cytoplasm are Important in Early Development

We will now examine how the first differences appear between cells of the early embryo. As we have seen in the first part of this

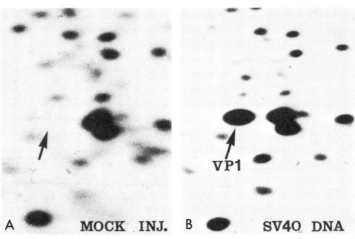

Figure 26–14 Radioactive proteins synthesized by *Xenopus* oöcytes injected into the nucleus with DNA of SV40 virus. The DNA was transcribed, and the mRNA produced was translated into the viral protein 1 (VP1), which is absent from mock injected controls. (From E. M. De Robertis.)

A   MOCK INJ.   B   SV40 DNA

chapter, the cytoplasm of oöcytes and other cells contains molecules that influence the activity of the nucleus. Eggs have in their cytoplasm such substances, and in some cases they are localized in specific regions of the egg. As development proceeds, these substances, which are called *determinants* of development, are unequally distributed into the cytoplasm of certain groups of cells, which become committed to a particular type of cell differentiation.

Eggs are in general very large cells which stockpile many of the molecules required for early development. For example, a *Xenopus* egg contains about 100,000 times more RNA polymerases, histones, mitochondria, and ribosomes than a normal adult *Xenopus* somatic cell.[40] One reason for accumulating these ready-made materials rather than making them *de novo* during early embryogenesis is the extraordinarily rapid rate of cell division during *cleavage*. At the mid-blastula stage, *Xenopus* replicates its DNA every 10 minutes, a process that takes 19 hours in the somatic cells of adults. Similarly, *Drosophila* blastulae double their DNA content in 3.4 minutes.[41] As shown for both amphibia[42] and *Drosophila*,[41] this very short S phase is due to an increase in the number of replication forks (Chap. 16).

This rapid pace allows little time for new RNA and protein synthesis, but it is during this period that the first differences between nuclei are established. At least in some cases these differences result from the presence of *determinants* synthesized during oögenesis and localized in some regions of the egg cytoplasm.

The best example of *determinants* in development is provided by the *germ plasm* (Fig. 26–15,C). Amphibian eggs contain in their vegetal (yolky) pole a specialized region of cytoplasm which can be recognized morphologically by the presence of special granules. *Drosophila* eggs also have an equivalent region located in the posterior pole of the egg (which is therefore called the *pole-plasm*). This cytoplasm has the property of inducing germ cell formation, i.e., those cells that contain the germ plasm will eventually become the germ cells of the new organism. When the posterior poles of eggs are irradiated with ultraviolet light, sterile (but otherwise normal) animals are obtained. However, if UV-treated eggs are injected with pole-plasm of normal eggs, fertile flies are obtained.[43] Furthermore, if cytoplasm containing the germ cell determinants is injected into the anterior part of a *Drosophila* egg, germ cells develop in an anterior position.[44]

Although determinants are undoubtedly very important in establishing early differences between cells, they cannot entirely explain development. For example, there is no evidence of cytoplasmic localization in mammalian eggs. Furthermore, as development advances, *cell interactions* become increasingly important. At gastrulation, extensive cell movements and migrations occur, and different types of cells interact with each other in the phenomenon known as *embryonic induction*. Notochord tissue induces the overlying ectoderm to become neural tissue, and the optic vesicles (an outgrowth of the brain) induce nearby ectoderm to become eye lens. These inductions are mediated by diffusible substances, but in spite of numerous attempts to isolate them, their chemical nature remains unknown.

It is possible that the same principles involved in the action of egg determinants could also apply to *adult cells* (reviewed by Gurdon, 1978). In the initial part of this chapter we saw that all cells contain in their cytoplasm molecules that can reprogram nuclear gene expression. Cells continually exchange information between nucleus and cytoplasm, so that gene products accumulated in the cytoplasm can subsequently modify nuclear activity. If these substances are localized in certain regions of the cytoplasm, upon cell division they can become unequally distributed between daughter cells, giving rise to two different cell types.

Figure 26–15 shows how this could occur in adult and embryonic cells. In differentiation of adult tissues it is frequently observed that only one of the daughter cells becomes specialized; the other one remains as a *stem cell*, which is able to divide again. This occurs in red blood cell differentiation and could occur in skin and intestinal epithelium, in which the dividing cells are located in certain regions of the epithelium (attached to the basal membrane or at the bottom of the intestinal crypts).

The hypothetical mechanism shown in Figure 26–15,B is supported by experimental evidence. During nerve cell differentiation in grasshoppers, some cell divisions result in the formation of a neuron (ganglion cell) and a stem cell (neuroblast) which are always in the same position and morphologically recognizable. By introducing a needle at mitosis, it is possible[45] to rotate the spindle and chromosomes 180 degrees; but in spite of this the resulting daughter cells still have the neuron and stem cell in the normal position. This shows that the ability to become a neuron does not depend on a particular chromosome set but rather on the type of cytoplasm inherited by the daughter cell.

(a) Propagation of the determined or specialized state

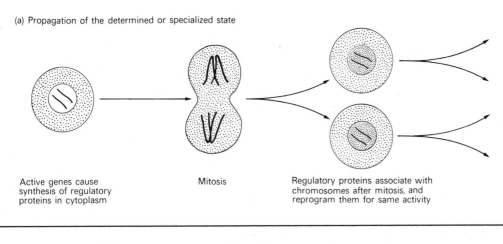

Active genes cause
synthesis of regulatory
proteins in cytoplasm

Mitosis

Regulatory proteins associate with
chromosomes after mitosis, and
reprogram them for same activity

(b) Unequal division following unequal distribution of cytoplasm

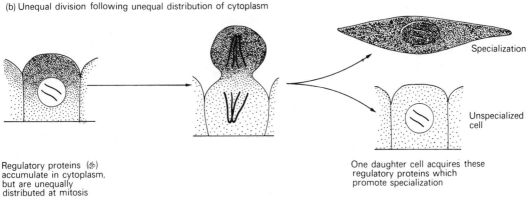

Specialization

Unspecialized
cell

Regulatory proteins (≛)
accumulate in cytoplasm,
but are unequally
distributed at mitosis

One daughter cell acquires these
regulatory proteins which
promote specialization

(c) Unequal distribution of materials in egg (e.g., germ plasm in Amphibia)

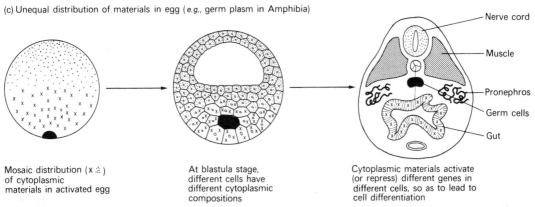

Nerve cord

Muscle

Pronephros

Germ cells

Gut

Mosaic distribution (x ⁞)
of cytoplasmic
materials in activated egg

At blastula stage,
different cells have
different cytoplasmic
compositions

Cytoplasmic materials activate
(or repress) different genes in
different cells, so as to lead to
cell differentiation

Figure 26–15 Possible mechanism for the cytoplasmic control of cell differentiation. **a,** under normal conditions the daughter cells have the same differentiated state as the parent cell. **b,** the parent's gene-regulating cytoplasmic molecules become unequally distributed, and one of the daughter cells differentiates, while the other remains as a stem cell. **c,** egg cells have unequally distributed determinants which cause embryonic cells to specialize in different ways, as in the case of the germ plasm, which induces germ cell differentiation. (Courtesy of J. B. Gurdon, from *Gene Expression During Cell Differentiation,* Carolina Biology Readers, Vol. 25, Carolina Biological, 1978.)

Cell Differentiation is Intimately Related to Cellular Interactions

Cell interactions are intimately related to the biology of cell differentiation. As development proceeds, cell movement and interactions become increasingly important. For example, as gastrulation, cell migrations produce an invagination of the embryo (Figs. 26–1 and 26–2), and eventually some of the invaginated cells (notochord) induce the ectoderm to become nerve tissue. In multicellular organisms the cells of the different tissues must cooperate in a harmonious way. Cells can interact with each other by way of long-range or short-range interactions. Interactions over longer distances are mediated by the diffusion of chemical substances, as in the case of the embryonic induction discussed earlier. In many cases the interaction depends on short-range action or *cell contact* (see Chapter 8–4). The importance of the recognition properties of the cell surface can be realized immediately by thinking of the development of the nervous system, in which millions of neurons must find their specific partners to establish the synaptic junctions that give rise to neuronal circuits of immense complexity.

Despite many attempts, the chemical nature of the diffusible substances involved in embryonic induction remains unknown. Long-range interactions are better understood in the case of the cellular slime mold *Dictyostelium discoideum* (Fig. 26–16). This organism lives in forests, under layers of decomposing leaves. While food is plentiful (*Dictyostelium* feeds on soil bacteria), the organism lives as unicellular amebae which are independent of each other and which multiply by mitosis. When food becomes scarce, some amebae start secreting a diffusible substance that attracts more amebae. A *nucleation center* is formed into which streams of amebae converge, until finally a *slug* or *pseudoplasmodium* is formed. The slug contains millions of cells and is able to migrate for considerable distances (towards light, where the spores will have a better chance to disperse). Eventually the slug differentiates, producing a stalk on top of which spores develop. By sporulating, *Dictyostelium* ensures survival during periods of low food supply.

The *Dictyostelium* life cycle has two distinct stages, one of unicellular life and a multicellular stage in which millions of cells move and differentiate in a coordinated way. This whole process is triggered by a substance that attracts the cells together. The attractant has been identified and is cyclic AMP.[46] Many cells use cyclic AMP as an intracellular signal to regulate metabolism (see Chapter 6). Cellular slime molds use cyclic AMP as an *extracellular messenger* to communicate positional information to distant cells.

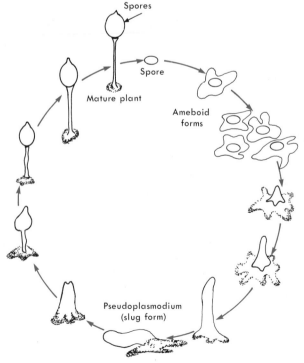

Spores

Spore

Mature plant

Ameboid forms

Pseudoplasmodium (slug form)

Figure 26–16 Life cycle of *Dictyostelium discoideum*. The slime mold can exist as single-cell amebae which aggregate when food is scarce, attracted by cAMP. The slug (pseudoplasmodium) thus formed migrates and eventually sporulates.

SUMMARY:

Mechanisms of Cell Differentiation

Experimental evidence indicates that the genome remains constant throughout cell differentiation; in other words, during differentiation *there is no loss* of genetic information. By transplanting a diploid nucleus from an intestinal cell into an egg in which the nucleus was previously destroyed, it was possible to obtain the development of a normal frog. This indicates that the donor nucleus had all the genes required for the complete life cycle of a frog. Since the genome remains constant, the differences between cells must be due to *differential transcription* or *translation* (protein synthesis) of genes.

The results of experiments in which foreign mRNAs have been injected into *Xenopus* oöcytes do not favor a difference in translation. For example, globin mRNA from reticulocytes is efficiently translated in oöcytes, and each molecule of mRNA produces up to 100,000 globin molecules. Other mRNAs of animal or plant origin are similarly translated. Translation of the heterologous mRNA is found not only in the oöcyte but also in specialized tissues. While a cell apparently cannot choose which kind of mRNA it will translate, it can regulate the amount of protein that the mRNAs synthesize. For example, in sea urchins after fertilization there is a ten-fold increase in protein synthesis, due to the translation of accumulated mRNAs. Protein synthesis is thus subjected to *quantitative control* but does not seem to be regulated qualitatively.

Gene expression can be regulated at the level of *transcription* (i.e., synthesis of mRNA). This is observed clearly in polytene chromosomes with the changes in the pattern of puffing in different tissues (see Chapter 25–3).

In any one cell only about 5 to 10 per cent of the total structural genes are transcribed. These genes may be classified into the so-called *housekeeping genes* (e.g., those coding for membrane proteins and ribosomal and mitochondrial proteins), which are common to all cells, and the genes for *luxury functions,* which are active only in differentiated cells (e.g., genes for ovalbumin, keratin, hemoglobin, and so forth). Labeled DNA complementary to mRNAs (cDNA) can be prepared by the use of reverse transcriptase. The cDNA can then be used as a probe to assay by hybridization the amount of a given mRNA. All luxury genes that have been assayed showed an increased transcription in the differentiated cells.

Regulation of transcription is difficult to study in eukaryotes. Recently progress has been made by the injection of DNA into the oöcyte nucleus. In this way the DNA can be accurately transcribed and translated (see Fig. 26–4).

The egg contains *cytoplasmic determinants* that have special localizations. These determinants are unequally distributed during development and may influence early cell differentiation. During oögenesis an amphibian oöcyte accumulates many molecules and organelles to be used later in development (e.g., RNA polymerases, histones, mitochondria, and ribosomes). These include cytoplasmic determinants, such as the *germ plasm* present in the yolky pole, which induces the formation of germ cells. It is possible that in adult cells similar principles may be active in cell differentiation. For example, when cytoplasmic gene-regulating molecules become unequally distributed between daughter cells, one cell may become differentiated, while the other remains as a *stem cell* (Fig. 26–15). As development progresses, *cellular interactions* become increasingly important. For example, the induction of the neural tissue by the notochord and of the lens by the optical vesicle are due to intercellular diffusion of substances. The example of the slime mold *Dictyostelium* is very interesting because it has two distinct stages: one unicellular and the other multicellular. In this organism cyclic AMP has been identified as the messenger that conveys positional information to attract the cells to form the multicellular organism.

REFERENCES

1. Yasumura, Y., Tashjian, A. H., and Sato, G. H. (1966) *Science, 154*:1186.
2. Augusti-Tocco, G., and Sato, G. (1969) *Proc. Natl. Acad. Sci. USA, 64*:311.
3. Fronbrune, P. (1949) *LaTechnique de Micromanipulation.* Masson & Cie., Editeurs, Paris.
4. Lorch, I. J., and Danielli, J. F. (1960) *Nature, 166*:329.
5. Haemmerling, J. (1963) *Annu. Rev. Plant Physiol., 14*:65.
6. Schweiger, H. (1969) *Curr. Top. Microbiol. Immunol., 50*:1.
7. Harris, H. (1965) *Nature, 206*:583.
8. Ringertz, N. R., Carlsson, S. A., Ege, T., and Bolund, L. (1971) Proc. Natl. Acad. Sci, USA, *68*:3228.
9. Ladda, R. L., and Estensen, R. D. (1970) *Proc. Natl. Acad. Sci., 67*:1528.
10. Ege, T., Zeuthen, J., and Ringertz, N. R. (1975) *1*:65.
11. Prescott, D. M., Myerson, D., and Wallace, J. (1972) *Exp. Cell Res., 71*:480.
12. Goldman, R. D., Pollack, R., and Hopkins, N. H. (1972) *Proc. Natl. Acad. Sci. USA, 70*:750.
13. Veomett, G., Prescott, D. M., Shay, J., and Porter, K. R. (1974) *Proc. Natl. Acad. Sci. USA, 71*:1999.
14. Krondhl, U., Bols, N., Ege, T., Linder, S., and Ringertz, N. (1977) *Proc. Natl. Acad. Sci. USA, 74*:606.
15. Malawista, S. E., and Weiss, M. C. (1974) *Proc. Natl. Acad. Sci. USA, 74*:1502.
16. Gurdon, J. B. (1968) *J. Embryol. Exp. Morphol., 20*:401.
17. Gurdon, J. B., De Robertis, E. M., and Partington, G. A. (1976) *Nature, 260*:116.
18. De Robertis, E. M., Gurdon, J. B., Partington, G. A., Mertz, J. E., and Laskey, R. A. (1977) *Biochem. Soc. Symp., 42*:181.
19. De Robertis, E. M., Partington, G. A., Longthorne, R. F., and Gurdon, J. B. (1977) *J. Embryol. Exp. Morphol., 40*:199.
20. Briggs, R., and King, T. J. (1959) In: *The Cell,* Vol. 2, Chap. 13. (Branchet, J., and Mirsky, A. E., eds.) Academic Press, Inc., New York.
21. Mirsky, A. E., and Ris, H. (1949) *Nature, 163*:166.
22. Gurdon, J. B. (1962) *J. Embryol. Exp. Morphol., 10*:622.
23. Gurdon, J. B. (1968) *Sci. Am., 219*:24.
24. Gurdon, J. B., Laskey, R. A., and Reeves, O. R. (1975) *J. Embryol. Exp. Morphol., 34*:93.
25. Whabl, M. R., Brun, R. B., and Du Pasquier, L. (1975) *Science, 190*:1310.
26. Steward, F. C. (1970) *Proc. R. Soc. London* [Biol.], *175*:1.
27. Pavan, C., and Da Cunha, A. B. (1969) *Annu. Rev. Genet., 3*:425.
28. Brack, C., Hirama, M., Lenhard-Schuller, R., and Tonegawa, S. (1978) *Cell, 15*:1.
29. Jeffrys, A. J., and Flavell, R. A. (1977) *Cell, 12*:429.
30. Breathnach, R., Mandel, J. L., and Chambon, P. (1977) *Nature, 270*:314.
31. Lane, C. D., Marbaix, G., and Gurdon, J. B. (1971) *J. Molec. Biol., 61*:73.
32. Gurdon, J. B., Lingrel, J. B., and Marbaix, G. (1973) *J. Molec. Biol., 80*:539.
33. Woodland, H. R., Gurdon, J. B., and Lingrel, J. B. (1974) *Dev. Biol., 39*:134.
34. Bradhorst, B. P. (1976) *Dev. Biol., 52*:310.
35. Davidson, E. H., and Britten, R. J. (1973) *Q. Rev. Biol., 48*:565.
36. Rhoads, R. E., McKnight, G. S., and Schimke, R. T. (1973) *J. Biol. Chem., 248*:2031.
37. Harris, S. E., Rosen, J. M., Means, A. R., and O'Malley, B. M. (1975) *Biochemistry, 14*:2072.
38. Mertz, J. E., and Gurdon, J. B. (1977) *Proc. Natl. Acad. Sci. USA, 74*:1502.
39. De Robertis, E. M., and Mertz, J. E. (1977) *Cell, 12*:175.
40. Laskey, R. A. (1979) In: *A companion to Biochemistry,* Vol. 2, Chap. 5. (Bull, A. T., ed.) Longman, Inc., New York.
41. Blumenthal, A. B., Kriegstein, H. J., and Hogness, D. S. (1973) *Cold Spring Harbor Symp. Quant. Biol., 38*:205.
42. Callan, H. G. (1973) *Cold Spring Harbor Symp. Quant. Biol., 38*:195.
43. Okada, M., Kleinman, I. A., and Scheiderman, H. A. (1974) *Dev. Biol., 37*:43.
44. Ilmensee, K., and Mahowald, A. (1974) *Proc. Natl. Acad. Sci. USA, 71*:1016.
45. Carlson, J. G. (1952) *Chromosoma, 5*:199.
46. Bonner, J. T. (1976) In: *The Developmental Biology of Plants and Animals.* (Graham, C. F., and Wareing, P. F., eds.) Blackwell Scientific Publications, Ltd., Oxford.
47. Moscona, A. A. (1974) In: *The Cell Surface in Development.* (Moscona, A. A., ed.) John Wiley & Sons, Inc., New York.

ADDITIONAL READING

Britten, R. J., and Davidson, E. H. (1969) Gene regulation for higher cells: A theory. *Science, 165*:349.

Briggs, R., and King, T. J. (1959) Nucleocytoplasmic interactions in eggs and embryos. In: *The Cell,* Vol. 2, Chap. 13. J. Brachet and A. E. Mirsky, eds. Academic Press, Inc., New York.

Davidson, E. H. (1977) *Gene Activity in Early Development.* Academic Press, New York.

Dawid, I. B., and Wahli, W. (1979) Application of recombinant DNA technology to questions of developmental biology: A review. *Dev. Biol., 69*:305.

De Robertis, E. M., and Gurdon, J. B. (1977) Gene activation in somatic nuclei after injection into amphibian oöcytes. *Proc. Natl. Acad. Sci. USA, 74*:2470.

De Robertis, E. M., Laskey, R. A., and Gurdon, J. B. (1977) Injected living cells as a biochemical test tube. *Trends Biochem. Sci. 2*:250.

De Robertis, E. M., and Gurdon, J. B. (1980) Gene transplantation and the analysis of development. *Sci. Am.* In press.

Graham, C. F., and Wareing, P. F., eds. (1976) *The Developmental Biology of Plants and Animals.* W. B. Saunders Co., Philadelphia.

Gehring, W. J. (1976) Developmental genetics in Drosophila. *Annu. Rev. Genet., 10*:209.

Gurdon, J. B. (1977) Egg cytoplasm and gene control in development. *Proc. R. Soc. London* [Biol.], *198*:211. Highly recommended.

Gurdon, J. B. (1978) *Gene Expression During Cell Differentiation.* Carolina Biology Readers, Vol. 25. Carolina Biological.

Gurdon, J. B., Laskey, R. A., De Robertis, E. M., and Partington, G. A. (1978) Reprogramming of transplanted nuclei in amphibia. *Int. Rev. Cytol., Suppl. 9*:161.

Gurdon, J. B., Wyllie, A. W., and De Robertis, E. M. (1978) Transcription and translation of DNA injected into oöcytes. *Philos. Trans. R. Soc. London,* [Biol. Sci.], *283*:367.

Gurdon, J. B., Melton, D. M., and De Robertis, E. M. (1979) Genetics in an oöcyte. In: Ciba Foundation Symposium, Elsevier North-Holland, Inc., New York.

Harris, H. (1974) *Nucleus and Cytoplasm.* Clarendon Press, Oxford. Highly recommended.

Kohler, G., and Milstein, C. (1975) Continuous cultures of fused cells secreting antibody of predefined specificity. *Nature, 256*:495.

Lane, C. D. (1976) Rabbit haemoglobin from frog eggs. *Sci. Am. 235*:60.

Pollack, R. (1973) *Readings in Mammalian Cell Culture.* Cold Spring Harbor Laboratory, Cold Spring Harbor, New York.

Potter, S. S., and Thomas, C A., Jr. (1978) The two-dimensional fractionation of *Drosophila* DNA. *Cold Spring Harbor Symp. Quant. Biol., 42*:1023.

Ringertz, N. R., and Savage, R. E. (1976) *Cell Hybrids.* Academic Press, New York. Recommended for study of cell fusion.

Schneiderman, H. A. (1976) New ways to probe pattern formation and determination in insects. In: *Insect Development.* (Lawrence, P. A., ed.) Blackwell Scientifc Publications, Ltd., Oxford.

Wilson, E. B. (1928) *The Cell in Development and Heredity.* 3rd Ed. MacMillan Pub. Co., Inc., New York. The best treatment of cytoplasmic localizations in development.

Part 7

CELL AND MOLECULAR BIOLOGY OF SPECIALIZED CELLS

The multiplicity of functions of a higher organism necessitates a greater degree of specialization of those functions, and this is accomplished by specially differentiated cells. These cells can be studied in textbooks of histology.

Here, we would like to include only two examples — muscle cells and neurons — in which the structural and functional adaptations represent a maximum degree of specialization. Both cell types will be considered at the cellular and molecular level and the mechanisms underlying their function will be interpreted.

In Chapter 27 the study of striated muscle presents an extraordinary example of macromolecular machinery adapted to the work of contraction. This phenomenon can be explained as the result of the interaction both of force-generating proteins, such as actin and myosin, and of regulatory and structural proteins, such as tropomyosin, the troponins, α-actinin, and others, which are organized in a very regular pattern within the intimate structure of the myofibril. Another exciting field of study is that of the coupling between excitation and contraction, which is based on a complex intracellular conducting system which carries the action potential to the deeper portion of the muscle fiber, producing the functional synchronization of the myofibrils. In this functional coupling, calcium ions and Ca^{2+}-activated ATPase play an essential role. Muscle is an admirable example of physiologic and structural integration at the macromolecular level comparable only to that of mitochondria and chloroplasts.

Chapter 28 introduces the cellular basis of nerve conduction and synaptic transmission. Both muscle and nerve fibers conduct impulses by way of action potentials, but the neuron is also specially adapted to receive stimuli, to transmit impulses across the synapse, and to induce a response at effector cells. The ultrastructure of the axon, the biosynthetic properties of the perikaryon, and the nature of axon flow both in orthograde and retrograde directions will be emphasized in the first part of this chapter. Synaptic transmission can be electrical, but it is more frequently chemically driven. This implies a neuro-chemical mechanism and the production of transmitter substances at the nerve ending.

Chemical synapses have a complex structure, both at the membranes and within the ending, in which the main components are the synaptic vesicles, the true quantal units of the transmitter. In general terms, in chemical synapses a localized process of neurosecretion takes place that is similar to the production of other neurohumors. Synaptic transmission is the result of the interaction between the neurotransmitter and specific receptor proteins present at the neuronal membrane. This interaction produces a translocation of ions and, in some cases, the production of the secondary messengers cyclic AMP and GMP.

CELLULAR AND MOLECULAR BIOLOGY OF MUSCLE

27–1 Structure of the Striated Muscle Fiber 586
    The Myofibril and Sarcomere are Structures Differentiated for Contraction
    Thick and Thin Myofilaments are the Macromolecular Contractile Components
    The Z-Line Shows a Woven-Basket Lattice and Contains α-Actinin
    The Sarcomere I-Band Shortens and the Banding Inverts During Contraction
    Smooth Muscles Lack the Z-Line
    Summary: *The Structure of Muscle*
27–2 Molecular Organization of the Contractile System 591
    The Thick Myofilament — Myosin Molecules; The Cross-Bridges — S_1 Subunits
    Actin, Tropomyosin, and the Troponins Constitute the Thin Myofilament
    Contractile and Regulatory Proteins — Localization by Fluorescent Antibodies
27–3 The Sliding Mechanism of Muscle Contraction 593
    Myosin and Actin have a Definite Polarization within the Sarcomere
    Muscle Contraction — A Cyclic Formation and Breakdown of Actin-Myosin Linkages

    Summary: *Molecular Organization and the Sliding Mechanism*
27–4 Regulation and Energetics of Contraction 597
    Molecular Regulation — Displacement of Tropomyosin After Binding of Ca^{2+} to Troponin-C
    Energy for Contraction Comes from Oxidative Phosphorylation and Glycolysis
    Summary: *Regulation and Energetics of Contraction*
27–5 Excitation-Contraction Coupling 598
    The Sarcoplasmic Reticulum — A Longitudinal Component with Terminal Cisternae Forms Part of the Triad
    The T-System is in Continuity with the Plasma Membrane and Conducts Impulses Inward
    Stimulation Releases Ca^{2+} from the Terminal Cisternae
    A Ca^{2+}-Activated ATPase is Present in the Sarcoplasmic Reticulum and Acts as a Ca^{2+} Pump
    Summary: *Excitation-Contraction Coupling and the Sarcoplasmic Reticulum*

In Chapter 9 we studied the general motility of cells and emphasized the fact that contraction was essentially the result of the interaction between the two contractile proteins *actin* and *myosin* present in the microfilamentous cytoskeleton.

Cell contractility reaches its highest development in the various types of muscular tissues. The structural organization of muscle is adapted to unidirectional shortening during contraction. Because of this, most muscle cells are elongated and spindle-shaped. The cytoplasmic matrix is considerably differentiated, and the major part of the cytoplasm is occupied by contractile fi-

brils. In smooth muscle these *myofibrils* are homogeneous and birefringent. In contrast, in cardiac and skeletal muscle the myofibrils are striated and have dark, birefringent (anisotropic) zones alternating with clear isotropic zones (Fig. 27–1). In muscle cells only a small part of the cytoplasm — the *sarcoplasm* — retains its embryonic characteristics. It lies between the myofibrils, particularly around the nucleus.

Some muscle cells are so highly differentiated that they are adapted to produce mechanical work equivalent to 1000 times their own weight and to contract 100 or more times per second.

The different types of muscle cells are includ-

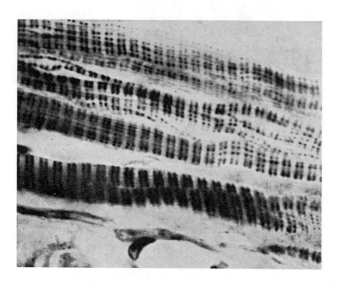

Figure 27-1 Myofibrils of a striated muscle fiber showing dark and clear bands. Iron hematoxylin stain.

ed in histology textbooks and the special types of contraction are in physiology textbooks. Here, the emphasis is on the macromolecular organization of the striated skeletal muscle and the coupling between excitation and contraction. The study of the molecular biology of muscle is one of the most rewarding examples of the intimate association between structure and function and of the way in which chemical energy is transformed into mechanical work.

27-1 STRUCTURE OF THE STRIATED MUSCLE FIBER

Striated skeletal muscles are composed of multinucleate cylindrical fibers, 10 to 100 μm in diameter and several millimeters or centimeters long. These enormous structures arise in the embryo by the fusion of several primordial cells, the so-called *myoblasts*. The entire fiber is surrounded by an electrically polarized membrane with an electrical potential of about −0.1 volt; the inner surface is negative with respect to the outer surface. This membrane, called the *sarcolemma*, becomes depolarized physiologically each time a nerve impulse that reaches the motor innervation of the muscle (*end plate*) activates the membrane. The final result is a coordinated contraction of the entire muscle fiber. Three cytoplasmic components are highly differentiated in the muscle fiber. One is represented by the contractile machinery, which is essentially made of protein myofilaments and is formed embryonically within the cytoplasmic matrix. The myofilaments are disposed in parallel to form the larger fibrillar structures, the *myofibrils* (Fig. 27–2).

The arrangement of the myofilaments deter-

mines the different classes of muscle that are now recognized. For example, in striated skeletal and heart muscle of vertebrates the filaments are longitudinally oriented, and there is also a transverse repeating organization. In other types of non-striated muscles the filaments are oriented in longitudinal or oblique arrays or have a more or less random distribution. The second component of striated muscle is a special differentiation of the endomembrane system, the so-called *sarcoplasmic reticulum,* which is involved with conduction inside the fiber and with coordination of the contractions of different myofibrils, in addition to being related to the relaxation of the muscle after a contraction.

The third component is represented by numerous mitochondria, the so-called *sarcosomes,* which in some cases may attain large dimensions. The abundance of mitochondria may be related to the constancy with which the muscle contracts; for example, there is a greater number in steadily active muscles, such as the heart.

The Myofibril and Sarcomere are Structures Differentiated for Contraction

Myofibrils are long cylindrical structures about 1 μm in diameter that have transverse striations. These striations consist of the repetition of a fundamental unit, the sarcomere, which is limited by a dense line called the Z-line or Z-disc. This line is located in the center of a less dense zone known as the I-band, which corresponds to the relatively *isotropic* disc (for a definition of *isotropy* and *anisotropy,* see Chapter 3–1) (Fig. 27–2). The A-band, which is *anisotropic* under polarized light, has a greater density

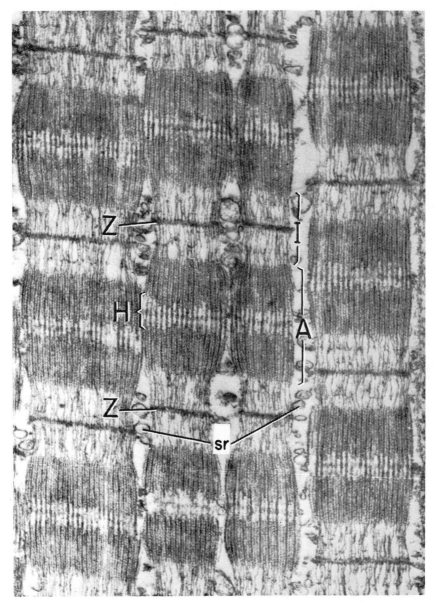

Figure 27–2 Electron micrograph of four myofibrils, showing the alternating sarcomeres with the Z lines and the H, A, and I bands; *sr,* sarcoplasmic reticulum situated between the myofibrils. The finer structure of the myofibril represented by the thin and thick myofilaments is also observed. ×60,000. (Courtesy of H. Huxley.)

than the I-band. Under certain conditions, a less dense zone may be observed in the center of the A-band, subdividing it into two dark semidiscs (Fig. 27–2). This zone consitutes the H-disc (Hensen's disc). In the middle of the H-disc an M-line can be observed.

In a relaxed mammalian muscle, the A-band is about 1.5 μm long, and the I-band, 0.8 μm. The striations of the myofibrils are the result of periodic variations in density, i.e., in concentration of the finer structures, the *myofilaments*, along the axis. These striations are in register in the different myofibrils, thus giving rise to the striation of the entire fiber.

Thick and Thin Myofilaments are the Macromolecular Contractile Components

Morphologically, there are two kinds of myofilaments, *thick myofilaments* that are about 1.5 μm long and 10 nm wide and that are separated by a 40 nm space, and *thin myofilaments* that are about 1.0 μm long and 5 nm in diameter. As will be mentioned later, thick myofilaments are made of *myosin*, and the thin myofilaments have a more complex structure containing several proteins (i.e., actin, tropomyosin, and troponins) of which *actin* is the most important.

As shown in Figure 27–3, these two types of filaments are disposed in register and overlap to an extent that depends on the degree of contraction of the sarcomere. In a relaxed condition, the I-band contains only thin filaments; the H-band contains only thick filaments; and within the A-band the thick and thin filaments overlap. In a cross section through the A-band, the regular disposition of the two types of filaments can be observed best (Fig. 27–3). In vertebrate muscle each thick filament is seen to be surrounded by six thin filaments, and each thin filament lies symmetrically among three thick ones (Fig. 27–3).[1] As a consequence of this geometry there are twice as many thin filaments as thick ones.

A cross section through the H-band shows only thick myofilaments and through the I-band, only thin myofilaments (Fig. 27–3).

Another interesting detail revealed by the electron microscope is that the two sets of filaments are linked together by a system of cross-bridges (Fig. 27–4). These arise from the thick filaments at intervals of about 7 nm. Each bridge is situated along the axis with an angular difference of 60 degrees. This means that the bridges describe a helix about every 43 nm. As a result of this arrangement one thick filament joins the six adjacent thin ones every 43 nm.

A repeat period corresponding to this distance can be observed in the A-band after staining with uranyl acetate. There are 11 stripes in each half an A-band starting from the M-line. Some of

these are due to the myosin cross-bridges, but others, to the presence of C-protein[2] (see later discussion). The complex fine structure of the A-band can also be revealed by negative staining of ultrathin sections of frozen muscle (i.e., cryosections).[3] In cross section the M-line shows that the thick myofilaments are joined, forming a triangular lattice (Fig. 27–3).

The Z-line Shows a Woven-basket Lattice and Contains α-Actinin

In cross section the Z-disc shows a woven-basket lattice which remains essentially unchanged during contraction. This lattice is made of Z-filaments which are connected to the thin filaments of the I-band. It is presumed that as one thin filament enters the Z-line, it is in continuity with three curved Z-filaments which unite it with three other thin filaments of the same sarcomere. According to this model the thin filaments of the opposite sarcomere are similarly arranged. The hexagonal lattice that is characteristic of the A-band (one thick filament surrounded by six thin ones) becomes compressed at the I-Z junction and is transformed into the square lattice characteristic of the woven-basket pattern (Fig. 27–3). This model, in which there is no interlooping of filaments from one sarcomere to the other, explains the splitting of the Z-line that may occur under certain conditions.[4] Later on we shall see that one of the main components of the Z-line is α-actinin, a protein that joins with the actin thin filament and may constitute the Z-filaments.

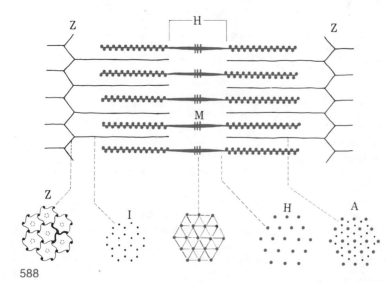

Figure 27–3 Diagram representing the structure of one sarcomere in striated muscle in longitudinal and transverse sections. Observe that the I band has only thin filaments, the H band only thick ones, and the A band both thick and thin filaments. Special structures are observed at the Z and M lines.

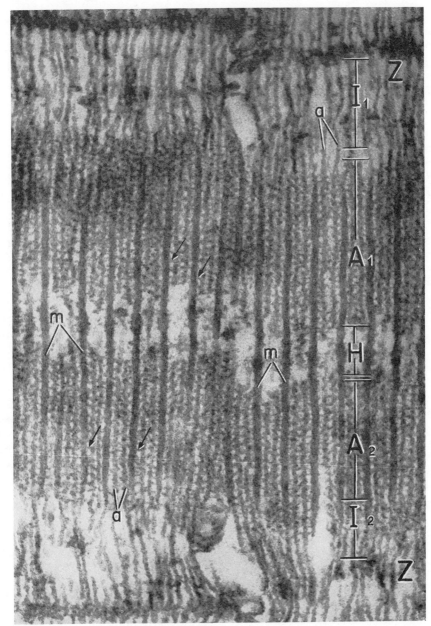

Figure 27–4   Electron micrograph of two sarcomeres in the adjacent myofibrils: Z, Z line; A_1 and A_2, aniso-tropic half bands; H, Hensen's band; I_1 and I_2, isotropic half bands; m, thick, and a, thin, filaments. The cross band between both types of filaments can be clearly seen. Some of them are indicated by arrows. ×175,000. (Courtesy of H. Huxley.)

The Sarcomere I-band Shortens and the Banding Inverts During Contraction

In a living muscle fiber, changes with contraction can be observed with the phase contrast and interference microscope. If this is done, one striking observation that can be made is that the A-band remains constant in a wide range of muscle lengths, whereas the I-band shortens in accordance with the contraction.

The shortening of the I-band is the result of the fact that the thin myofilaments slide farther and farther into the arrays of thick filaments (Fig. 27–5). With the progressing contraction, the thin filaments penetrate into the H-band and may even overlap, thereby producing a more dense band in the center of the sarcomere (*inversion of the banding*). Finally, the thick filaments make contact and are crumpled against the Z-lines.[5] These findings have been interpreted in

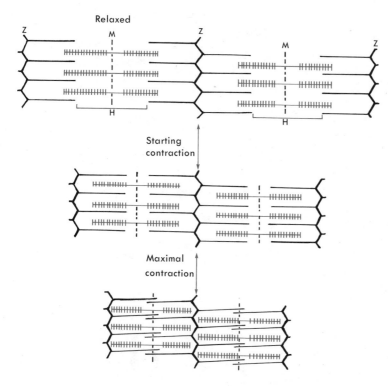

Figure 27–5 Diagram showing the sliding model of contraction. In the relaxed condition the H bands are wide and contain only thick filaments. With the beginning of contraction, the thin filaments slide toward the center of the sarcomere. With maximal contraction, the thin filaments penetrate into the H band and produce inversion of the banding.

the so-called *sliding filament mechanism* of contraction, which will be discussed in greater detail later. The degree of contraction thus achieved can be measured by determining the length of the sarcomere (i.e., the distance between Z-discs) at rest and when it has shortened. Note that insect muscle, in general, shortens only slightly (about 12 per cent), whereas the shortening in vertebrate muscle may be much greater (about 43 per cent).

Smooth Muscles Lack the Z-line

The electron microscope has revealed that "smooth" muscles may have a varied macromolecular organization. In many cases they contain thin and thick myofilaments, as do striated muscles, but the difference lies in the absence of the Z-line and the lack of periodicity. In mollusks and annelids there are muscles with a helical arrangement that have thin and thick myofilaments linked by cross bridges. In the adductor muscle of the oyster, the so-called paramyosin muscle, each thick filament is surrounded by 12 thin filaments.[6]

The smooth muscle of vertebrates apparently lacks these two types of myofilaments and even myofibrils are difficult to recognize. However, improvements in the preparative techniques have demonstrated coarse myofilaments.[7] Furthermore, thick and thin myofilaments have been isolated. In smooth muscles the contraction is very slow, but extreme degrees of shortening may be achieved.

SUMMARY

The Structure of Muscle

Muscle cells are adapted to mechanical work by unidirectional contraction. The functional unit is the *myofibril,* which may be either striated or smooth. In skeletal muscle, myofibrils fill most of the large muscle fiber, leaving small amounts of sarcoplasm, which contains the nuclei, the *sarcoplasmic reticulum,* and large mitochondria, or *sarcosomes.* A *sarcolemma,* with −0.1 volt polarization, surrounds the fiber. This is activated through the *end plate.*

Myofibrils result from the repetition of *sarcomeres.* These are limited by the Z-lines and contain the I-bands, the A-bands, and the H-band. The M-line may be observed in the middle of the sarcomere. The myofibril is composed of thick (10 nm) and thin (5 nm) myofilaments. In the relaxed condition, the I-band contains only thin myofilaments, the H-band only thick, and the A-band, both thick and thin. Myofilaments are organized in a paracrystalline hexagonal array. From the thick filaments, at 7 nm intervals, there are cross-bridges extending toward the thin filaments. One thick filament joins to six adjacent thin ones every 43 nm.

In each half an A-band there are about 11 transverse stripes, some of which are due to the cross-bridges, others to C-protein. The Z-disc shows a woven-basket square lattice made of Z-filaments of α-actinin which are in continuity with the thin filaments of each sarcomere.

During contraction the I-bands shorten as a result of the sliding of the thin filaments into the A- and H-bands. A further reversal of the band may be produced at the center of the sarcomere. In vertebrate muscle, the sarcomere may shorten up to 43 per cent.

27–2 MOLECULAR ORGANIZATION OF THE CONTRACTILE SYSTEM

In recent years knowledge about the molecular machinery involved in muscle contraction has become increasingly complex. To the classic force-generation proteins *myosin* and *actin* others having a regulatory function have been added, such as *tropomyosin* and the various *troponins.* Finally a series of other proteins, probably playing a structural role, such as α-*actinin* present in the Z-band, the *M-band proteins,* and the *C-protein,* have been described. Here, we shall only mention some of the main characteristics of these proteins in relation to the structure of the myofilaments and myofibrils (for further details, see Pepe, 1975; Lehman, 1976).

The Thick Myofilament — Myosin Molecules; The Cross-bridges — S_1 Subunits

Myosin can be extracted from muscles with a 0.3 M solution of KCl and purified by precipitation at lower ionic strength. This large molecule comprises about half of the total protein of the myofibril and has a molecular weight of about 500,000 daltons. It contains 2 polypeptide chains of about 200,000 daltons each and 4 smaller ones in the range of 20,000 daltons. As shown in the diagram of Figure 27–6, the 2 heavy chains are coiled around each other, forming a double helical α-helix about 140 nm long and 2 nm in diameter. Only about half the heavy chain is rod-like, the rest, together with some of the smaller chains, is folded into two globular regions at one end of the molecule.

Myosin can be fragmented by the action of proteolytic enzymes; *trypsin* breaks it into the long, rod-like *light meromyosin* (LMM) and *heavy meromyosin* (HMM). This last portion can be further subdivided by *papain* into the globular *subunit* S_1 and the *helical rod* S_2, which joins S_1 to the light meromyosin.[8] The most important part of the myosin molecule is HMM-S_1, since it contains the sites for the *ATPase* and for *binding to actin.* In fact it is the interaction between myosin and actin that results in contraction.[9]

At this point it is important to mention how the myosin molecules are integrated to constitute the *thick* or *myosin myofilament.* Each of these filaments has a smooth or bare portion in the middle (corresponding to the H-band) and two regions in which the surface is rough because of the presence of the *myosin cross-bridges.* It is now known that these cross-bridges represent the S_1 globular ends of myosin, while the shaft of the filament is formed by the rod-like portion of the molecule. As shown in Figure 27–7, *bottom,* one half of the molecules in the myofilament are oriented in the opposite direction from the other half. In other words, there is a definite polarization of the S_1 ends in each half of the myofilament.[1]

From the study of cross sections of the myosin filaments, which show a triangular shape, a model in which 12 parallel units are closely packed has been postulated.[10] Each one of the units, with a diameter of 4 nm, could be the result of the association of two myosin molecules.[11] Another interesting observation is that at the M-line a cross section of the myosin filament shows bridges which connect with the six neighboring thick filaments, forming a hexagonal lattice (Fig. 27–3). Such bridges repre-

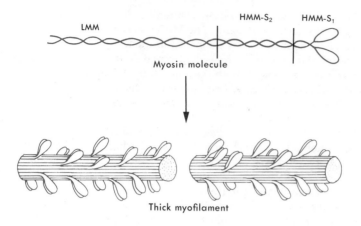

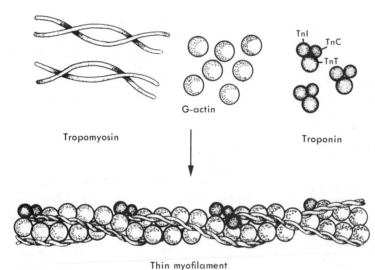

Figure 27-6 Diagram indicating the molecular structure of the thick myofilaments (**top**) and thin myofilaments (**bottom**) in the myofibril. The myosin molecule is made of two polypeptide chains 140 nm long, having at one end the S_1 head (LMM-light meromyosin, HMM-S_2, heavy meromyosin-S_2, HMM-S_1, heavy meromyosin-S_1). Observe that the thick filament results from the assembly of myosin molecules. The thin filament is the result of the association of G-actin monomers, tropomyosin and the troponins (TnT, troponin T; TnC, troponin C; and TnI, troponin I). Note that each tropomyosin molecule extends over seven actin molecules.

Figure 27-7 **A,** diagram illustrating the polarity of cross-bridges in the myosin filaments and in the actin myofilaments. The sliding forces tend to move the actin filaments toward the center of the sarcomere. (From Huxley, H. E.) **B,** arrangement of myosin molecules within the thick filaments. Each molecule has a tadpole shape with a globular head and a tail. The axis of the filament is formed by the assembly of the tails. Observe that in each half of the myofilament the molecules are polarized. (From Huxley, H. E., *Proc. Roy. Inst. Gr. Br.,* 44:274, 1970.)

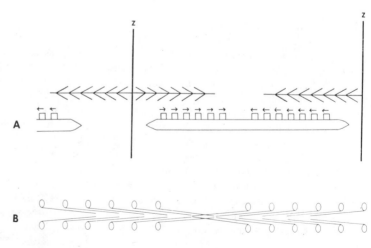

sent the attachment of the M-protein at the M-line of the sarcomere.[10]

Actin, Tropomyosin, and the Troponins Constitute the Thin Myofilament

Actin needs high concentrations of KCl (0.6 M) to be extracted. It is the second major protein and represents about one quarter of the protein of the myofibril. The molecular weight of actin is 42,000 daltons. The actin myofilament is made up of two helical strands which cross over every 36 to 37 nm. Each of these cross-over repeats contains between 13 and 14 *globular* monomers of *G-actin* about 5 nm in diameter (Fig. 27–6). In muscle, actin is present mainly as fibrous or *F-actin*, which is the polymerized form of *G-actin*.[12, 13]

The actin myofilament is more complex because it also contains *tropomyosin* and the *troponins*.

Tropomyosin. This protein represents between 5 and 10 per cent of the total and is extracted with 1 M KCl or weak acids. It has a molecular weight of 64,000 daltons and is an elongated molecule of 40 nm made of 2 parallel polypeptide chains in an α-helix. As shown in the diagram of Figure 27–6, tropomyosin stretches over seven actin monomers and forms two helixes that lie within the grooves of the actin double helix.

Troponin. Troponin is known to consist of 3 components forming a complex of proteins of 80,000 daltons. Each of these troponin complexes binds to tropomyosin every 40 nm, or every 7 actin monomers (Fig. 27–6). The three components of troponin that are present in equimolecular proportions are *troponin-C*, which binds Ca^{2+} specifically, *troponin-T*, which has a high affinity for tropomyosin, and *troponin-I*, which inhibits the ATPase of myosin[14] (see Lehman, 1976).

Contractile and Regulatory Proteins — Localization by Fluorescent Antibodies

Antibodies against myosin, actin, tropomyosin, and troponin that have been labeled with fluorescent dyes have permitted the localization of these proteins in relation to the macromolecular structure of the myofibril and myofilaments. For example, with *anti-troponin*, the fluorescent labeling of the thin filaments appears at a 40 nm interval, as shown in the model of Figure 27–6. Other minor muscle proteins may also be localized with fluorescent antibodies (see Chapter 4–6).

α-**Actinin.** This protein is a rod-shaped molecule of about 95,000 daltons localized exclusively at the Z-line, as can be demonstrated by anti-actinin antibodies.[15] The α-actinin molecule binds in vitro to actin filaments, forming regular cross-links. This protein apparently plays a structural role in the sarcomere.

C-Protein. This protein is present in the A-band along the middle portion of the myosin filament.[16] With electron microscopy the anti-C-protein antibody is observed to form between 7 and 9 transverse stripes in each half of the A-band.[2, 3]

M-Line Protein. Several proteins have been extracted from the M-line region; one of them is probably a structural protein, and the others are phosphorylase and creatine-kinase[17] (for further details, see Pepe, 1975).

27–3 THE SLIDING MECHANISM OF MUSCLE CONTRACTION

We mentioned earlier that contraction is currently explained by a *sliding mechanism* in which the thin actin filaments are displaced with respect to the thick myosin filaments during each contraction-relaxation cycle.[9, 12]

In the case of a frog sarcomere, 2.5 μm long, there is a shortening of 30 per cent at each contraction. This implies a sliding of 0.37 μm for each half sarcomere. The force generated seems to be proportional to the degree of overlap between thick and thin filaments. This implies that the force is of a short range and is produced directly by the cross-bridges between the filaments. A resting muscle is rather plastic and extensible because the cross-bridges are not attached and can slide upon an applied external force (Fig. 27–8,*A*).

Active sliding movement is thought to result from the repetitive interaction of the cross-bridges with the actin filaments. In the diagram of Figure 27–8 it is assumed that each cross-bridge represents the active end of the myosin molecule (i.e., the S_1 globular unit). The following mechanism takes place: (1) A cross-bridge binds to a specific site of the actin filament. (2) The cross-bridge undergoes a conformational change which displaces the

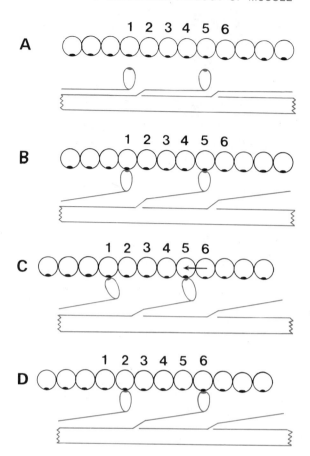

Figure 27–8 A very schematic interpretation of the mode of action of cross-bridges in the sliding of myofilaments. **A,** at relaxation there is no attachment of the S_1 cross-bridges to the actin filament; **B,** at the start of contraction interaction of S_1 and actin occurs; **C,** a conformational change in S_1 causes the actin myofilament to move toward the center of the A band; **D,** at the end of the stroke S_1 detaches, returns to the original configuration, and re-attaches to the following G actin.

point of attachment toward the center of the A-band, thereby pulling the acting filament; at the same time a second cross-bridge becomes attached. (3) At the end of the cycle the first bridge returns to the starting configuration in preparation for a new cycle. According to this theory, the actin filaments from each half sarcomere are pulled as ropes toward the center of the sarcomere by the myosin arms that move to and fro (Fig. 27–8,B through D).

The energy for this interaction is provided by the splitting of ATP because of the ATPase present in the cross-bridge. It is estimated that the splitting of one ATP accounts for a displacement of 5 to 10 nm for each myosin cross-bridge.

Myosin and Actin have a Definite Polarization within the Sarcomere

One of the most characteristic features of muscle is a molecular organization by which individual molecules may interact both spatially and temporally in a concerted fashion. Thus, the sliding mechanism depends on a very specific interaction between actin and myosin molecules and requires the existence of a *definite polarization* within the sarcomere.[1] In each half sarcomere actin and myosin should have a different polarization; furthermore, the forces developed must have opposite directions (Fig. 27–7).[18]

From electron microscopic studies of aggregates of myosin molecules, it is possible to obtain reconstituted myofilaments having the structure shown in the diagram of Figure 27–7,B. The molecules of each half of the filament are in two antiparallel sets with a reversal of structural polarity in the center. In the diagram of a sarcomere in Figure 27–7,A, the polarity of the cross-bridges is indicated by the arrows. This diagram also emphasizes that all the actin molecules should be polarized in opposite directions in each half sarcomere for the sliding process to take place.

Muscle Contraction — A Cyclic Formation and Breakdown of Actin-myosin Linkages

Since the classic experiments of Szent-Györgyi, it has been known that the mixing of myosin and actin in the test tube results in the formation of the complex *actomyosin*, which contracts in the presence of ATP.[19] Investigators following this interaction under the electron microscope have observed that the myosin molecules bind to the F-actin with a directional orientation. As shown in Figure 27–9, the complex can also be produced by using F-actin and heavy meromyosin. The actin filaments become "decorated" with the myosin, and the complex shows an arrow-like polarity. These findings demonstrate that in each filament of F-actin, the G-actin molecules are polarized with the same orientation.

Figure 27–10 shows a model illustrating the double helical filament of actin decorated with the S_1 myosin fragments.[20] The arrow-like configuration shown in the electron micrograph of Figure 27–9 results in the tilting and bending of the S_1 myosin along the helix of F_1 actin. This tilting of the S_1 myosin very likely corresponds to the configuration that the cross-bridge has at the end of the working stroke,

when ADP and Pi have been released and no further force is being exerted. On the other hand, there is electron microscopic and x-ray diffraction evidence that in the relaxed muscle the bridges are approximately perpendicular to the axis of the myofilaments. (For further details and x-ray diffraction data, see Huxley, 1976.)

In summary, the detailed ultrastructural and biochemical information obtained permits an interpretation of the macromolecular mechanisms involved in muscle contraction. It may be postulated that this is a cyclic event involving the repetitive formation and breakdown of actin-myosin linkages at the bridges between thick and thin filaments. At each bridge the following sequence of events is probably produced: (1) formation of a perpendicular linkage between a heavy meromyosin head and one G-actin (globular) unit; (2) rupture of this linkage by one ATP molecule; (3) hydrolysis of the ATP by the Ca^{2+}-activated ATPase of myosin; (4) formation of a new linkage between the same heavy meromyosin (bridge) and the next G-actin unit. The relative movement of the thin filament taking place in each sequence would be equivalent to the length of one G-actin unit (5.3 nm).

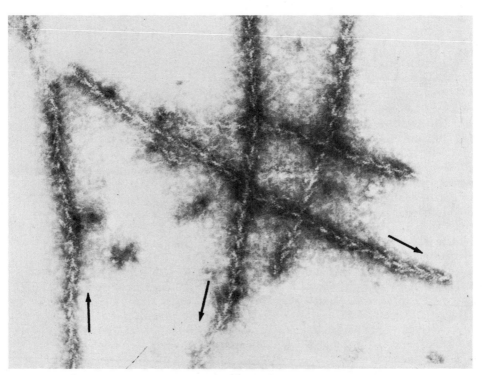

Figure 27–9 Filaments of actomyosin resulting from the interaction of actin and myosin (H-meromyosin). Observe the arrowlike polarity of the actomyosin complex (arrows). Negatively stained; ×155,000. (Courtesy of H. E. Huxley.)

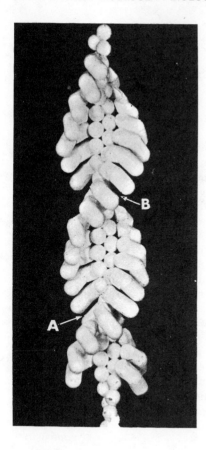

Figure 27–10 Three-dimensional model of an actomyosin complex. In the axis is the double helical filament of *F*-actin (see Fig. 23–6) made of G subunits. The actin has been treated with fragments of heavy meromyosin (fragment S_1, see Fig. 23–8). Points marked *A* and *B* are cross-over points in the actin helix. The arrowhead appearance is produced by the change in position of the point of attachment of the S_1 subunit to the actin molecules. (From Huxley, H. E., *Proc. Roy. Inst. Gr. Br.*, 44:274, 1970.)

SUMMARY:

Molecular Organization and the Sliding Mechanism

The molecular machinery involved in muscle contraction comprises (1) force-generating proteins (*myosin* and *actin*), (2) regulatory proteins (*tropomyosin* and *troponins*), and (3) structural proteins (α-actin in the Z-band, M-band proteins, and the C-protein).

Myosin represents 50 per cent of the myofibril protein, and its molecular weight is about 500,000 daltons, with 2 polypeptides of 200,000 daltons and 4 of 20,000. Trypsin separates the rodlike light meromyosin from the heavy meromyosin, which can be divided further by papain into the subunits S_1 and S_2. S_1 contains ATPase and the binding site for actin. Myosin is in the thick myofilament, and S_1, in the cross-bridge. There is a definite polarization of both the myosin molecule and the S_1 in each half sarcomere (Fig. 27–7).

Actin represents about 25 per cent of the myofibril protein, with a MW of 42,000 daltons. The monomeric or globular actin (G-actin) of about 5 nm polymerizes into fibrous actin (F-actin) in a double helical strand with cross-overs every 13 to 14 monomers. The actin filament also contains tropomyosin and the troponins.

Tropomyosin (5 to 10 per cent) is a molecule of 64,000 daltons and forms two elongated polypeptides 40 nm long that extend the length of 7 G-actins, within the grooves of the actin double helix.

Troponin has three components, troponin-C (binding Ca^{2+}), troponin-T (binding tropomyosin), and troponin-I (inhibiting the ATPase). Troponin binds to tropomyosin every seven G-actins.

The *sliding mechanism of muscle contraction* postulates that the thin actin filaments are displaced with respect to the thick filaments at each contraction-relaxation cycle. Short-range forces are generated at the cross-bridges between the filaments. An S_1 portion of the myosin molecules attaches to a binding site of F-actin, and by a conformational change, pulls it toward the center of the sarcomere. The splitting of ATP by the Ca^{2+}-activated ATPase provides the energy. One ATP may account for a 5 to 10 nm displacement of a cross-bridge. According to this theory, both the myosin and actin myofilaments should be polarized in each of the sarcomeres (Fig. 27–7). This polarization is actually demonstrated by the electron microscopic observation of reconstituted myosin filaments and of actin filaments treated with heavy meromyosin. The arrow-like configuration of these "decorated" filaments is the result of the F_1-actin helix and the tilting and binding of the S_1 heavy meromyosin.

27–4 REGULATION AND ENERGETICS OF CONTRACTION

We have seen earlier that in the *relaxed state* there is no attachment of the cross-bridges; the opposite condition could explain the state of *rigor* (i.e., of permanent contraction). In this state, the rigidity of the muscle could be due to the permanent attachment of the cross-bridges to the thin filaments. In order to have a contraction-relaxation cycle, a cyclic mechanism by which the cross-bridges become attached and detached must operate. It is now known that the transition between rest and activity is dependent on the *concentration of free calcium* in the vicinity of the contractile machinery[21] (see Murray and Weber, 1974).

The control of contraction by Ca^{2+} requires the presence of the regulatory proteins tropomyosin and troponin, which form part of the structure of the thin filament. We have seen earlier that *troponin-C* (Ca^{2+}-binding) and *troponin-I* and *troponin-T* are placed along the actin filament at intervals of about 40 nm and that such a periodicity corresponds to 7 G-actin monomers. Furthermore, we mentioned that tropomyosin also extends along the actin filament for the same distance and is in close relationship with the troponin complex (Fig. 27–6,*B*).

The mechanism by which Ca^{2+} regulates the contraction-relaxation cycle is thought to be the following: (1) In the absence of Ca^{2+} the regulatory proteins inhibit the interaction between actin and myosin. (2) When the Ca^{2+} concentration in the cytosol increases above 10^{-6} M, the muscle contraction is triggered by the binding of Ca^{2+} to troponin-C.[21] (3) Since the presence of tropomyosin is needed for this regulatory mechanism, it is supposed that the influence of troponin is transmitted by way of tropomyosin to the seven G-actin monomers with which it is associated. Thus, this mechanism is highly cooperative.

Molecular Regulation — Displacement of Tropomyosin After Binding of Ca^{2+} to Troponin-C

From electron microscopic and x-ray diffraction evidence a model has been constructed which may explain the regulatory action of the tropomyosin-troponin-Ca^{2+} system.[20, 22, 23]

In this model (Fig. 27–11), an end-on view of the relative positions of actin, the S_1-myosin, and tropomyosin is presented. The tropomyosin molecule is shown in two positions, one corresponding to the activated state and the other to the relaxed state. According to the model in the latter condition (dotted contour) tropomyosin is deep in the groove of F-actin and covers the actin binding site of the S_1-myosin.

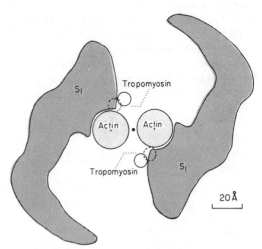

Figure 27–11 A possible molecular model of regulation of contraction. The diagram shows an end-on view of the relative positions of actin and the S_1-myosin. Tropomyosin is shown with a line contour in the activated condition, in which there is interaction between actin and myosin. The dotted contour corresponds to the relaxed condition, in which tropomyosin is deep in the groove of F-actin and covers the actin binding site of S_1-myosin (Courtesy of H. Huxley.)

597

When Ca^{2+} binds to troponin-C a displacement of tropomyosin away from the groove could be effected, exposing the actin binding site in S_1 for actin-myosin interaction.

Energy for Contraction Comes from Oxidative Phosphorylation and Glycolysis

Whereas the myofibrils constitute the mechanical machinery of the muscle, the fuel needed is produced mainly in the sarcoplasm. In all types of muscle, numerous mitochondria, called *sarcosomes,* provide the essential oxidative phosphorylation processes and the Krebs cycle system. These mitochondria are particularly prominent in size and number in heart muscle and in the flight muscles of birds and insects.

The sarcoplasmic matrix contains the glycolytic enzymes as well as other globular proteins, such as myoglobin, salts, and high phosphate compounds. Glycogen is present in the matrix as small granules observed under the electron microscope. There are about 1 per cent glycogen and 0.5 per cent creatine phosphate as sources of energy in muscle. Glycogen disappears with contraction through glycolysis, and lactic acid is formed, which can be transformed into pyruvic acid to enter the Krebs cycle (see Chapter 6).

The initial energy source for contraction is ATP. The ADP produced after the initial contraction is again recharged to ATP by glycolysis or from creatine phosphate. Oxidative phosphorylation is the last and most important source of ATP.

SUMMARY:

Regulation and Energetics of Contraction

During the contraction-relaxation cycle there is a cyclic attachment and detachment of cross-bridges to actin filaments. (In *relaxation* there is no attachment, and in the state of *rigor* the attachment is permanent.) Regulation of this cycle depends on the Ca^{2+} concentration and the presence of the regulatory proteins tropomyosin and the troponins. When Ca^{2+} increases above 10^{-6} M in the cytosol, contraction is triggered by the binding of Ca^{2+} to troponin-C. The influence of troponin is transmitted to seven G-actin monomers by way of tropomyosin. This regulatory action has been analyzed by electron microscopy and x-ray diffraction, and a model has been constructed. According to this, in the relaxed state tropomyosin is deep in the groove of F-actin and covers the actin binding site of S_1-myosin. When Ca^{2+} binds to troponin-C, tropomyosin is displaced from the groove, and the binding site in S_1 is exposed for the actin-myosin interaction.

The initial source of energy for contraction is provided by ATP. This energy is furnished by oxidative phosphorylation in mitochondria (*sarcosomes*), by glycolysis of glycogen, and by creatine phosphate.

27–5 EXCITATION-CONTRACTION COUPLING

Excitation-contraction coupling refers to the mechanism by which the electrical impulse is able to induce contraction in a muscle. It is known that with the arrival of the nerve impulse at the motor end plate a *chemical synaptic transmission* occurs (see Chapter 28). This in turn brings about a depolarization of the cell membrane which is conducted along the muscle fiber and penetrates into it to induce contraction.

We shall see that the essential components in this excitation-contraction coupling mechanism are (1) the *T-system*, which conducts the action potential to the interior of the muscle fiber; (2) the release of Ca^{2+} from the *sarcoplasmic reticulum;* (3) the induction of contraction by Ca^{2+} (explained earlier); and (4) the reaccumulation of Ca^{2+} into the sarcoplasmic reticulum by Ca^{2+}-*ATPase,* which results in *muscle relaxation.* The study of this mechanism involves the knowledge of each one of the elements involved (see Ebashi, 1976; Hasselbach, 1977).

Sarcoplasmic Reticulum — A Longitudinal Component with Terminal Cisternae Forms Part of the Triad

The sarcoplasmic reticulum found in skeletal and cardiac muscle fibers is one of the most interesting specializations of the endoplasmic reticulum. It was discovered by Veratti in 1902 as a reticulum present in the sarcoplasm of the muscle fiber and extending in between the myofibrils. It was completely neglected until 1953 when the first electron micrographs of this structure were published.[24, 25]

The sarcoplasmic reticulum is a membrane-limited reticular system whose organizational structure is regularly superimposed upon that of the sarcomeres. As shown in Figure 27–12, there are two main components in the *sarcoplasmic reticulum:* (1) a *longitudinal* component that is larger and represents the true endoplasmic reticulum and (2) a *transverse* component or *T-system* made of vesicles and tubules that are disposed at the level of the Z-lines and are connected to the sarcolemma.

The longitudinal reticulum is composed of wide anastomosing tubules that cover the surface of the sarcomere at the level of the A- and I-bands and that terminate by special terminal cisternae found at the end of the I-bands near the Z-line. In between the terminal cisternae of two consecutive sarcomeres is another flattened vesicle belonging to the T-system; the three together constitute the so-called *triad* (Fig. 27–12).

The T-system is in Continuity with the Plasma Membrane and Conducts Impulses Inward

The transverse component (i.e., the T-system) at certain points is continuous with the plasma membrane of the sarcolemma and is the structure organized to conduct impulses from the fiber surface into the deepest portions of the muscle fiber.[26]

The continuity between the T-system, the plasma membrane, and the extracellular space was first demonstrated indirectly. In frog muscle immersed for short periods in a solution containing ferritin molecules, it was found that the central vesicle of the triad had filled with these electron-opaque molecules. Ferritin was also found in certain tubules that were continuous with the central element of the triad. These findings suggested that at the plasma membrane there are a small number of open-

ings communicating directly with the transverse system of the sarcoplasmic reticulum through fine tubules. Interestingly, the lateral components of the triad never contained ferritin molecules;[27] these components, then, are not connected to the plasma membrane.

Direct observation of the connections between the T-system and the plasma membrane is difficult but has been achieved.[28, 29] Portions of the T-system have been observed to run longitudinally and even in a spiral arrangement (in Fig. 27–12 a contorted tubule that connects with the transverse component of the triad can be observed).

The function of the T-system is to transmit inward the electrical signal that will bring about the contraction of the individual myofibrils. The role of the system was suggested by experiments with microelectrodes in which stimulation of the sarcolemma, at the level of the Z-line, produced a localized contraction of adjacent sarcomeres.[30, 31] Apparently the T-system is not a passive channel but is actively involved in the conduction of action potentials. One interesting finding has been that in muscle immersed in glycerol there is a lesion of the T-system (i.e., *detubulation*), and the mechanism of excitation-coupling is abolished.[32]

The presence of this intracellular conducting system may explain the physiologic paradox that all the myofibrils of a fiber between 50 and 100 μm in diameter may contract quickly and synchronously, once the activating action potential has passed over the surface.

Stimulation Releases Ca^{2+} from the Terminal Cisternae

Within the muscle, Ca^{2+} is stored mainly in the longitudinal components of the sarcoplasmic reticulum, so that the Ca^{2+} concentration near the myofibrils is kept very low (i.e., about 10^{-7} M). Following stimulation, Ca^{2+} is rapidly released, so that the concentration increases (up to 10^{-5} M) and induces contraction. It is now assumed that the triad plays an essential role in this release of Ca^{2+}. It is thought that the action potential, arriving by way of the T-system, immediately opens Ca^{2+} channels in the *terminal cisternae* of the sarcoplasmic reticulum, producing a localized release of Ca^{2+}. Thus, the triad is the most important site in the excitation-contraction coupling (for further details, see Ebashi, 1976).

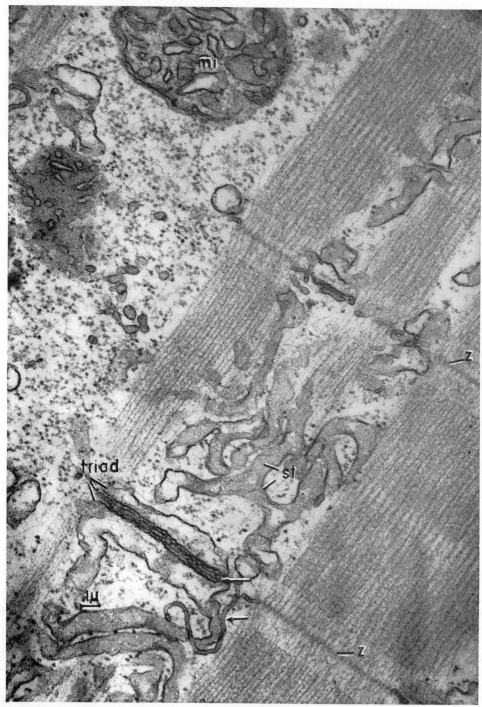

Figure 27-12 Electron micrograph of striated muscle showing two myofibrils, one of which is tangentially cut and shows the arrangement of the sarcoplasmic reticulum. The two components of this system can be seen clearly. The transverse component is represented by the *triad* and especially by the central cisternae of the triad, which continues in special tubules (arrows). Notice the relationship of the triad to the Z-line. The longitudinal component of the sarcoplasmic reticulum forms anastomosing tubules (*st*) on the surface of the sarcomere; *mi*, mitochondrion. (Courtesy of K. R. Porter.)

A Ca²⁺-activated ATPase is Present in the Sarcoplasmic Reticulum and Acts as a Ca²⁺ Pump

We have just seen that during excitation Ca^{2+} is liberated from special regions of the sarcoplasmic reticulum and that this causes the contraction. To produce relaxation the opposite phenomenon must take place. In other words, Ca^{2+} must be rapidly removed, and this is done by a *Ca²⁺-activated ATPase* that is present in the membrane of the sarcoplasmic reticulum. This acts as a Ca^{2+} pump that incorporates this cation under the action of ATP.

The reactions involved are the following:

$$2\ Ca^{2+}\ outside + ATP + E \rightleftharpoons \\ E\text{-}PCa_2^{2+} + ADP \qquad (1)$$

$$E\text{-}PCa^{2+} \rightleftharpoons E\text{-}Pi + 2Ca^{2+}\ inside \qquad (2)$$

The two equations show that the splitting of one molecule of ATP results in the translocation of two Ca^{2+} by the ATPase (*E*). This reaction is reversible[33] (i.e., Ca^{2+} inside can be used for the formation of ATP).

The sarcoplasmic reticulum can be isolated by cell fractionation and vesicular structures may be obtained. The vesicles contain 40 per cent lipid, most of which is in the form of phosphatidylcholine, and two main proteins: the *Ca²⁺ pump protein (ATPase)*, which, with a molecular weight of 105,000 daltons, represents about 80 per cent of the protein of the membrane, and a *Ca²⁺-binding protein* (i.e., *calsequestrin*), with a molecular weight of 65,000 daltons (see Meissner and Fleischer, 1974; Hasselbach, 1977). The Ca²⁺ATPase is thought to be a highly asymmetric molecule spanning the membrane, which needs the presence of a certain amount of phospholipid to function (see Metcalfe and Warren, 1977).

Since both the ATPase and the Ca^{2+} uptake are inhibited by agents that block —SH groups, an electron microscopic study was conducted using a cytochemical reagent in which ferritin molecules were attached to an —SH blocking molecule. It was demonstrated that the active groups of the membrane-bound ATPase were localized in the outer surface of the isolated sarcoplasmic reticulum.[34] All these findings suggest that the sarcoplasmic reticulum assumes the important role of relaxing the fiber after contraction and that this is accomplished by the binding of Ca^{2+} at the outer surface and its transport within the sarcoplasmic vesicles. *Calsequestrin*, on the other hand, could be responsible for the binding of Ca^{2+} within the sarcoplasmic reticulum.[35]

In summary, the series of events produced after the arrival of the electrical signal is as follows: (1) The signal is received at the level of the Z-line or the A-I junction by way of the T-system. (2) The coupling of the T-system with the terminal cisternae produces the release of Ca^{2+} (this may occur in a matter of milliseconds). (3) Ca^{2+} induces the contraction, affecting the regulatory proteins troponin and tropomyosin and thus enabling the interaction of actin and myosin. During this step ATP is used as an energy source. (4) Ca^{2+} is quickly removed and restored into the sarcoplasmic reticulum by way of the Ca^{2+} pump. Ca^{2+} uptake results in muscle relaxation.

All these data, as well as those related to the sliding mechanism of contraction, can be put together in a molecular theory of muscular contraction. This is one of the best examples, so far studied, of a tight coupling between the processes furnishing energy and the actual machinery involved in contraction. In this case, structure and function are so intimately related in the realm of molecular organization that they are an inseparable unit.

SUMMARY:

Excitation-Contraction Coupling and the Sarcoplasmic Reticulum

The essential components of the excitation-contraction coupling are (1) the T-system, which conducts the action potential to the myofibril; (2) the release of Ca^{2+} from the sarcoplasmic reticulum; (3) the induction of contraction by Ca^{2+}; (4) the reaccumulation of Ca^{2+} in the sarcoplasmic reticulum, which results in relaxation.

Within the muscle, Ca^{2+} is stored mainly in the longitudinal components of the sarcoplasmic reticulum, so that the concentration in the cytosol is kept at about 10^{-7} M. Following stimulation, Ca^{2+} is released and increases to 10^{-5} M, thus inducing contraction. This localized release occurs by the opening of Ca^{2+} channels in the terminal cisternae of the triad. *Relaxation* is caused by the removal of *Ca^{2+} by the Ca^{2+}-activated ATPase,* which is present in the membrane of the sarcoplasmic reticulum, under the action of ATP. The splitting of one molecule of ATP results in the translocation of two molecules of Ca^{2+}.

In the isolated sarcoplasmic reticulum the vesicles contain 40 per cent lipids, and 80 per cent of the protein consists of the Ca^{2+}-ATPase (or Ca^{2+} pump), which has a MW of 105,000 daltons. There is also *calsequestrin,* a Ca^{2+}-binding protein of 65,000 daltons. Muscle provides one of the best examples of the tight coupling between the processes furnishing energy and the actual machinery that produces contraction and relaxation. In muscle, structure and function constitute an inseparable unit.

REFERENCES

1. Huxley, H. E. (1966) *Harvey Lect., 60:*85.
2. Craig, R. (1977) *J. Molec. Biol., 109:*69.
3. Sjöström, M., and Squire, J. M. (1977) *J. Molec. Biol., 109:*49.
4. Ullrich, W. C., Toselli, P. A., Saide, J. D., and Phear, W. P. C. (1977) *J. Molec. Biol., 115:*61.
5. Huxley, H. E. (1970) *Proc. Roy. Inst. Gr. Br., 44:*274.
6. Hanson, J., and Lowy, J. (1961) *Proc. Roy. Soc. London [Biol.], 154:*173.
7. Pease, D. C. (1968) *J. Ultrastruct. Res., 23:*280.
8. Lowey, A., Slayter, S., Weeds, A. G., and Baker, H. (1969) *J. Molec. Biol., 42:*1.
9. Huxley, H. E. (1958) *Sci. Am., 199:*67.
10. Pepe, F. A. (1975) *J. Histochem. Cytochem., 23:*543.
11. Pepe, F. A., and Dowben, P. (1977) *J. Molec. Biol., 113:*199.
12. Davies, R. E. (1963) *Nature, 199:*1068.
13. Hanson, J., and Lowy, J. (1963) *J. Molec. Biol., 6:*46.
14. Potter, J. D., and Gergely, J. (1974) *Biochemistry, 13:*2697.
15. Masaki, T., Endo, M., and Ebashi, S. (1967) *J. Biochem. 62:*630.
16. Offer, G. (1972) *Cold Spring Harbor Symp. Quant. Biol., 37:*87.
17. Turner, D. C., Walliman, T., and Eppenberger, H. M. (1973) *Proc. Natl. Acad. Sci. USA, 70:*702.
18. Reedy, M. K. (1968) *J. Molec. Biol., 31:*155.
19. Szent-Györgi, A. (1947) *Chemistry of Muscular Contraction.* Academic Press, Inc., New York.
20. Moore, P. B., Huxley, H. E., and De Rosier, D. J. (1970) *J. Molec. Biol., 50:*279.
21. Ebashi, S. E., and Endo M. (1968) *Prog. Biophys. Molec. Biol., 18:*123.
22. Vibert, P. J., Halsegrove, J. C., Lowy, J., and Poulsen, F. R. (1972) *Nature, 236:*182.
23. Wakabayaski, T., Huxley, H. E., Amos, L., and Klug, A. (1975) *J. Molec. Biol., 93:*477.
24. Bennett, H. S., and Porter, K. R. (1953) *Am. J. Anat., 93:*1.
25. Porter, K. R. (1961) *J. Biophys. Biochem. Cytol., 10* (Suppl.):219.
26. Andersson-Cedergren, E. (1959) *J. Ultrastruct. Res.* Suppl. 1.
27. Huxley, H. E. (1964) *Nature, 202:*1067.
28. Franzini-Armstrong, C., Landmesser, L., and Pilar, G. (1974) *J. Cell Biol., 64:*493.
29. Zampighi, G., Vergara, J., and Ramón, F. (1974) *J. Cell Biol., 64:*734.
30. Huxley, A. F, and Taylor, R. E. (1958) *J. Physiol. London, 144:*426.
31. Huxley, A. F. (1971) *Proc. R. Soc. London [Biol.] . 178:*1.
32. Eisenberg, B., and Eisenberg, R. S. (1968) *J. Cell Biol., 39:*451.
33. Makinose, M., and Hasselbach, W. (1971) *FEBS Lett., 12:*271.
34. Hasselbach, W., and Elfvin, L. G. (1967) *J. Ultrastruct. Res., 17:*598.
35. MacLennan, D. H., and Wong, P. T. S. (1971) *Proc. Natl. Acad. Sci. USA, 68:*1231.

ADDITIONAL READING

Bourne, C., ed. (1960) *Structure and Function of Muscle,* 2 volumes. Academic Press, Inc., New York.

Craig, R. (1977) Structure of A-segments from frog and rabbit skeletal muscle. *J. Molec. Biol., 109:*69.

Curtin, N. A., and Woledge, R. C. (1978) Energy changes and muscular contraction. *Physiol. Rev., 58:*690.

Ebashi, S. E. (1976) Excitation-contraction coupling. *Annu. Rev. Physiol., 38:*293.

Ebashi, S. E., and Endo, M. (1968) Calcium ions and muscle contraction. *Prog. Biophys. Mol. Biol., 18:*123.

Hasselbach, W. (1977) The sarcoplasmic reticulum pump. *Biophys. Struct. Mech., 3:*43.

Huxley, H. E. (1973) Muscular contraction and cell motility. *Nature, 243:*445.

Huxley, H. E. (1976) The structural basis of contraction and regulation in skeletal muscle. In: *Molecular Basis of Motility,* p. 9. (Heilmeyer, L., et al., eds.) Springer-Verlag, Berlin.

Kelly, R. E., and Rice, R. V. (1968) Localization of myosin filaments in smooth muscle. *J. Cell Biol., 37*:105.

Lehman, W. (1976) Phylogenetic diversity of proteins regulating muscular contraction. *Int. Rev. Cytol., 44*:55.

Lowey, S. (1972) Protein interaction in the myofibril. In: *Polymerization in Biological Systems,* Ciba Foundation Symposium No. 7 (new series), p. 217. Elsevier North-Holland Co., Amsterdam.

Luther, P., and Squire, J. (1978) Three-dimensional structure of the vertebrate muscle M region. *J. Molec. Biol., 125*:313.

Margosian, S. S., and Lowey, S. (1978) Interaction of myosin subfragments with F-actin. *Biochemistry, 17*:5431.

Martonosi, A., et al. (1977) Development of sarcoplasmic reticulum in cultured chicken muscle. *J. Biol. Chem., 252*:318.

Meisser, G., and Fleischer, S. (1974) Characterization, dissociation and reconstitution of sarcoplasmic reticulum. In: *Calcium Binding Proteins,* p. 281. (Drabilowski, W., et al., eds.) Elsevier North-Holland Co., Amsterdam.

Metcalfe, J. C., and Warren, G. B. (1977) Lipid-protein interactions in a reconstituted calcium pump. In: *International Cell Biology.* (Brinkley, B. R., and Porter, K. R., eds.) The Rockefeller University Press, New York.

Murray, J. M., and Weber, A. (1974) The cooperative action of muscle protein. *Sci. Am., 230*(2):59.

Pepe, F. A. (1975) Structure of muscle filaments from immunohistochemical and ultrastructural studies. *J. Histochem. Cytochem., 23*:543.

Pepe, F. A. (1976) Detectability of antibody in fluorescence and electron microscopy. In: *Cell Motility,* p. 337. Cold Spring Harbor Laboratory, Cold Spring Harbor, New York.

Pepe, F. A., and Dowben, P. (1977) The myosin filament. *J. Molec. Biol., 113*:199.

Rowe, R. W. D. (1973) The ultrastructure of Z-disks from white intermediate and red fibers of mammalian striated muscles. *J. Cell Biol., 57*:261.

Seymour, J. (1978) Unfolding troponin-C. *Nature, 275*:177.

Sjöstrom, M., and Squire, J. M. (1977) Fine structure of the A-band in cryosections. *J. Molec. Biol., 109*:49.

Squire, J. (1977) Contractile filament organization mechanics. *Nature, 267*:753.

Trinick, J., and Lowey, S. (1977) M-protein from chicken pectoralis muscle. *J. Molec. Biol., 113*:343.

Ullrick, W. C., Toselli, P. A., Saide, J. D., and Phear, W. P. C. (1977) Fine structure of the vertebrate Z-disc. *J. Molec. Biol., 115*:61.

Wakabayashi, T., Huxley, H. E., Amos, L. A., and Klug, A. (1975) Three-dimensional image reconstruction of actin-tropomyosin complex and actin-tropomyosin-troponin T-troponin I complex. *J. Molec. Biol., 93*:477.

Wray, J. S. (1979) Structure of the backbone in myosin filament. *Nature, 277*:37.

28

CELLULAR AND MOLECULAR NEUROBIOLOGY

28–1 General Organization and Function of
Nerve Fibers 605
Axon Structure—Neurofibrils and
Neurotubules
The Biosynthetic Functions of the
Neuron Occur in the Perikaryon
Macromolecules are Transported Through
the Axon in an Orthograde Direction
Orthograde Axonal Flow may be Fast or
Slow
Macromolecules Such as NGF Undergo
Retrograde Axonal Transport
Tetanus Toxin and Some Neurotropic
Viruses may Undergo Retrograde and
Trans-Synaptic Transport
Nerve Conduction Velocity—Related to
Diameter, Myelin, and Internode
Distance
The Action Potential—Non-Decremental
and Propagated as a Depolarization
Wave
In Myelinated Nerve Fibers Conduction is
Saltatory
At the Physiologic Receptors and
Synapses Potentials are Graded and
Non-Propagated
Propagated Action Potentials Depend on
Na^+ and K^+ Channels in the Axon
Membrane
Summary: General Organization of the
Neuron and Function of the Nerve
Fiber

28–2 Synaptic Transmission and Structures
of the Synapse 619
Nerve Impulse Transmission—Possibly
Electrical but Mediated Mainly by a
Chemical Mechanism
Synaptic Transmission may be Excitatory
or Inhibitory
Several Thousand Synapses may Impinge
on a Single Neuron
The Number of Synapses is Related to
that of the Dendritic Spines
The Synapse Ultrastructure Suggests
Many Types of Functional Contacts
Lectin Receptors and Postsynaptic
Densities and the Formation and
Maintenance of Synapses
The Presynaptic Membrane Shows
Special Projections at Active Zones

The Postsynaptic Membrane Shows a
Complex Macromolecular Organization

28–3 Synaptic Vesicles and Quantal Release
of Neurotransmitter 626
Several Types of Synaptic Vesicles may
be Recognized by Morphology and
Cytochemistry
Neuronal Development—The Type of
Neurotransmitter and Synaptic Vesicle
may be Medium-Determined
Synaptosomal Membranes and Synaptic
Vesicles may be Isolated by Cell
Fractionation
Neurotransmitter Synthesis and Metabolism
is Exemplified by the Acetylcholine
System
The Transport of the Neurotransmitter
Involves the Synaptic Vesicles
Transmitter Release is Related to the Role
of Synaptic Vesicles in Nerve Trans-
mission
Transmitter Release Probably Involves
Exocytosis with Recycling of Synaptic
Vesicles
The Depolarization Transmitter-Secretion
Coupling is Mediated by Calcium Ions
Summary: Synaptic Transmission and
Structure of the Synapse

28–4 Synaptic Receptors and the Physio-
logic Response 636
Synaptic Receptors are Hydrophobic and
Embedded in the Lipoprotein Framework
of the Membrane
The Acetylcholine Receptor is Coupled to
the Translocation of Sodium and
Potassium Ions
Acetylcholine-Induced Fluctuations may be
Observed in a Reconstituted Cholinergic
Receptor
An Oligomeric Model for the Cholinergic
Receptor has been Postulated. Long-
Lasting Synaptic Functions Involve the
Use of a Second Messenger
Cyclic AMP is Involved in the Phosphoryla-
tion of Membrane and Other Proteins
Summary: Receptors and the Physiologic
Response

One of the most important functions of living organisms is reacting to an environmental change. Such a change, called a *stimulus*, generally elicits a *response*. In its most basic sense this general property is called *irritability*. For example, a unicellular protozoan may react to different stimuli, such as changes in heat or light or the presence of a food particle, by a mechanical response, such as ciliary motion, ameboid movement, etc. *Plants* react by slow responses, which produce differential growth, also called a *tropism*. For example, the responses to the gravitational field, temperature, light, touch, and chemicals are referred to, respectively, as *geotropism, thermotropism, phototropism, thigmotropism,* and *chemotropism*. Irritability reaches its maximal development in animals, and special cells forming the nerve tissue are differentiated to respond rapidly and specifically to the different stimuli.

In these organisms special physiologic *receptors* adapted to "receive" the various types of stimuli are differentiated. Receptors are made of special cells or of the distal endings of neurons, specialized to receive a particular stimulus. For example, the receptors of light, touch, taste, pressure, heat, cold, and others are characterized by their great sensitivity to the specific stimulus.

At the receptor the *threshold of excitation* is much lower than in any part of the nerve cell.

In an animal, the response to the stimulus may be of a varied nature. Most frequently the animal reacts with a rapid movement by contraction of muscle tissue. However, other types of reactions may be elicited. For example, a hungry dog in the presence of food reacts by secreting saliva; an electric fish, upon being touched, may produce an electrical discharge; and a firefly may give off light quanta. These different types of responses are produced in special tissue (e.g., muscles, glands, electric plates, luminous organs), called *effectors*, that are controlled by efferent neurons.

In an animal the simplest mechanism of nerve action is represented by the so-called monosynaptic reflex. This consists of a neuronal circuit formed by two *neurons*. One neuron is *sensorial* (afferent) and has a receptor at one end to receive the stimulus. At the other end the sensory neuron makes a special contact, also called a *synapsis*, with a *motor* (efferent) neuron, which in turn acts on the effector (i.e., muscle).

Figure 28–1 is a simplified diagram of the way in which the information *received* at the receptor is *conducted* along the sensory neuron and then *transmitted* at the synapse. Notice that a new wave of information starts in the second neuron, which finally reaches the effector, where the final response is elicited.

As will be shown later, in nerves and muscle information is propagated by changes in the

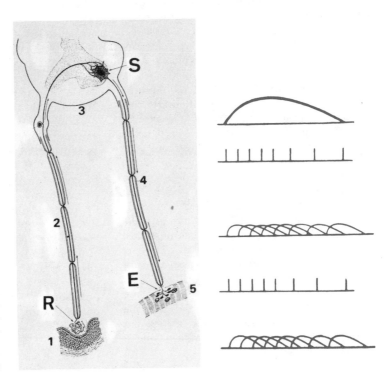

Figure 28–1 Diagram showing the monosynaptic reflex arc. **Left,** one sensory and one motor neuron with the synaptic junction. Notice the receptor and the effector. **Right,** different types of potentials produced at the portions of the reflex arc (1–5), indicated in the figure. *R,* receptor; *S,* synapsis; *E,* effector.

resting or *steady potential* at the surface membrane (see Chapter 8–3). This change originates a *wave of excitation,* which moves along the surface of the cell from one end to the other. In the nerve cell this *propagated* or *action potential* is also known as a *nerve impulse.*

Nerve impulses are *conducted* along the elongated parts of neurons (i.e., the nerve fibers) by way of action potentials. At the receptors, the synapse and the effectors, other electrical potentials having different characteristics are produced. The emphasis of this chapter will be on the *cellular bases* of *nerve conduction* and *synaptic transmission.*

28–1. GENERAL ORGANIZATION AND FUNCTION OF NERVE FIBERS

The neuron is differentiated for the conduction and transmission of nerve impulses. After embryonic life, neurons do not divide, but remain in a permanent interphase throughout the entire life of the organism. During this time a neuron undergoes changes in volume and in the number and complexity of its processes and functional contacts, but the number of neurons is not increased by cell division. This fact may be of paramount significance, since, in addition to conducting and transmitting impulses, nerve cells store instinctive and learned *information* (e.g., conditioned reflexes, memory) — a property that would be best served by a more permanent system of structures.

Neurons are adapted to their specialized functions by means of different types of outgrowths. The cell body (perikaryon) may emit one or more short outgrowths, or *dendrites,* which carry nerve impulses centripetally, and a longer one, the *axon,* which carries the impulse centrifugally to the next neuron or the effector. An axon is also called a *nerve fiber* when it is wrapped in the different sheaths after emerging from the neuron. The axon terminates, ramifying in the *telodendrons,* or endings. Some neurons have only one dendrite and the axon (i.e., *bipolars*) and others have only the axon (i.e., *monopolars*), in addition to the most common *multipolars.* In invertebrates most neurons are monopolar.

The different types of neurons and neuronal interconnections are discussed in histology and neuroanatomy textbooks. Only a few general considerations will be made here.

Axon Structure — Neurofibrils and Neurotubules

In fixed and stained preparations observed under the light microscope fine filaments, called *neurofibrils,* can be demonstrated in the cytoplasm of the neuron. These neurofibrils run in all directions and continue into the dendrites, axon, and nerve fiber. Better information about the structure of the axon can be obtained by polarization and electron microscopy.

The axoplasm of myelinated nerve fibers may be extruded and separated from the myelin sheath, permitting its study with the electron microscope and with polarization microscopy[1, 2] (Fig. 28–2,A). In these studies the partial volume occupied by the axially oriented material has been found to be less than 1 per cent. In the extruded axoplasm a fibrillar material was detected (Fig. 28–2,B).[1] In addition, mitochondria, strands of canaliculi, and vesicles of the endoplasmic reticulum have been observed.[2] The fibrillar material present in the axon, dendrites, and perikaryon is formed principally of long, tubular elements (the *neurotubules* or *neuronal microtubules*), 20 to 30 nm in diameter, which were previously seen in nerve homogenates[1, 3] (Fig. 28–2,C).

Unmyelinated nerve fibers show numerous neurotubules in cross section (Fig. 28–3,A). Among the neurotubules, there are also cross sections of neurofilaments. These two structural components of the axon are shown at higher magnification in Figure 28–3,B and C. The walls of the neurotubules show a subunit structure similar to that generally found in microtubules (see Chapter 9–2). Neurotubules containing dense cores (Fig. 28–3,C) have been described.[4] The electron-dense material of the cores has been interpreted as a result of endoluminal migration of some products. In the neurofilaments some lateral outgrowths or side-arms have been observed.[5] Neurotubules have been isolated from mammalian brain cells. Most of the biochemical and physiologic data described in Chapter 9–2 for the microtubules also apply to the neurotubules. Neurotubules and neurofilaments, when clumped together under the action of the fixatives and with the addition of colloid silver, form the neurofibrils of classic histology.

Although neurofibrils were described more than a century ago, their significance remained practically unknown. The hypothesis that they are involved in nerve conduction has been disproved. In Chapter 8–3 the experiments of ax-

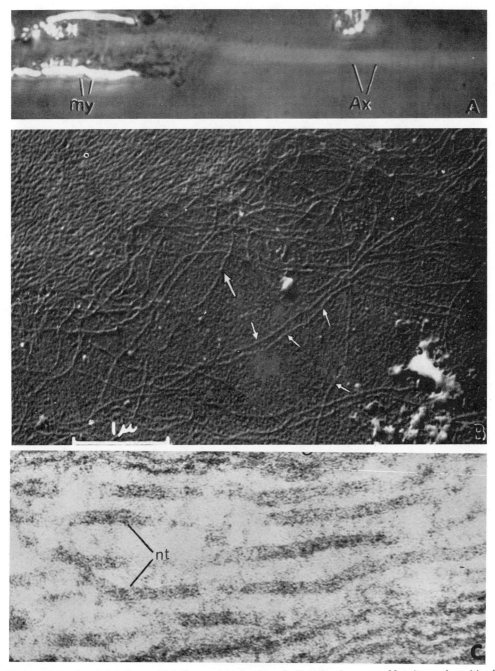

Figure 28-2 **A,** extruded axon (*Ax*) observed under the polarization microscope. Note its weak positive birefringence. After appropriate treatment, the axon exhibits both intrinsic and form birefringence; *my,* myelin with a strong birefringence. In the normal fiber, the axon birefringence is obscured by that of the myelin sheath. ×26,000. (From W. Thornburg and E. De Robertis, 1956.) **B,** fibrillar material (neurotubules and neurofilaments) observed under the electron microscope in an axon extruded from the myelin fibers and compressed. Preparation shadow cast with chromium. The arrows indicate some neurotubules. ×26,000. (From E. De Robertis and C. M. Franchi, 1953.) **C,** section of longitudinally oriented neurotubules (*nt*). ×120,000. (Courtesy of E. L. Rodríguez Echandía, R. S. Piezzi, and E. M. Rodríguez.)

oplasm extrusion and replacement by a saline solution with normal conduction of nerve impulses were mentioned. There is now no doubt that nerve conduction takes place at the surface membrane of the axon.

The Biosynthetic Functions of the Neuron Occur in the Perikaryon

The perikaryon is characterized by the presence of considerable amounts of basophilic ma-

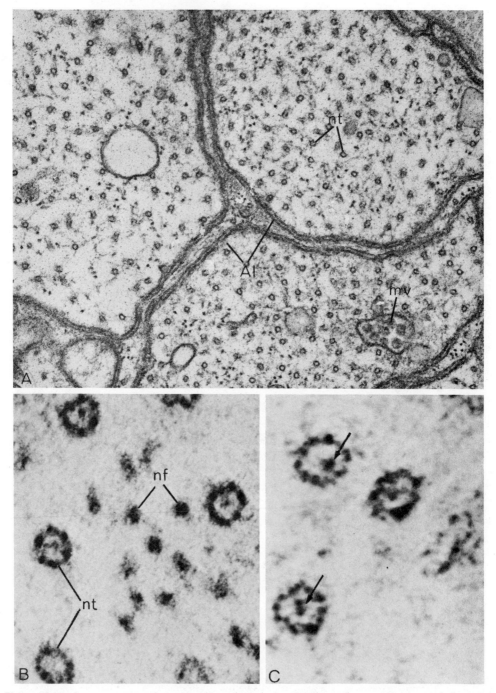

Figure 28–3 **A,** cross section of an unmyelinated nerve showing the axolemma (*Al*), neurotubules (*nt*), and a multivesicular body (*mv*). ×60,000. **B,** same as **A** at higher magnification; *nf,* neurofilaments; *nt,* neurotubules. ×400,000. **C,** neurotubules containing a dense granule (arrows). ×600,000. (Courtesy of E. L. Rodríguez Echandía, R. S. Piezzi, and E. M. Rodríguez.)

terial — the Nissl substance — which, as in other cells, is composed of ribosomes and endoplasmic reticulum. A well-developed Golgi complex is also characteristic of the neuron. The great abundance of ribosomes is related to the biosynthetic functions of the perikaryon, which

has to maintain a volume of cytoplasm in its outgrowths that may be considerably greater than its own. (In mammals axons may be as long as 1 meter or longer). If a nerve fiber is cut, the distal part degenerates (wallerian degeneration), and the proximal stump may regenerate later on by a growing process that is dependent on the perikaryon. There is also experimental evidence that the axon is continuously growing and being used at the endings.[6]

Macromolecules are Transported Through the Axon in an Orthograde Direction

The experiments just mentioned — on degeneration — should be correlated with the fact that the axon and the nerve endings are generally devoid of polysomes, and are unable to undertake a significant local protein synthesis. In fact, local synthesis by nerve endings may account for only 2 per cent of the rapidly renewed proteins.[7, 8] For this reason most axonal and synaptic proteins must be manufactured in the perikaryon and transported to the nerve endings.

The study of the axonal transport (*axon flow*) of macromolecules in the *orthograde direction* (i.e., from the perikaryon to the nerve endings) has become one of the most intense fields of research in neurobiology.[9-11] Macromolecules (such as proteins, glycoproteins, and enzymes) which are soluble in the axoplasm or integrate the various axonal and synaptic structures (i.e., axonal membrane, neurotubules, neurofilaments, mitochondria, synaptic vesicles, nerve-ending membranes, etc.), are first synthesized in the cell body and then find their way into the dendritic arborizations. They may also pass into the axon and nerve endings, at a distance that may vary from millimeters to meters. The proteins present in nerve endings have half-lives which range from 12 hours to 50 to 100 days;[12] to be replaced, therefore, new proteins must reach the nerve endings to compensate for their loss. Experimental work that has used colchicine and vinblastine has demonstrated that axoplasmic transport is related to the neurotubules.[13] In fact, the integrity of these axonal organelles is critical in maintaining a normal axoplasmic flow. One of the best approaches to the study of axonal transport is the use of radiolabeled precursors for proteins or glycoproteins (e.g., ^{3}H-lysine, ^{3}H-fucose) and the monitoring of the final product by radioautography at the electron microscopic level.

Experiments of this kind have been made on the ciliary ganglion of the chicken because it has definite advantages. In fact, the axons originate from the Edinger-Westphal nuclei, situated near the cerebral aqueduct, where the precursor can be injected. The axons are 10 mm long and terminate in a single giant nerve ending (the presynaptic calyx), which makes contact with the postsynaptic cell body of the ciliary ganglion

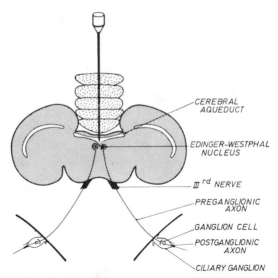

CEREBRAL AQUEDUCT

EDINGER-WESTPHAL NUCLEUS

III rd NERVE

PREGANGLIONIC AXON

GANGLION CELL

POSTGANGLIONIC AXON

CILIARY GANGLION

Figure 28–4 Diagrammatic representation of the administration of labeled precursors to the preganglionic neurons of the ciliary ganglion in chickens. The radioactive tracer injected into the cerebral aqueduct reaches the nerve cell bodies of the preganglionic neurons located in the Edinger-Westphal nucleus near the cerebral aqueduct. The preganglionic axons follow the pathway of the oculomotor nerve (III) and terminate in the ciliary ganglion by forming giant presynaptic calices. (From Droz, B., Koenig, H. L., and Di Giamberardino, L., *Brain. Res.*, 60:93, 1973.)

(Fig. 28–4).[14] Figure 28–5 shows a radioautograph obtained after the injection of ^{3}H-lysine. In the lower portion of the figure the postsynaptic ganglion cell is shown to be completely devoid of silver grains, indicating that there is no transynaptic migration of proteins. In the upper part, the presynaptic ending is covered with silver grains corresponding to proteins that have migrated from the perikaryon. The radioactive proteins are distributed over synaptic vesicles, mitochondria, and axoplasm.

Orthograde Axonal Transport may be Fast or Slow

The studies just described, as well as others, have revealed that the proteins may be transported in an orthograde direction at different rates, and that there are generally two definite groups of proteins — those of fast or those of slow transport. The diagram of Figure 28–6 gives a general idea of these two systems.

Fast Axonal Flow. Axonal flow of about 280 mm per day is found in proteins that are synthesized in polysomes bound to the endoplasmic reticulum (ergastoplasm), that are transferred to the Golgi complex, and then penetrate the axon

609

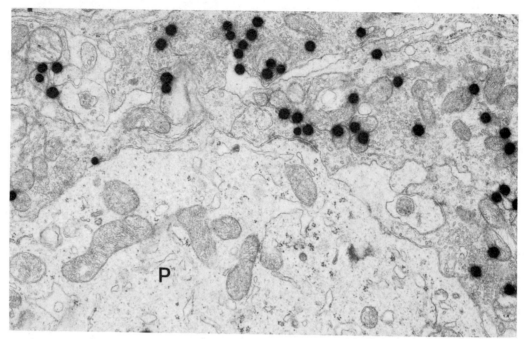

Figure 28–5 Radioautograph of the transport of a protein labeled with ^{3}H-lysine that was injected into the cerebral aqueduct observed under the electron microscope (see Fig. 28–4). The section corresponds to the ciliary ganglion of the chick. In the upper part of the figure the silver grains of the labeled protein are concentrated over the nerve endings, whereas there are none in the postsynaptic perikaryon (P). (Courtesy of B. Droz et al., 1973.)

and dendrites, mainly by way of tubules and vesicles of the smooth endoplasmic reticulum. On their way through the Golgi apparatus, some proteins are converted into glycoproteins by the addition of carbohydrates (see Chapter 11–3). Figure 28–6 indicates that the proteins of fast transport contribute to the renewal of the axon membrane, the turnover of macromolecules associated with synaptic vesicles, and the presynaptic membrane. Another fraction is related to axonal and nerve ending mitochondria. The fast moving proteins are thus assigned mainly to the renewal of various membrane components; by cell fractionation they are recovered in the particulate fractions.[15] Practically all glycoproteins are also transferred by this fast phase of axonal flow.[16] Since these fast moving proteins and glycoproteins can cover 280 mm per day, it is understandable that they begin to appear in the ciliary ganglion within one hour of the injection of the precursors into the aqueduct.

Slow Axonal Flow. In proteins present in the soluble fraction of the nervous tissue, and to a lesser degree in mitochondria, axonal flow is slow.[17] Figure 28–6 shows that these proteins are synthesized mainly by free polysomes and probably by-pass the Golgi complex. Soluble enzymes, such as choline acetyltransferase, protein subunits of tubulin that assemble to form the

neurotubules, or the protein of microfilaments, are transported by slow axonal flow. These enzymes enter the axon at a speed of 1 to 1.5 mm a day and exhibit maximal accumulation in the nerve ending at 6 days. In general, these proteins turn over at a slow rate in the axon, and only a minor fraction of them enters the nerve ending.

The fast and slow moving proteins that enter the axon and nerve endings may compensate for the local breakdown by proteolytic enzymes, as well as the release of proteins that may be related to the function of synaptic vesicles (see later discussion). A possible postsynaptic transfer of macromolecules of small magnitude should also be contemplated.

Macromolecules Such as NGF Undergo Retrograde Axonal Transport

We have described earlier the axon flow of macromolecules in the orthograde direction, but there is also an axon transport that goes the opposite way, i.e., a *retrograde axon flow*. Although by 1917 Levi, observing the axons of cultured nerve cells, had detected movement of particles in both orthograde and retrograde directions, the latter type of transport had been

610

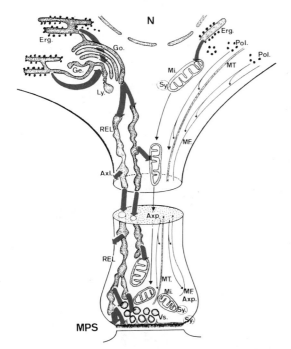

Figure 28-6 Diagram representing the ortho-grade axonal transport of macromolecules. **On the left,** the *fast transport;* **on the right,** the *slow transport.* In fast transport the proteins synthesized in the ergasto-plasm (*Erg.*) are transferred to the Golgi complex (*Go.* and *Ge.*) where carbohydrates may be added to make glycoproteins. Along the axon the proteins and glyco-proteins are transported by way of the smooth endo-plasmic reticulum (*REL*). At the rapid speed of 280 mm per day such proteins participate in the turnover of macromolecules associated with the axolemma (*Axl.*), the synaptic vesicles (*Vs.*), the presynaptic membrane (*MPS*), and mitochondria (*Mi.*). In slow transport, in-dicated by lightly drawn arrows, the proteins synthe-sized in free ribosomes (*Pol.*) are slowly transported with the axoplasm (*Axp.*) along the axon. The soluble proteins, as well as the microtubules (*MT.*) and micro-filaments (*MF.*), are transported at the slow speed of 1.5 mm per day. *N,* nucleus; *Ly,* lysosome, *Sy,* synaptic protein. (Courtesy of B. Droz et al., 1973.)

neglected until recently. By injecting high con-centrations of horseradish peroxidase or serum albumin, coupled to a fluorescent dye, in the vicinity of nerve terminals, it was found that these proteins were transferred through the axon to the corresponding cell bodies.[18-20] For exam-ple, if peroxidase is injected into the visual cor-tex, at the occipital pole of the brain, it is possi-ble to trace the neurons into the lateral geniculate nucleus. While this technique has be-come of great importance in tracing neuroana-tomic connections in the central nervous system and is now widely used,[18-21] the physiologic im-portance of the retrograde transport was ques-tionable. In fact, peroxidase is not a protein nor-

mally found in the nervous system and had to be injected at rather high concentrations.

The physiologic importance of this mechan-ism was recognized when it was found that the *nerve growth factor* (NGF) (a protein of low mo-lecular weight that is indispensable for the growth of adrenergic neurons)[22] is transported by a retrograde mechanism. For example, if ^{125}I-labeled NGF is injected into the anterior chamber of the eye of a rat, the radioactivity is taken up by the adrenergic terminals of the iris and transported to the superior cervical ganglia, where the adrenergic perikarya are present.[23, 24] These findings have suggested that NGF could act as a retrograde "trophic" messenger between the effector cells and the neurons.

At present the following two types of retro-grade transport are postulated: (1) a *non-specific* type which depends, as indicated above, on the injection of a macromolecule at high concentra-tion (e.g., peroxidase or serum albumin) and (2) a *specific retrograde axonal transport* which de-pends on the binding of the macromolecule to specific binding sites present at the nerve-ending membrane.[25] The typical example of spe-cific transport is that of the NGF. However, other macromolecules, such as tetanus and chol-era toxins and several lectins, may also be trans-ported by a selective mechanism. The transport velocity of NGF in sympathetic axons was es-timated at 2 to 3 mm/hr,[26] and in sensory neurons, at 10 to 13 mm/hr.[25] By coupling NGF to peroxidase it has been possible to follow the subcellular components that are involved in the retrograde axonal transport under the electron microscope. As shown in Figure 28–7, they fol-low essentially the same route as indicated in Figure 28–6 for fast orthograde transport. The macromolecules are found in the axon in tubules of smooth endoplasmic reticulum and then in the perikaryon, in tubules, multivesicular bodies, and dense bodies that may correspond to secondary lysosomes.[27] We mentioned that the NGF could be considered as a retrograde trophic factor, and it is well known that it is indispensable for the growth of the adrenergic neuron.[22] If newborn rats are treated with 6-hydroxydopamine (precursor analogue of the neurotransmitter norepinephrine) the adrener-gic nerve terminals and the cell bodies are de-stroyed. However, the degeneration of the cell bodies can be prevented with NGF.[28] Similar results are obtained by surgical axotomy or by the action of *vinblastine,* which acts on the as-sembly of microtubules and blocks both ortho-grade and retrograde axonal transport.

The NGF is indispensable at the early stages of

611

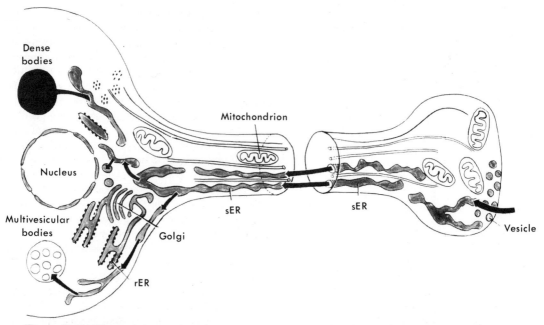

Figure 28–7 Diagram showing the retrograde axonal transport of macromolecules. The possible pathway includes endocytic vesicles, the smooth endoplasmic reticulum (SER), multivesicular and dense bodies. Notice that the macromolecules do not penetrate into the nucleus. (Courtesy of H. Thoenen.)

neuronal growth, and with increasing age (after 3 weeks) the neuronal perikaryon becomes resistant to axotomy, to the action of vinblastine, and to 6-hydroxydopamine. In the adult the last-mentioned neurotoxic drug affects only the adrenergic nerve endings. The NGF not only produces the growth of the adrenergic neuron but also stimulates the key enzymes involved in the metabolism of the neurotransmitter, *tyrosine hydroxylase* and *dopamine-β-hydroxylase*.[29]

Since the specific retrograde transport involves binding to specific receptor sites present at the nerve-ending membrane, by using different ligands (i.e., NGF, toxins, lectins, and others) it is possible to obtain qualitative and quantitative information about the composition of this membrane.

Tetanus Toxin and Some Neurotropic Viruses May Undergo Retrograde and Trans-synaptic Transport

Tetanus toxin, one of the most potent neurotoxins, is transported by the axons of the motoneurons to the anterior horn of the spinal cord by a retrograde transport similar to that of NGF. However, the toxin does not remain confined to the perikarya and dendrites of the motoneurons but is also found in the presynaptic nerve terminals that are attached to the neuron.[30] From

the medical viewpoint it is interesting that the trans-synaptic transfer of the toxin coincides with the first symptoms of local tetanus.

In addition to tetanus toxin, a series of neurotropic viruses (e.g., polio virus) reach the nervous system by binding to the peripheral nerve terminals and then undergoing retrograde transport, possibly followed by trans-synaptic transfer. In this way viruses may gain access to the central nervous system in spite of the blood-brain barrier that blocks entry from the bloodstream. More detailed information about the molecular mechanism of the interaction of toxins and viruses with the nerve-ending membrane would have medical importance in developing specific therapeutic approaches to cope with these generally lethal infections (see Thoenen and Schwab, 1976).

Nerve Conduction Velocity — Related to Diameter, Myelin, and Internode Distance

Nerve fibers are *non-myelinated* when wrapped only in Schwann cells. *Myelinated* nerve fibers also have a myelin sheath that consists of a multi-layer lipoprotein system (see Figure 8–10). In the autonomic system of vertebrates, most nerve fibers are unmyelinated and are contained within invaginations of the plasma membrane of the Schwann cells. The myelin sheath is inter-

TABLE 28-1 PROPERTIES OF NEURONS OF DIFFERENT SIZES (CAT AND RABBIT SAPHENOUS NERVES)*

Properties	Group		
	A	B	C
Diameter of fiber (μm)	20–1	3	—
Conduction velocity (m/sec)	100–5	14–3	2
Duration of action potential (msec)	0.4–0.5	1.2	2.0
Absolute refractory period (msec)	0.4–1.0	1.2	2.0

*Modified from Grundfest, H., *Ann. Rev. Physiol.*, 2:213–242, 1940.

rupted at the *nodes of Ranvier*. The distance between nodes varies with the diameter of the fiber. The *internode*, i.e., the distance between successive nodes, is the segment of myelin that is produced and contained within a single Schwann cell. The internode is 0.2 mm in a bull frog fiber of 4 μm, about 1.5 mm in a fiber of 12 μm, and 2.5 mm in one of 15 μm.

Within the internode, obliquitous (conic) *incisures* go across the myelin sheath where the myelin leaflets have a looser disposition. At the node the myelin lamellae are loosely arranged, and a small zone of axon is in direct contact with the extracellular fluid. The myelin sheath acts as an insulator and, as a consequence, myelinated fibers conduct nerve impulses at a much faster rate than unmyelinated fibers. The diameter of the fiber also influences the conduction rate. As shown in Table 28–1, nerve fibers can be classified according to their diameters into groups A, B, and C. C fibers are unmyelinated. The diameter may vary from 20 μm in A fibers to less than 1 μm in C fibers, and the conduction velocity varies from 100 to 2 meters or less per second. The rate of conduction of the nerve impulse follows a linear relationship with the fiber diameter in mammalian myelinated fibers, and it is also related to the internode distance.

The Action Potential – Non-decremental and Propagated as a Depolarization Wave

When a nerve fiber is stimulated, a profound change is produced in the electrical properties of the surface membrane and in the steady potential, and a wave of depolarization is conducted.

For the study of the physicochemical phenomena underlying the conduction of the nerve impulse, consult general physiology textbooks. Here, the subject will be discussed briefly and superficially as a continuation of the discussion of active transport and membrane potentials in Chapter 8–3 (see Figs. 8–14 and 8–15). As shown in Figure 28–8, using intracellular recording, it was demonstrated that upon excitation the resting potential suddenly changes. At the point of stimulation there is not only a depolarization, but also an overshoot, and the potential inside becomes positive.

With radioactive tracers it has been found that at the point of stimulation there is a sudden, and several hundredfold increase in permeability to Na^+, which reaches its peak in 100 microseconds.[31] At the end of this period the membrane again becomes essentially impermeable to Na^+, but the K^+ permeability increases, and this ion leaks out of the cell, repolarizing the nerve fiber. In other words, during the rising phase of the spike Na^+ enters, and in the descending phase K^+ is extruded. Complete restoration of the ionic balance takes a longer time after the electrical event.

The action potential that develops in the nerve fiber has several other characteristics: (1) The stimulus produces a slight local depolarization in the fiber, which, after reaching a certain *threshold of activation*, produces spikes of the same amplitude. If the intensity of the stimulus is increased, the height of the spike always remains the same. This is called an *all-or-none response*. (2) The nerve impulse is *non-decremental;* i.e., the amplitude of the spike does not decrease and is the same all along the course of the nerve fiber. The action potential is thus well adapted to conduction over long distances without losses (see Figure 28–1). (3) Once a nerve impulse has passed over any point of the fiber, there is a *refractory period* during which it cannot react to another stimulus.

The *propagation* of the nerve impulse is generally explained by the so-called *local circuit theory* (Fig. 28–9). At the point of stimulation the area becomes depolarized (negative outside) and acts as a sink toward which the current flows from the adjacent areas (Fig. 28–9,B and C). This wave of depolarization advances along the nerve fiber at a rate of conduction that is characteristic for each fiber (Table 24–1). While this wave of depolarization advances, repolarization is so rapid that only a fraction of the nerve fiber (a few millimeters or centimeters, depending on the conduction rate) is depolarized at a time. In the recovery period, sodium leaves the cell by the action of the sodium pump (Chapter 8–3) and potassium re-enters to restore the steady state.

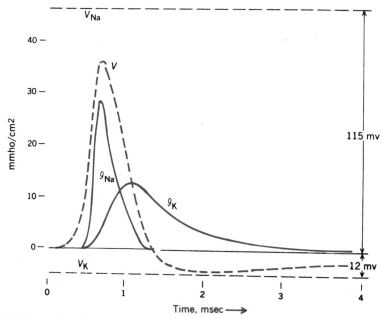

Figure 28–8 Diagram showing a propagated action potential (curve V) and the sodium (^gNa) and potassium (^gK) conductances. Observe that the entrance of sodium coincides with the rising phase of the spike. (From Hodgkin, A. L., and Huxley, A. F., *Cold Spring Harbor Symp. Quant. Biol.*, 17:43, 1952.)

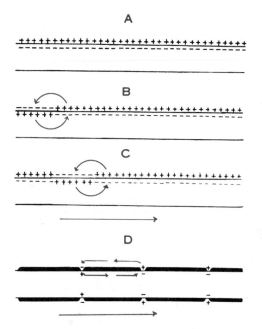

Figure 28–9 Diagram illustrating the local circuit theory of propagation of the action potential (**A, B, C**) in unmyelinated neurons and muscle fibers as compared to saltatory conduction in myelinated neurons (**D**). **A,** the membrane of an unexcited nerve (or muscle) fiber; **B,** the cell membrane excited at one end; **C,** the movement of the action potential, followed by recovery; **D,** node-to-node saltatory conduction. In large nerve fibers less than one hundredth as much ionic exchange occurs during an impulse in saltatory conduction as compared to conduction in an unmyelinated nerve fiber. The arrows in **C** and **D** show the direction of impulse propagation. (After Hodgkin, 1957, *Proc. Roy. Soc. London,* ser. B, *148*:1, 1957.)

This recovery is probably produced at the expense of high-energy phosphate bonds. However, impulses continue to discharge for some time in the absence of oxygen, and even when glycolysis is inhibited, which indicates that high energy bonds are stored at the membrane.

In Myelinated Nerve Fibers Conduction is Saltatory

The preceding theory of nerve conduction applies to unmyelinated nerve fibers. In myelinated fibers it is thought that the local circuits occur only at the nodules (Fig. 28–9,D). According to this so-called *saltatory theory*, at the internode the impulse is conducted electrotonically, and at each node the action potential is boosted to the same height by ionic mechanisms. In this way the amount of Na^+ and K^+ exchanged is greatly reduced and the net work required is much less. Stimulation of myelinated nerve fibers with fine electrodes has shown that at the nodes the threshold of stimulation is much lower than at the internode.[32] It has been found that the nerve impulse can jump across one anesthetized node, but not two of them.

At the Physiologic Receptors and Synapses Potentials are Graded and Non-propagated

Physiologic studies have demonstrated that in addition to the all-or-none response, there is another type of electrical activity in nervous tissue. This is, by far, the most frequent in the central nervous system and is referred to as a *graded response*. In the graded response the impulse is not *propagated* and the *amplitude* varies with the intensity of the stimulus. This type of response is characteristic of the physiologic receptors and synapses (see Figure 28–1). Both the *generator potentials* found at the receptors and the *synaptic potentials* are graded responses.

If a peripheral receptor, such as a Pacini corpuscle or a stretch receptor (neuromuscular spindle), is mechanically stimulated, a local, graded, and decremental potential is recorded, the amplitude and duration of which depend on the intensity and duration of the stimulus. In the Pacini corpuscle it is possible to remove most of the connective lamellae that surround the nerve ending without impairing the generator potential (Fig. 28–10). It appears that in this case the *biological transducer* capable of transforming the mechanical energy (pressure) into the electrical energy (generator potential) is localized at the sensory part of the ending (Fig. 28–10). Probably the mechanical deformation of the ending produces a change in permeability with entrance of ions and partial depolarization. In the Pacini corpuscle it has been observed that the nerve impulse starts at the first node of Ranvier (Fig. 28–10).[33]

The intensity of the sensory stimulus is reflected in the amplitude of the generator potential and this in turn in the *frequency* of the propagated signal — the stronger the generator potential, the higher the frequency. In this way the information received is coded for conduction along the nerve fiber in the form of a train or volley of impulses (see Figure 28–1).

Propagated Action Potentials Depend on Na^+ and K^+ Channels in the Axon Membrane

In Chapter 8–3 we gave an introduction to the concepts of Na^+ and K^+ channels, and we described the Na^+-K^+ pump (Na^+-K^+ ATPase) the function of which is to extrude Na^+ from the axon and to allow the entrance of K^+ (Fig. 8–14). These three components (i.e., Na^+ and K^+ channels and the Na^+-K^+ pump) are essential to explaining the molecular basis of the propagation of action potential along the axonal membranes, as well those of the muscle.

An ion channel can be envisioned as a hydrophilic channel formed by the interstice between protein subunits embedded in the axon membrane. It is now generally admitted that the ion channel consists of two functional elements: (1) a *selectivity filter*, which determines the kind of ion that will be translocated, and (2) *a gate*, which by opening and closing the channel, regulates the ion flow. In the Na^+ and K^+ channels of nerve (and muscle) membranes this gate is regulated by the membrane potential; in other words, the gating mechanism is electrically driven. In Figure 8–14,E the voltage dependence of Na^+ and K^+ channels is indicated diagrammatically. It may be observed that in the resting (polarized) condition both the Na^+ and K^+ channels are closed. With depolarization, the Na^+ channel is opened and during repolarization it closes again, and the K^+ channel opens. We shall see later that another type of channel, present at the postsynaptic membrane, is regulated chemically by the interaction of the neurotransmitter with specific receptor molecules.

Sodium Channel. The sodium channel is better known than the potassium channel because there are several drugs that affect it in a different manner. For example, *tetrodotoxin*

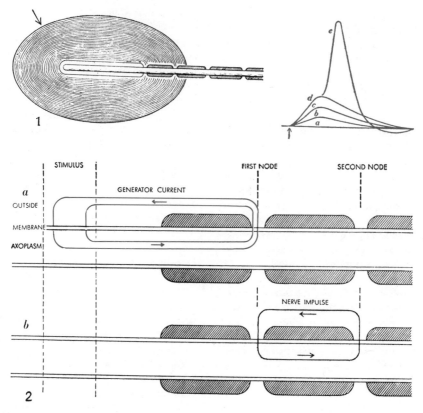

Figure 28–10 **1**, Pacini corpuscle with the nerve ending surrounded by multiple layers. Stimulation at the point marked by the arrow produces a generator potential (**right**) which increases in amplitude (*a–d*) until it fires an all-or-none nerve impulse (*e*). **2**, mechanism of the transducer. The stimulus produces a drop in the resistance of the membrane with ion transfer. Notice the generator current induced by the stimulus (*a*) and the nerve impulse (*b*) originating at the first node. (Courtesy of W. R. Loewenstein.)

(TTX), a toxin derived from the Japanese puffer fish, at concentrations well below 1 μM completely blocks the Na$^+$ channels. Other toxins act at different steps of channel function.[34, 35] By using ^{3}H-TTX the number of channels per μm^2 has been calculated.[36] In the axon membrane, this is a rather small number, i.e., between 50 and 500 per μm,[2] but there are many more at the nodes of Ranvier.

The TTX binding site appears to be a glycoprotein embedded in the lipid phase. The Na$^+$ channel is blocked by protons and, by comparing its permeability to various ions, a diameter of 0.3×0.5 nm has been estimated.[36] A single channel transports about 10^8 Na$^+$/sec, with a conductivity of 4×10^{-12} Ω^{-1}, which is about one third that of the K$^+$ channel.[37]

Since TTX has no effect when applied internally, it is generally accepted that the selectivity filter of the Na$^+$ channel is located on the outer side of the membrane.[38] This selectivity filter is independent of the gating mechanism, which operates even after blocking with TTX. This

mechanism has been explained by voltage- and time-dependent conformational changes in protein subunits, and at least three of these subunits are postulated.[38] A four-barrier model in which the limiting step is the dehydration of the cation has been postulated for the sodium channel.[39]

Potassium Channel. The potassium conductance is selectively inhibited by tetraethylammonium ions (TEA); however, the affinity is too low to carry out binding assays.[40] The affinity can be increased by replacing one of the ethyl groups of TEA with a more hydrophobic side chain. In the squid giant axon, TEA blocks only the K$^+$ channel from the inside. This molecule has the same diameter as the hydrated potassium ion. This and other findings have led to the postulation of a funnel shape for the axonal K$^+$ channel. As indicated in Figure 28–11 the large opening of the funnel is toward the interior, and it has a diameter of about 0.8 nm, which is large enough for the potassium ion, as well as TEA. The selectivity filter, which corresponds to the narrow end

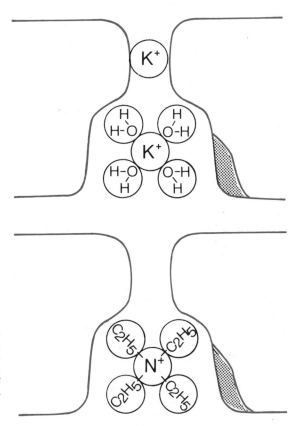

Figure 28–11 Model of an axonal potassium channel. The selectivity filter at the narrow portion of the funnel has a diameter only large enough for a dehydrated potassium ion. The large inner portion can accommodate a hydrated ion or TEA. The shaded portion corresponds to a possible hydrophobic region in the channel. (Courtesy of F. Hucho and W. Schiebler, from Hucho, et al., *Molec. Cell Biochem.*, 18:151, 1977.)

of the funnel, is only about 0.3 nm in diameter, i.e., large enough to let only the dehydrated potassium ion pass. A hydrophobic site is thought to be present at the large entrance of the channel.[41] The gating mechanism, sensitive to changes in the membrane potential, could be located on the inside of the channel.[40, 41] The number of sites for the squid axon has been calculated to be $70/\mu m^2$ [37] (see Hucho and Schiebler, 1977).

The Na^+-K^+ pump, which is the third component of the axon membrane involved in ion transport, was studied in Chapter 8–3.

SUMMARY:

General Organization of the Neuron and Function of the Nerve Fiber

The nervous tissue is specially differentiated to respond rapidly and specifically to different stimuli. Protozoa and plants contain no nerve differentiation but have more primitive mechanisms of *irritability*. The simplest mechanism in the nervous tissue is the circuit of the reflex arc, which is integrated by a sensorial (*afferent*) neuron and a motor (*efferent*) neuron. These two neurons are related, respectively, to a a physiological *receptor* and an *effector; they are connected by synapses.* The nerve impulse is *conducted* along the axon of the nerve fibers and is *transmitted* at the level of the synapse and the effector.

Neurons do not divide during postnatal life, but undergo an increase in volume and in the number and complexity of their processes and functional contacts. Neurofibrils can be observed in the axon and dendrites after fixation and staining of the perikaryon. They result from the clumping together of neurotubules and neurofilaments. Neurotubules are similar to the microtubules of other cells (see Chapter 9).

Neurofibrils are not involved in nerve conduction, since this takes place at the axonal membrane.

The perikaryon of neurons is rich in ribosomes and endoplasmic reticulum (Nissl substance). The great abundance of ribosomes is related to their biosynthetic functions. If the nerve fiber is cut, the distal stump degenerates (wallerian degeneration), and the proximal stump may regenerate later on. There is little or no local protein synthesis in the axon and nerve endings, which are devoid of polysomes. Proteins and glycoproteins are synthesized in the perikaryon and are transported to the axon and nerve endings as well as to the dendrites.

The orthograde axonal transport of macromolecules depends on the integrity of neurotubules. Colchicine and vinblastine, which destroy these organelles, also stop axonal transport. At the cellular level, this process can best be studied with radioautography. The ciliary ganglion of the chicken has definite advantages in such studies (Fig. 28–4). In the radioautograph the synthesized protein may be observed to enter the nerve ending, but it does not penetrate the postsynaptic ganglion cell.

Axonal transport may be *fast* (280 mm per day) or *slow* (1 to 1.5 mm per day). Fast transport is found in proteins integrating the axonal membrane, synaptic vesicles, presynaptic membrane, and mitochondria, i.e., in membrane-bound proteins. The glycoproteins studied with ^{3}H-fucose are also fast moving. Slow axonal flow is found in soluble proteins (e.g., enzymes, tubulin) that are in the axoplasm, or that integrate the neurotubules, neurofilaments and mitochondria. Few of these proteins reach the nerve ending. The fast and slow moving proteins compensate for the local breakdown and release of proteins taking place within the axon and at the nerve ending.

There is also a retrograde axonal transport of macromolecules that are taken up by the nerve terminals and transported to the perikaryon. This type of transport is *non-specific* when the macromolecule is an extraneous one and is injected at high concentration (e.g., peroxidase, bovine serum albumin), and *specific* when there are special binding sites at the nerve-terminal membrane. The nerve growth factor is transported in a retrograde direction at 2 to 13 mm/hr through the smooth endoplasmic reticulum of the axon to the perikaryon. Tetanus toxin is also transported by a specific mechanism, but after reaching the perikaryon of the motoneuron, it is transferred trans-synaptically to the nerve endings in contact with the motoneuron. It is thought that neurotropic viruses are transported by a similar mechanism.

Nerve fibers are *non-myelinated,* when they are wrapped only in Schwann cells, or they are *myelinated.* Myelin is interrupted at the nodes of Ranvier. The internode distance varies from 0.2 to 2.5 mm and is related to the conduction velocity. The diameter of the nerve fiber is also related to nerve conduction. Nerve conduction is faster with fibers of a larger diameter and which have longer distances between internodes. A, B, and C fibers can be distinguished. Nerve conduction is propagated along the axonal membrane by the *action potential.* This consists of a sudden depolarization with increased permeability to Na^+. The membrane potential may depolarize from -90 mV and may overshoot to $+50$ mV. In the ascending phase of the spike there is entrance of Na^+. In the descending phase, K^+ leaks out. The action potential has a threshold of activation, is an all-or-none response, is non-decremental, and has a refractory period. In unmyelinated fibers, *propagation* is accounted for by the *local circuit theory.* In myelinated fibers conduction is considered *saltatory,* from one node to the other. In the internode the impulse is electrotonically conducted.

In receptors and synapses there are *graded* responses that are *not propagated* (generator and synaptic potentials). A good example of a receptor is the Pacini corpuscle, which is a *biological transducer* capable of transforming mechanical energy into electrical energy.

Propagated action potentials depend on the presence of Na^+ and K^+ channels in the axon membrane. An ion channel has a kind of selectivity filter that is specific for the ion species and a gating mechanism that is electrically driven and regulates the ion flow. The Na^+ channel is blocked by tetrodotoxin, which acts externally at the selectivity filter. The voltage-dependent gate is explained by conformational changes in protein subunits caused by the rising action potential. The K^+ channel is blocked from inside by tetraethylammonium (TEA). A funnel shape has been postulated for the K^+ channel with a mouth that is 0.8 nm in diameter and that fits the hydrated K^+ or TEA. A selectivity filter 0.3 nm in diameter allows the passage of only the dehydrated K^+.

28–2 SYNAPTIC TRANSMISSION AND STRUCTURES OF THE SYNAPSE

The earliest knowledge of *synapses*, or *synaptic junctions*, came from the discoveries at the turn of the last century of the morphologic and physiologic organization of the nervous system. The so-called *neuron theory*, established mainly by Cajal, led to the assumption that the functional interactions between nerve cells was by way of contiguities or *functional contacts*. Different types of nerve terminals on dendrites or perikarya were described by the use of silver staining methods such as the characteristic boutons, the club endings, the so-called baskets or the contacts *en passant*.

In 1897, Sherrington coined the name *synapse* to explain the special properties of the reflex arc, which he considered to be dependent on the functional contact between neurons. He attributed to the synapse a valvelike action, which transmits the impulses in only one direction (see Figure 28–1). In his studies on reflex transmission he discovered some of the fundamental properties of synapses, such as the *synaptic delay* (the delay that the impulse experiences in traversing the junction), the fatigability of the synapse, and the greater sensitivity to reduced oxygen and anesthetics. He also pointed out that the many synapses situated on the surface of a motoneuron could interact, and that some would have an additive excitatory action, whereas others would be inhibitory and antagonize the excitatory ones.

Synapses are special zones of contact between two neurons, or a neuron and a nonneuronal element, such as those between a physiologic receptor and a neuron, or those with an effector cell, as in the case of the myoneural junction. Synapses can thus be defined as all the regions anatomically differentiated and functionally specialized for the transmission of excitations and inhibitions from one element to the following in an irreversible direction. However, a more modern definition of the synapse should also include the existence of a complex submicroscopic organization in both the presynaptic and postsynaptic parts of the junction and of the presence of a specific neurochemical mechanism in which transmitter, receptor protein, synthetic and hydrolytic enzymes, and so forth, are involved.

From Figure 28–1, it is clear that the main problem in synaptic transmission consists of finding out by what molecular mechanism the information carried by one neuron is transferred to the next. In other words, the problem is how the code of frequency conducted by one neuron originates a new code of frequency in the following neuron.

Nerve Impulse Transmission — Possibly Electrical but Mediated Mainly by a Chemical Mechanism

DuBois-Reymond (1877) was the first to suggest that transmission could be either *electrical* or *chemical*. These two types of mechanisms have been observed. However, chemical synapses seem to be by far the most frequent.

Electrical transmission was the first demonstrated in a giant synapse of the abdominal ganglion of the crayfish cord, and since then, in several other cases. In this type of synapse the membrane contact acts as an efficient rectifier, allowing current to pass relatively easily from the pre- to the postsynaptic element, but not in the reverse direction. In this case the arriving nerve impulse is passed without delay and can depolarize directly and excite the postsynaptic neuron. Here the one-way transmission is due to the valvelike resistance of the contacting synaptic membranes.

In addition to electrical transmission that is unidirectional there are regions of the adjacent neuronal membranes in which there is an *electrotonic coupling* due to the presence of gap junctions (see Chapter 8–4). In this type of transmission the electric flux is bidirectional.

Chemical transmission presupposes that a specific chemical transmitter is synthesized and stored at the nerve terminal and is liberated by the nerve impulse. The transmitter produces a change in ionic permeability at the postsynaptic component that causes a bioelectrical change. In 1904, Elliot suggested that sympathetic nerves act by liberating adrenalin at the junctions with smooth muscle. Later, von Euler demonstrated that *noradrenalin* was the true adrenergic transmitter. The studies of Dixon (1906) and particularly of Dale (1914) strongly supported chemical transmission in the parasympathetic system. This was finally proved in the heart by Loewi in 1921. Since then, *acetylcholine* has been demonstrated to act in sympathetic ganglia, neuromuscular junctions, and in many central synapses. Modern studies on chemical synaptic transmission have revealed that synapses are the sites of a transducing mechanism in which the electrical signals are converted into chemical signals, and these, in turn, again into electrical signals. It will be shown that active substances of low molecular weight (e.g., acetylcholine, noradrenalin, dopamine, glutamate, γ-aminobutyrate, and others) are produced at the nerve endings and packaged in special containers (i.e., synaptic vesicles) in multi-molecular quantities. These packages of the so-called *transmitters* are released when the activation is produced by the nerve impulse. The transmitter, in turn, reversibly reacts with special receptor proteins, present at the postsynaptic membrane. This transmitter-receptor interaction produces a change in permeability to certain ions, thereby creating a *synaptic potential* in the postsynatic cell.

Synaptic Transmission may be Excitatory or Inhibitory

Physiologic studies on synaptic transmission were greatly improved by the use of microelectrodes which could be implanted near the synaptic region or intracellularly in the pre- and postsynaptic neuron. The first synaptic potential to be recorded directly was the *end plate potential* of the myoneural junction.

With intracellular recordings in large nerve cells (e.g., motoneurons, pyramidal cells, invertebrate ganglion cells, etc.), it was also observed that the arrival of the presynaptic nerve impulse produces a local synaptic potential. *Synaptic potentials*, like the generator potentials studied above, are graded and decremental and do not propagate. They extend electrotonically only for a short distance with reduction in amplitude.[42]

A typical experiment, shown in Figure 28–12, involves two ganglion cells of *Aplysia* (a marine mollusk), one of which *(P)* acts synaptically upon the other *(F)*. Neuron P is impaled

with two microelectrodes, one of which is used for stimulation *(St)* and the other for recording *(R)*. Neuron F is impaled with one microelectrode *(R)* to register the synaptic potential. Two main types of cells can be found, one of which produces an excitatory synaptic potential and the other an inhibitory postsynaptic potential.[43]

Excitatory synapses induce a depolarization of the postsynaptic membrane, which, upon reaching a certain critical level, causes the neuron to discharge an impulse. The *excitatory postsynaptic potential* (EPSP) is due to the action of the transmitter released by the ending (Fig. 28–12,1). This causes a change in permeability of the subsynaptic membrane, allowing the free passage of small ions, such as Na^+, K^+, and Cl^-.

Similarly, *inhibitory synapses* affect the subsynaptic membrane. In these instances the transmitter causes a transient increase in membrane potential, the so-called *inhibitory postsynaptic potential* (IPSP) (Fig. 28–12,2). This hyperpolarizing effect induces a depression of the neuronal excitability and an inhibitory action.

The excitatory or inhibitory action is not dependent exclusively on the type of transmitter substance. For example, acetylcholine is excitatory in the myoneural junction, sympathetic ganglia, and so forth, but inhibitory in the vertebrate heart, in which it reduces the frequency of contraction.

Figure 28–13 shows that in the ganglion cells of *Aplysia* the injection of acetylcholine may have an excitatory synaptic effect in certain cells producing depolarization and increased frequency of discharges (1) or only a depolarization without firing' (2). In other cells the same treatment provokes a hyperpolarization and inhibition of spontaneous discharges (3).

These facts indicate that the nature of a synapse depends, in particular, on the mechanism of the receptor (i.e., the chemical reactivity of the membrane in the postsynaptic neuron). The use of intracellular recording has greatly contributed to the delineation of some of the basic mechanisms by which the code of signals is transmitted from one cell to the next. All the synaptic potentials from the different excitatory and inhibitory endings impinging upon a neuron are integrated. Both of these types of input will change the electrical properties of the membrane at a critical zone of the cell of low excitatory threshold, which is called the "pacemaker." In this region, which in the motoneurons is located at the initial segment of the axon, new impulses are fired.[44] In this re-

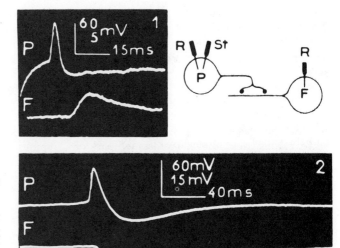

Figure 28-12 Diagram of the experiment in two ganglion cells of *Aplysia* that are related synaptically. (See the description in the text.) **1,** excitatory response. Depolarization of the membrane at *F* after arrival of the action potential from *P*. **2,** inhibitory response. Hyperpolarization of the membrane at *F* after arrival of the action potential from *P*. (Courtesy of L. Tauc and H. M. Gerschenfeld.)

spect a neuron acts in a way similar to an analogue computer, i.e., integrating all the inhibitory and excitatory messages that impinge on its surface before deciding whether or not to fire a new impulse.

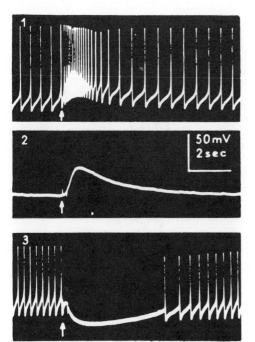

Figure 28-13 **1,** intracellular recording in a neuron of *Aplysia* (see Fig. 24–11) that is firing spontaneously. At the point marked by the arrow, acetylcholine is added, producing depolarization and increasing the frequency of discharges (excitatory synapse). **2,** same experiment on a neuron, without firing. Only depolarization is produced. **3,** neuron in which the action of acetylcholine induces hyperpolarization and inhibition of spontaneous discharges. (Courtesy of L. Tauc and H. M. Gerschenfeld.)

Several Thousand Synapses may Impinge on a Single Neuron

Morphologic studies with the light microscope revealed that the size, shape, and distribution of synapses of different regions of the central and peripheral nervous tissue vary considerably. Synapses are classically categorized as axodendritic, axosomatic, or axo-axonic, according to the relationship of the ending to the postsynaptic component. The endings may have different sizes and shapes, (e.g., bud, foot, or button ending, club ending, and calix (cup) ending). *The number* of synapses that terminate on the perikaryon and dendrites varies considerably. In the above-mentioned calix synapses of the ciliary ganglion there is a single huge synapse per cell (Fig. 28–5). A large motoneuron of the spinal cord may receive as many as 10,000 synaptic contacts, and a large Purkinje cell of the cerebellum may receive as many as 200,000. These figures give an idea of the extraordinary complexity of the nervous system. This immense number of synapses carries information from numerous other neurons, some of which have an excitatory effect, and others, an inhibitory. Thus, the neuron is a real computation center where all the information is integrated and from which new information is sent along the axon by nerve impulses.

The Number of Synapses is Related to that of the Dendritic Spines

In general the number of synapses relates to the number and length of the dendrites on which most of the synaptic contacts are produced. The dendrites have fine protrusions on the surface, the so-called *spines,* upon which synaptic contacts with the nerve endings are made.

A typical spine is pedunculated with an enlarged tip between 0.5 and 2.0 μm in diameter and a narrow stalk between 0.5 and 1 μm long. We shall see later that upon isolation of the nerve ending (i.e., the synaptosome) part of the spine is also separated (Fig. 28–15). The number and size of the dendritic spines, which can be easily studied in neurons stained by the Golgi method, allow one to judge the importance of the synaptic connections of a definite neuron (Fig. 28–14). Such studies are important in pathologic conditions, since the reduction in the number of spines may indicate a degeneration of afferent axons and terminals. Furthermore, the number and size of spines are related to the so-called plasticity of the central nervous system. It is well known that synaptic contacts are not completely stable and may change not only during development but also in relation to the influence of the

inputs upon a neuron. For example, sensory deprivation produces alteration and reduction in the number of spines.[45] Recently, it has been observed that fishes reared in community tanks have more dendritic branches and spines in interneurons of the deep tectal layers than those reared without visual-tactile contact with other fishes of the same species (Fig. 28–14). The socialized controls not only have more spines but also have shorter stems. These findings suggest that social stimulation induces formation and widening of the spines.[46] Such changes in plasticity may have a profound influence on the effectiveness of the synaptic transmission and thus may be at the base of learning processes.

The Synapse Ultrastructure Suggests Many Types of Functional Contacts

The discovery of the synaptic vesicles[47] and of special features of the synaptic membranes[48] with the electron microscope provided new morphologic landmarks for the study of syn-

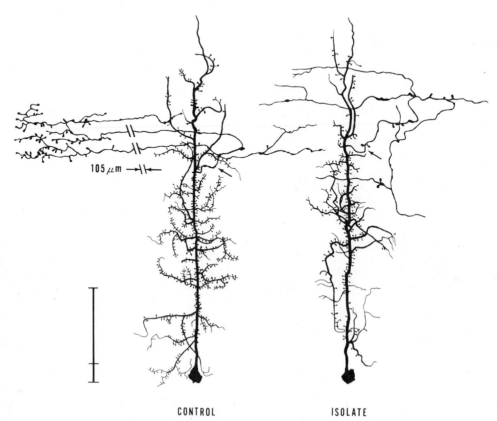

CONTROL   ISOLATE

Figure 28–14 Interneurons of the optic tectum of the jewel fish stained by the Golgi method. **Control,** numerous dendrites and spines; **Isolate,** a reduction in the dendrites and in the number of spines may be observed. These spines are longer and more slender than in the control. (From Goss, R. G. and Globus, A., *Science, 200:*787, 1978. Copyright 1978 by the American Association for the Advancement of Science.)

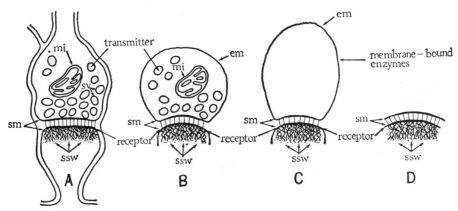

Figure 28–15 Diagram showing the systematic dissection of the synaptic region achieved by the use of cell fractionation methods. **A,** typical synapse of the cerebral cortex showing: *mi,* mitochondria; *sv,* synaptic vesicles; *sm,* synaptic membranes; and *ssw,* subsynaptic web. **B,** isolated nerve endings; *em,* nerve-ending membrane. **C,** after the osmotic shock, only the nerve-ending membrane remains. **D,** after treatment with a mild detergent, only the junctional complex remains. (From De Robertis, E., *Science, 171*:963, 1971. Copyright 1971 by the American Association for the Advancement of Science.)

apses. At the synaptic junction both synaptic membranes appear to be thicker and denser. The synaptic cleft between the synaptic membranes may be about 30 nm wide (up to 50 nm wide at the myoneural junction) and is occupied by a dense cleft material, which may show a system of fine *intersynaptic filaments* of about 5 nm that join both synaptic membranes (Fig. 28–15).

Another system of filaments has been observed to penetrate a variable distance into the postsynaptic cell. This is the so-called subsynaptic web[48] or *postsynaptic density.*[49] The demonstration of the intersynaptic material between the membranes suggests that there is greater adhesion at the junction and this was demonstrated by microdissection experiments. In fact, in an isolated cell the endings break their connection with the axons but remain attached to the cell. The discovery of the positions of the synaptic vesicles and the presynaptic and postsynaptic densities has suggested the possible functional polarity of the synapse and the recognition of a series of contacts that could not be demonstrated with histologic methods.

Synaptic contacts may be established between any portions of the neuronal membranes; in fact, the only parts of the neuron which never receive a synapse are the segments of nerve fibers covered by myelin. Furthermore, release of neurotransmitter does not seem to be restricted to axon terminals and it has been suggested recently that dendrites may also represent sites of release.

From the morphologic viewpoint the following types of synapses may be recognized

under the electron microscope: (1) *Axodendritic synapses* (Fig. 28–16,*a*), which may be directly on the dendrite but in most cases are established by way of a spine *(axospinous synapses),* (2) *Axosomatic synapses,* which terminate on the perikaryon and some of which may be at the axon-hillock of the neuron (i.e., the portion of the perikaryon from which the axon emerges). These synapses are sometimes referred to as *initial-segment synapses* and are assumed to exert an inhibitory action on the postsynaptic membrane. Such synapses are important because at this strategic point they may influence the firing of the nerve impulse. (3) *Axo-axonic* synapses are those in which a nerve ending terminates on another (Fig. 28–16,*d*). This type of synapse appears to be involved in *presynaptic inhibition,* since its effect may produce a reduction in the release of the neurotransmitter. (4) *Serial synapses,*[50] in which axo-axonic synapses are disposed serially, i.e., one process is postsynaptic to one synapse and presynaptic to another. (5) *Reciprocal synapses* correspond to the situation in which a neuronal process may be presynaptic at one point and postsynaptic at another (Fig. 28–16,*e*). (6) *Somatodendritic synapses* correspond to the special type of synapses between the perikaryon and a dendrite (Fig. 28–16,*c*). (7) *Dendrodendritic synapses,*[51] which are found in several regions of the central nervous system and in which dendrites contain cluster of synaptic vesicles opposing other dendrites. Some of these contacts may be of the reciprocal type (Fig. 28–16,*e*). These and other more complex types of synapses, such as the *glomerular* (Fig. 28–16,*b*), the *triad* (Fig. 28–16,*d*), and the *invaginated* syn-

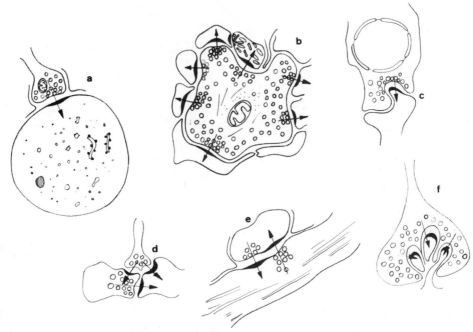

Figure 28–16 Diagram showing different types of synapses. The polarity in each case is indicated by an arrow. **a**, axodendritic; **b**, glomerular, showing several axodendritic and one axo-axonic synapse; **c**, somatodendritic; **d**, a triad, in which one terminal synapses with another and with a dendrite; **e**, a reciprocal synapse, which may involve two dendrites; **f**, an invaginated synapse. Notice that the polarity is determined by the synaptic vesicles and the postsynaptic density. (Courtesy of J. Pecci Saavedra, O. Vilar, and A. Pellegrino de Iraldi.)

apses (Fig. 28–16*f*), suggest that the functional interactions between neurons may be much more complex than previously imagined and that practically all regions of the neuronal membrane may play a synaptic role.

Lectin Receptors and Postsynaptic Densities May Play a Role in the Formation and Maintenance of Synapses

The cleft material and the postsynaptic density are also interesting because of the possibility that they may play a role in the formation and maintenance of the synaptic connections.[49] By using plant lectins, such as concanavalin A and others, it has been possible to detect the presence of specific glycoproteins on the neuronal membrane (see Chapter 8–5). It has been demonstrated that these so-called *lectin* receptors in the postsynaptic membrane are restricted in their mobility relative to similar receptors located outside. It is thought that the synaptic receptors are linked across the membrane and that the underlying postsynaptic density may be involved in this linkage (see Kelly et al., 1976).

The possibility that the postsynaptic density may play a role in the formation of synapses is

suggested by work on the formation of synapses in tissue culture. If neurons of the superior cervical ganglion of rat fetuses are grown in tissue culture, together with segments of the spinal cord, synapses are formed. *Growth cones* (i.e., enlargement of the axon terminal) of the spinal neurons come in contact with the ganglion cells by way of filopodia; then a more extended contact is made and from the Golgi region are formed numerous coated vesicles (see Chapter 13–2) which come in contact with the plasma membrane and fuse to it. In this way the cones apparently contribute to the formation of the postsynaptic density. Subsequent to this is the appearance of synaptic vesicles which become clustered in a position opposite the postsynaptic density, and the typical synapse is thus completed (see Rees et al., 1976).

The Presynaptic Membrane Shows Special Projections at Active Zones

From the initial electron microscopic studies of synapses it has been common knowledge that synaptic vesicles tend to converge on special regions of the presynaptic membrane that were called the *active zones*. At these points there are projections of a dense material that bear a special relationship with the synaptic vesicles.[52] In central synapses these *presynaptic projections* are about 50 nm in diameter and are arranged in precise hexagonal array on the presynaptic membrane.[53] In Figure 28–17 it may be observed that these projections form a hexagonal network and that the spaces in between accommodate the attachment sites of the vesicles. In freeze-fracturing studies the dense projections are not observed, since the plane of cleavage goes along the middle of the membrane (Fig. 14–9). In the inner membrane leaflet (P face) there are clusters of 10 nm particles; however, the most characteristic feature is the presence of small pits in the P face which correspond to protuberances on the outer leaflet (E face). These pits, which are geometrically arranged, correspond to the attachment sites of the vesicles (Fig. 28–17).[53]

In the neuromuscular junction of vertebrates the presynaptic dense projections appear as long bands that run perpendicular to the axis of the nerve terminal (Fig. 28–18). Facing these bands, on the postsynaptic side, are the typical folds of the postsynaptic membrane. While most of the synaptic vesicles of the neuromuscular junction are not in contact with the presynaptic membrane, those that are on both sides of the projections are in intimate relationship with that membrane.[54] In freeze-etching studies the outer leaflet (E face) shows double rows of 10 nm particles on both sides of the position corresponding to the presynaptic projection.[53, 55] In spite of the geometric dif-

ferences (i.e., a *hexagonal grid* in central synapses and *parallel bands* in the myoneural junction), the presynaptic projections on both synapses are essentially the same. There is, however, a quantitative difference; central synapses can accommodate up to eight times more vesicles near the membrane than the motor end plate. In synapses of photoreceptors there are special presynaptic projections, the *synaptic ribbons*, that appear as flattened bands upon which numerous synaptic vesicles converge.

The function of these various types of presynaptic projections is still unknown, but it seems possible that they may act as guides for transporting the vesicles to the sites of release. This hypothesis is supported by the finding that both actin and myosin are present within the nerve ending and are associated with the presynaptic membrane and the vesicles.[56] These proteins could provide the motive force that pulls the vesicles toward the presynaptic membrane (see Osborne, 1977).

The Postsynaptic Membrane Shows a Complex Macromolecular Organization

In addition to the subsynaptic web and the dense material that occupies the cleft with the intersynaptic filaments (see earlier discussion), electron microscopic studies using freeze-etching and negative staining have revealed a fine macromolecular organization in the postsynaptic membrane.

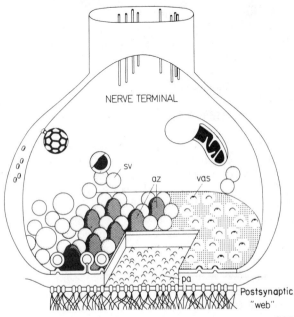

Figure 28–17 Diagram of a central synapse. *az,* active zone showing the presynaptic projection surrounded by vesicles attached to the membrane. Observe that the vesicles and the attachment sites (*vas*) are arranged to form a hexagonal pattern. *pa,* postsynaptic intramembrane particles restricted to the area facing the active zones; *sv,* synaptic vesicles. (Courtesy of K. Akert.)

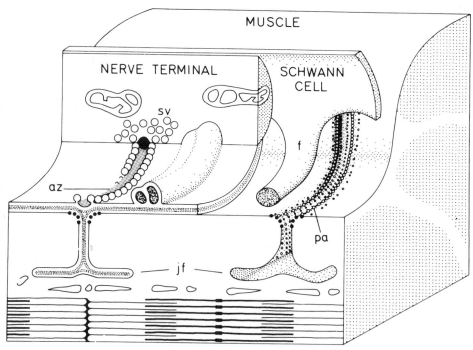

Figure 28–18 Diagram showing the main structural characteristics of the frog myoneural junction. *Az,* active zone showing the attachment of synaptic vesicles on both sides of the presynaptic band; *pa,* particles found in the postsynaptic membrane that may correspond to cholinergic receptors; *sv,* synaptic vesicles; *jf,* junctional folds; *f,* finger of a Schwann cell. Notice that on the right side of the figure the presynaptic membrane and the cleft are not represented. (Courtesy of K. Akert.)

In the myoneural junction, at the entrance of the postsynaptic folds, there are numerous 10 nm membrane particles that have a density on the order of $7500/\mu m^2$ and that are on both the P and E faces. As shown in Figure 28–18, such particles correspond, on the postsynaptic membrane, to regions opposed to presynaptic active zones.[45] It seems possible that these regions, at the entrance of the folds, correspond to the localization of the cholinergic receptors (see later discussion). In fact, radioautographic studies with ^{3}H-α-bungarotoxin (a toxin that binds specifically to nicotinic cholinergic receptors) have demonstrated that these particles are localized in the same region of the postsynaptic membrane.[57] On the other hand, acetylcholinesterase, the enzyme that hydrolyzes acetylcholine, is more widely distributed into the postsynaptic folds.[58] In the postsynaptic membrane of *Torpedo* electroplax, particles with a packing density of 10,000 to $15,000/\mu m^2$ have been observed. These particles have a hexagonal arrangement with a center-to-center distance of 8 to 9 nm.[59, 60]

A somewhat similar structure is observed in synapses of the central nervous system. Particles of 8 to 13 nm have been found in aggre-

gates confined roughly to the regions opposed to active sites in the presynaptic membrane. Some have interpreted such particles as anchoring devices for the filamentous material of the subsynaptic web, which is tightly attached to the postsynaptic membrane (Fig. 28–17).

28–3 SYNAPTIC VESICLES AND QUANTAL RELEASE OF NEUROTRANSMITTER

Since the discovery of the synaptic vesicles as subcellular organelles present in the presynaptic terminals, they have been implicated in the storage, transport, and release of the neurotransmitters. Following the demonstration by Fatt and Katz (1952) of the quantal nature of chemical transmission, De Robertis and Bennett postulated[47] that the synaptic vesicles are involved in the storage of the neurotransmitter. This was proved by the isolation of these organelles from synapses of the central nervous system and the demonstration that acetylcholine was concentrated in this fraction.[61] Later on, other neurotransmitters were localized in the vesicles (see De Robertis,

1964), and pure populations of the vesicles were isolated from cholinergic and adrenergic neurons.[62, 63] In this section we will study the morphology and structure of the synaptic vesicles in relation to their possible role in synaptic transmission.

Several Types of Synaptic Vesicles may be Recognized by Morphology and Cytochemistry

From the viewpoint of structure, five types of synaptic vesicles have been recognized:

Agranular or Electron-translucent Vesicles. These vesicles have a diameter of 40 to 50 nm with a limiting membrane 4 to 5 nm thick (Fig. 28–19). These vesicles are characteristic of peripheral cholinergic synapses and are those most frequently found in central synapses. They are distributed more or less uniformly throughout the axoplasm of the ending, but as mentioned earlier, a certain proportion make contact with the presynaptic membrane at regions generally called the *active points* of the synapse (Fig. 28–17). At the myoneural junction of the frog these active points are localized on both sides of the presynaptic band[54, 55] (Fig. 28–18). In this type of synapse there are about 1000 synaptic vesicles per μm^3 and a total of about 3×10^5. Of these about 20 per cent are near the membrane and readily available for release of the transmitter upon arrival of the nerve impulse.[64]

Large Dense-core or Large Granular Vesicles. These may be observed in central, as well as peripheral, synapses (Fig. 28–20). In peripheral cholinergic and adrenergic endings only a few per cent of the vesicles are large granular vesicles.

Flattened or Elliptical Vesicles. These vesicles are found in certain central and peripheral synapses (Fig. 28–20). They have been related to inhibitory synapses[66] and found in a fraction of isolated nerve ending (i.e., the synaptosomes) that is rich in the inhibitory transmitter γ-aminobutyrate.[67]

Small Granulated Vesicles. These are characteristic of the adrenergic nerve endings of the sympathetic system (Fig. 28–21). By using pharmacologic agents that release catecholamines, one may observe in the experimental animal, a depletion of the granulated vesicles. These vesicles increase in concentration in the presence of inhibitors of the enzyme monoamine oxidase or when the animal is given precursors of catecholamines.[69] All these results indicate that the small granulated vesicles contain the adrenergic transmitter. Specific cytochemical techniques have been used to demonstrate that the small granulated vesicles of sympathetic nerves are the storage site for norepinephrine, as well as for 5-hydroxytryptamine in some cases.[70]

Coated Vesicles. These are found in a small proportion in central synapses and were first recognized as complex vesicles in isolated fractions.[61] These coated vesicles are produced by the process of endocytosis and may be involved in the process of membrane recycling.[68] In myoneural junctions labeled with peroxidase that have produced a considerable amount of transmitter in response to stimulation, it has been found that these vesicles increase in number and contain the label. In Chapter 13–2 it was mentioned that the coat of these vesicles is made of the protein *clathrin*.

Neuronal Development — The Type of Neurotransmitter and Synaptic Vesicle may be Medium-Determined

The development of synaptic connections involves the synthesis of a specific neurotransmitter and a special type of vesicle in which the transmitter is stored. Elucidation of the mechanisms involved in this differentiation is extremely difficult in the central nervous system, in which more than a dozen neurotransmitters and at least five morphologic types of vesicles are observed (see Fig. 28–23). The problem can be more easily approached in the autonomic system, in which there are two main types of neurotransmitters, acetylcholine and norepinephrine, contained respectively in clear agranular vesicles and small granulated vesicles (Fig. 28–21). Both the adrenergic and the cholinergic neurons are embryologically derived from special regions of the neural crests. The neuroblasts migrate to the sites that they will occupy in the adult, cease dividing, and differentiate, producing one transmitter or the other.

Several experiments have demonstrated that this differentiation is influenced by the immediate environment. For example, cells of regions of the neural crests that normally provide adrenergic neurons, if transplanted to the rostral regions of the embryo, perform the cholinergic function, producing acetylcholine and clear vesicles; if neural crest cells from the rostral region are implanted in the mid-trunk, they develop into adrenergic neurons with terminals containing granular vesicles (see Le Douarin and Teillet, 1973).

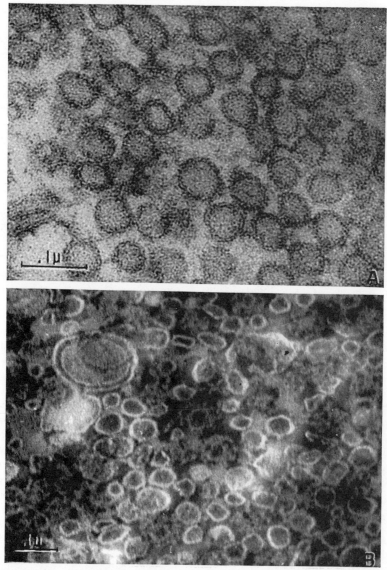

Figure 28–19 A, high resolution electron micrograph of synaptic vesicles in the hypothalamus of a rat, showing the fine structure of the vesicular membrane. ×180,000. **B,** isolated synaptic vesicles from rat brain after osmotic shock of the mitochondrial fraction. Negative staining with phosphotungstate. ×120,000. (From De Robertis, E., Rodríguez de Lores Arnaiz, G., Salganicoff, L., Pellegrino de Iraldi, A., and Zieher, L. M., *J. Neurochem., 10*:225, 1963.)

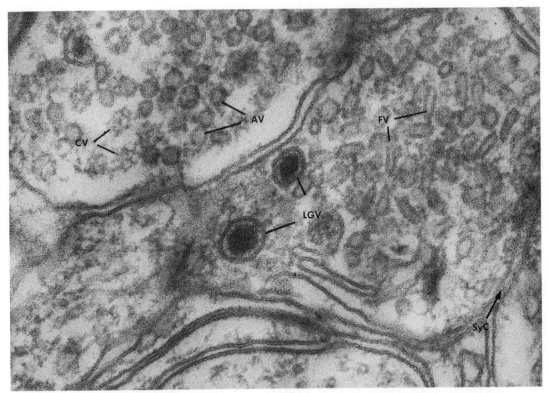

Figure 28-20 Four different types of synaptic vesicles are observed in two nerve endings of the rat hypothalamus. *AV*, agranular vesicle; *CV*, coated vesicle; *FV*, flattened vesicle; *LGV*, large granular vesicle. *SyC* points to the synaptic cleft. ×135,000. (From E. De Robertis, unpublished.)

A more direct experiment consists of culturing sympathetic neuronal cells from a newborn rat by a method that destroys all the non-neuronal (i.e., glial) cells. In this case, the neurons become adrenergic and produce norepinephrine. On the other hand, if the same neurons are grown in the presence of heart tissue most of them become cholinergic. This effect can also be obtained by adding the fluid in which heart tissue has been grown to the pure adrenergic cell culture. This is an indication that the heart cells produce a *cholinergic developmental factor* that is able to promote the differentiation of the neuron toward the synthesis of acetylcholine.

These and other experiments have led to the hypothesis that at an early stage neurons are still "plastic" and, under the influence of the medium, can change the neurotransmitter (and the type of vesicle) they produce. Work in which single neurons are cultured in the presence of heart cells has permitted the transformation from adrenergic to cholinergic to be followed. The most interesting finding is that there are neurons that have a dual function, i.e., that behave simultaneously as adrenergic and cholinergic and have the corresponding two types of synaptic vesicles.

These neurons could correspond to an intermediary stage in the differentiation between adrenergic and cholinergic. The hypothesis has been put forward that the chemical differentiation of a neuron (which determines the choice of neurotransmitter) is influenced by the non-neuronal cells by way of the release of special developmental factors which may enter the neuron by retrograde transport (see earlier discussion). In addition, the influence of the electrical activity of other neurons may modify the response to such developmental factors. (For more details on this very important problem, see Bunge et al., 1978; Patterson et al., 1978.)

Synaptosomal Membranes and Synaptic Vesicles may be Isolated by Cell Fractionation

The application of cell fractionation methods (see Chapter 4–4) to the study of the central nervous system has permitted a true dissection of the synaptic region, with the isolation of the synaptosome as a structural unit. As shown in Figure 28–15, a typical axodendritic synapse on a spine,

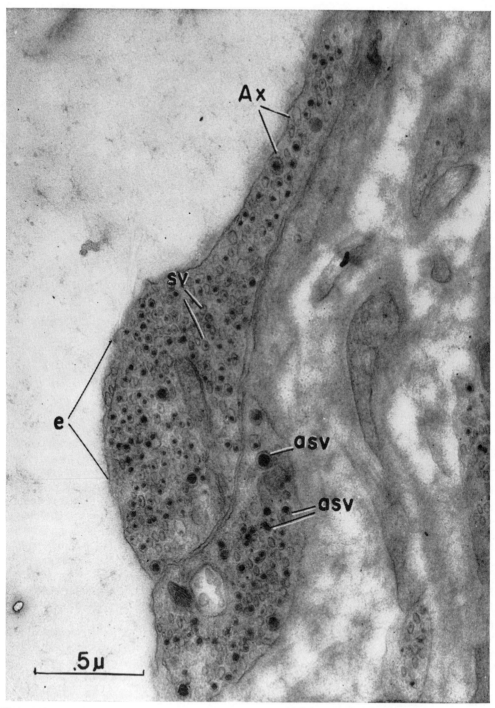

Figure 28–21 Electron micrograph of an adrenergic axon (*Ax*) and nerve ending (*e*) in the pineal gland of a rat. Both are filled with vesicles (*sv*), most of which contain a deposit of reduced osmium (*asv*). Observe the presence of a few large granulated vesicles. ×60,000. (From Pellegrino de Iraldi, A., and De Robertis, E., *Z. Zellforsch.*, *87*:330, 1968.)

under the action of shearing forces, breaks its connections with the axon and with the dendrite, and the nerve ending, or *synaptosome*, becomes isolated, together with the presynaptic and post-synaptic membranes. The synaptosomal membrane heals, and the synaptosome becomes a structural and functional unit in which many processes related to the metabolism of the neurotransmitters may be studied (see Rodríguez de Lores Arnaiz and De Robertis, 1972, 1973). The synaptosome is sensitive to changes in the osmotic pressure of the medium, and in a hypotonic solution it undergoes an osmotic shock, with breakage of the membrane and release of the synaptic vesicles, mitochondria, and the entire content of the nerve ending. At this point (Fig. 28–15,C), only the nerve-ending membrane remains. However, by a further treatment with a mild detergent, only the *junctional complex*, composed of the synaptic membranes and related structures, is found (Fig. 28–15,D).[71] Further treatment with detergents leads to the isolation of the subsynaptic web or postsynaptic density.[49] Such studies have demonstrated that the nerve-ending membrane contains important membrane-bound enzymes, such as acetylcholinesterase, Na^+-K^+ ATPase, and adenylate cyclase. They have also allowed the study of the binding of synaptic receptors to the neurotransmitters, which are found mainly at the postsynaptic membrane.[72] These cell fractionation studies also permitted the isolation of the synaptic vesicles[61] (Fig. 28–19) and demonstrated that they contain the neurotransmitters. In Table 28–2 it is shown that fraction M_2, corresponding to the synaptic vesicles, has the highest content of various neurotransmitters.

Neurotransmitter Synthesis and Metabolism is Exemplified by the Acetylcholine System

The various neurons integrating the nervous system can be differentiated not only by their morphology and synaptic relationships but also by the transmitter they synthesize, store, and release at the nerve endings. Thus, in the central nervous system there are nerve cells that produce various biogenic amines (acetylcholine, noradrenaline or dopamine, adrenaline, and serotonin or histamine) and amino acids (glutamate, aspartate, γ-aminobutyrate (GABA), and glycine) or polypeptides. Each of these neuronal types has its own biochemical mechanism which involves several enzymes for the synthesis and catabolism of the particular transmitter.

TABLE 28–2 CONTENT OF BIOGENIC AMINES IN THE BULK FRACTION (M_1), IN SYNAPTIC VESICLES (M_2), AND IN THE SOLUBLE FRACTION (M_3)

Biogenic Amines	*Fraction*		
	M_1	M_2	M_3
Acetylcholine	0.55	2.85	1.20
Noradrenaline	0.40	2.56	1.93
Dopamine	0.46	2.46	1.72
Histamine	0.39	2.24	2.27

The crude mitochondrial fraction of the brain was osmotically shocked and then centrifuged. The results are expressed in relative specific concentration. For literature, see De Robertis (1967).

In addition to the results obtained by cell fractionation, progress has been made on the identification of the various neuronal types. For example, neurons producing catecholamines (i.e., noradrenaline and dopamine) or indolamines (i.e., serotonin) may be identified by the use of fluorescence histochemical techniques.

Another powerful technique is based on the use of antibodies against polypeptide neurotransmitters such as substance P, enkephalin, and somatostatin) or of specific enzymes involved in the synthesis of the neurotransmitters (e.g., *glutamate decarboxylase* for γ-aminobutyrate, *cholineacetyltransferase* for acetylcholine or *tyrosine hydroxylase* for adrenergic neurotransmitters). In the acetylcholine system acetylcholine (ACh) is synthesized via a combination of choline and acetylcoenzyme A by the enzyme *cholineacetyltransferase* (Fig. 28–22). This process is most active in the motor nerve terminals of the myoneural junction. The enzyme is produced in the neuronal body and is transported to the nerve terminal by axonal flow. In some species, brain cholineacetyltransferase is associated with the synaptic vesicles, whereas in others, this enzyme is more soluble in the axoplasm.[73] Acetylcholine is stored mainly within the synaptic vesicles, although some ACh that is soluble in the cytoplasm may be present. After its release into the synaptic cleft, acetylcholine interacts with the cholinergic receptor present in the postsynaptic membrane and is degraded into choline and acetate by *acetylcholinesterase*, a hydrolytic enzyme present in nerve-ending membranes. As shown in Figure 28–22 choline is again taken up by a high-affinity transport mechanism into the nerve terminals. By cell fractionation it can be demonstrated that the three main components of the acetylcholine system (i.e., cholinea-

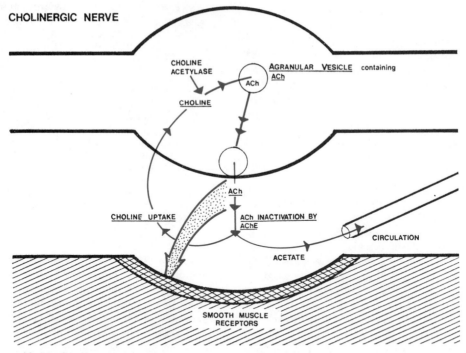

CHOLINERGIC NERVE

Figure 28–22 Diagram of a cholinergic synapse and the components of the acetylcholine system. Observe the storage of the acetylcholine (*ACh*) in the synaptic vesicle and its release. The mechanism of synthesis and inactivation by acetylcholinesterase (*AChE*) is indicated. (From Burnstock, N., *Pharmacol. Rev., 24*:509, 1972. © (1972) The Williams & Wilkins Company, Baltimore.)

cetyltransferase, acetylcholine, and acetylcholinesterase) are found in the synaptosomes (see Rodríguez de Lores Arnaiz and De Robertis, 1973).

The synthesis of the neurotransmitter in adrenergic neurons is better documented, and two of the key enzymes involved in the synthesis, *tyrosine hydroxylase* and *dopamine-β-hydroxylase*, are present, at least in part, in the vesicles.[74, 75]

The Transport of the Neurotransmitter Involves the Synaptic Vesicles

The synthesis of neurotransmitter, although occurring predominately at the nerve terminal, may take place in the perikaryon and the entire neuron. Thus, part of the transmitter must be transported to the sites of release by way of synaptic vesicles. The possible flux of the various kinds of vesicles is shown diagrammatically in Figure 28–23. The vesicles are transported by fast axoplasmic flow,[76] possibly with the involvement of neurotubules, which under certain conditions are even seen associated with the vesicles within the nerve terminal.[77] Drugs that depolymerize the neurotubules (i.e., colchicine and vinblas-

tine) produce a reduction or blockage of the transport of the vesicles along the axon. In Figure 28–23 it may be observed that while most types of synaptic vesicles are transported by orthograde transport, the coated vesicles, which arise through endocytosis, could travel by retrograde transport toward the perikaryon (see Chapter 28–1).

Transmitter Release is Related to the Role of Synaptic Vesicles in Nerve Transmission

The so-called *vesicle hypothesis* of synaptic transmission postulates that at the arrival of the nerve impulse at the nerve ending there is a synchronous release of a large number of quanta (i.e., multi-molecular units or packets) of neurotransmitter, which are liberated from the synaptic vesicles into the synaptic cleft. For example, the postsynaptic potential at the myoneural junction may be the result of the simultaneous release of 100 quanta of acetylcholine, which may represent the liberation of some 2×10^6 molecules of neurotransmitter.[78] In a resting synapse a spontaneous opening of some of the vesicles attached to the membrane may occur, and this produces the so-called *miniature end plate potentials*, each of which is attributed to the release of single quanta. Direct estimates of ACh content and the number of vesicles in nerve endings of *Torpedo* electroplax suggest that the content may be very concentrated. In fact, the solution of ACh inside the vesicle is probably isosmotic with plasma (i.e., 0.4 to 0.5 M).[62] Estimates of the ACh released in each quantum have varied in the literature; in the rat diaphragm a fair estimate could be from 12,000 to 21,000 ACh molecules, a number that could easily be packed within a vesicle.[64] The presence of ATP and some binding protein within synaptic vesicles has been dem-

onstrated, and it is possible that these substances may play some role in the ACh concentration within the vesicle.

Since the time of the discovery of the synaptic vesicles, experiments have been carried out to demonstrate their possible role in nerve transmission. Axotomy resulted in an early degeneration of the endings, with clumping and lysis of the vesicles.[79] These alterations coincide with the time in which nerve transmission is interrupted.[80] Electrical stimulation of the cholinergic nerve endings of the adrenal medulla at high frequencies resulted in a reduction in the number of vesicles.[81] This finding has now been con-

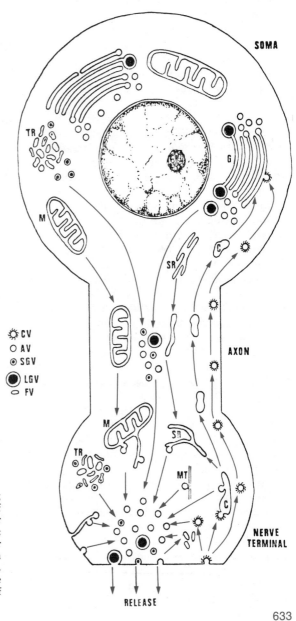

Figure 28–23 Composite diagram showing possible sites of origin, transport, and release of various types of synaptic vesicles. *AV,* agranular vesicle; *CV,* coated vesicles; *FV,* flattened vesicle; *LGV,* large granular vesicle; *SGV,* small granular vesicle. The SGV are shown to originate from a tubular reticulum (*TR*). *M,* mitochondria; *MT,* microtubule. Note that some coated vesicles are transported by retrograde axon flow. (Courtesy of M. P. Osborne, 1977.)

firmed by numerous authors in the myoneural junction and in electric tissue (see Osborne, 1977). In *Torpedo* electroplax stimulated to fatigue there is a loss of about 50 per cent of the vesicles;[62] at the same time the presynaptic membrane increases, forming finger-like protrusions, probably corresponding to the integration of the vesicle membranes.[82] An interesting finding is that a certain number of synaptic vesicles in the control synapse of the electroplax and in the myoneural junction, when fixed in the presence of Ca^{2+}, show a small granule attached to the membrane, which corresponds to calcium binding. When the synapse is intensely stimulated, the vesicles that carry this Ca^{2+} granule are preferentially released.[83] These findings are related to the role of calcium in transmitter release (see later discussion).

Transmitter Release Probably Involves Exocytosis with Recycling of Synaptic Vesicles

One of the most discussed problems of transmitter release is whether it involves exclusively a mechanism of exocytosis (Chapter 11–3) or whether the synaptic vesicles are reused again and again. In the latter case, the attachment of the vesicle to the releasing sites of the membrane would be temporary, and the vesicle could be recharged with transmitter for reuse. There are arguments that favor both sides of the discussion, and probably the two mechanisms are both at work. Quantitative work done on crayfish motor neurons has led to the proposal that there must be vesicle reuse (i.e., retention of the intact vesicle membrane within the axon terminal) or a mechanism by which incorporation of the vesicle membrane is balanced with an equivalent withdrawal.[84] Related to these mechanisms is the finding that newly synthesized transmitter or transmitter recaptured from the cleft, as occurs in adrenergic synapses, is preferentially released.[63, 85] In adrenergic neurons there is a large pool of transmitter (up to 99 per cent) that can be depleted with reserpine without failure in synaptic transmission, while a small pool is essential.[63] In synapses of *Torpedo* electroplax there are apparently two populations of cholinergic vesicles, one with a higher ratio of ACh/ATP, the so-called *supercharged vesicles*, and the other with a lower ratio.[86] It has been suggested that the supercharged vesicles are associated with the presynaptic membrane and upon stimulation are released first. As was mentioned earlier, it seems possible that the dense presynaptic projections

play a role in the transport of the vesicles to the sites of release.

It has been postulated that newly formed synaptic vesicles are capable of several cycles of secretion and reloading before they undergo exocytosis and are incorporated into the membrane.[82] However, the existence of a recycling mechanism for the vesicles is now generally accepted.[62, 82, 87] It is possible that after several rounds of reuse the vesicle is incorporated into the presynaptic membrane and subsequently recaptured. This could be done by direct endocytosis and formation of new vesicles or by intermediary membranous structures (e.g., by way of the smooth endoplasmic reticulum). Frequently, coated vesicles are re-formed in this process (Fig. 28–23). Recycling of vesicle membrane is an energy-saving mechanism that avoids the problem of constant synthesis, with all the overwork that this implies for the neuron.

The Depolarization Transmitter-Secretion Coupling is Mediated by Calcium Ions

The coupling between the depolarization caused by the arrival of the nerve impulse, and the secretion of acetylcholine is mediated by the influx of calcium ions. Extracellular Ca^{2+} is an absolute requirement for the release of ACh from the nerve terminal; in the absence of calcium, there is no release of ACh. When the nerve impulse arrives at the terminal, there is an increase in Ca^{2+} permeability resulting from the depolarization. The entry of calcium into the ending has been demonstrated by ^{45}Ca and the use of aequorin, a substance that fluoresces in the presence of Ca^{2+}.[88] It has been assumed that for the release of each quantum of ACh, one to four Ca^{2+} molecules must enter, and that in the release of ACh there is a cooperative action of Ca^{2+}.[89]

Once the Ca^{2+} has penetrated the terminal, the secretion of ACh is probably produced by the opening of the synaptic vesicles that were attached to the presynaptic membrane. Physiologic studies have demonstrated that the *synaptic delay* that takes place between the arrival of the nerve impulse and the production of the synaptic potential is due mainly to the interval between depolarization and actual secretion of the transmitter.

We mentioned earlier that synaptic vesicles carry a binding site for Ca^{2+} in the membrane. Such a site may be important in the actual release of the transmitter, since the entrance of Ca^{2+} at this point could determine the increased porosity of the vesicle.[82, 83]

634

SUMMARY:

Synaptic Transmission and Structure of the Synapse

Synapses are regions of contact between neurons, or between a neuron and a non-neuronal cell (i.e., a physiologic receptor or effector) at which excitatory or inhibitory actions are transmitted in one direction. They may be *electrical,* in places in which there is a direct contact of membranes, and in cases like this the nerve impulse passes from one neuron to the other without delay. However, most synapses involve *chemical transmission* based on the release of specific transmitters such as acetylcholine (ACh), noradrenalin, dopamine, and other biogenic amines, as well as amino acids such as glutamate, aspartate, γ-aminobutyrate (GABA), and glycine.

The microphysiologic study of synaptic transmission has demonstrated that the nerve impulse stops at the nerve ending, and that at the synapse a *synaptic potential* is generated in the postsynaptic cell. This potential may be *excitatory* (depolarization) or *inhibitory* (hyperpolarization); in both cases, it is graded and does not propagate. In some cases it may be observed that the same transmitter (i.e., ACh) may be excitatory or inhibitory; the different effect depends on the receptor protein with which the transmitter interacts. All the excitatory and inhibitory inputs acting on a neuron are added algebraically, and the neuron will fire new impulses in relation to these inputs as well as to its own spontaneous firing rhythm.

Morphologically, synapses may be *axodendritic, axosomatic,* or *axo-axonic.* With the electron microscope, *dendrodendritic, serial,* and *reciprocal* synapses may be distinguished by virtue of their presence on one side of synaptic vesicles and the differentiations of the synaptic membranes (i.e., presynaptic projections, intersynaptic filaments, and subsynaptic web).

Several thousand synapses may impinge on a single neuron, and in general, the number of synapses is related to that of the dendritic spines. Synaptic contacts are not completely stable and may change not only during development but also in relation to the stimulation of the neuron (e.g., in sensory deprivation the number of spines and synapses diminishes).

Special lectin receptors are present at the neuronal membrane, and these correspond to glycoproteins. Most of these are mobile within the membrane, but at the postsynaptic membrane they are less mobile. The postsynaptic density may play a role in this change in mobility, as well as in the development of the synapse. In cultured neurons it has been observed that the postsynaptic densities are made by coated vesicles coming from the Golgi, prior to the appearance of synaptic vesicles at the terminal.

A number of synaptic vesicles are attached to the presynaptic membrane at special sites arranged around presynaptic projections. These projections make a hexagonal grid in central synapses or parallel bands in the myoneural junction and may be related to the release of the vesicles.

At the postsynaptic membrane there are particles that in some cases correspond to the localization of receptor proteins.

The presence of *synaptic vesicles* is the main characteristic of synapses. Most of these structures are spherical and electron-translucent, and have a diameter of 40 to 50 nm and a membrane 4 to 5 nm thick. In the myoneural junction there are some 1000 vesicles per μm^3 and a total of 3×10^5; about 20 per cent of them are attached to the presynaptic membrane and are ready to discharge the transmitter. In many central and peripheral synapses there are *large dense-core* or *granular vesicles.*

Some synapses, thought to be inhibitory, contain *flattened* or *elliptical vesicles.* In sympathetic axons and endings there are small *granulated vesicles;* the granule contains the transmitter, noradrenalin. Another type is the *complex* or *coated vesicle,* which has a shell of material surrounding it. These vesicles are formed by endocytosis and may represent a mechanism of membrane recycling.

During synaptic development there is a critical period in which peripheral synapses may change from adrenergic (with small granular vesicles) into cholinergic (with clear vesicles). This change depends on the influence of the surrounding cells, which produce developmental factors. Cultured neurons with dual function and with both types of vesicles have been observed. By cell fractionation methods a true dissection of the synaptic region may be achieved. In this way, synaptosomes, synaptic vesicles, nerve-ending membranes, synaptic membranes, and postsynaptic densities may be isolated. A direct demonstration that synaptic vesicles are the sites of storage of transmitter was obtained after the isolation of these organelles (Table 28–2).

In the central nervous system, neurons produce different transmitters and have special biochemical mechanisms for their synthesis and catabolism. Neurons producing catecholamines or indolamines (i.e., serotonin) may be recognized by fluorescence. Recently, antibodies against specific enzymes have been used to identify a particular neuron. In the acetylcholine system, synthesis is produced by the enzyme cholineacetyltransferase in the following reaction:

$$\text{Choline} + \text{Acetyl CoA} \rightarrow \text{Acetylcholine} \qquad (2)$$

After its release, acetylcholine is hydrolyzed by acetylcholinesterase, thereby releasing acetyl groups and choline, which are again taken up by the nerve ending.

Transmitter release is related to the role of synaptic vesicles in nerve transmission. The vesicle hypothesis implies that the neurotransmitter is liberated in multi-molecular packets of quanta. Each quantum released produces a spontaneous miniature end plate potential. The ACh content of vesicles may be very high. In *Torpedo*, its concentration is about 0.4 to 0.5 M. Each quantum involves some 12,000 to 21,000 ACh molecules, which are packed within a vesicle. For the release, Ca^{2+} is indispensable, and the *synaptic delay* corresponds principally to the time taken by the actual secretion of the transmitter.

When nerve impulses arrive at the ending, 100 or more quanta are released, thereby producing the synaptic or end plate potential. The relationship between vesicles and quantal release has been demonstrated by a variety of experiments.

The mechanism of transmitter release probably involves reuse of the vesicles, followed by exocytosis and recycling of the membrane vesicle.

28–4 SYNAPTIC RECEPTORS AND THE PHYSIOLOGIC RESPONSE

Since the beginning of this century, through the work of Langley, Ehrlich, and others it has been postulated that the transmitter interacts with a specific receptor localized at the chemosensitive sites of the cell membrane. For many years, however, the knowledge of synaptic receptors has been principally indirect — based on the final response obtained from this interaction. For example, it was known that acetylcholine produced contraction of skeletal muscle, or the secretion of a gland, and that these two effects could be blocked with curare and atropine, respectively.

In 1955 Nachmanson had already postulated that the "acetylcholine receptor is a protein that upon the binding of ACh undergoes a conformational transition which results in changes in membrane permeability." Figure 28–24 shows that the primary transmitter-receptor interaction is coupled, probably by

way of a conformational change, to the translocation of ions across the cell membrane (i.e., the ionophore), and by several interposed mechanisms can produce a final response. This figure also shows the probable relationship between the primary interaction, the displacement of Ca^{2+}, and certain metabolic processes involving cyclic AMP and other substances.

Synaptic Receptors are Hydrophobic Proteins and Embedded in the Lipoprotein Framework of the Membrane

Only in recent years has it been possible to separate synaptic receptor proteins as chemical entities and to analyze, in a more direct way, their interaction with the transmitter. The separation of these substances has been difficult because they are intrinsic proteins (see Table 8–2) which require strong treatments to be separated from the lipoprotein framework of the membrane. Furthermore, they are present in

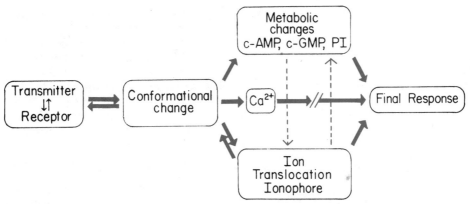

Figure 28–24 Diagram showing the primary interaction between transmitter and receptor and its consequences. (See the description in the text.)

extremely small concentrations. These properties imply that the membranes should first be separated by cell fractionation methods. The receptor molecules should then be isolated from the membrane. After isolation, the receptor protein should show a specific high-affinity binding for the neurotransmitter. Finally, it should be able to undergo conformational changes that are capable of inducing the translocation of ions and of producing a bioelectrical response.

Receptor proteins are highly hydrophobic and intimately related to lipids of the membrane. Two main procedures are used to isolate these proteins; extraction with organic solvents[91] and separation by detergents.[92, 93] Cholinergic receptor proteins have been isolated not only from the electroplax of *Torpedo* and *Electrophorus* tissues, which have the richest cholinergic innervation, but also from skeletal muscle, smooth muscle, and brain. Adrenergic receptor proteins have been separated from brain, from spleen capsule, and from heart. Receptor proteins for the amino acids glutamate and γ-aminobutyrate have also been isolated from muscle of *Crustacea* and from insects which have a double innervation: excitatory (via glutamate) and inhibitory (via γ-aminobutyrate). From synaptic membranes of the cerebral cortex, hydrophobic proteins binding L-glutamate, L-aspartate, and γ-aminobutyrate have been separated[94] (see De Robertis, 1975).

After extraction with an organic solvent or detergent, the receptor proteins are generally separated by column chromatography or, in the case of the cholinergic receptor, by affinity chromatography. In the latter technique, the gel used contains an active group attached by a long arm. This group specifically binds the receptor proteins and is then eluted by a pulse of the cor-

responding transmitter (see Hucho and Schiebler, 1977).

Figure 28–25 shows the two-step separation of the cholinergic receptor of skeletal muscle. The small second peak of protein in Figure 28–25,1 is the peak having a high affinity for ACh and other cholinergic drugs. In Figure 28–25,2 this protein has been further purified by affinity chromatography. In this case, the protein peak, which appears after the ACh pulse, is the specific receptor protein fraction. With this procedure a total purification of about 15,000 times has been achieved using rat diaphragm.[95] Complete purification of the nicotinic acetylcholine receptor has been obtained in the *Torpedo* electroplax which is the richest source of this protein. After affinity chromatography of detergent-extracted receptor, four polypeptide chains, ranging between 39,000 and 68,000 daltons, have been separated by SDS gel electrophoresis. These subunits form a complex of about 270,000 daltons. It is known that the small subunit of 39,000 daltons is the one that carries the binding site, corresponding to the true receptor.[96] Recently it has been postulated that a second subunit of about 43,000 daltons could represent the ionophore, i.e., the subunit involved in the translocation of ions.[97] (For further details, see Briley and Changeux, 1977; Eldefrawi and Eldefrawi, 1977.)

The Acetylcholine Receptor is Coupled to the Translocation of Sodium and Potassium Ions

Our previous consideration of the transmitter-receptor interaction suggests that the transduction of energy at the synapse is the result of a conformational change in the receptor macromolecule and that such a change is coupled with

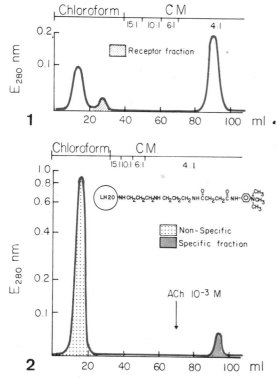

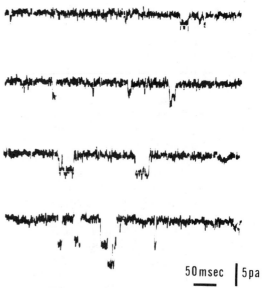

Figure 28–25 Separation of the cholinergic receptor from rat diaphragm. **1**, by conventional chromatography in Sephadex LH20; **2**, by affinity chromatography. Observe that the receptor protein fraction from **1** has been purified further (about fifteen times) in **2**. The specific receptor fraction appears only after a pulse of 10^{-3} M acetylcholine (*ACh*). The inset shows the chemical composition of the affinity column with the cholinergic end (p-aminophenyl-trimethylammonium) linked to the gel by a 1.4 nm "arm" (3'-3'-iminobispropylamine). (From Barrantes, F., Arbilla, S., de Carlin, C., and De Robertis, E., 1975.)

the translocation of ions across the membranes (Fig. 28–24). In the case of acetylcholine, these mechanisms can be expressed as shown in the diagram that follows. In this equation AChRc represents a reversible complex resulting from the binding of ACh to the receptor protein. In the second step, the ionophoric mechanism operates, producing the opening of the channel for the translocation of the ions (AChRo).

$$ACh + Rc \underset{K_{-1}}{\overset{K_{1}}{\rightleftharpoons}} AChRc \underset{K_{-2}}{\overset{K_{2}}{\rightleftharpoons}} AChRo$$

It is important to stress here the difference between the axonal channels we studied pre-

viously and the postsynaptic channels that are related to the cholinergic receptor. While in the axon there are separate Na^+ and K^+ channels which are regulated by the membrane potential, in the postsynaptic membrane both ions probably use a common channel, and the gating is driven by the chemical interaction of the neurotransmitter with the receptor protein. By the use of fine microelectrodes in the myoneural junction, some of the best evidence of the functioning of receptors at the molecular level has been obtained.[98] It was mentioned before that in this type of preparation it is possible to record the spontaneous miniature end plate potentials (mepp), which have an amplitude of about 0.1 mV and represent the discharge of single multi-molecular quanta of ACh. Katz and Miledi have observed that if a minimal but steady dose of acetylcholine is applied to the myoneural junction, there are minute fluctuations of the membrane potentials that are superimposed upon a small but steady depolarization of the membrane. The amplitude of these fluctuations is several hundred times smaller than the mepp (i.e., on the order of 0.3 μV) and constitute the so-called *membrane noise*.[98]

In denervated frog muscle, in which the acetylcholine receptors appear beyond the myo-

Figure 28–26 Conductance fluctuations recorded from the muscle membrane of a chronically denervated frog muscle in the presence of 10^{-7} M suberyldicholine and at different potentials (−60, −80, −110, and −150 mV). The discrete rectangular changes in conductance correspond to the opening and closing of single channels. When a great number of these changes are superimposed they produce the typical membrane noise. (Courtesy of E. Neher and C. F. Stevens.)

neural junction, it has been possible to record the actual opening and closing of individual channels[99] (Fig. 28–26) (see Neher and Stevens, 1977). The current that is produced is equivalent to the translocation of about 5×10^4 univalent ions. The enormous amplification that is produced at the chemical synapse is readily understandable; in fact, the interaction of one or two ACh molecules with a receptor translocated more than 10,000 ions!

Acetylcholine-induced Fluctuations may be Observed in a Reconstituted Cholinergic Receptor

Several attempts to reconstitute a functional excitable membrane from cholinergic receptor protein extracted by detergents have generally given negative results; i.e., in most cases the model systems were unreactive to the action of the neurotransmitter.[100]

The advantage of having a receptor isolated in organic solvent is that it can be directly incorporated into a planar lipid bilayer or into a liposome (Figs. 7–8 and 7–9). As shown in Figure 28–27, if the cholinergic receptor is incorporated into a lipid bilayer and acetylcholine is applied with a micropipette, there is a sudden increase in the level of conductance, which is accompanied by current jumps in the same range of amplitude as those observed in the muscle membrane (Fig. 28–26). These fluctuations disappear when the membrane voltage is reduced to zero or by addition of the blocking agent D-tubocurarine,[101] and may be considered as an increase in membrane noise. These and other experiments, in which the protein was incorporated into liposomes, suggest that the cholinergic receptor isolated by organic solvents may be reconstituted in the artificial membrane in a form that is "reactive" to cholinergic drugs.

An Oligomeric Model of the Cholinergic Receptor has been Postulated

Based on the hydrophobic properties of the synaptic receptor proteins and on studies of reconstitution in artificial membranes, a tentative model of the integration of the cholinergic receptor was postulated.[102] The most interesting features of this model are the oligomeric arrangement of the protein subunits and their disposition traversing the postsynaptic membrane. In Figure 28–28, the receptor, with the binding site for acetylcholine, is shown as a macromolecule different from the ionophore, as some recent studies suggest. However, a tight coupling between the two must exist. The interaction of the transmitter with the receptor produces a conformational change which could be transmitted to the coupled ionophore, leading to the opening of a hydrophilic channel through which Na^+ and K^+ could flow. It is possible that the ionophore or ion-conducting mechanism also resides in a hydrophobic protein (i.e., proteolipid), since this is the case for most channel-forming macromolecules.[103]

Since the ACh-receptor interaction is reversible, it may easily go back toward the closed position once the ACh has been removed from the site of binding. Such a reversible effect would be facilitated by the fact that the receptor molecules are held in place by hydrophobic interactions within the framework of the membrane.

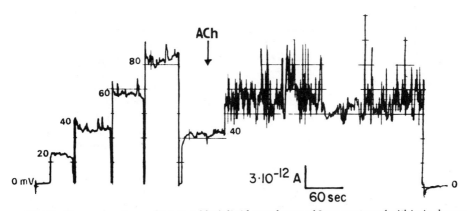

Figure 28–27 Time-current recording in a black lipid membrane of 2 per cent egg lecithin in decane containing, in the forming solution, 4 μg/ml of cholinergic proteolipid from *Torpedo*. At the time indicated, acetylcholine was applied to the membrane. Observe that the membrane is maintained at various voltages and that the transmitter is applied at 40 mV. (Courtesy of P. Schlieper and E. De Robertis.)

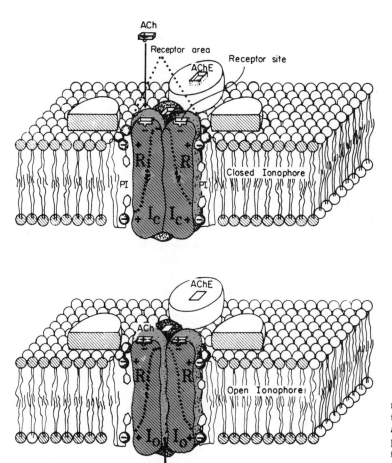

Figure 28–28 Oligomeric model of the cholinergic receptor. *Ach,* acetylcholine; *AchE,* acetylcholinesterase; *PI,* phosphatidylinositol; *R,* receptor protein; I_c, closed ionophore; I_o, open ionophore.

Long-lasting Synaptic Functions Involve the Use of a Second Messenger

The description given so far of the neurotransmitter-receptor interaction involves the direct translocation of ions. As we have seen in the case of the nicotinic cholinergic receptor of the myoneural junction, this translocation is a very fast phenomenon, occurring in a few milliseconds or less. However, in other cases, as indicated in Figure 28–24, the interaction may involve metabolic changes in the neuron which involve a transient but longer-lasting physiologic effect. In Chapter 6 a section was dedicated to discussion of the action of hormones on non-neuronal tissues, and this involves the intervention of specific receptors and the enzyme adenylate cyclase, discovered by Sutherland in 1956. As shown in Figure 6–8, the specific hormone receptor is coupled within the membrane with adenylate cyclase. When the hormone interacts with the receptor, the enzyme is activated and produces *3'-5' cyclic AMP (cAMP)* from ATP. This nucleotide is considered a *second messenger* that is able to

change the metabolism of the cell. In Figure 6–9 the classic example of the so-called *cascade effect* of cAMP on glucogenolysis is indicated. It may be observed that under the action of the hormones, epinephrine or glucagon adenylate cyclase is activated, increasing the levels of cAMP. This in turn activates *protein kinase,* a phosphorylating enzyme, which converts phosphorylase kinase from an inactive into an active form, which will finally degrade the glycogen. Adenylate cyclase was considered again in Chapter 8–2 when we mentioned that coupling between the receptor and the enzyme may be dependent on the fluidity of the cell membrane.

The first indication that adenylate cyclase could be involved in synaptic function was provided in 1967 with the demonstration that this enzyme was concentrated in synaptosomes and was found mainly in the synaptosomal membrane. It is interesting that after osmotic shock of the synaptosome, the enzyme activity was found to be increased, indicating that the active groups are facing toward the inside of the synaptosomal membrane.[104] In the same work it was found that the synaptosome also contains high levels of *cyclic phosphodiesterase*, the enzyme that degrades cAMP into inactive AMP. This enzyme was found to be partially soluble, but a high per cent was particulated and was also found in the synaptosomal membrane. Cytochemical studies demonstrated later that phosphodiesterase is localized mainly at the postsynaptic side of the synapse.[105]

At the present time it is thought that several neurotransmitter-receptor interactions that generate slow postsynaptic potentials (i.e., on the order of 100 to 500 msec) or that produce a modulation effect on synaptic function involve a second messenger mechanism, with the production of cAMP. More recently, another second messenger, *cyclic GMP*, was found in brain, and it is produced by the enzyme *guanylate cyclase*. In contrast to adenylate cyclase, this enzyme is usually soluble in the cytosol.

Recently, six putative neurotransmitters have been shown to stimulate adenylate cyclase specifically (i.e., dopamine, norepinephrine, serotonin, histamine, octopamine, and enkephalin). Cyclic GMP may mediate the effects of acetylcholine by way of muscarinic receptors, and of histamine and norepinephrine, acting on receptors that are different from those associated with adenylate cyclase (Fig. 28–29) (see Bartfai, 1978).

The levels of cAMP or cGMP depend not only on the activation of the corresponding enzymes but also on the action of phosphodiesterase, which apparently acts on both nucleotides. For example, if phosphodiesterase is inhibited by methylxanthines (such as caffeine) or other drugs, the levels of these cyclic nucleotides increase. Phosphodiesterase and adenylate cyclase are regulated by Ca^{2+} by way of a Ca^{2+}-binding protein.

The link between cAMP and synaptic transmission was more firmly established when it was found that the iontophoretic application of either norepinephrine or cAMP upon the Purkinje cells of the cerebellum slowed the firing of these neurons. Furthermore, it was shown the application of norepinephrine produced a large increase of cAMP within these cells.[106]

Cyclic AMP is Involved in the Phosphorylation of Membrane and Other Proteins

As in the control of glucogenolysis in liver, shown in Figure 6–9, cAMP phosphorylation of proteins also plays an important role in the central nervous system. In the synaptosomal fraction there is a complete system for the phosphorylation and dephosphorylation of proteins (i.e., the covalent attachment and detachment of phosphate in proteins).

This system is composed of a specific *cAMP-dependent protein kinase*, with the membrane-bound proteins acting as substrate and a specific *phosphatase* reversing the reaction.[107] The protein kinase is thought to be activated by the binding of cAMP to an inhibitory subunit, causing it to dissociate from the catalytic subunit. In this way the activated enzyme readily transfers a phosphate group to the substrate protein.

It was found that of the many proteins of the neuronal membrane only two or three were readily phosphorylated by the cAMP. One of these, designated *protein I*, has two subunits of 86,000 and 80,000 daltons, and it is present almost exclusively in the synaptic region, being absent from non-neuronal tissue and from fetal stages before the formation of the connections between neurons. Phosphorylation of protein I is extremely fast and may reach a maximum in five seconds.[107] It has been suggested that this special phosphorylation process, mediated by cAMP, could be responsible for the permeability of the membrane to ions; however, at the present time it is difficult to establish whether there is a real cause-and-effect relationship. In addition to the membrane-bound phosphorylating system, it should be noted that there are also soluble cAMP-dependent protein kinases which may phosphorylate other proteins in the cytosol, such as the tubulin of microtubules and tyrosine hydroxylase.

We mentioned earlier that in cyclic AMP-mediated synapses the physiologic events may last 100 milliseconds or more. There are indications in mollusc ganglion cells and in the

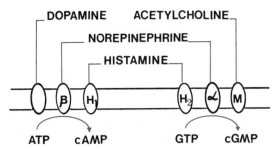

Figure 28–29 Diagram indicating various putative neurotransmitters that, by way of specific receptors, enhance the production of cyclic nucleotides. The following receptors are indicated: β, β-adrenergic; α, α-adrenergic; H_1 and H_2, histaminergic; and M, muscarinic.

641

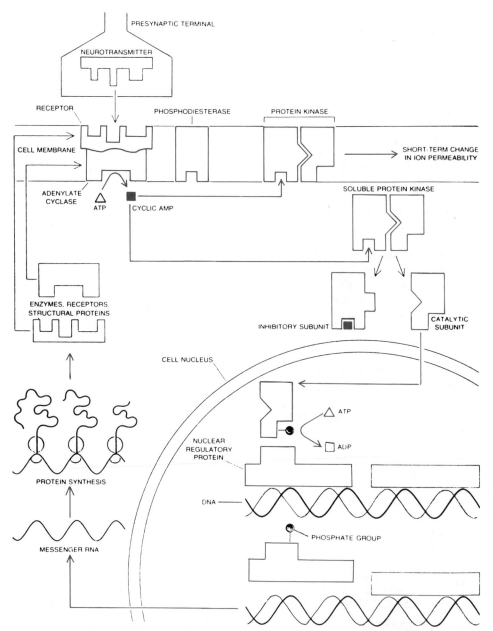

Figure 28–30 Diagram of events that may occur at a cyclic AMP–mediated synapse. The interaction of the neurotransmitter with the specific receptor activates adenylate cyclase, which produces cAMP from ATP. cAMP may be inactivated by phosphodiesterase, or it may act on a membrane-bound protein kinase which phosphory-lates proteins in the synaptic membrane. These proteins in turn could be involved in the translocation of ions. cAMP may also activate a soluble protein kinase which, by phosphorylating a regulatory protein in the nucleus, may affect gene expression, producing longer-lasting changes in the neuron. (Courtesy of J. A. Nathanson and P. Greengard.)

sympathetic system of vertebrates that these events may be much longer, lasting for hours. This has led to the hypothesis that synaptic events mediated by cyclic nucleotides could be related to long-term changes in the nervous system and could play a role in memory.

Phosphorylation of regulatory proteins within the nucleus of the neuron could be involved in gene expression by the transcription of RNA from certain DNA genes. This could lead to translation into important proteins, such as enzymes or receptors, and to more permanent changes in the neuron (see Nathanson and Greengard, 1977). A largely diagrammatic view of these processes is shown in Figure 28–30.

From the medical viewpoint these studies are important because some receptors using cyclic nucleotides (e.g., dopamine, β-adrenergic and H_2-histamine receptors) appear to be the targets of some of the psychoactive drugs used in the treatment of mental illness. Furthermore, some of these receptors are altered in neurologic diseases, such as parkinsonism and Huntington's chorea (see Bartfai, 1978).

SUMMARY:

Receptors and the Physiologic Response

Receptor proteins are localized at the chemosensitive sites of the postsynaptic membrane. They are in low concentration, highly hydrophobic, and embedded in the membrane.

They may be extracted with organic solvents or by the use of strong detergents and can be purified by column chromatography and affinity chromatography. Receptor proteins from electroplax, skeletal muscle, brain, smooth muscle, and crustacean heart muscle have been isolated. These proteins are either cholinergic, adrenergic, or related to the amino acids, glutamate, or GABA. The cholinergic receptor from electroplax is about 270,000 daltons and has several subunits ranging between 39,000 and 68,000 daltons. The smallest subunit carries the receptor site, and the ionophore is probably contained in a slightly larger subunit.

The essence of chemical transmission resides in the interaction of the transmitter with the receptor. This interaction produces a conformational change in the protein by which channels for the translocation of ions are opened (ionophoric response). In contrast to the axonal ionophores, in this type of response both Na^+ and K^+ are translocated. In the myoneural junction the application of a minimum dose of ACh produces fine fluctuations in the membrane potential. These are several hundred times smaller (i.e., 0.2 to 0.5 μV) than the mepp (700 to 1000 μV) and probably correspond to the opening of individual channels. Each of these elementary fluctuations corresponds to the passage of some 50,000 univalent ions across the membrane. In the extrajunctional receptors of denervated muscle the actual opening and closing of single channels has been observed.

The cholinergic receptor proteolipid from the electroplax has been reconstituted in black lipid membranes. In this case of reconstitution, under the influence of ACh, fluctuations of the same order of amplitude as those of the myoneural junction have been observed. An oligomeric model of the cholinergic receptor, in which the receptor and the ionophore traverse the membrane, has been postulated (Fig. 28–28).

In addition to the 1 millisecond events produced by the interaction of ACh with the nicotinic cholinergic receptor, there may be other cases in which these events are longer lasting. The physiologic effects involve the production of a second intracellular messenger represented by the cyclic nucleotides cAMP and cGMP. Cyclic AMP is produced by adenylate cyclase, which is concentrated in synaptosomal membranes, and cGMP, by guanylate cyclase, which is usually soluble. At least six neurotransmitters are known to stimulate adenylate cyclase, and three are known to stimulate guanylate cyclase (Fig. 29–29); the influence of cAMP in synaptic transmission has been demonstrated in certain cases.

Cyclic AMP acts on specific protein kinases that phosphorylate certain proteins in the synaptic membrane, and it has been suggested that this phenomenon may be at the basis of the permeability changes. Intracellular proteins may also be phosphorylated. It has been hypothesized that these mechanisms may be involved in long-lasting changes of the neuron (Fig. 28–30). These receptor–cyclic nucleotide interactions may have important medical implications.

REFERENCES

1. De Robertis, E., and Franchi, C. M. (1953) *J. Exp. Med., 98*:269.
2. Thornburg, W., and De Robertis, E. (1956) *J. Biophys. Biochem. Cytol., 2*:475.
3. De Robertis, E., and Schmitt, F. O. (1948) *J. Cell Comp. Physiol., 31*:1.
4. Rodríguez Echandia, E. L., Piezzi, R. S., and Rodríguez, E. M. (1968) *Am. J. Anat., 122*:157.
5. Wuerker, R. B., and Palay, S. L. (1968) Cited in: *Neurosci. Res. Program Bull., 6*:125.
6. Weiss, P., and Hiscoe, H. B. (1948) *J. Exp. Zool., 107*:315.
7. Ramirez, G., Levitan, I. B., and Mushynski, W. E. (1972) *J. Biol. Chem., 247*:5382.
8. Gilbert, J. M. (1972) *J. Biol. Chem., 247*:6541.
9. Grafstein, B. (1969) *Adv. Biochem. Psychopharmacol., 1*:11.
10. Ochs, S. (1972) *Science, 176*:252.
11. Droz, B., Koenig, H. L., and Di Giamberardino, L. (1973) *Brain Res., 60*:93.
12. Cuenod, M., and Schonbach, J. (1971) *J. Neurochem., 18*:809.
13. Fernandez, H. L., Burton, P. R., and Samson, F. E. (1971) *J. Cell Biol., 51*:176.
14. Droz, B., Koenig, H. L., and Di Giamberardino, L. (1973) *Brain Res., 60*:93.
15. Di Giamberardino, L., Bennett, G., Koenig, H. L., and Droz, B. (1973) *Brain Res., 60*:129.
16. Bennett, G., Di Giamberardino, L., Koenig, H. L., and Droz, B. (1973) *Brain Res., 60*:147.
17. Barondes, S. H. (1969) In: *Handbook of Neurochemistry*, Vol. 2, p. 435. (Lajtha, A., ed.) Plenum Publishing Corp., New York.
18. Kristensson, K. (1970) *Acta Neuropathol., 16*:293.
19. Kristensson, K. (1978) *Annu. Rev. Pharmacol. Toxicol., 18*:97.
20. La Vail, J. H., and La Vail, M. M. (1974) *J. Comp. Neurol., 157*:303.
21. Nauta, H. J. W., Pritz, M. B., and Lasek, R. J. (1974) *Brain Res., 67*:219.
22. Levi-Montalcini, R. (1966) *Harvey Lect., 60*:219.
23. Hendry, I. A., Stach, R., and Herrup, K. *Brain Res., 82*:117.
24. Stöckel, K., Schwab, M. E., and Thoenen, H. (1975) *Brain Res., 89*:1.
25. Schwab, M. E., and Thoenen, H. (1976) *Brain Res., 105*:213.
26. Hendry, I. A., Stöckel, K., Thoenen, H., and Iversen, L. L. (1974) *Brain Res., 68*:103.
27. Schwab, M. E. (1977) *Brain Res., 130*:190.
28. Levi-Montalcini, R., Aloe, L., Mugnaini, E., Oesch, F., and Thoenen, H. (1975) *Proc. Natl. Acad. Sci. USA, 72*:595.
29. Paravicini, U., Stöckel, K., and Thoenen, H. (1975) *Brain Res., 84*:279.
30. Schwab, M. E., and Thoenen, H. (1976) *Brain Res., 105*:213.
31. Hodgkin, A. L., and Huxley, A. F. (1952) *Cold Spring Harbor Symp. Quant. Biol., 17*:43.
32. Tasaki, I. (1953) *Nervous Transmission*. Charles C Thomas, Pub., Springfield, Ill.
33. Loewenstein, W. R. (1960) *Sci. Am., 203*:98.
34. Narahashi, T. (1974) *Physiol. Rev., 54*:813.
35. Beress, L., Beress, R., and Wunderer, G. (1975) *FEBS Letters, 50*:311.
36. Ritchie, J. M., and Rogart, R. B. (1977) *Proc. Natl. Acad. Sci. USA, 74*:211.
37. Conti, F., de Felice, L. J., and Wanke, E. (1975) *J. Physiol. (London), 248*:45.
38. Keynes, R. D. (1976) In: *The Structural Basis of Membrane Function*, p. 331. (Hatefi, Y., and Djavadi-Ohaniance, L., eds.) Academic Press, Inc., New York.
39. Hille, B. (1975) *J. Gen. Physiol., 66*:535.
40. Armstrong, C. M., and Hille, B. (1972) *J. Gen. Physiol., 59*:388.
41. Armstrong, C. M. (1975) In: *Membranes*. (Eisenman, G., ed.) Marcel Dekker, Inc., New York.
42. Furshpan, E. J., and Potter, D. D. (1957) *Nature, 180*:342.
43. Tauc, L., and Gerschenfeld, H. M. (1960) *C. R. Acad. Sci. (Paris), 257*:3076.
44. Eccles, J. C. (1964) *The Physiology of Synapses.* Springer-Verlag, Berlin.
45. Valverde, F. (1967) *Exp. Brain Res., 3*:337.
46. Coss, R. G., and Globus, A. (1978) *Science, 200*:787.
47. De Robertis, E., and Bennett, H. S. (1955) *J. Biophys. Biochem. Cytol., 2*:307.
48. De Robertis, E., Pellegrino de Iraldi, A., Rodríguez de Lores Arnaiz, G., and Salganicoff, L. (1961) *Anat. Rec., 139*:220.
49. Cotman, C. W., and Banker, G. A. (1974) In: *Review of Neurosciences*, Vol. 1, p. 2. (Ehrenpreis, S., and Kopin, I. J., eds.) Raven Press, New York.
50. Rall, W., and Shepherd, G. M. (1968) *J. Neurophysiol., 31*:884.
51. Dowling, J. E., and Boycott, B. B. (1966) *Proc. R. Soc. London [Biol.], 166*:80.
52. Gray, E. G. (1963) *J. Anat., 97*:101.
53. Akert, K., Peper, K., and Sandri, C. (1975) In: *Cholinergic Mechanisms*, p. 43. (Waser, P. G., ed.) Raven Press, New York.
54. Couteaux, R., and Pécot-Dechavassine, M. (1973) *Arch. Ital. Biol., 111*:231.
55. Heuser, J. E., Reese, T. S., and Landis, D. M. D. (1974) *J. Neurocytol., 3*:109.
56. Berl, S., Puzkin, S., and Nicklas, W. J. (1973) *Science, 179*:441.
57. Fertuck, H. C., and Salpeter, M. M. (1974) *Proc. Natl. Acad. Sci. USA, 71*:1376.
58. Davis, R., and Koelle, G. B. (1967) *J. Cell Biol., 34*:157.
59. Nickel, E., and Potter, L. T. (1973) *Brain Res., 57*:508.

60. Cartaud, J., Benedetti, E. L., Cohen, J. B., Meunier, J. C., and Changeux, J. P. (1973) *FEBS Letters, 33*:109.

61. De Robertis, E., Rodríguez de Lores Arnaiz, G., Salganicoff, L., Pellegrino de Iraldi, A., and Zieher, L. M. (1963) *J. Neurochem., 10*:225.

62. Whittaker, V. P., and Zimmerman, H. (1974) In: *Synaptic Transmission and Neuronal Interaction*, p. 217. (Bennett, M. U. L., ed.) Raven Press, New York.

63. Burnstock, G., and Costa, M. (1975) *Adrenergic Neurons*. Chapman & Hall, London.

64. Hubbard, J. I. (1973) *Physiol. Rev., 53*:674.

65. Pellegrino de Iraldi, A., Farini-Duggan, H., and De Robertis, E. (1963) *Anat. Rec., 145*:561.

66. Uchizono, K. (1965) *Nature, 207*:642.

67. De Robertis, E. (1968) In: *Structure and Function of Inhibitory Neuronal Mechanisms*, p. 511. (Von Euler, C., et al., eds.) Pergamon Press, Oxford.

68. Kanaseki, T., and Kadota, K. (1969) *J. Cell Biol., 42*:202.

69. Pellegrino de Iraldi, A., and De Robertis, E. (1963) *Int. J. Neuropharmacol., 2*:231.

70. Jaim-Etcheverry, J., and Zieher, L. M. (1971) In: *Advances in Cytopharmacology*, Vol. 1, p. 343. Raven Press, New York.

71. De Robertis, E. (1967) *Science, 156*:907.

72. De Robertis, E. (1971) *Science, 171*:963.

73. McCaman, R. E., Rodríguez de Lores Arnaiz, G., and De Robertis, E. (1965) *J. Neurochem., 12*:927.

74. Wooten, G. F., and Coyle, J. T. (1973) *J. Neurochem., 20*:1361.

75. Von Euler, U. S. (1972) In: *Catecholamines*, p. 186. (Blaschko, H., and Marshall, E., eds.) Springer-Verlag, Berlin.

76. Heslop, J. P. (1975) *Adv. Comp. Physiol. Biochem., 6*:75.

77. Gray, E. G. (1977) In: *Synapses*, p. 6. (Cottrell, G. A., and Usherwood, P. N., eds.) Blackie & Sons, Ltd., Glasgow.

78. Katz, B. (1971) *Science, 173*:123.

79. De Robertis, E. (1959) *Int. Rev. Cytol., 8*:61.

80. Pecci Saavedra, J., Vaccarezza, O. L., Reader, I. A., and Pasqualini, E. (1970) *Exp. Neurol., 26*:607.

81. De Robertis, E., and Vaz Ferreira, A. (1957) *J. Biophys. Biochem. Cytol., 3*:611.

82. Boyne, A. L., Bohan, T. P., and Williams, T. H. (1975) *J. Cell Biol., 67*:814.

83. Pappas, G. D., and Rose, S. (1976) *Brain Res., 103*:362.

84. Bittner, G. D., and Kennedy, D. (1970) *J. Cell Biol., 47*:485.

85. Jones, D. G. (1975) *Synapses and Synaptosomes: Morphological Aspects*. Chapman and Hall, London.

86. Dowall, M. J., and Zimmermann, H. (1974) *Brain Res., 71*:160.

87. Heuser, J. E., and Reese, T. S. (1974) In: *Synaptic Transmission and Neuronal Interaction*, p. 59. (Bennett, M. V. L., ed.) Raven Press, New York.

88. Llinas, R., and Nicholson, C. (1975) *Proc. Natl. Acad. Sci. USA, 72*:187.

89. Dodge, F. A., and Rahaminoff, R. (1967) *J. Physiol. London, 193*:419.

90. Nachmanson, D. (1955) *Harvey Lect., 41*:57.

91. De Robertis, E., Fiszer, S., and Soto, E. F. (1967) *Science, 158*:928.

92. Changeux, J. P., Kasai, M., Huchet, M., and Meunier, J. C. (1970) *C. R. Acad. Sci. Paris, 270*:2864.

93. Miledi, R., Molinoff, P., and Potter, L. T. (1971) *Nature, 229*:554.

94. De Robertis, E., and Fiszer de Plazas, S. (1976) *Nature, 260*:347.

95. Barrantes, F. J., Arbilla, S. De Carlin, M. C. L., and De Robertis, E. (1975) *Biochem. Biophys. Res. Commun., 63*:194.

96. Karlin, A., Weill, C., McNamee, M., and Valderrama, R. (1976) *Cold Spring Harbor Symp. Quant. Biol., 40*:203.

97. Sobel, A., Heidmann, T., Hofler, J., and Changeux, J. P. (1978) *Proc. Natl. Acad. Sci. USA, 75*:510.

98. Katz, B., and Miledi, R. (1972) *J. Physiol. (London), 224*:665.

99. Neher, E., and Sakmann, B. (1976) *Nature, 260*:799.

100. Briley, M. S., and Changeux, J. P. (1978) *Eur. J. Biochem., 84*:429.

101. Schlieper, P., and De Robertis, E. (1977) *Biochem. Biophys. Res. Commun., 75*:886.

102. De Robertis, E. (1971) *Science, 171*:963.

103. Racker, E. (1977) *J. Supramol. Struct., 6*:215.

104. De Robertis, E., Rodríguez de Lores Arnaiz, G., Alberici, M., Sutherland, E. W., and Butcher, R. W. (1967) *J. Biol. Chem., 242*:3487.

105. Florendo, N. T., Barnett, R. J., and Greengard, P. (1971) *Science, 173*:745.

106. Bloom, F. E. (1975) *Rev. Physiol. Biochem. Pharmacol., 74*:1.

107. Ueda, T., and Greengard, P. (1977) *J. Biol. Chem., 252*:5155.

ADDITIONAL READING

Adams, P. R. (1978) Molecular basis of synaptic transmission. *Trends Neurosci., 1*:14.

Barondes, S. H. (1969) Axoplasmic transport. In: *Handbook of Neurochemistry*, Vol. 2, p. 435. (Lajtha, A., ed.) Plenum Publishing Corp., New York.

Bartfai, T. (1978) Cyclic nucleotides in the central nervous system. *Trends Biochem. Sci., 3*:121.

Basbaum, C. B., and Heuser, J. E. (1979) Morphological studies of stimulated adrenergic axon varicosities in the mouse *vas deferens. J. Cell Biol., 80*:310.

Boyne, A. F. (1978) Neurosecretion—Integration of recent findings into vesicle hypothesis. *Life Sci., 22*:2057.

Breer, H., Morris, S. J., and Whittaker, V. P. (1978) A structural model of cholinergic synaptic vesicles from the electric organ of *Torpedo marmorata. Eur. J. Biochem., 87*:453.

Briley, M. S., and Changeux, J. P. (1977) Isolation and purification of the nicotinic receptor and its functional reconstitution into a membrane environment. *Int. Rev. Neurobiol., 20*:31.

Bunge, R., Johnson, M., and Ross, C. D. (1978) Nature and nurture in development of the autonomic neuron. *Science, 199*:1409.

Burnstock, N. (1972) Purinergic nerves. *Pharmacol. Rev., 24*:509.

Carlson, S. S., Wagner, J. A., and Kelly, R. B. (1978) Purification of synaptic vesicles from *Elasmobranch* electric organ. *Biochemistry, 17*:1188.

Ceccarelli, B. et al. (1979) Freeze-fracture studies of frog neuromuscular function during intense release of neurotransmitter I and II. *J. Cell Biol., 81*:163 and 178.

Cottrel, G. A., and Usherwood, P., eds. (1977) *Synapses.* Blackie & Sons, Ltd., London.

De Robertis, E. (1964) *Histophysiology of Synapses and Neurosecretion.* Pergamon Press, Oxford.

De Robertis, E. (1967) Ultrastructure and cytochemistry of the synaptic region. The macromolecular components involved in nerve transmission are being studied. *Science, 156*:907.

De Robertis, E. (1975) *Synaptic Receptors: Isolation and Molecular Biology,* pp. 1–387. Marcel Dekker, Inc., New York.

Eldefrawi, M. E., and Eldefrawi, A. T. (1977) Acetylcholine receptors. In: *Receptors and Recognition,* Vol. 4, p. 199. (Cuatrecases, P., and Greaves, M. F., eds.) Chapman & Hall, London.

Hille, B. (1976) Gating in sodium channels of nerve. *Annu. Rev. Physiol., 38*:139.

Hucho, F., and Schiebler, W. (1977) Biochemical investigations on ionic channels in excitable membranes. *Mol. Cell. Biochem., 18*:151.

Hydén, H. (1960) The neuron. In: *The Cell,* Vol. 4, p. 215. (Brachet, J., and Mirsky, A. E., eds.) Academic Press, Inc., New York.

Katz, B. (1966) *Nerve, Muscle, and Synapse.* McGraw-Hill Book Co., New York.

Kelly, P., Cotman, C. W., Gentry, C., and Nicolson, G. L. (1976) Distribution and mobility of lectin receptors on synaptic membranes of identified neurons in the central nervous system. *J. Cell Biol., 71*:487.

Kristensson, K. (1978) Retrograde transport of macromolecules in axons. *Annu. Rev. Pharmacol. Toxicol., 18*:97.

Le Douarin, N. M., and Teillet, M. A. (1973) The migration of neural crests to the wall of the digestive tract in avian embryo. *J. Embryol. Exp. Morphol., 30*:31.

Lester, H. A. (1977) The response to acetylcholine. *Sci. Am., 236*:106.

Lubinska, L. (1975) On axoplasmic flow. *Int. Rev. Neurobiol., 17*:241.

Miller, D. C., Weinstock, M. M., and Magleby, K. L. (1978) Is the quantum of transmitter released composed of subunits? *Nature, 274*:388.

Nathanson, J. A., and Greengard, P. (1977) Second messengers in the brain. *Sci. Am., 237*:109.

Neher, E., and Stevens, C. F. (1977) Conductance fluctuations and ionic pores in membranes. *Annu. Rev. Biophys. Bioeng., 6*:345.

Ochs, S. (1974) Energy metabolism and supply of ~P to the fast axoplasmic transport mechanism in nerve. *Fed. Proc., 33*:1049.

Ochs, S., Chan, S. Y., and Worth, R. (1977) Calcium and the mechanism of axoplasmic transport. In: *The Neurobiologic Mechanisms in Manipulative Therapy,* p. 359. (Korr, I. M., ed.) Plenum Publishing Corp., New York.

Osborne, M. P., and Monaghan, P. (1977) Effects of light and dark upon photoreceptor synapses in the retina of Xenopus laevis. *Cell Tissue Res., 173*:59.

Otsuka, M., and Takahashi, T. (1977) Putative peptide neurotransmitters. *Annu. Rev. Pharmacol. Toxicol., 17*:425.

Patterson, P. H., Potter, D. D., and Furshpan, E. J. (1978) The chemical differentiation of nerve cells. *Sci. Am., 239*:38.

Rees, R. P., Bunge, M. P., and Bunge, R. P. (1976) Morphological changes in the neuritic growth cone and target neuron during synaptic junction development in culture. *J. Cell Biol., 68*:240.

Rodríguez de Lores Arnaiz, G., and De Robertis, E. (1972) Properties of isolated nerve endings. In: *Current Topics in Membrane Transport,* Vol. 3, p. 237. Academic Press, Inc., New York.

Rodríguez de Lores Arnaiz, G., and De Robertis, E. (1973) Drugs affecting the synaptic components of the CNS. In: *Fundamentals of Cell Pharmacology,* p. 280 (Dikstein, S., ed.) Charles C Thomas, Springfield, Ill.

Shepherd, G. M. (1978) Microcircuits in the nervous system. *Sci. Am., 238*:92.

Suskiw, J. B., Zimmermann, H., and Whittaker, P. V. (1978) Vesicular storage and release of ACh in *Torpedo* electroplax synapses. *J. Neurochem., 30*:1269.

The Synapse. (1975) *Cold Spring Harbor Symposium Quantitative Biology, 40.*

Ulbricht, W. (1977) Ionic channels and gating currents in excitable membranes. *Annu. Rev. Biophys. Bioeng., 6*:7.

Williams, M. (1979) Protein phosphorylation in mammalian nervous system. *Trends Biochem. Sci., 4*:25.

Yamamura, H. I., Enna, S. J., and Kuhar, M. J. eds. (1978) Neurotransmitter receptor binding. Raven Press, New York.

INDEX

A-band, muscle contraction and, 588, 589
  striated muscle fiber and, 586–587
Acceptor end, of tRNA, 520
Acceptor site, for protein synthesis, 530
Acentric chromosome, 350
Acetabularia, nucleocytoplasmic interactions
  and, 563
Acetate, 264
Acetyl coenzyme A (acetyl CoA), 110, 111, 264,
  307, 308
Acetylcholine, fluctuations induced by, 639
Acetylcholine receptor, coupling of, 637–639
Acetylcholine system, neurotransmitter
  behavior in, 631–632
Acetylcholinesterase, 135, 139, 631
Acid(s)
  amino. See *Amino acid(s)*.
  citric, 264
  EDTA, 341
  fatty, 123
    evolution of, 128
    oxidation of, 262
  galacturonic, 302
  gibberellic, 307
  glycolic, 298
  hyaluronic, 166
  nitrous, 473
  nucleic evolution of, 128
    sequencing of, genetic code confirmed by,
      476, 478
  oxaloacetic, 264
  pantothenic, acetyl CoA and, 110
  phenylpyruvic, 464
  phosphatidic, 233, 292
  phosphoglyceric, 324
  phosphoric, 108
  sialic, 233, 235, 236
    cell coat and, 166, 167
    cell membrane and, 134
    glycophorin and, 138
  stearic, 123
  succinic, 264
  tricarboxylic, cycle of, 110–112
Acid phosphatase(s), 228, 234, 282
  acrosomes and, 295
Acid phosphomonoesterase, 138
Acid proteases, 296
Acid-base equilibrium, 70
Acidosis, 295
Aconitase, 307, 308
Acrocentric chromosome, 348, 349
Acrosome, 295
ACTH, 102, 245, 258
Actin, 182, 197
  chromosome movement and, 395
  contractile ring of cytokinesis and, 395
  endocytosis and, 292

Actin (*Continued*)
  muscle contractility and, 588, 595
  myofilaments and, 593
  polarization of within sarcomere, 594
  nonmuscle, 198–200
  nuclear, 339
  plant cells and, 200
  spectrin and, 142
α-actinin, 158, 198–200
  contractile proteins and, 593
  Z-line and, 588
Actinomycin D, 531
Action potential, 153
  in nerves, 613–615
  propagation of, 615–617
Activation
  allosteric, 102
  precursor, 102, 103
Active site, 96
  enzymes and, 141
  substrate binding to, 97
Active transport. See *Transport, active*.
Active zones, in presynaptic membranes,
  624–625
Actomyosin, 595
Adenine, 84, 85, 108
Adenosine, 84, 108
Adenosine diphosphate (ADP), 84, 108, 262,
  263
  mitochondrial transport and, 272, 274
  photochemical reaction and, 320, 321
Adenosine monophosphate (AMP), 84
  as extracellular messenger, 578
  cyclic (cAMP), 102–105, 143, 144, 640–643
    hormonal regulation of glycogen
      metabolism and, 104
    lac transcription and, 540–541
Adenosine triphosphatase (ATPase), 138, 139,
  263, 266, 269
  calcium-activated, 198, 601
  chloroplasts and, 321, 325
  dynein and, 192, 193
  F_2, 268
  mitochondrial, structure of, 268
    F_1 particles and, 253
  myosin and, 591
  Na$^+$ K$^+$ mediated, active transport and, 154,
    155
  proton transfer and, 270
Adenosine triphosphate (ATP), 84, 108–113,
  262, 263, 269
  active transport and, 150
  Calvin cycle and, 322, 323
  dynein and, 192
  endocytosis and, 292
  intracellular transport and, 242, 243
  microtubules and, 192

Adenosine triphosphate (ATP) (*Continued*)
mitochondrial transport and, 272, 274
photochemical reaction and, 320, 321
photosynthesis and, 322
protein synthesis and, 522–523
synthesis of, 268
Adenovirus, 119
Adenylate cyclase, 138, 139
coupling of receptors to, 143, 144
Adenylate kinase, 262
Adhesion, cellular, 161, 168, 169
ADP. See *Adenosine diphosphate.*
Adrenal cortex, mitochondrial crests and, 258
Adrenal gland, 258, 259
catecholamine granules in, 243
Adrenalin, 102
Adrenocorticotrophic hormone (ACTH), 102, 245, 258
Aequorin, 164
Aerobes, 129
respiration in, 109–113
Agglutinin, germ, 138, 167
Aggregates, cellular, 168, 169
supramolecular, 117
Agranular vesicles, 627
Aided assembly, 119
Albinism, 465
Albumin
bovine serum, 336
proalbumin and, 245
Alcian blue stain, 167
Aleurone body, 296, 306–309
Alga(ae)
blue-green, symbiont hypothesis and, 278
scales of, Golgi complex and, 302
Alkaptonuria, 465
Alleles, 421
Allolactose, 538
Allopolyploidy, 427
Allosomes, 444
Allosteric activator, 101, 102
Allosteric enzymes, 100, 101
Allosteric inhibitor, 101, 102
Allosteric modifier, 101
Altmann, R., 8, 250, 278
Amanita phalloides, 531
α-amanitine, RNA transcription and, 490
Ameba, cell coat of, 167
Ameboid movement, 182
microfilaments and, 202
Ames, B., 473
Amino acid(s)
codons and, 470
evolution of, 128
mutations and, 465–466
nucleotides as code for, 468
phenylalanine, tRNA and, 521
pinocytosis and, 289
properties of, 71
side-chains of, 71
tyrosine, in tRNA, 520–521
D-amino-acid oxidase, 297, 298
Aminoacyl site, protein synthesis and, 530
Aminoacyl-activating enzymes, protein synthesis and, 519–523
Amniocentesis, 448–449
Ammanita phalloides, 531
AMP. See *Adenosine monophosphate.*

AMP, cyclic (cAMP), 102–105, 143, 144, 640–643
lac transcription and, 540–541
Amphipathy, 140, 141
cell membrane phospholipids and, 81
Amphiuma, DNA content in, 340
Amphoteric molecules, 71
Ampicillin, 531
Amplification, gene, 550
α-amylase, 307
β-amylase, 296
Amylopectin, 122
in starch, 78
Amyloplast, 309
Amylose, in starch, 78
Amytal, 272
Anabolic systems, enzyme repression in, 537
Anabolism, 95
Anaerobe(s), 129
respiration and, 109–111
Anaphase
chromosomes and, 348, 349
meiotic, 410
mitotic, 385
chromosomal movement during, 394–395
tubulin and, 185
Anderson, 353
Anemia, hereditary, Cooley's, 258
sickle cell, 465–466
Aneuploidy, 426–427, 449–450
Angiosperms, C₄ pathway and, 324
Anions, diffusion and, 150
facilitated diffusion of, 138
membrane permeation channels and, 142
Anisotropism, 32
striated muscle fiber, 586–587
Annealing of DNA, 88
Annulus, of nuclear pores, 335
Anoxia, 295
Antibiotics, usefulness to medicine, 531
Antibody(ies)
anti-troponin, 593
anti-tubulin, 199
cell fusion and, 566–567
microtubules and, 185
mitotic apparatus and, 388
monoclonal, 566–567
secretion of, 237
Anticodon, 468
of tRNA, 519
Anticodon loop, tRNA, 520
Antigen(s), 143
ABO blood group, cell coat and, 168
antibody production and, 566–567
cell surface, 138
histocompatibility, cell coat and, 168
H-Y, 447
membrane, cancer cells and, 171
MN, cell coat and, 168
surface, 142
Antimycin, 269, 272
Antiparallel strands, of DNA, 86
Apoenzymes, 96, 97
Arabinose, 233, 302
Arthritis, rheumatoid, lysosomes and, 295
Aryl hydroxylases, 219, 220
Aryl sulfatase, 285
Asbestosis, lysosomes and, 295

Aspartate, 262, 272
Assembly, macromolecular, 118–123
  microtubular, 185
Asters, mitotic, 185, 384
Atmosphere, prebiotic, 127, 128
ATP. See *Adenosine triphosphate.*
ATPase. See *Adenosine triphosphatase.*
Attenuator, in enzyme repression, 541
Aurovertin, 269
Autolysis, 259
Autophagosomes, 287, 288
Autophagy, crinophagy and, 295
  lysosomes and, 294
Autoradiography, DNA fiber, 372
Autotrophs, 129
Avery, O., 467
Axo-axonic synapses, 623
Axodendritic synapses, 623
Axon(s), 144, 186
  giant, squid, 153
  macromolecular transport, 609
  membranes of, Na$^+$ and K$^+$ channels in,
    615–617
  microtubules in, 187
  mitochondria and, 275
  neurofilaments and, 200
  orthograde transport and, 609–610
  retrograde transport and, 610–612
  structure of, 606
Axoneme, microtubules and, 188
  sliding model of, 192, 193
Axosomatic synapses, 623

Bacteriophage(s), RNA, MS2, and hairpin
    loops, 91
  structure of, 119–121
Bacterium(a)
  lysosomes and, 295
  mitochondrial F$_1$ particles and, 253
  periplasmatic space in, 283
  sulfur, photosynthesis and, 319
  symbiont hypothesis and, 278
Balbiani, E., 333
Balbiani rings, 551
Banding, chromosomal, 358
Banding techniques, 438–439
  chromosomal identification and, 440, 442
  syndromes detected by, 454
Bands, chromosomal, 440
Bangham, A., 126
Barber, 9
Bärlund, H., 149
Barr, M., 444
Barr body, 444–445
  heterochromatin and, 354
Basal body(ies), ciliary, 184
  microtubules and, 190, 191
Basal plate, 190
Base(s), nucleic acid
  absorption of UV light and, 85
  evolution of, 128
  sequence of in DNA, 86
  tRNA and, 520
  types of, 85
  uracil in RNA, 91
Base-pairing, ribosomal, 504–505

Bateson, W., 9
Beadle, 464
Beerman, W., 551
Benda, 8, 250
Beneden, van, 8
Bensley, R. R., 10, 51, 250
Benson, 323
Benzopyrene, 219, 220
Benzoquinone, 265
Bertram, 444
Bicarbonate, 272
Bilayer(s), lipid, 132. See also *Membrane(s).*
  artificial, 123–125
  distribution of lipids in, 133, 134
  ER and, 209
  liposomes and, 270
  models of, 139, 140
  thylakoids and, 312–314
Binding proteins, primary, 506
Binding site, ribosome, in protein synthesis,
    526–527
Bioenergetics, 106–108
  muscle contractility and, 597–601
  photochemical reaction and, 320–322
Biosynthetic functions, of neurons, 606–609
Birefringence, 32, 33
  myelin sheath and, 144
  of liposomes, 126
Bivalents, lampbrush chromosomes and, 553
Blastula, 560
Blebs, cellular, 169–171
Blindness, color, 455
Blood, clotting of, 120, 121
Blood groups, ABO, 138
  cell coat and, 168
Blood-brain barrier, 161
Bond(s)
  covalent vs. non-covalent, 76
  disulfide, 74–75, 122
  1,6-α-glycosidic, 122
  hydrogen, in DNA, 86, 87
  in proteins, 75, 76
  molecular shape and, 118
  non-covalent, 140
  peptide, 71, 72
  phosphate, high energy, 108, 109
Boveri, T., 8
Bovine serum albumin, 336
Brachet, J., 181, 508
Brain, blood-brain barrier and, 161
BrdU, 351, 432
  chromosome replication visualized with, 371
Breuer, M., 552
Bridges, 9
5-bromo-4-chloroindoleacetate-esterase, 285
Bromodeoxyuridine (BrdU), 351, 432
  chromosome replication visualized with, 371
Brown, R., 8

C$_3$ cycle, photosynthesis and, 322, 323
C$_4$ pathway, Calvin cycle and, 324
Cajal, 236, 619
Calcium ions, 155, 274
  ameboid contraction and, 203
  ATPase activation and, 601
  cilia and, 192

Calcium ions (*Continued*)
 exocytosis and, 243
 flagella and, 192
 gap junctions and, 164, 165
 microtubules and, 184, 185
  assembly of, 393
 muscle contraction and, 597
 photoreceptors and, 147
 proalbumin and, 245
 terminal cisternal release of, 599
 transmitter-secretion coupling and, 634
Callan, H. G., 553
Callose, 305
Calvin, M., 323
Calvin cycle (C_3), photosynthesis and, 322, 323
cAMP, 102–105, 143, 144, 640–643
 hormonal regulation of glycogen metabolism
  and, 104
 lac transcription and, 540–541
Cancer
 cyclic AMP concentration and, 105
 glycosphingolipid loss in, 236
 microtubules and, 185
 mitochondria and, 252
 viral transformation and, 170, 171
CAP factor, 487
Capping, cellular, 142
Capsid, 119
Capsomere, 119
Carbamylcholine, 244
Carbohydrate(s)
 Calvin cycle and, 322, 323
 classification of, 78
 degradation of, 263, 264
 lysosomes and, 294
 membrane, 132, 134, 135
 photosynthesis and, 319
 as prosthetic groups, 235, 236
Carbon dioxide, Calvin cycle and, 322, 323
 fixation of, 322, 323
 photosynthesis and, 319, 320
Carbon reduction, photosynthesis and, 322,
 323
Carboxyatractylate, ^{35}S (^{35}S CAT), 272
Carboxydismutase, subunits of, 323
Carboxypeptidases, 296
Carcinogens, 219, 220
Cardiolipin, 134, 261, 262
Carotenes, 317
 β-carotene, structure of, 316
Carotenoids, 309, 315, 317, 325
Carrel, 45, 101
Carrier(s), electrons and, 320, 321
 protein, 219
  ATP and, 272
  mitochondrial, 262
 Q, 320
Carrier hypothesis, active transport and, 154
Cascade effect, cAMP, 640
Cascade mechanism, hormonal regulation of
 glycogen metabolism and, 104
Caspersson, T., 181, 438, 508
Catabolic systems, enzyme induction in, 537
Catabolism, 95. See also *Respiration, cellular*.
Catabolite repression, 541
Catalase(s), peroxisomal, 297, 298
Catalysis, control of, 101, 102
Catalysts, biological. See *Enzyme(s)*.
Catecholamine, granules of, 243

Cathepsins, 294
Cell(s)
 adhesion of, 161, 168, 169
 cancerous. See *Cancer*.
 chemical components of, 69–94
 contact between, differentiation and, 578
 crown, 193
 differentiation of. See *Differentiation,
  cellular*.
 energy use by, 107, 108
 ganglion, 576–577
 germ, lysosomes and, 295
  mutation of, 430
 haploid, 400
 hybrid, 165
 intercellular communications in, 158–166
 Lesch-Nyhan, 465
 oligodendroglial, 144
 recognition, 166–173
 reconstitution, 565
 red blood. See *Erythrocyte(s)*.
 Schwann, 144
 stem, 576
 types of molecular structures in, 117
 viral infection of, 119
 xeroderma pigmentosum, 465
Cell biology, history of, 7–10
Cell coat, 166–173
 pinocytosis and, 289, 290
Cell culture, types of, 45, 46
Cell cycle, 364–370. See also *Gap periods*.
 mitochondrial DNA and, 277
 nucleus and, 21, 22
Cell division, 182, 198, 364, 383, 395
Cell factor, 171
Cell fractionation, 53
Cell fusion, 142, 144, 165, 563–565
 antibodies from, 566–567
Cell injury, mitochondria and, 259
Cell plate, 302, 395, 397
 plasmodesmata and, 304
Cell respiration. See *Respiration, cellular*.
Cell surface, differentiations of, 158–166
 supramolecular structures of, 115–176
Cell theory, 8
Cell wall, 166, 300–304, 397
 osmotic pressure and, 148
Cellobiose, cell wall and, 301
 lysosomes and, 294
Cellular interactions, differentiation and, 578
Cellulose, 78, 166
 cell wall and, 301
 synthesis of, 302
Centric fusion, 431, 433
Centrifugation, differential, 53, 214
Centriole(s), 182, 184, 251
 cilia and, 194
 flagella and, 194
 kinetosomal vs. ciliary, 194, 195
 microtubules and, 190, 191
Centromere(s), 184, 349, 383
 chromosomal shape and, 348
 diffuse, 391
 homologous (sister), separation of in
  meiosis, 410
 kinetochore and, 391–392
Centrometric index, 439
Centrosome, 387
Centrosphere, 185

Cephalin, 259
Cerebrosides, 81
CF_0, 325
CF_1, 321, 325
cGMP, 105
Chain elongation, EFG factor and, 529
Chain termination, codon involvement,
    530–531
Chambers, R., 9
Chance, B., 265
Chance and Necessity, 127
Changeux, J. P., 101
Channels, gap junction, 161–163
    intercellular, 169
    membrane, protons and, 268
Chargaff's rules of DNA base composition, 86
Chelation, EDTA and, 341
Chemical-conformational hypothesis, 270
    chloroplast coupling and, 321
Chemiosmotic hypothesis, 269, 270
    chloroplasts and, 321, 327
Chemodynesis, 200
Chevremont, 276
Chiasmata, 409
Chitin, 78, 166
    cell wall and, 302
Chloramphenicol, 278, 328, 531
    ribosomal inhibition and, 507
Chlorine, ions of, electrical potential and, 148,
    149
Chlorophyll(s), 309, 315
Chlorophyll-*a*, structure of, 316
Chloroplast(s)
    characteristics of, 309–314
    membranes of, 324–327
    mitochondrial F_1 particles and, 253
    peroxisomes and, 298
    symbiont hypothesis and, 129, 278
    thylakoids in, 314–318
Chloroplast envelope, 310, 312
Cholera, adenylate cyclase activation and, 104
Cholesterol, 123, 133, 259, 261
    lecithin and, 144
    sphingomyelin and, 144
Cholinacetyltransferase, 631
Choline dehydrogenase, 262
Cholinergic development factor, 629
Cholinergic receptor, neurotransmitter
    fluctuation and, 639
Chondroitin sulfate, 78
Chromatid(s), 348, 349, 384
    bivalents and, in lampbrush chromosomes,
        553
    mononemic structure of, 428
    sister, 408
        exchange between, 428
        exchanges of, in diseases with altered
            DNA repair, 432–433
Chromatin, 333
    characteristics of, 340–348
    histones in, 84
    nuclear, at interphase, 444–445
    nucleolar, 359
    nucleolar-associated, 359, 360
    sex, 354, 445–446
    types of, 440
    X, 445–446
    Y, 446
Chromocenter(s), 340, 354

Chromogranins, 243
Chromomere(s), 349, 402
    chromosomal bands and, 440
    polytene chromosomes and, 550
Chromonemata, 348, 349
Chromoplasts, plant cells and, 309
Chromoproteins, types of, 72
Chromosome(s)
    aberrations of, 428–432
    accessory, 444
    banding technique for, 438–439
    bands of, 440
    cell cycle and, 368–369
    changes in, cytogenetics and, 426–433
    characteristics of, 349–354
    chromatin in types of, 440
    condensed, RNA synthesis and, 368–369
    daughter, 385
    eukaryotic, DNA replication in, 375–376
        bidirectionality of, 376, 377
        reinitiation prevented in, 377
    evolutionary role of, 433–435
    fibers of, 385
    genes localized in, 458
    giant, 548–557
    homologous, in meiosis, 402
        pairing of, 405–408
    human, aberrations of, mixed, 451–452
        in tumors, 454
        abnormalities of, 448–455
        in human syndromes, 450–454
        diploid number of, 438
        genetic map and, 455–458
    identification of, banding techniques and,
        440, 442
    lampbrush, 410
        oöcyte, 552–557
        RNA synthesis and, 553–556
    leptonemic, 402, 404
    mitotic and meiotic, distribution of, 410, 412
    movement of, in anaphase, 394–395
        in metaphase, 393–394
    mutation of, 379
    Philadelphia, 454
    polytene, 349
        chromomere alignment, 550
        DNA in, 548–550
        gene transcription in, 550–552
    radiation and, 429
    sex, 444–448
        true hermaphroditism and, 452
    structure of, banding techniques and, 440
    synapsis of, 405–408
    visualization of, G_1, S, and G_2 periods and,
        366–368
    X, 354
        inactivation of, 445–446
    Y, heterochromatic region of, 446
Chromosome mapping, 457–458
Chromosome number, diploid, 24
Ciliary apparatus, 188
Ciliary necklace, 190
Ciliary plate, 190
Cilium(a), 182
    development of, 195–197
    microtubules and, 183
    movement of, 191, 192
    photoreceptors and, 193
Cinematography, time-lapse, 250, 275

Cirri, 188
Cisterna(ae), 209
  cavities of, 210
  of Golgi dictyosomes, 226, 227
  nuclear pores and, 335, 336
  terminal, Ca^{2+} release from, 599
Cistrons, rDNA, repetitive, 510
Citrate synthetase, 307, 308
Citric acid, 264
Clathrin, coated vesicles and, 292
Claude, A., 10
Cleavage, cellular, 182, 198, 395
  of fertilized egg, 560
Cleocytoplasmic interaction, 561
Clone(s), cell culture and, 46
  development of, 571
  gene, 480
Clotting, of blood, 120, 121
Coacervates, 128
Coated vesicles, 627
Codons, 468
  amino acids and, 470
  AUG, 470
    ribosome binding site and, 526
  initial, signal peptides and, 221
  initiation, 468, 470
  nonsense, 470
  termination, 470–471, 530–531
  UAA, 470–471
  UAG, 470–471
  UGA, 470–471
Coenzyme(s), 96, 97
  A, 110, 111, 264, 307, 308
  NAD^+ and $NADP^+$, dehydrogenases and, 96, 97
  NADH, 112
  Q_{10} (CoQ), 265
  ubiquinone, 265
  vitamins and, 97
Cofactors, enzymatic, 96
Colchicine, 183, 185, 186, 200, 245
  spindle fibers and, 392
Collagen, 72
  structure of, 120
Collander, 149
Color blindness, 455
Concanavalin A, 167
Concentration gradient, 149, 150
Condensation, chromosome, chromatin and, 352–354
  premature, 366–368
Conductance, in artificial lipid bilayers, 124, 125
Conduction, in nerve impulses, myelinated, 615
Conduction velocity, nerves, 612–613
Cone(s), retinal, 193
  mitochondria in, 251
Conjugated proteins, 72
Conjugation, 468
Connexon, 163
Constitutive heterochromatin, 354, 356, 446
  centromeric, 440
Contact inhibition, cellular, 46, 168–171
Contact zones, mitochondrial, 263, 272
Continuous fibers, 385
Contraction
  chloroplasts and, 310
  ciliary, 191, 192

Contraction (Continued)
  microfilaments and, 197, 200
  mitochondrial, 250, 272–274
  muscle. See Muscle(s), contraction of.
Coomassie Blue stain, 135
Coons's technique, 61
Copper,
  enzyme-bound, 265
  plastoquinone and, 320
  in proteins, 267
Core polymerase, RNA, 486
Core proteins, ribosomal dissociation and, 506
Co-repressor, in enzyme repression, 541
Correns, 9
Cortisone, 295
Corynebacterium diphtheriae, 531
Costello, D. P., 351
Coupling
  ATPase and, 253
  cancer cells and, 165
  electron transport and, 267, 268
  intercellular, electrical, 161
  intercellular cooperation and, 165
Coupling factor (CF_1)
  chloroplast ATPase and, 325
  thylakoid membrane and, 321
C-protein, contractile proteins and, 593
Cretinism, goitrous, 465
Cri du chat syndrome, 454
Crick, F. H. C., 10, 86, 370, 463, 470, 496
Crinophagy, lysosomes and, 295
Crista(ae), 263
Critical temperature, 353
Critical-point drying method, 353
Cross-bridges, of dynein, 192
Crossing over, 409
  linkage broken by, 424
Crown region, ribosomal, 503
Crystalline birefringence, 33
CTP, 108
Culture, of cells, 45, 46
Cutin waxes, 302
Cyanide, 269, 272
Cyclic adenosine monophosphate (cAMP), 102–105, 143, 144, 640–643
  hormonal regulation of glycogen metabolism and, 104
  lac transcription and, 540–541
Cyclic AMP. See Cyclic adenosine monophosphate.
Cyclic guanosine monophosphate (cGMP), 105
Cycloheximide, 242, 244, 278, 328, 531
  ribosomal inhibition and, 507
Cyclosis, 181, 182, 198, 305
  chloroplasts and, 310
  plant cells and,
    microfilaments of, 200, 201
Cytochalasin B, 293
  cell contraction and, 395
  cell enucleation by, 565–566
  microfilaments and, 197–199
Cytochrome(s), 112
  b_5, 222
  b_6, 320, 321
  b559 (b_3), 320, 321
  b-c_1 complex, 278
  c, 135
  chain of, topography of, 264–269
  chlorophyll and, 315

Cytochrome(s) (*Continued*)
  f, 320, 321
  f-b$_6$ complex, 325
  P-450, 220
  photochemical reaction and, 320, 321
Cytochrome oxidase, 233, 265, 267, 269, 278
Cytochrome reductase, 266
Cytogenetics, 420. See also *Genetics.*
  chromosomal changes and, 426–433
  human, 438–439
Cytokinesis, 383, 387
  in animal cells, 395
  in plant cells, 395, 397
Cytolysosome, 259, 287, 288
Cytophotometry, principles of, 59
Cytoplasm
  basophilic, 181
  characteristics of, 18
  chromidial, 181
  oöcyte, importance of determinants in,
    575–577
  in plants, 305–309
  relation to nucleus, 562–570
Cytoplasmic determinants, importance in
  development, 575–577
Cytoplasmic matrix, 179
Cytoplasmic streaming, 181, 182, 198, 305
  chloroplasts and, 310
  plant cells and,
    microfilaments and, 200, 201
Cytoplasts, preparation by enucleation,
  565-566
Cytosine, 85
  mitochondrial, 277
Cytosine triphosphate (CTP), 108
Cytoskeleton, structure of, 179–183
Cytosol, constituents of, 179–183

D loop, tRNA, 520
DAB, 269, 297
Dale, 620
Daltonism, 455
Dan, 392
Danielli, J., 139
Dark reaction, photosynthetic, 320
Darnell, J. E., 496
Davson, 139
Decondensation, chromosomal, chromatin
  and, 352–354
De Duve, C., 282
De Robertis, E., 183, 295
De Robertis, E. M., 570
De Vries, 9
Defecation, cellular, 287
Dehydrogenase(s), 96, 112, 263, 264
  choline, 262
  glyceraldehyde-3-P, 137
  glycerol-phosphate, 262
  β-hydroxybutyrate, 263
  malate, 307, 308
  NAD$^+$ and NADP$^+$ as coenzymes of, 96, 97
  succinic, 263
Deletions, genetic, 428
  chromosomal material and, 430–431
  single base, 472

Deletions (*Continued*)
  of DNA, 88, 89
    mitochondrial, 277
    partial, in mapping, 89
  of protein, 74
Dendrites, 186
  relation of synapses to, 621–622
Dendrodendritic synapses, 623
Dense particles, ribosomes and, 501
Dense-core vesicles, large, synaptic, 627
Deoxynucleoside triphosphates, 374
Deoxyribonuclease (DNase), 262, 296
Deoxyribonucleic acid (DNA), 262
  annealing of, 88
  antiparallel strands of, 86
  assembly of, directed, 119
  bacteriophage, φ29 and, 119, 121
  basal bodies and, 193
  base composition of, 86
  Chargaff's rules of base composition in, 86
  chromatin and, 342
  chromosomal, packing of, 352–354
  circular, 89, 90
    mitochondrial, 253
  complementary, labeled, 493
  denaturation of, 88, 89
  gene composition and, 467
  hybridization of, 88, 173
  meiosis and, 401–402
  mitochondrial (mtDNA), 253, 275
    symbiont hypothesis and, 278
  mitosis and, 401–402
  molecule of, length of, 86
  non-coding, eukaryotic genes and, 494–495
  nuclear, intraspecies constancy of, 340
  nucleosomes and, 343–348
  pachynemic, 417
  polytene chromosomes and, 548–550
  primer, 374
  renaturation of, 88
  repair altered in, sister chromatid exchanges
    and, 432–433
  repetitive, 356, 545–547
    constitutive heterochromatin and, 357, 358
  replication of, 370–381
    eukaryotic, 375–376
      bidirectionality of, 376, 377
      reinitiation prevented in, 377
    in interphase, 365
    nucleosomes and, 377–378
    RNA synthesis during, 378
  ribosomal (rDNA), 359
    amplification in oöcytes, 511–513
    nucleolar organizer and, 508–509
    repetition in, 547
  5S, repetition in, 547
  satellite, 440
    repetitive sequences in, 546–547
    ribosomal, 511
  sequence of, in operator, 538–539
    restriction enzymes and, 479–480
  sequencing of, genetic code and, 474–478
  space-filling model of, 88
  spacer, ribosomal genes and, 509–511
  structure of, 86
  supercoiled, 89, 90
  synthesis of, enzymes and, 373–375
  as template, 374

Deoxyribonucleic acid (DNA) (*Continued*)
 transcription to RNA, 486–488
 transfer of, 467–468
 UV absorption of, and hyperchromic shift, 88
 viral, 173
 Watson-Crick model of, 86, 87
 zygonemic, 417
Deoxyribose, 78
Dephosphorylation, 102
 enzyme interconversion and, 102
Depolarization, of action potential, 153, 613–615
Desmosomes, 158–161
Detachment factor, 221
Determination, cell, differentiation and, 562
Detoxification, 219, 220
Development, early, importance of determinants, 575–577
 lysosomes and, 294
Dextran, lysosomes and, 294
Dextran sulfate, 354
Diakinesis, 410
3',3'-diaminobenzidine (DAB), 269, 297
Diazonium reaction, 55
Dicentric chromosome, 350
Dichroism, 33
Dictiotene, 415
Dictyosome(s), 226
 faces of, 226, 227
 polarization of, 228, 231
 secretion and, 307
 turnover rate of, 235
Differential centrifugation, 53
Differentiation
 cellular, 165
  endoplasmic reticulum and, 209
  general characteristics of, 562
  heterochromatin and, 357
  microtubules and, 186
  molecular mechanisms of, 571–579
  plants and, 301, 302
  plasmodesmata and, 304
  translational control and, 572–574
 cellular interactions and, 578
 of membranes, 218, 219
 morphological, determination and, 562
 structural, of myofibrils and sarcomeres, 586–587
Diffraction, x-ray, calculation of, 41
Diffusion, 149, 150
 coefficient of, 149, 150
 endoplasmic reticulum and, 219
 facilitated, 156
  anions and, 138
 selective, nuclear pores and, 335–337
Digestion, intracellular, 294–296
Digitonin, 259
Dimerization, 379
Dimethyl-sulfoxide, 119
2,4-dinitrophenol, 112, 321
Dipeptide, 71
Diphtheria toxin, 531
Diploid chromosome number, 24
 human, 438
Diplonema, 409–410
Diplotene stage, meiotic, lampbrush chromosomes and, 552–553
Dipole, 70

Directed assembly, 119
Disaccharides, 78
 lysosomes and, 294
Disc, isotropic, in striated muscle fiber, 586
Disease(s)
 genetic, 192
 lysosomes and, 294, 295
 sex-linked, 455–457
 storage, lysosomes and, 296
Dissociation, 433–434
 ribosomal proteins and, 506–507
Dissociation factors, protein synthesis and, 526
Dixon, 620
DNA. See *Deoxyribonucleic acid.*
DNA binding protein, 417
DNA fiber autoradiography, 372
DNA polymerases, 373–374, 474
DNA repair enzymes, 378–380
Dolichol, 79
 phosphates of, 236
Donnan, 150
Donnan equilibrium, 150, 152
Donor site, protein synthesis and, 530
Double helix, 86, 87
Drosophila melanogaster, 356
 DNA content of, 340
Down's syndrome, 438
Du Bois-Reymond, E., 9, 619
Dujardin, F., 8
Dutrochet, H. J., 8
Dynamic equilibrium hypothesis, 394–395
Dynein, arms of, 190
 crossbridges of, 192
 tubulin and, 192
Dysgenesis, gonadal, 451

E_s, 134
E. coli. See *Escherichia coli.*
Ecdysone, puff induction and, 551
 response of cells and, 575
Edidin, 142
EDTA, 119, 341
Edwards, 289
Effectors, 605
EFG factor, 529
EFs, 313
EFT, 527, 529
EFu, 313
Eggs. See *Oöcytes.*
Ehrlich, P., 49, 636
Elaioplast, 309
Electric charge
 cell coat and, 168
 membrane pores and, 153
 membranes and, 134
 mitochondrial membranes and, 263
 pH and, 168
 pinocytosis and, 290
Electric discharges, prebiotic evolution and, 128
Electrical gradient, 150
Electrical potential, membranes and, 148, 268–270
 nuclear pores and, 335, 337
Electrical transmission of nerve impulses, 619

Electron(s), carriers of, 320, 321
transfer of, 262, 264, 265
photochemical reaction and, 320–321
Electron microscope (EM), 34–39
high-voltage, 38
osmium tetroxide, use of with, 47
scanning, 38, 39
visualization of RNA transcription with, 491–492
Electron spin resonance (ESR), 142
Electron stains, 37
Electron transport system, 110–112. See also *Respiratory chain*.
phosphorylation and, 267
thylakoid membrane and, 318
Electron-translucent vesicles, 627
Electronic coupling, nerve impulses and, 619–620
Electrophoresis, gel, 233
polyacrylamide gel (PAGE), 135
of proteins, 76, 77
Electroplax, membranes of, 135
Elliptical vesicles, 627
Elongation, in protein synthesis, ribosomal role in, 530
RNA transcription and, 486–488
Elongation factor (EFT), protein synthesis and, 527–529
Embedding, 49
Embryonic induction, 576
Embryos, development of, lysosomes and, 294
End plate, 586
potential across, 620
Endocytosis, 186
exocytosis and, 244
lysosomes and, 288–294
Endomembrane system, 179
differentiations of, 20
lysosomes and, 288
morphology of, 206–213
subfractionation of, 213, 214
Endomitosis, DNA replication in polytenic chromosomes and, 548
Endonucleases, 262, 417
restriction, 478–482
Endopeptidases, 307
Endoplasmic reticulum (ER), 208
cellulose synthesis and, 302
functions of, 218–221
glyoxysomes and, 307
lysosomal enzymes and, 288
mitochondria and, 275
peroxisomal enzymes and, 297
plant cells and, 305
protein synthesis and, 221–224
rough (RER), 209, 211
nuclear pores and, 335
ribosomal binding to, 210
smooth (SER), 210, 211
β-endorphin, 245
Endosperm, 415
Endosperm nucleus, triploid, 415
Energetics, See *Bioenergetics*.
Energy, cellular requirements of, 250
free, 264, 265, 270
light, photoreceptors and, 147
Enucleation, in cytoplast and karyoplast preparation, 565, 566
Enzyme(s), 95–100. See also names of individual enzymes.

Enzyme(s) (*Continued*)
activating, tRNA and, 521–523
activation of, 101
active sites of, 96, 141
affinity of, apparent, 98, 99
allosteric, 100, 101
aminocyl-activating, in protein synthesis, 519–523
cancer cells and, 171
in cell coat, 168
cofactors of, 96
digestive, seed germination and, 296
distribution of in cell, 100
DNA repair, 378–380
dynein and, 190
of endoplasmic reticulum, carcinogens and, 219, 220
extracellular, lysosomes and, 294, 295
gene regulation of, repressors and, 539–540
glyoxylate cycle and, 297, 307, 308
hydrolytic, 283, 307
induced-fit model of, 97
inhibition of, 99–101
interconversions in, 102
kinetics of, 97–99
latency of, 283
Lineweaver-Burk plot and, 99
lock-and-key model of, 97
lysosomes and, 234, 245, 288
thyroid hormone and, 295
of membranes, 219
distribution in, 138, 139
microsomal, 216–218
mitochondrial, 260, 261, 263, 264, 276
modificaton, 480
peptidyltransferase, 530
peroxisomes and, 297, 298
phosphorylation and, 269
polymerases, 486, 490, 491
prohormone-converting, 245
protein kinase, eukaryotic initiation factors and, 529
reactions of, Michaelis-Menten equation for, 98, 99
regulation of, catalytic and genetic, 101, 102
respiratory, 251
restriction, 478–482
ribonuclease P, 521
substrate specificity of, 96, 97
synaptic function and, 640, 641
synthesis of, 101
thylakoid, CF_1 as, 325
Enzyme deficiencies, treatments for using liposomes, 126
Enzyme induction, 101
control of, 537
Enzyme repression, 101
control of, 537
lac operon as model for, 541
Epinephrine, 102
Epithelium(a), ciliated sheets of, 188
tonofilaments of, 199
5,6-epoxide, 220
Equatorial plate, 385
Equilibrium, acid-base, 70
ER. See *Endoplasmic reticulum (ER)*.
Ergastoplasm, 181, 209, 210
Erythrocyte(s)
anion transport in, 153, 154
formation of, 292

Erythrocyte(s) (*Continued*)
  freeze-fracturing studies of, 142
  hemolysis of, 132, 133
  membrane of, proteins in, 136–138
  reactivation by cell fusion, 563–565
  resealed ghosts of, 153
Erythrophores, 187
ES, 312
Escherichia coli (*E. coli*), 464
  anatomy of, 10
  chromosome replication in, 371–373
  DNA content of, 340
  lac operon and permease of, 156
  pyruvate dehydrogenase complex in, 118
ESR, 142
Ethylenediaminetetraacetate (EDTA), 119, 341
Ethylmethane sulfonate, 473
Euchromatin, 340, 354, 440
  RNA synthesis in, 545
Eukaryoste(s), 251
  DNA content of, 340
  evolution of, 127–129
  gene regulation in, 544–547
  mRNA in, 492–499
  protein synthesis and, 527
  repetitive DNA sequences in, 545
  ribosomal RNAs in, 503–505
  ribosomes in, 507
  transcription in, 490–492
  vs. prokaryotes, 10
Eukaryotic cell, shape of, 16
Eukaryotic genes, cloning of, 480–482
  plasmids and, 480
Euploidy, 426, 427
Evolution
  chemical and cellular, 127–129
  chromosomal role in, 433–435
  mitochondrial crests and, 256, 257
  photosynthesis and, 310
Excitation, coupling with contraction, 598–601
  threshold of, 605
Excitatory synaptic transmission, 620, 621
Exciton(s), 320
  photosynthesis and, 319
Exocytosis, secretion and, 243, 244
Exoskeleton, cell wall as, 301

F_1 particles, 253, 261, 263, 266, 268
F-actin, 595
Facultative heterochromatin, 446
FAD, 97, 264–266
$FADH_2$, 263
Fat body, 309
Fats, degradation of, 263, 264
Fatt, 626
Fatty acid(s), 123
  oxidation of, 262
Fatty acid coenzyme A ligase, 262
Feedback inhibition, 102, 103
Fermentation, 109–111
Ferredoxin, photochemcal reaction and, 320, 321
Ferritin
  hemoglobin and, 292
  intramitochondrial, 258
  labeling with, 143
  polycationic, 263

Fertilization, 400, 416
  lysosomes and, 295
  process of, 560
Fertilizin, 416
Feulgen, R., 2, 56, 467
Feulgen reaction, 56
  staining of, polytene chromosomes and, 550
Fibrillar region, nucleolar, 514–516
Fibrin, 121
Fibrinogen, blood clotting and, 120, 121
Fibroblast(s), 120, 165, 197
Fibrous proteins, 118
Film(s), balance for pressure measurement of, 123, 124
  lipid, monomolecular, 123, 124
Filmer, 101
Fingerprint, cellular, cell coat and, 168
Fischer, E., 9
Fission, cellular, 433, 434
Fixation, 46, 47
Fixative, effect of, 251
  types of, 46, 47
Flagellata, 188
Flagellum(a), 182
  microtubules and, 183
  movement of, 192
Flattened vesicles, 627
Flavin adenine dinucleotide (FAD), 97, 264–266
  vitamin B_2 and, 97
Flavoproteins, 262, 265
  ferredoxin-NADP oxireductase, 320
Flemming, W., 8, 333, 552
Fluid, interstitial, 148
  pinocytosis and, 289
Fluid mosaic model, of membranes, 132 140–142
Fluidity, membrane, 126, 142–144, 219
  Golgi complex and, 228
  mitochondrial, 265
  thylakoids and, 314
Fluorescein, 163
Fluorescence microscopy, 60
FMM, 265
Folch, J., 138
Fontana, F., 359
Form birefringence, 33
Fractionation, cell, 53
  mitochondrial membranes and, 260
  synaptic component isolation and, 629–631
  microsomal, 214
Franchi, C., 183
Franklin, R., 86
Freeze-drying, 48
Freeze-etching, 35
Freeze-fracturing, 35, 143
  gap junctions and, 162, 163
  membranes and, 137, 138, 140, 142, 143, 219
  chloroplasts and, 325, 326
  faces of 208, 209
  mitochondrial, 265, 267
  thylakoids and, 312–314
  mitochondrial, 261
Freeze-substitution, 48
Fructose, 78
Frye, L., 142
Functional contacts, synapse ultrastructure and, 622–624

Fucose, 233, 235, 236
Fusidic acid, 531
Fusion, cell, 142, 144, 165, 563–565
    antibodies from, 566–567

G₀ state, 365
G₁ period, 365, 366
  chromosomes visualized during, 366–368
G₂ period, 365, 366
  chromosomes visualized during, 366–368
G3PD, 137, 139
G-actin, 595
Galactolipids, 133
Galactose, 78, 233, 235, 236, 302
β-galactosidase, lac operon code for, 537, 538
Galactosyltransferase, 234
Galacturonic acid, 302
Ganglion cell, 576, 577
Ganglioside(s), 134, 135
  cancer cells and, 171
  cellular role of, 81
Gap junction, 158–166
  cancer cells and, 171
Gap periods, 365–368. See also *Cell cycle*.
Garnier, 181
Gastrulation, 560
Gated pore model, 272
Gel-sol, chloroplast stroma and, 312
Gene(s)
  5S, 511
  amplification of, 550, 571, 572
  DNA composition of, 467
  dominant vs. recessive, 421
  eukaryotic, cloning of, 480–482
    non-coding DNA insertions, 494, 495
    plasmids and, 480
  expression of, reprogramming and, 567–570
  histone, repetition in, 547
  housekeeping, 575
  inactive, heterochromatin and, 356, 357
  independent assortment of, 422
  linkage of, 422–424
  mutation of, 378–379, 428–430
  overlapping, of virus φX174, 475–476, 477
  prokaryotic, regulation of, 537–544
  radioactive probes for, 493
  regulation of, at chromosome level, 548–557
  repetitive, 547
  ribosomal, spacer DNA and, 509–511
  segregation of, 421, 422
  structural, 455, 537
  transcription of, polytene chromosomes and, 550–552
Gene cloning, 480
Gene mapping 455–458
Genetic code, 464–472
  deciphering of, artificial mRNAs and, 468–470
  degeneracy of, 470
  DNA sequencing and, 474–478
  mutations and, 472–474
  nucleic acid sequencing and, 476, 478
  synonyms in, 470
  universality of, 471
Genetic engineering, 10, 478–482
Genetic information, flow of, 484

Genetics, karyotypes and, 350, 439–443
  meiosis and, 413–417
  Neurospora used in study of, 424
Genome, constancy during differentiation, 571
  prokaryotic vs. eukaryotic, 536, 537
Genotype, 421, 422
GERL, 228, 244, 245
Germ cells, lysosomes and, 295
Germ plasm, determinants in, 576
Germinal vesicle, 567
Germination, seed, lysosomes in, 296
Ghost, red cell, 132
    resealed, 153
Gibberellic acid, 307
Giemsa stain, 440
Gilbert, W., 487
Gland(s), adrenal, 258, 259
  exocrine, secretion by, 238–244
  ionic transport and, 152
Glial cells, 197
Globular proteins, 118
Glomerular synapses, 623, 624
Glucagon, 102
  autophagy and, 288
Glucan, 302
Glucosamine, 233
Glucose, 167, 233
  active transport of, 155, 156
  β-cellobiose and, 301
  degradation of, 109, 110
Glucose phosphate, 322
Glucose-6-phosphatase, 233, 234
α-glucosidase, 295, 296
Glucosyl transferase, 302
β-glucuronidase, 285
Glutamate, 262, 272
Glutaraldehyde, fixation with, 144, 182
Glutathione, 274
Glyceraldehyde-3-P-dehydrogenase (G3PD), 137
Glycerol-phosphate dehydrogenase, 262
Glycocalyx, 166–173
  pinocytosis and, 289, 290
Glycogen, 78
  degradation and synthesis of, 104
  structure of, 78, 79, 122, 123
Glycogen phosphorylase, 102
Glycogenolysis, 220
Glycogenosis, lysosomes and, 296
Glycolic acid, 298
Glycolipid(s), 132, 134
  cancer cells and, 171
  classification of, 81, 82
  membrane, 134
  thylakoids and, 315
Glycolysis 264, 278. See also *Respiration, cellular*.
  anaerobic, 252
  contraction and, 598
Glycophorin, 138, 223
Glycoproteins, 72, 78, 79, 132
  anionic, 263
  antigenic, active sites of, 141
  cancer cells and, 171
  membrane, 134, 135
  secretion of, 237
  synthesis of, 167
    Golgi complex and, 235, 236

Glycosidase, 296
Glycosidation, Golgi complex and, 235
Glycosphingolipids, 233
    storage diseases and, 295
    synthesis of, Golgi complex and, 235, 236
Glycosyl transferases, 79, 234–236
    cancer and, 236
    Golgi complex and, 233, 234
Glyoxylate cycle, 307, 308
Glyoxysome(s), 297, 298
    plant cells and, 305
    triglycerides and, 307
Goitrous cretinism, 465
Golgi, C., 8, 226
Golgi complex, 208, 210, 220
    cell wall synthesis and, 302
    cytochemistry of, 232–234
    functions of, 235–246
    lysosomes and, 287, 288
    morphology of, 226–232
    secretion and, 307
Gonadal dysgenesis, 451
Gonocyte, 415
Gorter, 139
Gout, lysosomes and, 295
Graft rejection, 168
Granular region, nucleolar, 514–516
Granulated vesicles, synaptic, 627
Granule(s)
    hyalin, 259
    leukocytes and, 295
    microperoxisomes and, 297
    mitochondrial degeneration and, 259
    nuclear pores and, 335
    nucleolar, 359
    ribosomes and, 501
    starch, 296
    storage, 287
Granum(a), chloroplast and, 311–313
    development of, 314, 315
Green, 265
Grendell, D., 139
Grew, J., 8
Griffith, 467
Ground cytoplasm, 179
Guanine, 85
    mitochondrial, 277
Guanosine monophosphate, cyclic (cGMP), 105
Guanosine triphosphate (GTP), 108, 184
Guillermond, A., 307
Gurdon, J. B., 570, 571

H2A, 342
H2B, 342
H3, 342
H4, 342
Hadorn, E., 562
Hairpin loops, in mRNA, protein synthesis and, 527
    in RNA, 91
    ribosomal RNA, 504
Haldane, 128
Hardy, 9
Harris, H., 564
Harrison, 9
HAT, 457

H-band, muscle contractility and, 588
H-disc, striated muscle fiber, 587
Heat shock, puff induction and, 551
Heitz, E., 354
HeLa cells, cell fusion and, 564
    ribosomal membrane-binding, 502
Helix, coiling of, chromatin and, 349, 352–354
    α-soluble proteins and, 118
Heme, 72, 265
Hemoglobin, 466
    chlorophyll and, 315
    intramitochondrial, 258
Hemidesmosome, 161
Hemoglobin, self-assembly of, 118
Hemolysis, of erythrocytes, 132, 133
Hemophilia, 455, 456
Henneguy, 190
Hensen's disc, striated muscle fiber and, 587
Heparin, 78
Heredity, intermediary, 422
    Mendelian laws and, 420–426
    parental inheritance (bi- and uni-) and, 400
Hermaphroditism, true, sex chromosomes and, 452
Hertwig, O., 8
Heterochromatin, 340
    centrometric constitutive, 440
    characteristics of, 354–359
    intercalary, 440
    non-transcription of, 545
    nucleolar-associated, 359, 360
    satellite DNA in, 546
    types of, 446
Heterochromosomes, 444
Heterokaryon, cell fusion and, 563–565
Heterophagosome, 287, 293
Heteropyknosis, chromosomal, 354
Heterotroph, 129
Hexagonal phase, 124, 125
Hexose (fructose), carbon dioxide fixation and, 323
Hexose (fructose) diphosphate, 322
HF_0, 325
Hill, R., 320
Hill reaction, 320–322
Histocompatibility antigens, 566
Histone(s), 156
    chromatin and, 84, 342
    nucleosomes and, 343–348
    supercoiling of circular DNA and 90
Histone genes, repetition in, 547
hnRNA, 336
    as mRNA precursors, 497
Hoerr, L., 10, 250
Hofmeister, W., 10
Hogeboom, G. H., 10, 250, 259
Holley, R. W., 519
Holocentric chromosome, 350
Holoenzymes, 96
Hooke, R., 8
Hormones. See also names of individual hormones.
    adrenal, 258
    antidiuretic, 152
    cyclic AMP and, 102–105
    gibberellic acid, 307
    parathyroid, 294
    peptide, 143
    prohormone activation and, 245

Hormones (*Continued*)
  sex differentiation and, 446–448
  steroid, 219, 220
  thyroid, lysosomes and, 295
Huberman, J., 376
Hughes, 371
Hyalin, 259
Hyaline cap, 202
Hyaluronic acid, 78, 166
Hyaluronidase, 78
  acrosomes and, 295
Hybrid cell line, 563–565
Hybridization,
  of DNA, 88, 173
    radioactive probes for, 575
  in situ, 358, 458, 512
    chromosome mapping and, 550
  somatic cell, 457, 458
Hydrocortisone, 295
Hydrogen bond, in DNA, 86, 87
Hydrogen ions,
  chloroplasts and, 326, 327
  as electron carriers, 112
  proton "pump" and, 253
  transfer of, 269, 270
Hydrogen peroxide (H_2O_2), peroxisomes an,
  298
β-hydroxybutyrate dehydrogenase, 263
Hydroxylamine, 473
Hydroxylase aryl, 219, 220
Hydrolase(s), lysosomes and, 283, 294
α-hydroxylic acid oxidase, 298
Hyperchromic shift, 88
Hyperthyroidism, mitochondria and, 252
Hypophysectomy, 259
Hypoxanthine, 165

i gene, 537, 538
  code for repressor protein and, 538
I-band
  mitochondria and, 251
  muscle contractility and, 588–590
  striated muscle fiber and, 586, 587
Icosahedral symmetry, in viruses, 119
Imaginal disks, cell determination and, 562
  dipteran larval, 550
Immunocytochemistry
  cell fusion and, 142, 143
  cytochromes and, 267
  cytoskeletal proteins and, 182
  methods of, 61, 62
  microfilaments and, 197
  microtubules and, 185
  viral transformation and, 173
Immunofluorescence, probes using, 185
Inclusions, intramitochondrial, 258
Induced-fit model, 97
Inducer(s), of pinocytosis, 290
  protein binding to, 538
  repressor-operator binding and, 538
Induction, enzymes and, 102, 103
  embryonic, 576
Informosomes, 497, 498
Infusoria, cilia and, 188
Inheritance
  biparental, 400
  Mendelian laws of, 420–426
  uniparental, 400

Inhibition, allosteric, 102
  contact, 168–171
  feedback, 102, 103
  irreversible, 100
  reversible, competitive, 99, 100
    non-competitive, 100
Initial-segment synapses, 623
Initiation, of RNA transcription, 486–488
Initiation complex, protein synthesis and, 528
Initiation factors (IF), protein synthesis and,
    527–529
    prokaryotes vs. eukaryotes, 529
Inosinic diphosphatase, 307
Inositol, phosphatidyl, 261
Inoué, S., 394
Insulin, secretion of, 243, 244
Interbands, polytene chromosomes and, 550
Intercalary heterochromatin, 358, 440
Interconversion, in enzymes, 102
Interdoublet links, 190
Intermediary junctions, 158–161
Interphase, 364
  chromosomal arrangement and, 351, 352
  chromosomes and, 348, 349
  G_1, S, and G_2 phases of, 365, 366
  meiotic, 410
  tubulin and, 185
Intersynaptic filaments, 623
Intervening sequences, DNA, non-coding,
    495, 496
  mRNA precursors and, 497
Interzonal fibers, 385
Intrathylakoid space, 312
Intrinsic birefringence, 33
Inulin, lysosomes and, 294
Invaginated synapses, 623, 624
Inversion, of chromosomal material, 428,
    431, 432
    pericentric, 434
Ion(s)
  cellular role of, 70
  diffusion of, 150
  electrical potential and, 148, 149
  fluxes of, chloroplasts and, 326, 327
  gangliosides and, 135
  hydrated, diameters of, 153
  nuclear pores and, 335
  protein carriers of, 272
  transport of, endoplasmic reticulum and,
    219
Ionophore, 124
Iron
  cytochromes and, 112, 264, 265
  hemoglobin and, 292, 315
  non-heme, 325
  in proteins, 265–267
Irradiation, ultraviolet, microbeam, spindle
  fibers and, 392
Irritibility, 604
Insertions, single base, 472
Isochromosomes, 432
Isocitrate, 272
Isocitrate lyase, 298, 307, 308
Isoelectric focusing, in proteins, 76, 77
Isoelectric point, in proteins, 76
Isoenzyme(s), 100
  of dynein, 190
  of LDH, 100
Isotropic disc, 586
Isotropism, striated muscle fiber, 586, 587

Jacob, R., 485
Jacob-Monod model, lac transcription and, 540
Jagendorf, A., 321
Johanssen, 421
Junctions, intercellular, 158–166

K⁺. See *Potassium, ions of.*
Kartagener's syndrome, 192
Karyoplasts, preparation by enucleation, 565, 566
Karyotype(s), definition of, 350
    human, normal, 439–443
Katz, B., 626, 638
Keilin, D., 10, 264
Kendrew, J. C., 40
Keratin, 72
    fibers of, 122
Keratinization, 199
Khorana, H., 469
Kidney, 148
    brush border and, 158
    filtration by cell coats in, 168
    tubule of, mitochondria and, 251
Kinase(s)
    cyclic AMP-dependent, 185
    protein, 102–104, 139
        cAMP-dependent, 641
        eukaryotic initiation factors and, 529
        tubulin and, 184
Kinetics, of enzymes, 97–99
Kinetochore(s), 348–350, 383
    centromere and, 391, 392
    chromosome distribution and, 410, 412
    metaphase chromosomal motion and, 393, 394
Kinetosome(s). See *Basal body(ies).*
Kite, 9
Kleinschmidt, A. K., 35
Klinefelter's syndrome, 438, 450, 451
Km (Michaelis constant), as related to velocity and saturation, 98
    effects of inhibition on, 100
Kölliker, A., 8
Koltzoff, 181
Kornberg, A., 345
Kornberg, R., 345
Koshland, D. E., 101
Kossel, A., 10
Krebs tricarboxylic acid cycle, 110–112, 262, 264
Krogh, A., 152
Kynocilia, 188
Kynurenine hydroxylase, 262

Lac operon, 156
    codes of, 537, 538
Lac permease, 156
    lac operon code for, 537, 538
Lac transcription, 537–543
    CAP-cAMP complex and, 540, 541
    Jacob-Monod model and, 540
Lactate, 110
Lactic dehydrogenase (LDH), isoenzymes of, 100
Lactose, 78
    cleavage of, lac operon and, 537, 538
Lamarck, J. B., 8

Lamellae, annulated, nuclear pores and, 335, 336
Lamellar phase, 126
Lampbrush chromosomes. See *Chromosomes, lampbrush.*
Langley, 636
Langmuir, 123
Lanthanum, 167
Lateral loops, RNA synthesis and, 553–556
Laue, 40
Law of independent assortment, 422
Laws of inheritance, Mendelian, 420–426
Law of segregation, 421, 422
Lecithin, 134, 259, 275
    cholesterol and, 144
Lectins, 143, 167, 171
Lectin receptors, synapse formation and maintenance, 624
Leder, P., 469
Lees, M., 138
Leeuwenhoek, A. van, 8
Legumin, 306
Lejeune, T., 438
Leloir, 79
Lenhossek, 190
Leptonema, 402
Lesch-Nyhan syndrome, 446
Leukemia, myelogenous, chronic, 454
Leukocytes, polymorphonuclear, granules in, 295
    lysosomes and, 288
Leukoplast(s), 309, 310
Levan, 438
Levi, G., 9, 610
Lewis, 289
LH, 102
LHCP, 317
Ligands, 142
    active transport of, 155
    recognition of, 143
Light, wavelength of, photosynthesis and, 320
Light reaction, photosynthetic, 320
Light-harvesting chlorophyll-protein complex (LHCP), 317
Lignin, cell wall and, 301, 302
Lineweaver-Burk plot, 99
Linkage analysis, 457
Linker regions, nucleosomal, 347
Lipid(s)
    bilayer of, ER and, 209
    classification of, 80, 81
    distribution in bilayer, 133, 134
    glycosidation of, cancer cells and, 236
    membrane, 132
        mitochondrial, 265
        mobility of, 140, 142
    microsomal, 214
    mitochondrial, 258
        lipid/protein ratio of, 260
    molecule of, dimensions of, 123
    monolayer films of, 122, 124
    synthesis of, 220
    thylakoids and, 315
Lipmann, F., 529
Lipoproteins, 72, 220
    mitochondrial, 259
    multi-layered, 144, 145
    synaptic receptors and, 636, 637
    synthesis of, Golgi complex and, 231

Liposome(s), 125, 126
  bilayers in, 270
Local circuit theory, nerve impulse
  propagation, 613–615
Lock-and-key model, 97
Locus(i), 421
Loewi, 620
Lubrol, 259
Luteinizing hormone (LH), 102
Lycopene, 309
Lymphocytes, capping of, 142
  homing of, 169
Lyon, M., 356, 445
Lysis, auto, 259
Lysosome(s), 208, 209, 259
  GERL and, 228
  endocytosis and, 288–294
  functions of, 294–296
  glycoproteins and, 236
  primary, GERL region and, 244, 245
  types of, 282–288
Lysozyme, 336

Macrofibril(s), cell wall and, 301
Macromolecules, assembly of, 118–123
  structure of, 70
  transport of through axon, 609
Macrophages, endocytosis and, 292
  lysosomes and, 295
  phagocytosis and, 289
Macula adherens, 158–161
Magnesium ions,
  ATPase and, 138
  chlorophyll and, 315
  chloroplasts and, 326, 327
  histones and, 347
Malate, 272
Malate dehydrogenase, 307, 308
Malate synthetase, 307, 308
  peroxisomal, 298
Malpighi, M., 8
Maltose, 78
Mannose, 167, 233, 235, 236
MAPs, 185
Mapping, genetic, 89, 277, 455–458
Mattaei, H., 468
Matthey, R., 433
Mazia, D., 392
McClung, C. E., 9, 433, 444
Mealy bug, heterochromatin and, 356
Megaspores, 415
Meiosis, 26
  biochemistry of, 417, 418
  chromosome distribution in, 410, 412
  diplotene stage, lampbrush chromosomes
    and, 552, 553
  gametic, 400
  genetic consequences of, 413–417
  in human female, duration of, 415, 416
  in human male, 416
  intermediary, 400, 401
  vs. mitosis, 401, 402
  sporic, 400, 401
  stages of, 402–410, 417
  terminal, 400
  types of, 413–417
Meischer, 333
Melanin, 187
  GERL region and, 245

Melanocyte-stimulating hormone (MSH), 102
Membrane(s). See also *Bilayer, lipid.*
  artificial and elementary, 123–126
  axonal, Na⁺ and K⁺ channels, 615–617
  basement, 158
  biogenesis of, 218, 219
  cancer cells and, 171
  cellular,
    cell wall synthesis and, 302
    differentiations of, 158–166
    endocytosis and, 288–293
    excitation-contraction coupling and, 599
    microfilaments and, 199, 200
    molecular models of, 139–148
    molecular organization of, 132–139
    permeability of, 148–158
    structure and functions of, 132–173
  chloroplasts and, 312, 324–327
  ciliary, 188
  differentiation of, 228, 231
  endoplasmic reticulum and, 210, 222, 223
  fluidity of, 219
  freeze-fracturing of, nomenclature of, 312,
    313
  intracellular, faces of, 208, 209
    re-utilization of, 243
  lysosomeal, disease and, 295
  microsomal, 214
  mitochondrial, 271–275, 277, 278
  nuclear, 334–339
  pectocellulose, plasmodesmata and, 304
  phosphorylation and, cAMP and, 641–643
  postsynaptic, macromolecular organization
    of, 625, 626
  presynaptic, active zones and, 624, 625
  receptors in, and adenylate cyclase, 102–105
  respiratory, and multi-enzyme systems, 100
  ribosome binding to, 501, 502
  ruffling of, 186
  synaptic, receptor composition and, 636, 637
  synaptosomal, isolation of, 629–631
  thylakoid, 312–314, 317, 318
Membrane noise, 638
Mendel, G., 9, 420–421, 467
Menten, M., 98
Meromyosin, 198
  myosin and, 591
Merotomy, nucleocytoplasmic interactions
  and, 562, 563
Meselson, M., 371
Mesomorphism, thermotropic, 126
Mesophase, smetic, 125, 126
Mesosomes, symbiont hypothesis and, 278
Messenger(s), hormonal, 102
Messenger RNA. See *Ribonucleic acid,*
  messenger.
Metabolic cooperation, 165
Metabolic regulation, 101–105
Metabolism,
  cellular, 95
    regulation of, 101, 102
  gene regulation and, 544–547
  inborn errors of, 464
  intercellular cooperation and, 165
  neurotransmitter and, 631, 632
Metacentric chromosome, 348, 349
Metachromasia, 50, 51
Metafemales, 451
Metamorphosis, amphibian, lysosomes and,
  294

Metaphase, chromosomes and, 348, 349, 351
  I, meiotic, 410
  mitotic, 385
    chromosome motion during, 393, 394
  tubulin and, 185
Metazoa, phagocytosis and, 289
Methionine, formyl derivatives of, protein
    synthesis and, 527
7-methyl-G, blockage of mRNA, and, 493, 494
Metschnikoff, E., 289
Michaelis, L., 9, 98
Michaelis constant (Km), as related to velocity
    and saturation, 98
  effects of inhibition on, 100
Michaelis-Menten equation, 98
Microbodies, 208–210, 236, 297, 298
Microcentrum, 387
Microcinematography, 192
Microelectrodes, 148, 161, 335, 337
Microenvironment, cell coat and, 168
Microfibril(s), cell wall and, 301
  cell wall and, 305
  synthesis of, 302
Microfilaments, 137, 182
  cancer cells and, 171
  cytoplasmic motility and, 196–203
  endocytosis and, 292
Micromethods, 54
Microperoxisomes, nucleoids and, 297
Micropinocytosis, fluids and, 290
Microscopy
  darkfield, 31
  electron, 34–39
    artifacts of sample preparation and, 140
    glutareldehyde fixation and, 182
    high-voltage, 38
    membranes and, 139, 140
    osmium textraoxide and, 47
    sample preparation for, 144, 145
    scanning, 38, 39
  fluorescence, 60, 142
  light, light path in, 29
  Nomarski interference-contrast, 31
  phase contrast, 29–31
  polarization, 32, 33
  resolving power of, 28, 29
Microsomal fraction, biochemistry of, 213–218
Microspores, 415
Microsurgery, of cells, 46
Microtomes, 36
  freezing, 48, 49
Microtrabeculae, lattice of, 197, 198
Microtubular associated proteins, 185
Microtubule(s), 182, 198
  assembly of, 184, 185, 392, 393
  doublets of, sliding of, 192
  kinetochores and, 350, 392
  metaphase chromosomal motion and, 393, 394
  in organelles, 187–196
  plant cells and, 305
  properties of, 183–187
Microvesiculation, 244
Microvilli, 158, 186
  stress fibers in, 199, 200
Midbody, 388, 395
Miescher, F., 10
Miledi, R., 638
Miller, O., 378, 491–492
Miller, S., 128

Millon reaction, 55
Minerals, cellular role of, 70
Minimyosin, 198
Mirbel, C. F., 8
Mirsky, A., 467
Mitchell, 268–270, 321
Mitochondrial crests, 250, 252, 255, 256, 263, 268, 269
  mesosomes and, 278
  plant cells and, 307, 309
  viarations in, 256–258
Mitochondrial matrix, 252, 255, 276
Mitochondrion(a), 198
  chloroplasts and, 307, 309
  Krebs cycle in, 111, 112
  symbiont hypothesis and, 129
Mitomycin C, 342–433
Mitoplasts, 259, 260, 262
Mitoribosomes, 253, 254, 267, 275, 277
  symbiont hypothesis and, 278
Mitosis, 182
  asters and, 385
  cancer cells and, 169–171
  chromosome distribution in, 410, 412
  chromosome packing during, 352–354
  general description of, 383–387
  mitochondria and, 251, 275
  nucleolus and, 361
  stages of, 24, 25, 384, 385–387
  tubulin and, 185
  vs. meiosis, 401, 402
Mitotic apparatus, molecular organization and
    functional role of, 387–398
Mitotic selection, 366
Mitotic spindle, 383
  fibers of, 385
  microtubules of, 392, 393
M-line, muscle contractility and, 588
  protein of, contractility and, 593
  striated muscle fiber and, 587
MN antigens, 138
Mobile hypothesis, of membrane structure, 143
Mohl, H. von, 8
Mongolism, 438, 452, 454
Monoamine oxidase, 262
Monocentric chromosome, 350
Monoclonal antibody, 566, 567
Monocytes, lysosomes and, 295
Monod, J., 101, 127, 485
Monolayers, of lipids, 122, 124
Monolayer technique, 35
Monomer, 70
Monosaccharides, 78
18-monosomy, 454
21-monosomy, 452
Morgan, T. H., 9, 422
Mosaics, genetic, 450, 452
    X chromosome and, 356
Moses, M., 405
Motility,
  cancer cells and, 169–171
  cellular, 182, 188
    ameboid, 201–203
    non-muscular, 200
  chloroplasts and, 310
  mitochondrial, 250
MS2 bacteriophage, 91
MSH, 102
Mucilage, dictyosomes and, 307

Mucin, 166
Mucopolysaccharides, 236
    acidic, 78
Multi-enzyme complex(es), 95, 100
    respiratory membranes and, 100
    self assembly of, 118
Muscle
    cardiac, gap junctions and, 165
    contraction of, 118, 192
        actin-myosin linkages in, 595
        excitation coupling to, 598–601
        non-muscular contraction and, 200
        myofilaments and, 587–588
        proteins and, 593
        regulation and energetics in, 597–601
        sliding mechanism and, 593–594
        structural differentiation for, 586–587
    mitochondria and, 251, 252
    sarcomere of, 199
    sarcoplasmic reticulum and, 219
    skeletal, myofibril of, 182
Muscle fiber, contraction of, molecular
        organization for, 591–593
    striated, contraction of, 587–588
        structure of, 586–591
Mutagens, chemical, DNA molecule and, 429,
        430
    specificity of, 473, 474
Mutation(s), 378–379, 428–430
    amino acids changed by, 465, 466
    auxotrophic, 457
    genetic code and, 472–474
    germ cell, 430
    somatic, 430
    storage diseases and, 296
    types of, 472, 473
Mycoplasms, 16
Myelin, figures of, 126
    nerve conduction velocity and, 612, 613
    nerve fibers with, conduction in, 615
    protein/lipid ratio in, 132, 134
    sheath of, 144
    storage diseases and, 295
Myoblasts, 586
Myofibrils, 585
    mitochondria and, 251
    of skeletal muscle, 182
    of striated muscle fiber, 586
    structural differentiation of, 586, 587
Myofilament(s), contractile system and, 591
    of striated muscle fiber, 587
        contractility and, 587, 588
    thick, function of, 591–593
Myosin, 182
    actin linkage and, muscle contractility and,
        595
    chromosome movement and, 395
    contractile ring of cytokinesis and, 395
    endocytosis and, 293
    non-muscle, 198–200
    polarization of within sarcomere, 594
    structure and function of, 591–593
Myxovirus, 173

Na$^+$. See Sodium ions.
Na$^+$-K$^+$ activated–Mg^{2+} ATPase, 138
Na$^+$ K$^+$ ATPase, active transport and, 154, 155

N-acetyl-β-glucosaminidase, 285
N-acetylglucosamine, 167, 235, 236
Nachmanson, D., 636
NAD, 110, 112, 264
NAD$^+$, 96, 97
NADH, 263, 268, 269
NADH-cytochrome-b$_5$-reductase, 234
NADH-cytochrome-c-reductase, 234, 262, 263
NADH-diphosphorase, 139
NADH-Q-reductase, 265–267
NADP$^+$, 96, 97, 320
NADPH, Calvin cycle and, 322, 323
NADPH$_2$, 320–322
    Calvin cycle and, 322, 323
Nägeli, 6
Nass, M., 276
Nass, S., 276
Navashin, S., 433
Negative staining, 36, 37
Nemethy, 101
Neoplasms, 171, 172
Nernst equation, 150
Nerves, conduction velocity in, 612, 613
Nerves, impulse transmission in, 619, 620
        synaptic vesicles and, 632–634
Nerve fibers, 144
    myelinated, saltatory conduction in, 615
    organization and function of, 604–619
    stimulation of, action potential and, 613–615
Nerve growth factor (NGF), 610–612
Neuraminidase, 166, 167
Neuroblast, 197, 576, 577
Neurofibrils, axon structure and, 606
Neurofilaments, 199, 200
Neurohypophysis, 152
Neurons
    biosynthetic functions of, 606–609
    differentiation of, 605, 606
    gangliosides and, 134, 135
    molecular recognition and, 168
    synapse impingement on, 621
Neuron theory, 619
Neuronal development, medium
        determination and, 627–629
Neuronal microtubules, 606
Neurospora, genetic study and, 275, 424
Neurotransmitter(s), 143
    coupling with secretions, mediation of, 634
    neuronal development and, 627–629
    release of, 626, 627, 632, 633
    synthesis and metabolism of, 631, 632
    transport of, synaptic vesicles and, 632
Neurotropic virus, transport of, 612
Neurotubules, 183
    axon structure and, 606
Nexin, microtubular doublets and, 190
Nexus, 158–166
Nicolson, G., 314
Nicotinamide, 97
Nicotinamide-adenine dinucleotidase, 139
Nicotine adenine dinucleotide (NAD), 96, 97,
        110, 112, 264
Niemann-Pick disease, 295
Nirenberg, N., 468, 469
Nissl substance, 606
Nitrogen mustard, 432, 433
Nitrous acid, 473
NMR, 142
Nodes, nerve, distance between and
        conduction velocity, 612, 613

Nodes (*Continued*)
  of Ranvier, nerve conduction velocity and, 612–613
Non-disjunction, 427, 449, 450
Norepinephrine, 102
Novikoff, 245
Nuclear envelope, 208, 334–339
Nuclear magnetic resonance (NMR), 142
Nuclear transplantation, constancy of genome and, 571
Nucleases, 296
Nucleation center, differentiation and, 578
Nucleic acid(s), 82–94. See also *Deoxyribonucleic acid* and *Ribonucleic acid.*
  major types of, 85
  ribose in, 90
  RNA, general description of, 91
  sequence of, in DNA, 86
  sequencing of, genetic code confirmed by, 476, 478
  viroids and, 91, 92
Nucleocytoplasmic interaction, 562–570
  use of *Acetabularia* and, 563
Nucleoids, peroxisomes and, 297
Nucleolar organizer, 350, 361
  ribosomal DNA in, 508–509
  5S genes outside, 511
Nucleolar-associated chromatin, 359, 360
Nucleolus(i), characteristics of, 359–361
  function of, ribosomal biogenesis and, 508–517
  ribosomal RNA processing and, 513–514
Nucleoplasmic index, equation for, 21
Nucleoproteins, 72
  nuclear export of, 219
Nucleoside, 84
Nucleoside diphosphokinase, 262
Nucleosomes, DNA replication and, 377, 378
  histones and, 343–348
5′-nucleotidase, 138, 233, 234
Nucleotides
  adenosine monophosphate, 108
  amino acid coded by, 468
  metabolic cooperation and, 164, 165
  sequence of, mitochondria and, 276, 277
  structure of, 84
Nucleotide modifications, RNA processing and, 484
Nucleus(i)
  characteristics of, 22
  endosperm, triploid, 415
  envelope of, 208, 218, 219
  interphase, X chromatin in, 444, 445
  mitochondrial biogenesis and, 276
  relation to cytoplasm, 562–570
  somatic, gene expression and, 567–570

Ochoa, S., 469, 527
Okasaki fragments, 375
Oken, 8
Oleic acid, 80
Oligomycin, 269
Oligomycin-sensitive conferring protein (OSCP), 268
Oleosome, 309
Oligomeric model, of cholinergic receptor, 639
Oligopeptide, 71

Oligosaccharide(s), membrane, 132
  distribution of, 134
One gene–one enzyme hypothesis, 464
One gene–one polypeptide chain hypothesis, 464
Oöcytes, cytoplasm of, importance of determinants in, 575–577
  differentiation experiments and, 567–570
  fertilization of, 295
  lampbrush chromosomes in, 552–557
  mitochondria in, 252
  ribosomal DNA amplification in, 511–513
  tRNA processing and, 521
Oögenesis, 340, 400, 401
Oparin, A., 128
Operators, gene expression and, 487
  gene regulation and, 537
  symmetry of, 538, 539
Operon, tryptophan, 541–543
Operon codes, lac, 537, 538
Orthograde transport, axonal, 609
  speed of, 609, 610
OSCP, 268
Osmium tetroxide, and electron microscopy, 47
  as a stain for lipid, 57
Osmotic equilibrium, active transport and, 152
Osmotic pressure, 70
  cell wall and, 301
Osteoblasts, 274
Osteoclasts, 294
Ostergren, 394, 412
Ostwald, W., 10
Ouabain, 155
  membrane binding site for, 139
Ovalbumin, 336
Overton, H., 9, 139, 149
Ovum, aneuploid, 449
Oxaloacetate, 110
Oxaloacetic acid, 264
Oxidase(s),
  D-amino-acid, 297, 298
  cytochrome, 250
  glycolic acid, 298
  α-hydroxylic acid, 298
  urate, 297, 298
Oxidation, 112, 253
  decreased, cancer and, 252
  of fatty acids, 262
  of foodstuffs, 108
  mitochondrial, 263, 264
  phosphorylation and, 112
  photorespiration and, 298
Oxidation-reduction, 111
  potential of, 264, 265
Oxidative phosphorylation, photosynthesis and, 319
Oxygen, active transport and, 150

P680, 317
P700, 315, 317
PF, 312
PFs, 313
PFu, 312, 313
P_s, 134
PS, 313
Pachynema, 408–409, 417

Pachytene stage, DNA accumulation during, 512
PAGE, 135, 216
Palade, G. E., 239, 501
Palmitic acid, 80
Pancreas, secretory process in, 238–244
Pantothenic acid, 97
    acetyl-CoA and, 110
Parasites, viral, 119
Parathyroid hormone, 102, 294
Particle(s), F_1, 253, 256, 261, 263, 265, 268
    glycogen, types of, 122
    granal, chloroplast shape and, 326
    intramembrane, chloroplasts and, 325
    nucleosomal, 347
    protein, thylakoid membranes and, 312, 314
Partition coefficient, 149, 150
PAS, 135, 167
PAS reaction, 56
Pauling, L., 40
Pavan, C., 552
Pectin, 166
    cell wall formation and, 397
    cell wall matrix and, 302
Penicillin, 531
Pentoses, types of, 84, 85
Peptidase, signal, 221, 245
Peptide, signal, hydrophobic, secretory
        proteins and, 531, 532
    initial codons for, 221
Peptide bond, 71, 72
Peptide synthetase, 530
Peptidoglycan, sulfated, secretion granules
    and, 242
Peptidyl site, protein synthesis and, 530
Peptidyl transferase, 530
Pericanicular dense body, 286
Pericentric inversion, 434
Perikaryon, neuron functions in, 606–609
Periodic Acid–Schiff (PAS) reaction, 56
Permeability, 148–158. See also Transport.
    anionic permeation channels and, 142
    calcium and, 164, 165
    gap junctions and, 163–165
    liposome studies of, 126
    mitochondrial, 271–275
    nuclear pores and, 335–337
Permeases, 219
    active transport and, 156
    ATP and, 272
Peroxidase, 285, 297
    labeling with, 143
Peroxide, hydrogen (H_2O_2), peroxisomes and, 298
Peroxisome(s), 208–210, 236, 297–298
Perry, R. P., 498
Perutz, M., 40
Peterfi, T., 9
pH, membranes and, 268–270
    photophosphorylation and, 321
    proteins and, 76
PHA, 138, 439
Phagocytosis, 156, 219
    characteristics of, 289
Phagosomes, 293
Phases, in bulk phospholipid-water systems, 124–126
Phenobarbital, 219
Phenotype, 421, 422

Phenylalanine, tRNA and, 521
Phenylketonuria, 464, 465
Phloem, differentiation of, 301, 302
Phosphatase, acid, 228, 282, 285
    acrosomes and, 295
    alkaline, 138
Phosphate, 262, 263, 268
    cellular role of, 70
    dolichol, 236
    inorganic (Pi), mitochondrial transport and, 272, 274
    mitochondria and, 274, 275
Phosphatides, 259
    classification of, 82
    structure of, 80, 81
Phosphatidic acid, 233, 292
Phosphatidylcholine, 133, 233
Phosphatidylethanolamine, 133, 134
Phosphatidylglycerol, 134, 233
Phosphatidylinositol, 133, 261, 292
Phosphatidylserine, 133, 134
Phosphoenzymes, as covalent intermediate, active transport and, 155
3-phosphoglyceraldehyde, 323
3-phosphoglycerate, 322, 323
Phospholipases, 134, 262
Phospholipid(s), 123, 133, 325
    acidic, 134
    of Golgi complex, 233
    of membranes, mitochondrial, 265
    mitochondria and, 262
    structure of, 80, 81
    thylakoids and, 315
    vesicle of, 125, 126
    with water, bulk systems of, 124–126
Phosphoric acid, 81, 108
Phosphorylase, glycogen, 102
Phosphorylation, 253
    cAMP and, 641–643
    contraction and, 598
    initiation factors and, 529
    mitochondrial, 268
    oxidative, 112, 250, 262. See also Respiration, cellular.
        mitochondrial conformation and, 272
        photosynthesis and, 319
        of tubulin, 185
Photodynesis, 200
Photon(s), 320
    photosynthesis and, 319
Photophosphorylation, photosystem II and, 321, 322
    thylakoid membrane and, 318
Photopigment(s), 147, 317
Photoreceptor(s), cilia and, 193
    membranes of, 144, 145
Photorespiration, peroxisomes and, 298
Photosynthesis, evolution and, 310
    pigments and, 309
    reactions of, 319–324
    thylakoids and, 315
Photosystem(s)
    I, complex of, 317, 318
    II, complex of, 317, 318
    inter-relationship of I and II, 320, 321
    thylakoid membrane and, 318
    thylakoids and, 315
Phragmoplast, 395, 397
Phycocyanin, 309

Phycoerythrin, 309
Physiologic response, synaptic receptors and, 636–644
Physiologic receptors, synapse potentials and, 615
Phytohemagglutinin, 138, 439
Pigment(s), P700, 315, 317
  photosynthesis and, 309, 317
  respiratory, 264, 265
Pinocytosis, 156, 158, 219
  characteristics of, 289
  plant cells and, 305
Planococcus citri (mealy bug), heterochromatin and, 356
Plant cells
  chloroplasts in, membranes of, 324–327
  cytokinesis in, 395
  cytoplasm of, 305–309
  endoplasmic reticulum in, 219
  lysosomes and, 296
  microfilaments and, 200, 201
  mitochondria in, 252
  peroxisomes and, 298
  photosynthesis in, 319–324
  symbiont hypothesis and, 278
  walls of, 300–304, 397
Plasm, germ, determinants in, 576
Plasm, pole-, 576
Plasmal reaction, 57
Plasmids, and genetic engineering, 10
  eukaryotic genes and, 480
Plasmin, 171
Plasminogen, 171
Plasmodesmata, 169, 397
  cell wall and, 302–304
Plasmodium, 395
Plastids, characteristics of, 309–314
Plastocyanine, 320
Plastoquinone, 315, 320
Pleuropneumonia-like organism, 16
Polarization, actin and myosin, within sarcomere, 594
Pole-plasm, 576
Pollister, 467
Polyacrylamide gel electrophoresis (PAGE), 135, 216
Polyadenylic acid, mRNAs and, 494
Polymer, 70
Polymerase(s)
  DNA, 373, 374
  mitochondrial, 276
  RNA, repressors and, 539, 540
  RNA transcription and, prokaryotic, 486
Polymerization, chemical evolution and, 127, 128
  microtubules and, 184, 185
  of proteins, 117
Polymorphism, 440
Polynucleotide ligase, 374, 375
Polynucleotide phosphorylase, 469
Polyoma virus, 171, 173
Polypeptide(s), 71
  membrane, particles of 8 to 9 nm and, 154
  erythrocytes and, 137
Polypeptide chain, termination of, protein synthesis and, 530
Polyploidy, 427
Polyribosomes, 210, 212, 503
  protein synthesis and, 524–526

Polysaccharide(s), 71
  cellulose and, 301, 302
  complex of, types of, 78
  lysosomes and, 294
Polysomes, 210, 212, 503
  protein synthesis and, 524–526
Polysomy, X, females with, 451
Polytene chromosomes. See Chromosomes, polytene.
Pore(s)
  fixed, 156
  gated, anionic, 272
  ionic transport and, 153
  nuclear, 208, 334, 335
  sieve, 305
Porphyrin, chlorophyll and, 315, 316
Postsynaptic density, 623
  synapse formation and maintenance, 624
Potassium ions
  acetylcholine receptor and, 637–639
  active transport and, 152
    channels for, action potentials and, 615–617
  chloroplasts and, 325, 326
  electrical potential and, 148, 149
  mitochondria and, 274
Potentials, action, propagation of, 615–617
PPLO, 16
Precursor(s), protein, 245
  radiolabeled, 239
Precursor activation, 102, 103
Prekeratin, filaments of, 199
Preleptonema, 402
Premature chromosome condensation, 366–368
Pre-proalbumin, 222
Pre-proinsulin, 245
Pre-protein(s), 221, 222
Preregion, 245
Presecretory granules, 228
Pressure, osmotic, 70
  surface, of monomolecular film, 123, 124
Presynaptic projections, 624, 625
Prezygonema, 417
Primary binding proteins, 506
Primary constriction, chromosomal, 349
Proalbumin, 245
Procentriole(s), 190, 194
Proenzymes, 100, 237
Prohormones, 245
Prokaryote(s), 251
  DNA content of, 340
  evolution of, 127–129
  gene regulation in, 537–544
  mRNA in, 484–490
    ribosome binding sites and, 527
  protein synthesis and, 527
  ribosomes in, 502, 503, 507
  symbiont hypothesis and, 278
  transcription in, protein synthesis and, 489
  vs. eukaryotes, 10
Prolamellar body, 314, 315
Prometaphase, mitotic, 385
Prometaphase I, meiotic, 410
Promoters, repressors binding within, 539, 540
  structural genes of, 537
Pronucleus, female, 416, 560
Pronucleus, male, 416
Propagation, of nerve impulses, 613
Prophase, chromosomes and, 348, 349

Prophase (*Continued*)
  mitotic, 384, 385
  tubulin and, 185
Proplastids, 309
  chloroplast development and, 314, 315
Proprotein(s), 222, 245
Prostaglandins, 143
Prosthetic group, 96
Protamine, 156
Protease(s)
  acid, 296
  acrosomes and, 295
  cancer cells and, 171
  lysosomes and, 295
Proteins, 71–77
  carrier, 154, 155, 262
  basic, 156
  bonds in, types of, 75, 76
  ciliary, synthesis of, 194
  classification of, 72–73
Protein(s), contractile. See *Contraction.*
  cyclic-AMP binding (CAP), 539
  cytoskeletal, 182
  degradation of, 263
  denaturation and renaturation of, 74
  DNA-binding, 417
  eukaryotic mRNAs and, 497, 498
  gene cloning and, 480
  glycosidation of, cancer cells and, 236
  histones as, 342
  hydrophobic, 76, 277, 278, 636, 637
  iron-sulfur, 265–267
  isoelectric focusing in, 76, 77
  lysosomes and, 294
  membrane, 132, 219
    chloroplasts and, 317, 325
    distribution of, 135, 137–139
    of the Golgi complex, 233
    integral, ER and, 209
    mitochondrial, 260, 262, 265
    particles of 8 to 9 nm in, 142, 162, 163
    peripheral vs. integral, 135
  microsomal, 214, 216
  microtubular associated, 185
  mitochondrial, 276
  molecular shape of, 118
  non-histone, chromosomal scaffold and, 354–356
  nuclear, accumulation of, 337–339
  nuclear pores and, 336, 337
  of chloroplast stroma, carboxydismutase and, 323
  oligomycin-sensitive conferring, 268
  particles of, nuclear pores and, 335
  pH and, 76
  phosphorylation, cAMP and, 641–643
  pinocytosis and, 289, 292, 293
  reassociation of, 417
  receptor, hormonal, 102, 103
  repressor, i gene and, 538
  ribosomal, 506–508
  secretion and, 78, 79, 214, 221–224, 231
    hydrophobic signal peptides and, 531, 532
  structure of, 72–76
  synthesis of, 221–224, 489
    chloroplasts and, 328
    fidelity of, 521–523
    mitochondrial, 277

Protein(s), synthesis of (*Continued*)
  nucleoli and, 359
  peroxisomal enzymes and, 297, 298
  protein factors and, 527–529
  ribosomes and, 209, 210, 505, 524–534
  symbiont hypothesis and, 279
  transmembrane, 141
Protein body, 296, 306–309
Protein coat, viral, 119
Proteinoid droplets, 128
Proteinoplast, 296, 306–309
Proteolipid(s), 268, 269
  characteristics of, 138
  mitochondrial, 278
Proton(s), transport of, 253, 266–268, 285
Protozoa, cilia and, 188
  phagocytosis and, 289
Prozymogen granules, 239, 242
PS I complex, 317, 318. See also *Photosystem(s).*
PS II complex, 317, 318. See also *Photosystem(s).*
Pseudohermaphroditism, feminine-masculine, 447
Pseudoplasmodium, differentiation and, 578
Pseudopodium(a), 187
  phagocytosis and, 289
  types of, 201
Puffs, chromosomal, 550–552
Pulse-chase radiolabeling, pancreatic cells and, 239–241
Purines, 85
Purkinje, J. E., 8
Puromycin, 531
Pyridoxal, 97
Pyrimidines, 85
Pyrrole rings, chlorophyll and, 315, 316
Pyruvate, 110
Pyruvate dehydrogenase complex, self-assembly of, 118

QH_2-cytochrome-c-reductase, 266, 269
Quinacrine mustard, 440
Quinones, 320

Racker, E., 268
Radial spokes, of axoneme, 190
Radiation(s), DNA molecule and, 429, 430
  ultraviolet, prebiotic evolution and, 128
Radicals, of amino acids, 71
Radioactive probes, hybridization and, 575
Radioautography, 62, 63
Radioisotope(s), labeling with, 142, 169
  "pulse-chase" method and, 239–242
Reaction(s)
  condensation, and peptide bond, 71
  dark, photosynthetic, 320
  endergonic, 95, 107, 108
  enzymatic, 95, 96
    Michaelis-Menten equation for, 98, 99
  exergonic, 95, 107, 108
  light, photosynthetic, 320
  oxidation-reduction, 111
    and NAD^+ and $NADP^+$, 96, 97
  photochemical (Hill), 320–322
  photosynthetic, 319–324
Reaggregation, cellular, 169

Receptor(s), 168
  acetylcholine, coupling of to sodium and
      potassium ions, 637–639
  β-adrenergic
  cancer cells and, 171
  coupling to enzymes and, 143, 144
  cholinergic, 142
    neurotransmitter fluctuation and, 639
    oligomeric model of, 639
  hormonal, 168
  physiologic, synapse potentials and, 615
  ribosomal, endoplasmic reticulum and, 221,
      222
  synaptic, composition of, 636, 637
    physiologic response and, 636–644
Response, physiologic, synaptic receptors and,
    636–644
Reciprocal synapses, 623
Recognition, intercellular, 166–173
Recombination, 409
  linkage broken by, 424
Recombination index, 424
Recombination nodule, 409
Reconstituted cells, 565
Reconstitution, ribosomal proteins and, 506,
    507
Reconstitution intermediate, formation of, 507
Recruitment zone, 201, 202
Red blood cell. See Erythrocyte.
Redox loops, 269
Redox potential, 264, 265
Reduction, 112
  of carbon, photosynthesis and, 322, 323
  photochemical reaction and, 320, 321
Reduction division. See Meiosis.
Refractory period, nerve impulse propagation
    and, 613–615
Regression, organs and, 294
Regulation, of contractility, 597–601
  metabolic, enzyme activity and, 101, 102
Regulator gene, 537, 538
Regulatory enzymes, 100, 101
Remak, R., 8
Renaturation, of proteins, 74
  of DNA, 88
Replication, heterochromatin and, 356
  semiconservative, chromatids and, 350, 351
Replicons, 376
Repolarization, of action potential, 153
Repression, 102, 103
Repressors, binding of within promoters, 539,
    540
  gene expression and, 487
  gene regulation and, 537
RER, 211
  nuclear pores and, 335
  ribosomal binding to, 210
Resealing, of red cell ghosts, 132
Residual bodies, lysosomes and, 287
Resolving power, of microscopes, 28, 29
Respiration
  cellular, 109–113, 263
    active transport and, 150
    aerobic, 109–113
    anaerobic, 109–111
    electron transport and, 112
    intracellular transport and, 242
    phosphorylation and,
        topography of, 268, 269

Respiratory chain, 112, 262
  symbiont hypothesis and, 278
  multi-molecular complexes of, 265–267
Resting potential, 148, 152
Restriction enzymes, 478–482
Restriction-modification phenomenon, 480
Reticulocytes, polysome study and, 524
Reticuloendothelial system, 289
Retina, rods and cones of, 144–147, 193, 194
Retinenes, 147
Retinoblastoma, 454
Retrograde transports, axonal, 610–612
  tetanus toxin and, 612
  NGF and, 610–612
Reverse transcriptase, 88, 375
  DNA preparation and, 495, 496
  gene probes and, 493
  mRNA labeling and, 575
R-groups, of amino acids, 71
  rho factor, RNA transcription and, 489
Rhodopsin, 147
Riboflavin, 97
Ribonuclease (RNase), 138, 156, 181, 296
Ribonuclease P, 521
Ribonucleic acid (RNA), 209, 259, 262
  basal bodies and, 193
  base composition of, 90, 91
  chloroplasts and, 328
  chromatin and, 342
  circular, in virioids, 91, 92
  directed assembly of, 119
  hairpin loops in, 91
  heterogenous nuclear (hnRNA), 336, 497
  in situ hybridization of, 458
  major classes of, 91
  messenger, (mRNA), eukaryotic, 492–499
    genetic code deciphered by, 468–470
    globin, 494
    insulin and, 245
    labeling and, 575
    prokaryotes and, 484–490
  protein synthesis and, 524–526
    ribosome binding site and, 526, 527
    signal peptides and, 221
  mitochondrial, 276, 277
  nuclear pores and, 335
  nuclear export of, 219
  nucleolus and, 359, 361
  polymerase, 486
    repressors and, 539, 540
    transcription and, 490, 491
  protein synthesis and, fidelity of, 521–523
  ribose in, 90
  ribosomal, 503–505
    nucleolar organizers and, 350
  nucleolar processing, 513–514
  synthesis of, condensed chromosomes and,
      368, 369
    DNA replication and, 378
    lateral loops as sites, 553–556
  tertiary structure of, 91
  tobacco mosaic virus and, 119
  transcription stages of, 486–488
  transfer (tRNA), 181, 468
    transcription of, 521
  viral, 173
Ribonucleoprotein matrix, asymmetry of, 554
Ribonucleoprotein particles (RNP), eukaryotic,
    497, 498

Ribophorins, 210, 221, 222
Ribose, 78, 84, 108
  in RNA, 90
Ribosome(s), 208, 262, 468, 501–505
  aleurone bodies and, 306, 307
  binding of, mRNA and, 526, 527
    RER and, 210, 212, 213
  biogenesis of, 508–517
  chloroplasts and, 327, 328
  composition of, 502, 503
  cytoplasmic, mitochondria and, 263
  DNA in, nucleolar organizer and, 508, 509
  endoplasmic reticulum and, 221–223
  eukaryotic, synthesis of, 508
  genes of, spacer DNA and, 509–511
  mitochondrial, 253, 254, 267, 275, 277
    symbiont hypothesis and, 278
  nucleolus and, 359
  peroxisomal enzymes and, 297
  protein synthesis and, 524–534
  proteins in, 506–508
  rough endoplasmic reticulum and, 209, 210
  subunit of, involvement in elongation, 530
Ribosome-polysome cycle, protein synthesis
    and, 524–526
Ribulose, 78
Ribulose 1, 5-diphosphate, regeneration of,
    323
Ribulose 1, 5-diphosphate carboxylase
    (carboxydismutase), subunits of, 323
Rifampicin, 531
  RNA transcription and, 490, 491
Riggs, 376
RNA. See Ribonucleic acid (RNA).
Robertson, J., 140
Rods, retinal, 144–147, 193, 194
    mitochondria in, 251
Rotenone, 269
Rough endoplasmic reticulum. See
    Endoplasmic reticulum, rough.
Rous, P., 173
Roux, W., 9
RNase, 138, 156, 181, 296,
RNase P, 521
RNP, eukaryotic, 497, 498
Ruckert, 552
Ruffles, cellular, 169–171
Ruthenium red stain, 167

S period, 365, 366
  chromosomes visualized during, 366–368
Sakano, H., 497
Salts, cellular role of, 70
Sanger, F., 474
Sarcomere, I-band in, 589, 590
  polarization within, actin and myosin and,
    594
  structural differentiation of, 586, 587
Sarcoplasm, 585
Sarcoplasmic reticulum, 219
  Ca²⁺ transport in, 601
  excitation-contraction coupling and, 598, 599
  of striated muscle fiber, 586
Satellite DNA, 358
Satellites, chromosomal, 350
Sato, 394
Saturation, of membrane fatty acids, 142

Scales, algal, Golgi formation of, 231, 302
Schiff's reagent, 55
Schimper, 278, 309
Schmidt, W. J., 6
Schleicher, 8
Schleiden, M., 8
Schouten, 9
Schrader, 412
Schultze, M., 8
Schwann, T., 8
Schwann cells, 144
Scurvy, 259
SDS, 135
  in electrophoresis of proteins, 77
Secondary constriction, 361
  chromosomal, 350
    nucleoli formation and, 508
Secretion, 208, 218, 219, 226
  cell coat and, 167
  continuous vs. discontinuous, 236, 237
  dictyosomes and, 307
  Golgi complex and, 235, 236
  granules of, 208
  lysosomes and, 288
  pancreatic, 238–244
Secretory granules, 242, 243
Secretory proteins, hydrophobic signal
    peptides in, 531, 532
Seeds, development of, 306
  germination of, 307
Self-assembly, 75, 118
Semiautonomous organelles, 275
  chloroplasts as, 327, 328
Semiconservative replication and, 350, 351
Seminiferous tubules, 416
Sendai virus, 142
SER, 210, 211
Serial synapses, 623
Sex, determination of, 444–448
Sex chromatin, 354
Sex chromosomes, 444–448
  true hermaphroditism and, 452
Sex-linked diseases, 455–457
Shadow casting, 36
Shear zone, 202
Shock, heat, puff induction and, 551
Sialic acid, 233, 235, 236
  cell membrane and, 134
Sickle cell anemia, 465, 466
Side-chains
  of animo acids, 71
  of coenzyme Q₁₀, 265
  glycoproteins and, 236
  membrane surface and, 166
Sieve tubes, 302
Signal hypothesis, 221, 531, 532
Signal peptidase, 221, 245
Signal peptide, insulin mRNA and, 245
Silicosis, lysosomes and, 295
Simian virus 40 (SV40), 171, 173
Singer, 314
Sliding mechanism hypothesis, 395
Smooth endoplasmic reticulum (SER), 210, 211
Sodium ions
  acetylcholine receptor and, 637–639
  active transport and, 150–153, 155
  channels for, action potentials and, 615–617
  electrical potential and, 148, 149
  mitochondria and, 274

Sodium dodecyl sulfate (SDS), 135
  in electrophoresis of proteins, 77
Sol-gel transformation, 197, 198
Somatic cell hybridization, 457, 458
Somatic nuclei, gene expression in, 567–570
Somatic pairing, 550
Somatodentritic synapses, 623
Spacer deoxyribonucleic acid, 509–511
Sparsomycin, 531
Specificity, of enzyme activity, 96
  of enzyme inhibitors, 99, 100
Specimen, preparation of, 35, 36
Spectrin, 135
  actin and, 142
Sperm, fertilization and, 295
  flagella of, 192
Spermatogenesis, 340, 400, 401
Spermatozoa, 188
Spherocytosis, hereditary, 137
Sphingolipids, classification of, 81
  storage diseases and, 295
Sphingomyelins, 81, 133, 134
  cholesterol and, 144
Sphingosine, in sphingolipids, 81
Spindle, 185
  centrioles and, 190
  microtubules and, 183
Spirin, A., 497, 498
Split proteins, ribosomal dissociation and, 506
Sporophyte, 415
Stahl, F., 371
Stains, electron, 37
Stain, Feulgen, polytene chromosomes and,
  550
  fluorescent, 250
  Giemsa, 440
  Janus green, 250
Staining, chromatin and, 341, 342
  chromosomal, 348
  cytochemical methods of, 49, 50, 55–59
  of cytosol, 181
  of the Golgi, 228, 231
  histochemical methods of, for enzymes, 57,
    58
  negative, 253, 256
    ribosomes and, 503
  polyacrylamide gels and, 135
Starch, 78
Starch granules, 296
Steady potential, 148, 152
Stearic acid, 80
  film of, surface pressure of, 123
Stem body, 394
Stem cell, 576
Stereocilia, 188
Stevens, 9
"Sticky" ends, chromosomes and, 350
Storage granules, 287
Strain birefringence, 33
Strasburger, E., 8
Streptomycin, 531
Stress fibers, 199, 200
Striated border, 158
Stroma, chloroplast and, 311, 312
Structural differentiation, of myofibrils and
  sarcomeres, 586, 587

Structural genes, 537
Subfibers, microtubular, 188–190, 192
Subfractionation, 233, 234
Sturtevant, 9
Submetacentric chromosome, 348, 349
Substrate(s), enzymatic binding of, induced-fit
  model of, 97
    lock-and-key model of, 97
  enzymatic specificity for, 96
Subunits, ribosomal, 503
Succinate, 272
Succinate dehydrogenase, 263–265, 269
Succinate-Q-reductase, 265, 266
Succinic acid, 264
Sucrose, 78
  lysosomes and, 294
Sudan stains, 57
Sugars, evolution of, 128
Sulfatase, aryl, 285
Sulfate, 236
Sulfatides, 81
Sulfolipid(s), 134
  thylakoids and, 315
Sulfur, photosynthesis and, 319
  in protein, 265, 266, 268
Supercharged vesicles, synaptic, 634
Supercoiling, of circular DNA, 89, 90
  and histones, 90
Sutherland, E. W., 102, 640
SV40 virus, DNA supercoiling and, 90
  overlapping genes of, 476
Symbiont hypothesis, 129, 278
Symbiosis, intracellular, 279
Synapses, formation and maintenance of, 624
  gap junctions and, 165
  structure of, transmission and, 610–626
  types of, 623
  ultrastructure of, 622–624
Synaptic functions, long-lasting, second
  messenger for, 640, 641
Synaptic junctions, 619–626
Synaptic potentials, 620, 621
  physiologic receptors and, 615
Synaptic receptors, 636–644
Synaptic transmission, nature of, 620, 621
  synapse structure and, 619–626
Synaptonemal complex, formation of, 405–408
  functions of, 409
Synaptic vesicles. See Vesicle(s) synaptic.
Synaptosomal membranes, isolation of,
  629–631
Synchronization, of cell growth, 466
Synkaryon, cell fusion and, 563–565
Synthetases
  AA-tRNA, protein synthesis and, 521–523
  citrate, 307, 308
  glycogen, 104
  malate, 298, 307, 308
Synthetic period, 365–368
Systematics, cytogenetics and, 433, 434
Szent-Györgi, A., 595

Tadpole, tails of, lysosomes and, 294
Tatum, E., 464

Taylor, J., 371
Tay-Sachs disease, 295
Telocentric chromosome, 348, 349
Telomere, 350
Telophase, mitotic, 385–387
Telophase I, meiotic, 410
Template DNA, 374
Terminal addition, of nucleotides, 484
Terminal bar, 158–161
Terminal cisternae, Ca^{2+} release from, 599
Terminalization, 410
Termination, of RNA transcription, 486–488
Termination factor, RNA transcription and, 489
Testosterone, action of, 447
Tetanus toxin, retrograde transport and, 612
Tetracycline, 531
Tetraethylammonium ions (TEA), 616, 617
Tetrazole, 269
Tetrodotoxin (TTX), 615, 616
Thiamine-pyrophosphatase, 234, 307
Threshold of activation, 613
Thrombin, blood clotting and, 120, 121
Thylakoid(s), membrane of, 253
    molecular organization of, 314–318
    structure of, 311–314
Thymidine, tritiated (^{3}H-), 277
Thymine, 85
    dimers of, DNA repair enzymes and, 378–380
Thymidine kinase (TK), 457
Thyroglobulin, 295
Thyroid-stimulating hormone (TSH), 102
Thyroxine, 102, 295
    mitochondria and, 252
    mitochondrial swelling and, 274
Tight junction, 158–161
Tijo, J., 438
Tissue culture, types of, 45, 46
T loop, of tRNA, 520
TMV, 119
Tobacco mosaic virus (TMV), assembly of, 119
Toluidine blue, visualization of ribosomes and, 503
Tonofilaments, 161, 199
Tortoiseshell cats, heterochromatin and, 356, 357
Toxin(s), bacterial, 143
    usefulness to medicine, 531
    tetanus, retrograde transport of, 612
Tracers, electron opaque, 37, 38
Track radioautography, 63
Transacetylase, lac operon code for, 537, 538
Transcriptase, reverse, 88, 375
    DNA preparation and, 495, 496
    gene probes and, 493
    mRNA labeling and, 575
Transcription, 463
    cell differentiation and, 574, 575
    eukaryotic, 490–492
    gene, polytene chromosomes and, 550–552
    heterochromatin and, 545
    lac, 537–543
    prokaryotic, protein synthesis and, 489
    RNA, in prokaryotes, 486
        stages of, 486–488

Transcription, RNA (Continued)
    tRNA, precursors and, 521
    visualization of, 491, 492
Transduction, 468
Transfer, of electrons. See Electrons, transfer of.
    of protons, 270
Transfer RNA, 181, 468
    transcription of, 521
Transferase, UDPG, 220
Transformation, viral, 156, 172, 173, 467
        cancer and, 132
        microfilaments and, 200
        microtubules and, 185
Transitional elements, Golgi complex and, 239
Transitions, 472
Translation, 463
Translational control, cell differentiation and, 572–574
Translocase, 529
Translocation, 428
    ADP-ATP, 272
    balanced, 450
    of chromosomal material, 431
    of protons, mitochondria and, 266–268
    reciprocal, 450
    of sodium and potassium ions, acetylcholine receptors and, 637–639
    Transmission, synaptic, nature of, 620, 621
        synapse structure and, 619–626
Transmitter, neuro-. See Neurotransmitter.
Transplantation, nuclear, constancy of genome and, 571
    active, 153, 156
        endoplasmic reticulum and, 219
        energy use by, 108
        ions and, 150
        mitochondria and, 251
    of ATP, mitochondrial, 272
    axonal, 187
    cancer cells and, 171
    electrons and, thylakoid membrane and, 318
        respiratory chain and, 110–112
    intracellular, microtubules and, 186
        respiration and, 242
    ionic, 152
    neurotransmitters, synaptic vesicles and, 632
    orthograde, axonal, 609, 610
    passive, 152
    of protons, 278
Trans-synaptic transport, tetanus toxin and, 612
Transversions, 472
Triad synapses, 623, 624
Tricarboxylic acid cycle. See Krebs tricarboxylic acid cycle.
Triglyceride(s), 220
    glyoxysomes and, 307
    structure of, 80
Triiodothyronine, 295
Trillium, DNA content in, 340
Triose phosphate, 322
Tripeptide, 71
Triphosphates, deoxynucleoside, 374
Trisomy, 438
13-Trisomy, 454
18-Trisomy, 452

21-Trisomy, 438, 452, 454
Tritium, use of, in radioautography, 63
tRNA, 181, 468
    transcription of, 521
Tropocollagen, assembly of, in collagen, 120
Tropomyosin, 182, 198, 199
    displacement in contractility mechanism, 597
    thin myofilaments and, 593
Troponin, 198
    thin myofilaments and, 593
Tryptophan operon, 541–543
Tschermack, 9
TSH, 102
T-system, excitation-contraction coupling and, 599
Tubulin, 182
    antibodies and, 388
    dimers of, 183
    dynein and, 192
    mitosis and, 185
    monomers of, 392
Tubules, seminiferous, 416
Tumor(s), 173
    chromosomal aberrations and, 454
Turgidity, plant cells and, 148
Turner's syndrome, 438, 451
Turpin, 8
Tyrosine, tRNA of, 520, 521
Tyrosinosis

Ultramicromethods, 54
Ultramicroscopy, 31
Ultraphagocytosis, 289
Ultraviolet absorption, by DNA, and
        hyperchromic shift, 88
    by nucleic acid bases, 85
Ultraviolet light, mutations and, 379, 380
Uncoupling agents, 112
Unineme theory, 350, 351
Unit membrane, mitochondrial, 253, 255
    nuclear, 334
UDPG-glycogen transferase, 220
Uracil, 85
Uranyl acetate, 340
Urate oxidase, 297, 298
Urea, 268
Uric acid oxidase, 234
Uridine diphosphate glucose (UDPG), 220
Uridine triphosphate (UTP), 108
Uroid, 202
Uterus, regression of, 294

Vacuolar system. See Endomembrane system.
Vacuole(s), 236
    autophagy and, 287, 288
    condensing, 239, 242
    contractile, 148
    digestive, 287
    of the Golgi, 227, 228, 231
    mitochondrial degeneration and, 259
    plant cells and, 304, 305

Vasopressin, 102
Velocity, maximum (Vmax), of an enzymatic
        reaction, 98
Vesicle(s)
    coated, micropinocytosis and, 292, 293
    fluid, 289
    germinal, 567
    micropinocytosis and, 290–293
    mitochondrial, 268
    phospholipid, 125, 126
    secretory, 226–228, 231, 236
    synaptic, isolation of, 629–631
        neuronal development and, 627–629
        neurotransmitter release and, 626–627, 632–634
        neurotransmitter transport and, 632
        types of, 627
Vesicle hypothesis, synapses and, 632–634
Vibrio cholerae, adenylate cyclase activation
    and, 104
Vicilin, 306
Vinblastine, tubulin and, 184
Virchow, R., 8
Virioids, circular RNA in, 91, 92
Virus(es), 119, 120. See also Bacteriophage(s).
    φ29, 119, 120, 121
    φX174, 119
        overlapping genes of, 475, 476, 477
    influenza, 138
    oncogenic, 171
    properties of, 10
    Sendai, 142
    SV40, and DNA supercoiling, 90
        overlapping genes of, 476
Vmax (maximum velocity), and enzyme
    inhibition, 100
    of an enzymatic reaction, 98
Vitamin(s),
    A, lysosomal enzymes and, 294, 295
    B₂, 97
    B₅, 97
    B₆, 97
    coenzymes and, 97, 110
    K₁, 320
    liposoluble, lysosomes and, 295

Waldeyer, W., 8, 333
Wall, cell. See Cell wall.
Warburg, O., 10, 250, 264
Water
    as component of cell, properties of, 70
    heavy, spindle fibers and, 392
    with phospholipids, bulk systems of, 124–126
    photosynthesis and, 319, 320
Watson, J. D., 10, 86, 370
Watson-Crick base pairing, RNA transcription
    and, 486
Watson-Crick model of DNA, 86, 87
Wavelength, of light, photosynthesis and, 320
Waxes, cutin, 302
Weissman, G., 9
Wieland, H., 10
Wilkins, M., 40, 86
Wilson, E. B., 9

Woods, P., 371
Wyman, J., 101

Yeast, 252
Yok bodies, mitochondria and, 258, 259
Yeast, mitochondria in, 275

φX174 virus, overlapping genes of, 475–477
X chromosome(s), heterochromatin and, 354, 356
    inactivation of, 545
    tortoiseshell cats and, 356
X polysomy, females with, 451
XYY syndrome, 451
Xanthophyll, 317
Xenopus, oöcytes of, 336, 337
Xeroderma pigmentosum, 379, 380
X-ray diffraction, 142
    of lipid monolayer, 123
    principles of, 39–41
Xylem, differentiation of, 301, 302
Xylose, 233

Z, as photosynthetic electron acceptor, 320, 321
Z-line, lack of, smooth muscles and, 590
Z-line, muscle contraction, behavior during, 589, 590
Z-line, woven-basket lattice in, 588
Zein, 306
Zone of exclusion, 227
Zonula occludens, 158–161
Zwitterions, 71
Zygonema, 405–408, 417
Zygote, diploid, 415
Zymogen granules, 237–239, 242, 243, 245